EIGHTH EDITION

Astronomy

A Beginner's Guide to the Universe

Eric Chaisson

Harvard University

Steve McMillan

Drexel University

Editor-in-Chief: Jeanne Zalesky
Executive Editor: Nancy Whilton
Project Managers: Tema Goodwin/Lizette Faraji
Program Manager: Mary Ripley
Development Manager: Cathy Murphy
Content Producers: Ziki Dekel, Kristin Mayo
Director of Product Management Services: Erin Gregg
Program/Project Management Team Lead: Kristen Flathman
Production Management: Thistle Hill Publishing Services
Copyeditor: Jane Loftus

Compositor: Cenveo® Publisher Services
Design and Cover Manager: Marilyn Perry
Interior Designer: Jerilyn Bockorick
Cover Designer: Jeff Puda
Illustrations: Rolin Graphics Inc.
Photo Permissions Project Manager: Maya Gomez
Photo Researcher: Cordes McMahan Hoffman/QBS Learning
Manufacturing Buyer: Maura Zaldivar-Garcia
VP Marketing: Christy Lesko
Printer/Binder/Cover Printer: RR Donnelley Owensville/Phoenix Color
Cover Photo Credit: Substantial Coronal Hole: NASA/SDO

Library of Congress Control Number: 2015956919

ISBN 10: 0-134-08770-4; ISBN 13: 978-0-134-08770-2 (Student edition)

www.pearsonhighered.com

1 2 3 4 5 6 7 8 9 10—V0N4—18 17 16 15

PART 2 OUR PLANETARY SYSTEM

4 The Solar System
Interplanetary Matter and the Birth of the Planets 102

5 Earth and Its Moon
Our Cosmic Backyard 138

PART 3   THE STARS

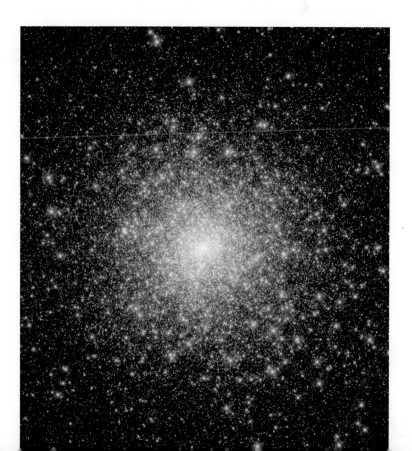

PART 4   GALAXIES AND THE UNIVERSE

We are pleased to have the opportunity to present in this book a representative sample of the known facts, evolving ideas, and frontier discoveries in astronomy today.

Astronomy: A Beginner's Guide to the Universe has been written and designed for students who have taken no previous college science courses and who will likely not major in physics or astronomy. We present a broad view of astronomy, straightforwardly descriptive and without complex mathematics. The absence of sophisticated mathematics, however, in no way prevents discussion of important concepts. Rather, we rely on qualitative reasoning as well as analogies with objects and phenomena familiar to the student to explain the complexities of the subject without oversimplification. We have tried to communicate the excitement that we feel about astronomy and to awaken students to the marvelous universe around us.

We are very gratified that the first seven editions of this text have been so well received by many in the astronomy education community. In using those earlier texts, many of you—teachers and students alike—have given us helpful feedback and constructive criticisms. From these, we have learned to communicate better both the fundamentals and the excitement of astronomy. Many improvements inspired by your comments have been incorporated into this edition.

Organization and Approach

As in previous editions, our organization follows the popular and effective "Earth out" progression. We have found that most students, especially those with little scientific background, are much more comfortable studying the relatively familiar solar system before tackling stars and galaxies. With Earth and Moon as our initial planetary models, we move through the solar system. Integral to our coverage of the solar system is a discussion of its formation. This line of investigation leads directly into a study of the Sun.

With the Sun as our model star, we broaden the scope of our discussion to include stars in general—their properties, their evolutionary histories, and their varied fates. This journey naturally leads us to coverage of the Milky Way Galaxy, which in turn serves as an introduction to our treatment of other galaxies. Finally, we reach cosmology and the large-scale structure and dynamics of the universe as a whole. Throughout, we strive to emphasize the dynamic nature of the cosmos—virtually every major topic, from planets to quasars, includes a discussion of how those objects formed and how they evolve.

We place much of the needed physics in the early chapters—an approach derived from years of experience teaching thousands of students. Additional physical principles are developed as needed later, both in the text narrative and in the *Discovery* and *More Precisely* boxes (described on p. xiv). We have made the treatment of

physics, as well as the more quantitative discussions, as modular as possible, so that these topics can be deferred if desired. Instructors presenting this material in a one-quarter course, who wish to (or have time to) cover only the essentials of the solar system before proceeding on to the study of stars and the rest of the universe, may want to teach only Chapter 4 (the solar system) and then move directly to Chapter 9 (the Sun).

What's New in This Edition

Astronomy is a rapidly evolving field, and almost every chapter in the eighth edition has been updated with new and late-breaking information. Several chapters have also seen significant internal reorganization in order to streamline the overall presentation, strengthen our focus on the process of science, and reflect new understanding and emphases in contemporary astronomy. Among the many improvements are the following:

- New chapter-opening images reflecting the latest astronomical discoveries.
- Updated astronomical imagery throughout.
- Streamlined art program providing more direct and accurate representations of astronomical objects.
- Increased use of annotations to clarify figure content.
- Updates in Chapter 3 on the *Hubble Space Telescope* and its successor, the *James Webb Space Telescope;* new material on new very large ground-based telescopes now under construction.
- New imagery throughout from the recently completed *ALMA* interferometric array.
- Updated coverage in Chapter 4 of the *Dawn* mission to asteroids Vesta and Ceres and the *Rosetta* mission to comet 67 P/Churyumov–Gerasimenko.
- Updates in Chapter 4 on exoplanet properties, with a new focus on Earths and super-Earths; revised discussion of habitable zones and Earth-like worlds.
- Updates on global CO_2 levels and global warming in Chapter 5.
- Updated data in Chapter 5 on lunar interior structure following the *LCROSS* and *GRAIL* missions; new *Lunar Reconnaissance Orbiter* imagery of lunar surface features.
- Updated discussion in Chapter 6 of Mercury's surface and internal structure in light of the findings of the *Messenger* probe.
- Updated discussion of Mars in Chapter 6, including results from the *Curiosity* mission.
- Updates in Chapter 7 on storm systems on the outer planets, Jupiter's shrinking Great Red Spot, and Saturn's polar vortices.
- New *Discovery* feature in Chapter 7 on solar system exploration.

- Updated discussion in Chapter 8 of Ganymede's magnetism, subsurface water, and aurora; updated material on Enceladus and its internal ocean.
- New material and rewritten discussion in Chapter 8 of the Pluto system following the *New Horizons* flyby; updated discussion of trans-Neptunian objects.
- New higher-resolution *Solar Dynamics Observatory* imagery in Chapter 9 of the corona, sunspots, and coronal mass ejections.
- Improved diagrams and updated discussion of the sunspot cycle in Chapter 9.
- Updated text and *ALMA* imagery on star-forming regions in Chapter 11.
- New supernova imagery and discussion in Chapter 12.
- Updated discussion in Chapter 13 of gamma-ray bursts and hypernovae.
- Added new art to the discussion of Special Relativity in *More Precisely* 13-1.
- Updated discussion in Chapter 14 of Milky Way formation.
- Updated discussion in Chapter 14 of stellar orbits around the central supermassive black hole.
- New discussion in Chapter 14 of energetic outflows from the Galactic center.
- Updated discussion in Chapter 16 of hot gas in galaxy clusters.
- New discussion in Chapter 16 of the star formation history of the universe.
- Updated discussion in Chapter 16 of galactic cannibalism.
- Integrated treatment in Chapter 16 of tidal streams in the Milky Way.
- Updated discussion in Chapter 17 of the cosmic microwave background; new discussion of acoustic oscillations and their relevance to cosmology.
- Expanded discussion of extremophilic life in Chapter 18.

The Illustration Program

Visualization plays an important role in both the teaching and the practice of astronomy, and we continue to place strong emphasis on this aspect of our book. We have tried to combine aesthetic beauty with scientific accuracy in the artist's conceptions that enrich the text, and we have sought to present the best and latest imagery of a wide range of cosmic objects. Each illustration has been carefully crafted to enhance student learning; each is pedagogically sound and tightly tied to nearby discussion of important scientific facts and ideas.

Full-Spectrum Coverage and Spectrum Icons

Increasingly, astronomers are exploiting the full range of the electromagnetic spectrum to gather information about the cosmos. Throughout this book, images taken at radio, infrared, ultraviolet, X-ray, or gamma-ray wavelengths are used to supplement visible-light images. As it is sometimes difficult (even for a professional) to tell at a glance which images are visible-light photographs and which are false-color images created with other wavelengths, each photo in the text is accompanied by an icon that identifies the wavelength of electromagnetic radiation used to capture the image.

Other Pedagogical Features

As with many other parts of our text, instructors have helped guide us toward what is most helpful for effective student learning.

Learning Outcomes Studies indicate that beginning students often have trouble prioritizing textual material. For this reason, a few (typically five or six) well-defined learning outcomes are provided at the start of each chapter. These help students determine what mastery they should be able to demonstrate after reading the chapter and then structure their reading accordingly. The outcomes are numbered and keyed to the items in the Chapter Summary and the Review and Discussion section, which in turn refer back to passages in the text. This highlights the most important aspects of the chapter, helping students prioritize information and aiding their review of the material. The outcomes are organized and phrased in such a way as to make them objectively testable, affording students a means of gauging their own progress.

The Big Picture and The Big Question Each chapter begins with a Big Picture overview of not only the content of the chapter, but also the critical issues facing the topic covered in the chapter; its purpose is to help students see how chapter content is connected to a broad understanding of the universe.

The Big Picture is bookended with a similar feature called The Big Question at the end of each chapter; the Big Question poses a cosmic, open-ended question about the chapter's topic area that is intended to ignite students' curiosity about the still-unanswered questions at the forefront of astronomical knowledge and research.

THE BIG PICTURE

Light collected tonight from the most distant galaxies was emitted by those objects long before Earth even formed. Racing for billions of years across the darkened realms of the cosmos, a minute fraction of their radiation is now intercepted by our telescopes and spacecraft. Captured in the many images of this book, that radiation tells us not only about the properties of faraway galaxies, but also about the history of our Galaxy and the universe in which we live.

THE BIG QUESTION

Galactic research lags stellar research by about 50 years. That's because galaxies were discovered only in the 20th century, and we are still learning about them. How did they form, and how do they evolve? Those are the biggest questions regarding galaxies, and they will not be answered until more and better data accumulate, especially regarding the most distant systems. Will the much larger galaxy surveys now on the horizon help solve these important issues?

Data Points Data Points sidebars in each chapter, based on data captured from thousands of students, alert students to the statistically most common mistakes made when working problems on related topics in MasteringAstronomy.

∞**Concept Links** The connection between the astronomical material and the physical principles set forth early in the text is crucial. It is important that students, when they encounter, say, Hubble's law in Chapter 15, recall what they learned about spectral lines and the Doppler shift in Chapter 2. Similarly, the discussions of the masses of binary star components (Chapter 10) and of galactic rotation (Chapter 14) both depend on the discussion of Kepler's and Newton's laws in Chapter 1. Throughout, discussions of new astronomical objects and concepts rely heavily on comparison with topics introduced earlier in the text.

It is important to remind students of these links so that they can recall the principles on which later discussions rest and, if necessary, review them. To this end, we have inserted "Concept Links" throughout the text—symbols that mark key intellectual bridges between material in different chapters. The links, denoted by the symbol ∞, signal students that the topic under discussion is related in some significant way to ideas developed earlier and direct them to material that they might wish to review before proceeding.

(MA) **Interactive Figures and Photos** Icons throughout the text direct students to dynamic versions of art and photos on MasteringAstronomy™. Using online applets, students can manipulate factors such as time, wavelength, scale, and perspective to increase their understanding of these figures.

Key Terms Like all subjects, astronomy has its own specialized vocabulary. To aid student learning, the most important astronomical terms are boldfaced at their first appearance in the text. Boldfaced key terms in the Chapter Summary are linked with the page number where the term was defined. In addition, a full alphabetical glossary, defining each key term and locating its first use in the text, appears at the end of the book.

Concept Checks We incorporate into each chapter a number of Concept Checks—key questions that require the reader to reconsider some of the material just presented or attempt to place it into a broader context. Answers to these in-chapter questions are provided at the back of the book.

CONCEPT CHECK

Why does a star get brighter as it runs out of fuel in its core?

Process of Science Checks Similar to Concept Checks, Process of Science Checks are aimed specifically at clarifying how science is done and how scientists reach the conclusions they do. Answers to these in-chapter questions are also provided at the back of the book.

PROCESS OF SCIENCE CHECK

Why are observations of star clusters so important to the theory of stellar evolution?

Visual Analogies Revised for clarity, visual analogies link art and analogy, explaining complex astronomical concepts with references to everyday experience.

A good analogy to a tangled solar magnetic field is a garden hose with loops and kinks.

▼ **Compound Art** It is rare that a single image, be it a photograph or an artist's conception, can capture all aspects of a complex subject. Wherever possible, multiple-part figures are used in an attempt to convey the greatest amount of information in the most vivid way:

- Visible images are often presented along with their counterparts captured at other wavelengths.
- Interpretive line drawings are often superimposed on or juxtaposed with real astronomical photographs, helping students to really "see" what the photographs reveal.
- Breakouts—often multiple ones—are used to zoom in from wide-field shots to close-ups, so that detailed images can be understood in their larger context.

Figure Annotations The eighth edition incorporates the research-proven technique of strategically placing annotations (which always appear in blue type) within key pieces of art, fostering students' ability to read and interpret complex figures, focus on the most relevant information, and integrate verbal and visual knowledge.

H–R Diagrams All of the book's H–R diagrams are drawn in a uniform format, using real data. H–R diagrams help us organize our information about the stars and track their evolutionary histories.

◄ **FIGURE 13.4 Crab Pulsar** In the core of the Crab Nebula (a), the Crab pulsar (c) blinks on and off about 30 times each second. In the top frame, the pulsar is off; in the bottom frame, it is on (arrow). (b) This *Chandra* X-ray image of the Crab, superimposed on a *Hubble* optical image, shows the central pulsar, as well as rings of hot X-ray-emitting gas in the equatorial plane, driven outward by the pulsar wind. Also visible in the image is a jet of hot gas escaping perpendicular to the equatorial plane. *(ESO; NASA; UC/Lick Observatory)*

▶ **MORE PRECISELY Boxes** These boxes provide more quantitative treatments of subjects discussed qualitatively in the text. Removing these more challenging topics from the main flow of the narrative and placing them within a separate modular element of the chapter design allow instructors greater flexibility in setting the level of their coverage.

13.2 MORE PRECISELY
Tests of General Relativity

Special relativity is the most thoroughly tested and most accurately verified theory in the history of science. General relativity, however, is on somewhat less firm experimental ground. The problem with verifying general relativity is that its effects on Earth and in the solar system—the places where we can most easily perform tests—are very small. Just as special relativity produces major departures from Newtonian mechanics only when speeds approach the speed of light, general relativity predicts large departures from Newtonian gravity only when extremely strong gravitational fields are involved.

Here we consider just two "classical" tests of the theory—solar system experiments that helped ensure acceptance of Einstein's theory. Bear in mind, however, that there are no known tests of general relativity in the "strong-field" regime—that part of the theory that predicts black holes, for example—so the full theory has never been experimentally tested.

At the heart of general relativity is the premise that everything, including light, is affected by gravity because of the curvature of spacetime. Shortly after he published his theory in 1915, Einstein noted that light from a star should be deflected by a measurable amount as it passes the Sun. The closer to the Sun the light comes, the more it is deflected. Thus, the maximum deflection should occur for a ray that just grazes the solar surface. Einstein calculated that the deflection angle should be 1.75, a small but detectable amount.

Of course, it is normally impossible to see stars so close to the Sun. During a solar eclipse, however, when the Moon blocks the Sun's light, the observation does become possible, as illustrated in the figure at lower left. In 1919 a team of observers led by British astronomer Sir Arthur Eddington succeeded in meas-

14.2 DISCOVERY
Density Waves

In the late 1960s, American astrophysicists C. C. Lin and Frank Shu proposed a way in which spiral arms in the Galaxy could persist for many Galactic rotations. They argued that the arms themselves contain no "permanent" matter. They should not be viewed as assemblages of stars, gas, and dust moving intact through the disk because such structures would quickly be destroyed by differential rotation. Instead, as described in the text, a spiral arm should be envisaged as a *density wave*—a wave of alternating compression and expansion sweeping through the Galaxy.

A wave in water builds up material temporarily in some places (crests) and lets it down in others (troughs). Similarly, as Galactic matter encounters a spiral density wave, the matter is compressed to form a region of higher than normal density. The matter enters the wave, is temporarily slowed down (by gravity) and compressed as it passes through, then continues on its way. This compression triggers the formation of new stars and nebulae. In this way, the spiral arms are formed and reformed repeatedly, without wrapping up.

The accompanying figure illustrates the formation of a density wave in a more familiar context—a traffic jam triggered by the presence of a repair crew moving slowly down the road. Cars slow down temporarily as they approach the crew, then speed up again as they pass the worksite and continue on their way. The result observed by a traffic helicopter flying overhead is a region of high traffic density concentrated around the work crew and

moving with it. An observer on the side of the road, however, sees that the jam never contains the same cars for very long. Cars constantly catch up to the bottleneck, move slowly through it, then speed up again, only to be replaced by more cars arriving from behind.

The traffic jam is analogous to the region of high stellar density in a Galactic spiral arm. Just as the traffic density wave is not tied to any particular group of cars, the spiral arms are not attached to any particular piece of disk material. Stars and gas enter a spiral arm, slow down for a while, then exit the arm and continue on their way around the Galactic center. The result is a moving region of high stellar and gas density, involving different parts of the disk at different times. Notice also that, just as in our Galaxy, the traffic jam wave moves more slowly than, and independently of, the overall traffic flow.

We can extend our traffic analogy a little further. Most drivers are well aware that the effects of such a tie-up can persist long after the road crew has stopped work and gone home for the night. Similarly, spiral density waves can continue to move through the disk even after the disturbance that originally produced them has long since subsided. According to spiral density wave theory, this is precisely what has happened in the Milky Way Galaxy. Some disturbance in the past—an encounter with a satellite galaxy, perhaps, or the effect of the central bar—produced a wave that has b

Traffic speed

Traffic density   Density wave

Location

◀ **DISCOVERY Boxes** Exploring a wide variety of interesting topics, *Discovery* boxes provide the reader with insight into how scientific knowledge evolves and emphasizes the process of science.

CHAPTER REVIEW

SUMMARY

LO1 A core-collapse supernova may leave behind an ultracompressed ball of material called a **neutron star** (p. 362). This is the remnant of the inner core that rebounded and blew the rest of the star apart. Neutron stars are extremely dense and, at formation, are predicted to be extremely hot, strongly magnetized, and rapidly rotating. They cool down, lose much of their magnetism, and slow down as they age.

LO2 According to the **lighthouse model** (p. 364), neutron stars, because they are magnetized and rotating, send regular bursts of electromagnetic energy into space. The beams are produced by charged particles confined by the strong magnetic fields. When we can see the beams from Earth, we call the source neutron star a **pulsar** (p. 362). The pulse period is the rotation period of the neutron star.

LO3 A neutron star that is a member of a binary system can draw matter from its companion, forming an accretion disk, which is usually a strong source of X-rays. As gas builds up on the star's surface, it eventually becomes hot enough to fuse hydrogen. When hydrogen burning starts on a neutron star, it does so explosively, and an **X-ray burster** (p. 366) results. The rapid rotation of the inner part of the accretion disk causes the neutron star to spin faster as new gas arrives on its surface. The eventual result is a very rapidly rotating neutron star—a **millisecond pulsar** (p. 366). Many millisecond pulsars are found in the hearts of old globular clusters. They cannot have formed

explosions involve the violent merger of neutron stars in a distant binary system, or the recollapse and subsequent violent explosion following a "failed" supernova in a very massive star.

LO5 Einstein's special theory of relativity deals with the behavior of particles moving at speeds comparable to the speed of light. It agrees with Newton's theory at low velocities, but makes many very different predictions for high-speed motion. All of its predictions have been repeatedly verified by experiment. The modern replacement for Newtonian gravity is Einstein's **general theory of relativity** (p. 373), which describes gravity in terms of the warping, or bending, of space by the presence of mass. The more mass, the greater the warping. All particles—including photons—respond to that warping by moving along curved paths.

LO6 The upper limit on the mass of a neutron star is about three solar masses. Beyond that mass, the star can no longer support itself against its own gravity, and it collapses to form a **black hole** (p. 373), a region of space from which nothing can escape. The most massive stars, after exploding in a supernova, form black holes rather than neutron stars. Conditions in and near black holes can only be described by general relativity. The radius at which the escape speed from a collapsing star equals the speed of light is called the **Schwarzschild radius** (p. 374). The surface of an imaginary sphere centered on the collapsing star and having a radius equal to the star's Schwarzschild radius is called the **event horizon** (p. 374).

▶ **Chapter Summaries** Key terms introduced in each chapter are listed again, in context and in boldface, along with key figures and page references to the text discussion. Summary items are keyed to the Learning Outcomes presented at the start of the chapter.

End-of-Chapter Questions and Problems Many elements of the end-of-chapter material have seen substantial reorganization:

- Each chapter has 15 Review and Discussion questions, which may be used for in-class review or for assignment. The material needed to answer Review and Discussion questions can be found within the chapter. The Review and Discussion questions explore particular topics more deeply, often asking for opinions, not just facts. As with all discussions, these questions usually have no single "correct" answer. A few (2–4) questions per chapter are marked as directly relevant to the Process of Science theme of the book, and each Learning Outcome is reflected in one of the Review and Discussion questions.

- Each chapter incorporates 15 Self-Test questions, roughly divided between true/false and multiple choice formats, designed to allow students to assess their understanding of the chapter material. As with the Review and Discussion questions, the Self-Test questions can be answered based on material presented in the chapter. Two of the multiple choice questions in each chapter are tied directly to a specific figure or diagram in the text to test students' comprehension of the visual material presented there. Answers to all these questions appear at the end of the book.

- The end-of-chapter material also includes 10 Problems based on the chapter contents and requiring some numerical calculation. In many cases, the Problems are tied directly to quantitative statements made (but not worked out in detail) in the text. The solutions to the Problems are not contained verbatim within the chapter, but the information necessary to solve them has been presented in the text. Answers appear at the end of the book.

- The end-of-chapter material now includes a number of astronomical Activities relevant to the material presented in the text. Activities include both group and individual projects, ranging from basic naked-eye and telescopic observing exercises, to opinion polls, surveys and group discussion, and astronomical research on the Web.

Instructor Resources

MasteringAstronomy™

www.masteringastronomy.com

MasteringAstronomy is the most widely used and most advanced astronomy tutorial and assessment system in the world. By capturing the step-by-step work of students nationally, MasteringAstronomy has established an unparalleled database of learning challenges and patterns. Using this student data, a team of astronomy education researchers has refined every activity and problem. The result is a library of activities of unique educational effectiveness and assessment accuracy. MasteringAstronomy provides students with two learning systems in one: a dynamic self-study area and the ability to participate in online assignments.

MasteringAstronomy, now easier to use than ever, provides instructors with a fast and effective way to assign uncompromising, wide-ranging online homework assignments of just the right difficulty and duration. The tutorials coach 90 percent of students to the correct answer with specific wrong-answer feedback. The powerful post-diagnostics allow instructors to assess the progress of their class as a whole or to quickly identify individual student's areas of difficulty. Tutorials built around text content and all the end-of-chapter problems from the text are also available.

MasteringAstronomy has been revised in this edition with extensive new interactive opportunities in the item library for assignment and in the media-rich study area for open-ended student exploration, which students can use whether the instructor assigns homework or not.

Instructor Resource Manual Updated for the eighth edition, this manual provides sample syllabi and course schedules, an overview of each chapter, pedagogical tips, useful analogies, suggestions for classroom demonstrations, writing questions, answers to the end-of-chapter Review and Discussion questions, Conceptual Self-Test, and Problems, selected readings, and additional references and resources. The Instructor Resource Manual is available for download on the Pearson Instructor Resource Center (www.pearsonhighered.com/educator) and in the Mastering-Astronomy Instructor Resource Area.

Test Bank This extensive file of approximately 2500 multiple-choice, true/false, fill-in-the-blank, short-answer and essay questions is newly updated and revised for the eighth edition. The questions are organized and referenced by chapter section and by question type. The Test Bank is available within TestGen® as well as electronically on the Pearson Instructor Resource Center (www.pearsonhighered.com/educator) and in MasteringAstronomy.

Instructor Resource DVD This DVD provides virtually every electronic asset you'll need in and out of the classroom. The DVD is organized by chapter and contains all text illustrations, tables, and photos in jpeg and PowerPoint® formats, as well as animations, videos, and self-guided tutorials from the self-study section of MasteringAstronomy.
ISBN 0-134-24163-0

Learner-Centered Astronomy Teaching: Strategies for ASTRO 101

Timothy F. Slater, University of Wyoming; Jeffrey P. Adams, Millersville University

Strategies for ASTRO 101 is a guide for instructors of the introductory astronomy course for nonscience majors. Written by two leaders in astronomy education research, this book details various techniques instructors can use to increase students' understanding and retention of astronomy topics, with an emphasis on making the lecture a forum for active student participation. Drawing from the large body of recent research to discover how students learn, this guide describes the application of multiple classroom-tested

techniques to the task of teaching astronomy to predominantly nonscience students.
ISBN 0-13-046630-1

Peer Instruction for Astronomy

Paul Green, Harvard Smithsonian Center for Astrophysics
Peer instruction is a simple yet effective method for teaching science. Techniques of peer instruction for introductory physics were developed primarily at Harvard and have aroused interest and excitement in the physics education community. This approach involves students in the teaching process, making science more accessible to them. Peer instruction is a new trend in astronomy that is finding strong interest and is ideally suited to introductory astronomy classes. This book is an important vehicle for providing a large number of thought-provoking, conceptual short-answer questions aimed at a variety of class levels. While significant numbers of such questions have been published for use in physics, *Peer Instruction for Astronomy* provides the first such compilation for astronomy.
ISBN 0-13-026310-9

Student Resources

MasteringAstronomy™

www.masteringastronomy.com
This homework, tutorial, and assessment system is uniquely able to tutor each student individually by providing students with instantaneous feedback specific to their wrong answers, simpler sub-problems upon request when they get stuck, and partial credit for their method(s) used. Students also have access to a self-study area that contains practice quizzes, self-guided tutorials, animations, videos, and more.

Pearson eText 2.0 is available through MasteringAstronomy, either automatically when MasteringAstronomy is packaged with new books, or available as a purchased upgrade online. Allowing the students to access the text wherever they have access to the Internet, Pearson eText comprises the full text, including figures that can be enlarged for better viewing. Within Pearson eText 2.0, students are also able to pop up definitions and terms to help with vocabulary and reading. Students also can take notes in Pearson eText using the annotation feature.

Starry Night® College Student Access Code Card (Simulation Curriculum) This access kit provides a one-time download of Starry Night College, the best-selling planetarium software that lets you escape the Milky Way and travel across 700 million light-years of space. Hailed for its breathtaking realism, powerful features, and intuitive interface, Starry Night College is available to be packaged (for a minimal charge) with new copies of introductory astronomy textbooks. This access kit also enables users to download *Starry Night College Activities & Observation and Research Projects* by Erin O'Connor and Steve McMillan.
ISBN 0-321-71295-1

Starry Night® College Activities & Observation and Research Projects This online booklet contains over 35 activities and 70 observation and research projects written by Erin O'Connor and Steve McMillan and is based on Starry Night College planetarium software.
ISBN 0-321-75307-0

SkyGazer 5.0 Student Access Code Card (Carina Software) This access kit provides a one-time download of SkyGazer 5.0 that combines exceptional planetarium software with informative pre-packaged tutorials. Based on the popular Voyager software, this access kit is available to be packaged at no additional charge with new copies of introductory astronomy textbooks. This access kit also enables users to download the *Astronomy Media Workbook* by Michael LoPresto.
ISBN 0-321-76518-4

***Astronomy Media Workbook*, 7th Edition** *Astronomy Media Workbook* by Michael LoPresto includes a wide selection of in-depth activities based on *Voyager*: SkyGazer 5.0 planetarium software, and the Interactive FiguresTM and RSS Feeds in MasteringAstronomy. These thought-provoking projects are suitable for labs or homework assignments. It is downloadable with a SkyGazer access code.
ISBN 0-321-74124-2

Observation Exercises in Astronomy This workbook by Lauren Jones contains a series of astronomy exercises that integrate technology from planetarium software such as Stellarium, Starry Night College, WorldWide Telescope, and SkyGazer. Using these online products adds an interactive dimension to students' learning.
ISBN: 0-321-63812-3

Edmund Scientific Star and Planet Locator The famous rotating roadmap of the heavens shows the location of the stars, constellations, and planets relative to the horizon for the exact hour and date you determine. This 8-inch square star chart was plotted by the late astronomer and cartographer George Lovi. The reverse side of the locator is packed with additional data on the planets, meteor showers, and bright stars. Included with each star chart is a 16-page, fully-illustrated, pocket-size instruction booklet.

***Norton's Star Atlas and Reference Handbook*, 20th edition** Now in a superbly redesigned, two-color landmark 20th edition, this combination star atlas and reference work by Ian Ridpath has no match in the field. First published in 1910, *Norton's* owes much of its legendary success to its unique maps, arranged in slices known as gores, each covering approximately one-fifth of the sky. Every star visible to the naked eye under the clearest skies—down to magnitude 6.5—is charted, along with star clusters, nebulae, and galaxies. Extensive tables of data on interesting objects for observation accompany each of the precision-drawn maps. Preceding the maps is the unique and authoritative reference

handbook covering timekeeping and positional measurements on the celestial sphere; the Sun, Moon, and other bodies of the solar system; telescopes and other equipment for observing and imaging the sky; and stars, nebulae, and galaxies. Throughout, succinct fundamental principles and practical tips guide the reader into the night sky. The appendices Units and Notation, Astronomical Constants, Symbols and Abbreviations, and Useful Addresses complete what has long been the only essential reference for the stargazer.
ISBN 0-13-145164-2

***Lecture-Tutorials for Introductory Astronomy,* 3rd Edition** Edward E. Prather, University of Arizona; Timothy F. Slater, University of Wyoming; Jeffrey P. Adams, Millersville University; Gina Brissenden, University of Arizona

Funded by the National Science Foundation, *Lecture-Tutorials for Introductory Astronomy* is designed to help instructors bring interactive teaching strategies into general education astronomy courses of all sizes. The third edition features new lecture-tutorials entitled the Greenhouse Effect; Dark Matter; Making Sense of the Universe and Expansion; Hubble's Law; Expansion, Lookback Times, and Distances; and Big Bang. Each of the 44 lecture-tutorials is presented in a classroom-ready format, asks students to collaborate in groups of two to three, takes approximately 15 minutes, challenges students with a series of carefully designed questions that spark student discussions, engages students in critical reasoning, and requires no equipment.
ISBN 0-321-82046-0

Sky and Telescope This supplement, edited by Evan Skillman, contains nine articles that originally appeared in the popular amateur astronomy magazine, plus a summary and four question sets focusing on the issues professors most want to address: general review, process of science, scale of the universe, and our place in the universe.
ISBN 0-321-70620-X

Acknowledgments

Throughout the many drafts that have led to this book, we have relied on the critical analysis of many colleagues. Their suggestions ranged from the macroscopic issue of the book's overall organization to the minutiae of the technical accuracy of each and every sentence. We have also benefited from much good advice and feedback from users of the first seven editions of the text and our more comprehensive text, *Astronomy Today*. To these many helpful colleagues, we offer our sincerest thanks.

Reviewers of the Eighth Edition

Ron Armale, *Cypress College*
Sandra Doty, *Ohio University, Lancaster*
Jacqueline Dunn, *Midwestern State University*
Saeed Safaie, *Rockland Community College*
Gene Tracy, *William and Mary College*

Anastasia Phillis, *Richard J. Daley College*
Charles Hakes, *Fort Lewis College*

Reviewers of Previous Editions

Todd Adams, *Florida State University*
Stephen G. Alexander, *Miami University*
Ron Armale, *Cypress College*
Nadine G. Barlow, *Northern Arizona University*
Michael L. Broyles, *Collin County Community College*
Juan Cabanela, *Haverford College*
Erik Christensen, *South Florida Community College*
Michael L. Cobb, *Southeast Missouri State University*
Anne Cowley, *Arizona State University*
Manfred Cuntz, *University of Texas at Arlington*
James R. Dire, *U.S. Coast Guard Academy*
Jacqueline Dunn, *Midwestern State University*
John Dykla, *Loyola University–Chicago*
Tina Fanetti, *Western Iowa Tech Community College*
Doug Franklin, *Western Illinois University*
Michael Frey, *Cypress College*
Alina Gabryzewska-Kukawa, *Delta State University*
Peter Garnavich, *University of Notre Dame*
Richard Gelderman, *Western Kentucky University*
Martin Goodson, *Delta College*
David J. Griffiths, *Oregon State University*
Martin Hackworth, *Idaho State University*
Susan Hartley, *University of Minnesota–Duluth*
David Hudgins, *Rockhurst University*
Doug Ingram, *Texas Christian University*
Marvin Kemple, *Indiana University–Purdue University, Indianapolis*
Linda Khandro, *Shoreline Community College*
Mario Klaric, *Midlands Technical College*
Patrick Koehn, *Eastern Michigan University*
Andrew R. Lazarewicz, *Boston College*
Paul Lee, *Middle Tennessee University*
M.A. K. Keohn Lodhi, *Texas Tech University*
Andrei Kolmakov, *Southern Illinois University*
F. Bary Malik, *Southern Illinois University-Carbondale*
Fred Marschak, *Santa Barbara College*
John Mattox, *Francis Marion University*
Chris McCarthy, *San Francisco State University*
George E. McCluskey, *Jr., Leigh University*
Scott Miller, *Pennsylvania State University*
L. Kent Morrison, *University of New Mexico*
George Nock, *Northeast Mississippi Community College*
Richard Nolthenius, *Cabrillo College*
Edward Oberhofer, *University of North Carolina–Charlotte*
Robert S. Patterson, *Southwest Missouri State University*
Jon Pedicino, *College of the Redwoods*
Cynthia W. Peterson, *University of Connecticut*
Robert Potter, *Bowling Green State University*
Heather L. Preston, *U.S. Air Force Academy*
Andreas Quirrenbach, *University of California–San Diego*
James Regas, *California State University, Chico*
Tim Rich, *Rust College*

Frederick A. Ringwald, *California State University, Fresno*
Gerald Royce, *Mary Washington College*
Louis Rubbo, *Coastal Carolina University*
Rihab Sawah, *Moberly Area Community College*
John C. Schneider, *Catonsville Community College*
C. Ian Short, *Florida Atlantic University*
Philip J. Siemens, *Oregon State University*
Earl F. Skelton, *George Washington University*
Don Sparks, *Los Angeles Pierce College*
Angela Speck, *University of Missouri–Columbia*
Phillip E. Stallworth, *Hunter College of CUNY*
Peter Stine, *Bloomsburg University of Pennsylvania*
Irina Struganova, *Barry University*
Jack W. Sulentic, *University of Alabama*
Jonathan Tan, *University of Florida*
Gregory Taylor, *University of New Mexico*
Gregory R. Taylor, *California State University, Chico*
George Tremberger, Jr., *Queensborough Community College*
Craig Tyler, *Fort Lewis College*
Robert K. Tyson, *University of North Carolina*
Alex Umantsev, *Fayetteville State University*
Michael Vaughn, *Northeastern University*
Jimmy Westlake, *Colorado Mountain College*
Donald Witt, *Nassau Community College*
J. Wayne Wooten, *University of West Florida*
Garett Yoder, *Eastern Kentucky University*
David C. Ziegler, *Hannibal-LaGrange College*

We would also like to express our gratitude to those that have contributed to our accompanying media products and print supplements, as well as Lola Judith Chaisson for assembling and drawing all the H-R diagrams for this edition.

The publishing team at Pearson has assisted us at every step along the way in creating this text. Much of the credit must go to our project editor, Tema Goodwin, for her superb editorial judgment, fine attention to detail, dogged determination to bring the project in on schedule, and unflappable skill in keeping things on an even keel, despite the authors' best efforts to the contrary. We have worked with her on several editions of our texts and will sorely miss her now that she has moved on to new professional challenges. Thanks also to our editor, Nancy Whilton, who has once again successfully navigated us through the many unfathomable twists and turns of the publishing world. Program manager Mary Ripley has done an excellent job of tying together the threads of this very complex project, made all the more complex by the necessity of combining text, art, and electronic media into a coherent whole.

Special thanks are also in order to the project management team at Thistle Hill Publishing Services, Andrea Archer and Angela Williams Urquhart, and to interior and cover designer Jerilyn Bockorick for making the eighth edition look spectacular.

We are always interested in your feedback on this text. Please email us at aw.astronomy@pearson.com if you find any errors or have comments.

Eric Chaisson
Steve McMillan

ABOUT THE AUTHORS

ERIC CHAISSON

Eric holds a doctorate in astrophysics from Harvard University, where he spent 10 years on the faculty of Arts and Sciences. For more than two decades thereafter, he served on the senior science staff at the Space Telescope Science Institute and held various professorships at Johns Hopkins and Tufts universities. He is now back at Harvard, where he teaches natural science and conducts research at the Harvard-Smithsonian Center for Astrophysics. Eric has written 12 books on astronomy and has published nearly 200 scientific papers in professional journals.

STEVE MCMILLAN

Steve holds a bachelor's and master's degree in mathematics from Cambridge University and a doctorate in astronomy from Harvard University. He held post-doctoral positions at the University of Illinois and Northwestern University, where he continued his research in theoretical astrophysics, star clusters, and numerical modeling. Steve is currently Distinguished Professor of Physics at Drexel University and a frequent visiting researcher at Princeton's Institute for Advanced Study and Leiden University in the Netherlands. He has published more than 100 scientific papers in professional journals.

Astronomy

A Beginner's Guide to the Universe

Foundations

Astronomy is the study of the universe—the totality of all space, time, matter, and energy. It is a subject like no other, for it requires us to profoundly change our perspective and to consider sizes, scales, and times unfamiliar to us from everyday experience. To appreciate astronomy, we must broaden our view and expand our minds. We must think big!

Part 1 presents the basic methods used by astronomers to chart the space around us. We describe the progress of scientific knowledge, from stories of chariots and gods to today's well-tested ideas of planetary motion and quantum physics. We also delve into the microscopic realm of atoms and molecules, whose properties hold keys to understanding the universe on macroscopic scales.

The images here illustrate the range of scales encountered in Part 1, from atoms to humans to Earth itself.

Atoms ~10^{-10} m

Cells ~10^{-5} m

Humans ~2 m

Earth ~10^{-7} m

Mountains ~10^4 m

Earth is neither central nor special;
we inhabit no unique place in the universe.

0

Charting the Heavens

The Foundations of Astronomy

Nature offers no greater splendor than the starry sky on a clear, dark night. Silent and jeweled with the constellations of ancient myth and legend, the night sky has inspired wonder throughout the ages—a wonder that leads our imaginations far from the confines of Earth and out into the distant reaches of space and time. Astronomy, born in response to that wonder, is built on two basic traits of human nature: the need to explore and the need to understand. People have sought answers to questions about the universe since the earliest times. Astronomy is the oldest of all the sciences, yet never has it been more exciting than it is today.

LEARNING OUTCOMES

Studying this introductory chapter will enable you to:

LO1 Arrange the basic levels of structure in the universe in order of increasing size.

LO2 Describe the celestial sphere, and explain how astronomers use constellations and angular measurement to locate objects in the sky.

LO3 Account for the apparent motions of the Sun and the stars in terms of Earth's actual motion, and explain how the axial tilt of our planet causes the seasons.

LO4 Explain the changing appearance of the Moon, and describe how the relative motions of Earth, the Sun, and the Moon lead to eclipses.

LO5 Give an example of how simple geometric reasoning can be used to measure the distances and sizes of otherwise inaccessible objects.

LO6 Distinguish between scientific theories, hypotheses, and observations, and describe how scientists combine observation, theory, and testing to understand the universe.

THE BIG PICTURE

Although early astronomy was defined by just the few thousand stars visible to the unaided eye, stars exist everywhere throughout the universe. Roughly as many stars reside in the observable universe as there are grains of sand in all Earth's beaches—about a hundred sextillion, or 10^{23}.

◄ High overhead on a clear, dark night, we can see a rich band of stars known as the Milky Way— so called for its resemblance to a milky band of countless stars stretching across the sky. All these stars (and more) are part of a much larger system called the Milky Way Galaxy, of which our star, the Sun, is one member. This image shows the awesome splendor of the Milky Way shining above some of the large telescopes of the European Southern Observatory, a major astronomical facility located high in the Chilean Andes. (ESO/Y. Beletsky)

MasteringAstronomy®

Visit www.masteringastronomy.com for quizzes, animations, videos, interactive figures, and self-guided tutorials.

0.1 The "Obvious" View

Our Place in Space

Of all the scientific insights achieved to date, one stands out boldly: Earth is neither central nor special. We inhabit no unique place in the universe. We live on an ordinary rocky **planet** called Earth, one of eight known planets orbiting an average **star** called the Sun, a middle-aged star near the edge of a huge collection of stars called the Milky Way Galaxy, one **galaxy** among countless billions of others spread throughout the observable **universe.** Figure 0.1 illustrates the enormous range of scales of these very different objects.

Today, our scientific understanding of the cosmos extends as far as we can see, spanning the vast reaches of intergalactic space. But the modern view of the universe depicted here is in many ways the "punch line" of the story presented in this text. It is the culmination of countless scientific discoveries, large and small—the work of generations of astronomers. How have astronomers achieved the perspective sketched in Figure 0.1? Our study of the universe, the science of **astronomy,** begins by examining the sky.

Constellations in the Sky

Between sunset and sunrise on a clear night, we can see some 3000 points of light. Include the view from the opposite side of Earth, and nearly 6000 stars are visible to the unaided eye. A natural human tendency is to see patterns and relationships between objects even when no true connection exists, and people long ago connected the brightest stars into configurations called **constellations.** In the Northern Hemisphere, most constellations were named after mythological heroes and animals. Figure 0.2 shows a constellation especially prominent in the northern night sky from October through March: the hunter Orion, named for a mythical Greek hero famed, among other things, for his amorous pursuit of the

▼ **FIGURE 0.1 Size and Scale in the Universe**   The bottom right of this figure shows humans on Earth, a view that widens progressively in the other four scenes from bottom to top—Earth, the solar system, a galaxy, and truly deep space. The numbers within the dashed zooms indicate approximately the increase in scale between successive images: Earth is 10 million times larger than humans, our solar system in turn is some million times larger than Earth, and so on. (See the Preface for an explanation of the icon at the bottom, which here indicates that these images were made in visible light.) *(NASA; J. Lodrigus; NOAA)*

Universe

$\times 10^4$

Galaxy

$\times 10^8$

Planetary system

$\times 10^6$

$\times 10^7$

R I V U X G

Earth

Humans

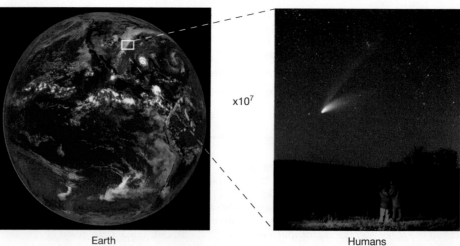

▶ **FIGURE 0.2 Constellation Orion** INTERACTIVE (a) A photograph of the group of bright stars that make up the constellation Orion. (b) The stars connected to show the pattern visualized by the Greeks: the outline of a hunter. The Greek letters serve to identify some of the brighter stars in the constellation (see Figure 0.3). You can easily find Orion in the northern winter sky by identifying the line of three bright stars in the hunter's "belt." *(P. Sanz/Alamy)*

This is a real photo of the Orion constellation . . .

. . . and this is a mapped interpretation, to exactly the same scale.

16°

(a)

R I V U X G

(b)

Pleiades, the seven daughters of the giant Atlas. According to Greek mythology, the gods placed the Pleiades among the stars to protect them from Orion, who still stalks them nightly across the sky. Many other constellations have similarly fabulous connections with ancient cultures.

The stars making up a particular constellation are generally not close together in space. They merely are bright enough to observe with the naked eye and happen to lie in the same direction in the sky as seen from Earth. Figure 0.3 illustrates this point for Orion, showing the true relationships between that constellation's brightest stars. Although constellation patterns have no real significance, the terminology is still used today. Constellations provide a convenient means for astronomers to specify large areas of the sky, much as geologists use continents or politicians use voting precincts to identify certain localities on Earth. In all, there are 88 constellations, most of them visible from North America at some time during the year.

The Celestial Sphere

Over the course of a night, the constellations appear to move across the sky from east to west. However, ancient sky-watchers noted that the *relative* positions of stars (to each other) remained unchanged as this nightly march took place. It was natural for early astronomers to conclude that the stars were attached to a **celestial sphere** surrounding Earth—a canopy of stars like an astronomical painting on a vast heavenly ceiling. Figure 0.4 shows how early astronomers pictured the stars as moving with this celestial sphere as it turned around a fixed Earth. Figure 0.5 shows how stars appear to move in circles around a point in the sky very close to the star Polaris (better known as the Pole Star or the North Star). To early astronomers, this point represented the axis around which the celestial sphere turned.

From our modern standpoint, the apparent motion of the stars is the result of the spin, or **rotation,** not of the celestial sphere, but of Earth. Even though we now know that a revolving celestial sphere is an incorrect

▶ **FIGURE 0.3 Orion in 3D** The true three-dimensional relationships among the most prominent stars in Orion. The distances in light-years were measured by the European *Hipparcos* satellite in the 1990s. (See Section 10.1.)

What we see projected on sky

|←—————— 1000 light-years ——————→|

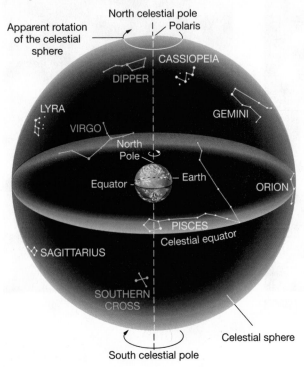

Imagine yourself at the center of this sphere, looking out at the whole sky around you.

Apparent rotation of the celestial sphere

North celestial pole

Polaris

CASSIOPEIA

DIPPER

LYRA

GEMINI

VIRGO

North Pole

Equator — Earth

ORION

PISCES

Celestial equator

SAGITTARIUS

SOUTHERN CROSS

Celestial sphere

South celestial pole

▲ FIGURE 0.4 **The Celestial Sphere** INTERACTIVE   Planet Earth sits fixed at the hub of the celestial sphere. This is one of the simplest possible models of the universe, but it doesn't agree with the facts that astronomers now know about the universe.

CONCEPT CHECK

Earth isn't really enclosed in a sphere with stars attached. Why then do astronomers find it convenient to retain the fiction of the celestial sphere? What vital piece of information about stars is lost when we talk about their positions "on" the sky?

---

description of the heavens, astronomers still use the idea as a convenient fiction that helps us visualize the positions of stars in the sky. The point where Earth's rotation axis (the line through the center around which the planet rotates) intersects the celestial sphere in the Northern Hemisphere is known as the **north celestial pole;** it is directly above Earth's North Pole. The star Polaris happens to lie close to the north celestial pole, which is why its direction indicates due north. In the Southern Hemisphere, the extension of Earth's axis in the opposite direction defines the **south celestial pole.** (There are no bright stars conveniently located near the south celestial pole and hence no "southern Pole Star.") Midway between the north and south celestial poles lies the **celestial equator,** representing the intersection of Earth's equatorial plane (the plane through Earth's center, perpendicular to the rotation axis) with the celestial sphere.

Celestial Coordinates

The simplest method of locating stars in the sky is to specify their constellation and then rank the stars in that constellation in order of brightness. The brightest star is denoted by the Greek letter α (alpha), the second brightest by β (beta), and so on. For example, Betelgeuse and Rigel, the two brightest stars in the constellation Orion, are also known as α Orionis and β Orionis, respectively (see Figures 0.2 and 0.3). (Precise observations show that Rigel is actually brighter than Betelgeuse, but the names are now permanent.) Because there are many more stars in any given constellation than there are letters in the Greek alphabet, this method is of limited use. However, for naked-eye astronomy, where only bright stars are involved, it is quite satisfactory.

For more precise measurements, astronomers find it helpful to use a system of **celestial coordinates** on the sky. If we think of the stars as being attached to the celestial sphere centered on Earth, then the familiar system of angular measurement on Earth's surface—latitude and longitude (Figure 0.6a)—extends quite naturally to the sky. The celestial analogs of latitude and longitude are called **declination** and **right ascension,** respectively (Figure 0.6b). Just as latitude and longitude are tied to Earth, right ascension and declination are fixed on the celestial sphere. Although the stars appear to move across the sky because of Earth's rotation, their celestial coordinates remain constant over the course of a night.

Declination (dec) is measured in *degrees* (°) north or south of the celestial equator, just as latitude is measured in degrees north or south of Earth's equator (see *More Precisely 0-1*). The celestial equator is at a declination of 0°, the north celestial pole is at +90°, and the south celestial pole is at −90° (the plus sign here just means "north of the celestial equator"; minus means "south"). Right ascension (RA) is measured in angular units called *hours, minutes,* and *seconds,* and it increases in the eastward direction. Like the choice of the Greenwich Meridian as the zero-point of longitude on Earth, the choice of zero right ascension is quite arbitrary—it is conventionally taken to be the position of the Sun in the sky at the instant of the vernal equinox (to be discussed in the next section).

Polaris

The duration of this exposure is about 5 hours, . . .

. . . since each star traces out approximately 20 percent of a circle.

R I V U X G

◀ FIGURE 0.5 **The Northern Sky** INTERACTIVE Time-lapse photograph of the northern sky. Each trail traces the path of a single star across the night sky. The concentric circles are centered near the North Star, Polaris. *(AURA)*

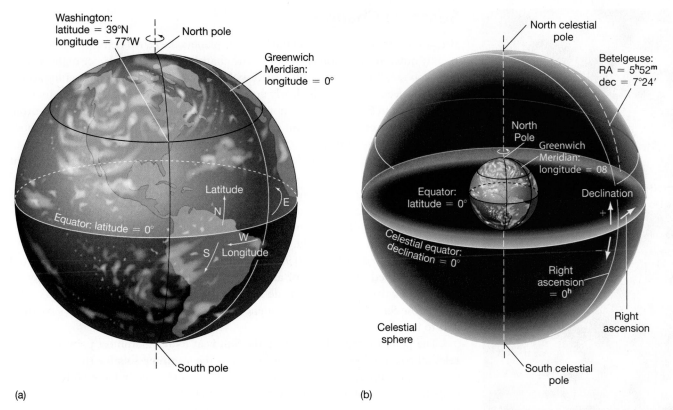

(a)

(b)

▲ FIGURE 0.6 Right Ascension and Declination
(a) Longitude and latitude allow us to locate a point on the surface of Earth by specifying its distance (as an angle) east or west of the Greenwich Meridian, and north or south of the equator. For example, to find Washington, D.C., on Earth, look 77° west of Greenwich and 39° north of the equator.
(b) Similarly, right ascension and declination specify locations on the sky. To locate the star Betelgeuse on the celestial sphere; look 5^{h}52^m east of the vernal equinox (the line on the sky with a right ascension of zero) and 7°24′ north of the celestial equator.

0.2 Earth's Orbital Motion

Day-to-Day Changes

We measure time by the Sun. The rhythm of day and night is central to our lives, so it is not surprising that the period of time from one sunrise (or noon, or sunset) to the next, the 24-hour **solar day,** is our basic social time unit. As we have just seen, this apparent daily progress of the Sun and other stars across the sky, known as **diurnal motion** is a consequence of Earth's rotation. But the stars' positions in the sky do not repeat themselves exactly from one night to the next. Each night, the whole celestial sphere appears shifted a little compared with the night before—you can confirm this for yourself by noting over the course of a week or two which stars are visible near the horizon just after sunset or just before dawn. Because of this shift, a day measured by the stars—called a **sidereal day** after the Latin word *sidus*, meaning "star"—differs in length from a solar day.

The reason for the difference in length between a solar day and a sidereal day is sketched in Figure 0.7. Earth moves in two ways simultaneously: it rotates on its central axis while at the same time **revolving** around the Sun. Each time Earth rotates once on its axis, it also moves a small distance along its orbit. Therefore, each day Earth has to rotate through slightly more than 360° in order for the Sun to return to the same apparent location in the sky. As a result, the interval of time between noon one day and noon the next (a solar day) is slightly greater than the true rotation period (one sidereal day). Our planet takes 365 days to orbit the Sun, so the additional angle is 360°/365 = 0.986°. Because Earth takes about 3.9 minutes to rotate through this angle, the solar day is 3.9 minutes longer than the sidereal day.

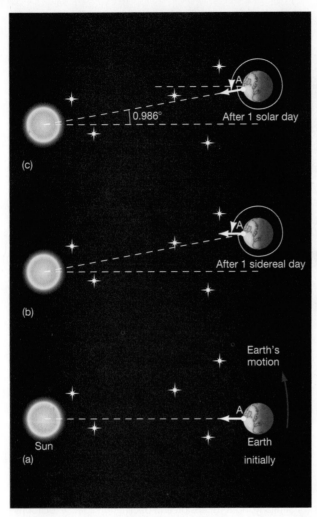

▲ **FIGURE 0.7 Solar and Sidereal Days**   A sidereal day is Earth's true rotation period—the time taken for our planet to return to the same orientation in space relative to the distant stars. A solar day is the time from one noon to the next. The difference in duration between the two is easily explained because Earth revolves around the Sun at the same time it rotates on its axis. Frames (a) and (b) are one sidereal day apart, when Earth rotates exactly once on its axis and also moves a little in its solar orbit—approximately 1°. Consequently, between noon at point A on one day and noon at the same point the next day, Earth actually rotates through about 361° (c), and the solar day exceeds the sidereal day by about 4 minutes. Note that the diagrams are not drawn to scale; the 1° angle is actually much smaller than that shown here.

▶ **FIGURE 0.8 The Zodiac INTERACTIVE**   The night side of Earth faces a different set of constellations at different times of the year. The 12 constellations named here make up the astrological zodiac. The arrows indicate the most prominent zodiacal constellations in the night sky at various times of year. For example, in June, when the Sun is "in" Gemini, Sagittarius and Capricornus are visible at night.

Seasonal Changes

Because Earth revolves around the Sun, our planet's darkened hemisphere faces in a slightly different direction each night. The change is only about 1° per night (Figure 0.7)—too small to be easily discerned with the naked eye from one evening to the next. However, the change is clearly noticeable over the course of weeks and months, as illustrated in Figure 0.8. In 6 months, Earth moves to the opposite side of its orbit, and we face an entirely different group of stars and constellations at night. Because of this motion, the Sun appears, to an observer on Earth, to move slowly (at a rate of 1° per day) relative to the background stars over the course of a year. This apparent motion of the Sun on the sky traces out a path on the celestial sphere known as the **ecliptic.** The 12 constellations through which the Sun passes during the year as it moves along the ecliptic—that is, the constellations we would see looking in the direction of the Sun if they weren't overwhelmed by the Sun's light—had special significance for astrologers of old. They are collectively known as the **zodiac.**

As illustrated in Figure 0.9, the ecliptic forms a great circle on the celestial sphere, inclined at an angle of about 23.5° to the celestial equator. In reality, as shown in Figure 0.10, the plane defined by the ecliptic is *the plane of Earth's orbit around the Sun.* Its tilt is a consequence of the *inclination* of our planet's rotation axis to its orbital plane.

The point on the ecliptic where the Sun is at its northernmost point above the celestial equator (see Figure 0.9) is known as the **summer solstice** (from the Latin words *sol*, meaning "sun," and *stare*, "to stand"). As indicated in Figure 0.10, it represents the point on Earth's orbit where our planet's North Pole is oriented closest to the Sun. This occurs on or near June 21—the exact date varies slightly from year to year because the actual length of a year is not a whole number of days. As Earth rotates on that date, points north of the equator spend the greatest fraction of their time in sunlight, so the summer solstice corresponds to the longest day of the year (that is, the greatest number of daylight hours—Earth's rotation period doesn't change!) in Earth's Northern Hemisphere and the shortest day in Earth's Southern Hemisphere.

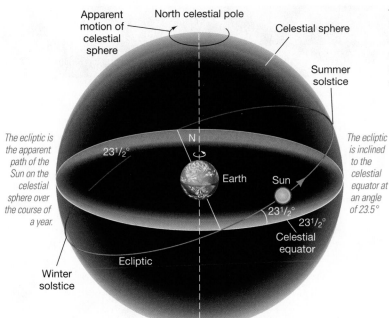

Apparent motion of celestial sphere

North celestial pole

Celestial sphere

Summer solstice

The ecliptic is the apparent path of the Sun on the celestial sphere over the course of a year.

N

23½°

Earth

Sun

23½°

23½°

The ecliptic is inclined to the celestial equator at an angle of 23.5°

Celestial equator

Ecliptic

Winter solstice

◀ **FIGURE 0.9 Ecliptic** The seasons result from the changing height of the Sun above the celestial equator. At the summer solstice, the Sun is at its northernmost point on its path around the ecliptic. It is therefore highest in the sky, as seen from Earth's Northern Hemisphere, and the days are longest. The reverse is true at the winter solstice. At the vernal and autumnal equinoxes, when the Sun crosses the celestial equator, day and night are of equal length.

Six months later, the Sun is at its southernmost point below the celestial equator (Figure 0.9)—or, equivalently, Earth's North Pole is oriented farthest from the Sun (Figure 0.10). We have reached the **winter solstice** (December 21), the shortest day in Earth's Northern Hemisphere and the longest in the Southern Hemisphere.

The tilt of Earth's rotation axis relative to the ecliptic is responsible for the **seasons** we experience—the marked difference in temperature between the hot summer and cold winter months. As illustrated in Figure 0.10, two factors combine to cause this variation. First, there are more hours of daylight during the summer than in winter. To see why this is, look at the yellow lines on the surfaces of the Earths in the figure. (For definiteness, they correspond to a latitude of 45° north—roughly that of the Great Lakes or the south of France.) A much larger fraction of the line is sunlit in the summertime, and more daylight means more solar heating. Second, as illustrated in the insets in Figure 0.10, when the Sun is high in the sky in summer, rays of sunlight striking Earth's surface are more concentrated—spread out over a smaller area—than in winter. As a result, the Sun feels hotter. Therefore summer, when the Sun is highest above the horizon and the days are longest, is generally much warmer than winter, when the Sun is low and the days are short.

A popular misconception is that the seasons have something to do with Earth's distance from the Sun. Figure 0.11 illustrates why this is not the case. It shows

ANIMATION/VIDEO
Summer Solstice

ANIMATION/VIDEO
Winter Solstice

▶ **FIGURE 0.10 Seasons INTERACTIVE** Earth's seasons result from the inclination of our planet's rotation axis with respect to its orbit plane. The summer solstice corresponds to the point on Earth's orbit where our planet's North Pole points most nearly toward the Sun. The opposite is true of the winter solstice. The vernal and autumnal equinoxes correspond to the points in Earth's orbit where our planet's axis is perpendicular to the line joining Earth and the Sun. The insets show how rays of sunlight striking the ground at an angle (during northern winter) are spread over a larger area than rays coming nearly straight down (e.g., during northern summer). As a result, the amount of solar heat delivered to a given area of Earth's surface is greatest when the Sun is high in the sky.

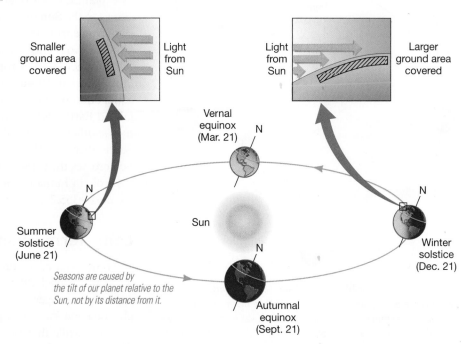

Smaller ground area covered

Light from Sun

Light from Sun

Larger ground area covered

Vernal equinox (Mar. 21)
N

N

Sun

Summer solstice (June 21)

N

Winter solstice (Dec. 21)

Seasons are caused by the tilt of our planet relative to the Sun, not by its distance from it.

Autumnal equinox (Sept. 21)

0.1 MORE PRECISELY

Angular Measure

The size and scale of astronomical objects are often specified by measuring lengths and angles. The concept of length measurement is fairly intuitive. The concept of angular measurement may be less familiar, but it too can become second nature if you remember a few simple facts:

- A full circle contains 360 arc degrees (or 360°). Therefore, the half circle that stretches from horizon to horizon, passing directly overhead and spanning the portion of the sky visible to one person at any one time, contains 180°.
- Each 1° increment can be further subdivided into fractions of an arc degree, called arc minutes; there are 60 arc minutes (60′) in 1 arc degree. Both the Sun and the Moon project an angular size of 30 arc minutes on the sky. Your little finger, held at arm's length, does about the same, covering about a 40-arc minute slice of the 180° horizon-to-horizon arc.
- An arc minute can be divided into 60 arc seconds (60″). Put another way, an arc minute is 1/60 of an arc degree, and an arc second is 1/60 × 1/60 = 1/3600 of an arc degree. An arc second is an extremely small unit of angular measure—it is the angular size of a centimeter-size object (a dime, say) at a distance of about 2 kilometers (a little over a mile).

The above figure illustrates this subdivision of the circle into progressively smaller units.

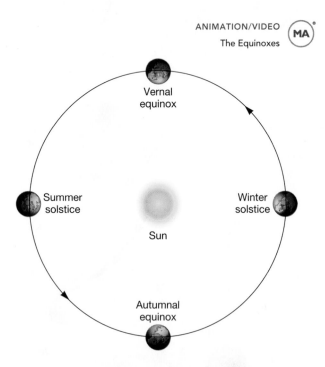

ANIMATION/VIDEO **MA** The Equinoxes

▲ **FIGURE 0.11 Earth's Orbit**   Seen face on, Earth's orbit around the Sun is almost indistinguishable from a perfect circle. The distance from Earth to the Sun varies only slightly over the course of a year and is *not* the cause of the seasonal temperature variations we experience on our planet.

Earth's orbit face on, instead of almost edge-on, as in Figure 0.10. Notice that the orbit is almost perfectly circular, so the distance from Earth to the Sun varies very little (in fact, by only about 3 percent) over the course of a year—not nearly enough to explain the seasonal changes in temperature. What's more, Earth is actually *closest* to the Sun in early January, the dead of winter in the Northern Hemisphere, so distance from the Sun cannot be the main factor controlling our climate.

The two points where the ecliptic intersects the celestial equator (Figure 0.9)—that is, where Earth's rotation axis is perpendicular to the line joining Earth to the Sun (Figure 0.10)—are known as **equinoxes.** On those dates, day and night are of equal duration. (The word *equinox* derives from the Latin for "equal night.") In the fall (in Earth's Northern Hemisphere), as the Sun crosses from the northern into the southern celestial hemisphere, we have the **autumnal equinox** (on September 21). The **vernal equinox** occurs in spring, on or near March 21, as the Sun crosses the celestial equator moving north. The vernal equinox plays an important role in human timekeeping. The interval of time from one vernal equinox to the next—365.242 solar days—is known as one **tropical year.**

Long-Term Changes

The time required for Earth to complete exactly one orbit around the Sun, relative to the stars, is called a **sidereal year.** One sidereal year is 365.256 solar days long, about 20 minutes longer than a tropical year. The reason for this slight difference is a phenomenon known as **precession.** Like a spinning top that rotates rapidly on its own axis while that axis slowly revolves about the vertical, Earth's axis changes its direction over the course of time, although the angle between the axis and a line

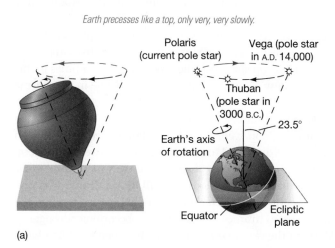

Earth precesses like a top, only very, very slowly.

Polaris (current pole star)

Vega (pole star in A.D. 14,000)

Thuban (pole star in 3000 B.C.)

23.5°

Earth's axis of rotation

Equator

Ecliptic plane

(a)

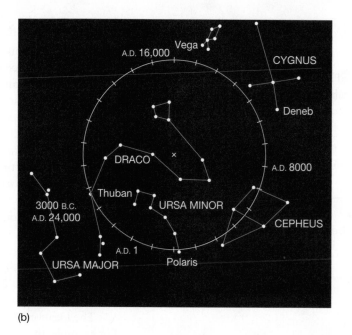

(b)

perpendicular to the plane of the ecliptic always remains close to 23.5°. Figure 0.12 illustrates Earth's precession, which is caused by the combined gravitational pulls of the Moon and the Sun (see Chapter 1). During a complete cycle of precession, taking about 26,000 years, Earth's axis traces out a cone. Because of this slow shift in the orientation of Earth's rotation axis, the vernal equinox, which defines the tropical year, drifts slowly around the ecliptic over the course of the precession cycle. Notice in Figure 0.12(b) that most of the time there is *no* bright "Pole Star" marking due north.

The tropical year is the year our calendars measure. If our timekeeping were tied to the sidereal year, the seasons would slowly march around the calendar as Earth precessed—13,000 years from now, summer in the Northern Hemisphere would be at its height in mid-February! By using the tropical year instead, we ensure that July and August will always be (northern) summer months. However, in 13,000 years' time, Orion, now a prominent feature of the northern winter sky, will be a summer constellation.

▲ **FIGURE 0.12 Precession** **INTERACTIVE** (a) Earth's axis currently points nearly toward the star Polaris. Some 12,000 years from now, Earth's axis will point toward a star called Vega, which will then be the "North Star." Five thousand years ago, the North Star was a star named Thuban in the constellation Draco. (b) The yellow circle shows the precessional path of the north celestial pole among some prominent northern stars and depicts the direction toward which Earth's pole points on the sky. Each tick mark represents 1000 years.

CONCEPT CHECK

Earth is actually farthest from the Sun during northern summer. Why, then, is it hottest in North America during this season?

0.3 The Motion of the Moon

Early astronomers had very practical reasons for studying the sky. Some stars (such as Polaris) served as navigational guides, while others served as primitive calendars to predict planting and harvesting seasons. By observing repeating patterns in the sky and associating them with events on Earth, astronomers began to establish concrete connections between celestial events and everyday life and took the first steps toward true scientific understanding of the heavens. In a real sense, then, human survival depended on astronomical knowledge. The ability to predict and even explain astronomical events was undoubtedly a highly prized, and perhaps jealously guarded, skill.

The Moon also played an important role in ancient astronomy. Calendars and religious observances were often tied to its phases and cycles, and even today the calendars of most of the world's major religions are still based wholly or partly on the lunar orbit. The Moon's regularly changing appearance (as well as its less regular, but much more spectacular, eclipses) was an integral part of the framework within which ancient astronomers sought to understand the universe. We will study the Moon's physical properties in more detail in Chapter 5. Here we continue our inventory of the sky with a brief description of the motion of our nearest neighbor in space.

SELF-GUIDED TUTORIAL **MA**
Phases of the Moon

Lunar Phases

Apart from the Sun, the Moon is by far the brightest object in the sky. Like the Sun, the Moon appears to move relative to the background stars. Unlike the Sun, however, the explanation for this motion is the obvious one—the Moon really does revolve around Earth.

The Moon's appearance undergoes a regular cycle of changes, or **phases,** taking a little more than 29 days to complete. (The word *month* is derived from the word *Moon.*) Figure 0.13(a) illustrates the appearance of the Moon at different times in this monthly cycle. Starting from the **new Moon,** which is all but invisible in the sky, the Moon appears to *wax* (grow) a little each night and is visible as a growing *crescent* (frame 1 of Figure 0.13a). One week after new Moon, half of the lunar disk (the circular face we would see if the Moon were completely illuminated) can be seen (frame 2). This phase is known as a **quarter Moon.** During the next week, the Moon continues to wax, passing through the *gibbous* phase (more than half of the lunar disk visible, frame 3), until 2 weeks after new Moon the **full Moon** (frame 4) is visible. During the next 2 weeks, the Moon *wanes* (shrinks), passing in turn through the gibbous, quarter, and crescent phases (frames 5–7), eventually becoming new again.

The location of the Moon in the sky, as seen from Earth, depends on its phase. For example, the full Moon rises in the east as the Sun sets in the west, while the first quarter Moon actually rises at noon, but often only becomes visible late in the day as the Sun's light fades. By this time the Moon is already high in the sky. These connections between lunar phase and rising/setting times are indicated on Figure 0.13(a).

Unlike the Sun and the other stars, the Moon emits no light of its own. Instead, it shines by reflected sunlight, giving rise to the phases we see. As indicated in Figure 0.13(a), half of the Moon's surface is illuminated by the Sun at any moment, but not all of the Moon's sunlit face can be seen because of the Moon's position with respect to Earth and the Sun. When the Moon is full, we see the entire "day lit" face because the Sun and the Moon are in opposite directions from Earth in the sky. The Sun's light is not blocked by Earth at the full phase because, as shown in Figure 0.13(b), the Moon's orbit is inclined at a small angle (5.2°) to the plane of the ecliptic, so the alignment of the three bodies is not perfect. (The sizes of Earth and the Moon are greatly exaggerated in these figures.) In the case of a new Moon, the Moon and the Sun are in nearly the same part of the sky, and the sunlit side of the Moon is oriented away from us—at new Moon the Sun is almost behind the Moon, from our perspective.

Notice, by the way, that the Moon always keeps the same face toward Earth—as indicated on the figure, it rotates on its axis in exactly the same time it takes to orbit Earth. This is called *synchronous rotation.* We will discuss the reason for it in Chapter 5.

As it revolves around Earth, the Moon's position in the sky changes with respect to the stars. In one **sidereal month** (27.3 days), the Moon completes one revolution and returns to its starting point on the celestial sphere, having traced out a great circle in the sky. The time required for the Moon to complete a full cycle of phases, one **synodic month,** is a little longer—about 29.5 days. The synodic month is a little longer than the sidereal month for the same basic reason that a solar day is slightly longer than a sidereal day (Figure 0.7): Because of Earth's motion around the Sun, the Moon must complete slightly more than one full revolution to return to the same phase in its orbit.

Eclipses

From time to time—but only at new or full Moon—the Sun, Earth, and the Moon line up precisely, and we observe the spectacular phenomenon known as an **eclipse.** When the Sun and the Moon are in exactly *opposite* directions as seen from Earth, Earth's shadow sweeps across the Moon, temporarily blocking the Sun's light and darkening the Moon in a **lunar eclipse,** as illustrated in

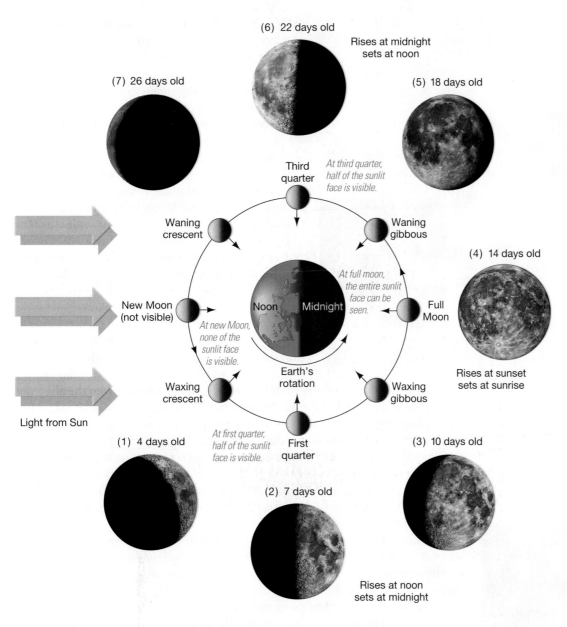

(a) Top-down view

◀ **FIGURE 0.13 Lunar Phases**   (a) Because the Moon orbits Earth, this top-down view shows how the visible fraction of the lunar sunlit face varies from night to night, although the Moon always keeps the same face toward our planet. (Note the location of the small, straight arrows that mark the same point on the lunar surface at each phase shown.) The complete cycle of lunar phases takes about 29 days to complete. Rising and setting times for some phases are also indicated. (b) A side view shows how the Moon's orbit is inclined at about 5° to the ecliptic, so not all orbital configurations produce an eclipse. *(Insets: UC/Lick Observatory)*

(b) Side view

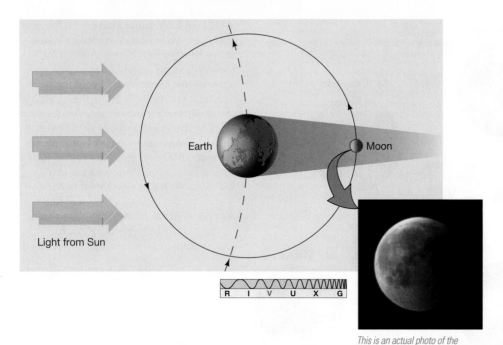

▲ **FIGURE 0.14 Lunar Eclipse INTERACTIVE** When the Moon passes through Earth's shadow, we see a darkened, copper-colored Moon, as shown by the partial eclipse in the inset photograph. The red coloration is caused by sunlight deflected by Earth's atmosphere onto the Moon's surface. *(Inset: G. Schneider)*

This is an actual photo of the eclipsed Moon, one of the great light shows visible to the naked eye.

ANIMATION/VIDEO (MA)

Solar Eclipse in Indiana

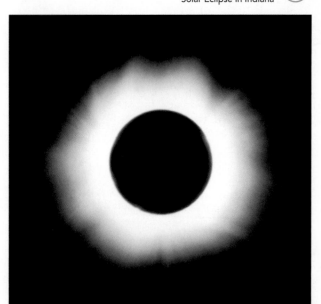

▲ **FIGURE 0.15 Solar Eclipse** During a total solar eclipse, the Sun's corona becomes visible as an irregularly shaped halo surrounding the blotted-out disk of the Sun. This was the August 1999 eclipse, as seen from the banks of the Danube River near Sofia, Bulgaria. *(Bencho Angelov)*

Figure 0.14. From Earth we see the curved edge of Earth's shadow begin to cut across the face of the full Moon and slowly "eat" its way into the circular lunar disk.

Usually the alignment of the Sun, Earth, and the Moon is imperfect, so the shadow never completely covers the Moon. Such an occurrence is known as a **partial eclipse.** Occasionally, however, the entire lunar surface is obscured in a **total eclipse** (such as that shown in the inset in Figure 0.14). Total lunar eclipses last only as long as is needed for the Moon to pass through Earth's shadow—no more than about 100 minutes. During that time, the Moon often acquires an eerie, deep red coloration, the result of a small amount of sunlight being refracted (bent) by Earth's atmosphere onto the lunar surface, preventing the shadow from being completely black.

When the Moon and the Sun are in exactly the *same* direction as seen from Earth, an even more awe-inspiring event occurs. The Moon passes directly in front of the Sun, briefly turning day into night in a **solar eclipse** (Figure 0.15). In a *total solar eclipse,* when the alignment is perfect, planets and some stars become visible in the daytime as the Sun's light is blocked. By pure chance the Sun and Moon have almost exactly the same angular size as seen from Earth—the Sun is much bigger than the Moon, but it also lies much farther away (see *More Precisely 0-1*). Consequently, during a total solar eclipse, we can often see the Sun's ghostly outer atmosphere, or *corona,* its faint glow becoming temporarily visible when the rest of the Sun's glare is obscured. In a *partial solar eclipse,* the Moon's path is slightly off center, and only a portion of the Sun's face is covered.

Unlike a lunar eclipse, which is simultaneously visible from all locations on Earth's night side, a total solar eclipse can be seen from only a small portion of the daytime side. The Moon's shadow on Earth's surface is about 7000 kilometers wide—roughly twice the diameter of the Moon. Outside that shadow, no eclipse is seen. However, only within the central region of the shadow, in the **umbra,** is the eclipse total. The umbra is the part of a shadow (in this case, the Moon's) where all light from the source is blocked. Within the shadow but outside the umbra, in the **penumbra,** some but not all of the Sun's light is blocked and the eclipse is partial, with less and less of the Sun being obscured the farther one travels from the shadow's center.

The connections between the umbra, the penumbra, and the relative locations of Earth, Sun, and Moon are illustrated in Figure 0.16. The umbra is always very small—even under the most favorable circumstances, its diameter never exceeds 270 kilometers. Because the Moon's shadow sweeps across Earth's surface at a speed of more than 1700 kilometers per hour, the duration of a total solar eclipse at any given point can never exceed 7.5 minutes (270 km divided by 1700 km/hour, times 60 minutes per hour).

The Moon's orbit around Earth is not exactly circular. As a result, the Moon may be far enough from Earth at the moment of an eclipse that its disk fails to cover the disk of the Sun completely, even though their centers coincide. In that case, there is no *region of totality*—the umbra never reaches Earth at all, and a thin ring of sunlight can still be seen surrounding the Moon. Such an occurrence, called an **annular eclipse** (from the word *annulus,* meaning "ring"), is depicted at the bottom of Figure 0.16. Roughly half of all solar eclipses are annular.

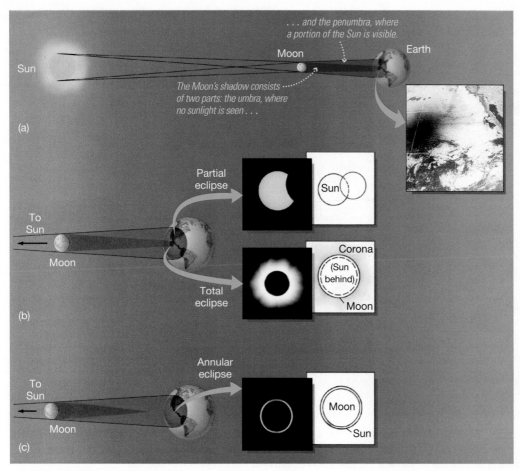

... and the penumbra, where a portion of the Sun is visible.

Sun

Moon

Earth

The Moon's shadow consists of two parts: the umbra, where no sunlight is seen ...

(a)

To Sun

Moon

Partial eclipse

Sun

Total eclipse

Corona (Sun behind)

Moon

(b)

To Sun

Moon

Annular eclipse

Moon

Sun

(c)

◄ **FIGURE 0.16 Solar Eclipse Types** INTERACTIVE (a) The Moon's shadow consists of two parts: the umbra, where no sunlight is seen, and the penumbra, where a portion of the Sun is visible. (b) Situated in the umbra, we see a total eclipse; in the penumbra, we see a partial eclipse. (c) If the Moon is too far from Earth at the moment of the eclipse, the umbra does not reach Earth and there is no region of totality; instead, an annular eclipse is seen. (Note that these figures are not drawn to scale.) *(Insets: NOAA; G. Schneider)*

CONCEPT CHECK

If Earth's distance from the Sun were to double, would you still expect to see total solar eclipses? What if the distance became half its present value?

Because the Moon's orbit is slightly inclined to the plane of the ecliptic (Figure 0.13b), we don't see a lunar eclipse at every full Moon. The Moon is usually above or below the ecliptic when a full Moon occurs and, hence, is untouched by Earth's shadow. Similarly, most new Moons do not result in solar eclipses. The chance that a new (or full) Moon will occur just as the Moon happens to cross the ecliptic plane (so Earth, Moon, and Sun are perfectly aligned, as illustrated in Figure 0.17) is quite low. As a result, eclipses are relatively infrequent events. On average, there

▼ **FIGURE 0.17 Eclipse Geometry**   An eclipse occurs when Earth, Moon, and Sun are precisely aligned. If the Moon's orbital plane lay in exactly the plane of the ecliptic, this alignment would occur once a month. However, the Moon's orbit is inclined at about 5° to the ecliptic, so not all configurations are actually favorable for producing an eclipse. For an eclipse to occur, the line of intersection of the two planes must lie along the Earth–Sun line. For clarity, only the umbra of each shadow is shown (see Figure 0.16).

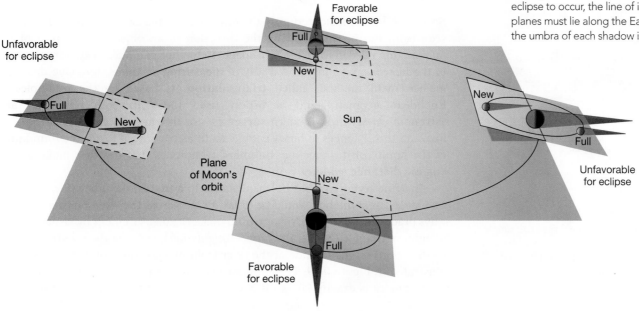

Favorable for eclipse

Full

New

Unfavorable for eclipse

Full

New

New

Sun

New

Full

Unfavorable for eclipse

Plane of Moon's orbit

New

Full

Favorable for eclipse

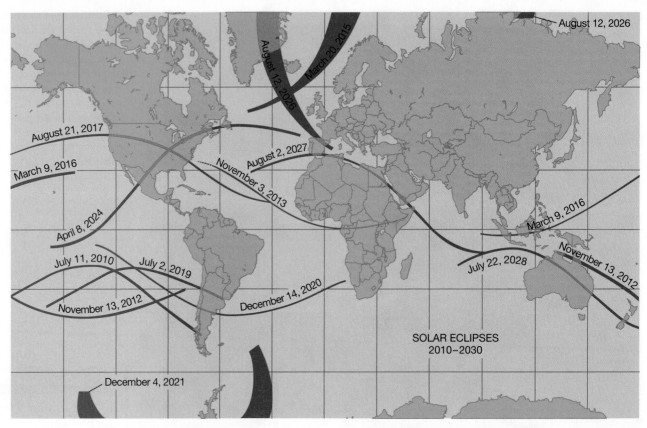

▲ **FIGURE 0.18 Eclipse Tracks**   Regions of Earth that saw or will see total solar eclipses between the years 2009 and 2030. Each track represents the path of the Moon's umbra across Earth's surface during an eclipse. High-latitude tracks are broader because sunlight strikes Earth's surface at an oblique angle near the poles and because of the projection of the map.

▲ **FIGURE 0.19 Triangulation**   Surveyors often use simple geometry to estimate the distance to a faraway object by triangulation. By measuring the angles at A and B and the length of the baseline, the distance can be calculated without the need for direct measurement (or getting wet!).

are 7 total lunar eclipses and 15 total or annular solar eclipses each decade. Because we know the orbits of Earth and the Moon to great accuracy, we can predict eclipses far into the future. Figure 0.18 shows the location and duration of all total and annular eclipses of the Sun between 2010 and 2030.

0.4 The Measurement of Distance

So far we have considered only the *directions* to the Sun, Moon, and stars, as seen from Earth. But knowing the direction to an object is only part of the information needed to locate it in space. Before we can make a systematic study of the heavens, we must find a way of measuring *distances,* too. One distance-measurement method, called **triangulation,** is based on the principles of Euclidean geometry and is widely used today in both terrestrial and astronomical settings. Modern surveyors use this age-old geometrical idea to measure the distances to faraway objects. In astronomy, it forms the foundation of the family of distance-measurement techniques that together make up the **cosmic distance scale.**

Imagine trying to measure the distance to a tree on the other side of a river. The most direct method would be to lay a tape across the river, but that's not always practical. A smart surveyor would make the measurement by visualizing an imaginary triangle (hence, the term *triangulation*), sighting the tree (measuring its direction) on the far side of the river from two positions on the near side, as illustrated in Figure 0.19. The simplest triangle is a right triangle, in which one of the angles is exactly 90°, so it is often convenient to set up one observation position directly opposite the object, as at point A, although this isn't necessary for the method to work. The surveyor then moves to another observation position at point B, noting the distance covered between A and B. This distance is called the **baseline** of the imaginary triangle. Finally, the surveyor, standing at point B, sights toward the tree and notes the angle formed at point B by the intersection of this sight line and the baseline. No further observations are required. Knowing the length of one side (AB) and two angles (the right angle at A and the angle at B) of the triangle, and using elementary trigonometry, the surveyor can construct the remaining sides and angles and so establish the distance from A to the tree.

Obviously, for a fixed baseline, the triangle becomes longer and narrower as the tree's distance from A increases. Narrow triangles cause problems because it is hard to measure the angles at A and B with sufficient accuracy. A surveyor on Earth can "fatten" the triangle by lengthening the baseline, but in astronomy there are limits on how long a baseline we can choose. For example, consider an imaginary triangle extending from Earth to a nearby object in space, perhaps a neighboring planet. The triangle is now extremely long and narrow, even for a relatively nearby object (by cosmic standards). Figure 0.20(a) illustrates a case in which the longest baseline possible on Earth—Earth's diameter, measured from point A to point B—is used. In principle, we could sight the planet from opposite sides of Earth, measuring the angles at A and B. However, in practice it is easier to measure the third angle of the imaginary triangle. Here's how.

The observers each sight toward the planet, taking note of its position *relative to some distant stars* seen on the plane of the sky. The observer at A sees the planet at apparent location A′ (pronounced "A prime") relative to those stars, as indicated in Figure 0.20(a). The observer at B sees the planet at location B′. If each observer takes a photograph of the same region of the sky, the planet will appear at slightly different places in the two images, as shown in Figure 0.20(b). (The positions of the background stars appear unchanged because of their much greater distance from the observer.) This apparent shift of a foreground object relative to the background as the observer's location changes is known as **parallax.** The size of the shift in Figure 0.20(b), measured as an angle on the celestial sphere, is equal to the third angle of the imaginary triangle in Figure 0.20(a).

The closer an object is to the observer, the larger the parallax. To see this for yourself, hold a pencil vertically just in front of your nose (see Figure 0.21). Look at some far-off object—a distant wall, say. Close one eye, then open it while closing the other. You should see a large shift in the apparent position of the pencil relative to the wall—a large parallax. In this example, one eye corresponds to point A in Figure 0.20, the other eye to point B. The distance between your eyeballs is the baseline, the pencil represents the planet, and the distant wall the remote field of stars. Now hold the pencil at arm's length, corresponding to a more distant object (but still not as far away as the distant stars). The apparent shift of the pencil will be smaller. By moving the pencil farther away, you are narrowing the triangle and decreasing the parallax. If you were to paste the pencil to the wall, corresponding to the case where the object of interest is as far away as the background star field, blinking would produce no apparent shift of the pencil at all.

The amount of parallax is inversely proportional to an object's distance. Small parallax implies large distance. Conversely, large parallax implies small distance. Knowing the amount of parallax (as an angle) and the length of the baseline, we can easily derive the distance through triangulation.

Surveyors of the land routinely use these simple geometric techniques to map out planet Earth (*Discovery 0-1* presents an early example). As surveyors of the sky, astronomers use the same basic principles to chart the universe.

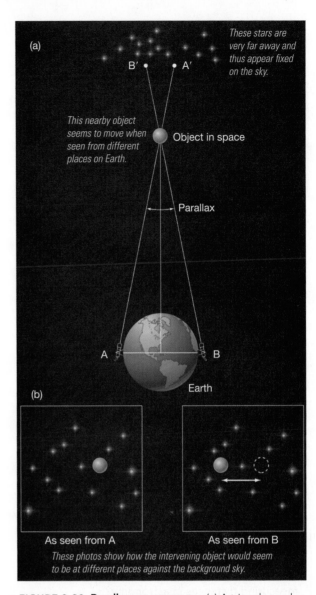

▲ **FIGURE 0.20 Parallax INTERACTIVE**   (a) A triangle can be imagined to extend from Earth to a nearby object in space. The group of stars at the top represents a background field of very distant stars. (b) Hypothetical photographs of the same star field showing the nearby object's apparent shift, relative to the distant unshifted stars.

Try this experiment: Move a pencil to and fro away from your nose while blinking your eyes.

Parallax for nearby pencil

Parallax for distant pencil

Baseline

SELF-GUIDED TUTORIAL **MA**
Stellar Parallax

CONCEPT CHECK

Why is elementary geometry essential for measuring distances in astronomy?

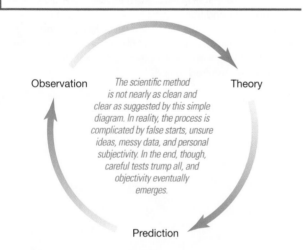

Observation

The scientific method is not nearly as clean and clear as suggested by this simple diagram. In reality, the process is complicated by false starts, unsure ideas, messy data, and personal subjectivity. In the end, though, careful tests trump all, and objectivity eventually emerges.

Theory

Prediction

▲ **FIGURE 0.22 Scientific Method** Scientific theories evolve through a combination of observation, theory, and prediction, which in turn suggests new observations. The process can begin at any point in the cycle, and it continues forever—or until the theory fails to explain an observation or makes a demonstrably false prediction.

◄ **FIGURE 0.21 Parallax Geometry** Parallax is inversely proportional to an object's distance. An object near your nose has a much larger parallax than an object held at arm's length.

0.5 Science and the Scientific Method

Science is a step-by-step process for investigating the physical world, based on natural laws and observed phenomena. However, the scientific facts just presented did not come easily or quickly. Progress in science is often slow and intermittent and may require a great deal of patience before significant progress is made. The earliest known descriptions of the universe were based largely on imagination and mythology and made little attempt to explain the workings of the heavens in terms of testable earthly experience. However, history shows that some early scientists did come to realize the importance of careful observation and testing to the formulation of their ideas. The success of their approach changed, slowly but surely, the way science was done and opened the door to a fuller understanding of nature. Experimentation and observation became central parts of the process of inquiry.

Theories and Models

To be effective, a **theory**—the framework of ideas and assumptions used to explain some set of observations and make predictions about the real world—must be continually tested. Scientists accomplish this by using a theory to construct a **theoretical model** of a physical object (such as a planet or a star) or phenomenon (such as gravity or light), accounting for its known properties. The model then makes further predictions about the object's properties or perhaps how it might behave or change under new circumstances. If experiments and observations favor those predictions, the theory can be further developed and refined. If they do not, the theory must be reformulated or rejected, no matter how appealing it originally seemed. The process is illustrated schematically in Figure 0.22. This approach to investigation, combining thinking and doing—that is, theory and experiment—is known as the **scientific method.** It lies at the heart of modern science, separating science from pseudoscience, fact from fiction.

Notice that there is no end point to the process depicted in Figure 0.22. A theory can be invalidated by a single wrong prediction, but no amount of observation or experimentation can ever prove it "correct." Theories simply become more and more widely accepted as their predictions are repeatedly confirmed. The process can fail at any point in the cycle. If a theory cannot explain an experimental result or observation, or if its predictions are demonstrated to be untrue, it must be discarded or amended. And if it makes no predictions at all, then it has no scientific value.

Scientific theories share several important defining characteristics:

■ They must be *testable*; that is, they must admit the possibility that both their underlying assumptions and their predictions can be exposed to experimental verification. This feature separates science from, for example, religion, since ultimately divine revelations or scriptures cannot be challenged within a religious framework: We can't design an experiment to "understand the mind of God." Testability also distinguishes science from a pseudoscience such as astrology, whose underlying assumptions and predictions have been repeatedly tested and never verified, with no apparent impact on the views of those who continue to believe in it!

■ They must continually be *tested*, and their consequences tested, too. This is the basic circle of scientific progress depicted in Figure 0.22.

■ They should be *simple*. This is a practical outcome of centuries of scientific experience—the most successful theories tend to be the simplest ones that fit the facts. This view is often encapsulated in a principle known as *Occam's razor* (after the 14th-century English philosopher William of Ockham):

0.1 DISCOVERY

Sizing Up Planet Earth

In about 200 B.C. a Greek philosopher named Eratosthenes (276–194 B.C.) used simple geometric reasoning to calculate the size of our planet. He knew that at noon on the first day of summer, observers in the city of Syene (now called Aswan), in Egypt, saw the Sun pass directly overhead. This was evident from the fact that vertical objects cast no shadows and sunlight reached to the very bottoms of deep wells, as shown in the insets in the accompanying figure. However, at noon of the same day in Alexandria, a city 5000 *stadia* to the north, the Sun was seen to be displaced slightly from the vertical. The *stadium* was a Greek unit of length, roughly equal to 0.16 km—the modern town of Aswan lies about 780 (5000 × 0.16) km south of Alexandria. Using the simple technique of measuring the length of the shadow of a vertical stick and applying elementary geometry, Eratosthenes determined the angular displacement of the Sun from the vertical at Alexandria to be 7.2°.

What could have caused this discrepancy between the two measurements? As illustrated in the figure, the explanation is simply that Earth's surface is not flat, but *curved*. Our planet is a sphere. Eratosthenes was not the first to realize that Earth is spherical—the philosopher Aristotle had done that over 100 years earlier (Section 0.5)—but he was apparently the first to build on this knowledge, combining geometry with direct measurement to infer our planet's size. Here's how he did it.

Rays of light reaching Earth from a very distant object, such as the Sun, travel almost parallel to one another. Consequently, as shown in the figure, the angle measured at Alexandria between the Sun's rays and the vertical (that is, the line joining Alexandria to the center of Earth) is equal to the angle between Syene and Alexandria, as seen from Earth's center. (For the sake of clarity, this angle has been exaggerated in the drawing.) The size of this angle in turn is proportional to the fraction of Earth's circumference that lies between Syene and Alexandria:

$$\frac{7.2°}{360°} = \frac{5000 \text{ stadia}}{\text{Earth's circumference}}.$$

Earth's circumference is therefore 50 × 5000 or 250,000 stadia, or about 40,000 km. Earth's radius is therefore $250{,}000/2\pi$ stadia, or 6366 km. The correct values for Earth's circumference and radius, now measured accurately by orbiting spacecraft, are 40,070 km and 6378 km, respectively.

Eratosthenes' reasoning was a remarkable accomplishment. More than 20 centuries ago he estimated the circumference of Earth to within 1 percent accuracy, using only simple geometry. A person making measurements on only a small portion of Earth's surface was able to compute the size of the entire planet on the basis of observation and pure logic—an early triumph of scientific reasoning.

If two competing theories both explain the facts and make the same predictions, then the simpler one is better. A good theory should contain no more complexity than is absolutely necessary.

- Finally, most scientists have the additional prejudice that a theory should in some sense be *elegant*. When a clearly stated simple principle naturally explains several phenomena previously thought to be distinct, this is widely regarded as a strong point in favor of the new theory.

You may find it instructive to apply these criteria to the many physical theories—some old and well established, others much more recent and still developing—we will encounter throughout the text.

The notion that theories must be tested and may be proven wrong sometimes leads people to minimize their importance. We have all heard the expression, "Of course, it's only a theory," used to deride or dismiss an idea that someone finds unacceptable. Don't be fooled! Gravity (see Section 1.4) is "only" a theory, but calculations based on it have guided human spacecraft throughout the solar system. Electromagnetism and quantum mechanics (Chapter 2) are theories, too, yet they form the foundation for most of 20th- (and 21st-) century technology. Facts about the universe are a dime a dozen. Theories are the intellectual "glue" that combine seemingly unrelated facts into a coherent and interconnected whole.

An Early Application

The birth of modern science is usually associated with the Renaissance, the historical period from the late 14th to the mid-17th century that saw a rebirth (*renaissance* in French) of artistic, literary, and scientific inquiry in European culture following the chaos of the Dark Ages. However, one of the first documented uses of the scientific method in an astronomical context was by Aristotle (384–322 B.C.) some 17 centuries earlier. Aristotle is not normally remembered as a strong proponent of this approach—many of his best-known ideas were based on pure thought, with no attempt at experimental test or verification. Still, his brilliance extended into many areas now thought of as modern science. He noted that, during a lunar eclipse (Section 0.3), Earth casts a curved shadow onto the surface of the Moon. Figure 0.23 shows a series of photographs taken during a recent lunar eclipse. Earth's shadow, projected onto the Moon's surface, is indeed slightly curved (as indicated by the dashed line). This is what Aristotle must have seen and recorded so long ago.

Because the observed shadow seemed always to be an arc of the same circle, Aristotle theorized that Earth, the cause of the shadow, must be round. Don't underestimate the scope of this apparently simple statement. Aristotle also had to reason that the dark region was indeed a shadow and that Earth was its cause—facts we regard as obvious today, but far from clear 25 centuries ago. On the basis of this *hypothesis*—one possible explanation of the observed facts—he then predicted that any and all future lunar eclipses would show Earth's shadow to be curved, regardless of our planet's orientation. That prediction has been tested every time a lunar eclipse has occurred. It has yet to be proved wrong. Aristotle was not the first person to argue that Earth is round, but he was apparently the first to offer observational proof using the lunar eclipse method.

Today, scientists worldwide use an approach that relies heavily on testing ideas. They gather data, form a working hypothesis that explains the data, and then explore the implications of the hypothesis using experiment and observation. Eventually, one or more "well-tested" hypotheses may be elevated to the stature of a physical law and come to form the basis of a theory of even broader applicability. The new predictions of the theory will in turn be tested, as scientific knowledge continues to grow.

Don't think that the scientific method is perfect. We will see many examples in this book of good scientists making mistakes by following incorrect lines of reasoning, or simply placing too much faith in faulty observations. Nevertheless, used properly over a period of time, this rational, methodical approach enables us to arrive at conclusions that are mostly free of the personal biases and human failings of any one scientist. The scientific method is designed to yield—eventually—an objective view of the universe we inhabit.

The Universe Today

Our conception of the cosmos has changed a lot since ancient times. The modern universe is much larger, far more complex, and infinitely stranger than anything early astronomers ever imagined. In sharp contrast to the predictable and orderly heavens of the ancients, the universe we inhabit today is dynamic, expanding, evolving, and yet (we must admit) apparently dominated by fundamental forces that still lie beyond our understanding. Despite this, scientific inquiry today is guided by the same fundamental principles that led our ancestors to uncover the

R　I　V　U　X　G

▲ **FIGURE 0.23　A Lunar Eclipse**　These photographs show Earth's shadow sweeping across the Moon during an eclipse. Aristotle reasoned that Earth was the cause of the shadow and concluded that Earth must be round. (*G. Schneider*)

basic workings of the universe—gravity, light, relativity, quantum physics, and the Big Bang that brought our cosmos into being.

Experiment and observation are integral parts of the process of modern science. Untestable theories, or theories unsupported by experimental evidence, rarely gain any measure of acceptance in scientific circles. Observation, theory, and testing are cornerstones of the scientific method, a technique whose power will be demonstrated again and again throughout our text.

THE BIG QUESTION

Take another look at the spectacular photo at the start of this chapter (p. 4). Think about all those stars—about 100 billion in our Galaxy alone. We cannot help but wonder: Are there planets around some of those stars and perhaps intelligent beings on some of those planets? This grandest of all unsolved questions about the universe now lies at the heart of modern astronomy.

CHAPTER REVIEW

SUMMARY

LO1 The **universe** (p. 6) is the totality of all space, time, matter, and energy. **Astronomy** (p. 6) is the study of the universe. In order of increasing size, the basic constituents of the cosmos are **planets** (p. 6), **stars** (p. 6), **galaxies** (p. 6), and the universe itself. They differ enormously in scale—a factor of a billion billion from planet Earth to the entire observable universe.

LO2 Early observers grouped the stars visible to the naked eye into patterns called **constellations** (p. 6), which they imagined were attached to a vast **celestial sphere** (p. 7) centered on Earth. Constellations have no physical significance, but are still used to label regions of the sky. **Celestial coordinates** (p. 8) are a more precise way of specifying a star's location on the celestial sphere.

LO3 The nightly motion of the stars across the sky is due to Earth's **rotation** (p. 7) on its axis. Because of Earth's **revolution** (p. 9) around the Sun, we see different stars at night at

different times of the year. The Sun's apparent yearly path around the celestial sphere (or the plane of Earth's orbit around the Sun) is called the **ecliptic** (p. 10). We experience **seasons** (p. 11) because Earth's rotation axis is inclined to the ecliptic plane. At the **summer solstice** (p. 10), the Sun is highest in the sky, and the length of the day is greatest. At the **winter solstice** (p. 11), the Sun is lowest and the day is shortest. Because of **precession** (p. 12), the orientation of Earth's axis changes slowly over thousands of years.

LO4 As the Moon orbits Earth, it keeps one face permanently turned toward our planet. We see lunar **phases** (p. 14) as the fraction of the Moon's sunlit face visible to us varies. A **lunar eclipse** (p. 14) occurs when the Moon enters Earth's shadow. A **solar eclipse** (p. 16) occurs when the Moon passes between Earth and the Sun. An eclipse may be **total** (p. 16) if the body in question (Moon or Sun) is completely obscured, or **partial** (p. 16) if only part of the surface is affected. If the Moon is too far from Earth for its disk to completely hide the Sun, an **annular eclipse** (p. 16) occurs. Because the Moon's orbit around Earth is slightly inclined to the ecliptic, solar and lunar eclipses are relatively rare events.

LO5 Astronomers use **triangulation** (p. 18) to measure the distances to planets and stars, forming the foundation of the **cosmic distance scale** (p. 18), the family of distance-measurement techniques used to chart the universe. **Parallax** (p. 19) is the apparent motion of a foreground object relative to a distant background as the observer's position changes. The larger the **baseline** (p. 19)—the distance between the two observation points—the greater the parallax.

LO6 **Science** (p. 20) is a step-by-step process for investigating the physical world. The **scientific method** (p. 20) is a methodical approach employed by scientists to explore the universe around us in an objective manner. A **theory** (p. 20) is a framework of ideas and assumptions used to explain some set of observations and make predictions about the real world. These predictions in turn are amenable to further observational testing. In this way, the theory expands and science advances.

MasteringAstronomy®   For instructor-assigned homework go to www.masteringastronomy.com

Problems labeled **POS** explore the process of science. **VIS** problems focus on reading and interpreting visual information. **LO** connects to the introduction's numbered Learning Outcomes.

REVIEW AND DISCUSSION

1. **LO1** Compare the size of Earth with that of the Sun, the Milky Way Galaxy, and the entire universe.

2. **LO2** What is a constellation? Why are constellations useful for mapping the sky?

3. Why does the Sun rise in the east and set in the west each day? Does the Moon also rise in the east and set in the west? Why? Do stars do the same? Why?

4. How and why does a day measured by the Sun differ from a day measured by the stars?

5. How many times in your life have you orbited the Sun?

6. Why do we see different stars at different times of the year?

7. **LO3** Why are there seasons on Earth?

8. What is precession, and what is its cause?

9. **LO4** If one complete hemisphere of the Moon is always lit by the sun, why do we see different phases of the Moon?

10. What causes a lunar eclipse? A solar eclipse? Why aren't there lunar and solar eclipses every month?

11. **POS** Do you think an observer on another planet might see eclipses? Why or why not?

12. What is parallax? Give an everyday example.

13. Why is it necessary to have a long baseline when using triangulation to measure the distances to objects in space?

14. **LO5** What two pieces of information are needed to determine the diameter of a faraway object?

15. **LO6 VIS** What is the scientific method? In what ways does science differ from religion?

CONCEPTUAL SELF-TEST: TRUE OR FALSE?/MULTIPLE CHOICE

1. Our Galaxy is about 1 million times larger than Earth. (T/F)

2. The stars in a constellation are physically close to one another. (T/F)

3. The solar day is longer than the sidereal day. (T/F)

4. The seasons are caused by the precession of Earth's axis. (T/F)

5. A lunar eclipse can occur only during the full phase. (T/F)

6. The angular diameter of an object is inversely proportional to its distance from the observer. (T/F)

7. If we know the distance of an object from Earth, we can determine the object's size by measuring its parallax. (T/F)

8. If Earth rotated twice as fast as it currently does, but its motion around the Sun stayed the same, then (a) the night would be twice as long; (b) the night would be half as long; (c) the year would be half as long; (d) the length of the day would be unchanged.

9. A long, thin cloud that stretched from directly overhead to the western horizon would have an angular size of (a) 45°; (b) 90°; (c) 180°; (d) 360°.

10. When a thin crescent of the Moon is visible just before sunrise, the Moon is in its (a) waxing phase; (b) new phase; (c) waning phase; (d) quarter phase.

11. If the Moon's orbit were a little larger, solar eclipses would be (a) more likely to be annular; (b) more likely to be total; (c) more frequent; (d) unchanged in appearance.

12. If the Moon orbited Earth twice as fast, but in the same orbit, the frequency of solar eclipses would (a) double; (b) be cut in half; (c) stay the same.

13. **VIS** According to Figure 0.8 (The Zodiac), in January the Sun is in the constellation (a) Cancer; (b) Gemini; (c) Leo; (d) Aquarius.

14. **VIS** In Figure 0.19 (Triangulation), using a longer baseline would result in (a) a less accurate distance to the tree; (b) a more accurate distance to the tree; (c) a smaller angle at point B; (d) a greater distance across the river.

15. **VIS** In Figure 0.20 (Parallax), a smaller Earth would result in (a) a smaller parallax angle; (b) a shorter distance measured to the object; (c) a larger apparent displacement; (d) stars appearing closer together.

PROBLEMS

The number of squares preceding each problem indicates its approximate level of difficulty.

1. ■ The vernal equinox is now just entering the constellation Aquarius. In what constellation will it lie in the year A.D. 10,000?

2. ■ Given that Earth orbits at 150,000,000 km from the Sun, through what distance does Earth move in a second? An hour? A day?

3. ■■■ How, and by roughly how much, would the length of the solar day change if Earth's rotation were suddenly to reverse direction?

4. ■■ How long does it take for the Moon to move a distance equal to its own diameter of angular 0.5°?

5. ■■ Given that the distance to the Moon is 384,000 km, and taking the Moon's orbit around Earth to be circular, estimate the speed (in kilometers per second) at which the Moon orbits Earth.

6. ■■■ Use reasoning similar to that illustrated in Figure 0.7 to verify that the length of the synodic month (the time from one full Moon to the next; Section 0.3) is 29.5 days.

7. ■ The baseline in Figure 0.19 is 100 m and the angle at B is 60°. By constructing the triangle on a piece of graph paper, determine the distance from A to the tree.

8. ■■ Use reasoning similar to that in *Discovery 0-1* (but now using a circle centered on the object and containing the baseline) to determine the distance to an object if its parallax, as measured from either end of a 1000-km baseline, is (a) 1°; (b) 1′; (c) 1″.

9. ■ What would the measured angle in *Discovery 0-1* have been if Earth's circumference were 100,000 km instead of 40,000 km?

10. ■ What angle would Eratosthenes have measured (see *Discovery 0-1*) had Earth been flat?

ACTIVITIES

Collaborative

1. On a clear night, sketch a 10°-wide patch of the sky containing the Moon, with the Moon initially toward the west side of the patch. (See Individual Activity 3 below for how to estimate angles on the sky.) Repeat the observation of the same collection of stars every hour over the course of a night (take turns!). The Moon's position relative to the stars will change noticeably even in a few hours. What is the Moon's angular speed, in degrees per hour? Now observe the Moon at the *same* time each night over the course of a month. Sketch its appearance and note its position on the sky each night. Can you interpret its changing phase in terms of the relative positions of Earth, the Sun, and the Moon? (See Figure 0.13.)

2. Consider Figure 0.18, which shows solar eclipse paths on a world map. Write a description of which two eclipses you would most like to observe together and where and when you would go to observe them. Explain why you chose the dates and sites you did.

Individual

1. Find the star Polaris, also known as the North Star, in the evening sky. Identify any separate pattern of stars in the same general vicinity of the sky. Wait several hours, at least until after midnight, and then locate Polaris again. Has Polaris moved? What has happened to the nearby pattern of stars? Why?

2. Consider the curved star trails shown in Figure 0.5, a time-lapse photograph of the northern sky. They are arcs of circles centered on Polaris. What was the exposure time used for the photograph? How long would you need to take a similar picture from (a) the planet Mercury? (b) Jupiter's moon Europa? (Jump ahead to Chapters 6 or 8 to find the rotation periods of these bodies.)

3. Hold your little finger out at arm's length. Can you cover the disk of the Moon? The Moon projects an angular size of 30′ (half a degree); your finger should more than cover it. You can use this fact to make some basic sky measurements. As a simple rule, your little finger at arm's length is about 1° across, your middle three fingers are about 4° across, and your clenched fist is about 10° across. If the constellation Orion is visible, use this information to estimate the angular size of Orion's belt and the angular distance between Betelgeuse and Rigel. Compare your findings with Figure 0.2(a).

1

The Copernican Revolution
The Birth of Modern Science

Living in the Space Age, we have become accustomed to the modern view of our place in the universe. Images of Earth taken from space leave little doubt that we inhabit just one planet among several, and no one seriously questions the idea that we orbit the Sun. Yet there was a time, not so long ago, when our ancestors maintained that Earth had a special role in the cosmos and lay at the center of all things. Our view of the universe—and of ourselves—has undergone a radical transformation since those early days. Humankind has been torn from its throne at the center of the cosmos and relegated to an unremarkable position on the periphery of the Milky Way Galaxy. But in return we have gained a wealth of scientific knowledge. The story of how this came about is the story of the rise of science and the genesis of modern astronomy.

LEARNING OUTCOMES

Studying this chapter will enable you to:

LO1 Explain how geocentric models of the solar system accounted for the apparent retrograde motion of the planets.

LO2 Explain how retrograde motion actually arises from the orbital motions of the planets around the Sun.

LO3 Describe the scientific developments that led to the modern view of the solar system, and identify the main contributions of Copernicus, Tycho, Galileo, and Kepler.

LO4 State Kepler's laws of planetary motion.

LO5 Explain how astronomers have measured the true size of the solar system.

LO6 State Newton's laws of motion and his law of universal gravitation, and explain how they enable us to measure the masses of astronomical bodies.

THE BIG PICTURE

Exploration is at the heart of the modern scientific method used by all scientists around the world. Ideas must be tested against what is observed in nature, and those ideas that fail the test must be discarded. In this way, astronomers progressively generate, not "truth," but a better and better understanding of reality.

◄ Exploration is at the heart of modern science. This engraving from a 19th-century popular science text presents a stylized depiction of the medieval worldview: the sky as a domed firmament arcing over Earth's surface. Here, a traveler finds the place where Earth and Sky meet and, pushing through the gap between them, sees the wonders of the larger universe. The image was intended primarily as a counterpoint to the then-modern view of the universe, but it also serves as a metaphor for the scientific transformation that occurred during the Renaissance—when scientists broke through the artificial constraints of cultural orthodoxy to discover the intellectual riches that lay beyond.

MasteringAstronomy®

Visit www.masteringastronomy.com for quizzes, animations, videos, interactive figures, and self-guided tutorials.

1.1 The Motions of the Planets

Over the course of a night, the stars slide smoothly across the sky. ∞ *(Sec. 0.1)* Over the course of a month, the Moon moves smoothly and steadily along its path in the sky relative to the stars, passing through its familiar cycle of phases. ∞ *(Sec. 0.3)* Over the course of a year, the Sun progresses along the ecliptic at an almost constant rate, varying little in brightness from day to day. ∞ *(Sec. 0.2)* This basic predictability of the night sky provided ancient cultures with a means of tracking the seasons and organizing their activities (see *Discovery 1-1*).

But ancient astronomers were also aware of five other bodies in the sky—the planets Mercury, Venus, Mars, Jupiter, and Saturn—whose behavior was not so easy to grasp. The explanation for their motion would change forever our perception of the universe and our place in it.

Wanderers in the Heavens

Planets do not behave in nearly as orderly a fashion as do the Sun, Moon, and stars. They vary in brightness and don't maintain fixed positions in the sky. Planets never stray far from the ecliptic and generally traverse the celestial sphere from west to east, like the Sun. But, they speed up and slow down as they go, and at times even appear to loop back and forth relative to the stars, as shown in Figure 1.1. Indeed, the word *planet* derives from the Greek word *planetes,* meaning "wanderer." Astronomers refer to the planets' normal eastward motion as *prograde* motion. The backward (westward) loops are **retrograde motion.**

Unlike the Sun and stars, but like the Moon, the planets produce no visible light of their own. Instead, they shine by reflected sunlight. Ancient astronomers correctly reasoned that the apparent brightness of a planet in the night sky is related to its distance from Earth—a planet appears brightest when closest to us. However, Mars, Jupiter, and Saturn are always brightest during the retrograde portions of their orbits. The challenge facing astronomers was to explain the observed motions of the planets and to relate those motions to the variations in planetary brightness.

The Geocentric Universe

The earliest models of the solar system followed the teachings of the Greek philosopher Aristotle (384–322 B.C.) and were **geocentric,** meaning that Earth lay at the center of the universe and all other bodies moved around it. (Figures 0.4 and 0.9 illustrate the basic geocentric view.) ∞ *(Sec. 0.2)* These models employed what Aristotle had taught was the perfect form: the circle. The simplest possible description— uniform motion around a circle having Earth at its center—provided a fairly good approximation to the orbits of the Sun and the Moon, but it could not account for the variations in planetary brightness, nor for their retrograde motion. A better model was needed.

Motions of the planets relative to the stars produce continuous streaks on a planetarium "sky."

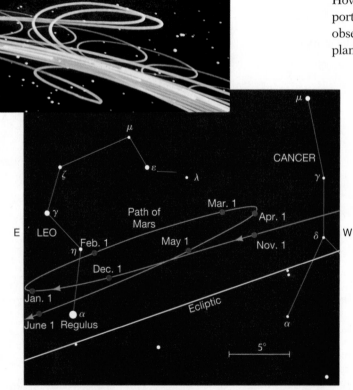

Observed planet motions can be complicated because each planet travels at a different speed around the Sun.

◀ **FIGURE 1.1 Planetary Motions**   Planets normally move from west to east relative to the background stars. Occasionally—roughly once per year—they change direction and temporarily undergo retrograde motion (east to west). The main illustration shows a retrograde loop in the motion of the planet Mars. The inset depicts the movements of several planets over the course of several years, as reproduced on the inside dome of a planetarium. *(Inset: Museum of Science, Boston)*

In the first step toward this new model, each planet was taken to move uniformly around a small circle, called an **epicycle,** whose *center* moved uniformly around Earth on a second and larger circle, the **deferent** (Figure 1.2). The motion was therefore composed of two separate circular orbits, creating the possibility of retrograde at certain times. Also, the distance from the planet to Earth would vary, accounting for changes in brightness. By tinkering with the relative sizes of epicycle and deferent and with the speeds at which the planet and epicycle moved, early astronomers managed to bring this model into good agreement with the observed paths of planets in the sky. It predicted quite well the positions of the known planets, according to the accuracy of observations at the time.

However, as the number and the quality of observations increased, astronomers had to introduce small corrections into the simple epicyclic model to bring it into line with new observations. The center of the deferents had to be shifted away from Earth's center, and the motion of the epicycles had to be imagined uniform with respect not to Earth, but to yet another point in space. Around A.D. 140 a Greek astronomer named Claudius Ptolemaeus (known today as Ptolemy) constructed perhaps the best geocentric model of all time. Illustrated in simplified form in Figure 1.3, it described remarkably well the observed paths of the five planets then known, as well as the paths of the Sun and the Moon. However, to achieve its explanatory and predictive power, the full **Ptolemaic model** required a series of no fewer than 80 circles. To account for the paths of the Sun, Moon, and all eight planets known today would require a vastly more complex set.

Today our scientific training leads us to seek simplicity, because simplicity in science has so often proved to be an indicator of truth. ∞ *(Sec. 0.5)* The intricacy of a model as complicated as the Ptolemaic system is a clear sign of a fundamentally flawed theory. We now recognize that the major error lay in the assumption of a geocentric universe, compounded by the insistence on uniform circular motion, the basis of which was largely philosophical, rather than scientific, in nature.

Actually, history records that some ancient Greek astronomers reasoned differently about the motions of heavenly bodies. Foremost among them was Aristarchus of Samos (310–230 B.C.), who proposed that all the planets, including Earth, revolve around the Sun and that Earth rotates on its axis once each day. This, he argued, would create an apparent motion of the sky—a simple idea that is familiar to anyone who has ridden on a merry-go-round and watched the landscape appear to move past in the opposite direction. However, this description of the heavens, though essentially correct, did not gain widespread acceptance. Aristotle's influence was too strong, his followers too numerous, his writings too comprehensive.

The Aristotelian school did present several arguments in favor of their views, the strongest asking—in fact, quite correctly—If the vantage point from which we view the stars changes over the course of a year, then why don't we see stellar parallax? ∞ *(Sec. 0.4)* We now know that there is stellar parallax as Earth orbits the Sun, but because the stars are so distant, it is less than 1 arc second, even for the closest stars. Early astronomers simply would not have noticed it (it was conclusively measured only in the latter half of the 19th century). We will encounter many instances in astronomy where correct reasoning led to the wrong conclusions because it was based on inadequate data.

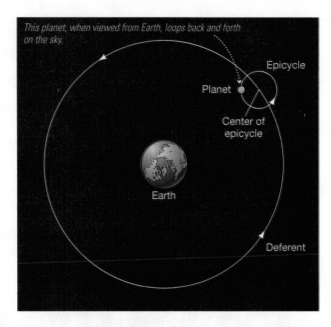

▲ **FIGURE 1.2 Geocentric Model**   To explain retrograde motion, in the geocentric model of the solar system, each planet was thought to follow a small circular orbit (the epicycle) about an imaginary point that itself traveled in a large, circular orbit (the deferent) about Earth.

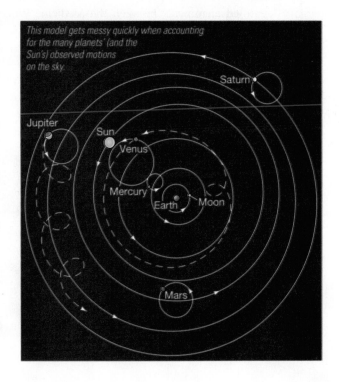

▶ **FIGURE 1.3 Ptolemaic Model INTERACTIVE**   The basic features, drawn roughly to scale, of Ptolemy's geocentric model of the inner solar system, a model that enjoyed widespread popularity prior to the Renaissance. To avoid confusion, partial paths (dashed) of only two planets, Venus and Jupiter, are drawn here.

1.1 DISCOVERY

Ancient Astronomy

Many ancient societies took a keen interest in the changing night-time sky. Seafarers needed to navigate their ships and farmers had to know when to plant their crops. Cultures all over the world built elaborate structures to serve, at least in part, as primitive calendars to predict celestial events. Often, the keepers of the secrets of the sky enshrined their knowledge in myth and ritual, and these astronomical sites were also used for religious ceremonies.

(M. Boulton)

Perhaps the best known such site is Stonehenge, in England, shown above. This ancient stone circle dates from the Stone Age and probably served as a kind of three-dimensional almanac. Many of the stones are aligned (within a degree or so) with important astronomical events, such as the rising Sun on the summer solstice and the rising and setting of the Sun and the Moon at other key times of the year, allowing its builders to track the seasons.

The Big Horn Medicine Wheel in Wyoming (below) is similar to Stonehenge in design, and perhaps also in intent. Some researchers have identified alignments between the Medicine Wheel's spokes and the rising and setting Sun at solstices and equinoxes, and with some bright stars, suggesting that its builders—the Plains Indians—were familiar with the changing night sky. Others dispute the accuracy of the alignments, however, suggesting that the wheel's purpose was more symbolic than practical. A similar controversy swirls around the Caracol temple (bottom right of the same figure) in the Mayan city of Chitzen Itza, built around A.D. 1000 on Mexico's Yucatán peninsula. Were its windows aligned with astronomical events, such as the rising of Venus, or did it have some different purpose? Experts disagree. Researchers do seem to agree that the Sun Dagger (bottom right), in Chaco Canyon, New Mexico, is a genuine astronomical calendar that probably aided agriculture. The rock is sculpted so that a thin streak of light passes precisely through the center of a carved spiral pattern at noon on the summer solstice.

The ancient Chinese also observed the heavens. Their astrology attached particular importance to "omens" such as comets and "guest stars" that appeared suddenly in the sky and then slowly faded away, and they kept careful records of such events. Perhaps the best-known guest star was one that appeared in A.D. 1054 and was visible in the daytime sky for many months. We now know that the event was a *supernova*, the explosive death of a giant star (see Chapter 13). It left behind a remnant that is still detectable today, nine centuries later. The Chinese data are a prime source of historical information for supernova research.

A vital link between the astronomy of ancient Greece and that of modern Europe was provided by astronomers in the Muslim world. This 16th-century illustration shows Turkish astronomers at work in the Istanbul Observatory. From the depths of the Dark Ages to the beginning of the Renaissance, Islamic astronomy flourished, preserving and adding to the knowledge of the Greeks. Its influence on modern astronomy is widespread. Many techniques used in trigonometry were developed by Islamic astronomers in response to practical problems, such as determining the precise dates of holy days or the direction of Mecca at any given location on Earth. Many astronomical terms, such as *azimuth, zenith,* and the names of many stars—for example, Rigel, Betelgeuse, and Vega—all bear witness to this extended period of Muslim scholarship.

(G. Gerster)

(H. Lapahie Jr.)

(J. Cornell)

(Istanbul University Library)

The Heliocentric Model of the Solar System

The Ptolemaic picture of the universe survived, more or less intact, for almost 13 centuries until a 16th-century Polish cleric, Nicholas Copernicus (Figure 1.4), rediscovered Aristarchus's **heliocentric** (Sun-centered) model. Copernicus asserted that Earth spins on its axis and, like all other planets, orbits the Sun. Not only does this model explain the observed daily and seasonal changes in the heavens, as we have seen, but it also naturally accounts for planetary **retrograde motion** and brightness variations. The critical realization that Earth is not at the center of the universe is now known as the **Copernican revolution.**

Figure 1.5 shows how the Copernican view explains both the changing brightness of a planet (in this case, Mars) and its apparent looping motions. If we suppose that Earth moves faster than Mars, then every so often Earth "overtakes" that planet. Each time this happens, Mars appears to move backward in the sky, in much the same way as a car we overtake on the highway seems to slip backward relative to us. Furthermore, at these times Earth is closest to Mars, so Mars appears brightest, as observed. Notice that in the Copernican picture the planet's looping motions are only apparent—in the Ptolemaic view, they were real.

Copernicus's major motivation for introducing the heliocentric model was simplicity. Even so, he still clung to the idea of circles to model the planets' motions. To bring his theory into agreement with observations, he was forced to retain the idea of epicyclic motion, though with the deferent centered on the Sun rather than on Earth and with smaller epicycles than in the Ptolemaic picture. Thus, he retained unnecessary complexity and actually gained little in accuracy over the geocentric model. For Copernicus the primary attraction of heliocentricity was its simplicity, its being "more pleasing to the mind." To this

---

PROCESS OF SCIENCE CHECK

In terms of the scientific method presented in Chapter 0, what were the principal advantages of the heliocentric theory over the geocentric model? ∞ *(Sec. 0.5)*

▲ **FIGURE 1.4 Nicholas Copernicus (1473–1543)**
(E. Lessing/Art Resource, NY)

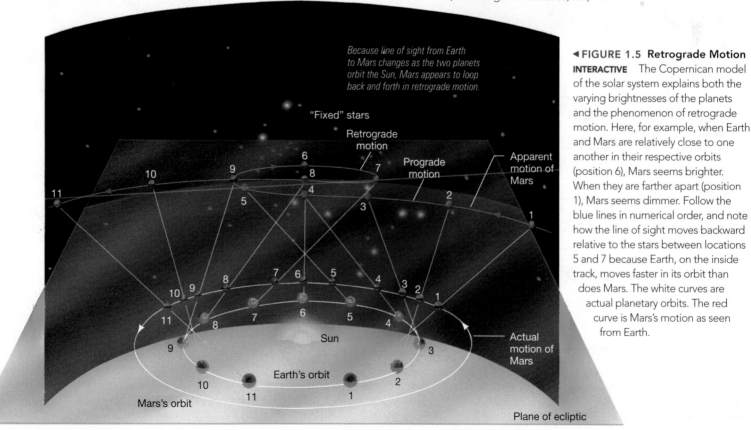

Because line of sight from Earth to Mars changes as the two planets orbit the Sun, Mars appears to loop back and forth in retrograde motion.

◀ **FIGURE 1.5 Retrograde Motion**
INTERACTIVE The Copernican model of the solar system explains both the varying brightnesses of the planets and the phenomenon of retrograde motion. Here, for example, when Earth and Mars are relatively close to one another in their respective orbits (position 6), Mars seems brighter. When they are farther apart (position 1), Mars seems dimmer. Follow the blue lines in numerical order, and note how the line of sight moves backward relative to the stars between locations 5 and 7 because Earth, on the inside track, moves faster in its orbit than does Mars. The white curves are actual planetary orbits. The red curve is Mars's motion as seen from Earth.

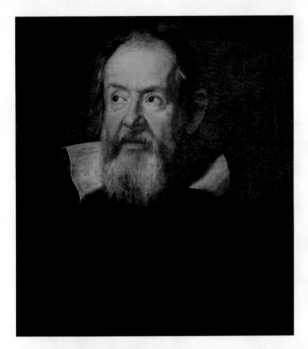

▲ FIGURE 1.6 Galileo Galilei (1564–1642)   *(Scala/Art Resource, NY)*

CONCEPT CHECK

How do the geocentric and heliocentric models of the solar system differ in their explanations of planetary retrograde motion?

The asterisks show the positions of the moons, now called Io, Europa, Ganymede, and Callisto, around Jupiter (open circle).

▲ FIGURE 1.7 **Galilean Moons**   The four large moons of Jupiter, as sketched by Galileo on seven nights between January 7 and 15, 1610. They are now known as the Galilean moons. *(From Sidereus Nuncius)*

day, scientists still are guided by simplicity, symmetry, and elegance in modeling the universe. ∞ *(Sec. 0.5)*

Copernicus's ideas were never widely accepted during his lifetime. By relegating Earth to a noncentral and undistinguished place within the solar system, heliocentricity contradicted the conventional wisdom of the time and violated the religious doctrine of the Roman Catholic Church. Copernicus surely discussed and debated his theory with his fellow scholars, but, possibly because he wished to avoid direct conflict with the Church, his book *On the Revolution of the Celestial Spheres* was not published until 1543, the year he died. Only later, when others extended and popularized the heliocentric model—and as supporting observational evidence began to mount—did the Copernican theory gain widespread recognition.

1.2 The Birth of Modern Astronomy

In the century following the death of Copernicus, two scientists—Galileo Galilei and Johannes Kepler—made indelible imprints on the study of astronomy. Each achieved fame for his discoveries and made great strides in popularizing the Copernican viewpoint.

Galileo's Historic Observations

Galileo Galilei (Figure 1.6) was an Italian mathematician and philosopher. By his willingness to perform experiments to test his ideas—a radical approach in those days—and by embracing the brand-new technology of the telescope, he revolutionized the way science was done, so much so that he is now widely regarded as the father of experimental science. The telescope was invented in Holland in 1608. Having heard of the invention (but without having seen one), Galileo built one for himself in 1609 and aimed it at the sky. What he saw conflicted greatly with the philosophy of Aristotle and strongly supported the ideas of Copernicus.

Using his telescope, Galileo discovered the following:

- The Moon has mountains, valleys, and craters—terrain in many ways reminiscent of that on Earth.
- The Sun has imperfections—dark blemishes now known as *sunspots* (see Chapter 9). By noting the changing appearance of these sunspots from day to day, Galileo inferred that the Sun *rotates* approximately once per month around an axis roughly perpendicular to the ecliptic plane.
- Four small points of light, invisible to the naked eye, orbit the planet Jupiter. He realized that they were *moons* (Figure 1.7) circling that planet just as our Moon orbits Earth. To Galileo, the fact that another planet had moons provided the strongest support for the Copernican model. Clearly, Earth was not the center of all things.
- Venus shows a complete cycle of *phases* (Figure 1.8), much like the familiar monthly changes exhibited by our own Moon. This finding can be explained only by the planet's motion around the Sun.

All these observations ran directly counter to the accepted scientific beliefs of the day. They showed that Earth is not the center of all things and that at least one planet orbited the Sun.

In 1610 Galileo published his observational findings and his controversial conclusions supporting the Copernican theory, challenging both scientific orthodoxy and the religious dogma of the day. In 1616 his ideas were judged heretical, both his works and those of Copernicus were banned by the Church, and Galileo

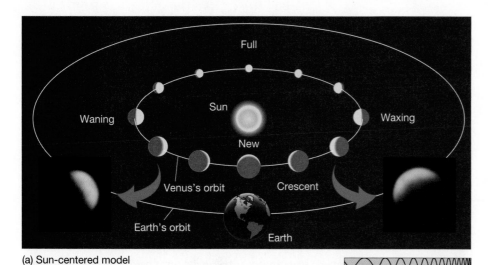

(a) Sun-centered model

R I V U X G

(b) Ptolemy's model

◄ **FIGURE 1.8 Venus Phases INTERACTIVE** Both the Ptolemaic and the Copernican models of the solar system predict that Venus should show phases as it moves in its orbit. (a) In the Copernican picture, when Venus is directly between Earth and the Sun, its unlit side faces us and the planet is invisible to us. As Venus moves in its orbit, progressively more of its illuminated face is visible from Earth. Note the connection between the orbital phase and the apparent size of the planet: Venus seems much larger in its crescent phase than when it is full because it is much closer to us during its crescent phase. This is the behavior actually observed. The insets at bottom left and right are photographs of Venus taken at two of its crescent phases (*Courtesy of New Mexico State University*). (b) In the Ptolemaic model, the full phase of the planet cannot be explained. Seen from Earth, Venus reaches only a "fat crescent," yet never a full phase, then begins to wane as it nears the Sun. (Note that both these views are from a sideways perspective; from overhead, both orbits are very nearly circular, as shown in Figure 1.14.)

was instructed to abandon his astronomical pursuits. Instead, he continued to collect and publish data supporting the heliocentric view. These actions brought Galileo into direct conflict with the Church. The Inquisition forced him, under threat of torture, to retract his claim that Earth orbits the Sun, and he was placed under house arrest in 1633. He remained imprisoned for the rest of his life. Not until 1992 did the Church publicly forgive Galileo's "crimes."

Ascendancy of the Copernican System

The Copernican revolution illustrates how the scientific method, although affected at any given time by the subjective whims, human biases, and even sheer luck of individual researchers, does ultimately lead to objectivity. ∞ *(Sec. 0.5)* Over time, groups of scientists checking, confirming, and refining experimental tests can neutralize the subjective attitudes of individuals. Heliocentricity was only confirmed observationally some three centuries after Copernicus published his work and more than 2000 years after Aristarchus had proposed the concept. Nonetheless, objectivity *did in fact* eventually prevail.

CONCEPT CHECK

In what ways did Galileo's observations of Venus and Jupiter conflict with the prevailing view at the time?

▲ **FIGURE 1.9 Johannes Kepler (1571–1630)**
(E. Lessing/Art Resource, NY)

▲ **FIGURE 1.10 Tycho Brahe**   The astronomer in his observatory Uraniborg, on the island of Hveen, in Denmark. *(Joan Bleau/ Newberry Library/SuperStock)*

1.3 The Laws of Planetary Motion

At about the time Galileo was becoming famous for his telescopic observations, Johannes Kepler (Figure 1.9), a German mathematician and astronomer, announced his discovery of a set of simple laws that accurately described the motions of the planets. While Galileo was the first "modern" observer who used telescopic observations of the skies to confront and refine his theories, Kepler was a pure theorist. He based his work almost entirely on the observations of another scientist. Those observations, which predated the telescope by several decades, had been made by Kepler's employer, Tycho Brahe (1546–1601), arguably one of the greatest observational astronomers who ever lived.

Brahe's Complex Data

Tycho, as he is often called, was an eccentric aristocrat and a skillful observer. Born in Denmark, he was educated at some of the best universities in Europe, where he studied astrology, alchemy, and medicine. Most of his observations, which predated the invention of the telescope by several decades, were made at his own observatory, named Uraniborg, in Denmark (Figure 1.10). There, using instruments of his own design, Tycho maintained meticulous and accurate records of the stars, the planets, and noteworthy celestial events.

In 1597, having fallen out of favor with the Danish court, Tycho moved to Prague. Kepler joined Tycho in Prague in 1600 and was put to work trying to find a theory to explain Brahe's planetary data. When Tycho died a year later, Kepler inherited not only Brahe's position as Imperial Mathematician of the Holy Roman Empire, but also his most priceless possession: the accumulated observations of the planets, spanning several decades. Tycho's observations, though made with the naked eye, were nevertheless of very high quality. Kepler set to work seeking a unifying principle to explain the motions of the planets without the need for epicycles. The effort occupied much of the remaining 29 years of his life.

Kepler's goal was to find a simple description of the solar system, within the basic framework of the Copernican model, that fit Tycho's complex mass of detailed observations. In the end, he had to abandon Copernicus's original simple notion of circular planetary orbits, but even greater simplicity emerged as a result. Kepler determined the shapes and relative sizes of each planet's orbit by triangulation, not from different points on Earth, but from different points on Earth's orbit, using observations made at many different times of the year. ∞ *(Sec. 0.4)* Noting where the planets were on successive nights, he inferred the speeds at which they moved. After long years working with Brahe's data and after many false starts and blind alleys, Kepler succeeded in summarizing the motions of all the known planets, including Earth, in the three **laws of planetary motion** that now bear his name.

Kepler's Simple Laws

Kepler's first law addresses the *shapes* of the planetary orbits:

   I. The orbital paths of the planets are elliptical (*not* necessarily circular), with the Sun at one focus.

An **ellipse** is simply a flattened circle. Figure 1.11 illustrates a means of constructing an ellipse using a piece of string and two thumbtacks. Each point at which the string is pinned is called a **focus** (plural: foci) of the ellipse. The long axis of the ellipse, containing the two foci, is known as the *major axis*. Half the length of this long axis is referred to as the **semimajor axis,** a conventional measure of the ellipse's size.

The **eccentricity** of the ellipse is equal to the distance between the foci divided by the length of the major axis. The length of the semimajor axis and the eccentricity are all we need to describe the size and shape of a planet's orbital path. (A circle is an ellipse in which the two foci happen to coincide, so the eccentricity is zero. The semimajor axis of a circle is simply its radius.) Figure 1.12 illustrates how two other useful quantities—the planet's **perihelion** (its point of closest approach to the Sun) and its **aphelion** (greatest distance from the Sun)—can be computed from its orbital semimajor axis and eccentricity.

In fact, no planet's elliptical orbit is nearly as elongated as the one shown in Figure 1.11. With one exception (Mercury), the planets' orbits have such small eccentricities that our eyes would have trouble distinguishing them from true circles. Only because the orbits are so nearly circular were the Ptolemaic and Copernican models able to come as close as they did to describing reality.

Kepler's second law, illustrated in Figure 1.13, addresses the *speed* at which a planet traverses different parts of its orbit:

II. An imaginary line connecting the Sun to any planet sweeps out equal areas of the ellipse in equal intervals of time.

While orbiting the Sun, a planet traces the arcs labeled A, B, and C in Figure 1.13 in equal times. Notice, however, that the distance traveled along arc A is greater than the distance traveled along arc B or arc C. Because the time is the same and the distance is different, the speed must vary: When a planet is close to the Sun, as in sector A, it moves much faster than when farther away, as in sector C.

These laws are not restricted to planets. They apply to *any* orbiting object. Spy satellites, for example, move very rapidly as they swoop close to Earth's surface, not because they are propelled by powerful onboard rockets, but because their highly eccentric orbits are governed by Kepler's laws.

Kepler published his first two laws in 1609, stating that he had proved them only for the orbit of Mars. Ten years later he extended them to all the known planets (Mercury, Venus, Earth, Mars, Jupiter, and Saturn) and added a third law relating the size of a planet's orbit to its sidereal orbital **period,** defined as the time needed for the planet to complete one circuit around the Sun. *Kepler's third law* states:

III. The square of a planet's orbital period is proportional to the cube of its semimajor axis.

This law becomes particularly simple when we choose the (Earth) year as our unit of time and the **astronomical unit** as our unit of length. One **astronomical unit** (AU) is the semimajor axis of Earth's orbit around the Sun—the average distance between Earth and the Sun. Using these units for time and distance, we can conveniently write Kepler's third law for any planet in the form

$$P^2 \text{ (in Earth years)} = a^3 \text{ (in astronomical units)},$$

where *P* is the planet's sidereal orbital period and *a* is its semimajor axis.

Table 1.1 presents some basic data describing the orbits of the eight major planets of the solar system. The semimajor axis of each planet's orbit is measured in astronomical units—that is, relative to the size of Earth's orbit—and all orbital periods are measured in years. Renaissance astronomers knew these properties for the innermost six planets only. However, *all* the known bodies orbiting the Sun obey Kepler's laws, *not just the six planets on which he based his conclusions*. The rightmost column lists the ratio P^2/a^3. As we have just seen, in the units used in the table, Kepler's third law implies that this number should equal 1 in all cases. The small deviations of P^2/a^3 from 1 in the cases of Uranus and Neptune are caused by the gravitational attraction between those two planets (see Chapter 7).

CONCEPT CHECK

In what ways did Galileo and Kepler differ in their approach to science? In what ways did each advance the Copernican view of the universe?

The wider the separation of the foci, the more elongated, or eccentric, the ellipse.

When the two foci are at the same place, the drawn curve is a circle.

Focus   Focus

Major axis   Semimajor axis

▲ FIGURE 1.11 Ellipse **INTERACTIVE** An ellipse can be drawn using a string, a pencil, and two thumbtacks.

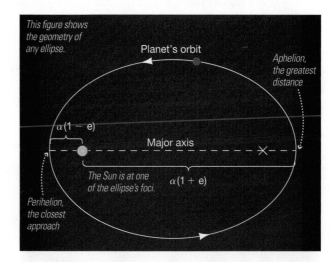

This figure shows the geometry of any ellipse.

Planet's orbit

Aphelion, the greatest distance

$a(1 - e)$

Major axis

The Sun is at one of the ellipse's foci. $a(1 + e)$

Perihelion, the closest approach

▲ FIGURE 1.12 Orbital Properties **INTERACTIVE** A planet's perihelion and aphelion are related in a simple way to the orbital semimajor axis *a* and eccentricity *e*. Note that while the Sun resides at one focus, the other focus (depicted by an X) is empty and has no particular significance. No planet in our solar system has an orbital eccentricity as large as shown here (see Table 1.1), but some meteorites and all comets do (see Chapter 4).

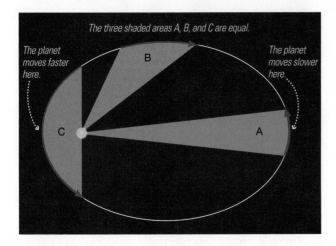

TABLE 1.1	Some Planetary Properties			
Planet	Orbital Semimajor Axis, a (AU)	Orbital Period, P (Earth Years)	Orbital Eccentricity, e	P^2/a^3
Mercury	0.387	0.241	0.206	1.002
Venus	0.723	0.615	0.007	1.001
Earth	1.000	1.000	0.017	1.000
Mars	1.524	1.881	0.093	1.000
Jupiter	5.203	11.86	0.048	0.999
Saturn	9.537	29.42	0.054	0.998
Uranus	19.19	83.75	0.047	0.993
Neptune	30.07	163.7	0.009	0.986

▲ FIGURE 1.13 Kepler's Second Law INTERACTIVE   A line between a planet and the Sun sweeps out equal areas in equal intervals of time. Any object traveling along the elliptical path would take the same amount of time to cover the distance indicated by the three red arrows. Planets move faster when closer to the Sun.

CONCEPT CHECK

Why is it significant that Kepler's laws apply to Uranus and Neptune?

CONCEPT CHECK

Why don't Kepler's laws tell us the value of the astronomical unit?

Kepler's laws were far more than mere fits to existing data. They made testable predictions about the future locations of the planets. Those predictions have been borne out to high accuracy every time they have been tested by observation—the hallmark of any credible scientific theory. ∞ *(Sec. 0.5)*

The Dimensions of the Solar System

Kepler's laws allow us to construct a scale model of the solar system, with the correct shapes and *relative* sizes of all the planetary orbits, but they do not tell us the *actual* size of any orbit. Because Kepler's measurements were based on triangulation using Earth's orbit as a baseline, his distances were expressed relative to the size of that orbit—the astronomical unit—which was not itself determined.

Our model of the solar system would be analogous to a road map of the United States showing the *relative* positions of cities and towns but lacking the all-important scale marker indicating distances in kilometers or miles. For example, we would know that Kansas City is about three times farther from New York than it is from Chicago, but we would not know the actual mileage between any two points on the map. If we could somehow determine the value of the astronomical unit—in kilometers, say—we would be able to add the vital scale marker to our map and compute the exact distances between the Sun and each of the planets.

The modern method for deriving the absolute scale of the solar system uses a technique called *radar ranging*. The word **radar** is an acronym for **ra**dio **d**etection **a**nd **r**anging. Radio waves are transmitted toward an astronomical body, such as a planet. Their returning echo indicates the body's direction and distance, in absolute terms—in other words, in kilometers rather than in astronomical units. Multiplying the round-trip travel time of the radar signal (the time elapsed between transmission of the signal and reception of the echo) by the speed of light (300,000 km/s, which is also the speed of radio waves), we obtain twice the distance to the planet.

We cannot use radar ranging to measure the distance to the Sun directly, because radio signals are absorbed at the solar surface and are not reflected back to Earth. Instead, the planet Venus, whose orbit brings it closest to Earth, is the most common target for this technique. Figure 1.14 is an idealized diagram of the

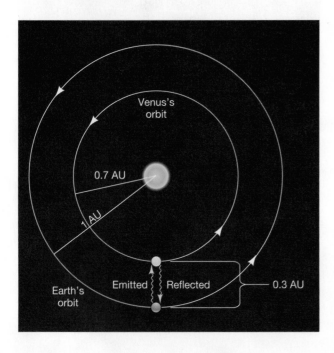

◀ FIGURE 1.14 Astronomical Unit INTERACTIVE   The wavy blue lines represent the paths along which radar signals are transmitted toward Venus and received back at Earth when Venus is at its minimum distance from Earth. Because the radius of Earth's orbit is 1 AU and that of Venus is about 0.7 AU, the one-way distance covered by the signal is 0.3 AU. Thus, we can calibrate the astronomical unit in kilometers.

Sun–Earth–Venus orbital geometry. Neglecting for simplicity the small eccentricities of the two planets' orbits, we see from Table 1.1 that the distance from Venus to the Sun is roughly 0.7 AU. Hence (from Figure 1.14), the distance from Earth to Venus at closest approach is approximately 0.3 AU. Radar signals bounced off Venus at that instant return to Earth in about 300 seconds, indicating that Venus lies 300,000 km/s × 300 s ÷ 2 (dividing by 2 for a one-way trip) = 45,000,000 km from Earth. Since 0.3 AU is 45,000,000 km, it follows that 1 AU is 45,000,000 km/0.3, or 150,000,000 km.

Through precise radar ranging, the astronomical unit is now known to be 149,597,870 km. In this text, we will round this value off to 1.5×10^8 km. Having determined the value of the astronomical unit, we can re-express the sizes of the other planetary orbits in more familiar units, such as miles or kilometers. The entire scale of the solar system can then be calibrated to high precision.

1.4 Newton's Laws

Kepler's three laws, which so simplified the solar system, were discovered *empirically*. In other words, they resulted solely from the analysis of observational data, rather than being derived from a theory or mathematical model. Indeed, Copernicus, Galileo, and Kepler had no real understanding of *why* the planets orbit the Sun. Only by considering something more fundamental can we truly comprehend planetary motion. The critical test of any complete scientific theory is its ability to explain physical phenomena, not simply describe them.

The Laws of Motion

In the 17th century, the British physicist and mathematician Isaac Newton (Figure 1.15) developed a deeper understanding of the way all objects move and interact with one another. Newton's theories form the basis for what today is known as **Newtonian mechanics.** Three basic laws of motion, the law of universal gravitation, and a little calculus (which Newton also developed) are sufficient to explain virtually all of the complex dynamic behavior we see throughout the universe.

Figure 1.16 illustrates *Newton's first law of motion*, which states:

   I. An object at rest remains at rest, and a moving object continues to move forever in a straight line with constant speed, unless some external **force** changes their state of motion.

An example of an external force would be the force exerted by, say, a brick wall when a rolling ball glances off it, or the force exerted on a pitched ball by a baseball bat. In either case, a force changes the original motion of the object. The

DATA POINTS

Kepler's Laws

Almost half of all students had difficulty applying Kepler's laws to the orbits of objects in the solar system. Remember:

- According to Kepler's second law ("equal areas"), a planet moves fastest in its orbit when it is closest to the Sun.

- Kepler's third law tells us that planets on larger orbits take longer to complete one trip around the Sun. The orbital period increases faster than the size of the orbit—doubling the size increases the period by a factor of $2\sqrt{2}$, or approximately 2.8.

▲ **FIGURE 1.15 Isaac Newton (1642–1727)**
(S. Terry)

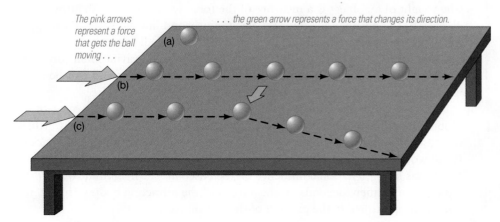

◄ **FIGURE 1.16 Newton's First Law** An object at rest will remain at rest (a) until some force acts on it. When a force (represented by the red arrow) does act (b), the object will remain in that state of uniform motion until another force acts on it. When a second force (green arrow) acts in a direction different from the first force (c), the object changes its direction of motion.

tendency of an object to keep moving at the same speed and in the same direction unless acted upon by a force is known as **inertia.** A familiar measure of an object's inertia is its **mass**—loosely speaking, the total amount of matter the object contains. The greater an object's mass, the more inertia it has and the greater is the force needed to change its motion.

Newton's first law contrasts sharply with the view of Aristotle, who maintained (incorrectly) that the natural state of an object was to be at *rest*—most probably an opinion based on Aristotle's observations of the effect of friction. To simplify our discussion, we will neglect friction—the force that slows balls rolling along the ground, blocks sliding across tabletops, and baseballs moving through the air. In any case, this is not an issue for the planets, as there is no appreciable friction in outer space.

The rate of change of the velocity of an object—speeding up, slowing down, or simply changing direction—is called its **acceleration.** *Newton's second law* states:

II. The acceleration of an object is directly proportional to the net applied force and inversely proportional to the object's mass.

In other words, the greater the force acting on the object, or the smaller the mass of the object, the greater its acceleration. Thus, if two objects are pulled with the same force, the more massive one accelerates less. If two identical objects are pulled with different forces, the one experiencing the greater force accelerates more.

Finally, *Newton's third law* tells us that forces always occur in pairs:

III. To every action there is an equal and opposite reaction.

If body A exerts a force on body B, then body B necessarily exerts a force on body A that is equal in magnitude but oppositely directed.

Gravity

Forces can act either *instantaneously* or *continuously.* The force from a baseball bat that hits a home run can reasonably be thought of as instantaneous. A good example of a continuous force is the one that prevents the baseball from zooming off into space—**gravity,** the phenomenon that started Newton on the path to the discovery of his laws. Newton hypothesized that any object having mass exerts an attractive **gravitational force** on all other massive objects. The more massive an object, the stronger its gravitational pull.

Consider a baseball thrown upward from Earth's surface, as illustrated in Figure 1.17. In accordance with Newton's first law, the downward force of Earth's gravity steadily modifies the baseball's velocity, slowing its initial upward motion and eventually causing the ball to fall back to the ground. Of course, the baseball, having some mass of its own, also exerts a gravitational pull on Earth. By Newton's third law, this force is equal in magnitude to the weight of the ball (the weight of an object is a measure of the force with which Earth attracts that object), but oppositely directed. By Newton's second law, however, Earth has a much greater effect on the light baseball than the baseball has on the much more massive Earth. The ball and Earth each feel the same gravitational force, but Earth's *acceleration* as a result of this force is much smaller and can be safely ignored.

Now consider the trajectory of a baseball batted from the surface of the Moon. The pull of gravity is about one-sixth as great on the Moon as on Earth, so the baseball's velocity changes more slowly—a typical home run in a ballpark on Earth would travel nearly half a mile on the Moon. The Moon, less massive than Earth, has less gravitational influence on the baseball. The magnitude of the gravitational force, then, depends on the *masses* of the attracting bodies. In fact, it is directly proportional to the product of the two masses.

▲ FIGURE 1.17 Gravity A ball thrown up from the surface of a massive object, such as a planet, is pulled continuously down (arrows) by the gravity of that planet—and, conversely, the gravity of the ball continuously pulls the planet (although very, very little).

Surface of Earth

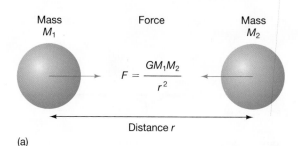

▶ **FIGURE 1.18 Gravitational Force** (a) The gravitational force between two bodies is proportional to the mass of each and is inversely proportional to the square of the distance between them. (b) Inverse-square forces weaken rapidly with distance from their source, never quite reaching zero no matter how far away.

Studying the motions of the planets reveals a second aspect of the gravitational force. At locations equidistant from the Sun's center, the gravitational force has the same strength and is always directed toward the Sun. Furthermore, detailed calculation of the planets' accelerations as they orbit the Sun reveals that the strength of the Sun's gravitational pull decreases in proportion to the *square* of the distance from the Sun. (Newton is said to have first realized this fact by comparing the accelerations not of the planets, but of the Moon and an apple falling to the ground—the basic reasoning is the same in either case.) The force of gravity is said to obey an **inverse-square law** (Figure 1.18a).

We can combine the preceding statements about mass and distance to form a law of gravity that dictates the way in which *all* massive objects (i.e., objects having some mass) attract one another:

Every particle of matter in the universe attracts every other particle with a force that is directly proportional to the product of the masses of the particles and inversely proportional to the square of the distance between them.

As shown in Figure 1.18(b), inverse-square forces decrease rapidly with distance from their source. For example, tripling the distance makes the force $3^2 = 9$ times weaker, while multiplying the distance by five results in a force that is $5^2 = 25$ times weaker. Despite this rapid decrease, the force never quite reaches zero. The gravitational pull of an object having some mass can never be completely extinguished.

Orbital Motion

The mutual gravitational attraction of the Sun and the planets, as expressed by Newton's law of gravity, is responsible for the observed planetary orbits. As depicted in Figure 1.19, this gravitational force continuously pulls each planet toward the Sun, deflecting its forward motion into a curved orbital path. In the solar system, at this very moment, Earth is moving under the combined influence of these two effects: the competition between gravity and inertia. Because the Sun is much more massive than any of the planets, it dominates the interaction. We might say that the Sun "controls" the planets, not the other way around.

The planet–Sun interaction sketched here is analogous to what occurs when you whirl a rock at the end of a string above your head. The Sun's gravitational pull is your hand and the string, and the planet is the rock at the end of that string. The tension in the string provides the force necessary for the rock to move in a circular path. If you were suddenly to release the string—which would be like eliminating the Sun's gravity—the rock would fly away along a tangent to the circle, in accordance with Newton's first law.

Kepler's Laws Reconsidered

Newton's laws of motion and his law of universal gravitation provided a theoretical explanation for Kepler's empirical laws of planetary motion. Just as Kepler modified the Copernican model by introducing ellipses in place of circles, so too did Newton make corrections to Kepler's first and third laws. Because the Sun and a planet feel

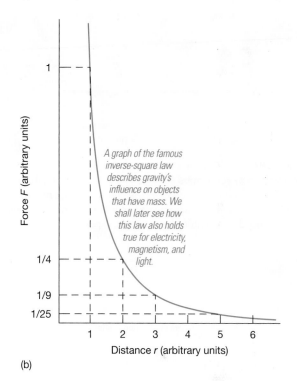

A graph of the famous inverse-square law describes gravity's influence on objects that have mass. We shall later see how this law also holds true for electricity, magnetism, and light.

(b)

INTERACTIVE Center of Mass of a Binary Star (MA)

CONCEPT CHECK

Explain, in terms of Newton's laws of motion and gravity, why planets orbit the Sun.

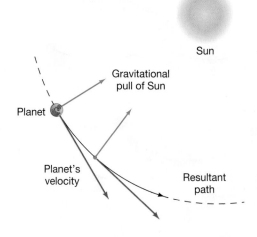

▶ **FIGURE 1.19 Sun's Gravity** The Sun's inward pull of gravity on a planet competes with the planet's tendency to continue moving in a straight line. These two effects combine, causing the planet to move smoothly along an intermediate path, which continuously "falls around" the Sun.

(a) Equal masses   Center of mass

(b) Unequal masses   Center of mass

▲ **FIGURE 1.20 Center of Mass**   (a) The center of mass of two bodies of equal mass lies midway between them. (b) As the mass of one body increases, the center of mass moves toward it. Experienced seesawers know that when both sides are balanced, the center of mass is at the pivot point.

ANIMATION / VIDEO   **(MA)**

Gravity Demonstration on the Moon

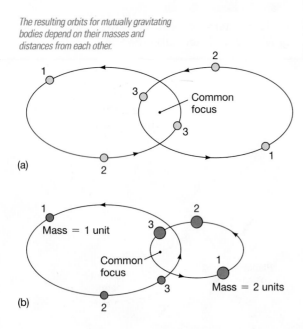

The resulting orbits for mutually gravitating bodies depend on their masses and distances from each other.

equal and opposite gravitational forces (by Newton's third law), the Sun must also move (by Newton's first law) due to the planet's gravitational pull. As a result, the planet does not orbit the exact center of the Sun but, instead, both the planet and the Sun orbit their common **center of mass**—the "average" position of the matter comprising the two bodies (see Figure 1.20)—and Kepler's first law becomes:

> I. The orbit of a planet around the Sun is an ellipse having the center of mass of the planet–Sun system at one focus.

As illustrated in Figure 1.20, the center of mass of a system consisting of two objects of comparable mass does not lie within either object. For identical masses orbiting one another (Figure 1.21a), the orbits are identical ellipses with a common focus located midway between the two objects. For unequal masses (Figure 1.21b), the elliptical orbits still share a focus and have the same eccentricity, but the more massive object moves more slowly and on a tighter orbit. (Note that Kepler's second law continues to apply without modification to each orbit separately, but the rates at which the two orbits sweep out area are different.)

The change to Kepler's third law is small in the case of a planet orbiting the Sun but may be very important in other circumstances, such as the orbits of two stars that are gravitationally bound to one another. Following through the mathematics of Newton's theory, we find the true relationship between the semimajor axis a (measured in astronomical units) of the planet's orbit relative to the Sun and its orbital period P (in Earth years) is

$$P^2 \text{ (in Earth years)} = \frac{a^3 \text{ (in astronomical units)}}{M_{\text{total}} \text{ (in solar units)}},$$

where M_{total} is the *combined* mass of the two objects, expressed in terms of the mass of the Sun. Notice that this restatement of Kepler's third law preserves the proportionality between P^2 and a^3, but now the proportionality also includes M_{total} so it is *not* quite the same for all the planets. The Sun's mass is so great, however, that the differences in M_{total} among the various combinations of the Sun and the other planets are almost unnoticeable. Therefore, Kepler's third law, as originally stated, is a very good approximation.

This modified form of Kepler's third law is true in all circumstances, inside or *outside* the solar system. Most importantly, it provides a means of measuring mass *anywhere* in the universe, so long as we can determine the orbital properties—separation and period—of the bodies involved. In fact, this is how *all* masses are measured in astronomy. When we need to know an object's mass, we always look for its gravitational influence on something else. This principle applies to planets, stars, galaxies, even clusters of galaxies—very different objects but all subject to the same fundamental physical laws. Newton's laws extend our intellectual reach far beyond the tiny fraction of the universe we can actually visit or touch.

The Circle of Scientific Progress

The progression from the complex Ptolemaic model of the universe to the elegant simplicity of Newton's laws is a case study in the scientific method. *(Sec.0.5)* Copernicus made a radical conceptual leap away from the Ptolemaic view, gaining much in insight but little in predictive power. Kepler made critical changes to

◀ **FIGURE 1.21 Orbits**   (a) The orbits of two bodies with equal masses, under the influence of their mutual gravity, are identical ellipses with a common focus. The pairs of numbers (e.g., the two 2s in each orbit) indicate the positions of the two bodies at three different times. (b) The orbits of two bodies, one twice as massive as the other, are again elliptical and with the same eccentricity, but according to Newton's laws, the more massive body moves more slowly and in a smaller orbit.

the Copernican picture and gained both accuracy and predictive power, but still fell short of a true physical explanation of planetary motion within the solar system, or of orbital motion in general. Eventually, Newton showed how all known planetary motion could be explained in detail by the application of four simple, fundamental laws—the three laws of motion and the law of gravity. The process was slow, with many starts and stops and a few wrong turns, but it worked!

In a sense, the development of Newton's laws and their application to planetary motion represent the end of the first "loop" around the schematic diagram shown in Figure 0.22. The practical and conceptual questions raised by ancient observations of retrograde motion were finally resolved, and new predictions, themselves amenable to observational testing, became possible. But Newton's laws went much further. Unlike the essentially descriptive models of Ptolemy, Copernicus, and Kepler, Newtonian mechanics is not limited to the motions of planets. They apply to moons, comets, spacecraft, stars, and even the most distant galaxies, extending the range of our scientific inquiries across the observable universe.

Newton's laws are still being tested today. Every time a comet appears in the night sky right on schedule or a spacecraft reaches the end of a billion-kilometer journey within meters of its target and seconds of the predicted arrival time, our confidence in the laws is further strengthened.

Yet even Newton's laws are just an approximation to reality. They break down in extreme circumstances—although this fact was only realized in the early 20th century, when Albert Einstein's theories of relativity (Chapter 13) again revolutionized our view of gravity and the universe. The modern conception of an expanding universe dominated on the largest scales by non-gravitational forces (Chapter 17) differs radically from Newton's conception of the cosmos, but it too is the result of the same basic scientific methodology followed by Galileo and Newton.

PROCESS OF SCIENCE CHECK

Describe some ways in which Newtonian mechanics superseded Kepler's laws as a model of the solar system.

THE BIG QUESTION

Newtonian mechanics and the law of gravity work well for small masses and low velocities, and hence for almost every application on or near Earth. But in the early 20th century, Albert Einstein overthrew these ideas with a better one—relativity theory, which deals with fast-moving, often massive objects in curved space. Now, in the 21st century, scientists are beginning to grapple with the notion that other, still unknown, forces may dominate the universe on the largest scales. Will relativity in turn be replaced by a new paradigm? No one knows, but the scientific method is designed to help us find out.

CHAPTER REVIEW

SUMMARY

LO1 **Geocentric** (p. 28) models of the universe, such as the **Ptolemaic model** (p. 29), have the Sun, Moon, and planets all orbiting Earth. They explain **retrograde motion** (p. 31) as a real backward motion of a planet as it moves along its epicyclic path.

LO2 The **heliocentric** (p. 31) view of the solar system, due to Aristarchus and later Copernicus, holds that Earth, like all the other planets, orbits the Sun. This model explains both retrograde motion as Earth overtakes other planets in its orbit and the observed brightness variations of the planets. The

realization that the solar system is Sun centered, not Earth centered, is known as the **Copernican revolution** (p. 31).

LO3 Copernicus suggested that if the planets orbited the Sun, rather than Earth, this would provide a simpler explanation of their observed motion. Galileo Galilei was the first experimental scientist. His telescopic observations of the Sun, Moon, and planets provided experimental evidence against the geocentric theory and supporting Copernicus's heliocentric model. Johannes Kepler constructed a set of three simple laws describing the motions of the planets around the Sun, explaining Tycho Brahe's detailed observational data.

LO4 Kepler's three **laws of planetary motion** (p. 34) state that (1) planetary orbits are **ellipses** (p. 34) having the Sun as one **focus** (p. 34), (2) a planet moves faster as its orbit takes it closer to the Sun, and (3)

the orbital **semimajor axis** (p. 34) is simply related to the planet's orbital **period** (p. 35).

LO5 The average distance from Earth to the Sun is one **astronomical unit** (p. 35), today precisely determined by bouncing **radar** (p. 36) signals off the planet Venus. Once this is known, the distances to all other planets can be inferred from Kepler's laws.

LO6 To change a body's velocity, a **force** (p. 34) must be applied. The rate of change of velocity, called the **acceleration** (p. 38), is equal to the applied force divided by the body's mass. **Gravity** (p. 38) attracts the planets to the Sun. Every object having any mass exerts a **gravitational force** (p. 38) on all other objects, and the strength of this force decreases with distance according to an **inverse-square law** (p. 39).

MasteringAstronomy®   For instructor-assigned homework go to www.masteringastronomy.com

Problems labeled **POS** explore the process of science. **VIS** problems focus on reading and interpreting visual information. **LO** connects to the introduction's numbered Learning Outcomes.

REVIEW AND DISCUSSION

1. **LO1** Describe the strengths and weaknesses of the geocentric model of the universe.

2. **POS** The benefit of our current knowledge lets us see flaws in the Ptolemaic model of the universe. What is its basic flaw?

3. What was the great contribution of Copernicus to our knowledge of the solar system?

4. **LO2** How did the heliocentric model explain planetary motions and brightness variations?

5. **LO3 POS** How did Galileo help confirm the views of Copernicus?

6. Why is Galileo often thought of as the first experimental scientist?

7. **LO4** Briefly describe Kepler's three laws of planetary motion.

8. **LO3** How did Tycho Brahe contribute to Kepler's laws?

9. **POS** What does it mean to say Kepler's laws are empirical?

10. **LO5** If radio waves cannot be reflected from the Sun, how can radar be used to find the distance from Earth to the Sun?

11. List the two modifications made by Newton to Kepler's laws.

12. Why do we say that a baseball falls toward Earth, and not Earth toward the baseball?

13. Why would a baseball thrown from the surface of the Moon go higher than one thrown with the same velocity from Earth's surface?

14. **LO6** According to Newton, why does Earth orbit the Sun?

15. What would happen to Earth if the Sun's gravity were suddenly "turned off"?

CONCEPTUAL SELF-TEST: TRUE OR FALSE?/MULTIPLE CHOICE

1. Aristotle proposed that all planets revolve around the Sun. (T/F)

2. During retrograde motion, planets actually stop and move backward in space. (T/F)

3. The heliocentric model of the universe holds that Earth is at the center and everything else moves around it. (T/F)

4. Copernicus's theories gained widespread scientific acceptance during his lifetime. (T/F)

5. Galileo's observations of the sky were made with the naked eye. (T/F)

6. The speed of a planet orbiting the Sun is independent of the planet's position in its orbit. (T/F)

7. Kepler's laws hold only for the six planets known in his time. (T/F)

8. You throw a baseball to someone; before the ball is caught, it is temporarily in orbit around Earth's center. (T/F)

9. A major flaw in Copernicus's model was that it still had (a) the Sun at the center; (b) Earth at the center; (c) retrograde loops; (d) circular orbits.

10. VIS An accurate sketch of Mars's orbit around the Sun would show (a) the Sun far off center; (b) an oval twice as long as it is wide; (c) a nearly perfect circle; (d) phases.

11. A calculation of how long it takes a planet to orbit the Sun would be most closely related to Kepler's (a) first law of orbital shapes; (b) second law of orbital speeds; (c) third law of planetary distances; (d) first law of inertia.

12. An asteroid with an orbit lying entirely inside Earth's (a) has an orbital semimajor axis of less than 1 AU; (b) has a longer orbital period than Earth's; (c) moves more slowly than Earth; (d) has a highly eccentric orbit.

13. If Earth's orbit around the Sun were to double in size, the new "year" would be (a) less than 2; (b) 2; (c) more than 2 current Earth years.

14. VIS As shown in Figure 1.8 (Venus Phases), Galileo's observations demonstrated that Venus must be (a) orbiting Earth; (b) orbiting the Sun; (c) larger than Earth; (d) similar to the Moon.

15. VIS Figure 1.17 (Gravity), showing the motion of a ball near Earth's surface, depicts how gravity (a) increases with altitude; (b) causes the ball to accelerate downward; (c) causes the ball to accelerate upward; (d) has no effect on the ball.

PROBLEMS

The number of squares preceding each problem indicates its approximate level of difficulty.

1. ■■ Tycho's observations were accurate to about 1 arc minute (1′). To what distance does this correspond at the distance of (a) the Moon, (b) the Sun, (c) Saturn at closest approach? (∞*Problem 0.8*)

2. ■ How long would a radar signal take to complete a round-trip between Earth and Mars when the two planets are 0.7 AU apart?

3. ■■ Seen from Earth, through what angle will Mars appear to move relative to the stars over the course of 24 hours when the two planets are at closest approach? Assume that Earth and Mars move on circular orbits of radii 1.0 AU and 1.5 AU, respectively, in the same plane. Will the apparent motion be prograde or retrograde?

4. ■■ An asteroid has a perihelion distance of 2.0 AU and an aphelion distance of 4.0 AU. Calculate its orbital semimajor axis, eccentricity, and period (see Figure 1.12).

5. ■■ Halley's Comet has a perihelion distance of 0.6 AU and an orbital period of 76 years. What is its aphelion distance from the Sun?

6. ■■ Using the data in Table 1.1, calculate how much farther Mercury is from the Sun at aphelion than at perihelion.

7. ■■■ Jupiter's moon Callisto orbits the planet at a distance of 1.88 million kilometer. Callisto's orbital period about Jupiter is 16.7 days. What is the mass of Jupiter? [Assume that Callisto's mass is negligible compared with that of Jupiter, and use the modified version of Kepler's third law (Section 1.4).]

8. ■■ The acceleration due to gravity at Earth's surface is 9.80 m/s². What is the acceleration at altitudes of (a) 100 km? (b) 1000 km? (c) 10,000 km? Take Earth's radius to be 6400 km.

9. ■ Use Newton's law of gravity to calculate the force of gravity between you and Earth. Convert your answer, which will be in newtons, to pounds using the conversion 4.45 N = 1 pound. What do you normally call this force?

10. ■■ The Moon's mass is 7.4×10^{22} kg, and its radius is 1700 km. What would be the period and the speed of a spacecraft moving in a circular orbit just above the lunar surface?

ACTIVITIES

Collaborative

1. You have been asked to arbitrate a dispute between a tour bus company and a nearby Native American tribe concerning a recently discovered ancient medicine wheel. Using sketches as necessary, compose a document describing what a medicine wheel is designed to do astronomically, and summarize the pros and cons of letting the public have unrestricted access to the site.

2. Select what you believe to be Galileo's single most important astronomical observation, state why you think it was most important, and explain using sketches what he observed.

Individual

3. Look in an almanac for the dates of opposition of Mars, Jupiter, and Saturn. At opposition, these planets are at their closest points to Earth and are at their largest and brightest in the night sky. Observe these planets. How long before opposition does each planet's retrograde motion begin? How long afterward does it end?

4. Use a small telescope to replicate Galileo's observations of Jupiter's largest moons. Note the brightness and location of the moons relative to Jupiter. If you watch over a period of several nights, draw what you see; the moons' positions will change as they orbit the planet.

5. Draw an ellipse (see Figure 1.11). You'll need two pins, a piece of string, and a pencil. Tie the string in a loop and place around the pins. Place the pencil inside the loop and run it around the inside of the string, holding the loop taut. The two pins will be at the foci of the ellipse. What is the eccentricity of the ellipse you have drawn? How does its shape change as you vary the distance between the pins?

2

Light and Matter
The Inner Workings of the Cosmos

Astronomical objects are more than just things of beauty in the night sky. Planets, stars, and galaxies are of vital significance if we are to understand fully the big picture—the grand design of the universe. Every object is a source of information about the universe—its temperature, its chemical composition, its state of motion, its past history. The starlight we see tonight began its journey to Earth decades, centuries, even millennia ago. The faint rays from the most distant galaxies have taken billions of years to reach us. The stars and galaxies in the night sky show us not just the far away, but also the long ago. In this chapter, we begin our study of how astronomers extract information from the light emitted by astronomical objects. The observational and theoretical techniques that enable researchers to determine the nature of distant atoms by the way they emit and absorb light are the indispensable foundation of modern astronomy.

THE BIG PICTURE

The human eye sees just a tiny fraction of the radiation that pervades the universe. A much larger cosmos lies unseen beyond the visible spectrum. Space is filled with many different kinds of invisible radiation, ranging from radio waves to gamma rays. Detailed spectroscopic study of this broad range of visible and invisible information is the primary way in which astronomers study stars and other distant objects.

LEARNING OUTCOMES

Studying this chapter will enable you to:

LO1 Describe the nature of electromagnetic radiation, and tell how that radiation transfers energy and information through interstellar space.

LO2 List the major regions of the electromagnetic spectrum, and arrange them in order of wavelength, frequency, or photon energy.

LO3 Explain how we can determine an object's temperature by observing the radiation it emits.

LO4 Describe the characteristics of continuous, emission, and absorption spectra and the conditions under which each is produced.

LO5 Specify the basic components of the atom, and describe our modern conception of atomic and molecular structure.

LO6 Explain how electron transitions within atoms produce unique emission and absorption spectra and how an object's composition can be determined from its spectrum.

LO7 Describe how the relative motion of an object can be determined from shifts in the wavelengths or frequencies of its spectral lines.

◄ This remarkable composite image of the nearby galaxy M106 was created by combining observations made in the radio (purple), infrared (red), visible (yellow/white), and X-ray (blue) parts of the spectrum. In the visible and infrared, M106 appears to be a rather normal disk galaxy, much like millions of others observed throughout the universe (see Chapter 15). But at radio and X-ray wavelengths, an entirely different picture emerges.

The data show huge "bubbles" of superheated gas apparently blasted out of the disk by intense jets of radiation emanating from a supermassive black hole at the galaxy's center. These observations illustrate a key mechanism by which galaxies are thought to evolve, as well as the importance of multiwavelength observations in our understanding of the workings of the cosmos. (*Chandra X-Ray Observatory*)

MasteringAstronomy®

Visit www.masteringastronomy.com for quizzes, animations, videos, interactive figures, and self-guided tutorials.

2.1 Information from the Skies

Figure 2.1 shows our nearest large galactic neighbor, which lies in the constellation Andromeda. On a dark, clear night, far from cities or other sources of light, the Andromeda galaxy, as it is generally called, can be seen with the naked eye as a faint, fuzzy patch on the sky, comparable in diameter to the full Moon. Yet the fact that it is visible from Earth belies this galaxy's enormous distance from us. It lies roughly 2.5 million light-years away. An object at such a distance is truly inaccessible in any realistic human sense.

Light and Radiation

How do astronomers know anything about objects far from Earth? The answer is that we use the laws of physics, as we know them here on Earth, to interpret the light, or **electromagnetic radiation** emitted by these objects. *Radiation* is any way in which energy is transmitted through space from one point to another without the need for any physical connection between those two locations. The term *electromagnetic* refers to the fact that the energy in a beam of light is actually carried in the form of rapidly changing *electric* and *magnetic* fields, as discussed in Section 2.2. Virtually all we know about the universe beyond Earth's atmosphere has been learned from analysis of electromagnetic radiation received from afar.

Visible light is the particular type of electromagnetic radiation to which the human eye happens to be sensitive. But, just as there are sound waves that humans can't hear, there is also *invisible* electromagnetic radiation, which goes completely undetected by our eyes. **Radio, infrared,** and **ultraviolet** waves, as well as **X-rays** and **gamma rays,** all fall into this category. Recognize that, despite the different names, these terms really all refer to the same thing. The names are just historical accidents, reflecting the fact that it took many years for scientists to realize that these apparently very different types of radiation are in reality one and the same physical phenomenon. Throughout this text, we will use the general terms *light* and *electromagnetic radiation* more or less interchangeably.

Wave Motion

All types of electromagnetic radiation travel through space in the form of **waves.** Simply stated, a wave is a way in which energy is transferred from place to place without physical movement of material from one location to another. In wave motion, the energy is carried by a *disturbance* of some sort that occurs in a distinctive, repeating pattern. Ripples on the surface of a pond, sound waves in air, and electromagnetic waves in space, despite their many obvious differences, all share this basic defining property.

As a familiar example, imagine a twig floating in a pond (Figure 2.2). A pebble thrown into the pond at some distance from the twig disturbs the surface of the water, setting it into up-and-down motion. This disturbance propagates outward from the point of impact in the form of waves. When the waves reach the twig, some of the pebble's energy is transferred to it, causing the twig to bob up and down. In this way, both energy and *information*—the fact that the pebble entered the water—are transferred from where the pebble landed to the location of the twig. We could tell just by

◀ **FIGURE 2.1 Andromeda Galaxy** The pancake-shaped Andromeda galaxy is about 2.5 million light-years away and contains a few hundred billion stars. *(R. Gendler)*

observing the twig that a pebble (or some small object) had entered the water. With a little additional physics, we could even estimate the pebble's energy.

A wave is *not* a physical object. No water traveled from the point of impact of the pebble to the twig; at any location on the surface, the water surface simply moved up and down as the wave passed. What, then, *does* move across the pond surface? The answer is that the wave is the *pattern* of up-and-down motion, and it is this pattern that is transmitted from one point to the next as the disturbance moves across the water.

Figure 2.3 shows how wave properties are quantified. The **wave period** is the number of seconds needed for the wave to repeat itself at some point in space. The **wavelength** is the number of meters needed for the wave to repeat itself at a given moment in time. It can be measured as the distance between two adjacent wave *crests*, two adjacent wave *troughs*, or any other two similar points on adjacent wave cycles (for example, the points marked "×" in Figure 2.3). The maximum departure of the wave from the undisturbed state—still air, say, or a flat pond surface—is called its **amplitude.** The number of wave crests passing any given point per unit time is called the wave's **frequency.** If a wave of a given wavelength moves at high speed, then many crests pass by per second and the frequency is high. Conversely, if the same wave moves slowly, then its frequency is low. The frequency of a wave is just one divided by the wave's period:

$$\text{frequency} = \frac{1}{\text{period}}.$$

Since frequency counts events per unit time, it is expressed in units of inverse time (cycles per second), called hertz (Hz) in honor of the 19th-century German scientist Heinrich Hertz, who studied the properties of radio waves. For example, a wave with a period of 5 seconds (5 s) has a frequency of (1/5) cycles/s = 0.2 Hz, meaning that one wave crest passes a given point in space every 5 seconds.

A wave moves a distance equal to one wavelength in one wave period. The product of wavelength and frequency therefore equals the *wave velocity:*

$$\text{wavelength} \times \text{frequency} = \text{velocity}.$$

Thus, if the wave in our earlier example had a wavelength of 0.5 m, its velocity is (0.5 m) × (0.2 Hz) = 0.1 m/s. Wavelength and wave frequency are *inversely* related—doubling one halves the other.

Waves ripple out from where a pebble hit the water . . .

. . . to where a twig is floating.

This insert shows a series of "snapshots" of the pond surface as the wave passes by.

Undisturbed pond surface

Direction of wave motion

▲ **FIGURE 2.2 Water Wave INTERACTIVE** The passage of a wave across a pond causes the surface of the water to bob up and down, but there is no movement of water from one part of the pond to another.

CONCEPT CHECK

What is a wave? What four basic properties describe a wave, and what relationships, if any, exist among them?

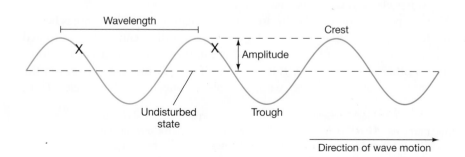

Wavelength

Crest

Amplitude

X   X

Undisturbed state

Trough

Direction of wave motion

◄ **FIGURE 2.3 Wave Properties INTERACTIVE** A typical wave has a direction of motion, wavelength, and amplitude. In one wave period, the entire pattern shown here moves one wavelength to the right.

In diffraction, a wave bends around an obstacle, such as a breakwater.

Motion

Breakwater

(a)

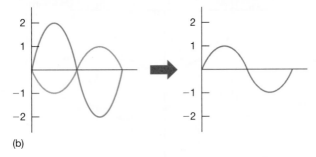

In interference, two waves combine to produce a wave equal to the sum of the two.

(b)

▲ **FIGURE 2.4 Wave Behavior**   Diffraction (a) and interference (b) are two basic types of behavior shared by all waves, including electromagnetic radiation.

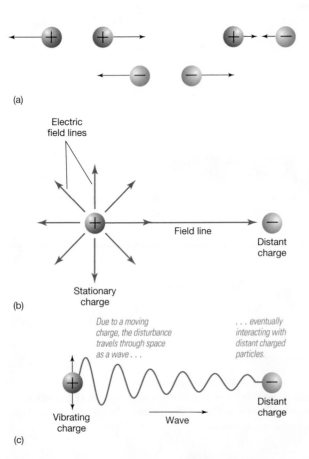

(a)

Electric field lines

Field line

Distant charge

Stationary charge

(b)

Due to a moving charge, the disturbance travels through space as a wave . . .

. . . eventually interacting with distant charged particles.

Vibrating charge

Wave

Distant charge

(c)

◀ **FIGURE 2.5 Charged Particles**   (a) Particles carrying like electrical charges repel one another; particles with unlike charges attract. (b) A charged particle is surrounded by an electric field, which determines the particle's influence on other charged particles. (c) If a charged particle begins to vibrate, its electric field changes.

2.2 Waves in What?

Scientists know that radiation travels as a wave because light displays characteristic behavior common to *all* kinds of wave motion. For example, light waves tend to "bend around corners," just like sound waves or ocean waves passing a breakwater (Figure 2.4a). This is called **diffraction.** In addition, the crests and troughs of waves coming from different sources can reinforce or partly cancel one another (Figure 2.4b). This is known as **interference.** We are not normally aware of this behavior in the case of electromagnetic radiation because the effects are generally too small to be noticeable in everyday life. However, they are easily measured in the laboratory and are very important considerations when designing and building telescopes, as we will see in Chapter 3. Their discovery early in the 19th century provided strong evidence in favor of the theory that light moves as a wave.

Waves of radiation differ in one fundamental respect from water waves, sound waves, or any other waves that travel through a material medium—radiation needs *no* medium. When light travels from a distant cosmic object, it moves through the vacuum of space. Sound waves, by contrast, cannot do this, despite what you hear in sci-fi movies! If we were to remove all the air from a room, conversation would be impossible because sound waves cannot exist without air or some other physical medium to support them. Communication by flashlight or radio, however, would be entirely feasible.

The ability of light to travel through empty space was once a great mystery. The idea that radiation could move as a wave through nothing at all seemed to violate common sense, yet it is now a cornerstone of modern physics.

Interactions Between Charged Particles

To understand more about the nature of light, consider an *electrically charged* particle, such as an **electron** or a **proton.** Electrons and protons are elementary particles—fundamental components of matter—that carry the basic unit of charge. Electrons are said to carry a *negative* charge, while protons carry an equal and opposite *positive* charge. Just as a massive object exerts a gravitational force on any other massive object, an electrically charged particle exerts an *electrical* force on every other charged particle in the universe. ∞ *(Sec. 1.4)* Unlike gravity, however, which is always attractive, electrical forces can be either attractive or repulsive. Particles having like charges (both negative or both positive) repel one another; particles having unlike charges attract (Figure 2.5a).

Extending outward in all directions from our charged particle is an **electric field,** which determines the electric force exerted by the particle on other charged particles (Figure 2.5b). The strength of the electric field, like that of the gravitational field, decreases with increasing distance from the source according to an inverse-square law—doubling the distance reduces the strength of the field by a factor of 4. ∞ *(Sec. 1.4)* By means of the electric field, the particle's presence is "felt" by other charged particles, near and far.

Now suppose our particle begins to vibrate, perhaps because it becomes heated or collides with some other particle. Its changing position causes its associated electric field to change, and this changing field in turn causes the electrical force exerted on other charges to vary (Figure 2.5c). If we measure the changes in the forces on these other charges, we learn about our original particle. Thus, *information about our particle's motion is transmitted through space via a changing electric field.* This *disturbance* in the particle's electric field travels through space as a wave.

Electromagnetism

The laws of physics tell us that a **magnetic field** must accompany every changing electric field. Magnetic fields govern the influence of *magnetized* objects on one another, much as electric fields govern interactions between charged particles. The fact that one end of a compass needle always points to magnetic north is the result of the interaction between the magnetized needle and Earth's magnetic field (Figure 2.6). Magnetic fields also exert forces on moving electric charges (that is, electric currents)—electric meters and motors rely on this basic fact. Conversely, moving charges create magnetic fields (the electromagnets found in loudspeakers and electric motors are familiar examples).

Electric and magnetic fields are inextricably linked to one another. A change in either one *necessarily* creates the other. For this reason, the disturbance produced by our moving charge actually consists of oscillating electric *and* magnetic fields, always oriented perpendicular to one another and moving together through space (Figure 2.7). These fields do not exist as independent entities. Rather, they are different aspects of a single physical phenomenon: **electromagnetism.** Together they constitute an *electromagnetic wave* that carries energy and information from one part of the universe to another.

Now consider a distant cosmic object—a star. It is made up of charged particles, mainly protons and electrons, in constant motion. As these charged contents move around, their electric fields change, and electromagnetic waves are produced. These waves travel outward into space, and eventually some reach Earth. Other charged particles, either in the molecules in our eyes or in our experimental apparatus (a telescope or radio receiver, for example), respond to the electromagnetic field changes by vibrating in tune with the received radiation. This response is how we "see" the radiation—with our eyes or with our detectors.

How fast does an electromagnetic wave travel? Both theory and experiment tell us that all electromagnetic waves move at a very specific speed—the **speed of light** (always denoted by the letter c). Its value is 299,792.458 km/s in a vacuum (and somewhat less in material substances, such as air or water). In this text, we round this value off to $c = 3.00 \times 10^5$ km/s. This is an extremely high speed. In the time needed to snap your fingers—about a tenth of a second—light can travel three quarters of the way around our planet! According to the theory of relativity (see *More Precisely 13-1*), the speed of light is the fastest speed possible.

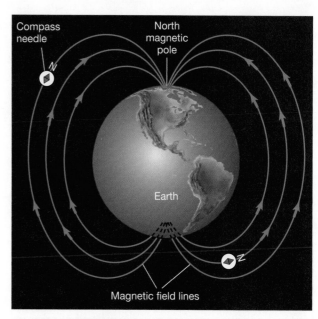

▲ **FIGURE 2.6 Magnetism** Earth's magnetic field interacts with a magnetic compass needle, causing the needle to become aligned with the field—that is, to point toward Earth's north (magnetic) pole.

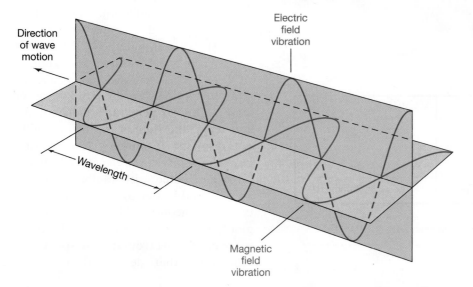

◀ **FIGURE 2.7 Electromagnetic Wave** Electric and magnetic fields vibrate perpendicular to each other. Together they form an electromagnetic wave that moves through space at the speed of light.

2.3 The Electromagnetic Spectrum

White light is a mixture of colors, which we conventionally divide into six major hues—red, orange, yellow, green, blue, and violet. As shown in Figure 2.8, we can separate a beam of white light into a rainbow of these basic colors—called a *spectrum* (plural: spectra)—by passing it through a prism. This experiment was first reported by Isaac Newton more than 300 years ago. In principle, the original beam of white light could be recovered by passing the spectrum through a second prism to recombine the colored beams.

Components of Visible Light

What determines the color of a beam of light? The answer is its frequency (or, equivalently, its wavelength). We see different colors because our eyes react differently to electromagnetic waves of different frequencies. Red light has a frequency of roughly 4.3×10^{14} Hz, corresponding to a wavelength of about 7.0×10^{-7} m. Violet light, at the other end of the visible range, has nearly double the frequency—7.5×10^{14} Hz—and just over half the wavelength—4.0×10^{-7} m. The other colors we see have frequencies and wavelengths intermediate between these two extremes.

Because the wavelength of visible light is so small, scientists often use a unit called the *nanometer* (nm) when describing it (see Appendix 2). A nanometer is one-billionth of a meter, or 10^{-9} m. An older unit called the *angstrom* ($1\text{Å} = 10^{-10}$ m = 0.1 nm) is also widely used by many astronomers and atomic physicists. Thus, the visible spectrum covers the wavelength range from 400 to 700 nm. The radiation to which our eyes are most sensitive has a wavelength near the middle of this range, at about 550 nm, in the yellow-green region of the spectrum.

The Full Range of Radiation

Figure 2.9 plots the entire range of electromagnetic radiation. To the low-frequency, long-wavelength side of visible light lies radio and infrared radiation. Radio frequencies include radar, microwave radiation, and the familiar AM, FM, and TV bands. We perceive infrared radiation as heat. To the high-frequency, short-wavelength side of visible light lies ultraviolet, X-ray, and gamma-ray radiation. Ultraviolet radiation, lying just beyond the violet end of the visible spectrum, is responsible for suntans and sunburns. X-rays are best known for their ability to penetrate human tissue and reveal the state of our insides without resorting to surgery. Gamma rays have the shortest wavelengths. They are often associated with radioactivity and are invariably damaging to any living cells they encounter. All these spectral regions, including the visible, collectively make up the **electromagnetic spectrum.** Remember that despite their greatly

▼ FIGURE 2.8 **Visible Spectrum**   When passed through a prism, white light splits into its component colors, spanning red to violet in the visible part of the electromagnetic spectrum. The slit narrows the beam of radiation. The "rainbow" of colors projected on the screen is just a series of different-colored images of the slit.

▲ **FIGURE 2.9 Electromagnetic Spectrum INTERACTIVE**
The entire electromagnetic spectrum, running from long-wavelength, low-frequency radio waves to short-wavelength, high-frequency gamma rays. Note how small the visible portion of the spectrum is.

differing wavelengths and the very different roles they play in everyday life on Earth, all types of electromagnetic radiation are basically the same and all move at the same speed—the speed of light c.

Note that wave frequency (in hertz) in Figure 2.9 increases from left to right, while wavelength (in meters) increases from right to left. Scientists often disagree on the "correct" way to display wavelengths and frequencies in diagrams of this type, and it is common for astronomers working in different parts of the spectrum to adopt opposite conventions. Throughout this book we will consistently display frequency increasing toward the *right*.

The wavelength and frequency scales in Figure 2.9 do not increase by equal increments of 10. Instead, successive values on the horizontal axis increase by *factors* of 10—each successive value is 10 times greater than its neighbor. This type of scale (called a *logarithmic* scale) is often used in science to condense a very large range of data into a manageable size. We will use such scales frequently throughout this book.

Many objects emit only a tiny fraction of their total energy in the visible range. A wealth of knowledge can be gained by studying the invisible regions of the electromagnetic spectrum. To remind you of this important fact and to identify the region of the electromagnetic spectrum in which a particular observation was made, we have attached a spectrum icon—an idealized version of the wavelength scale in Figure 2.9—to every astronomical image presented in this text.

Only a small fraction of the radiation arriving at our planet from space actually reaches Earth's surface because of the *opacity* of Earth's atmosphere. **Opacity** is the extent to which radiation is blocked by the material through which it is passing. The more opaque an object is, the less radiation gets through. For example, a pane of glass has low opacity (i.e., is *transparent*) to visible light, while a sheet of paper or the air on a foggy day has high opacity. Earth's atmospheric opacity is plotted along the bottom of Figure 2.9. Where the shading is greatest, no radiation can get in or out. Where there is no shading at all, our atmosphere is almost totally transparent.

Note that there are just a few *windows* (the unshaded regions) at well-defined locations in the electromagnetic spectrum, where Earth's atmosphere is transparent. In much of the radio and all of the visible portion of the spectrum, the opacity is low, and we can study the universe at those wavelengths from ground level. In parts of the infrared range, the atmosphere is partially

2.1 MORE PRECISELY

The Kelvin Temperature Scale

The atoms and molecules that make up any piece of matter are in constant random motion. This motion represents a form of energy known as *thermal energy*. The quantity we call temperature is a direct measure of this internal motion: the higher an object's temperature, the faster the random motion of its constituent particles. More precisely, the temperature of a piece of matter specifies the average thermal energy of the particles it contains.

The temperature scale probably most familiar to you, the Fahrenheit scale, is now a peculiarity of American society. Most of the rest of the world uses the Celsius temperature scale, in which water freezes at 0 degrees (0°C) and boils at 100 degrees (100°C), as illustrated in the accompanying figure.

There are, of course, temperatures below the freezing point of water. Temperatures can in principle reach as low as –273.15°C (although we know of no matter anywhere in the universe that is actually this cold); this is the temperature at which, theoretically, all atomic and molecular motion effectively ceases. It is convenient to construct a temperature scale based on this lowest possible temperature, which is called *absolute zero*. Scientists commonly use such a scale, called the Kelvin scale in honor of the 19th-century British physicist Lord Kelvin. Since it takes absolute zero as its starting point, the Kelvin scale differs from the Celsius scale by 273.15°. In this book, we round off the decimal places and simply use the relationship

kelvins = degrees Celsius + 273.

Thus,

- Thermal motion of atoms and molecules ceases at 0 kelvins (0 K).
- Water freezes at 273 kelvins (273 K).
- Water boils at 373 kelvins (373 K).

Note, by the way, that the unit is "kelvins," or "K," not "degrees kelvin" or "°K."

	Fahrenheit	Celsius	Kelvin
Hydrogen fuses	18,000,032	10,000,000	10,000,273
Water boils	212	100	373
Water freezes	32	0	273
All thermal motion stops	–459	–273	0

transparent, so we can make some infrared observations from the ground. Over the rest of the spectrum, however, the atmosphere is opaque. As a result, most infrared, and all ultraviolet, X-ray, and gamma-ray observations, can be made only from high-flying balloons or (more commonly) above the atmosphere, from orbiting satellites (see Section 3.5).

2.4 Thermal Radiation

All macroscopic objects—fires, ice cubes, people, stars—emit radiation at all times. They radiate because the microscopic charged particles in them are in constant random motion, and whenever charges change their state of motion, electromagnetic radiation is emitted. The **temperature** of an object is a direct measure of the amount of microscopic motion within it (see *More Precisely 2-1*). The hotter the object, the higher its temperature, the faster its constituent particles move and the more energy they radiate.

The Blackbody Spectrum

Intensity is a term often used to specify the amount or strength of radiation at any point in space. Like frequency and wavelength, intensity is a basic property of electromagnetic radiation. No natural object emits all of its radiation at just one frequency. Instead, the energy is often spread out over a range of frequencies. By studying the way in which the intensity of this radiation is distributed across the electromagnetic spectrum, we can learn much about the object's properties.

Figure 2.10 (a) illustrates schematically the distribution of radiation emitted by any object. Note that the curve peaks at a single, well-defined frequency (marked) and falls off to lesser values above and below that frequency. However, the curve is not symmetrical about the peak—the falloff is more rapid on the high-frequency side of the peak than it is toward lower frequencies. This overall shape is characteristic of the electromagnetic radiation emitted by *any* object, regardless of its size, shape, composition, or temperature.

The curve drawn in Figure 2.10(a) refers to a mathematical idealization known as a *blackbody*—an object that absorbs all radiation falling upon it. In a steady state (that is, if the temperature remains constant), a blackbody must reemit the same amount of energy as it absorbs. The **blackbody curve** shown in the figure describes the distribution of that reemitted radiation. No real object absorbs or radiates as a perfect blackbody. For example, the Sun's actual emission is shown in Figure 2.10(b). However, in many cases the blackbody curve is a very good approximation to reality, and the properties of blackbodies provide important insights into the behavior of real objects, including stars.

 SELF-GUIDED TUTORIAL
Continuous Spectra and Blackbody Radiation

(a)

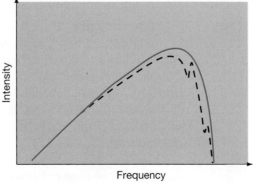

(b)

◄ **FIGURE 2.10 Blackbody Curves, Ideal vs. Reality**   The blackbody curve represents the spread of the intensity of radiation emitted across the electromagnetic spectrum. Note the contrast between the perfect "textbook" case (a) and a real graph (dashed) of the Sun's emission (b). The difference is due to absorption in the atmospheres of the Sun and Earth.

The Radiation Laws

As illustrated in Figure 2.11, the entire blackbody curve shifts toward higher frequencies (shorter wavelengths) and greater intensities as an object's temperature increases. Even so, the *shape* of the curve remains the same. This shifting of radiation's peak frequency with temperature is familiar to us all. Very hot glowing objects, such as lightbulb filaments or stars, emit visible light because their blackbody curves peak in or near the visible range. Cooler objects, such as warm rocks or household radiators, produce invisible radiation—they are warm to the touch but are not glowing hot to the eye. These objects emit most of their radiation in the lower-frequency infrared portion of the electromagnetic spectrum.

There is a very simple connection between the frequency or wavelength at which most radiation is emitted and the absolute temperature (that is, temperature measured in kelvins—see *More Precisely 2-1*) of the emitting object: The peak frequency is *directly proportional* to the temperature. This relationship, known as **Wien's law** after Wilhelm Wien, the German scientist who formulated it in 1897, is usually expressed as:

$$\text{wavelength of peak emission} \propto \frac{1}{\text{temperature}}.$$

The symbol "$\propto$" just means "is proportional to." Wien's law tells us that the hotter the object, the bluer its radiation. For example, an object with a temperature of 6000 K (Figure 2.11c) emits most of its energy in the visible part of the spectrum, with a peak wavelength of 480 nm. At 600 K (Figure 2.11b), the object's emission would peak at 4800 nm, well into the infrared. At a temperature of 60,000 K (Figure 2.11d), the peak would move all the way through the visible range to a wavelength of 48 nm, in the ultraviolet.

It is also a matter of everyday experience that as the temperature of an object increases, the *total* amount of energy it radiates (summed over all frequencies) increases rapidly. For example, the heat given off by an electric heater increases sharply as the heater warms up and begins to emit visible light. In fact, the total amount of energy radiated per unit time is proportional to the *fourth power* of an object's temperature:

$$\text{total energy radiated per second} \propto \text{temperature}^4.$$

This relationship is called **Stefan's law,** after the 19th-century Austrian physicist Josef Stefan. It implies that the energy emitted by a body rises dramatically as the body's temperature increases. Doubling the temperature, for example, causes the total energy radiated to increase by a factor of 16. The radiation laws are presented in more detail in *More Precisely 2-2*.

Astronomical Applications

Astronomers use blackbody curves as thermometers to determine the temperatures of distant objects. For example, study of the solar spectrum makes it possible to measure the temperature of the Sun's surface. Observations of the electromagnetic radiation from the Sun at many frequencies yield a curve shaped somewhat like that shown in Figure 2.10. The Sun's curve peaks in the visible part of the electromagnetic spectrum. The Sun also emits a lot of infrared and a little ultraviolet radiation. Applying Wien's law to the blackbody curve that best fits the solar spectrum, we find that the temperature of the Sun's surface is approximately 6000 K (Figure 2.11c). A more precise measurement, based on the detailed solar spectrum (Section 2.6), yields a temperature of 5800 K.

Other cosmic objects have surfaces very much cooler or hotter than the Sun's, emitting most of their radiation in invisible parts of the spectrum. For example,

---

MA ANIMATION/VIDEO
Solar Eclipse Viewed from Space, in X-rays

MA ANIMATION/VIDEO
Shoemaker-Levy 9 Impact at 2.15 Microns

MA ANIMATION/VIDEO
Multispectral View of Orion Nebula

MA ANIMATION/VIDEO
Earth Aurora in X-rays

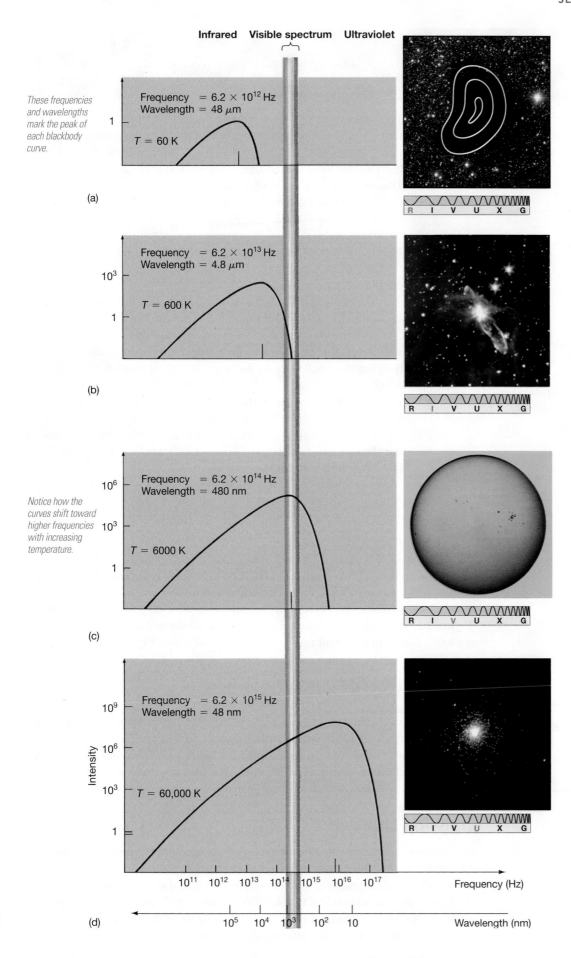

These frequencies and wavelengths mark the peak of each blackbody curve.

Infrared   Visible spectrum   Ultraviolet

(a)

Frequency = 6.2×10^{12} Hz
Wavelength = 48 μm

T = 60 K

(b)

Frequency = 6.2×10^{13} Hz
Wavelength = 4.8 μm

T = 600 K

Notice how the curves shift toward higher frequencies with increasing temperature.

(c)

Frequency = 6.2×10^{14} Hz
Wavelength = 480 nm

T = 6000 K

(d)

Frequency = 6.2×10^{15} Hz
Wavelength = 48 nm

T = 60,000 K

Intensity

Frequency (Hz)

10^{11} 10^{12} 10^{13} 10^{14} 10^{15} 10^{16} 10^{17}

10^5 10^4 10^3 10^2 10

Wavelength (nm)

◄ **FIGURE 2.11 Blackbody Curves**

INTERACTIVE Comparison of blackbody curves for four cosmic objects. (a) A cool, dark galactic cloud called Barnard 68. At a temperature of 60 K, it emits mostly radio radiation, shown here as overlaid contours. (b) A dim, young star (shown white in the inset photograph) called Herbig-Haro 46. The star's atmosphere, at 600 K, radiates mainly in the infrared. (c) The Sun's surface, at approximately 6000 K, is brightest in the visible region of the electromagnetic spectrum. (d) Some very hot, bright stars in a cluster called Messier 2, as observed by an orbiting space telescope above Earth's atmosphere. At a temperature of 60,000 K, these stars radiate strongly in the ultraviolet. *(ESO; AURA; NASA; Kennedy Space Center)*

2.2 MORE PRECISELY

More on the Radiation Laws

Wien's law relates the temperature T of an object to the wavelength λ_{max} at which the object's blackbody radiation spectrum peaks. (The Greek letter λ—lambda—is conventionally used to denote wavelength.) Mathematically, if we measure T in kelvins, we find that

$$\lambda_{max} = \frac{0.29 \text{ cm}}{T}.$$

Thus, at 6000 K (the approximate surface temperature of the Sun), the wavelength of maximum intensity is 0.29/6000 cm, or 480 nm, corresponding to the yellow-green part of the visible spectrum. A cooler star with a temperature of 3000 K has a peak wavelength of 970 nm, just longward of the red end of the visible spectrum, in the near infrared. The blackbody curve of a star with a temperature of 12,000 K peaks at 242 nm, in the near ultraviolet, and so on.

We can also give Stefan's law a more precise mathematical formulation. With T again measured in kelvins, the total amount of energy emitted per square meter of surface per second (a quantity known as the *energy flux F*) is given by

$$F = \sigma T^4.$$

energy per unit area | constant | temperature to the fourth power

This equation is usually referred to as the *Stefan-Boltzmann* equation. (Stefan's student, Ludwig Boltzmann, was an Austrian physicist who played a central role in the development of the laws of thermodynamics during the late 19th and early 20th centuries.) The constant σ (the Greek letter sigma) is known as the Stefan-Boltzmann constant. In SI units, the Stefan-Boltzmann constant has the value $\sigma = 5.67 \times 10^{-8}$ W/m^2 K^4.

Thus, red-hot metal in a furnace, at a temperature of 3500 K, radiates energy at a rate of about 850 W for every square centimeter of its surface area. Doubling its temperature to 7000 K (so that it becomes yellow to white hot, by Wien's law) increases the energy emitted by a factor of 16 (four "doublings"), to 13.6 *kilo*watts (kW) (13,600 W) per square centimeter.

Finally, note that the law relates to energy emitted *per unit area*. The flame of a blowtorch is considerably hotter than a bonfire, but the bonfire emits far more energy in total, because it is much larger.

Thus, in computing the total energy emitted from a hot object, both the object's temperature and its surface area must be taken into account. This fact is of great importance in determining the "energy budget" of planets and stars, as we will see in later chapters.

CONCEPT CHECK

Describe, in terms of the radiation laws, how the appearance of an incandescent lightbulb changes as you turn a dimmer switch to increase its brightness from "off" to "maximum."

the relatively cool surface of a very young star might measure 600 K and emit mostly infrared radiation (Figure 2.11b). Cooler still is the interstellar gas cloud from which the star formed. At a temperature of 60 K, such a cloud would emit mainly long-wavelength radiation in the radio and infrared parts of the spectrum (Figure 2.11a). The brightest stars, by contrast, have surface temperatures as high as 60,000 K and, hence, emit mostly ultraviolet radiation (Figure 2.11d).

2.5 Spectroscopy

Radiation can be analyzed with an instrument known as a **spectroscope.** In its most basic form, this device consists of an opaque barrier with a slit in it (to form a narrow beam of light), a prism (to split the beam into its component colors), and either a detector or a screen (to allow the user to view the resulting spectrum). Figure 2.12 shows such an arrangement.

Emission Lines

The spectra encountered in the previous section are examples of **continuous spectra.** A lightbulb, for instance, emits radiation of all wavelengths (but mostly in the visible and near-infrared ranges), with an intensity distribution that is well described by the blackbody curve corresponding to the bulb's temperature. Viewed through a spectroscope, the spectrum of the light from the bulb would show the familiar rainbow running from red to violet without interruption, as presented in Figure 2.12.

Not all spectra are continuous, however. For instance, if we took a glass jar containing pure hydrogen gas and passed an electrical discharge through it (a little like a lightning bolt arcing through Earth's atmosphere), the gas would begin to glow—that is, it would emit radiation. If we were to examine that radiation with our spectroscope, we would find that its spectrum consisted of only a few bright lines on an otherwise dark background, quite unlike the continuous spectrum of the lightbulb. Figure 2.13 shows this schematically (the lenses have been removed for clarity), and a more detailed rendering of the spectrum of hydrogen appears in the top panel of Figure 2.14. The light produced by the hydrogen in this experiment does *not* consist of all possible colors but instead includes only a few narrow, well-defined **emission lines**, narrow "slices" of the continuous spectrum. The black background represents all the wavelengths *not* emitted by hydrogen.

After some experimentation, we would also find that although we could alter the intensity of the lines (for example, by changing the amount of hydrogen in the jar or the strength of the electrical discharge), we could not alter their color (in other words, their frequency or wavelength). The particular pattern of spectral emission lines shown in Figure 2.13 is a property of the element hydrogen— whenever we perform this experiment, the same characteristic **emission spectrum** is the result. Other elements yield different emission spectra. Depending on which element is involved, the pattern of lines can be fairly simple or very complex. Always, though, it is *unique* to that element. The emission spectrum of a gas thus provides a "fingerprint" that allows scientists to deduce its presence by spectroscopic means. An analogy is a bar code that specifies uniquely the cost of a supermarket item. Examples of the emission spectra of some common substances are shown in Figure 2.14.

Absorption Lines

When sunlight is split by a prism, at first glance it appears to produce a continuous spectrum. However, closer scrutiny shows that the solar spectrum is interrupted by a large number of narrow dark lines, as shown in Figure 2.15. These lines represent wavelengths of light that have been removed (absorbed) by gases present either in the outer layers of the Sun or in Earth's atmosphere. These gaps in the spectrum are called **absorption lines**. The absorption lines in the solar spectrum are referred to collectively as *Fraunhofer lines*, after the 19th-century German physicist Joseph Fraunhofer, who measured and cataloged more than 600 of them.

▲ FIGURE 2.12   **Spectroscope** A simple spectroscope allows a narrow beam of light to pass through a thin slit and then into a prism where it is split into its component colors. A lens then focuses the light into a sharp image that is either projected onto a screen, as shown here, or analyzed as it strikes the detector.

(MA) **SELF-GUIDED TUTORIAL**
Emission Spectra

CONCEPT CHECK

What are absorption and emission lines? What do they tell us about the properties of the gas producing them?

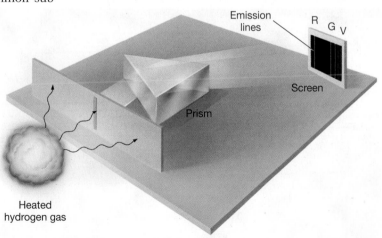

▶ FIGURE 2.13 **Emission Spectrum** INTERACTIVE   Instead of a continuous spectrum, the light from excited (heated) hydrogen gas consists of a series of distinct spectral lines. (For simplicity, the focusing lens has been omitted.)

Spectra are like supermarket bar codes that uniquely specify the type and cost of a product.

US $24.00 / $36.50 CAN
ISBN 0-465-07835-4
52400
9 780465 078356

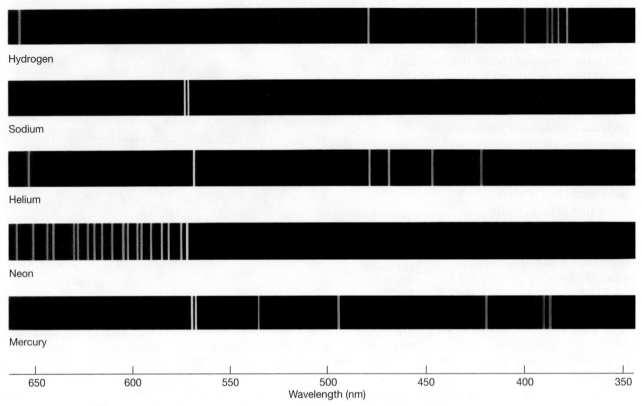

▲ **FIGURE 2.14 Elemental Emission** The emission spectra of some well-known elements. Note that wavelengths shorter than 400 nm, shown here in shades of purple, are in the ultraviolet part of the spectrum and thus invisible to the human eye. *(Wabash Instrument Corporation)*

At around the time solar absorption lines were discovered, scientists found that absorption lines could also be produced in the laboratory by passing a beam of light from a continuous source through a cool gas, as shown in Figure 2.16. They quickly observed a connection between emission and absorption lines: The absorption lines associated with a given gas occur at precisely the *same* wavelengths as the emission lines produced when the gas is heated. Both sets of lines therefore contain the *same* information about the composition of the gas.

The study of the ways in which matter emits and absorbs radiation is called **spectroscopy.** The observed relationships between the three types of spectra—continuous, emission line, and absorption line—are illustrated in Figure 2.17 and may be summarized as follows:

The scale of this spectrum extends from long wavelengths in red . . .

. . . to short wavelengths in blue.

◄ **FIGURE 2.15 Solar Spectrum** The Sun's visible spectrum shows hundreds of dark absorption lines superimposed on a bright continuous spectrum. This high-resolution spectrum is displayed in a series of 48 horizontal strips stacked vertically; each strip covers a small portion of the entire spectrum from left to right. If the strips were placed side by side, the full spectrum would be some 6 meters (20 feet) across! (AURA)

1. A luminous solid or liquid, or a sufficiently dense gas, emits light of all wavelengths and so produces a continuous spectrum of radiation (Figure 2.12).

2. A low-density hot gas emits light whose spectrum consists of a series of bright emission lines. These lines are characteristic of the chemical composition of the gas (Figure 2.13).

3. A low-density cool gas absorbs certain wavelengths from a continuous spectrum, leaving dark absorption lines in their place, superimposed on the continuous spectrum. These lines are characteristic of the composition of the intervening gas. They occur at precisely the same wavelengths as the emission lines produced by the gas at higher temperatures (Figure 2.16).

These rules are collectively known as **Kirchhoff's laws,** after the German physicist Gustav Kirchhoff, who published them in 1859.

Astronomical Applications

Once astronomers realized that spectral lines were indicators of chemical composition, they set about identifying the observed lines in the Sun's spectrum. Almost all of the lines observed from extraterrestrial sources could be attributed to known elements. For example, many of the Fraunhofer lines in sunlight are associated with the element iron. However, some unfamiliar lines also appeared in the solar spectrum. In 1868 astronomers realized that those lines must correspond to a previously unknown element. It was given the name helium, after the Greek word *helios*, meaning "Sun." Only in 1895 was helium discovered on Earth.

The development of spectroscopy is another example of the scientific method in action. ∞ *(Sec. 0.5)* As technology evolved and experimental measurement techniques improved, scientists realized that spectra were unique fingerprints of matter. They then went on to use this knowledge as a means of determining the composition of objects that were otherwise impossible to reach.

Yet for all the information that 19th-century astronomers could extract from observations of stellar spectra, they still lacked a theory explaining *how* those spectra arose. Despite their sophisticated spectroscopic equipment, they knew little more about the physics of stars than did Galileo or Newton. The next step in the circle of scientific progress was an explanation of Kirchhoff's empirical laws in terms of fundamental physical principles, just as Newtonian mechanics explained Kepler's empirical laws of planetary motion. ∞ *(Sec. 1.4)* And as with Newton's laws, the new theory—called *quantum mechanics*—opened the door to an explosion of scientific understanding far beyond the context in which it was originally conceived.

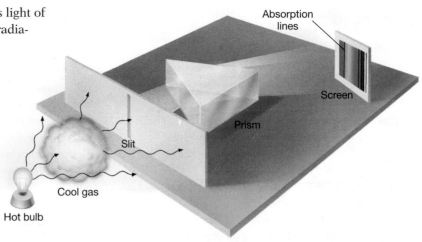

▲ **FIGURE 2.16 Absorption Spectrum INTERACTIVE** When a cool gas is placed between a source of continuous radiation (such as a hot lightbulb) and the detector/screen, the resulting color spectrum is crossed by a series of dark absorption lines. These lines are formed when the intervening cool gas absorbs certain wavelengths (colors) from the original beam of light. The absorption lines appear at precisely the same wavelengths as the emission lines that would be produced if the gas were heated to high temperatures (see Figure 2.13).

MA SELF-GUIDED TUTORIAL
Absorption Spectra

► **FIGURE 2.17 Kirchhoff's Laws** A source of continuous radiation, here represented by a lightbulb, is used to illustrate Kirchhoff's laws of spectroscopy. (a) The unimpeded beam shows the familiar continuous spectrum of colors. (b) When the source is viewed through a cloud of hydrogen gas, a series of dark hydrogen absorption lines appear in the continuous spectrum. These lines are formed when the gas absorbs some of the bulb's radiation and reemits it in random directions. (c) When the gas is viewed from the side, a fainter hydrogen emission spectrum is seen, consisting of reemitted radiation.

Atomic Structure and Spectroscopy

Fifty-seven percent of students had problems connecting atomic structure to an atom's emission and absorption of energy. Remember the following:

- Electrons in atoms can occupy only certain orbits, with precisely defined energies; the allowed energies increase as we move outward from the nucleus.

- When an electron moves from one orbit to another, the law of conservation of energy says that the change must be accompanied by the emission (if the electron drops from a high-energy orbit to a lower one) or absorption (low to high) of a photon of radiation.

- The energy of the photon is exactly equal to the change in orbital energy. High-energy photons have shorter wavelength (higher frequency).

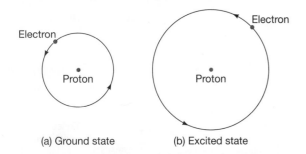

(a) Ground state   (b) Excited state

▲ **FIGURE 2.18 Classical Atom** An early-20th-century conception of the hydrogen atom—the Bohr model—pictured its electron orbiting the central proton in a well-defined orbit, much like a planet orbiting the Sun. Two electron orbits of different energies are shown: (a) the ground state and (b) an excited state.

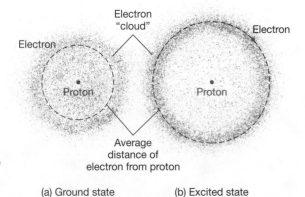

(a) Ground state   (b) Excited state

▲ **FIGURE 2.19   Modern Atom** The modern view of the hydrogen atom sees the electron as a "cloud" surrounding the nucleus.

2.6 Formation of Spectral Lines

By the start of the 20th century, physicists knew that light sometimes behaves in a manner that cannot be explained by the wave theory of radiation. Absorption and emission lines are produced only at certain, very specific wavelengths. This would not be the case if light behaved only as a continuous wave and matter always obeyed the laws of Newtonian mechanics. Clearly, when light interacts with matter on very small scales, it does so not in a smooth, continuous way but in a discontinuous, stepwise manner. The challenge was to explain this unexpected behavior. The solution revolutionized our view of nature and now forms the foundation not just for physics and astronomy, but for all of modern science.

Atomic Structure

To explain the formation of spectral lines, we must understand not just the nature of light but also something of the structure of **atoms**—the microscopic building blocks of matter. Let's start with the simplest atom, hydrogen, which consists of a single negatively charged electron orbiting a positively charged proton. The proton forms the central **nucleus** (plural: nuclei) of the atom. Because the positive charge on the proton exactly cancels the negative charge on the electron, the hydrogen atom as a whole is electrically neutral. If an atom absorbs some energy in the form of radiation, that energy must cause some internal change. And if the atom emits energy, it must come from somewhere within the atom. The energy absorbed or emitted by the atom is associated with changes in the motion of the orbiting electron.

In 1913, Danish physicist Niels Bohr developed a model of the atom that provided the first explanation of hydrogen's observed spectral lines. His work earned him the 1922 Nobel Prize in Physics. The essential features of the **Bohr model** of the atom are as follows. First, there is a state of lowest energy—the **ground state**—which represents the "normal" condition of the electron as it orbits the nucleus. Second, there is a maximum energy that the electron can have and still be part of the atom. If the electron acquires more than that maximum energy, it is no longer bound to the nucleus, and the atom is said to be *ionized*. An atom having fewer (or more) than its normal complement of electrons, and, hence, a net electrical charge, is called an **ion.** Third, and most important (and also least intuitive), between those two energy levels, the electron can exist only in certain sharply defined energy states, often referred to as *orbitals*.

An atom is said to be in an **excited state** when an electron occupies an orbital other than the ground state. The electron then lies at a greater than normal distance from its parent nucleus, and the atom has a greater than normal amount of energy. The excited state with the lowest energy (that is, the one closest to the ground state) is called the *first excited state*, that with the second-lowest energy the *second excited state*, and so on.

In Bohr's model, each electron orbital was pictured as having a specific radius, like a planetary orbit in the solar system, as shown in Figure 2.18. The modern view is not so simple. Although each orbital *does* have a precise energy, the electron is now envisioned as being smeared out in an *electron cloud* surrounding the nucleus, as illustrated in Figure 2.19. It is common to speak of the average distance from the cloud to the nucleus as the radius of the electron's orbital. When a hydrogen atom is in its ground state, the radius of the orbital is about 0.05 nm. As the orbital energy increases, the radius increases, too. For clarity, Figures 2.20 and 2.21 below use solid lines to represent electron orbitals, but bear in mind that the fuzziness in Figure 2.19 is a more accurate depiction of reality.

An atom can become excited by absorbing some light energy from a source of electromagnetic radiation or by colliding with some other particle—another atom, for example. However, the atom cannot stay in that state forever. After about 10^{-8} s, it returns to its ground state.

The Particle Nature of Radiation

Here is the crucial point that links atoms to radiation and allows us to interpret atomic spectra. Because electrons may exist only in orbitals having specific energies, atoms can absorb only specific amounts of energy as their electrons are boosted into excited states. Likewise, atoms can emit only specific amounts of energy as their electrons fall back to lower energy states. Thus, the amount of light energy absorbed or emitted in these processes *must correspond precisely to the energy difference between two orbitals*. This requires that light must be absorbed and emitted in the form of little "packets" of electromagnetic radiation, each carrying a very specific amount of energy. We call these packets **photons.** A photon is, in effect, a "particle" of electromagnetic radiation.

The idea that light sometimes behaves not as a continuous wave but as a stream of particles was first suggested by Albert Einstein in 1905. By then, experimental evidence clearly indicated that, on microscopic scales, electromagnetic radiation often displayed particle properties. Scientists understood that the notion of radiation as a wave was incomplete, but they did not know how to reconcile its

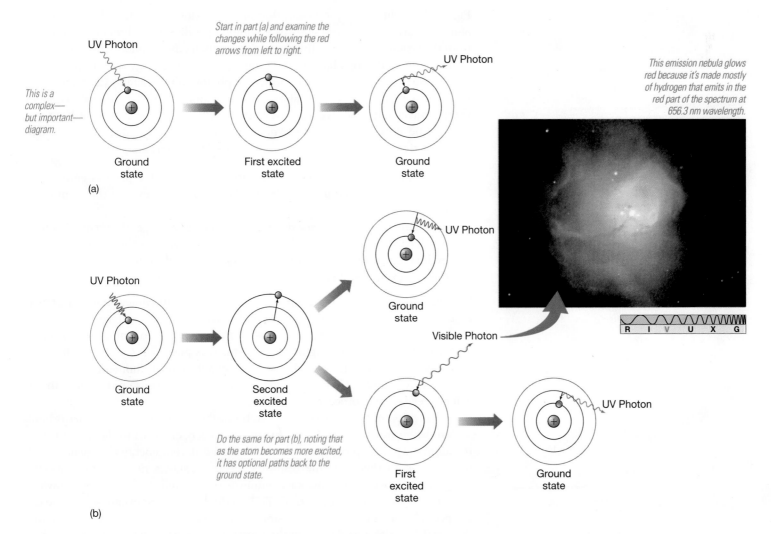

▲ **FIGURE 2.20 Atomic Excitation INTERACTIVE**   (a) Absorption of an ultraviolet (UV) photon (left) by a hydrogen atom causes the momentary excitation of the atom into its first excited state (center). Eventually, the atom returns to its ground state (right), in the process emitting a photon having exactly the same energy as the original photon. (b) Absorption of a higher-energy UV photon may boost the atom into a higher excited state, from which there are several possible paths back to the ground state. At the top, the electron falls immediately back to the ground state, emitting a photon identical to the one it absorbed. At the bottom, the electron initially falls into the first excited state, producing visible radiation of wavelength 656.3 nm—the characteristic red glow of excited hydrogen. (*Inset:* NASA)

ANIMATION/VIDEO
Classical Hydrogen Atom I

ANIMATION/VIDEO
Classical Hydrogen Atom II

CONCEPT CHECK

In what ways do electron orbits in the Bohr atom differ from planetary orbits around the Sun?

ANIMATION/VIDEO
Photon Emission

two seemingly contradictory natures. Einstein's great breakthrough (for which he won the 1919 physics Nobel Prize) came when he realized that all of the puzzling experiments could be explained by a simple, but critically important, connection between the particle and wave aspects of light. He found that the energy contained within a photon had to be proportional to the frequency of the radiation:

$$\text{photon energy} \propto \text{radiation frequency.}$$

Thus, for example, a "red" photon having a frequency of 4×10^{14} Hz (corresponding to a wavelength of about 750 nm) has 4/7 the energy of a "blue" photon, of frequency of 7×10^{14} Hz. Because it connects the *energy* of a photon with the *color* of the light it represents, this relationship is the final piece in the puzzle of how to understand the spectra we see.

Environmental conditions ultimately determine which description—wave or stream of particles—better fits the behavior of electromagnetic radiation. As a general rule of thumb, in the macroscopic realm of everyday experience, radiation is more usefully described as a wave, and in the microscopic domain of atoms, it is best characterized as a stream of particles.

The Spectrum of Hydrogen

Figure 2.20 illustrates the absorption and emission of photons by a hydrogen atom. In part (a), the atom absorbs a photon of radiation and makes a transition from the ground state to the first excited state, then emits a photon of precisely the same energy and drops back to the ground state. The energy difference between the two states corresponds to an ultraviolet photon of wavelength 121.6 nm.

Figure 2.20(b) depicts the absorption of a more energetic (higher frequency, shorter wavelength) ultraviolet photon, this one having a wavelength of 102.6 nm, causing the atom to jump to the *second* excited state. From that state, the electron may return to the ground state via either one of two alternate paths:

1. It can proceed directly back to the ground state, in the process emitting an ultraviolet 102.6-nm photon identical to the one that excited the atom in the first place.
2. Alternatively, it can *cascade* down one orbital at a time, emitting *two* photons: one having an energy equal to the difference between the second and first excited states and the other having an energy equal to the difference between the first excited state and the ground state.

The second step in the cascade produces a 121.6-nm ultraviolet photon, just as in Figure 2.20(a). However, the first part of the cascade—the one from the second to the first excited state—produces a photon of wavelength 656.3 nm, which is in the visible part of the electromagnetic spectrum. This photon is seen as red light. An individual atom—if one could be isolated—would emit a momentary red flash. The inset in Figure 2.20 shows an astronomical object whose red color is the result of precisely this process.

Absorption of still more energy can boost the electron to even higher orbitals within the atom. As the excited electron cascades back down to the ground state, the atom may emit many photons, each with a different energy and, hence, a different color. In this case, the resulting spectrum shows many distinct spectral lines. For hydrogen, all transitions ending at the ground state produce ultraviolet photons. However, downward transitions ending at the *first* excited state give rise to spectral lines in or near the visible portion of the electromagnetic spectrum (Figure 2.14). Because they form the most easily observable part of the hydrogen spectrum and were the first to be discovered, these lines (also known as *Balmer lines*) are often referred to simply as the "hydrogen series" and are denoted by the letter H. Individual transitions are labeled with Greek letters, in order of increasing energy (decreasing wavelength): The Hα (H alpha) line corresponds to the transition from the second to the first excited state and has a wavelength of 656.3 nm (red); Hβ (H beta; third to first) has wavelength 486.1 nm (green); Hγ (H gamma; fourth to first) has wavelength 434.1 nm (blue); and so on.

Kirchhoff's Laws Explained

Let's reconsider our earlier discussion of emission and absorption lines in terms of the model just presented. In Figure 2.17(b) a beam of radiation shines through a cloud of gas. The beam contains photons of all energies, but most of them do not interact with the gas because the gas can absorb only photons having precisely the right energy to cause an electron to jump from one orbital to another. Photons having energies that cannot produce such a jump pass through the gas unhindered. Photons having the right energies are absorbed, excite the gas, and are removed from the beam. This is the cause of the dark absorption lines in the spectrum. These lines are direct indicators of the energy differences between orbitals in the atoms making up the gas.

The excited gas atoms rapidly return to their original states, each emitting one or more photons in the process. Most of these reemitted photons leave at angles that do *not* take them through the slit and on to the detector. A second detector looking at the cloud from the side (Figure 2.17c) would record the reemitted energy as an emission spectrum. (This is what we are seeing in the inset to Figure 2.20.) Like the absorption spectrum, the emission spectrum is characteristic of the gas, not of the original beam.

The case of a *continuous* spectrum in Figure 2.17(c) is shown for emitted photons escaping from the bulb without further interaction with matter. Actually, the situation in a denser source of radiation (a thick gas cloud or in a liquid or solid body) is more complex. There, a photon is likely to interact with atoms, free electrons, and ions in the body many times before finally escaping, exchanging some energy with the matter at each encounter. The net result is that the emitted radiation displays a continuous spectrum, in accordance with Kirchhoff's first law. The spectrum is approximately that of a blackbody with the same temperature as the source.

More Complex Spectra

All hydrogen atoms have the same structure—a single electron orbiting a single proton—but, of course, there are many other kinds of atoms, each having a unique internal structure. The number of protons in the nucleus of an atom determines which **element** the atom represents. That is, just as all hydrogen atoms have a single proton, all oxygen atoms have 8 protons, all iron atoms have 26 protons, and so on.

The next simplest element after hydrogen is helium. Its nucleus is made up of two protons and two **neutrons** (another elementary particle with a slightly larger mass than a proton, but carrying no electrical charge). Two electrons orbit the nucleus. As with hydrogen and all other atoms, the "normal" condition for helium is to be electrically neutral, with the negative charge of the orbiting electrons exactly canceling the positive charge of the nucleus (Figure 2.21a).

More complex atoms contain more protons (and neutrons) in the nucleus and have correspondingly more orbiting electrons. For example, an atom of carbon (Figure 2.21b) consists of six electrons orbiting a nucleus containing six protons and six neutrons. As we progress to heavier and heavier elements, the number of orbiting electrons increases, and consequently the number of possible electronic transitions rises rapidly. The number increases further as the temperature increases and atoms become excited or even ionized. The result is that very complicated spectra can be produced. The complexity of atomic spectra generally reflects the complexity of the source atoms.

Even more complex spectra are produced by **molecules.** A molecule is a tightly bound group of atoms held together by interactions among their orbiting

PROCESS OF SCIENCE CHECK

In what ways was the wave theory of radiation unable to account for detailed observations of atomic spectra? Describe some key aspects of the theory of quantum mechanics, and explain how the new theory explains spectral lines.

(a)

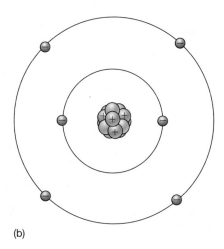

(b)

▲ FIGURE 2.21 **Helium and Carbon** (a) A helium atom in its ground state has two electrons within the lowest-energy orbital around a nucleus containing two protons and two neutrons. (b) A carbon atom in its ground state has six electrons orbiting around a six-proton, six-neutron nucleus—two of the electrons in an inner orbital and the other four at a greater distance from the center.

This spectrum is for molecular hydrogen, H₂.

This much simpler spectrum is for atomic hydrogen, H.

▲ **FIGURE 2.22 Hydrogen Spectra** The emission spectrum of molecular hydrogen (a) differs greatly from that of simpler atomic hydrogen (b). (Bausch & Lomb Inc.)

CONCEPT CHECK

How does the structure of an atom determine the atom's emission and absorption spectra?

electrons—interactions called *chemical bonds*. Much like atoms, molecules can exist only in certain well-defined energy states, and they produce emission or absorption spectral lines when they make a transition from one state to another. Because molecules are more complex than atoms, the rules of molecular physics are also much more complex. Nevertheless, as with atomic spectral lines, painstaking experimental work over many decades has determined the precise frequencies at which millions of molecules emit and absorb radiation. These lines are molecular fingerprints, just like their atomic counterparts, enabling researchers to identify and study one kind of molecule to the exclusion of all others.

Molecular lines usually bear little resemblance to the spectral lines associated with their component atoms. For example, Figure 2.22(a) shows the emission spectrum of the simplest molecule known—molecular hydrogen. Notice how different it is from the spectrum of atomic hydrogen shown in part (b).

2.7 The Doppler Effect

Most of us have had the experience of hearing the pitch of a train whistle change from high shrill (high frequency, short wavelength) to low blare (low frequency, long wavelength) as the train approaches and then recedes. This motion-induced change in the observed frequency of a wave is known as the **Doppler effect,** in honor of Christian Doppler, the 19th-century Austrian physicist who first explained it. Applied to cosmic sources of electromagnetic radiation, it has become one of the most important observational tools in all of 20th-century astronomy. Here's how it works.

Imagine a wave moving from the place where it is generated toward an observer who is not moving with respect to the wave source (Figure 2.23a). By noting the distances between successive wave crests, the observer can determine the wavelength of the emitted wave. Now suppose that the wave source begins to move (Figure 2.23b). Because the source moves between the times of emission of one wave crest and the next, wave crests in the direction of motion of the source are seen to be *closer together* than normal, while crests behind the source are more widely spaced. Thus, an observer in front of the source measures a *shorter* wavelength than normal, while one behind sees a *longer* wavelength.

The greater the relative speed of source and observer, the greater the observed shift. In terms of the net velocity of *recession* between source and observer, the apparent wavelength and frequency (measured by the observer) are related to the true quantities (emitted by the source) by

$$\frac{\text{apparent wavelength}}{\text{true wavelength}} = \frac{\text{true frequency}}{\text{apparent frequency}} = 1 + \frac{\text{recession velocity}}{\text{wave speed}}.$$

A positive recession velocity means that the source and the observer are moving apart; a negative value means that they are approaching. The wave speed is the

► FIGURE 2.23 **Doppler Effect INTERACTIVE** (a) Wave motion from a source toward an observer at rest with respect to the source. As seen by the observer, the source is not moving, so the wave crests are just concentric spheres (shown here as circles). (b) Waves from a moving source tend to "pile up" in the direction of motion and be "stretched out" on the other side. As a result, an observer situated in front of the source measures a shorter-than-normal wavelength—a blueshift—while an observer behind the source sees a redshift.

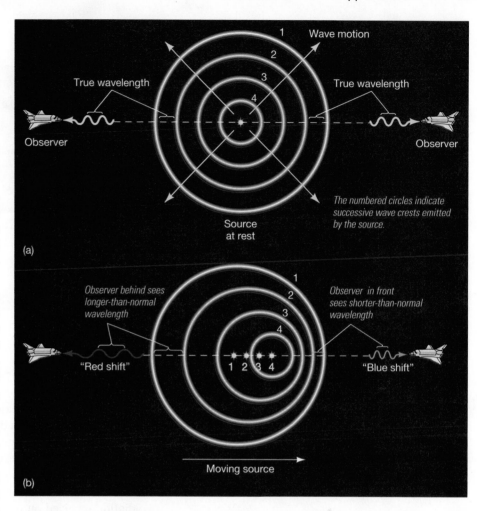

speed of light c in the case of electromagnetic radiation. For most of this text, the recession velocity will be small compared to the speed of light. Only when we come to discuss the properties of black holes (Chapter 13) and the structure of the universe on the largest scales (Chapter 17) will we have to reconsider this formula.

Note that, in the figure, the source is shown in motion. However, the same general statements hold whenever there is any *relative* motion between source and observer. Note also that only motion along the line joining source and observer, known as *radial* motion, appears in the above equation. Motion *transverse* (perpendicular) to the line of sight has no significant effect.

In astronomical jargon, the wave measured by an observer situated in front of a moving source is said to be *blueshifted*, because blue light has a shorter wavelength than red light. Similarly, an observer situated behind the source will measure a longer-than-normal wavelength—the radiation is said to be *redshifted*. This terminology holds even for invisible radiation, for which "red" and "blue" have no meaning. Any shift toward shorter wavelengths is called a blueshift, and any shift toward longer wavelengths is called a redshift.

The Doppler effect is not normally noticeable in visible light on Earth—the speed of light is so large that the wavelength change is far too small to be noticeable for everyday terrestrial velocities. However, using spectroscopic techniques, astronomers routinely use the Doppler effect to measure the line-of-sight velocity of cosmic objects by determining the extent to which known spectral lines are shifted to longer or shorter wavelengths. For example, suppose that an astronomer observes the red Hα line in the spectrum of a star to have a wavelength of 657 nm instead of the 656.3 nm measured in the laboratory. (How does she know it is the same line? Because, as illustrated in Figure 2.24, she realizes that *all* the hydrogen lines are shifted by the same fractional amount—the characteristic pattern of lines identifies hydrogen as the source.) Using the above equation, she calculates that the star's radial velocity is 657/656.3 − 1 = 0.0056 times the speed of light. In other words, the star is receding from Earth at a rate of 320 km/s.

The motions of nearby stars and distant galaxies—even the expansion of the universe itself—have all been measured in this way. Motorists stopped for speeding on the highway have experienced another, much more down-to-earth, application. As illustrated in the inset to Figure 2.23, police radar measures speed by means of the Doppler effect, as do the radar guns used to clock the velocity of a pitcher's fastball or a tennis player's serve.

Police use handheld radar guns—and the Doppler effect—to catch speeders.

Reflected radiation (blue crests) from the oncoming car shifts toward shorter wavelengths.

Motion

SELF-GUIDED TUTORIAL
Doppler Effect

CONCEPT CHECK

How might the Doppler effect be used in determining the mass of a distant star?

These are realistic spectra emitted by objects moving at different speeds.

The middle spectrum is unshifted for an object at rest.

▲ FIGURE 2.24 **Doppler Shift** INTERACTIVE The Doppler effect shifts the entire spectrum of a moving object to higher or lower frequencies. The spectrum at top shows the redshift of the hydrogen lines from an object moving at a speed of 300 km/s away from the observer; that at bottom shows those same lines blueshifted while the object moves at double that speed toward the observer.

2.8 Spectral-Line Analysis

Astronomers apply the laws of spectroscopy in analyzing radiation from beyond Earth. A nearby star or a distant galaxy takes the place of the lightbulb in our previous examples, an interstellar cloud or a stellar (or even planetary) atmosphere plays the role of the intervening cool gas, and a spectrograph attached to a telescope replaces our simple prism and detector. We list below some of the properties of emitters and absorbers that can be determined by careful analysis of radiation received on (or near) Earth. We will encounter other important examples as our study of the cosmos unfolds.

1. The *composition* of an object is determined by matching its spectral lines with the laboratory spectra of known atoms and molecules.
2. The *temperature* of an object emitting a continuous spectrum can be measured by matching the overall distribution of radiation with a blackbody curve. Temperature may also be determined from detailed studies of spectral lines. In Section 10.3 we will see how stellar temperatures are most accurately measured by spectroscopic means.
3. The *(line-of-sight) velocity* of an object is measured by determining the Doppler shift of its spectral lines.
4. An object's *rotation rate* can be determined by measuring the *broadening* (smearing out over a range of wavelengths) produced by the Doppler effect in emitted or reflected spectral lines.
5. The *pressure* of the gas in the emitting region of an object can be measured by its tendency to broaden spectral lines. The greater the pressure, the broader the line.
6. The *magnetic field* of an object can be inferred from a characteristic splitting it produces in many spectral lines, when a single line divides into two. (This is known as the *Zeeman effect*.)

   Given sufficiently sensitive equipment, there is almost no end to the wealth of data contained in starlight. The challenge facing astronomers is to unravel the extent to which each of the above mechanisms, in a complex mix of atoms and molecules, contributes to spectral line profiles, and thereby obtain meaningful information about the source of the lines.

CONCEPT CHECK

Why is it important for astronomers to analyze spectral lines in detail?

THE BIG QUESTION

Atoms are the basic components of normal matter—the stuff from which stars, planets, and we ourselves are built. In turn, all atoms are made of smaller elementary particles: protons, neutrons, and electrons. But even those particles are not fundamental, for physicists know that protons and neutrons are made of quarks, and many theorists think that quarks and electrons may have even deeper structure. How far does this hierarchy continue, and what new windows on the cosmos will be opened by our understanding of these unknown levels of nature?

CHAPTER REVIEW

SUMMARY

LO1 **Electromagnetic radiation** (p. 46) travels through space in the form of a **wave** (p. 46). Any electrically charged object is surrounded by an **electric field** (p. 48) that determines the force it exerts on other charged objects. When a charged particle moves, information about that motion is transmitted via the particle's changing **electric** and **magnetic fields** (pp. 48, 49). The information travels at the **speed of light** (p. 47) as an electromagnetic wave.

LO2 The **electromagnetic spectrum** (p. 50) consists of (in order of decreasing wavelength, or increasing frequency) **radio waves**, **infrared radiation**, **visible light**, **ultraviolet radiation**, **X-rays**, and **gamma rays** (p. 46). The **opacity** (p. 52) of Earth's atmosphere—the extent to which it absorbs radiation—varies greatly with the wavelength of the radiation. Only radio waves, some infrared wavelengths, and visible light can penetrate the atmosphere and reach the ground.

LO3 The **temperature** (p. 53) of an object is a measure of the speed with which its constituent particles move. The **intensity** (p. 53) of radiation emitted by an object has a characteristic distribution, called a **blackbody**

curve (p. 53), that depends only on the temperature of the object. **Wien's law** (p. 54) tells us that the wavelength at which the object's radiation peaks is inversely proportional to its temperature. **Stefan's law** (p. 54) states that the total amount of energy radiated is proportional to the fourth power of the temperature.

LO4 Many hot objects emit a **continuous spectrum** (p. 56) of radiation, containing light of all wavelengths. A hot gas may instead produce an **emission spectrum** (p. 57), consisting only of a few well-defined **emission lines** (p. 57) of specific frequencies, or colors. Passing a continuous beam of radiation through cool gas will produce **absorption lines** (p. 57) at precisely the same frequencies as would be present in the gas's emission spectrum.

LO5 **Atoms** (p. 60) are made up of negatively charged electrons orbiting a positively charged **nucleus** (p. 60) consisting of positively charged protons and electrically neutral **neutrons** (p. 63). The number of protons determines the type of **element** (p. 63) the atom represents. In the **Bohr model** (p. 60), a hydrogen atom has a minimum-energy **ground state** (p. 60), representing its "normal" condition. When the electron has a higher-than-normal energy, the atom is in an **excited state** (p. 60). For any given atom, only certain, well-defined energies are possible. In the modern view, the electron is envisaged as being spread out in a "cloud" around the nucleus but still having a sharply defined energy.

LO6 When an electron moves from one energy state to another in an atom, the difference in the energy between the states is emitted or absorbed in the form of "packets" of electromagnetic radiation—**photons** (p. 61). Because the energy levels have definite energies, the photons also have definite energies that are characteristic of the type of atom involved. The energy of a photon determines the frequency, and hence the color, of the light emitted or absorbed.

LO7 Our perception of the wavelength of a beam of light can be altered by the source's velocity relative to us. This motion-induced change in the observed frequency of a wave is called the **Doppler effect** (p. 64). Any net motion of the source away from the observer causes a redshift—a shift to lower frequencies—in the received beam. Motion toward the observer causes a blueshift. The extent of the shift is directly proportional to the source's radial velocity relative to the observer.

MasteringAstronomy®   For instructor-assigned homework go to www.masteringastronomy.com

Problems labeled POS explore the process of science. VIS problems focus on reading and interpreting visual information. LO connects to the introduction's numbered Learning Outcomes.

REVIEW AND DISCUSSION

1. Define the following wave properties: period, wavelength, amplitude, frequency.

2. Compare and contrast the gravitational and electric forces.

3. **LO1** Describe how light radiation leaves a star, travels through the vacuum of space, and finally is seen by someone on Earth.

4. **LO2** What do radio waves, infrared radiation, visible light, ultraviolet radiation, X-rays, and gamma rays have in common? How do they differ?

5. In what regions of the electromagnetic spectrum is the atmosphere transparent enough to allow observations from the ground?

6. What is a blackbody? Describe the radiation it emits.

7. **POS** If Earth were completely blanketed with clouds and we couldn't see the sky, could we learn about the realm beyond the clouds? What forms of radiation might penetrate the clouds and reach the ground?

8. **LO3** In terms of its blackbody curve, describe what happens as a red-hot glowing coal cools.

9. **LO6 POS** Explain how astronomers might use spectroscopy to determine the composition and temperature of a star.

10. **LO4** What are continuous, emission, and absorption spectra? How are they produced?

11. What is the normal condition for atoms? What is an excited atom? What are orbitals?

12. **LO5** Why do atoms absorb and reemit radiation at characteristic frequencies?

13. **POS** Suppose a luminous cloud of gas is discovered emitting an emission spectrum. What can be learned about the cloud from this observation?

14. What is the Doppler effect, and how does it alter the way in which we perceive radiation?

15. **LO7** How do astronomers use the Doppler effect to determine the velocities of astronomical objects?

CONCEPTUAL SELF-TEST: TRUE OR FALSE?/MULTIPLE CHOICE

1. The wavelength of green light is about the size of an atom. (T/F)

2. Two otherwise identical objects have temperatures of 1000 K and 1200 K, respectively. The object at 1200 K emits roughly twice as much radiation as the object at 1000 K. (T/F)

3. As you drive away from a radio transmitter, the radio signal you receive from the station is shifted to longer wavelengths. (T/F)

4. Imagine an emission spectrum produced by a container of hydrogen gas. Changing the amount of hydrogen in the container will change the colors of the lines in the spectrum. (T/F)

5. In the previous question, changing the gas in the container from hydrogen to helium will change the colors of the lines occurring in the spectrum. (T/F)

6. The energy of a photon is inversely proportional to the wavelength of the radiation. (T/F)

7. An electron moves to a higher energy level in an atom after absorbing a photon of a specific energy. (T/F)

8. Compared with ultraviolet radiation, infrared radiation has a greater (a) wavelength; (b) amplitude; (c) frequency; (d) energy.

9. An X-ray telescope located in Antarctica would not work well because of (a) the extreme cold; (b) the ozone hole; (c) continuous daylight; (d) Earth's atmosphere.

10. A star much cooler than the Sun would appear (a) red; (b) blue; (c) smaller; (d) larger.

11. The blackbody curve of a star moving toward Earth would have its peak shifted (a) to lower intensity; (b) toward higher energies; (c) toward longer wavelengths; (d) toward lower energies.

12. The visible spectrum of sunlight reflected from Saturn's cold moon Titan would be expected to be (a) continuous; (b) an emission spectrum; (c) an absorption spectrum.

13. Astronomers analyze starlight to determine a star's (a) temperature; (b) composition; (c) motion; (d) all of the above.

14. VIS According to Figure 2.11 (Blackbody Curves), an object having a temperature of 1000 K emits mostly (a) infrared light; (b) visible light; (c) ultraviolet light; (d) X-rays.

15. VIS In Figure 2.20 (Atomic Excitation), the total energy of the two photons emitted in the branch at the lower right is (a) greater than, (b) less than, (c) approximately equal to, (d) exactly equal to, the energy of the UV photon absorbed at the left.

PROBLEMS

The number of squares preceding each problem indicates its approximate level of difficulty.

1. ■ A sound wave moving through water has a frequency of 256 Hz and a wavelength of 5.77 m. What is the speed of sound in water?

2. ■ What is the wavelength of a 100-MHz (FM 100) radio signal?

3. ■ What would be the frequency of an electromagnetic wave having a wavelength equal to Earth's diameter (12,800 km)? In what part of the electromagnetic spectrum would such a wave lie?

4. ■■ The blackbody emission spectrum of object A peaks in the ultraviolet region of the electromagnetic spectrum at a wavelength of 200 nm. That of object B peaks in the red region, at 650 nm. Which object is hotter, and, according to Wien's law, how many times hotter is it? According to Stefan's law, how many times more energy per unit area does the hotter body radiate per second?

5. ■ Normal human body temperature is about 37°C. What is this temperature in kelvins? What is the peak wavelength emitted by a person with this temperature? In what part of the spectrum does this lie?

6. ■■ According to the Stefan-Boltzmann law, how much energy is radiated into space per unit time by each square meter of the Sun's surface (see *More Precisely 2-2*)? If the Sun's radius is 696,000 km, what is the total power output of the Sun?

7. ■ By what factor does the energy of a 1-nm X-ray photon exceed that of a 10-MHz radio photon? How many times more energy has a 1-nm gamma ray than a 10-MHz radio photon?

8. ■■ How many different photons (that is, photons of different frequencies) can be emitted as a hydrogen atom in the second excited state falls back, directly or indirectly, to the ground state? What are their wavelengths? What about a hydrogen atom in the third excited state?

9. ■ The Hα line (Section 2.6) of a star is received on Earth at a wavelength of 656 nm. What is the star's radial velocity relative to Earth?

10. ■■■ You are observing a spacecraft moving in a circular orbit of radius 100,000 km around a distant planet. You happen to be located in the plane of the spacecraft's orbit. You find that the spacecraft's radio signal varies periodically in wavelength between 2.99964 m and 3.00036 m. Assuming that the radio is broadcasting at a constant wavelength, what is the mass of the planet?

ACTIVITIES

Collaborative

1. Find a spectrum of the Sun with a wavelength scale on it. Google is a good place to start. Select some absorption lines and determine their wavelengths by interpolation. Now, try to identify the element that produced these lines. Use a reference such as Moore's *A Multiplet Table of Astrophysical Interest*, available on the NASA Astrophysics Data System. Work with the darkest lines before trying the fainter ones. How many elements can you find?

2. Stand near (but not too near!) a train track or busy highway and wait for a train or traffic to pass by. Can you notice the Doppler effect in the pitch of the engine noise or whistle blowing? How does the sound frequency depend on the train's (a) speed and (b) motion toward or away from you? Divide your group into two. One subgroup should time the train's motion and hence calculate approximately its speed. The other (consisting of the more musically inclined!) should estimate the perceived frequency change of the whistle when the train is moving first toward you and then away from you.

Individual

1. Locate the constellation Orion. Its two brightest stars are Betelgeuse and Rigel. Which is hotter? How can you tell? Which of the other stars scattered across the night sky are hot, and which are cool?

2. Obtain a handheld spectroscope, available from your school science lab or online. In the shade, point the spectroscope at a white cloud or white piece of paper that is in direct sunlight. Look for the absorption lines in the Sun's spectrum. Note their wavelength from the scale inside the spectroscope. How many of the lines can you identify by comparing your list with the Fraunhofer lines given in many astronomy reference books or on Wikipedia?

3

Telescopes
The Tools of Astronomy

At its heart, astronomy is an observational science. Painstaking observations of cosmic phenomena almost always precede any clear theoretical understanding of their nature. As a result, our detecting instruments—our telescopes—have evolved to observe as broad a range of wavelengths as possible. Until the middle of the 20th century, telescopes were limited to visible light. Since then, however, technological advances have expanded our view of the universe to all parts of the electromagnetic spectrum. Some telescopes are sited on Earth, others must be placed in space, and designs vary widely from one region of the spectrum to another. Whatever the details of its construction, however, the basic purpose of a telescope is to collect electromagnetic radiation and deliver it to a detector for detailed study.

THE BIG PICTURE

Telescopes are time machines, and astronomers, in a sense, are historians. Telescopes enhance our senses, enabling us to look far out in space—and hence far back in time. They allow us to explore objects at a much farther distance than would be possible with our unaided eyes and to perceive radiation at wavelengths far beyond human vision. Almost everything in this book would be unknown without these powerful and versatile instruments.

◄ Astronomers like to think big, and really big telescopes are now on the drawing board. This artist's conception for the European Southern Observatory shows E-ELT—the European Extremely Large Telescope. Its mirror diameter of nearly 40 m means that E-ELT will combine unrivaled light-gathering power with the ability to examine cosmic objects at unprecedented resolution. This largest telescope in the world will be built on Cerro Armazones, a 3000-m mountaintop in Chile's Atacama Desert. (*ESO/L. Calcada*)

▲ **FIGURE 3.1 Reflecting Mirror**   Curved mirrors focus to a single point all rays of light arriving parallel to the mirror axis. The arrows indicate the directions of the incoming and reflected rays.

3.1 Optical Telescopes

Simply put, a **telescope** is a "light bucket"—a device whose primary function is to capture as much radiation as possible from a given region of the sky and concentrate it into a focused beam for analysis. We begin our study of astronomical hardware with *optical* telescopes, designed to collect wavelengths visible to the human eye. Later, we will look at telescopes designed to capture and analyze radiation in other, *invisible* regions of the electromagnetic spectrum. Throughout this text we will see how advances in our understanding of the universe have always gone hand in hand with technological improvements in both the sensitivity and the spectral range of our detectors. This combination of technology and science is as vital today as it was when Galileo first turned his telescope toward the skies. ∞ *(Sec. 1.2)*

Reflecting and Refracting Telescopes

Optical telescopes fall into two basic categories—*reflectors* and *refractors*. Figure 3.1 shows how a **reflecting telescope** uses a curved mirror to gather and concentrate a beam of light. This mirror, usually called the *primary mirror* because telescopes generally contain more than one, is constructed so that all light rays arriving parallel to its axis (the imaginary line through the center of and perpendicular to the mirror) are reflected to pass through a single point, called the *focus*. The distance between the primary mirror and the focus is the *focal length*. In astronomical contexts, the focus of the primary mirror is referred to as the **prime focus.**

A **refracting telescope** uses a lens instead of a mirror to focus the incoming light, relying on refraction rather than reflection to achieve its purpose. **Refraction** is the bending of a beam of light as it passes from one transparent medium (for example, air) into another (such as glass). Figure 3.2 (a) illustrates how a prism can be used to change the direction of a beam of light. As illustrated in Figure 3.2 (b), we can think of a lens as a series of prisms combined in such a way that all light rays striking the lens parallel to the axis are refracted to pass through the focus.

Astronomers often use telescopes to make **images** of their fields of view. Figure 3.3 illustrates how this is accomplished, in this case by the primary mirror in a reflecting telescope. Light from a distant object reaches the telescope as nearly parallel rays. A ray of light entering the instrument parallel to the mirror axis is reflected through the focus. Light from a slightly different direction—that is, at a slight angle to the axis—is focused to a slightly different point. In this way, an image is formed near the focus. Each point in the image corresponds to a different angle in the field of view. Often, a lens known as an *eyepiece* is used to magnify the image before it is viewed by eye or recorded using a

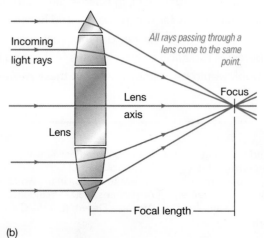

◀ **FIGURE 3.2 Refracting Lens** (a) Refraction by a prism changes the direction of a light ray. (b) A lens can be thought of as a series of prisms.

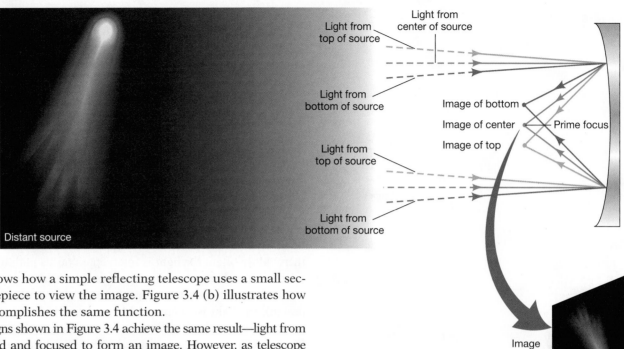

camera. Figure 3.4 (a) shows how a simple reflecting telescope uses a small secondary mirror and an eyepiece to view the image. Figure 3.4 (b) illustrates how a refracting telescope accomplishes the same function.

The two telescope designs shown in Figure 3.4 achieve the same result—light from a distant object is captured and focused to form an image. However, as telescope *size* has steadily increased over the years (as discussed in Section 3.2) astronomers have come to favor reflecting instruments over refractor for a number of reasons:

1. Just as a prism separates white light into its component colors, the lens in a refracting telescope focuses red and blue light differently (the blue focus lying closer to the lens). This deficiency is known as *chromatic aberration*. Careful design and choice of materials can largely correct this problem, but it is difficult to eliminate it entirely, and it requires the use of very high-quality glass for the body of the lens. Mirrors do not suffer from this defect.

2. As light passes through the lens, some of it is absorbed by the glass. This absorption is a relatively minor problem for visible radiation, but it can be severe for infrared and ultraviolet observations because glass blocks most of the radiation in those regions of the electromagnetic spectrum. This problem obviously does not affect mirrors.

▲ FIGURE 3.3 **Image Formation**   An image is formed by a mirror as rays of light coming from different points on a distant object focus to slightly different locations. Notice that the image is inverted (that is, upside down).

MA SELF-GUIDED TUTORIAL

Chromatic Aberration

(a) Reflector

(b) Refractor

◀ FIGURE 3.4 **Reflectors and Refractors**   Comparison of (a) reflecting and (b) refracting telescopes. Both types are used to gather and focus electromagnetic radiation. The image is viewed with a small magnifying lens called an *eyepiece*.

3. A large lens can be quite heavy. Because it can be supported only around its edge (so as not to block the incoming radiation), the lens tends to deform under its own weight. A mirror does not have this drawback because it can be supported over its entire back surface.

4. A lens has two surfaces that must be accurately machined and polished, which can be a difficult task. A mirror has only one surface.

For these reasons, *all* large modern optical telescopes are reflectors.

Types of Reflecting Telescopes

SELF-GUIDED TUTORIAL
Reflecting Telescopes

SELF-GUIDED TUTORIAL
The Optics of a Simple Lens

Figure 3.5 shows some basic reflecting telescope designs. Radiation from a star enters the instrument, passes down the main tube, strikes the primary mirror, and is reflected back toward the prime focus, near the top of the tube. Sometimes astronomers place their recording instruments at the prime focus. However, it can be very inconvenient, or even impossible, to suspend bulky pieces of equipment there. More often, the light is intercepted on its path to the focus by a secondary mirror and redirected to a more convenient location, as in Figure 3.5 (b–d).

In a **Newtonian telescope** (named after Isaac Newton, who invented this design), the light is intercepted by a flat secondary mirror before it reaches the prime focus and deflected 90°, usually to an eyepiece at the side of the instrument (Figure 3.5b). This is a popular design for smaller reflecting telescopes, such as those used by amateur astronomers.

Alternatively, astronomers may choose to work on a rear platform where they can use detecting equipment too heavy or delicate to hoist to the prime focus. In this case, the light reflected by the primary mirror toward the prime focus is intercepted by a convex secondary mirror, which reflects the light back down the tube and through a small hole at the center of the primary mirror (Figure 3.5c). This arrangement is known as a **Cassegrain telescope** (after Guillaume Cassegrain, a French lensmaker). The point behind the primary mirror where the light from the star finally converges is called the *Cassegrain focus*. A well-known example of a Cassegrain telescope is the *Hubble Space Telescope (HST; see Discovery 3-1)*, named for one of America's most notable astronomers, Edwin Hubble. The telescope's detectors all lie directly behind the primary mirror and are capable of making measurements in the optical, infrared, and ultraviolet parts of the spectrum.

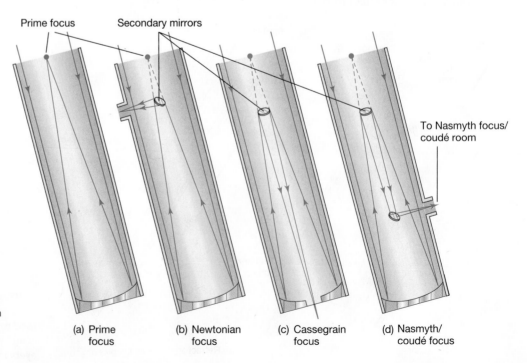

▶ FIGURE 3.5 **Reflecting Telescopes** Four reflecting telescope designs: (a) prime focus, (b) Newtonian focus, (c) Cassegrain focus, and (d) Nasmyth or coudé focus. Each design uses a primary mirror at the bottom of the telescope to capture radiation, which is then directed along different paths for analysis.

(a) Prime focus   (b) Newtonian focus   (c) Cassegrain focus   (d) Nasmyth/coudé focus

A more complex observational configuration requires starlight to be reflected by several mirrors. As in the Cassegrain design, light is first reflected by the primary mirror toward the prime focus and is then reflected back down the tube by a secondary mirror. Next, a third, much smaller, mirror reflects the light out of the telescope, where (depending on the details of the telescope's construction) the beam may be analyzed by a detector mounted alongside, at the *Nasmyth focus*, or it may be directed via a further series of mirrors into an environmentally controlled laboratory known as the *coudé* room (from the French word for "bent"). This laboratory is separate from the telescope itself, enabling astronomers to use very heavy and finely tuned equipment that cannot be placed at any of the other foci (all of which necessarily move with the telescope). The arrangement of mirrors is such that the light path to the coudé room does not change as the telescope tracks objects across the sky.

Figure 3.6 (a) shows the twin 10-m-diameter optical/infrared telescopes of the Keck Observatory on Mauna Kea in Hawaii, operated jointly by the California Institute of Technology and the University of California. The diagram in part (b) illustrates the light paths and some of the foci. Observations may be made at the Cassegrain, Nasmyth, or coudé focus, depending on the needs of the user. As the size of the person in part (c) indicates, this is indeed a very large telescope—in fact, the two mirrors are currently among the very largest on Earth. We will see numerous examples throughout this text of Keck's many important discoveries.

◄ **FIGURE 3.6 Keck Telescope** (a) The two 10-m telescopes of the Keck Observatory atop a dormant volcano in Hawaii, high above the surrounding clouds. (b) Artist's illustration of the telescope; the blue arrows show some of the paths starlight can take within the telescope and some of the locations where instruments may be placed. (c) One of the 10-m mirrors. Note the technician in orange coveralls at center. (*W. M. Keck Observatory*)

3.1 DISCOVERY

The Hubble Space Telescope

The *Hubble Space Telescope (HST)* is the largest, most complex, most sensitive observatory ever deployed in space. At more than $8 billion (including the cost of several missions to service and refurbish the system), it is also the most expensive scientific instrument ever built and operated. Built jointly by NASA and the European Space Agency, *HST* was designed to allow astronomers to probe the universe with at least 10 times finer resolution and with some 30 times greater sensitivity to light than existing Earth-based devices. The first figure shows the telescope being lifted out of the cargo bay of the space shuttle *Discovery* in the spring of 1990.

The accompanying "see-through" illustration displays *HST*'s main features. The telescope's Cassegrain design (Section 3.1) reflects light from its 2.4-m-diameter primary mirror (the large bluish disk at center) back to a smaller, 0.3-m secondary mirror, which sends the light through a small opening in the center of the primary mirror, from where it can be directed to any of several instruments (shown here in various colors at left) arrayed in the aft bay of the spacecraft. The large red objects are sensors that guide the pointing of the telescope, and the huge blue panels collect light from the Sun to power everything onboard. For scale, each of the instruments is about the size of a refrigerator. They were designed to be maintained by NASA astronauts, and indeed most of the telescope's original instruments have been upgraded or replaced since *HST* was launched. The current detectors on the telescope span the visible, near-infrared, and near-ultraviolet regions of the electromagnetic spectrum, from about 100 nm (UV) to 2200 nm (IR).

Soon after launch, astronomers discovered that the telescope's primary mirror had been polished to the wrong shape. The mirror is too flat by 2 μm, about 1/50 the width of a human hair, making it impossible to focus light as well as expected. In 1993, shuttle astronauts replaced *Hubble*'s gyroscopes to help the telescope point more accurately, installed sturdier versions of the solar panels that power the telescope's electronics, and—most important—inserted an intricate set of small mirrors to compensate for the faulty primary mirror. *Hubble*'s resolution is now close to the original design specifications, and the telescope has regained much of its lost sensitivity. Additional service missions were performed in 1997, 1999, 2002, and 2009 to replace instruments and repair faulty systems.

A good example of *Hubble*'s scientific capabilities can be seen by comparing the two images of the spiral galaxy M100, shown below.

(NASA)

ANIMATION/VIDEO **MA**
Hubble Space Telescope in Orbit

ANIMATION/VIDEO **MA**
Deployment of the *James Webb Space Telescope*

Guidance sensors

Primary mirror

Light enters here . . .

Solar panels

Detectors

. . . and is captured by instruments here.

(D. Berry)

On the left is one of the best ground-based photographs of this beautiful galaxy at the time of launch, showing rich detail and color in its spiral arms. On the right is an exquisitely detailed *HST* image of the galaxy's core, showing great improvements in both resolution and sensitivity. However, its small coverage reminds us that the trade-off for *Hubble*'s highest resolution is a relatively limited field of view.

During its 25 years of operation, *Hubble* has revolutionized our view of the sky, helping to rewrite some theories of the universe along the way. By measuring the properties of stars and supernovae within distant galaxies, it has helped establish the size and the expansion rate of the cosmos and has provided insights into the past and future evolution of the universe. *Hubble* has studied newborn galaxies almost at the limit of the observable universe with unprecedented clarity, allowing astronomers to see the interactions and collisions that may have shaped the evolution of our own Milky Way. Turning its gaze to the hearts of galaxies closer to home, *Hubble* has provided strong evidence for supermassive black holes in their cores. Within our own Galaxy, it has given astronomers stunning new insights into the physics of star formation and the evolution of stellar systems and stars of all sizes, from superluminous giants to objects barely more massive than planets. Finally, in our solar system, *Hubble* has given scientists new views of both the planets and their moons, and of the tiny fragments from which they formed long ago. Many spectacular examples of the telescope's remarkable capabilities appear throughout this book.

With *Hubble* now well into its third decade of highly productive service, NASA's plans for the telescope's successor are well underway. The *James Webb Space Telescope* (*JWST*), named after the administrator who led NASA's *Apollo* program during the 1960s and 1970s, will dwarf *Hubble* in both scale and capability. Sporting a 6.5-m segmented mirror with seven times *Hubble*'s collecting area and containing a formidable array of detectors optimized for use at visible and infrared wavelengths, *JWST* will orbit the Sun some 1.5 million km outside Earth's orbit, far beyond the Moon. NASA plans to launch the telescope in 2018. Its primary mission is to study the formation of the first stars and galaxies, measure the large-scale structure of the universe, and investigate the evolution of planets, stars, and galaxies.

Note that the *JWST* launch schedule likely means an interruption of a few years in space-based astronomical observations at near-optical wavelengths. The 2009 servicing mission extended *Hubble*'s operating lifetime at least into 2015, but few astronomers expect that the telescope will survive until *JWST* becomes fully operational. *HST* was designed to be retrieved by a shuttle and one day returned to Earth for placement in a museum, but the termination of the shuttle program in 2011 now means a much less glamorous end for the mission. Once the satellite is declared nonfunctional, it will be most likely be taken out of orbit by a robot tug and simply dropped into the ocean.

R I V U X G

(European Southern Observatory/VLT)

ANIMATION/VIDEO   **MA**
Gemini Control Room

ANIMATION/VIDEO **MA**
Speckle Imaging

Detectors and Image Processing

Most modern telescopes use electronic detectors known as **charge-coupled devices,** or **CCDs,** to record and store their data. A CCD (Figure 3.7) consists of a wafer of silicon divided into a two-dimensional grid of many tiny picture elements called *pixels.* When light strikes a pixel, an electric charge builds up on the device. The amount of charge is directly proportional to the number of photons striking each pixel—in other words, to the intensity of the light at that point. The charge buildup is monitored electronically, and a two-dimensional image is obtained (see Figures 3.7c and d). This is the same basic technology used in consumer video and digital cameras. A CCD is typically a few square centimeters in area and may contain several million pixels. As technology continually improves, both the areas of CCDs and the number of pixels they contain are steadily increasing.

CCDs have two important advantages over photographic plates, which were the staple of astronomers for over a century. First, CCDs are much more *efficient* than photographic plates, recording as many as 75 percent of the photons striking them, compared with less than 5 percent for photographic methods. This means that a CCD instrument can image objects 10 to 20 times fainter—or the same object 10 to 20 times faster—than can a photographic plate. Second, CCDs produce a faithful representation of an image in a digital format that can be manipulated by software, stored on disk, or sent across the Internet as needed.

Astronomers use computer processing on digital CCD images to compensate for known instrumental defects and even partly remove unwanted background noise in the signal, allowing them to see features in their data that would otherwise remain hidden. In addition, the computer can often carry out many of the tedious and time-consuming chores that must be performed before an image or spectrum reaches its final "clean" form. Figure 3.8 illustrates how computerized image-processing techniques were used to correct for known instrumental problems in the *Hubble Space Telescope,* allowing much of the planned resolution of the telescope to be recovered even before its repair in 1993.

(a)   (b)

(c)

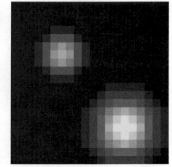

(d)

▲ **FIGURE 3.7 CCD Chip**   A charge-coupled device (CCD) consists of millions of tiny light-sensitive cells called pixels. Light striking a pixel causes an electrical charge to build up on it. By electronically reading out the charge on each pixel, a computer can reconstruct the pattern of light—the image—falling on the chip. (a) Detail of a CCD array. (b) A CCD chip mounted for use at the focus of a telescope. (c) Typical data from the chip consist of an array of numbers, running from 0 to 9 in this simplified example. Each number represents the intensity of the radiation striking that particular pixel. (d) The resulting image. *(MIT Lincoln Lab; AURA)*

(a)   (b)   (c)   (d)

▲ **FIGURE 3.8 Image processing** (a) Ground-based view of the star cluster R136, a group of stars in the Large Magellanic Cloud (a nearby galaxy). (b) The "raw" image of this same region as seen by the *Hubble Space Telescope* in 1990, before the first repair mission. (c) The same image after computer processing that partly compensated for imperfections in the mirror. (d) The same region as seen by the repaired *HST* in 1994, here observed at a somewhat bluer wavelength. *(AURA/NASA)*

3.2 Telescope Size

Large telescopes have two main advantages over small ones—*light-gathering power*, the ability to capture light, and *resolving power*, the ability to distinguish fine detail.

Light-Gathering Power

Astronomers spend much of their time observing very distant—and, hence, very *faint*—cosmic sources. In addition, they often want to obtain *spectra*, which entail spreading the incoming light out into its component colors. ∞ *(Sec. 2.5)* In order to make such detailed observations, astronomers must collect as much light as possible for analysis by their detectors.

The two factors determining a telescope's ability to collect light are *exposure time* and *collecting area*. **Exposure time** is simply the time spent gathering light from a source. Doubling the exposure time doubles the total amount of radiation captured. In Chapter 16 we will see an extreme example of a long-exposure image: constructed in 2003 from more than 300 hours of observations of the same small region of the sky, the so-called Hubble Ultra-Deep Field continues to provide astronomers with volumes of information on some of the most distant galaxies in the universe. However, such long exposures are rare. Access to telescopes and the sophisticated instruments attached to them is a precious commodity, and observatories are reluctant to allocate large amounts of time to a single observation.

The second factor controlling a telescope's light-gathering power is **collecting area,** the area capable of intercepting and focusing radiation—the "size of the bucket" (recall Figure 3.1). The larger the telescope's reflecting mirror (or refracting lens), the more light it collects and the easier it is to measure and study an object's radiative properties. The observed brightness of an astronomical object is directly proportional to the area of the telescope's mirror and, hence, to the *square* of the mirror diameter (Figure 3.9). For example, a 5-m telescope produces an image 25 times brighter than a 1-m instrument because a 5-m mirror

(a)

(b)

▶ **FIGURE 3.9 Sensitivity** Telescope size affects the image of a cosmic source, in this case the Andromeda galaxy. Both of these photographs had the same exposure time, but image (b) was taken with a telescope twice the size of that used to make image (a). Fainter detail can be seen as the diameter of the telescope mirror increases because larger telescopes are able to collect more photons per unit time. *(Adapted from AURA)*

▶ **FIGURE 3.10 Mauna Kea Observatory** (a) The world's highest ground-based observatory, at Mauna Kea, Hawaii, sits atop a dormant volcano some 4 km above sea level. Among the domes visible here are those housing the Canada–France–Hawaii 3.6-m telescope, the 8.1-m Gemini North instrument, the 2.2-m telescope of the University of Hawaii, the UK's 3.8-m infrared facility, and the twin 10-m Keck telescopes. To the right of the Kecks is the Japanese 8.3-m Subaru telescope. (b) The Subaru mirror. *(R. Wainscoat; NAOJ)*

Canada–France–Hawaii telescope
Gemini North telescope
United Kingdom infrared telescope
NASA infrared telescope
Keck 2
Keck 1
Subaru telescope

(a)

(b)

CONCEPT CHECK

Why do the largest modern telescopes use mirrors to gather and focus light?

has $5^2 = 25$ times the collecting area of a 1-m mirror. We can also think of this relationship in terms of the *time* required for a telescope to collect enough energy to create a recognizable image. A 5-m telescope produces an image 25 times faster than a 1-m device because it gathers energy at a rate 25 times greater. Put another way, a 1-hour time exposure with a 1-m telescope is roughly equivalent to a 2.4-minute (1/25 of an hour) time exposure with a 5-m instrument.

▶ **FIGURE 3.11 VLT Observatory** Located at the Paranal Observatory in Atacama, Chile, the European Southern Observatory's Very Large Telescope (VLT) is the world's largest optical telescope. Four 8.2-m reflecting telescopes are used in tandem to create the effective area of a single 16-m mirror. *(ESO)*

The twin Keck telescopes, shown in Figure 3.6 and in a larger view in Figure 3.10(a), are a case in point. Each telescope combines 36 hexagonal 1.8-m mirrors into the equivalent collecting area of a single 10-m reflector. The high altitude and large size of these telescopes makes them particularly well suited for detailed spectroscopic studies of very faint objects. To the right of the Keck domes in Figure 3.10(a) is the 8.3-m Subaru (the Japanese name for the Pleiades) telescope. Its mirror, shown in Figure 3.10 (b), is one of the largest single mirrors (as opposed to the segmented design used in Keck) yet built. In terms of total available collecting area, the largest telescope system now operating is the European Southern Observatory's Very Large Telescope (VLT), located at Cerro Paranal, in Chile (Figure 3.11). Its four separate 8.2-m mirrors can function as a single instrument of equivalent diameter 16.4 m. Both Keck and the VLT are designed to operate in the optical and near-infrared parts of the electromagnetic spectrum.

With 10-meter-class telescopes now almost routine, astronomers are turning their attention to the next generation of ground-based systems, the so-called *extremely large telescopes*, with mirrors tens of meters in diameter. All employ multimirror designs and real-time image enhancement technology; the first should come online early in the 2020s. The largest (so far) is the European Extremely Large Telescope, shown in the chapter-opening image on p. 70.

Resolving Power

The second advantage of large telescopes over smaller instruments is their superior **angular resolution.** In general, *resolution* refers to the ability of any device, such as a camera or a telescope, to form distinct, separate images of objects lying close together in the field of view. The finer the resolution, the better we can distinguish the objects and the more detail we can see. In astronomy, where we are always concerned with angular measurement, "close together" means "separated by a small angle on the sky," so angular resolution is the factor that determines our ability to see fine structure (see *More Precisely 0-1* for more detail on angular measure). Astronomers typically need to resolve objects only a few arc seconds (″) across. Figure 3.12 illustrates the result of increasing resolving power with views of the Andromeda galaxy at several different resolutions.

One important factor limiting a telescope's resolution is *diffraction*, the tendency of light—and all waves, for that matter—to "bend" around corners. ∞ *(Sec. 2.2)* Diffraction introduces a certain fuzziness, or loss of resolution, into the system. The degree of fuzziness—the minimum angular separation that can be distinguished—determines the angular resolution of the telescope.

As discussed in *More Precisely 3-1*, the amount of diffraction is directly proportional to the wavelength of the radiation and inversely proportional to the diameter of the telescope mirror. For light of any given wavelength, large telescopes produce less diffraction than small ones. The resolution given by the formula in *More Precisely 3-1* is known as the **diffraction-limited resolution** of the telescope. Thus, a 5-m telescope observing in blue light has a diffraction-limited resolution of about 0.02″. A 1-m telescope would have a diffraction limit of 0.1″ at the same wavelength, and so on. For comparison, the angular resolution of the human eye in the middle of the visual range is about 0.5′.

(a)

(b)

(c)

(d)

▶ **FIGURE 3.12 Resolution**   Detail becomes clearer in the Andromeda galaxy as the angular resolution is improved some 600 times, from (a) 10′, to (b) 1′, (c) 5″, and (d) 1″. (*Adapted from AURA*)

3.1 MORE PRECISELY

Diffraction and Telescope Resolution

The resolution of a telescope is ultimately controlled by *diffraction*, the process where light spreads out as it passes a corner or through an opening. ∞ *(Sec. 2.2)* Because of diffraction, it is impossible to focus a parallel beam of light to a sharp point, even with a perfectly constructed mirror. As illustrated in the accompanying figure, a wave passing through a gap is diffracted, creating a "fuzzy" shadow on the screen at right. Light and dark gradations represent crests and troughs of the wave, which define the wavelength (see Figure 2.3). In the absence of any diffraction, the shadow would be perfectly sharp—but that never happens in reality.

The amount of diffraction—that is, the amount of "fuzziness" introduced into an instrument—depends both on the *wavelength* of the radiation and the *size* of the opening (in our case, the *diameter* of the primary mirror). For a circular mirror and otherwise perfect optics, the angular resolution of a telescope (in convenient units) is

$$\text{angular resolution (arc seconds)} = 0.25 \frac{\text{wavelength } (\mu\text{m})}{\text{mirror diameter (m)}},$$

where 1 µm (micrometer) = 10^{-6} m = 1000 nm. (Recall that the symbol µ is the Greek letter mu.)

Thus, the best possible angular resolution that can be obtained using a 1-m telescope in blue light (with a wavelength of 400 nm = 0.4 µm) is about 0.25 (0.4/1)″ = 0.1″. But if we were to use our 1-m telescope to make observations in the near infrared, at a wavelength of 10 µm (10,000 nm), the best resolution we could obtain would be only 0.25 (10/1)″ = 2.5″. Observations in the infrared or radio range are often limited by the effects of diffraction. A 1-m radio telescope operating at a wavelength of 1 cm would have an angular resolution of just under 1°.

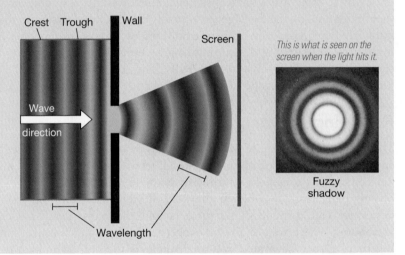

This is what is seen on the screen when the light hits it.

Fuzzy shadow

3.3 High-Resolution Astronomy

Even large telescopes have limitations. For example, according to the discussion in the preceding section, the 10-m Keck telescope should have an angular resolution of around 0.01″ in blue light. In practice, however, without the technological advances discussed in this section, it could not do better than about 1″. In fact, apart from instruments using special techniques developed to examine some particularly bright stars, *no* ground-based optical telescope built before 1990 can resolve astronomical objects to much better than 1″. The reason is *turbulence* in Earth's atmosphere—small-scale eddies of swirling air all along the line of sight, which blur the image of a star even before the light reaches our instruments.

Atmospheric Blurring

As we observe a star, atmospheric turbulence produces continuous small changes in the optical properties of the air between the star and our telescope (or eye). The light from the star is refracted slightly, again and again, and the stellar image dances around on the detector (or retina). This is the cause of the well-known twinkling of stars. The same basic process makes objects appear to shimmer when viewed across a hot roadway on a summer day because turbulent hot air just above the road surface constantly deflects and distorts the light rays reaching us.

On a typical night at a good observing site, the maximum deflection produced by the atmosphere is around 1″. Consider taking a photograph of a

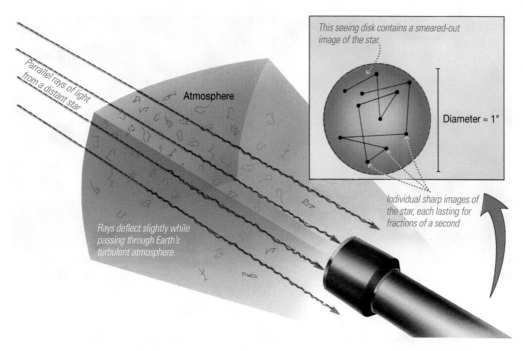

This seeing disk contains a smeared-out image of the star.

Atmosphere

Parallel rays of light from a distant star

Diameter ≈ 1"

Rays deflect slightly while passing through Earth's turbulent atmosphere.

Individual sharp images of the star, each lasting for fractions of a second

◀ FIGURE 3.13 Atmospheric Turbulence Light rays from a distant star strike a telescope detector at slightly different locations because of turbulence in Earth's atmosphere. Over time, the light covers a roughly circular region on the detector, and even the point-like image of a star is recorded as a small disk, called the seeing disk.

star under such conditions. After a few minutes' exposure time (long enough for the intervening atmosphere to have undergone many small, random changes), the dancing sharp image of the star has been smeared out over a roughly circular region 1″ or so in diameter (Figure 3.13). Astronomers use the term **seeing** to describe the effects of atmospheric turbulence. The circle over which a star's light is spread is called the **seeing disk.** In this jargon, "good seeing" means that the air is relatively stable and the seeing disk is small—as little as a few tenths of an arc second across in some exceptional cases.

To achieve the best possible seeing, telescopes are sited on mountaintops (to get above as much of the atmosphere as possible) in locations where the atmosphere is known to be fairly stable and relatively free of dust, moisture, and light pollution from cities. In the continental United States, these sites tend to be in the desert Southwest. The U.S. National Observatory for optical astronomy in the Northern Hemisphere is located high on Kitt Peak near Tucson, Arizona. The site was chosen because of its many dry, clear nights, and typical seeing of around 1″. Even better conditions are found on Mauna Kea in Hawaii (Figure 3.10) and in the Andes Mountains of Chile (Figure 3.11), which is why so many large telescopes have recently been constructed at those two exceptionally clear sites.

A telescope placed above the atmosphere, in Earth orbit, can achieve resolution close to the diffraction limit, subject only to the engineering restrictions of building and placing large structures in space. The *Hubble Space Telescope* (*Discovery 3-1*) has a 2.4-m mirror and a diffraction limit of 0.05″ (in blue light).

New Telescope Design

In addition to using high-quality optics and selecting the best observing sites, astronomers have developed other techniques to produce the sharpest possible images. By analyzing the image *while the light is still being collected*, it is possible to adjust the telescope from moment to moment to reduce the effects of mirror distortion, temperature changes, and bad seeing. By these means, an increasing number of large telescopes, including Keck and the VLT, can now achieve resolutions close to their theoretical diffraction limits.

The collection of techniques aimed at controlling environmental and mechanical fluctuations in the properties of the telescope itself is known as

CONCEPT CHECK

Give two reasons why astronomers need to build very large telescopes.

  ANIMATION/VIDEO

Adaptive Optics

▶ FIGURE 3.14 **Active Optics**   Infrared images of part of the star cluster R136 (see also Figure 3.8) contrast the resolution obtained (a) without and (b) with an active-optics system. *(ESO)*

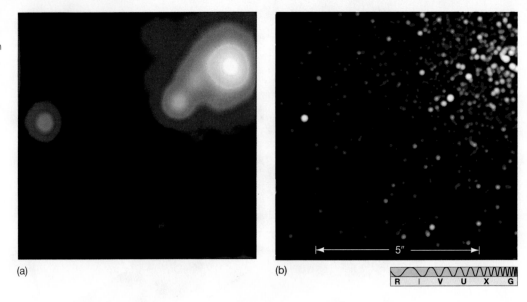

(a)   (b)

| R | I | V | U | X | G |

active optics. Active optics systems often include improved dome design to control airflow, precise control of the mirror temperature, and the use of actuators (pistons) behind the mirror to maintain its precise shape at all times (Figure 3.14a). Figure 3.14 (b) illustrates dramatically how active optics can improve image resolution.

An even more ambitious approach is known as **adaptive optics.** This technique deforms the shape of a mirror's surface under computer control while the image is being exposed in order to undo the effects of atmospheric turbulence. In the system shown in Figure 3.15 (a), lasers probe the atmosphere above the telescope, returning information about the air's swirling motion to a computer that modifies the mirror thousands of times per second to compensate for poor seeing. Other adaptive optics systems monitor standard stars in the field of view, constantly adjusting the mirror's shape to preserve those stars' appearance. Figure 3.15 (b) presents an example of the improvement in image quality that can be obtained by these means.

Many of the world's largest telescopes now incorporate sophisticated adaptive optics systems, and resolutions as fine as a few hundredths of an arc second have been reported for observations in the near-infrared, significantly better than the resolution of the much smaller *HST* at the same wavelengths. Adaptive optics systems are giving astronomers the best of both worlds, achieving with large ground-based optical telescopes the kind of resolution once attainable only from space.

CONCEPT CHECK

Why is Earth's atmosphere a problem for optical astronomers? What can they do about it?

3.4 Radio Astronomy

In addition to the visible radiation that penetrates Earth's atmosphere on a clear day, radio radiation also reaches the ground, and astronomers have built many ground-based **radio telescopes** capable of detecting cosmic radio waves. ∞ *(Sec. 2.3)* These devices have all been constructed since the 1950s—radio astronomy is a much younger field than optical astronomy.

Essentials of Radio Telescopes

Figure 3.16 shows the world's largest steerable radio telescope, the large 105-m-diameter (340-foot-diameter) telescope located at the National Radio Astronomy Observatory in West Virginia. Conceptually, the operation of a radio telescope is the same as a prime-focus optical reflector (Figure 3.5a). It

(a)

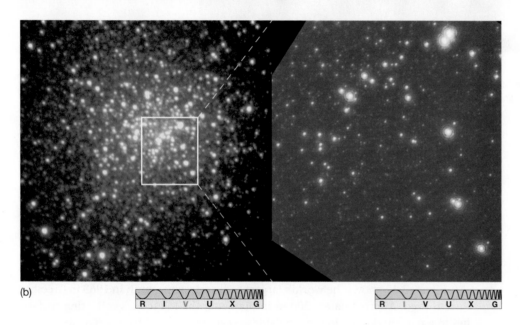

(b)

◄ **FIGURE 3.15 Adaptive Optics**   (a) In this test at the Lick Observatory's 3-m Shane telescope in California, a laser is used to create an "artificial star" to improve guiding. The laser beam probes the atmosphere above the telescope, allowing tiny computer-controlled changes to be made to the shape of the mirror surface thousands of times each second. (b) Clarity of view is very important in astronomy. The uncorrected, visible-light image (left) of the star cluster NGC 6934 is resolved to a little less than 1″. With adaptive optics applied (right), the resolution in the infrared is improved to nearly 10 times better, and more stars are seen more clearly. *(L. Hatch/Lick Observatory; NOAO)*

has a large, horseshoe-shaped mount supporting a huge, curved metal dish that serves as the collecting area. The dish captures cosmic radio waves and reflects them to the focus, where a receiver detects the signals and channels them to a computer.

Radio telescopes are built large in part because cosmic radio sources are extremely faint. Many sources simply don't emit many radio photons (see for example Figure 2.11), the photons don't carry much energy, and the sources themselves are often very distant. ∞ *(Sec. 2.6)* In fact, the total amount of radio energy received by Earth's entire surface is less than a trillionth of a watt. Compare this with the roughly 10 *million* watts our planet's surface receives in the form of infrared and visible light from any of the bright stars visible in the night sky. In order to capture enough radio energy to allow detailed measurements to be made, a large collecting area is essential.

Because of diffraction, the angular resolution of radio telescopes is generally quite poor compared with that of their optical counterparts. Typical wavelengths of radio waves are about a million times longer than those of visible light, and

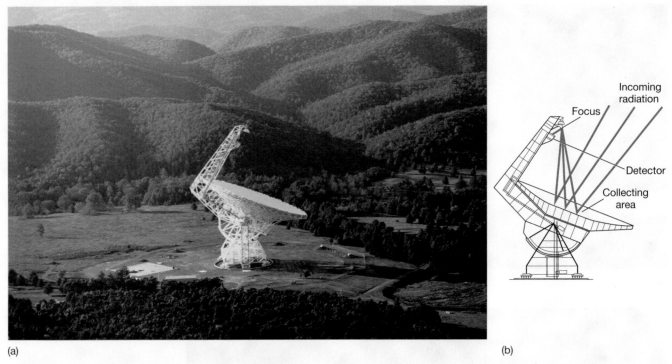

(a)   (b)

▲ **FIGURE 3.16 Radio Telescope**   (a) The world's largest fully steerable radio telescope, the 105-m-diameter device at the
National Radio Astronomy Observatory in Green Bank, West Virginia, is 150 m tall—taller than the Statue of Liberty and nearly as
tall as the Washington Monument. (b) A schematic diagram shows the path taken by an incoming beam of radio radiation
(colored blue). *(NRAO)*

even the enormous sizes of radio dishes only partly offset this effect. The 105-m
telescope shown in Figure 3.16 can achieve resolution of about 1' at a wavelength
of 3 cm, although the instrument was designed to be most sensitive to wavelengths around 1 cm, where the resolution is approximately 20'. The best angular
resolution obtainable with a single radio telescope is about 10" (for the largest
instruments operating at millimeter wavelengths), about 100 times coarser than
the capabilities of the best optical mirrors.

Figure 3.17 shows the world's largest radio telescope, built in 1963 in Arecibo,
Puerto Rico. Approximately 300 m in diameter, the telescope's reflecting surface
lies in a natural depression in a hillside and spans nearly 20 acres. The receiver is
strung among several limestone hills. Its enormous collecting area makes this telescope the most sensitive on Earth. The huge but fixed dish creates one distinct
disadvantage, however. The Arecibo telescope cannot be pointed to follow cosmic
objects across the sky. Its observations are limited to those objects that happen to
pass within about 20° of overhead as Earth rotates.

The Value of Radio Astronomy

Despite the inherent disadvantage of relatively poor angular resolution, radio
astronomy enjoys some advantages too. Unlike visible light, radio waves are not
deflected or scattered by Earth's atmosphere, and the Sun itself is a relatively
weak source of radio energy. As a result, radio observations can cover almost the
entire sky, 24 hours a day. Only within a few degrees of the Sun does solar
radiation swamp radio observations of more distant objects. In addition, radio
observations can often be made through cloudy skies, and radio telescopes can
detect the longest-wavelength radio waves even during rain or snowstorms.

However, perhaps the greatest value of radio astronomy—and all invisible
astronomies—is that it opens up a whole new window on the universe. There are

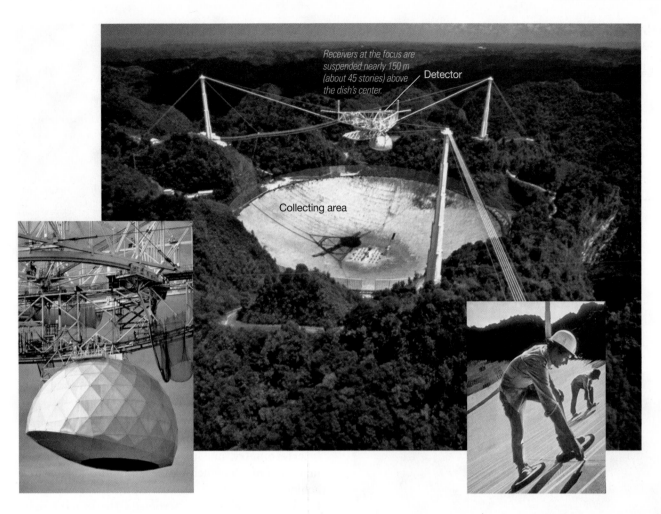

Receivers at the focus are suspended nearly 150 m (about 45 stories) above the dish's center.

Detector

Collecting area

▲ **FIGURE 3.17 Arecibo Observatory** The 300-m-diameter dish at the National Astronomy and Ionospheric Center near Arecibo, Puerto Rico. The two insets show close-ups of the receivers hanging high above the dish (left) and technicians adjusting the dish surface to make it smoother (right). *(D. Parker/T. Acevedo/NAIC; Cornell University)*

two main reasons for this. First, just as objects that are bright in the visible part of the spectrum (the Sun, for example) are not necessarily strong radio emitters, many of the strongest radio sources in the universe emit little or no visible light. Second, visible light may be strongly absorbed by interstellar dust along the line of sight to a source. Radio waves, on the other hand, are generally unaffected by intervening matter. Many parts of the universe cannot be seen at all by optical means but are easily detectable at radio wavelengths. Radio observations therefore allow us to see whole new classes of objects that would otherwise be completely unknown.

Figure 3.18 shows a visible-light photograph of a distant galaxy called Centaurus A. Superimposed on the optical image is a radio map of the same region. The radio image is represented in **false color,** a technique commonly used for displaying images taken in nonvisible light. The colors do not represent the actual wavelength of the radiation emitted, but instead some other property of the source, in this case intensity, descending from red to yellow, green, and blue. Notice just how different the galaxy looks at radio and optical wavelengths. The visible galaxy emits little or no radio energy, while the large blobs (called *radio lobes,* and thought to have been ejected from the center of the galaxy by explosive events long ago) are completely invisible. Centaurus A is an example of a *radio galaxy* (to be studied in more detail in Chapter 15). It actually emits far more energy in the form of radio waves than it does as visible light, but without radio astronomy we would be entirely ignorant of this fact, and of the galaxy's violent past.

CONCEPT CHECK

Cosmic radio waves are very weak, and the resolution of radio telescopes is often poor, so what can astronomers hope to learn from radio astronomy?

▶ **FIGURE 3.18 Radio Galaxy** This is a composite image of the Centaurus galaxy, showing its optical view at center and its radio emission in lobes well beyond (here shown in false color, with red indicating greatest radio intensity, blue the least). *(J. Burns)*

(a)

(b)

Interferometry

Radio astronomers can sometimes improve angular resolution by using a technique known as **interferometry,** making it possible to produce radio images of much higher angular resolution than can be achieved with even the best optical telescopes, on Earth or in space. In interferometry, two or more radio telescopes are used in tandem to observe the same object at the same wavelength and at the same time. The combined instruments together make up an **interferometer** (Figure 3.19). By means of electronic cables or radio links, the signals received by each antenna in the array making up the interferometer are sent to a central computer that combines and stores the data as the antennas track their target.

Interferometry works by analyzing how the signals *interfere* with each other when added together. ∞ *(Sec. 2.2)* Consider an incoming wave striking two detectors (Figure 3.20). Because the detectors lie at different distances from the source, the signals they record will, in general, be out of step with one another and, when combined, will interfere destructively, partly canceling each other out. Notice that the amount of interference depends on the *direction* in which the wave is traveling relative to the line joining the detectors. As Earth rotates and the antennae track their target, careful computer analysis of the combined signal results in a detailed image of the distant object.

Figure 3.21 shows the most powerful interferometer yet constructed. The Atacama Large Millimeter Array (ALMA) is the largest astronomical instrument

◀ **FIGURE 3.19 VLA interferometer** (a) This large interferometer, called the Very Large Array, or VLA for short, is located on the plain of San Augustin in New Mexico. It comprises 27 dishes spread along a Y-shaped pattern about 30 km across. (b) The dishes are mounted on railroad tracks so that they can be repositioned easily. (Another powerful interferometer is shown in the Chapter Opener photo on page 70.) *(NRAO)*

◀ **FIGURE 3.20 Interferometry** Two telescopes, A and B, record different signals from the same incoming wave because of the time it takes the radiation to traverse the distance between them. When the signals are combined, the amount of interference depends on the wave's direction of motion, providing a means of measuring the position of the source in the sky. Here, the dark blue waves come from a source high in the sky and are out of phase, thus causing destructive interference. But when the same source has moved because of Earth's rotation (light-blue waves), the interference can be constructive.

A ⟋⟍⟋⟍⟋⟍   Waves arrive out of step
B ⟋⟍⟋⟍⟋⟍   = destructive interference

A ⟋⟍⟋⟍⟋⟍   Waves arrive in step
B ⟋⟍⟋⟍⟋⟍   = constructive interference

on Earth, sited at an altitude of about 5000 m in the Atacama Desert in northern Chile. Completed in 2014, the system's 66 antennas, most of them 12 m in diameter, operate at millimeter wavelengths, between 0.3 and 10 mm. ALMA is expected to be a telescopic workhorse for the next generation of astronomers, opening a whole new window on the universe and capturing never-before-seen details about the stars, galaxies, and planetary systems beyond our own solar system.

◀ **FIGURE 3.21 The ALMA Array.** Built by an international consortium of astronomers and engineers from the United States, Canada, Europe, East Asia, and Chile, and completed in 2014, the ALMA interferometer scans the universe at millimeter wavelengths from one of the remotest and driest places on our planet, without clouds, radio interference, or light pollution. This single most powerful telescope ever built is designed to probe one of astronomy's last observational frontiers. The inset shows the instrument's control room, where one of the fastest supercomputers on the planet is used to analyze the incoming data and reconstruct images of the field of view. *(ESO/NAOJ/NRAO; STScI)*

(a)

|←———— 100,000 light-years ————→|

(b)

◄ **FIGURE 3.22 Radio—Optical Comparison** (a) This ALMA image of the colliding Antennae galaxies was made using just 12 of the system's antennae operating close together and has an angular resolution of a few arc seconds. The full ALMA system is capable of resolution some 10 times higher than shown here. (b) A visible-light photograph of that same galaxy, made with the HST and displayed on the same scale as (a). *(ESO NAOJ/ NRAO; STScI)*

As far as resolving power is concerned, the effective diameter of an interferometer is the distance between its outermost dishes. In other words, two small dishes can act as opposite ends of an imaginary but huge single radio telescope, dramatically improving the angular resolution. Large interferometers made up of many dishes, like the Very Large Array (VLA) shown in Figure 3.19, can attain radio resolution of a few arc seconds, comparable to that of many ground-based optical instruments (without adaptive optics). ALMA's best resolution is about 0.1″, similar to that of *HST* or VLT. Figure 3.22 compares an ALMA radio map of a nearby galaxy with a photograph of that same galaxy made using a large optical telescope. (The radio image in Figure 3.18 was also obtained by interferometric means.) The larger the distance (or *baseline*) separating the telescopes, the better the resolution attainable. Very-long-baseline interferometry (VLBI), using instruments separated by thousands of kilometers, can achieve resolution in the milli-arcsecond (0.001″) range.

Although the technique was originally developed by radio astronomers, interferometry is no longer restricted to the long-wavelength domain. Radio interferometry became feasible when electronic equipment and computers achieved speeds great enough to combine and analyze radio signals from separate radio detectors without loss of data. As the technology has improved, it has become possible to apply the same methods to higher-frequency radiation. Even before ALMA, millimeter-wavelength interferometry had already become an established and important observational technique. Both the Keck and VLT telescopes are now used for both infrared and optical interferometric work.

3.5 Space-Based Astronomy

As we saw in Chapter 2, Earth's atmosphere is opaque to electromagnetic radiation outside of the radio, infrared, and optical windows. ∞ *(Sec. 2.3)* Most other wavelengths can only be studied from space. The rise of these "other astronomies" has therefore been closely tied to the development of the space program.

Infrared and Ultraviolet Astronomy

Infrared studies are a very important component of modern observational astronomy. Much of the gas between the stars has a temperature between a few tens and a few hundreds of kelvins, and Wien's law tells us that the infrared domain is the natural portion of the electromagnetic spectrum in which to study this material. ∞ *(Sec. 2.4)* Generally, **infrared telescopes** resemble optical telescopes, but the infrared detectors are sensitive to longer-wavelength radiation. Although most infrared radiation is absorbed by the atmosphere (primarily by water vapor), there are a few windows in the high-frequency part of the infrared spectrum where the opacity is low enough to permit ground-based observations. ∞ *(Fig. 2.9)* As we have seen, some of the most useful infrared observing is done from the ground using large telescopes equipped with adaptive optics systems.

For most infrared observations, astronomers must place their instruments above most or all of Earth's atmosphere. Improvements in balloon-, aircraft-,

(a)

(b)

◄ **FIGURE 3.23 Infrared Telescopes** (a) A gondola containing a 1-m infrared telescope is readied for its balloon-borne ascent to an altitude of about 30 km, where it will capture infrared radiation that cannot penetrate Earth's atmosphere. (b) This artist's conception shows the *Spitzer Space Telescope* orbiting in an Earth-trailing orbit around the Sun while surveying the infrared sky at wavelengths ranging from 3 to 200 μm. Note the large baffle at left protecting the telescope from the Sun's heat. *(SAO; JPL)*

rocket-, and satellite-based telescope technologies have made infrared research a powerful tool for studying the universe (Figure 3.23). As might be expected, the infrared telescopes that can be carried above the atmosphere are considerably smaller than the massive instruments found in ground-based observatories. Nevertheless, their infrared view penetrates the clouds of dust and gas surrounding many cosmic objects, allowing astronomers to study regions of space that are completely obscured at visible wavelengths and revolutionizing our understanding of the universe. Figure 3.24 shows some terrestrial and astronomical examples of advantages gained by viewing the infrared part of the spectrum.

In 2003, NASA launched the 0.85-m *Spitzer Space Telescope (SST,* shown in Figure 3.23b), named in honor of Lyman Spitzer, Jr., a renowned astrophysicist and the first person to propose (in 1946) that a large telescope be located in space. The facility's detectors are designed to operate at wavelengths between 3.6 and 160 μm. Unlike previous space-based observatories, *SST* does not orbit Earth, but instead follows our planet in its orbit around the Sun, trailing millions of kilometers behind to minimize Earth's heating effect on its detectors. The spacecraft is currently drifting away from Earth at the rate of 0.1 AU per year. The detectors were cooled to near absolute zero in order to observe infrared signals from space without interference from the telescope's own heat. Figure 3.25 shows some examples of the spectacular imagery obtained from NASA's latest eye on the universe; many more of its images are seen throughout this book.

▶ **FIGURE 3.24 Infrared Images**
(a) An optical photograph taken near San Jose, California, and (b) an infrared photo of the same area taken at the same time. Infrared radiation can penetrate smog much better than short-wavelength visible light. The same advantage pertains to astronomi-cal observations (c). An optical view of an especially dusty part of the central region of the Orion Nebula is more clearly revealed (d) in this infrared im-age showing a cluster of stars behind the obscuring dust. (*Lick Observatory; NASA*)

These foggy optical views . . .

. . . are much clearer when viewed in the infrared.

(a) R I V U X G

(b) R I V U X G

(c) R I V U X G

(d) R I V U X G

Unfortunately, the liquid helium keeping *SST*'s detectors cool could not be confined indefinitely, and it slowly leaked away into space. In early 2009, *Spitzer* entered a new "warm" phase of operation as its temperature increased to roughly 30K—still very cool by Earth standards, but warm enough for the telescope's own thermal emission to overwhelm the long-wavelength detectors on board. (Recall how Wien's law tells us that the thermal emission of a 30-K object peaks at a wavelength of roughly 100 μm.) ∞ *(Sec. 2.4)* However, the craft's shorter-wave-length detectors (at around 3.6 and 4.5 μm) are still operational and will continue to be an important astronomical resource until the telescope drifts out of range of Earth controllers.

To the short-wavelength, higher-energy side of the visible spectrum lies the ultraviolet domain. This region of the spectrum, extending in wavelength from 400 nm (blue light) down to a few nanometers (the soft, or low-energy, end of the X-ray region), has only recently begun to be explored. Because Earth's atmosphere is partially opaque to radiation below 400 nm and is totally opaque to radiation below about 300 nm, astronomers cannot conduct any useful ultraviolet observations from the ground, not even from the highest mountaintop. Rockets, balloons, or satellites are therefore essential to any **ultraviolet telescope**—a device designed to capture and analyze this high-frequency radiation.

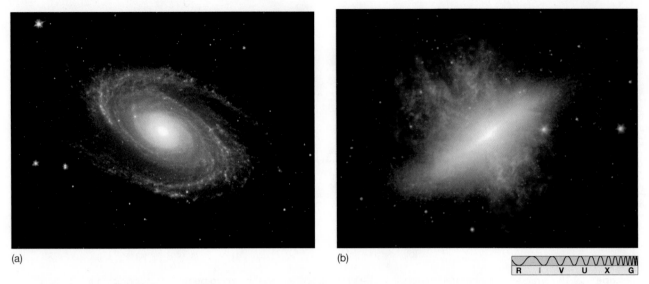

(a)

(b)

R I V U X G

▲ FIGURE 3.25 **Spitzer Images**   Images from the *Spitzer Space Telescope* show its superb capabilities. (a) The magnificent spiral galaxy M81 is about 12 million light-years away. (b) Its companion, M82, is not so serene, rather resembling a smoking hot cigar. *(JPL)*

Figure 3.26 (a) shows an image of a supernova remnant—the remains of a violent stellar explosion that occurred some 12,000 years ago (see Chapter 12)—obtained by the *Extreme Ultraviolet Explorer (EUVE)* satellite, launched in 1992. Since its launch, *EUVE* has mapped out our local cosmic neighborhood as it presents itself in the far ultraviolet and has radically changed astronomers' conception of interstellar space in the vicinity of the Sun. The *Hubble Space Telescope*, best known as an optical telescope, is also a superb ultraviolet instrument, as is the *Galaxy Evolution Explorer (GALEX)* satellite, launched in 2003, one of whose images is seen in Figure 3.26 (b).

PROCESS OF SCIENCE CHECK

Describe some of the kinds of observations that astronomers make across the entire electromagnetic spectrum. Why are such observations necessary?

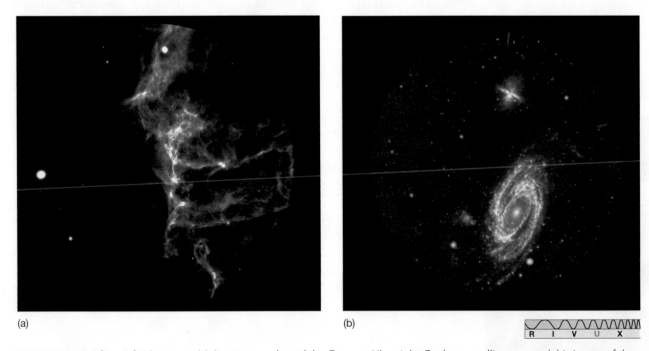

(a)

(b)

R I V U X

▲ FIGURE 3.26 **Ultraviolet Images**   (a) A camera on board the *Extreme Ultraviolet Explorer* satellite captured this image of the Cygnus Loop supernova remnant, the result of a massive star having exploded. The glowing field of debris lies some 1500 light-years from Earth. Based on the velocity of the outflowing debris, astronomers estimate that the explosion occurred about 12,000 years ago. (b) This false-color image of the galaxies M81 and M82 (the same ones as in Figure 3.25), made here by the *Galaxy Evolution Explorer* satellite, reveals stars forming in the blue arms well away from the galaxy's center. *(NASA/GALEX)*

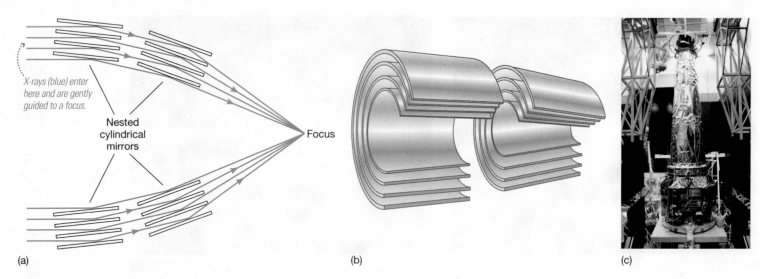

X-rays (blue) enter here and are gently guided to a focus.

Nested cylindrical mirrors

Focus

(a)

(b)

(c)

▲ FIGURE 3.27 X-Ray Telescope   (a) The arrangement of nested mirrors in an X-ray telescope allows the rays to be reflected at grazing angles and focused to form an image. (b) A cutaway 3D rendition of the mirrors, showing their shape more clearly. (c) The *Chandra* X-ray telescope, shown here during the final stages of its construction, has an effective angular resolution of 1″, allowing it to produce images of quality comparable to those of optical photographs. *(NASA)*

High-Energy Astronomy

ANIMATION/VIDEO
Chandra Light and Data Paths

High-energy astronomy studies the universe as it presents itself to us in X-rays and gamma rays—the types of radiation whose photons have the highest frequencies and, hence, the greatest energies. How do we detect radiation of such short wavelengths? First, it must be captured high above Earth's atmosphere because none of it reaches the ground. Second, its detection requires the use of equipment basically different from that used to capture the relatively low-energy radiation discussed up to this point.

The difference in the design of **high-energy telescopes** comes about because X-rays and gamma rays cannot be reflected easily by any kind of surface. Rather, these rays tend to either pass straight through or else be absorbed by any material they strike. When X-rays barely graze a surface, however, they can be reflected from it in a way that yields an image, although the mirror design is fairly complex (Figure 3.27). High-quality data from imaging X-ray telescopes have driven major advances in our understanding of high-energy phenomena throughout the universe.

In 1999, NASA launched the *Chandra X-Ray Observatory* (named in honor of the Indian astrophysicist Subramanyan Chandrasekhar and shown in Figure 3.27c). With greater sensitivity, a wider field of view, and better resolution than any previous X-ray telescope, *Chandra* is providing high-energy astronomers with new levels of observational detail. Figure 3.28 shows the first image returned by *Chandra*—a supernova remnant known as Cas A, all that now remains of a star in the constellation Cassiopeia that was observed to explode about 320 years ago. The false-color image shows 50 million-kelvin gas in the wisps of ejected stellar material; the bright white point at the very center of the debris may be a black hole.

The colors here indicate different intensities of X-rays and also different amounts of heavy elements.

R I V U X G

◀ FIGURE 3.28 X-Ray Image   A false-color *Chandra* X-ray image of the supernova remnant Cassiopeia A, a debris field of scattered, hot gases that were once part of a massive star. Roughly 10,000 light-years from Earth and barely visible in the optical part of the spectrum, Cas A is awash in brilliantly glowing X-rays spread across some 10 light-years. *(CXC/SAO)*

▶ FIGURE 3.29 **Gamma-Ray Image** A false-color gamma-ray image from the *Fermi* telescope, showing the remains of a violent event (a supernova) in a region named IC443. The gamma rays are shown mainly in magenta and are superimposed on infrared (blue/green) and optical (yellow) images of the region. *(NASA)*

R I V U X G

The European *X-Ray Multi-Mirror* satellite (now known as *XMM-Newton*) was launched in 1999. *XMM-Newton* is more sensitive than *Chandra* (that is, it can detect fainter X-ray sources), but it has significantly poorer angular resolution (5′, compared with 0.5″ for *Chandra*), making the two missions complementary to one another.

Gamma-ray astronomy is the youngest entrant into the observational arena. For gamma rays, no method of producing a true image (in the sense of Section 3.1) has yet been devised—present-day gamma-ray telescopes simply point in a specified direction and count photons received. As a result, only fairly coarse (1° resolution) observations can be made. Nevertheless, even at this resolution, there is much to be learned. Cosmic gamma rays were originally detected in the 1960s by the U.S. *Vela* series of satellites, whose primary mission was to monitor illegal nuclear detonations on Earth. Since then, several X-ray telescopes have also been equipped with gamma-ray detectors.

Gamma-ray astronomy traces the most violent events in the universe, on scales ranging from stars to galaxies. NASA's *Compton Gamma-Ray Observatory (CGRO)*, placed in orbit in 1991, scanned the sky and studied individual objects in much greater detail than ever before. The *CGRO* mission ended in 2000 following a failure of one of the spacecraft's three gyroscopes. In 2008, NASA launched the *Fermi Gamma-Ray Space Telescope*. With greater sensitivity to a broader range of gamma-ray energies than *CGRO*, *Fermi's* capabilities have greatly expanded astronomers' view of the high-energy universe. Figure 3.29

CONCEPT CHECK

List two scientific benefits of placing telescopes in space. What might be some drawbacks?

TABLE 3.1	Astronomy at Many Wavelengths	
Radiation	**General Considerations**	**Common Applications (Chapter Reference)**
Radio	Can penetrate dusty regions of interstellar space. Earth's atmosphere largely transparent to radio wavelengths. Can be observed day or night. High resolution at long wavelengths requires very large telescopes.	Radar studies of planets (1, 6) Planetary magnetic fields (7) Interstellar gas clouds (11) Center of Milky Way Galaxy (14) Galactic structure (14, 15) Active galaxies (15) Cosmic background radiation (17)
Infrared	Can penetrate dusty regions of interstellar space. Earth's atmosphere only partially transparent to infrared radiation, so some observations must be made from space.	Star formation (11) Cool stars (11, 12) Center of Milky Way Galaxy (14) Active galaxies (15) Large-scale structure of the universe (16, 17)
Visible	Earth's atmosphere transparent to visible light.	Planets (6, 7) Stars and stellar evolution (9, 10, 12) Galactic structure (14, 15) Large-scale structure of the universe (16, 17)
Ultraviolet	Earth's atmosphere opaque to UV radiation, so observations must be made from space.	Interstellar medium (11) Hot stars (12)
X-ray	Earth's atmosphere opaque to X-rays, so observations must be made from space. Special mirror configurations needed to form images.	Stellar atmospheres (9) Neutron stars and black holes (13) Active galactic nuclei (15) Hot gas in galaxy clusters (16)
Gamma ray	Earth's atmosphere opaque to gamma rays, so observations must be made from space. Cannot form images.	Neutron stars (13) Active galactic nuclei (15)

(a)

(b)

(c)

(d)

(e)

◀ **FIGURE 3.30** **Multiple Wavelengths**   The Milky Way Galaxy as it appears at (a) radio, (b) infrared, (c) visible, (d) X-ray, and (e) gamma-ray wavelengths. Each frame is a panoramic view covering the entire sky, with the center of our Galaxy at the center of each map. *ESA; UMass/Caltech; A. Mellinger; MPI; NASA*

shows a Fermi image of the aftermath of a vast stellar explosion, the kind of violent event that emits much of its energy in the form of gamma rays. An all-sky image from *Fermi* appears in Figure 3.30 (e).

Full-Spectrum Coverage

Table 3.1 lists the basic regions of the electromagnetic spectrum and describes objects typically studied in each frequency range. Bear in mind that the list is far from exhaustive and that many astronomical objects are now routinely observed at many different electromagnetic wavelengths. As we proceed through the text, we will discuss more fully the wealth of information that high-precision astronomical instruments can provide us.

As an illustration of the sort of comparison that full-spectrum coverage allows, Figure 3.30 shows a series of images of the Milky Way Galaxy. They were made by several instruments, at wavelengths ranging from radio to gamma ray, over a period of about 5 years. By comparing the features visible in each, we immediately see how multiwavelength observations can complement each other, greatly extending our perception of the universe around us.

THE BIG QUESTION

Astronomy is a data-driven science. The most stunning discoveries are often made as new telescopes come online; some instruments are bigger, others operate in orbit, and almost all are better than anything available previously. The biggest advances have been made by instruments built to sense new domains of the electromagnetic spectrum. Today, astronomers are opening new, nonelectromagnetic windows, in the form of neutrinos (Chapter 9) and gravitational radiation (Chapter 13). What will we learn from them, and are there other, as yet unknown, channels still awaiting discovery?

CHAPTER REVIEW

SUMMARY

LO1 A **telescope** (p. 72) is a device designed to collect as much light as possible from some distant source and deliver it to a detector for detailed study. Astronomers prefer **reflecting telescopes** (p. 72) because large mirrors are lighter and much easier to construct than large lenses, and they also suffer from fewer optical defects. Instruments may be placed inside the telescope at the **prime focus** (p. 72), or a secondary mirror may be used to reflect the light to an external detector. Most modern telescopes use **charge-coupled devices** (p. 78) to collect and store data in digital form for later analysis.

LO2 The light-gathering power of a telescope depends on its **collecting area** (p. 79), which is proportional to the square of the mirror diameter. To study the faintest sources of radiation, astronomers must use large telescopes. Large telescopes also suffer least from the effects of diffraction and, hence, can achieve better **angular resolution** (p. 81) once the blurring effects of Earth's atmosphere are overcome. The amount of diffraction is proportional to the wavelength of the radiation under study and inversely proportional to the size of the mirror.

LO3 The resolution of most ground-based optical telescopes is limited by **seeing** (p. 83)—the blurring effect of Earth's turbulent atmosphere, which smears the point-like images of a star into a **seeing disk** (p. 83) a few arc seconds in diameter. Astronomers can improve a telescope's resolution by using **active optics** (p. 84), in which a telescope's environment and focus are carefully monitored and controlled, and **adaptive optics** (p. 84), in which the blurring effects of atmospheric turbulence are corrected for in real time.

LO4 **Radio telescopes** (p. 84) are conceptually similar in construction to optical reflecting telescopes, although they are generally much larger than optical instruments, in part because so little radio energy reaches Earth from space. Their main disadvantage is that diffraction of long-wavelength radio

waves limits their resolution. Their principal advantage is that they allow astronomers to explore a new part of the electromagnetic spectrum and of the universe—many astronomical radio emitters are completely undetectable in visible light. In addition, radio observations are largely unaffected by Earth's atmosphere, weather, and the location of the Sun.

LO5 To increase the effective area of a telescope, and, hence, improve its resolution, several instruments may be combined into an **interferometer** (p. 88), in which the interference pattern of radiation received by two or more detectors is used to reconstruct a high-precision map of the source. Using **interferometry** (p. 88), radio telescopes can produce images much sharper than those from the best optical equipment.

LO6 **Infrared** (p. 90) and **ultraviolet telescopes** (p. 92) are similar in basic design to optical systems. **High-energy telescopes** (p. 94) study the X-ray and gamma-ray regions of the electromagnetic spectrum. X-ray telescopes can form images of their field of view, although the mirror design is more complex than for lower-energy instruments. Gamma-ray telescopes simply point in a certain direction and count photons received. Infrared studies in some parts of the infrared range can be carried out using large ground-based systems. However, because of atmospheric opacity, most infrared, and all ultraviolet, X-ray, and gamma-ray observations must be made from space.

LO7 Different physical processes can produce very different kinds of electromagnetic radiation, and the image of a given object in long-wavelength, low-energy radio waves may bear little resemblance to its appearance in high-energy X-rays or gamma rays. Observations at wavelengths spanning the electromagnetic spectrum are essential to a complete understanding of astronomical events.

(a)

(b)

MasteringAstronomy®   For instructor-assigned homework go to www.masteringastronomy.com

Problems labeled **POS** explore the process of science. **VIS** problems focus on reading and interpreting visual information. **LO** connects to the introduction's numbered Learning Outcomes.

REVIEW AND DISCUSSION

1. **LO1** List three advantages of reflecting telescopes over refracting telescopes.

2. **LO2** What are the largest optical telescopes in use today? Why do astronomers want their telescopes to be as large as possible?

3. How does Earth's atmosphere affect what is seen through an optical telescope?

4. What advantages does the *Hubble Space Telescope* have over ground-based telescopes? List some disadvantages.

5. What are the advantages of a CCD over a photograph?

6. **LO3** Is the resolution of a 2-m telescope on Earth's surface limited more by atmospheric turbulence or by the effects of diffraction?

7. Why do radio telescopes have to be very large?

8. **LO4** Which astronomical objects are best studied with radio techniques?

9. **LO5** What is interferometry, and what problem in radio astronomy does it address?

10. Compare the highest resolution attainable with optical telescopes to the highest resolution attainable with radio telescopes (including interferometers).

11. What special conditions are required to conduct observations in the infrared?

12. In what ways do the mirrors in X-ray telescopes differ from those found in optical instruments?

13. **LO6** Why was *Fermi* placed in space, rather than on the ground?

14. **LO7 POS** What are the main advantages of studying objects at many different wavelengths of radiation?

15. **POS** Our eyes can see light with an angular resolution of about 1′—equivalent to about a third of a millimeter at arm's length. Suppose our eyes detected only infrared radiation, with 1° angular resolution. Would we be able to make our way around on Earth's surface? To read? To sculpt? To create technology?

CONCEPTUAL SELF-TEST: TRUE OR FALSE/MULTIPLE CHOICE

1. The main advantage to using the *HST* is the increased amount of "nighttime" viewing it affords. (T/F)

2. The term "seeing" is used to describe how faint an object can be detected by a telescope. (T/F)

3. One of the primary advantages of CCDs over photographic plates is the former's high efficiency in detecting light. (T/F)

4. Radio telescopes are large in part to improve their angular resolution, which is poor because of the long wavelengths at which they are used to observe the skies. (T/F)

5. Infrared astronomy can only be done from space. (T/F)

6. Gamma-ray telescopes employ the same basic design that optical instruments use. (T/F)

7. Because gamma rays have very short wavelengths, gamma-ray telescopes can achieve extremely high angular resolution. (T/F)

8. The main reason that most professional research telescopes are reflectors is that (a) mirrors produce sharper images than lenses do; (b) their images are inverted; (c) they do not suffer from the effects of seeing; (d) large mirrors are easier to build than large lenses.

9. If telescope mirrors could be made in odd sizes, the one with the *most* light-gathering power would be a triangle with 1-m sides; a square with 1-m sides; a circle 1 m in diameter; a rectangle with two 1-m sides and two 2-m sides.

10. The primary reason professional observatories are built on the highest mountaintops is to (a) get away from city lights; (b) be above the rain clouds; (c) reduce atmospheric blurring; (d) improve chromatic aberration.

11. When multiple radio telescopes are used for interferometry, resolving power is most improved by increasing (a) the distance between telescopes; (b) the number of telescopes in a given area; (c) the diameter of each telescope; (d) the electrical power supplied to each telescope.

12. The *Spitzer Space Telescope* is stationed far from Earth because (a) this increases the telescope's field of view; (b) the telescope is sensitive to electromagnetic interference from terrestrial radio stations; (c) doing so avoids the obscuring effects of Earth's atmosphere; (d) Earth is a heat source and the telescope must be kept very cool.

13. The best way to study young stars hidden behind interstellar dust clouds would be to use (a) X-rays; (b) infrared light; (c) ultraviolet light; (d) blue light.

14. **VIS** The image shown in Figure 3.12 (Resolution) is sharpest when the ratio of wavelength to telescope size is (a) large; (b) small; (c) close to 1; (d) none of these.

15. **VIS** Table 3.1 (Astronomy at Many Wavelengths) suggests that the best frequency range in which to study the hot (million-kelvin) gas found among the galaxies in the Virgo cluster would be (a) at radio frequencies; (b) in the infrared; (c) in X-rays; (d) in gamma rays.

PROBLEMS

The number of squares preceding each problem indicates its approximate level of difficulty.

1. ■ A certain telescope has a 10′ × 10′ field of view that is recorded using a CCD chip having 2048 × 2048 pixels. What angle on the sky corresponds to 1 pixel? What would be the diameter of a typical seeing disk (1″ radius), in pixels?

2. ■ The *SST*'s planned operating temperature is 5.5 K. At what wavelength (in micrometers, μm) does the telescope's own blackbody emission peak? How does this wavelength compare with the wavelength range in which the telescope is designed to operate? (∞ *More Precisely 2-2*)

3. ■ A 2-m telescope can collect a given amount of light in 1 hour. Under the same observing conditions, how much time would be required for a 6-m telescope to perform the same task? A 12-m telescope?

4. ■ A space-based telescope can achieve a diffraction-limited angular resolution of 0.05″ for red light (wavelength 700 nm). What would the resolution of the instrument be (a) in the infrared, at wavelength 3.5 μm, and (b) in the ultraviolet, at wavelength 140 nm?

5. ■■ Two identical stars are moving in a circular orbit around one another with an orbital separation of 2 AU. The system lies 200 light-years from Earth. If we happen to view the orbit head-on, how large a telescope would we need to resolve the stars, assuming diffraction-limited optics at a wavelength of 2 μm?

6. ■■ What is the greatest distance at which *HST*, operating in blue light (400 nm), could resolve the stars in the previous question?

7. ■ The photographic equipment on a telescope is replaced by a CCD. If the photographic plate records 5 percent of the light reaching it, but the CCD records 90 percent, how much time will the new system take to collect as much information as the old detector recorded in a 1-hour exposure?

8. ■ The Andromeda galaxy lies about 2.5 million light-years away. To what distances do the angular resolutions of *SST* (3″), *HST* (0.05″), and a radio interferometer (0.001″) correspond at that distance?

9. ■ What would be the equivalent single-mirror diameter of a telescope constructed from two separate 10-m mirrors? Four separate 8-m mirrors?

10. ■ Estimate the angular resolutions of (a) a radio interferometer with a 5000-km baseline, operating at a frequency of 5 GHz, and (b) an infrared interferometer with a baseline of 50 m, operating at a wavelength of 1 μm.

ACTIVITIES

Collaborative

1. Determine the maximum size interferometer your group could build if you placed 2-m radio telescopes at each of your homes. What would be its resolution at a wavelength of 1 cm?

2. Your group has been assigned to observe the region of the sky around Orion to look for hot, bright young stars hidden in molecular clouds. Explain which of the telescopes described in the text would be your best choice, and estimate the level of detail you might expect to see.

Individual

1. Take some photographs of the night sky. You will need a location with a clear, dark sky, a good digital camera that lets you control the exposure time, a tripod and cable release, and a watch with a seconds display visible in the dark. Set your camera to the "manual" setting for the exposure and attach the cable release so you can control it. Set the focus on infinity. Point the camera at the desired constellation, seen through the viewfinder, and take a 20- to 30-second exposure. Don't touch any part of the camera or hold on to the cable release during the exposure to minimize all vibration. Keep a log of your shots.

2. For some variation, vary your exposure times, take hours-long exposures for star trails, use different lenses such as wide-angle or telephoto, or place the camera piggyback on a telescope tracking an object in the sky so you can take exposures that are a few minutes long. Experiment and have fun!

3. Which image of the Milky Way Galaxy in Figure 3.30 (Multiple Wavelengths) provides the most interesting information? Explain your reasoning.

Our Planetary System

As our study of astronomy expands, we embark on the next in a series of increasingly larger steps that will ultimately take us to the limits of the observable universe. This journey begins with Earth, progresses to the study of other planets, and continues on to the magnificent Sun. The dimension of our planetary family is so large that more than a million Earths could be stacked side by side across the whole solar system.

Part 2 relates our present understanding of our extended, complex home in space. That understanding is still a work in progress, but the results thus far have revolutionized our knowledge both of our present cosmic neighborhood and of Earth's rich natural history.

The images here illustrate the range of sizes encountered in Part 2, from the scattered debris of asteroids and comets to the central Sun.

Venus ~10^7 m

Moon ~10^6 m

Asteroid ~10^4 m

Sun ~10⁹ m

Saturn ~10⁸ m

Much of what we know about the planets
was discovered during the last few decades.

4

The Solar System

Interplanetary Matter and the Birth of the Planets

In less than a generation, we have learned more about our solar system—the Sun and everything that orbits it—than in all previous centuries. By studying the planets, their moons, and the countless fragments of material that orbit in interplanetary space, astronomers have gained a richer outlook on our own home in space. The discoveries of the past few decades have revolutionized our understanding not only of the present state of our cosmic neighborhood, but also of its history, for our solar system is filled with clues to its own origin and evolution. Paradoxically, the richest sources of information about the earliest days of the solar system are not the large bodies orbiting the Sun, but rather the asteroids, comets, and meteoroids that pepper interplanetary space. Far more than the planets themselves, these fragments hold a record of the formative stages of our planetary system and have much to teach us about the origin of our world.

THE BIG PICTURE

Our own solar system formed some 4.5 billion years ago, a time so ancient that it is virtually impossible to reconstruct the details of that singular event. Ironically, studies of other planetary systems far beyond our own are now helping us decipher, like Rosetta stones, our own origins. Just as comparative planetology of our own eight planets guides our understanding of Earth's formation and history, extrasolar planets have much to teach us about how planetary systems form and evolve.

◄ Comets are small but vital components of our planetary system, dating back to the earliest times and in many cases containing pristine material largely unchanged since the solar system formed. They are one of the few direct links we have with the processes that formed the Sun and the planets. This image shows comet 67 P/Churyumov–Gerasimenko, a 4-km-wide chunk of ice that orbits between Earth and Jupiter. It was taken by the European *Rosetta* spacecraft, which went into orbit around the comet in 2014 and remained with it during its closest approach to the Sun in 2015. *Rosetta* carried out detailed measurements of the comet's interior and environment, and soft-landed a small probe on the surface, providing by far the best view of a comet ever obtained. (*ESA/Rosetta/MPS*)

LEARNING OUTCOMES

Studying this chapter will enable you to:

LO1 Describe the scale and structure of the solar system, and list the basic differences between terrestrial and jovian planets.

LO2 Summarize the orbital and physical properties of asteroids.

LO3 Describe the composition and structure of a typical comet, and explain what a comet's orbit tells us about its probable origin.

LO4 Summarize the orbital and physical properties of meteoroids, and explain how these bodies are related to asteroids and comets.

LO5 List the major facts and exceptions that any theory of solar system formation must explain.

LO6 Outline the condensation theory of planetary formation, and indicate how it accounts for the major features of the solar system.

LO7 Explain how the condensation theory accounts for the terrestrial and jovian planets, as well as the smaller bodies scattered throughout the solar system.

LO8 Describe the main methods by which astronomers have detected planets beyond the solar system.

LO9 Outline the properties of the known extrasolar planets and relate to current theories of solar system formation.

MasteringAstronomy®

Visit www.masteringastronomy.com for quizzes, animations, videos, interactive figures, and self-guided tutorials.

4.1 An Inventory of the Solar System

Our **solar system** contains one star (the Sun), eight planets orbiting that star, 173 moons (at last count) orbiting those planets, five *dwarf planets*, seven *asteroids* and more than 100 *Kuiper belt objects* larger than 300 km (200 miles) in diameter, tens of thousands of smaller (but well-studied) asteroids and Kuiper belt objects, myriad *comets* a few kilometers in diameter, and countless *meteoroids* less than 100 m across. This list will undoubtedly grow as we continue to explore our cosmic neighborhood.

ANIMATION/VIDEO (MA)
An Astronomical Ruler

(a)   10 AU

(b)

▲ **FIGURE 4.1 Solar System**   Major bodies of the solar system include the Sun, planets, and asteroids. Except for Mercury, the orbits of the planets are almost circular (a) and lie nearly in the same plane (b). The entire solar system, including the Kuiper belt, spans nearly 100 AU.

Planetary Properties

The overall arrangement of the solar system is shown in Figure 4.1. Mercury is the planet closest to the Sun, followed by Venus, Earth, Mars, Jupiter, Saturn, Uranus, and Neptune. The asteroids lie mainly in a broad belt between the orbits of Mars and Jupiter, the Kuiper belt beyond Neptune.

The planets' distances from the Sun are known from their periods and Kepler's laws once the scale of the solar system is set by radar ranging on Venus. ∞ *(Sec. 1.3)* The distance from the Sun to Neptune is about 30 AU, some 700,000 times Earth's radius and roughly 12,000 times the distance from Earth to the Moon. Yet despite this vast extent, the planets all lie very close to the Sun, astronomically speaking. The diameter of Neptune's orbit is less than 1/3000 of a light-year, while the next nearest star to the Sun is several light-years distant. Note that the planetary orbits are not evenly spaced. Roughly speaking, the spacing between adjacent orbits doubles as we move outward from the Sun.

All the planets orbit the Sun counterclockwise as seen from above Earth's North Pole and in nearly the same plane as Earth (the ecliptic plane). ∞ *(Sec. 1.1)* Mercury deviates slightly from this rule—its orbital plane lies at 7° to the ecliptic. Still, as illustrated in Figure 4.1(b), we can think of the solar system as being remarkably flat. Figure 4.2 is a photograph of the planets Mercury, Venus, Mars, Jupiter, and Saturn, taken during an April 2002 planetary alignment. These five planets can (occasionally) be found in the same region of the sky in large part because their orbits lie nearly in the same plane in space.

Table 4.1 lists some basic orbital and physical properties of the eight planets, with a few other solar system objects included for comparison.

Earth's radius has long been known through conventional surveying techniques and, more recently, by satellite measurements. ∞ *(Discovery 0-1)* Other planetary radii are found by measuring the planets' angular sizes and then employing elementary geometry, as discussed in *More Precisely 4-1*. Figure 4.3 illustrates the sizes of the planets relative to the Sun.

A planet's mass is determined by observing its gravitational influence on some nearby object and applying Newton's laws of motion and gravity. ∞ *(More Precisely 1-1)* Before the Space Age, astronomers calculated planetary masses either by tracking the orbits of the planets' moons (if any) or by measuring the small but detectable distortions the planets produce in each other's orbits. Today astronomers can accurately determine the masses of most of the objects listed in Table 4.1 through their gravitational effects on artificial satellites and space probes launched from Earth. With nearly 99.9 percent of the total mass, the Sun is clearly the "senior partner" in the solar system. Its gravity dominates the motion of everything else.

The final column in Table 4.1 lists a quantity called **density,** a measure of the "compactness" of an object, computed by dividing the object's mass (in kilograms) by its volume (in cubic meters). Earth's average density is 5500 kg/m^3. For comparison, the density of water is 1000 kg/m^3, rocks on Earth's surface have densities in the range 2000–3000 kg/m^3, and iron has a density of about 8000 kg/m^3. Earth's atmosphere at sea level has a density of about 1 kg/m^3.

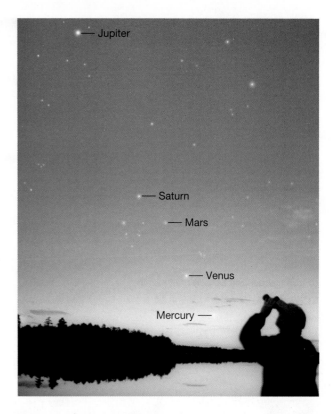

▶ **FIGURE 4.2 Planetary Alignment** Six planets—Mercury, Venus, Mars, Jupiter, Saturn, and Earth—are shown during a planetary alignment in April 2002. The Sun and Moon are just below the horizon. *(J. Lodriguss)*

4.1 MORE PRECISELY

Measuring Sizes with Geometry

We saw in Chapter 1 how astronomers use parallax, radar ranging, and Kepler's laws to determine the distances to objects in the solar system. ∞ *(Sec. 1.3)* Knowing the distance, we can convert a body's angular size into its physical size using a simple argument first made by the Greek geometer Euclid.

The accompanying figure shows an observer measuring the angular diameter of an object of known distance; we have added a large circle centered on the observer and passing through the object. To compute the object's size, we note that the ratio of its actual diameter to the circumference of the circle (2π times the distance to the object) must be equal to the ratio of the observed angular diameter to one full revolution, 360°:

$$\frac{\text{diameter}}{2\pi \times \text{distance}} = \frac{\text{angular diameter}}{360°}.$$

Note the basic similarity of this reasoning to that presented in Chapter 0. ∞ *(Discovery 0-1)* From the above equation we find

$$\text{diameter} = \text{distance} \times \frac{\text{angular diameter}}{57.3°}.$$

For example, radar ranging on the Moon tells us that its distance is 384,000 km. The Moon's angular diameter is measured to be about 31 arc minutes—a little over half a degree. It therefore follows that the Moon's actual diameter is 384,000 km × (31/60)° 57.3° = 3460 km. A more precise measurement gives 3476 km.

Although the observations are straightforward and the geometrical reasoning elementary, simple measurements such as these form the basis for almost every statement made in this book about size and scale in the universe.

TABLE 4.1 Properties of Some Solar System Objects

Object	Orbital Semimajor Axis (AU)	Orbital Period (Earth Years)	Mass (Earth Masses)	Radius (Earth Radii)	Number of Known Moons	Average Density (kg/m³)	Average Density (Earth = 1)
Mercury	0.39	0.24	0.055	0.38	0	5400	0.98
Venus	0.72	0.62	0.82	0.95	0	5200	0.95
Earth	1.0	1.0	1.0	1.0	1	5500	1.00
Moon	—	—	0.012	0.27	—	3300	0.60
Mars	1.5	1.9	0.11	0.53	2	3900	0.71
Ceres (asteroid)	2.8	4.7	0.00015	0.073	0	2700	0.49
Jupiter	5.2	11.9	318	11.2	67	1300	0.24
Saturn	9.5	29.4	95	9.5	62	700	0.13
Uranus	19.2	84	15	4.0	27	1300	0.24
Neptune	30.1	164	17	3.9	14	1600	0.29
Pluto (Kuiper-belt object)	39.5	249	0.002	0.2	5	2100	0.38
Comet Hale-Bopp	180	2400	1.0×10^{-9}	0.004	—	100	0.02
Sun	—	—	332,000	109	—	1400	0.25

Terrestrial and Jovian Planets

Astronomers can infer a planet's bulk composition from a combination of spacecraft data and theoretical calculations (see Chapters 5–8). Based on these, we can draw a clear distinction between the inner and outer members of our planetary system. Simply put, the inner planets—Mercury, Venus, Earth, and Mars—are small, dense, and *rocky* in composition. The outer worlds—Jupiter, Saturn, Uranus, and Neptune—are large, of low density, and *gaseous*.

Because the physical and chemical properties of Mercury, Venus, and Mars are somewhat similar to Earth's, the four innermost planets are called the **terrestrial planets.** (The word *terrestrial* derives from the Latin word *terra*, meaning "land" or "earth.") The larger outer planets—Jupiter, Saturn, Uranus, and Neptune—are all similar to one another chemically and physically (and very different from the terrestrial worlds). They are labeled the **jovian planets,** after Jupiter, the largest member of the group. The Roman god Jupiter was also known as Jove. The jovian worlds are all much larger than the terrestrials and quite

ANIMATION/VIDEO
Size and Scale of the Terrestrial Planets I

ANIMATION/VIDEO
The Gas Giants

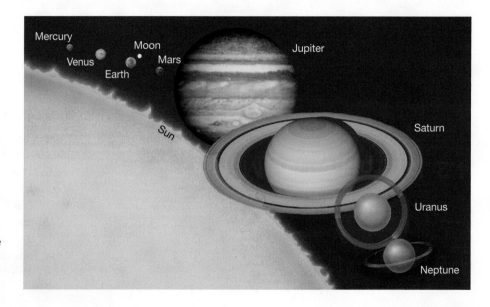

▶ **FIGURE 4.3 Sun and Planets** Relative sizes of the planets and our Sun. Notice that Jupiter, Saturn, Uranus, and Neptune are much larger than Earth and the other inner planets. However, even these large planets are dwarfed by the still larger Sun.

different from them in both composition and structure. Table 4.2 compares and contrasts some key properties of these two distinct planetary classes.

There are important variations within each category. For example, when we correct for how gravity compresses their interiors, we find that the *uncompressed densities* of the terrestrial planets decrease steadily with increasing distance from the Sun, so these planets' compositions depend on their location in the solar system. Similarly, all four jovian worlds have large, dense "terrestrial" cores of up to 20 times the mass of Earth, but these cores account for an increasing fraction of the jovian planets' total mass as we move outward.

Solar System Debris

One of the principal goals of planetary science is to understand how the solar system formed and to explain the physical conditions now found on Earth and elsewhere in our planetary system. Ironically, studies of the most accessible planet—Earth itself—do not help us much in our quest because information about our planet's early stages was obliterated long ago by atmospheric erosion and geological activity, such as earthquakes and volcanoes. Similar statements hold for the other planets. The problem is that the large bodies of the solar system have all *evolved* significantly since they formed, making it difficult to decipher the circumstances of their births.

A much better place to look for hints of conditions in the early solar system is on its *small* bodies—the planetary moons and the asteroids, meteoroids, Kuiper belt objects, and comets that make up interplanetary debris—for nearly all such fragments contain traces of solid and gaseous matter from the earliest times. They represent truly ancient material—many have scarcely changed since they formed along the rest of the solar system billions of years ago.

4.2 Interplanetary Matter

In the vast space among the eight known planets move countless small chunks of matter, ranging in size from a few hundred kilometers in diameter down to tiny grains of dust. The major constituents of this cosmic debris are asteroids, comets, Kuiper belt objects, and meteoroids. **Asteroids** and **meteoroids** are fragments of rocky material, somewhat similar in composition to the outer layers of the terrestrial planets. The distinction between the two is simply their size—anything larger than 100 m in diameter (corresponding to a mass exceeding 10,000 tons) is conventionally called an asteroid; anything smaller is a meteoroid. **Comets** are predominantly icy rather than rocky in composition (although they do contain some rocky material) and have typical diameters in the 1- to 10-km range. The **Kuiper belt** is an "outer asteroid belt" of sorts, also consisting of icy bodies, including the former planet Pluto. Comets and Kuiper belt objects are quite similar in chemical makeup to some of the icy moons of the outer planets—and may very well be the progenitors of those small bodies.

In 2006, the International Astronomical Union, the organization that oversees the rules for astronomical terminology, introduced a new category of solar system object: A **dwarf planet** is a body that orbits the Sun and is massive enough for its gravity to have pulled it into a spherical shape, but not so massive that it has cleared the region around its orbit of smaller bodies. By this definition, the three largest "small bodies" in the solar system—the asteroid Ceres, three Kuiper belt objects, including Pluto, and the "trans-Neptunian object" Eris—are all dwarf planets.

Together, these small bodies account for a negligible fraction of the total mass of the solar system—just a few millionths the mass of the Sun—and they are irrelevant to the present-day workings of the planets and their moons. Yet they are are key to answering some very fundamental questions about our planetary environment.

TABLE 4.2	Comparison Between the Terrestrial and Jovian Planets
Terrestrial	**Jovian**
close to the Sun	far from the Sun
closely spaced orbits	widely spaced orbits
small masses	large masses
small radii	large radii
predominantly rocky	predominantly gaseous
solid surface	no solid surface
high density	low density
slower rotation	faster rotation
weak magnetic fields	strong magnetic fields
no rings	many rings
few moons	many moons

CONCEPT CHECK

Why do astronomers draw such a clear distinction between the inner and the outer planets?

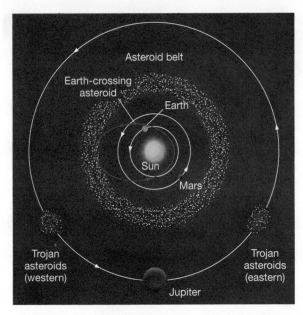

▲ **FIGURE 4.4 Inner Solar System**   The main asteroid belt, along with the orbits of Earth, Mars, and Jupiter. Note the Trojan asteroids clumped at two locations in Jupiter's orbit. The orbits of the main-belt asteroids generally lie within the belt region of the figure. The red ellipse shows the orbit of an Earth-crossing asteroid.

▲ **FIGURE 4.5 Asteroids, from Earth INTERACTIVE**   The Earth-crossing asteroid Icarus has an orbit that passes within 0.2 AU of the Sun, well within Earth's orbit. Icarus occasionally comes close to Earth, making it one of the best-studied asteroids in the solar system. Its motion relative to the stars makes it appear as a streak (marked) in this long-exposure photograph. (*Palomar Observatory/Caltech*)

Asteroid Orbits

The name *asteroid* comes from the Greek word *asteroeides*, which means "like a star," but asteroids are definitely not stars. They are too small even to be classified as planets. Astronomers often refer to them as minor planets. Researchers have so far cataloged well over 400,000 asteroids with well-determined orbits. The total number of known asteroids (that is, including those whose orbits are not yet known with sufficient accuracy to make them "official") is now approaching 1 million. As sketched in Figure 4.4, the vast majority are found in a region of the solar system known as the **asteroid belt,** located between 2.1 and 3.3 AU from the Sun, roughly midway between the orbits of Mars (at 1.5 AU) and Jupiter (at 5.2 AU). All but one of the known asteroids revolve about the Sun in prograde orbits—in the same sense (direction) as Earth and the other planets. Again like the planets, most asteroids have orbits fairly close to the ecliptic plane (with inclinations less than 10–20°). Unlike the almost circular paths of the major planets, however, asteroid orbits are generally noticeably elliptical in shape.

In addition to the main-belt asteroids, a few hundred **Trojan asteroids** share an orbit with Jupiter, remaining a constant 60° ahead of or behind that planet as it circles the Sun. This peculiar orbital behavior is not a matter of chance—the Trojan asteroids are held in place by a stable balance between the gravitational fields of Jupiter and the Sun. Calculations first performed by the 18th-century French mathematician Joseph Louis Lagrange show that interplanetary matter that strays into one of the two regions of space now occupied by the Trojan asteroids can remain there indefinitely, perfectly synchronized with Jupiter's motion.

The orbits of most asteroids have eccentricities lying in the range 0.05–0.3, ensuring that they always remain between the orbits of Mars and Jupiter. However, some **Earth-crossing asteroids** move on orbits that intersect Earth's orbit, most likely because the gravitational field of Mars or Jupiter has deflected these small bodies into the inner solar system. Currently, some 12,000 Earth-crossing asteroids are known. Most have been discovered since the late 1990s, when systematic searches for such objects began. The mile-wide Earth-crossing asteroid Icarus is shown (as the white streak) in Figure 4.5.

Roughly 1600 Earth crossers are officially designated "potentially hazardous," meaning that they are more than about 150 m in diameter (three times the size of the impactor responsible for the Barringer crater shown in Figure 4.19) and move in orbits that could bring them within 0.05 AU (7.5 million kilometer) of our planet. All told, between 2005 and 2015, some 200 asteroids (that we know of!) passed within 0.05 AU of our planet, a couple of them coming within the orbit of the Moon. Similar numbers are expected during the next decade. None of the currently known potentially hazardous asteroids is expected to impact Earth during the next century—the closest predicted near miss will occur in April 2029, when the 350-m asteroid Apophis will pass approximately 30,000 km above our planet's surface.

Calculations indicate that most Earth-crossing asteroids *will* eventually collide with Earth and that during any given million-year period, roughly three asteroids strike our planet. Several dozen large basins and eroded craters on Earth are thought to be sites of ancient asteroid collisions. The many large impact craters on the Moon, Venus, and Mars are direct evidence of similar events on other worlds. Such an impact could be catastrophic by human standards. Even a 1-km object carries more than 100 times more energy than all the nuclear weapons currently in existence on Earth and would devastate an area hundreds of kilometers in diameter. Should a sufficiently large asteroid hit our planet, it might even cause the extinction of an entire species. Indeed, many scientists think that the extinction of the dinosaurs was the result of just such an impact. ∞ *(Discovery 4-1)*

Asteroid Properties

Most asteroids are too small to resolve with even the largest Earth-based telescopes. ∞ *(Sec. 3.2)* Astronomers generally estimate asteroid sizes from the amount of sunlight they reflect and the heat they radiate. Occasionally, an asteroid passes directly in front of a star, allowing astronomers to determine its size and shape with great accuracy. The largest asteroid, the dwarf planet Ceres, has a diameter of 940 km. Only about two dozen asteroids are larger than 200 km across; most are far smaller. Large asteroids are roughly spherical because (as with planets) gravity is the dominant force determining their shape. Smaller bodies can be highly irregular in shape. The gravitational effect of an asteroid on its neighbors is very small and hard to measure accurately, so astronomers have measured the masses of only a few of them. The mass of Ceres is just 1/10,000 the mass of Earth. The total mass of all the known asteroids probably amounts to less than one-tenth the mass of the Moon.

Asteroid compositions are inferred from spectroscopic and other measurements in the infrared, visible, and ultraviolet parts of the spectrum. The darkest (least reflective) carbonaceous asteroids contain significant amounts of water ice and other volatile substances and are rich in organic (carbon-based) molecules. The more reflective silicate asteroids are composed primarily of rocky material. Silicate asteroids predominate in the inner parts of the asteroid belt, but carbonaceous asteroids are more common overall, and their fraction increases as we move outward.

Several asteroids have been visited by space probes. Figure 4.6(a) shows asteroid Ida, as seen by the *Galileo* probe en route to Jupiter in 1993. Ida is an irregularly shaped, cratered body some 30 km across, covered with layers of dust of variable thickness. It is thought to be a fragment of a larger object that broke up following a violent collision hundreds of millions of years ago. It also has a small moon (named Dactyl) orbiting it. By studying Dactyl's orbit and applying Newton's laws, astronomers were able to estimate Ida's mass at about $4\text{–}5 \times 10^{16}$ kg and infer a density of 2100–3100 kg/m^3. ∞ *(Sec. 1.4)* For comparison, the density of surface rock on Earth is around 3000 kg/m^3.

In 1997 the *Near Earth Asteroid Rendezvous* (*NEAR*) spacecraft flew by the 60-km wide asteroid Mathilde (Figure 4.6b). By sensing its gravitational pull, *NEAR* measured Mathilde's mass at about 10^{17} kg, implying a density of just 1300 kg/m^3. To account for this low density, scientists speculate that the asteroid's interior must be quite porous. Indeed, many asteroids seem to be more like

(a)

|← 30 km →|

(b)

|← 30 km →|

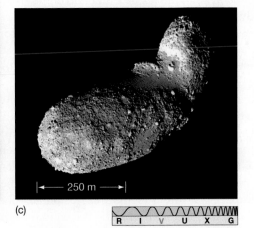

(c)

|← 250 m →|

R I V U X G

▲ **FIGURE 4.6 Asteroids, Close-up**   (a) Asteroid Ida photographed by the space probe *Galileo* from a distance of 3400 km. (Ida's moon, Dactyl, is visible at right.) Resolution is about 100 m. (b) Asteroid Mathilde, imaged by the *NEAR* spacecraft on its way to the asteroid Eros. The largest craters in this image are about 20 km across—much larger than those seen on Ida. The reason may be Mathilde's low density and rather soft composition. (c) A typical "rubble pile" asteroid, Itokawa, as seen by the Japanese *Hayabusa* spacecraft. It has few craters and is thought to have formed from fragments that coalesced over time. *(NASA; JAXA)*

4.1 DISCOVERY

What Killed the Dinosaurs?

The name *dinosaur* derives from the Greek words *deinos* (terrible) and *sauros* (lizard), but dinosaurs were no ordinary reptiles. In their prime, they were the rulers of Earth. Their fossilized remains have been found on every continent. They dominated Earth for well over 100 million years (for comparison, humans have been around for a little over 2 million years). Yet according to the fossil record, these creatures vanished from Earth quite suddenly about 65 million years ago. What happened to them?

Many explanations have been offered for the extinction of these creatures. Devastating plagues, magnetic field reversals, increased geological activity, severe climate change, and supernova explosions (see Chapter 12) have all been proposed. In the 1980s, it was suggested that a large extraterrestrial object—a 10- to 15-km-wide asteroid or comet—struck Earth 65 million years ago, and this is now (arguably) the leading explanation for the demise of the dinosaurs. Sketched in the first figure, the impact released millions of times more energy than the largest nuclear bombs ever constructed by humans and kicked huge quantities of dust high into the atmosphere. The dust may have shrouded our planet for years, virtually extinguishing the Sun's rays. On the darkened surface, plants could not survive. The entire food chain was disrupted, and the dinosaurs, at the top of that chain, became extinct.

(D. Hardy)

Although we have no direct astronomical evidence for or against this idea, we can estimate the chances that a large asteroid or comet will strike Earth today on the basis of the number of objects presently on Earth-crossing orbits. The second figure shows the likelihood of an impact as a function of the size of the impacting body. The horizontal scale indicates the energy released by the collision, measured in *megatons* of TNT. The megaton—4.2×10^{16} joules, the explosive yield of a large nuclear warhead—is the only common terrestrial measure of energy adequate to describe the violence of these events. ∞ *(More Precisely 2-2)*

We see that 100 million-megaton impacts, like the planet-wide catastrophe thought to have wiped out the dinosaurs, are very rare, occurring only once every 10 million years or so. However, smaller impacts, equivalent to "only" a few tens of kilotons of TNT (roughly equivalent to the bomb that destroyed Hiroshima in 1945), could happen every few years. The most recent large impact was the Tunguska explosion in Siberia in 1908, which packed a roughly 1-megaton punch (see Figure 4.21).

The main geological evidence supporting this theory is a layer of clay enriched with the element iridium, found in 65 million-year-old rocky sediments all around our planet. Iridium is rare on Earth's surface because most of it sank into our planet's interior long ago. The abundance of iridium in this clay layer is about 10 times greater than in other terrestrial rocks, but it matches closely the abundance of iridium found in meteorites (and, we assume, in asteroids and comets too). The site of the catastrophic impact has also been tentatively identified near Chicxulub, in the Yucatan Peninsula in Mexico, where evidence of a heavily eroded crater of just the right size and age has been found.

The idea that extraterrestrial events could cause catastrophic change on Earth was rapidly accepted by most astronomers, but it was controversial among paleontologists and geologists. Opponents argued that the amount of iridium in the clay layer varies greatly from place to place across the globe, and there is no complete explanation of why that should be so. Perhaps, they suggested, the iridium was produced by volcanoes and had nothing to do with an extraterrestrial impact at all.

Still, in the decades since the idea was first suggested, the focus of the debate seems to have shifted. As is often the case in science, the debate has evolved, sometimes erratically, as new data have been obtained, but the reality of a major impact 65 million years ago has become widely accepted. ∞ *(Sec. 0.5)* Much of the argument now revolves around the question of whether that event actually caused the extinction of the dinosaurs or merely accelerated a process already under way. The realization that catastrophic impacts can and do occur marked an important milestone in our understanding of planetary evolution (see also *Discovery 7-2*).

As a general rule, we can expect that global catastrophes are bad for the dominant species on a planet. As the dominant species on Earth, we are the ones who now stand to lose the most.

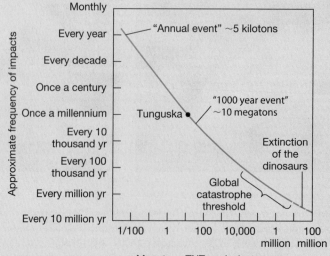

► FIGURE 4.7 **Asteroid Eros** The asteroid Eros, as seen by the *NEAR-Shoemaker* probe. Craters of all sizes, ranging from 50 m (the resolution of the image) to 5 km, pit the surface. The inset shows a close-up image of a "younger" section of the surface, where loose material from recent impacts has apparently filled in and erased all trace of older craters. (*JHU/NASA*)

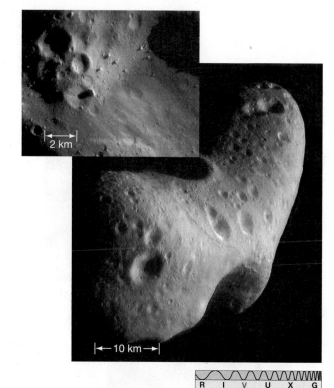

loosely bound rubble piles than pieces of solid rock. Figure 4.6(c) shows another rubble pile, the tiny (0.5-km long) asteroid Itokawa, which was visited in 2005 by the Japanese spacecraft *Hayabusa*. Itokawa's average density is 1900 kg/m³, but ground-based measurements indicate that the density of the left end is significantly greater, about 2900 kg/m³, suggesting that the asteroid may be the result of a collision in the relatively recent past. *Hayabusa* soft-landed on Itokawa, scooped up some rocky debris, and returned it to Earth in 2010. Analysis of the debris provided strong evidence that asteroids like Itokawa are the source of most meteorites—the oldest matter in the solar system.

On February 14, 2000, the *NEAR* spacecraft (by then renamed *NEAR-Shoemaker*) went into orbit around the asteroid Eros, sending back high-resolution images and making detailed measurements of the asteroid's size, shape, gravitational and magnetic fields, composition, and structure (Figure 4.7). Eros is a heavily cratered, rocky body, with mass 7×10^{15} kg and a roughly uniform density of around 2400 kg/m³. It is extensively fractured, apparently due to innumerable impacts in its past.

In 2011, NASA's *Dawn* probe entered orbit around Vesta, the second largest asteroid in the solar system. Vesta's most striking surface features (Figure 4.8a) are a set of deep troughs girdling the equator and one of the largest mountains in the solar system—a peak some 22 km high near the south pole. Age estimates based on the extent of cratering (see Section 5.6) indicate that Vesta's southern hemisphere is much younger than the north, suggesting that the major surface features may have been formed by an impact with another large body 1–2 billion years ago. The solar system is a violent place!

Dawn's next target was the dwarf planet Ceres, where it arrived in 2015. This small body appears to have a mainly rocky central core surrounded by an icy outer mantle and an icy/rocky surface. Figure 4.8(b) shows a portion of Ceres's southern hemisphere. The dwarf planet's heavily cratered surface actually shows fewer craters than had been expected, given its orbit in the crowded asteroid belt, perhaps indicating that past surface activity has obliterated some older features. The surface is also punctuated by a dozen or so bright white regions (one of them shown in the inset to Figure 4.8b), of unknown origin and composition.

MA ANIMATION/VIDEO
Orbiting Eros

CONCEPT CHECK

Describe some basic similarities and differences between asteroids and the inner planets.

◄ FIGURE 4.8 **Vesta and Ceres** (a) Vesta, the solar system's second largest asteroid, as seen here by NASA's *Dawn* spacecraft, is 500 km across and orbits between Mars and Jupiter. Note the deep grooves at top that span much of this rocky body and the towering mountain near the bottom. (b) Part of Ceres's heavily cratered southern hemisphere, imaged by *Dawn* in 2015. The inset shows crater Occator, home to a collection of an intriguing but poorly understood complex of bright spots. (*NASA*)

(a)

50,000 km

R I V U X G

10 km

R I V U X G

Hydrogen
envelope

Tail

Coma

Nucleus

To Sun

(b)

▲ **FIGURE 4.9 Halley's Comet** (a) Halley's Comet in 1986, about 1 month before it rounded the Sun. (b) Diagram of a typical comet, showing the nucleus, coma, hydrogen envelope, and tail, approximately to the same scale as in part (a). (c) The inset is Halley's nucleus, as seen by the European *Giotto* spacecraft, showing it to be very dark. The Sun is toward the left in this image. The brightest areas are jets of evaporated gas and dust spewing from the comet's nucleus. *(NOAO ESA/Max Planck Institute)*

Comets

Comets are usually discovered as faint, fuzzy patches of light on the sky while still several astronomical units away from the Sun. Traveling in a highly elliptical orbit, a comet brightens and develops an extended **tail** as it nears the Sun. (The name *comet* derives from the Greek word *kome*, meaning "hair.") As the comet departs from the Sun's vicinity, its brightness and its tail diminish until it once again becomes a faint point of light receding into the distance.

Probably the most famous comet of all is Halley's Comet (Figure 4.9a), whose appearance at 76-year intervals has been documented at every passage since 240 B.C. Its most recent appearance was in 1986. A spectacular show, the tail of Halley's Comet can reach almost a full astronomical unit in length, stretching many tens of degrees across the sky. More recently, in 1997, comet Hale-Bopp, an unusually large and bright comet with a 40°-long tail, was almost certainly the most-watched comet in history (Figure 4.10).

The various parts of a typical comet are shown in Figure 4.9(b). Like the planets, comets emit no visible light of their own—they shine by reflected (or reemitted) sunlight. The **nucleus,** or main solid body, of a comet is only a few kilometers in diameter. During most of the comet's orbit, far from the Sun, only this frozen nucleus exists. But if a comet comes within a few astronomical units of the Sun, its icy surface becomes too warm to remain stable. Part of it becomes gaseous and expands into space, forming a diffuse **coma** (halo) of dust and evaporated gas around the nucleus. The inset to Figure 4.9(a) shows the nucleus of Halley's comet, as seen by the European *Giotto* spacecraft, which approached within 600 km of the nucleus in 1986. The coma gets larger and brighter as the comet nears the Sun. At maximum size, it can measure 100,000 km in diameter—as large as Jupiter or Saturn. Engulfing the coma, an invisible **hydrogen envelope** stretches across millions of kilometers of space. The comet's tail, most pronounced when the comet is closest to the Sun, is larger still, sometimes spanning as much as 1 AU. From Earth, only the coma and tail of a comet are visible to the naked eye. Most of the comet's light comes from the coma. However, most of the mass resides in the nucleus.

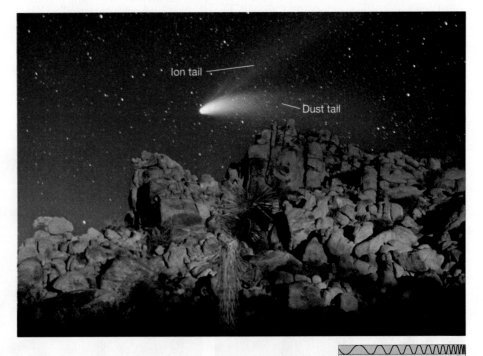

Ion tail

Dust tail

R I V U X G

▲ **FIGURE 4.10 Comet Tails** In 1997, comet Hale-Bopp displayed both an ion tail (dark blue) and a dust tail (white blue), showing the gentle curvature and inherent fuzziness of the dust. At the comet's closest approach to the Sun, its tail stretched nearly 40° across the sky. *(W. Pacholka)*

Comets can have two types of tail. An **ion tail** is approximately straight, and often made of glowing, linear streamers like those shown toward the top of Figure 4.10. Its emission spectrum indicates numerous ionized atoms and molecules that have lost some of their normal complement of electrons. ∞ *(Sec. 2.6)* A **dust tail** is usually broad, diffuse, and gently curved, as shown clearly at the center of Figure 4.10. It is rich in microscopic dust particles that reflect sunlight, making the tail visible from afar.

Both types of tails are in all cases directed away from the Sun by the **solar wind,** an invisible stream of matter and radiation escaping from the Sun. (In fact, astronomers first inferred the existence of the solar wind from observations of comet tails.) Consequently, as depicted in Figure 4.11, the tail, be it ion or dust, always lies outside a comet's orbit and actually *leads* the comet when it is outbound from the Sun. The light particles that make up the ion tails are more strongly influenced by the solar wind than by the Sun's gravity, so those tails always point directly away from the Sun. The heavier dust particles are less affected by the pressure of the solar wind and tend to follow the comet's orbit around the Sun, giving rise to the slight curvature of the dust tails.

The nucleus of a comet is composed of dust particles and small rocky fragments all trapped within a loosely packed mixture of methane, ammonia, carbon dioxide, and ordinary water ice, the density of the mixture being about 100 kg/m^3. Comets are often described as "dirty snowballs." Even as atoms, molecules, and dust particles boil off into space, creating a comet's coma and tail, the nucleus remains frozen at a temperature of only a few tens of kelvins. Estimates of typical cometary masses range from 10^{12} to 10^{16} kg, comparable to the masses of small asteroids.

In 2004, NASA's *Stardust* mission approached within 150 km of the nucleus of comet P/Wild 2, collecting cometary particles in a specially designed foam-like aerogel dust detector (see Figure 4.12). The comet was chosen because it is a relative newcomer to the inner solar system, having been deflected onto its present orbit by an encounter with Jupiter in 1974. It has therefore not undergone significant solar heating or evaporation and hasn't changed significantly since our solar system formed. On its return to Earth in 2006, *Stardust* provided researchers with the first-ever samples of cometary material. Detailed chemical analysis revealed organic material apparently formed in deep space, as well as the unexpected presence of silicate (rocky) materials that should only have formed at high temperatures, challenging astronomers' models of solar system formation.

The next year saw a much more violent encounter, when a 400-kg projectile from NASA's *Deep Impact* spacecraft crashed into comet Tempel 1 at more than 10 km/s (23,000 mph), blasting gas and debris from the comet's interior out into interplanetary space, while the spacecraft itself watched from a distance of 500 km.

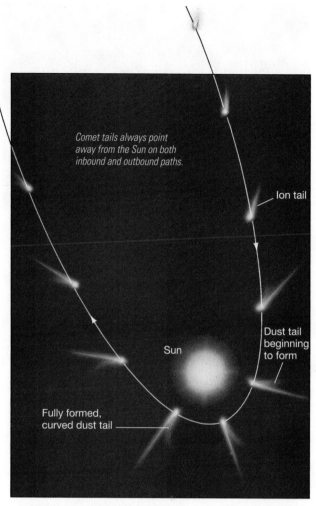

Comet tails always point away from the Sun on both inbound and outbound paths.

Ion tail

Dust tail beginning to form

Sun

Fully formed, curved dust tail

▲ **FIGURE 4.11 Comet Trajectory**   As it approaches the Sun, a comet develops an ion tail, which is directed away from the Sun. Closer in, the dust tail displays marked curvature and tends to lag behind the ion tail. Compare this to comet Hale-Bopp (Figure 4.10).

MA ANIMATION/VIDEO

Comet Hale-Bopp Nucleus Animation

(a) |← 5 km →|

(b)

Cometary dust particles

|← 1 cm →|

(c)

R I V U X G

▲ **FIGURE 4.12 Stardust at Wild-2 INTERACTIVE**   (a) The *Stardust* spacecraft captured this image (a) of comet P/Wild-2 in 2004, just before the craft passed through the comet's coma. (b) Onboard is a detector made of a foam-like jelly (called *aerogel*) that is 99.8% air, yet is strong enough to stop and store cometary dust particles as they hit the spacecraft. (c) Upon return of the craft to Earth in 2006, analysis began of the minute tracks in the aerogel, the ends of which contain captured comet dust fragments. *(NASA)*

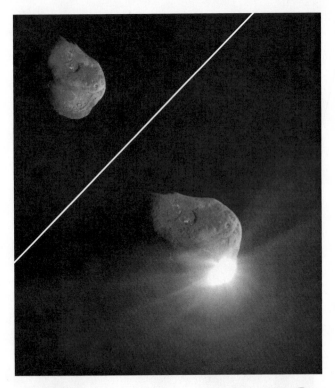

ANIMATION/VIDEO   (MA)
Deep Impact Simulation

ANIMATION/VIDEO   (MA)
Anatomy of a Comet Part I

◀ **FIGURE 4.13 Deep Impact**   The 5-km-sized nucleus of Comet Tempel 1 is shown at top before impact in 2005 and at bottom shortly after collision with a small projectile launched from the *Deep Impact* robot spacecraft.*(NASA)*

Figure 4.13 shows the mothership's spectacular view of the explosion about 1 minute after impact. Spectroscopic analysis of the ejected gas has provided scientists with their clearest view yet of the internal composition of a comet, confirming the presence of water ice and many organic molecules. ∞ *(Sec. 2.8)* Observations of the crater suggest a low-density "fluffy" internal composition, consistent with the "snowball" picture of cometary structure just described.

In August 2014, after a 10-year journey that took it almost as far as the orbit of Jupiter, the European *Rosetta* spacecraft rendezvoused with, and went into orbit around, a 4-km-wide comet called 67 P/Churyumov–Gerasimenko while it was still several astronomical units from the Sun. Like comet P/Wild 2, comet 67P was chosen because it is a relative newcomer to the inner solar system, having been placed on its current 3.5 AU orbit by a series of kicks from the planet Jupiter over the last 2 centuries. The chapter-opening image shows a detailed view of the comet from *Rosetta* shortly after arrival.

Rosetta carried an array of 12 sensors designed to study the comet's surface, interior, atmosphere, and chemical and magnetic environment. Already (see Section 4.3), some of its measurements of water on and near the comet may have upended some long-held theories of planet formation. However, the star of the show was a 100-kg lander called *Philae*, which made the first ever soft landing on a comet in November 2014 and returned data from the surface for several days before its power supply was depleted. Figure 4.14(a) shows another view of comet 67P, with

▶ **FIGURE 4.14 Comet Close-Up**   (a) This *Rosetta* image of comet 67 P/Churyumov–Gerasimenko shows the comet almost end-on (the view is roughly right to left along the chapter-opening image). The inset highlights the landing site of the *Philae* lander. (b) *Philae*'s view of the surface of this alien world. One of the lander's feet can be seen at the bottom of this panoramic image. *(ESA/Rosetta/MPS)*

(b)

(a)

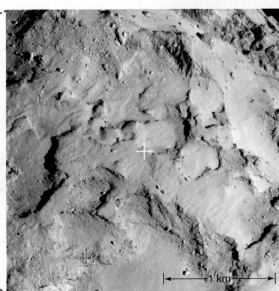

1 km

the landing site enlarged; Figure 4.14(b) shows *Philae's* view from the surface. Following the landing, *Rosetta* continued to orbit the comet as it rounded the Sun in August 2015, studying the evolving surface as the Sun's heat warmed and vaporized it.

Comet Orbits

Their highly elliptical orbits take most comets far beyond Pluto, where, in accordance with Kepler's second law, they spend most of their time. ∞ *(Sec. 1.3)* The majority take hundreds of thousands, even millions, of years to complete a single orbit around the Sun. Unlike the orbits of other solar system objects, most comet orbits are not confined to within a few degrees of the ecliptic plane. Rather, they exhibit all inclinations and orientations, roughly uniformly distributed in all directions from the Sun. Only a tiny portion of a cometary orbit can lie within the inner solar system, so it follows that for every comet we see, there must be many more similar objects far from the Sun.

Thus, astronomers reason, there must be a huge cloud of comets lying far beyond the orbit of Pluto, completely surrounding the Sun in all directions (Figure 4.15a). It is named the **Oort cloud,** after the Dutch astronomer Jan Oort, who first wrote in the 1950s of the possibility of such a vast reservoir of inactive, frozen comets. Researchers think that the Oort cloud may be up to 100,000 AU (0.5 light-years) in diameter. However, most Oort cloud comets never come anywhere near the Sun. Indeed, Oort cloud comets rarely approach even the orbit of Pluto. Only when the gravity of a passing star happens to deflect a comet into an extremely eccentric orbit that passes through the inner solar system do we get to see it at all.

A few *short-period comets* (conventionally defined as those having orbital periods of less than 200 years) are exceptions to the above general statements. According to Kepler's third law, these bodies never venture far beyond the orbit of Pluto. Their orbital orientations also differ from the norm—they tend to be prograde and lie close to the ecliptic plane, like the orbits of the planets and most asteroids.

Short-period comets originate beyond the orbit of Neptune in or near a region of the outer solar system called the **Kuiper belt** (after Gerard Kuiper, a pioneer in infrared and planetary astronomy). A little like the asteroids in the inner solar system, most Kuiper belt objects move in roughly circular orbits between about 30 and 100 AU from the Sun, always remaining outside the orbits of the jovian planets (Figure 4.15b). Occasionally, either a chance encounter between two objects or (more likely) the gravitational influence of Neptune "kicks" a Kuiper belt object into an eccentric orbit that brings it into the inner solar system and into our view. The observed orbits of these comets reflect the flattened structure of the Kuiper belt.

More than 1000 Kuiper belt objects are currently known. Because they are so small and distant, only a tiny fraction of the total has so far been observed. Researchers estimate that the total mass of the Kuiper belt may exceed that of the asteroid belt by a factor of 100 or more (although it is probably still less than the mass of Earth). The best-known Kuiper belt object is the former planet Pluto, but larger bodies orbit beyond Neptune. In 2005, astronomers discovered an object (since named Eris) some 10 percent larger and 30 percent more massive than Pluto, moving on an eccentric orbit with a semimajor axis of 70 AU. The discovery of Eris was a major factor in Pluto's demotion from major planet status in 2006.

Meteoroids

On a clear night it is possible to see a few *meteors*—shooting stars—every hour. A **meteor** is a sudden streak of light in the night sky caused by friction between air molecules in Earth's atmosphere and an incoming piece of asteroid, meteoroid, or comet. This friction heats and excites the air molecules, which then emit light as they return to their ground states, producing the characteristic bright streak

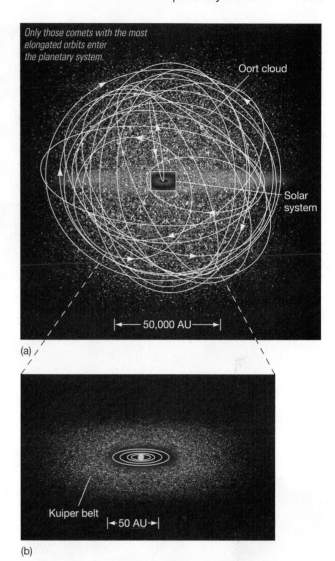

Only those comets with the most elongated orbits enter the planetary system.

Oort cloud

Solar system

|← 50,000 AU →|

(a)

Kuiper belt |← 50 AU →|

(b)

▲ **FIGURE 4.15 Comet Reservoirs**   (a) Diagram of the Oort cloud, showing a few cometary orbits. The solar system is smaller than the overlaid box at the center of the figure. (b) The Kuiper belt, the source of short-period comets, whose orbits hug the ecliptic plane.

MA ANIMATION/VIDEO
Anatomy of a Comet Part II

CONCEPT CHECK

In what sense are the comets we see very unrepresentative of comets in general?

▶ **FIGURE 4.16 Meteor Trails** A bright streak of light called a meteor is produced when a fragment of interplanetary debris plunges into the atmosphere, heating the air to incandescence. (a) A small meteor photographed against a backdrop of stars and the Northern Lights. (b) These meteors (one with a red smoke trail) streaked across the sky during the height of the Leonid meteor storm in November 2001. (*P. Parviainen, J. Lodriguss*)

(a)

(b)

R I V U X G

shown in Figure 4.16. Note that the brief flash that is a meteor is in no way similar to the broad, steady swath of light associated with a comet's tail. A meteor is a fleeting event in Earth's atmosphere, whereas a comet tail exists in deep space and can remain visible in the sky for weeks or even months.

Before encountering the atmosphere, the chunk of debris causing a meteor was almost certainly a meteoroid, simply because these small interplanetary fragments are far more common than either asteroids or comets. Any piece of interplanetary debris that survives its fiery passage through our atmosphere and finds its way to the ground is called a **meteorite.**

The smallest meteoroids are mainly the rocky remains of broken-up comets. Each time a comet passes near the Sun, some fragments dislodge from the main body. The fragments initially travel in a tightly knit group of dust or pebble-sized objects called a **meteoroid swarm,** moving in nearly the same orbit as the parent comet. Over the course of time, the swarm gradually disperses along the orbit, so that eventually the **micrometeoroids,** as these small meteoroids are known, become more or less smoothly spread around the parent comet's orbit. If Earth's orbit happens to intersect the orbit of such a young cluster of meteoroids, a spectacular *meteor shower* can result. Earth's motion takes it across a given comet's orbit at most twice a year (depending on the precise orbit of each body). Because the intersection occurs at the same time each year (Figure 4.17), the appearance of certain meteor showers is a regular and (fairly) predictable event (Table 4.3).

Meteor showers are named for their *radiant*, the constellation from whose direction they appear to come (Figure 4.18). For example, the Perseid shower appears to emanate from the constellation Perseus. It can last for several days but reaches its maximum every year on the morning of August 12, when upward of 50 meteors per hour can be observed. Astronomers use the speed and direction of a meteor's flight to compute its interplanetary trajectory. This is how some meteoroid swarms have come to be identified with well-known cometary orbits.

Larger meteoroids—those more than a few centimeters in diameter—are usually *not* associated with comets. They are more likely small bodies that have strayed from the asteroid belt, possibly as the result of asteroid collisions. Their

◀ **FIGURE 4.17 Meteor Showers** A meteoroid swarm associated with a given comet intersects Earth's orbit at specific locations, giving rise to meteor showers at specific times of the year. If the comet's path happens to intersect Earth's, the result is a meteor shower each time Earth passes through the intersection (point 4).

TABLE 4.3	Some Prominent Meteor Showers		
Morning of Maximum Activity	Shower Name/Radiant	Rough Hourly Count	Parent Comet
Jan. 3	Quadrantid/Bootes	40	—
Apr. 21	Lyrid/Lyra	10	1861I (Thatcher)
May 4	Eta Aquarid/Aquarius	20	Halley
June 30	Beta Taurid/Taurus	25	Encke
July 30	Delta Aquarid/Aquarius/Capricorn	20	—
Aug. 12	Perseid/Perseus	50	1862III (Swift-Tuttle)
Oct. 9	Draconid/Draco	up to 500	Giacobini-Zimmer
Oct. 20	Orionid/Orion	30	Halley
Nov. 7	Taurid/Taurus	10	Encke
Nov. 16	Leonid/Leo	12[1]	1866I (Tuttle)
Dec. 13	Geminid/Gemini	50	3200 Phaeton[2]

[1]Every 33 years (most recently in 1999 and 2000), as Earth passes through the densest region of this meteoroid swarm, we see intense showers, reaching 1000 meteors per minute for brief periods of time.
[2]Phaeton is an asteroid with no cometary activity, but its orbit matches the meteoroid paths very well.

orbits can sometimes be reconstructed in a manner similar to that used to determine the orbits of meteor showers. In most cases, their computed orbits do indeed intersect the asteroid belt, providing the strongest evidence we have that this is where they originated. These objects are responsible for most of the cratering on the surfaces of the Moon, Mercury, Venus, Mars, and some of the moons of the jovian planets. A few meteorites have been identified as originating on the Moon or Mars, blasted off the surfaces of those bodies by some impact long ago.

Meteoroids smaller than about a meter across (roughly a ton in mass) generally burn up in Earth's atmosphere. Larger bodies reach the surface, where they can cause significant damage, such as the kilometer-wide Barringer Crater shown in Figure 4.19. From the size of this crater, we can estimate that the meteoroid responsible must have had a mass of about 200,000 tons and a diameter of perhaps 50 m. Only 25 tons of iron meteorite fragments have been found at the crash site. The remaining mass must have been scattered by the explosion at impact, broken down by subsequent erosion or buried in the ground. Currently, Earth is scarred with nearly 100 craters larger than 0.1 km in diameter. Most of these are so heavily eroded by weather and geological activity that they can be identified only in satellite photography, as in Figure 4.20. Fortunately, such major collisions between Earth and large meteoroids are thought to be rare events now. On average, they occur only once every few hundred thousand years (see *Discovery 4-1*).

One of the most recent documented meteoritic events occurred in central Siberia on June 30, 1908 (Figure 4.21). The presence of only a shallow depression as well as a complete lack of fragments implies that this intruder exploded several kilometers above the ground, leaving a blasted depression at ground level but no well-formed crater. Recent calculations suggest that the object in question

Observer   *The view from the side*

(a)

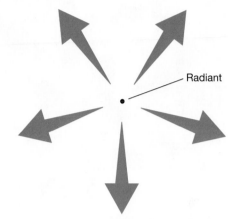

Radiant

Meteors coming at you face-on

(b)

Meteor swarms coming at you face-on are analogous to parallel train tracks that seem to converge in the distance.

▲ **FIGURE 4.18 Radiant** (a) A group of meteoroids approaches an observer, all of them moving in the same direction at the same speed. (b) From an observer's viewpoint, the trajectories of the meteoroids (and the meteor shower they produce) appear to spread out from a central point, called the radiant.

Access road

1 km

R I V U X G

▶ **FIGURE 4.19 Barringer Crater** The Barringer Meteor Crater, near Winslow, Arizona, 1.2 km in diameter and 0.2 km deep, resulted from a meteorite impact about 25,000 years ago. The meteoroid was probably about 50 m across and likely weighed around 200,000 tons. *(U.S. Geological Survey)*

R I V U X G

▲ **FIGURE 4.20 Manicouagan Reservoir** This photograph, taken from orbit by the U.S. *Skylab* space station, shows the ancient impact basin that forms Quebec's Manicouagan Reservoir. A large meteorite landed there about 200 million years ago. *(NASA)*

CONCEPT CHECK

Why are astronomers so interested in interplanetary matter?

◄ **FIGURE 4.21 Tunguska Debris** The Tunguska event of 1908 leveled trees over a vast area. Although the impact of the blast was tremendous and its sound audible for hundreds of kilometers, this Siberian site is so remote that little was known about the event until scientific expeditions arrived to study it many years later. *(Sovfoto/Eastfoto)*

was a rocky meteoroid about 30 m across. The explosion, estimated to have been equal in energy to a 10-megaton nuclear detonation, was heard hundreds of kilometers away and produced measurable increases in atmospheric dust levels all across the Northern Hemisphere.

Most meteorites are rocky (Figure 4.22a), although a few are composed mainly of iron and nickel (Figure 4.22b). Their composition is much like that of the rocky inner planets or the Moon, except that some of their lighter elements—such as hydrogen and oxygen—are depleted. Scientists think that the rocky meteorites are associated with the rocky, silicate asteroids and that their light elements boiled away when the bodies on which they originated were wholly or partially molten. Some meteorites do show evidence of strong heating in the past, most likely during the collisions that liberated them from their parent asteroids. Others show no evidence of past heating and may date back to the formation of the solar system. Oldest of all are the *carbonaceous* meteorites, black or dark gray in color and very likely related to the carbonaceous asteroids.

Finally, almost all meteorites are *old*. Radioactive dating shows most of them to be between 4.4 and 4.6 billion years old—roughly the age of the oldest Moon rocks brought back to Earth by *Apollo* astronauts (Chapter 5). Meteorites, along with asteroids, comets, and some lunar rocks, provide essential clues to the original state of matter in the solar neighborhood and to the birth of our planetary system.

► **FIGURE 4.22 Meteorite Samples** (a) A stony (silicate) meteorite often has a dark crust, created when its surface is melted by the tremendous heat generated during its passage through the atmosphere. The coin at the bottom is for scale. (b) Iron meteorites are much rarer than the stony variety. Most, like the one in this photo, show characteristic crystalline patterns when their surfaces are cut, polished, and etched with acid. *(Science Graphics)*

(a)

(b)

4.3 Formation of the Solar System

You might be struck by the vast range of physical and chemical properties found in the solar system. Our astronomical neighborhood may seem more like a great junkyard than a smoothly running planetary system. Is there some underlying principle that unifies the facts just outlined? Remarkably, the answer is yes. The origin of the solar system is a complex and as yet incompletely solved puzzle, but (we think!) the basic outlines are at last becoming understood.

A comprehensive theory of the formation of the solar system has been a dream of astronomers for centuries. Its development is a case study in the scientific method, on a par with the sweep of ideas presented in Chapters 1 and 2, although in this case the final form of the theory remains to be determined. ∞ *(Sec. 0.5)* As we will see, competing hypotheses have risen, fallen, and in some cases risen again in response to improved observations of our planetary environment. Nevertheless, we now seem to have a coherent picture of how our solar system formed.

Until the mid-1990s, theories of planetary system formation concentrated almost exclusively on our own solar system, for the very good reason that astronomers had no other examples of planetary systems on which to test their ideas. However, all that has now changed. We currently know of about 700 **extrasolar planets**—planets orbiting stars other than the Sun (Figure 4.23)—to challenge our theories. And challenge them they do! As we will see in Section 4.4, the new planetary systems discovered so far seem to have properties quite different from our own and may require us to rethink in some radical ways our concept of how stars and planets form.

Still, we currently have only limited information on these systems—little more than orbits and mass estimates for the largest planets. Accordingly, we begin our study by outlining the comprehensive theory that accounts, in detail, for most of the observed properties of our own planetary system: the solar system. Bear in mind, though, that no part of the scenario we describe here is in any way unique to our own system. The same basic processes should have occurred during the formative stages of many of the stars in our Galaxy. Later we will see how our theories hold up in the face of the new extrasolar data.

Model Requirements

Based on the measured ages of the oldest meteorites, as well as the ages of Earth and lunar rocks, planetary scientists think that the age of the solar system is 4.6 billion years—an enormous period of time (although still far less than the 14 billion-year age of the universe—see Chapter 17). What happened so long ago to create the planetary system we see today? And what evidence survives from those early times to constrain our theories? Any theory of the origin and architecture of our planetary system must adhere to these 10 known facts:

1. *Each planet is relatively isolated in space.* The planets orbit at progressively larger distances from the central Sun; they are not bunched together.
2. *The orbits of the planets are nearly circular.* With the exception of Mercury, which we will argue is a special case, each planetary orbit closely describes a perfect circle.
3. *The orbits of the planets all lie in nearly the same plane.* The planes swept out by the planets' orbits are accurately aligned to within a few degrees. Again, Mercury is a slight exception.
4. *The planets all orbit the Sun in the same direction (counterclockwise as viewed from above Earth's North Pole).* Virtually all the large-scale motions in the solar

▲ **FIGURE 4.23 Extrasolar Planet** Most known extrasolar planets are too faint to be detectable against the glare of their parent stars. However, in this system, called 2M1207, the parent itself (centered) is very faint—a so-called *brown dwarf* (see Chapter 12)—allowing the planet (lower left) to be detected in the infrared. This planet has a mass about five times that of Jupiter and orbits 55 AU from the star, which is 230 light-years away. *(VLT/ESO)*

system (other than cometary orbits) are in the same plane and in the same sense. The plane is that of the Sun's equator, and the sense is that of the Sun's rotation.

5. *Most planets rotate on their axis in roughly the same sense as the Sun.* Exceptions are Venus and Uranus.

6. *Most of the known moons revolve about their parent planet in the same direction as the planet rotates on its axis.* Of the large moons in the solar system, only Neptune's moon Triton is an exception.

7. *Our planetary system is highly differentiated.* The terrestrial planets are characterized by high densities, moderate atmospheres, slow rotation rates, and few or no moons. The jovian planets have low densities, thick atmospheres, rapid rotation rates, and many moons.

8. *Asteroids are very old and exhibit a range of properties not characteristic of either the terrestrial or the jovian planets or their moons.* Asteroids share, in rough terms, the bulk orbital properties of the planets. However, they appear to be made of primitive, unevolved material, and the meteorites that strike Earth are the oldest rocks known.

9. *The Kuiper belt is a collection of asteroid-sized icy bodies orbiting beyond Neptune.* Pluto is now considered such a Kuiper belt object.

10. *The Oort cloud comets are primitive, icy fragments that do not orbit in the plane of the ecliptic and reside primarily at large distances from the Sun.* While similar to the Kuiper belt in composition, the Oort cloud is a completely distinct part of the outer solar system.

These 10 facts strongly suggest a high degree of order in our solar system. The large-scale architecture is too neat, and the ages of the components too uniform, to be the result of random chaotic events. The overall organization points toward a single formation, an ancient but one-time event 4.6 billion years ago.

In the next few chapters we will see that some planetary properties (such as atmospheric composition and interior structure) have gradually *evolved* into their present states during the billions of years since the planets formed. However, *no* such evolutionary explanation exists for the items in the preceding list. For example, Newton's laws imply that the planets must move in elliptical orbits with the Sun at one focus, but they offer no explanation of why the observed orbits should be roughly circular, coplanar, and prograde. We know of no way in which the planets could have started off in random paths, then later evolved into the orbits we see today. Their basic orbital properties must have been established at the outset.

In addition to its many regularities, our solar system also has many notable *irregularities*, some of which we have already mentioned. Far from threatening our theory, however, these irregularities are important facts for us to consider in shaping our explanations. For example, any theory explaining solar system formation must not insist that *all* planets rotate in the same sense or have only prograde moons, because that is not what we observe. Instead, the theory should provide strong reasons for the observed planetary characteristics yet be flexible enough to allow for and explain the deviations, too. And, of course, the existence of the asteroids and comets that tell us so much about our past must be an integral part of the picture. That's a tall order, yet many researchers now think we are close to that level of understanding.

PROCESS OF SCIENCE CHECK

Why is it important that a theory of solar system formation make clear statements about how planets arose yet not be too rigid in its predictions?

Nebular Contraction

Modern theory holds that planets are by-products of the process of star formation (Chapter 11). Imagine a large cloud of interstellar dust and gas—called a **nebula**—a light-year or so across. Now suppose that due to some external influence, such as a collision with another interstellar cloud or perhaps the explosion of a nearby star, the nebula starts to contract under the influence of its own gravity. As it contracts, it becomes denser and hotter, eventually forming a star—the Sun—at its center.

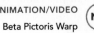

ANIMATION/VIDEO   **MA**
Beta Pictoris Warp

▶ **FIGURE 4.24 Angular Momentum INTERACTIVE** (a) Conservation of angular momentum demands that a contracting, rotating cloud must spin faster as its size decreases. (b) Eventually, a small part of it destined to become the solar system came to resemble a gigantic pancake. The large blob at the center ultimately became the Sun.

As a cloud contracts . . .

(a)

. . . it spins up and forms a disk.

(b)

In 1796 the French mathematician-astronomer Pierre Simon de Laplace showed mathematically that conservation of angular momentum (objects spin faster as they shrink—see *More Precisely 4-2*) demands that our hypothetical nebula must spin faster as it contracts. The increase in rotation speed, in turn, causes the nebula's *shape* to change as it shrinks. Centrifugal forces (the outward "push" due to rotation) tend to oppose the contraction in directions perpendicular to the rotation axis, with the result that the nebula collapses most rapidly along the rotation axis. As shown in Figure 4.24, by the time it has shrunk to about 100 AU, the cloud has flattened into a pancake-shaped disk. This swirling mass destined to become our solar system is usually referred to as the **solar nebula.**

If we now suppose that planets form out of this spinning disk, we can begin to understand the origin of much of the architecture observed in our planetary system today, such as the near-circularity of the planets' orbits and the fact that they move in the same sense in almost the same plane. The idea that planets form from such a disk is called the **nebular theory.**

Astronomers are fairly confident that the solar nebula formed such a disk because we can see similar disks around other stars. Figure 4.25(a) shows a visible-light image of the region around the star Beta Pictoris, which lies about 50 light-years from the Sun. When the light from Beta Pictoris itself is removed and the resulting image enhanced by a computer, a faint disk of matter (viewed almost edge-on here) can be seen. This particular disk is roughly 1000 AU across—about 10 times the diameter of the Kuiper belt. Astronomers think that Beta Pictoris is a very young star, perhaps only 100 million years old, and that we are witnessing it pass through an evolutionary stage similar to that experienced by our own Sun 4.6 billion years ago. Figure 4.25(b) shows an artist's conception of the disk.

As we saw in Chapter 0, scientific theories must continually be tested and refined as new data become available. ∞ *(Sec. 1.5)* Unfortunately for the original nebular theory, while Laplace's description of the collapse and flattening of the

(MA)° ANIMATION/VIDEO
The Formation of the Solar System

Size of Kuiper belt

Solar system to scale

(a)

R I V U X G

(b)

▲ **FIGURE 4.25 Beta Pictoris** (a) A computer-enhanced view of a disk of warm matter surrounding the star Beta Pictoris. False color is used here to accentuate the details. The bottom image is a close-up of the inner part of the disk, whose warp is possibly caused by the gravitational pull of unseen companions. In both images, the overwhelmingly bright central star has been covered to let us see the much fainter disk surrounding it. (b) An artist's conception of the disk of clumped matter, showing the warm disk with a young star at the center and several comet-sized or larger bodies already forming. Mottled dust is seen throughout—such protoplanetary regions are probably very "dirty." (*NASA; D. Berry*)

4.2 MORE PRECISELY

The Concept of Angular Momentum

Most celestial objects rotate. Planets, moons, stars, and galaxies all have some *angular momentum*, which we may define loosely as the tendency of a body to keep spinning or moving in a circle, or, equivalently, how much effort must be expended to stop it. Angular momentum is a fundamental property of any rotating object, as important a quantity as its mass or its energy.

Intuitively, we know that the more massive an object, or the larger it is, or the faster it spins, the harder it is to stop. In fact, angular momentum depends on the object's *mass, rotation rate* (measured in, say, revolutions per second), and *radius*, in a very specific way:

$$\text{angular momentum} \propto \text{mass} \times \text{rotation rate} \times \text{radius}^2.$$

Recall that the symbol $\propto$ means "is proportional to." The constant of proportionality depends on the details of how the object's mass is distributed.

According to Newton's laws of motion, angular momentum is *conserved* at all times—that is, it must remain constant before, during, and after a physical change in any object, so long as no external forces act. This allows us to relate changes in size to changes in rotation rate. As illustrated in the accompanying figure, if a spherical object having some spin begins to contract, the above relationship demands that it spin faster, so that the product mass $\times$ angular speed $\times$ radius^2 remains constant. The sphere's mass does not change during the contraction, but its radius clearly decreases. The rotation speed must therefore increase in order to keep the total angular momentum unchanged. Figure skaters use this principle to spin faster by drawing in their arms and slow down by extending them (as shown in the second accompanying figure). The mass of the human body remains the same, but its overall radius changes, causing the body's rotation rate to change in order to keep the angular momentum constant.

To take an example from the text, suppose that the interstellar gas cloud from which our solar system formed was initially 1 light-year across and rotating very slowly—once every 10 million years, say. The cloud's mass doesn't change (we assume) during the contraction, but its radius decreases. The rotation rate must therefore *increase* in order to keep the total angular momentum unchanged. By the time it has collapsed to a radius of 100 AU, its radius has shrunk by a factor of (1 light-year/100 AU) $\approx 9.5 \times 10^{12}$ km/$1.5 \times 10^{10} \approx 630$. Conservation of angular momentum then implies that its (average) spin rate increases by a factor of $630^2 \approx 400,000$, to roughly one revolution every 25 years, close to the orbital period of Saturn.

Large radius

Small radius

Slow rotation

Rapid rotation

D. Lefranc/Gamma-Rapho/Getty Images

solar nebula was essentially correct, we now know that a disk of gas would *not* form clumps of matter that would subsequently evolve into planets. In fact, modern computer calculations predict just the opposite: Any clumps in the gas would tend to disperse, not contract further. However, the model currently favored by most astronomers, known as the **condensation theory,** rests squarely on the old nebular theory, combining its basic physical reasoning with new information about interstellar chemistry to avoid most of the original theory's problems.

The key new ingredient is *interstellar dust* in the solar nebula. Astronomers now recognize that the space between the stars is strewn with microscopic dust grains, an accumulation of the ejected matter of many long-dead stars. These dust particles probably formed in the cool atmospheres of old stars, then grew by accumulating more atoms and molecules from the interstellar gas. The end result is

▶ **FIGURE 4.26 Dark Clouds** Glowing interstellar gas and dark dust clouds mark this region of star formation. These opaque dense dust clouds, called globules, are silhouetted here against bright emission and newborn stars in the star-forming region IC 2944. *(NASA)*

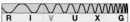

that interstellar space is littered with tiny chunks of icy and rocky matter, having typical diameters of about 10^{-5} m. Figure 4.26 shows one of many such dusty regions in the vicinity of the Sun.

Dust grains play an important role in the evolution of any gas cloud. Dust helps to cool warm matter by efficiently radiating heat away in the form of infrared radiation. When the cloud cools, its molecules move around more slowly, reducing the internal pressure and allowing the nebula to collapse more easily. ∞ *(More Precisely 2-1)* Furthermore, the dust grains greatly speed up the process of collecting enough atoms to form a planet. They act as **condensation nuclei**—microscopic platforms to which other atoms can attach, forming larger and larger balls of matter. This is similar to the way raindrops form in Earth's atmosphere; dust and soot in the air act as condensation nuclei around which water molecules cluster.

Planet Formation

According to the condensation theory, the planets formed from the solar nebula (Figure 4.27a) in three distinct stages. The first two apply to all planets, the third only to the giant jovian worlds.

The first stage of planet formation began when dust grains in the solar nebula formed condensation nuclei around which matter began to accumulate (Figure 4.27b). This vital step greatly hastened the critical process of forming the first small clumps of matter. Once these clumps formed, they grew rapidly by sticking to other clumps. (Imagine a snowball thrown through a fierce snowstorm, growing bigger as it encounters more snowflakes.)

As the clumps grew larger, their surface areas increased, and consequently, the rate at which they swept up new material accelerated. They gradually formed larger and larger objects, the size of pebbles, baseballs, basketballs, and boulders. Figure 4.28 shows radio and infrared views of two young stars whose protostellar disks are thought to be in just this state. Eventually, this process of **accretion**—the gradual growth of small objects by collision and sticking—created objects a few hundred kilometers across.

At the end of the first stage of planet formation, the solar system was made up of hydrogen and helium gas and millions of **planetesimals**—objects the size of small moons, having gravitational fields just strong enough to affect their neighbors. By then, their gravity was strong enough to sweep up material that would otherwise not have collided with them, so their rate of growth became faster still. During this second phase of planet formation, gravitational forces between planetesimals caused them to collide and merge, forming larger and larger objects. Because larger bodies exert stronger gravitational pulls, eventually almost all the planetesimal material was swept up into a few large **protoplanets**—accumulations of matter that would eventually evolve into the planets we know today.

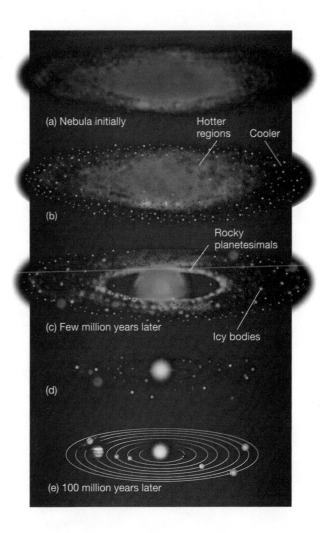

▶ **FIGURE 4.27 Solar System Formation** The condensation theory of planet formation. (a) The solar nebula after it has contracted and flattened to form a spinning disk (Figure 4.24b). The large red blob in the center will become the Sun. Smaller blobs in the outer regions may become jovian planets (see Figure 4.29). (b) Dust grains act as condensation nuclei, forming clumps of matter that collide, stick together, and grow into moon-sized planetesimals. (c) After a few million years, strong winds from the still-forming Sun begin expelling nebular gas, and some massive planetesimals in the outer solar system have already captured gas from the nebula. (d) With the gas ejected, planetesimals continue to collide and grow. (e) Over the course of a hundred million years or so, planetesimals are accreted or ejected, leaving a few large planets that travel in roughly circular orbits.

▶ **FIGURE 4.28 Newborn Solar Systems?** (a) This infrared image, taken by the *Spitzer Space Telescope,* of the bright star Fomalhaut, some 25 light-years from Earth, shows a circumstellar disk in which the process of accretion is underway. The star itself is well inside the yellowish blob at center. (b) This ALMA image of the disk around the star HL Tauri, 450 light-years away, shows fine structure in the gas thought to be caused by still-forming planets. The angular resolution is a remarkable 35 milliarcseconds—finer than available even using *HST. (NASA)*

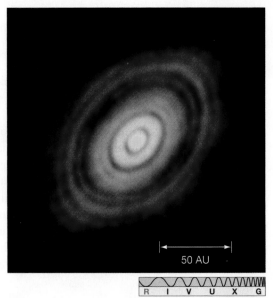

100 AU

50 AU

R I V U X G

R I V U X G

ANIMATION/VIDEO **MA**
Protoplanetary Disk Destruction

ANIMATION/VIDEO **MA**
Protoplanetary Disks in the Orion Nebula

As the protoplanets grew, a competing process became important. Their strong gravitational fields produced many high-speed collisions between planetesimals and protoplanets. These collisions led to **fragmentation** as small objects broke into still smaller chunks that were then swept up by the protoplanets. Only a relatively small number of 10- to 100-km fragments escaped capture by a planet or a moon and became the asteroids and comets.

After about 100 million years, the primitive solar system had evolved into eight protoplanets, dozens of protomoons, and a glowing **protosun** at the center. Roughly a billion more years were required to sweep the system clear of interplanetary trash. This was a period of intense meteoritic bombardment whose effects on the Moon and elsewhere are still evident today (see Chapter 5).

Making the Jovian Planets

The two-stage accretion picture just described has become the accepted model for the formation of the terrestrial planets in the inner solar system. However, the origin of the giant jovian worlds is less clear. Two somewhat different views, with important consequences for our understanding of extrasolar planets, have emerged.

In the first, more conventional scenario depicted in Figures 4.27(c and d), the four largest protoplanets in the outer solar system grew rapidly and became massive enough to enter a third phase of planetary development—their strong gravitational fields swept up large amounts of gas directly from the solar nebula. As we will see, there was a lot more raw material available for planet building in the outer solar system, so protoplanets grew much faster there. In this scenario, the cores of the jovian planets reached the point where they could capture nebular gas in less than a few million years. Compare this to the inner solar system, where the smaller, terrestrial protoplanets never reached this stage. Their growth was slow, taking 100 million years to form the planets we know today, and their masses remained relatively low.

In the second scenario, the giant planets formed through instabilities in the cool outer regions of the solar nebula—not so different from Laplace's original idea—mimicking on small scales the collapse of the initial interstellar cloud. In this view, the jovian protoplanets formed directly and very rapidly, skipping the initial accretion stage and perhaps taking less than a thousand years to acquire much of their mass. These first protoplanets had gravitational fields strong enough to scoop up gas and dust from the solar nebula, allowing them to grow into the giants we see today. Figure 4.29 illustrates this alternative formation path.

▶ **FIGURE 4.29 Jovian Condensation INTERACTIVE**   As an alternative to the growth of massive protoplanetary cores followed by accretion of nebular gas, some or all of the giant planets may have formed directly by instabilities in the cool gas of the outer solar nebula. Part (a) shows the same instant as Figure 4.27(a). (b) Only a few thousand years later, four gas giants have formed (red blobs), circumventing the accretion process sketched in Figure 4.27. With the nebula gone (c), the giant planets have taken their place in the outer solar system.

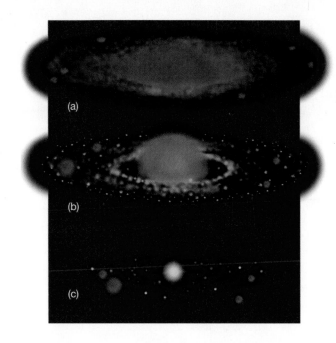

Ongoing studies of planetary composition and internal structure (for example, the size of the protoplanetary core), as well as observations of extrasolar planets, may one day allow us to distinguish between these competing theories. In either case, once the jovian protoplanets reached the critical size at which they could capture nebular gas, they grew rapidly. Their growth rate increased further as their gravitational fields intensified, and they reached their present masses in just a few million years.

Many of the jovian moons probably also formed by accretion, but on a smaller scale, in the gravitational fields of their parent planets. Once nebular gas began to flow onto the large jovian protoplanets, conditions probably resembled a miniature solar nebula, with condensation and accretion continuing to occur. The large moons of the outer planets almost certainly formed in this way. Some of the smaller moons may be captured planetesimals.

What of the gas that made up most of the original cloud? Why don't we see it today throughout the planetary system? All young stars apparently experience a highly active evolutionary stage known as the *T-Tauri phase* (see Chapter 11), during which their radiation emission and stellar winds are very intense. When our Sun entered this phase, a few million years after the solar nebula formed, any gas remaining between the planets was blown away into interstellar space by the solar wind and the Sun's radiation pressure. Afterward, all that remained were protoplanets and planetesimal fragments, ready to continue their long evolution into the solar system we know today. Note that the evolution of the Sun sets the time frame for the formation of the jovian planets. The outer planets *must* have formed before the nebula dispersed.

Differentiation of the Solar System

The condensation theory explains the basic composition differences among the terrestrial and jovian planets and the smaller bodies that constitute the solar system. Indeed, it is in this context that the term *condensation* derives its true meaning.

As the solar nebula contracted under the influence of gravity, it heated up as it flattened into a disk. The density and temperature were greatest near the central protosun and much lower in the outlying regions. In the hot inner regions, dust grains broke apart into molecules, which in turn split into atoms. Most of the original dust in the inner solar system disappeared at this stage, although the grains in the outermost parts probably remained largely intact.

The destruction of the dust in the inner solar nebula introduced an important new ingredient into the theoretical mix, one that we omitted from our earlier discussion. With the passage of time, the gas radiated away its heat, and the temperature decreased at all locations except in the very core, where the Sun was forming. Everywhere beyond the protosun new dust grains began to condense out, much as raindrops, snowflakes, and hailstones condense from moist, cooling air here on Earth. It may seem strange that although there was plenty of interstellar dust early on, it was mostly destroyed, only to form again later. However, a critical change had occurred. Initially, the nebular gas was uniformly peppered with dust grains. When the dust re-formed later, the distribution of grains was very different.

Figure 4.30 shows the temperature in various parts of the primitive solar system just before the onset of accretion. At any given location, the only materials to

(a)

(b)

▲ **FIGURE 4.30 Temperature in the Early Solar Nebula**
(a) Theoretically computed variation of temperature across the primitive solar nebula illustrated in part (b), which shows half of the disk in Figure 4.27(b).

condense out were those able to survive the temperature there. In the innermost regions, around Mercury's present orbit, only metallic grains could form—it was simply too hot for anything else to exist. Farther out, at about 1 AU, it was possible for rocky, silicate grains to form, too. Beyond about 3 or 4 AU, water ice could exist, and so on. More and more material could condense out at greater and greater distances from the Sun. The composition of the grains at any given distance from the Sun determined the type of planetesimal—and ultimately planet—that formed there. The present-day structure of the asteroid belt, with rocky silicates more common in the inner regions and carbonaceous bodies, descendents of icier planetesimals, most prevalent at larger radii, still reflects these early conditions.

Beyond about 5 AU from the center, the temperature was low enough to allow several abundant gases—water vapor, ammonia, and methane—to condense into solid form. As we will see, these compounds are still important constituents of jovian atmospheres. Consequently, the planetesimals destined to become the cores of the jovian planets were formed under cold conditions out of low-density, icy material. Because more material could condense out of the solar nebula at these radii than in the inner regions near the protosun, accretion began sooner, with more resources to draw on. If they hadn't already formed through instabilities in the cold nebular gas, the outer planets grew rapidly to the point where they could accrete not just grains, but nebular gas also, and the eventual result was the hydrogen-rich jovian worlds we see today.

In the inner regions of the primitive solar system, the environment was too hot for ices to survive. Many of the abundant heavier elements, such as silicon, iron, magnesium, and aluminum, combined with oxygen to produce a variety of rocky materials. Planetesimals in the inner solar system were therefore rocky or metallic, as were the protoplanets and planets they ultimately formed. Here is an additional reason why the jovian planets grew so much bigger than the terrestrial worlds. The inner regions of the nebula had to wait for the temperature to drop so that a few rocky grains could appear and begin the accretion process, whereas accretion in the outer solar system began almost as soon as the solar nebula collapsed to a disk.

The heavier materials condensed into grains in the outer solar system, too, of course. However, there they would have been vastly outnumbered by the far more abundant light elements. The outer solar system is not deficient in heavy elements—rather, the inner solar system is *underrepresented in light material.*

Asteroids and Comets

In the inner solar system, most rocky planetesimals collided with or were ejected by the growing terrestrial planets. Only a few remain today as the asteroid belt. Planetesimals beyond the orbit of Mars failed to accumulate into a protoplanet because the huge gravitational field of nearby Jupiter continuously disturbed their motion, nudging and pulling at them, preventing them from aggregating into a planet. Many of the Trojan asteroids have likely been locked in their odd orbits by the combined gravitational pulls of Jupiter and the Sun since those earliest times.

Once the jovian planets formed, they exerted strong gravitational forces on the planetesimals in the outer solar system. Over a period of hundreds of millions of years, and after repeated gravitational "kicks" from the giant planets, particularly Jupiter and Saturn, most of the interplanetary fragments in the outer solar system were flung into orbits taking them far from the Sun (Figure 4.31). Those fragments now make up the Oort cloud. Interactions with Uranus and Neptune were gentler, and generally did not eject planetesimals to large distances. Rather, these encounters tended to deflect small bodies into eccentric orbits that carried them into the inner solar system, where they collided with a planet or were kicked into the Oort cloud by Jupiter or Saturn. Most of the original planetesimals formed beyond the orbit of Neptune are still there, making up the Kuiper belt.

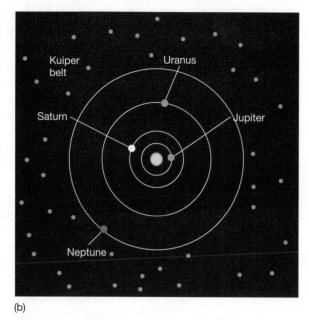

(a)   (b)

▲ **FIGURE 4.31 Planetesimal Ejection** **INTERACTIVE**   The ejection of icy planetesimals to form the Oort cloud and Kuiper belt. (a) Initially, once the giant planets had formed, leftover planetesimals were found throughout the solar system. Interactions with Jupiter and Saturn apparently "kicked" planetesimals out to very large radii (the Oort cloud). (b) After hundreds of millions of years and as a result of the inward and outward "traffic," the orbits of all four giant planets were significantly modified by the time the planetesimals inside Neptune's orbit had been ejected. As depicted here, Neptune was affected most and may have moved outward by as much as 10 AU.

The interactions that cleared the outer solar system of comets also caused significant changes in the orbits of the planets themselves. Computer simulations indicate that, during the ejection process, Jupiter moved slightly closer to the Sun, its orbital semimajor axis decreasing by a few tenths of an astronomical unit, while the other giant planets migrated outward—possibly by as much as 10 AU in the case of Neptune (Figure 4.31b).

The deflection of icy planetesimals into the inner solar system may have played an important role in the evolution of the terrestrial planets too. A long-standing puzzle in the condensation theory's account of the formation of the inner planets was the origin of the water and other volatile gases on Earth and elsewhere. At formation, the inner planets' surface temperatures were far too high, and their gravity too low, to capture and retain those gases. The answer may be that the comets from the outer solar system bombarded the newly born inner planets, supplying them with water *after* their formation.

This theory has been a leading, if controversial, explanation for Earth's water for some time, but recent data from *Rosetta* has cast doubt on it. *Rosetta's* sensors found that the amount of deuterium (a heavy form of hydrogen) in the ice and water vapor surrounding comet 67P is three times greater than is found in Earth's oceans, implying that comets like 67P from the outer solar system could not be responsible for our planet's water. Meteorites from the asteroid belt do match Earth's deuterium levels, however, suggesting that bombardment by asteroids, rather than comets, could have been the source of water on Earth.

Solar System Regularities and Irregularities

The condensation theory accounts for the 10 facts listed at the start of this section. The growth of planetesimals throughout the solar nebula, with each protoplanet ultimately sweeping up the material near it, accounts for the planets' wide spacing (point 1, although the theory does not adequately explain the regularity of the

CONCEPT CHECK

Would you expect to find asteroids and comets orbiting other stars?

spacing). That the planets' orbits are circular (2), in the same plane (3), and in the same direction as the Sun's rotation on its axis (4) is a direct consequence of the solar nebula's shape and rotation. The rotation of the planets (5) and the orbits of the moon systems (6) are due to the tendency of smaller structures to inherit the overall sense of rotation of their parent. The heating of the nebula and the Sun's ignition resulted in the observed differentiation (7), while the debris from the accretion–fragmentation stage naturally accounts for the asteroids (8) and comets (9).

We mentioned earlier that an important aspect of any theory of solar system formation is its ability to accommodate deviations. In the condensation theory, that capacity is provided by the randomness inherent in the encounters that combined planetesimals into protoplanets. As the numbers of large bodies decreased and their masses increased, individual collisions acquired greater and greater importance. The effects of these collisions can still be seen today in many parts of the solar system.

We will discuss many solar system irregularities in the next few chapters; most can be explained as the results of random collisions. A case in point is the anomalously slow and retrograde rotation of Venus (Chapter 6), which can be explained if we assume that the last major collision in the formation history of that planet just happened to involve a near head-on encounter between two protoplanets of comparable mass. A similar encounter, this time involving Earth, may have formed our Moon (Chapter 5). Scientists usually do not like to invoke random events to explain observations. However, there seem to be many instances where pure chance has played an important role in determining the present state of the universe.

4.4 Planets Beyond the Solar System

The condensation theory was developed to explain just one planetary system—our own. But a critical test of any scientific theory is its applicability and predictive power outside the context in which it was originally conceived. ∞ *(Sec. 0.5)* The discovery in recent years of numerous planets orbiting other stars presents astronomers with the opportunity—indeed, the scientific obligation—to confront their theories of solar system formation against a new body of observational data.

Detecting Extrasolar Planets

The past few years have seen enormous strides in the search for extrasolar planets. These advances have been achieved through steady improvements in both telescope and detector technology and computerized data analysis. With few exceptions it is not yet possible to obtain images of these newly discovered worlds. The techniques used to find them are generally indirect, based on analysis of light from the parent star, not from the unseen planet.

To date, some 600 planets have been discovered via their gravitational effect on their parent star. As a planet orbits a star, gravitationally pulling one way and then the other, the star "wobbles" slightly. The more massive the planet or the closer its orbit to the star, the greater its gravitational pull and hence the star's movement. If the wobble happens to occur along our line of sight to the star, then we see small fluctuations in the star's radial velocity, which astronomers can measure using the Doppler effect. ∞ *(Sec. 2.7)*

Figure 4.32 shows two sets of radial velocity data that betray the presence of planets orbiting other stars. Part (a) shows the line-of-sight velocity of the star 51 Pegasi, a near-twin to our Sun some 40 light-years away. These data were acquired in 1994 by Swiss astronomers using the 1.9-m telescope at Haute-Provence Observatory in France, and they were the first firm evidence for an extrasolar planet orbiting a Sun-like star. The regular 50 m/s fluctuations in the star's velocity have since been

▶ **FIGURE 4.32 Planets Revealed INTERACTIVE** (a) *Discovery* data of the Doppler shift of the star 51 Pegasi reveal a clear periodic signal indicating the presence of a planetary companion of mass at least half the mass of Jupiter. (b) Radial velocity data for Upsilon Andromedae are more complex, but are well fit (solid line) by a three-planet system orbiting the star. (c) A sketch of the inferred orbits of the three planets in the Upsilon Andromedae system (in orange) with the orbits of the terrestrial planets superimposed for comparison (in white).

(a)

The blue lines show the maximum possible signal produced by Jupiter orbiting the Sun.

confirmed by several groups of astronomers and imply that a planet of at least half the mass of Jupiter orbits 51 Pegasi in a circular orbit with a period of 4.2 days.

Figure 4.32(b) shows another set of Doppler data, this time revealing a much more complex planetary system—a triple-planet system orbiting another nearby Sun-like star named Upsilon Andromedae. The three planets have minimum masses of 0.7, 2.1, and 4.3 times the mass of Jupiter, and orbital semimajor axes of 0.06, 0.83, and 2.6 AU, respectively. Figure 4.32(c) sketches their orbits, with the orbits of the solar system terrestrial planets shown for scale.

The Doppler technique suffers from the limitation that the angle between the line of sight and the planet's orbital plane cannot be determined. Simply put, we cannot distinguish between a low-speed orbit seen edge-on and a high-speed orbit seen almost face-on (so only a small component of the orbital motion contributes to the line-of-sight Doppler effect). However, in some systems this is not the case. For example, observations of the solar-type star HD 209458, lying some 150 light-years from Earth, reveal a small (1.7 percent) but clear drop in brightness each time its 0.6-Jupiter-mass companion, orbiting just 7 million km (0.05 AU) away, passes between the star and Earth (Figure 4.33). The drop in brightness occurs every 3.5 days—the orbital period inferred from radial-velocity measurements. Such **planetary transits** are relatively rare because they require us to see the orbit almost exactly edge-on, but when they do occur, taken in conjunction with radial velocity measurements, they allow an unambiguous determination of the planet's mass and radius. In this case, the resulting density of the planet is just 200 kg/m³, consistent with a high-temperature gas giant planet orbiting very close to its parent star.

Only a small fraction of planetary systems are oriented in just the right way to show transits, so planet hunters adopt the strategy of repeatedly surveying thousands of stars in the hope of detecting a transit should one occur. Space-based telescopes are particularly well suited to this task, as they can stare continuously at a given region of the sky, making simultaneous, high-precision observations of the target stars. Orbiting instruments can measure the tiny brightness changes (less than 1 part in 10⁴) needed to detect an Earth-like planet orbiting a Sun-like star. NASA's *Kepler* mission, launched in 2009, has been spectacularly successful in finding exoplanets among the hundreds of thousands of stars it has monitored. It has produced more than 4000 exoplanet candidates, of which more than 1000 have so far been confirmed, including some of the smallest (i.e., Earth-sized) exoplanets yet discovered. Suspended in 2013 following the failure of a gyroscope needed to accurately point the spacecraft, the mission was redefined in 2014 to monitor planets orbiting low-mass stars.

In total, some 1200 confirmed extrasolar planets have been detected by the transit method. Unfortunately, in most cases, radial velocity measurements are not available, so we know these planets' orbits and radii, but not their masses.

(b)

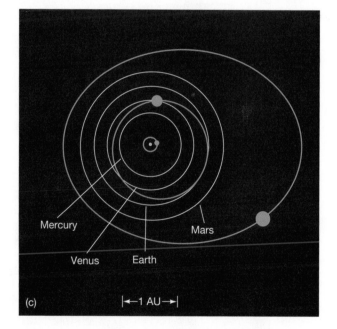

(c) |←—1 AU—→|

Exoplanet Properties

To date (as of mid-2015), astronomers have confirmed almost 2000 extrasolar planets orbiting more than 1000 stars. Most lie within 500 light years of the Sun, although many of the planets discovered by *Kepler* lie at much greater distances. About 10 percent of the nearby stars surveyed to date have been found to host planets. In most cases only a single planet has been detected, but roughly 500 *multiple-planet systems* are also known. Most astronomers expect both the

▶ **FIGURE 4.33 Extrasolar Transit INTERACTIVE** As an extrasolar planet passes between us and its parent star, the light from the star dims in a characteristic way. The planet orbiting a Sun-like star known as HD 209458 is 200,000 km across and transits every 3.5 days, blocking about 2 percent of the star's light each time it does so.

The dark planet cannot actually be seen against the bright star . . .

. . . but the total light from the two objects dims as in this plot.

Time

Ⓜ ANIMATION/VIDEO

Survey for Transiting Extrasolar Planets

▲ **FIGURE 4.34 Extrasolar Planetary Parameters** Masses and orbital semimajor axes of roughly 1000 extrasolar planets. Each point represents one planetary orbit. Corresponding points for Earth, Jupiter, and Neptune in our solar system are also shown. Planets are classified by familiar solar system names, depending on mass, and as hot or cold, depending on distance from their parent star.

fraction of stars with planets and the number of planets per star to increase as detection and data analysis technology continue to improve.

Figure 4.34 presents the observed masses and semimajor axes of about 1000 extrasolar planets. Each dot in the figure represents a planet, and we have added points corresponding to Earth, Neptune, and Jupiter in our own solar system. Massive exoplanets are often referred to as *Jupiters,* while less massive, but still "jovian" planets are called *Neptunes.* The dividing line between Jupiters and Neptunes is somewhat arbitrary, but is often taken at about twice the mass of Neptune, or 0.1 Jupiter masses. The terminology is intended to distinguish between planets that are mostly gas, like Jupiter, and those that have substantial rocky cores, like Neptune, but bear in mind that the division is based only on mass—we have no information on most of these planets' composition or internal structure.

Planets having masses between roughly 2 and 10 Earth masses (about half the mass of Neptune) are known as **super-Earths.** The upper limit in this case is significant, as theorists think that 10 Earth masses represents the minimum mass of a planetary core needed for it to accrete large amounts of nebular gas and begin to form a gas giant (Section 4.3). Below 2 Earth masses, exoplanets are simply called *Earths.*

As just mentioned, many transiting exoplanets have known radii but do not have measured masses. Making reasonable assumptions about planet properties, astronomers conventionally extend the above definitions as follows: Jupiters have radii greater than 5 Earth radii, Neptunes are between 2 and 5 Earth radii, super-Earths between 1.25 and 2 Earth radii, and Earths less than 1.25 Earth radii.

Exoplanets are further subdivided depending on their distances from their parent stars. Planets with orbital semimajor axes less than 0.1 AU are said to be *hot,* while those on wider orbits are called *cold.* Again, the dividing line is arbitrary—the actual temperature of a planet depends not just on its orbit, but also on the composition of the planet's atmosphere and the temperature and brightness of the central star.

Most of the planets observed so far fall into the "cold Jupiter" or "cold Neptune" categories, like the jovian planets in our own solar system, although their orbits are generally somewhat smaller than those of the jovian planets—less than a few astronomical units across—and much more eccentric. Fewer than 20 percent have eccentricities less than 0.1, whereas no jovian planet in our solar system has an eccentricity greater than 0.06. Figure 4.35 plots the actual orbits of some of these planets, with Earth's orbit superimposed for comparison. A sizable minority—about a third—of all observed exoplanets move in "hot" orbits very close to their parent stars and have surface temperatures as high as 1000–2000 K. The most massive ones were the first to be discovered, and they were quickly dubbed **hot Jupiters.** They represent a new class of planet and have no counterparts in our own solar system.

More than 300 confirmed super-Earths are currently known. They are found in both hot and cold orbits. Some, especially the lower-mass ones, might be large terrestrial planets. Others could be icy planetary cores that never managed to accrete significant amounts of nebular gas. Still others may have substantial atmospheres of light gases but never grew to "Neptune" status—they are sometimes referred to as *gas dwarfs*. These latter two categories, if real, would be new classes of planet unknown in the solar system. So far, some 150 confirmed exo-Earths have been found. Most move on hot orbits close to their parent stars and are unlikely to resemble anything we'd want to call home. However, a few Earths and more than a dozen super-Earths are of great interest, as their orbits may lead to relatively balmy surface conditions possibly conducive to life. We discuss them in more detail below.

The fact that we see relatively few low-mass planets on wide orbits (i.e., toward the right or lower part of Figure 4.34) is not too surprising. It is what astronomers call a **selection effect.** Lightweight and/or distant planets simply don't produce large enough velocity fluctuations for them to be easily detectable. Generally speaking, the methods currently employed are biased toward finding large and/or massive objects orbiting close to their parent stars.

Exoplanet Composition

As mentioned earlier, if we have radial velocity data on a transiting exoplanet, we can determine both its mass and its radius, and hence estimate its density and possibly even its composition. More than 200 transiting hot Jupiters have been measured in this way. But when astronomers calculate their densities, the numbers they obtain are generally much lower than predicted by theory. The computed densities range from around 1500 kg/m^3 (slightly more than the density of Jupiter) down to just 200 kg/m^3 (roughly the density of styrofoam) and are generally *inconsistent,* by a wide margin, with theoretical models, even assuming the lightest possible composition of pure hydrogen and helium. The leading explanation for this discrepancy is that the heat of the nearby parent star has puffed these planets up far beyond their normal sizes. However, no current model can account for the observed range in densities, and for now a complete explanation of this phenomenon eludes astronomers.

Currently, some 30 Earths and super-Earths have masses and radii measured well enough for approximate densities to be derived. The results range from roughly 500 to more than 9000 kg/m^3. The low end of this range suggests planets harboring large amounts of light gases, probably with rocky/icy cores and hydrogen/helium atmosphere—gas dwarfs. The high end suggests mainly rocky composition—compressed Earths. The intermediate densities suggest planets composed of water and/or other ices. Figure 4.36 presents a visual comparison of Earth, Neptune, and two super-Earths whose physical properties are relatively well known. CoRoT 7b is 4.8 times more massive and 1.7 times larger than Earth, implying a mean density of 5300 kg/m^3—very similar to that of Earth, although since this planet orbits just 0.02 AU from its parent Sun-like star, surface conditions are much more extreme than those on Earth. Meanwhile, the planet GJ 1214b has a mass 5.7 times that of Earth and a radius of 2.7 Earth radii, for a mean density of 1600 kg/m^3. Definitely not rocky, this planet may be composed mainly of water and/or ice, possibly surrounding a small rocky core, with an atmosphere of hydrogen and helium.

The composition, structure, and history of Earths and super-Earths are of great interest to astronomers as they seek to understand how these bodies form—and why they are absent in our solar system. Transit times, once measured, can be accurately predicted, and astronomers can then time their observations to obtain spectroscopic and other information on the starlit face of the planet during its "quarter" phase,

▲ **FIGURE 4.35 Extrasolar Orbits**   The orbits of many extrasolar planets residing more than 0.15 AU from their parent star, superimposed on a single plot, with Earth's orbit shown for comparison (as dashed white). All the extrasolar planets shown are comparable in mass to Jupiter. A plot of all known extrasolar planets would be too cluttered to show.

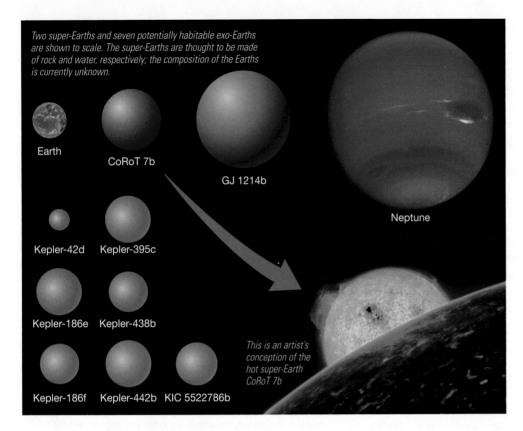

Two super-Earths and seven potentially habitable exo-Earths are shown to scale. The super-Earths are thought to be made of rock and water, respectively; the composition of the Earths is currently unknown.

Earth

CoRoT 7b

GJ 1214b

Neptune

Kepler-42d   Kepler-395c

Kepler-186e   Kepler-438b

This is an artist's conception of the hot super-Earth CoRoT 7b

Kepler-186f   Kepler-442b   KIC 5522786b

▲ **FIGURE 4.36** **Super-Earth Comparison** Two transiting super-Earths whose masses and radii are accurately known are shown here compared with Earth and Neptune. Based on their average densities, these two new worlds seem to be very different from one another—one is rocky, somewhat like Earth or Neptune's core, but the other may be composed predominantly of water and ice. Seven potentially habitable exo-Earths (see Figure 4.39) are also shown, to scale.

allowing them to probe the planet's atmospheric composition and dynamics. ∞ *(Sec. 2.5)* Since planets—even the hot ones—are cooler than their parent stars, their reflected light is most easily distinguished from the background starlight at infrared wavelengths, and the *Spitzer Space Telescope* has played a vital role in these studies. ∞ *(Secs. 2.4, 3.5)* So far, hydrogen, sodium, methane (CH_4), carbon dioxide (CO_2), and water vapor have been detected. Observations such as these are critical in pinning down the content and structure of exoplanet atmospheres.

Spectroscopic observations of the parent stars reveal what may be a crucial piece in the puzzle of extrasolar planet formation. Stars having compositions similar to that of the Sun are statistically *much more likely* to have planets orbiting them than are stars containing smaller fractions of the key elements carbon, nitrogen, oxygen, silicon, and iron. Because the elements found in a star reflect the composition of the nebula from which it formed, and the elements just listed are the main ingredients of interstellar dust, this finding provides strong support for the condensation theory (Section 4.3). Dusty disks really are more likely to form planets.

Is Our Solar System Unusual?

Not so long ago, many astronomers argued that the condensation scenario described earlier in this chapter was in no way unique to our own system. Today we know that planetary systems *are* common, but, by and large, the ones we see don't look much like ours! We can legitimately ask whether our solar system really is unusual, and whether the observations described above undermine the current theory of solar system formation.

Most cold extrasolar Jupiters and Neptunes move in orbits that are considerably more eccentric than the orbits of the jovian planets in the solar system. Does this make our system fundamentally different from the others? Probably not. To a large extent, this discrepancy can be explained by the selection effect just described—eccentric orbits tend to produce larger velocities and hence are more readily discovered. As search techniques improve, astronomers are finding more and more Jupiter-mass (and lower-mass) planets on wider and less eccentric orbits. Figure 4.37 shows evidence for one of the most "Jupiter-like" planets yet detected: a 0.95-Jupiter-mass object moving on a roughly circular orbit around a near-twin of our own Sun. The planet's period is 9.1 years. It is too early to say whether cold Jupiters on nearly circular orbits will turn out to be unusual or common among exoplanets, but they clearly exist among the systems already observed.

Are the eccentric extrasolar orbits we see consistent with the condensation theory? The answer is yes. The theory actually allows many ways in which massive planets can end up in eccentric orbits. Indeed, an important aspect of solar system formation not mentioned in our earlier discussion is the fact that many theorists worried about how Jupiter could have remained in a circular orbit after it formed in the protosolar disk! Jupiter-sized planets may be knocked into eccentric orbits by interactions with other Jupiter-sized planets or by the effects of nearby stars. And if they formed by gravitational instability, they could have had eccentric

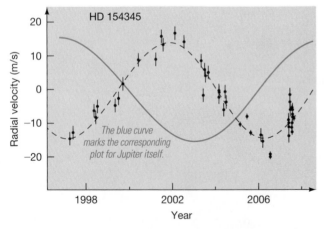

HD 154345

The blue curve marks the corresponding plot for Jupiter itself.

1998   2002   2006

Year

▲ **FIGURE 4.37** **An Exo-Jupiter?** Velocity "wobbles" in the star HD 154345 reveal the presence of the extrasolar planet with the most "Jupiter-like" orbit yet discovered. The parent star is almost identical to the Sun, and the 0.95-Jupiter-mass planet orbits at a distance of 4.2 AU with an orbital eccentricity of 0.04.

orbits right from the start, and then we must explain how those orbits circularized in the case of the solar system.

What of the hot Jupiters, which have no counterparts among the planets of the solar system? Here, too, there is an explanation that fits within the condensation theory! In the mid-1980s theorists realized that friction between the giant planets and the nebula in which they moved should have caused the planets to drift inward quite rapidly. Even before the first hot Jupiter was observed, theorists knew that, depending on how long the disk survived before being dispersed by the newborn Sun, the process could easily have deposited planets in an orbit very close to the parent star, as illustrated in Figure 4.38. The theorists were right, and the observed hot Jupiters may provide a much-needed connection between extrasolar planetary systems and our own.

Curiously, a Jupiter-sized planet sinking inward through the disk in which planetesimals are forming is not detrimental to the formation of terrestrial planets. The hot Jupiters must have reached their scorched orbits before the gas disk dispersed—that is, within a few million years of the formation of the solar nebula and long *before* the formation of the terrestrial planets. Computer simulations indicate that the inward migration stirs up the planetesimals as the giant planet moves through, but for the most part does not disrupt or eject them. The main consequence is to mix in more icy material from farther out, possibly resulting in planets more massive and more water-rich than would otherwise be the case.

Searching for Earth

Giant planets are interesting objects, but to many astronomers the real goal of extrasolar planet research is the detection of terrestrial planets with conditions similar to those found on Earth—and, by extension, the potential discovery of life elsewhere in the universe. This latter goal is increasingly becoming the motivation behind solar system exploration, too.

If Earth-like bodies are our ultimate goal, what parameters are of greatest interest to astronomers? Many astronomers think that a key requirement for the development of life as we know it is the existence of liquid water on (or under) a planet's surface, implying a surface temperature roughly in the 0–100°C range. The planet's temperature depends both on its distance from its parent star and on the star's intrinsic brightness. As sketched in Figure 4.39, any given star is surrounded by a **habitable zone,** within which liquid water can in principle exist on the planet, making it a possible abode for life. For low-mass, faint stars, the habitable zone is small and lies close to the star. For more massive, brighter stars, it lies much farther out and can be 1 AU or more wide. Three terrestrial planets—Venus, Earth, and Mars—lie in or near the Sun's habitable zone. Any or all of them might have seen the development of life, given the right circumstances (see Chapter 6). Earth-like planets orbiting within the habitable zone of their parent stars are regarded as the best candidates for harboring life as we know it beyond the solar system.

▶ **FIGURE 4.39 Habitable Zones** Every star is surrounded by a habitable zone, within which an Earth-like planet could have liquid water on its surface. The planets of the solar system, numerous extrasolar super-Earths, and seven possible exo-Earths lying in or near their habitable zones are marked.

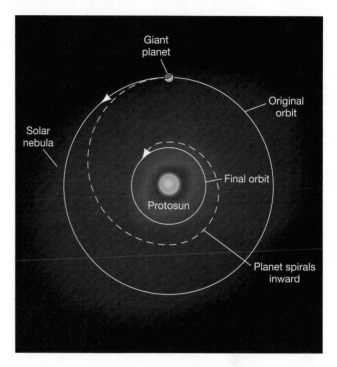

▲ **FIGURE 4.38 Sinking Planet** Friction between a giant planet and the nebular disk in which it formed tends to make the planet spiral inward. The process continues until the disk is dispersed by the wind from the central star, possibly leaving the planet in a "hot" orbit.

(MA) ANIMATION/VIDEO
Hot Jupiter Extrasolar Planet Evaporating

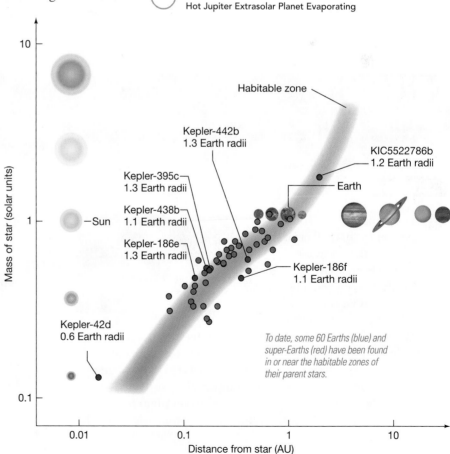

At the time of writing, 13 super-Earths and 4 Earths (excluding our own) are known to orbit in or near the habitable zones of their parent stars. They are indicated on Figure 4.39. Unfortunately, none of these planets has had both its mass and its radius reliably measured, so in all cases the densities, and hence the compositions, of these worlds are unknown. Most of these bodies orbit close to the "hot" edge of the habitable zone, but this is yet another aspect of the selection effect described earlier—of all small planets, those in close orbits are the most likely to be detected. The *Kepler* candidate list contains a further 27 Earths and 40 super-Earths that may soon be added to this figure.

Many planet hunters are confident that within the next decade (or sooner), observational techniques will reach the level of sophistication at which jovian and terrestrial planets similar to those in our solar system should be readily detectable—if they exist. Advances during the next decade will either bring numerous detections of extrasolar planets in "solar system" orbits, or allow astronomers to conclude that systems like our own really are a small minority. Either way, the consequences are profound.

CONCEPT CHECK

Why is it not too surprising that the extrasolar planetary systems detected thus far have properties quite different from our solar system?

THE BIG QUESTION

People from all walks of life—not just astronomers—eagerly await the discovery of a true Earth-like planet orbiting another star. When will Earth's twin be found, and will it have blue skies, deep oceans, and livable lands? Most intriguing of all, will it be inhabited? We live in a remarkable time, when we can actually address—with a reasonable expectation of an answer—some of the most profound questions that humans have pondered for thousands of years.

CHAPTER REVIEW

SUMMARY

LO1 The planets of our **solar system** (p. 104) all orbit the Sun counterclockwise, as viewed from above Earth's North Pole, on roughly circular orbits that lie close to the ecliptic plane. The orbit of Mercury is the most eccentric and also has the greatest inclination. The orbital spacing increases as we move outward from the Sun. The inner **terrestrial planets** (p. 106)—Mercury, Venus, Earth, Mars—are all of comparable density and generally rocky, whereas all the outer **jovian planets** (p. 106)—Jupiter, Saturn, Uranus, and Neptune—have much lower densities and are composed mostly of gaseous or liquid hydrogen and helium.

LO2 Most **asteroids** (p. 107) orbit in a broad band called the **asteroid belt** (p. 108), between the orbits of Mars and Jupiter. The **Trojan asteroids** (p. 108) share Jupiter's orbit, remaining 60° ahead or behind the planet as it circles the Sun. A few **Earth-crossing asteroids** (p. 108) have orbits that intersect Earth's orbit and will eventually collide with our planet. The largest asteroids are a few hundred kilometers across. Most are much smaller. Asteroids in the inner part of the asteroid belt are predominantly rocky; asteroids farther out contain larger fractions of water ice and organic material.

LO3 **Comets** (p. 107) are icy and rocky fragments that normally orbit far from the Sun. We see a comet by the sunlight reflected from the dust and vapor released if its orbit happens to bring it near the Sun. Stretching behind the kilometer-size **nucleus** (p. 112) of a comet is a long **tail** (p. 112) formed by the interaction between the cometary material and the **solar wind** (p. 113). Most comets reside in the **Oort cloud** (p. 115), a vast reservoir of cometary material tens of thousands of astronomical units across, completely surrounding the Sun. Short-period comets, having orbital periods of less than about 200 years, originate in the **Kuiper belt** (p. 107), a broad band of icy material beyond the orbit of Neptune.

LO4 A **meteoroid** (p. 107) is a piece of rocky interplanetary debris smaller than 100 m across. A meteoroid that enters Earth's atmosphere produces a **meteor** (p. 115), a bright streak of light across the sky. If part of it reaches the ground, the remnant is called a **meteorite** (p. 116). Each time a comet rounds the Sun, some cometary material becomes dislodged, forming a **meteoroid swarm** (p. 116)—a group of small **micrometeoroids** (p. 116) that travel in the comet's orbit. Larger meteoroids are pieces of material chipped off asteroids following collisions in the asteroid belt. Most meteorites are between 4.4 and 4.6 billion years old.

LO5 Any theory of solar system formation must explain the planets' roughly circular, coplanar, widely spaced orbits in the same direction around the Sun, with terrestrial planets near the Sun and jovian planets farther out—as well as the existence of the asteroids, Kuiper belt, and comets, and irregularities such as retrograde rotation of planets or moons.

LO6 According to the **condensation theory,** as the **solar nebula** (p. 121) contracted under its own gravity, it began to spin faster, eventually forming a disk. **Protoplanets** (p. 123) formed in the disk and became planets, while the central **protosun** (p. 124) became the Sun. Particles of interstellar dust helped cool the solar nebula and acted as **condensation nuclei** (p. 123) that began the planet-building process. Small clumps of matter grew by **accretion** (p. 123), sticking together and growing into moon-size **planetesimals** (p. 123) whose gravitational fields accelerated the accretion. As planetesimals collided and merged, a few planet-size objects remained.

LO7 Planets in the outer solar system became so large that they captured hydrogen and helium from the solar nebula, forming the jovian worlds. When the Sun became a star, its strong winds blew away any remaining nebular gas. The terrestrial planets are rocky because they formed in the hot inner regions of the solar nebula. Farther out, the nebula was cooler, so water and ammonia ice could also form. Many leftover planetesimals were ejected into the Oort cloud by the outer planets, leaving the Kuiper belt behind. The planetesimals in the asteroid belt never formed a planet, probably because of Jupiter's gravity.

LO8 Astronomers have confirmed nearly 2000 **extrasolar planets** (p. 119), with thousands more candidates awaiting confirmation. Several hundred have been discovered by observing their parent star wobble back and forth as the planet orbits. However, thanks largely to the Kepler mission, most known exoplanets have been discovered via **planetary transits** (p. 129), where the planet is observed passing in front of the star, slightly reducing its brightness.

LO9 Known exoplanetary masses range from that of Earth to many times that of Jupiter. Some planets move on "hot" orbits close to their parent star, while others move on wide, "cold" orbits similar to those of the jovian planets in the solar system. **Hot Jupiters** (p. 130) and **super-Earths** (p. 130) are new classes of planet not known in the solar system. It is not known whether our solar system is unusual among planetary systems. The new exoplanetary classes are compatible with the condensation theory. A few exo-Earths and numerous super-Earths have been found in or near the **habitable zones** (p. 133) of their parent stars.

Problems labeled POS explore the process of science. VIS problems focus on reading and interpreting visual information. LO connects to the introduction's numbered Learning Outcomes.

REVIEW AND DISCUSSION

1. LO1 Name three important differences between the terrestrial planets and the jovian planets.

2. LO2 POS Why are asteroids, comets, and meteoroids important to planetary scientists?

3. Are all asteroids found in the asteroid belt?

4. Describe the consequences of a 10-km asteroid striking Earth?

5. LO3 What are comets like when they are far from the Sun? What happens when they enter the inner solar system?

6. Why can comets approach the Sun from any direction but asteroids generally orbit close to the plane of the elliptic?

7. LO4 Distinguish between a meteor, a meteoroid, and a meteorite.

8. What causes a meteor shower?

9. LO5 POS Describe the basic features of the nebular theory of solar system formation, and give three examples of how this theory explains some observed features of the present-day solar system.

10. LO6 POS What key ingredient in the modern condensation theory was missing or unknown in the nebula theory?

11. Why are jovian planets so much larger than terrestrial planets?

12. LO7 How did the temperature structure of the solar nebula determine planetary composition?

13. Where did Earth's water come from?

14. LO8 How do astronomers set about looking for extrasolar planets?

15. LO9 POS In what ways do observed extrasolar planetary systems differ from our own solar system?

CONCEPTUAL SELF-TEST: TRUE OR FALSE?/MULTIPLE CHOICE

1. The largest planets also have the largest densities. (T/F)

2. The total mass of all the planets is much less than the mass of the Sun. (T/F)

3. Asteroids were recently formed by the collision and breakup of an object orbiting in the asteroid belt. (T/F)

4. Most comets have short periods and orbit close to the ecliptic plane. (T/F)

5. The solar system is of largely uniform composition. (T/F)

6. Asteroids, meteoroids, and comets are remnants of the early solar system. (T/F)

7. Astronomers have no theoretical explanation for the "hot Jupiters" observed orbiting some other stars. (T/F)

8. If we were to construct an accurate scale model of the solar system on a football field with the Sun at one end and Pluto at the other, the planet closest to the center of the field would be (a) Earth; (b) Jupiter; (c) Saturn; (d) Uranus.

9. In the leading theory of solar system formation, the planets (a) were ejected from the Sun following a close encounter with another star; (b) formed from the same flattened, swirling gas cloud that formed the Sun; (c) are much younger than the Sun; (d) are much older than the Sun.

10. The solar system is differentiated because (a) the heavy elements in the outer solar system sank to the center; (b) the light elements in the inner solar system fell into the Sun; (c) the light elements in the inner solar system were carried off as comets; (d) only rocky and metallic particles could form close to the Sun.

11. Water on Earth was (a) transported here by comets; (b) accreted from the solar nebula; (c) produced by volcanoes in the form of steam; (d) created by chemical reactions involving hydrogen and oxygen shortly after Earth formed.

12. The number of confirmed exoplanets is (a) less than 10; (b) roughly 50; (c) more than 500; (d) more than 5000.

13. Astronomers have not yet reported any Earth-like planets orbiting other stars because (a) there are none; (b) they are not detectable with current technology; (c) no nearby stars are of the type expected to have Earth-like planets; (d) the government is preventing us from reporting their discovery.

14. VIS In Figure 4.39 (Habitable Zones), the habitable zone of a star twice as massive as the Sun (a) is centered at roughly 3 AU from the star; (b) is more than 10 AU wide; (c) lies entirely within 1 AU of the star; (d) is the same size as the Sun's habitable zone.

15. VIS According to Figure 4.30, the temperature in the solar nebula at the location now at the center of the asteroid belt is (a) 2000 K; (b) 900 K; (c) 400 K; (d) 100 K.

PROBLEMS

The number of squares preceding each problem indicates its approximate level of difficulty.

1. ■ Suppose the average mass of each of 20,000 asteroids in the solar system is 10^{17} kg. Compare the total mass of these asteroids to the mass of Earth. Assuming a spherical shape and a density of 3000 kg/m³, estimate the diameter of an asteroid having this average mass.

2. ■■ The asteroid Icarus (Figure 4.5) has a perihelion distance of 0.19 AU and an orbital eccentricity of 0.83. What is its semimajor axis and aphelion distance from the Sun?

3. ■ The largest asteroid, Ceres, has a radius 0.073 times the radius of Earth and a mass of 0.0002 Earth masses. How much would an 80-kg astronaut weigh on Ceres?

4. ■■ It is observed that the number of asteroids (or meteoroids) of a given diameter is roughly inversely proportional to the square of the diameter. Assuming a density of 3000 kg/m³ and approximating the actual distribution of asteroids and meteoroids as a single 1000-km body (Ceres), one hundred 100-km bodies, 10 thousand 10-km bodies, and so on, calculate the total mass in the form of 1000-km bodies, 100-km bodies, 10-km bodies, and 1-km bodies.

5. ■■ (a) Using the version of Kepler's laws of planetary motion from Section 1.5, calculate the orbital period of an Oort cloud comet if the semimajor axis of the comet's orbit is 50,000 AU. (b) What is the maximum possible aphelion distance for a short-period comet with an orbital period of 125 years?

6. ■■ The planet orbiting star HD187123 has a semimajor axis of 0.042 AU. If the star's mass is 1.06 times the mass of the Sun, calculate how many times the planet has orbited its star since the paper announcing its discovery was published on December 1, 1998.

7. ■ Given the data provided in the text, calculate the gravitational acceleration at the surfaces of the two transiting super-Earths (CoRoT 7b and GJ 1214B) discussed in Section 4.4.

8. ■■ Stellar radial velocity variations as small as 1 m/s can be detected with current technology. For a Jupiter-mass planet orbiting a Sun-like star, this corresponds to a planetary orbital velocity of approximately 1 km/s. What is the radius of the widest circular orbit on which Jupiter could currently be detected orbiting the Sun?

9. ■ A typical comet contains 10^{13} kg of water ice. How many comets would be needed to account for the 2×10^{21} kg of water presently found on our planet? If this amount of water accumulated over a period of 0.5 billion years, how frequently must Earth have been struck by comets during that time?

10. ■■■ The amount of energy reaching a planet's surface per unit time is proportional to the luminosity of its parent star divided by the square of the planet's distance from the star. If the luminosity is proportional to the fourth power of the star's mass, estimate the orbital distance (in AU) at which an Earth-like exoplanet orbiting a 0.5-solar-mass star would receive the same amount of energy from its star as Earth does from the Sun.

ACTIVITIES

Collaborative

1. Which of the "irregular" characteristics of the solar system listed in Section 4.3. do you think are most and least likely to have occurred by pure chance? Can you think of other irregular features not on the list?

2. Do you think our solar system is unusual? Justify your opinion using data obtained from *The Extrasolar Planets Encyclopedia*, which you can find online at http://exoplanet.eu. Use the *Encyclopedia* to find examples of (1) a hot Jupiter, (2) a cold Jupiter on an orbit like Jupiter's, (3) a hot super-Earth, (4) a super-Earth in the habitable zone.

3. What should be the U.S. government's policy on the mining of minerals from asteroids? Justify the policy.

Individual

1. You can begin to visualize the ecliptic—the plane of the planets' orbits—just by noticing the path of the Sun throughout the day and of the full Moon in the course of a single night. It helps if you watch from one spot, such as your backyard or a rooftop. It's also good to have a general notion of direction. (West is where the Sun sets!) The movements of the Sun, Moon, and planets are confined to a narrow pathway across our sky. This pathway reflects the plane of the solar system, the ecliptic.

2. The only way to tell an asteroid from a star is to watch it over several nights. The magazines *Sky & Telescope* and *Astronomy* often publish charts for especially prominent asteroids. Look for Ceres, Pallas, or Vesta, the brightest asteroids. Use the chart to locate the appropriate star field and aim binoculars at that location in the sky. You may be able to pick out the asteroid from the chart. If you can't, make a rough drawing of the entire field. Come back a night or two later and look again. The "star" that has moved is the asteroid.

3. There are a number of major meteor showers every year, but if you plan to watch one, be sure to note the phase of the Moon. Bright moonlight (or city lights) can obliterate a meteor shower. A common misconception is that most meteors are seen in the direction of the shower's radiant. If you trace the paths of the meteors backward in the sky, they do all come from the radiant, but most meteors don't become visible until they are 20° or 30° from the radiant—they can appear in all parts of the sky! Just relax and let your eyes rove among the stars. You will generally see many more meteors in the hours before dawn than in the hours after sunset. Why do meteors have different brightnesses? Can you detect their variety of colors? Watch for meteors that appear to "explode" as they fall and vapor trails that linger after the meteor itself has disappeared.

5

Earth and Its Moon

Our Cosmic Backyard

If we are to appreciate the universe, we must first come to know our own neighborhood. By cataloging Earth's properties and attempting to explain them, we set the stage for a comparative study of all the other planets. From an astronomical perspective, this is a compelling reason to study the structure and history of our own world. What about the Moon? It is by far our closest neighbor in space, yet despite its nearness, it is a world very different from our own. It has no air, no sound, no water, no weather. Boulders and pulverized dust litter the landscape. Why, then, do we study it? In part, simply because it is our nearest neighbor and dominates our night sky. Beyond that, however, we study this body because it holds important clues to our own past. The very fact that it hasn't changed much since its formation means the Moon is a crucial key to unlocking the secrets of the solar system.

THE BIG PICTURE

Earth and its small neighbor, the Moon, are natural starting points for our investigation of the cosmos. Both bodies are very well studied—to the point that they are sometimes viewed as the purview of geology, rather than astronomy. They differ in many ways, yet closer scrutiny reveals key similarities, too. Their familiarity lets us frame some critical questions for the study of other planets and beyond. The science of astronomy begins at home.

◄ This iconic photograph, taken from NASA's *Apollo 8* spacecraft orbiting the Moon in 1968, shows Earth seen from afar, hanging in space like a "blue marble." Earth appears as a complex, fragile environment, very different from the rock-solid planet we know from everyday experience. Perhaps more than any other, this view of our world helps us appreciate our existence on our planet—and our place in the universe.

Studying this chapter will enable you to:

LO1 Summarize and compare the basic structural properties of Earth and the Moon.

LO2 Explain how the gravitational interaction between Earth, the Moon, and the Sun leads to tides in Earth's oceans.

LO3 Describe how Earth's atmosphere helps heat and protect us, and explain how greenhouse gases trap heat and increase Earth's surface temperature.

LO4 Outline the current model of Earth's interior structure, and describe some experimental techniques used to establish it.

LO5 Summarize the evidence for continental drift, and identify the physical processes driving it.

LO6 Describe how dynamic events early in the Moon's history formed its major surface features, and explain how cratering can be used to estimate the age of the lunar surface.

LO7 Describe the nature and origin of Earth's magnetosphere.

LO8 Outline the leading theories of how the Moon formed and evolved.

MasteringAstronomy®

Visit www.masteringastronomy.com for quizzes, animations, videos, interactive figures, and self-guided tutorials.

5.1 Earth and the Moon in Bulk

ANIMATION/VIDEO MA®

Earth as Seen by Galileo

In Chapter 0 we discussed some characteristics of Earth and the Moon known to ancient astronomers—the phases of the Moon, eclipses, and some basic geometrical inferences about the size of our planet. ∞ *(Secs. 0.3, 0.4)* Today we have direct and much more detailed measurements of many physical properties of both Earth and the Moon. Let's start by cataloging a few facts about the Earth–Moon system.

Physical Properties

Table 5.1 presents some basic data on Earth and the Moon. The first five columns list mass, radius, and average density for Earth and the Moon, determined as described in Chapter 4. ∞ *(Sec. 4.1)* The next two columns present important measures of a body's gravitational field. **Surface gravity** is the strength of the gravitational force at the body's surface. ∞ *(More Precisely 1-1)* **Escape speed** is the speed required for any object—an atom, a baseball, or a spaceship—to escape forever from the body's gravitational pull (see *More Precisely 5-1*, p. 147). By either measure, the Moon's gravitational pull is much weaker than Earth's. As we will see, this fact has played an important role in determining the very different evolutionary paths followed by the two worlds. The final column in Table 5.1 lists (sidereal) rotation periods, which have long been accurately known from Earth-based observations.

One important quantity not listed in Table 5.1 is the *distance* from Earth to the Parallax, with Earth's diameter as a baseline, has long provided astronomers with reasonably accurate measurements of the Earth–Moon distance. ∞ *(Sec. 0.4)* However, radar and laser ranging (using lasers instead of radar bounced from the Moon's surface) now yield far more accurate measurements of the Moon's orbit. ∞ *(Sec. 1.3)* Today the Moon's precise distance from Earth at any instant is known to within about 2 cm. The semimajor axis of the Moon's orbit around Earth is 384,000 km.

Overall Structure

Figure 5.1 compares and contrasts the main regions of these two very dissimilar worlds. As indicated in Figure 5.1(a), our planet may be divided into six main regions. In Earth's interior a thick **mantle** surrounds a smaller, two-part **core.** At the surface we have a relatively thin **crust,** comprising the solid continents and the seafloor, and the **hydrosphere,** comprising rivers, lakes, and the liquid oceans. An **atmosphere** of air lies just above the surface. At much greater altitudes a zone of charged particles trapped by our planet's magnetic field forms Earth's **magnetosphere.**

The Moon lacks a hydrosphere, an atmosphere, and a magnetosphere. Its internal structure (Figure 5.1b) is not as well studied as that of Earth, for the very good reason that the Moon is much less accessible. Nevertheless, as we will see, the basic interior regions on Earth—*crust, mantle,* and *core*—also exist on the Moon, although their properties differ somewhat from those of their Earthly counterparts.

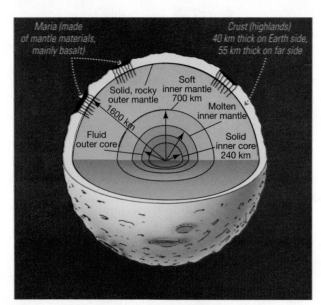

(a)

(b)

◀ **FIGURE 5.1 Earth and Moon** (a) Earth's inner core of radius 1300 km is surrounded by a 2200-km-thick liquid outer core. Most of the rest of Earth's interior is taken up by the mantle, topped by a thin crust only a few tens of kilometers thick. The liquid portions of Earth's surface make up the hydrosphere. Above the hydrosphere and solid crust lies the atmosphere, most of it within 50 km of the surface. Earth's outermost region is the magnetosphere, extending thousands of kilometers into space. (b) The Moon's rocky outer mantle is about 900 km thick. Its inner mantle is a semisolid layer somewhat similar to the upper regions of Earth's mantle.

TABLE 5.1	Some Properties of Earth and the Moon						
	Mass		Radius		Average Density	Surface Gravity	Escape Speed	
	(kg)	(Earth = 1)	(km)	(Earth = 1)	(kg/m³)	(Earth = 1)	(km/s)	Rotation Period
Earth	6.0×10^{24}	1.00	6400	1.00	5500	1.00	11	$23^h\,56^m$
Moon	7.4×10^{22}	0.012	1700	0.27	3300	0.17	2.4	27.3 days

5.2 The Tides

Earth is unique among the planets in having large quantities of liquid water on its surface. Approximately three-quarters of Earth's surface is covered by water, to an average depth of some 3.6 km.

A familiar hydrospheric phenomenon is the daily fluctuation in ocean level known as the **tides.** At most coastal locations on Earth, there are two low tides and two high tides each day. The "height" of the tides—the magnitude of the variation in sea level—can range from a few centimeters to many meters, depending on location and time of year, averaging about a meter on the open sea. A huge amount of energy is contained in this daily tidal motion.

Gravitational Deformation

What causes the tides? A clue comes from the fact that they exhibit daily, monthly, and yearly cycles. In fact, the tides are a direct result of the gravitational influence of the Moon and the Sun on Earth. We have already seen how gravity keeps Earth and the Moon in orbit about one another, and both in orbit around the Sun. Let's first consider just the interaction between Earth and the Moon.

Recall that the strength of the gravitational force depends on the inverse square of the distance separating any two objects. ∞ *(Sec. 1.4)* Thus, the Moon's gravitational attraction is greater on the side of Earth that faces the Moon than on the opposite side, some 12,800 km (Earth's diameter) farther away. This difference in the gravitational force is small, only about 3 percent, but it produces a noticeable deformation—a stretching along the line joining Earth to the Moon called a **tidal bulge.** The effect is greatest in Earth's oceans, because liquid can most easily move around on our planet's surface. As illustrated in Figure 5.2, the ocean becomes a little deeper in some places (along the Earth–Moon line) and shallower in others (in the perpendicular direction). The daily tides we see result as Earth rotates beneath this deformation.

This *differential force,* the variation of the Moon's (or the Sun's) gravity across Earth, is called a **tidal force.** The *average* gravitational interaction between two bodies determines their orbit around one another. However, the tidal force, superimposed on that average, tends to deform the bodies. We will see many situations in this book where such forces are critically important in understanding astronomical phenomena. Note that the tidal force is not in any sense a "new"

DATA POINTS

Tidal Forces

Fifty-four percent of students had difficulty describing the tidal interaction between Earth and the Moon. Some points to remember:

- Tides occur because the inverse-square gravitational force varies with location, so Earth exerts a greater force on the lunar near side than on the far side.

- This differential force tends to stretch the Moon along the Earth–Moon line, creating a tidal bulge.

- Earth's gravitational force on the lunar tidal bulge has synchronized the Moon's rotation with its orbit so that the Moon now always presents the same face toward Earth.

▶ **FIGURE 5.2 Lunar Tides**   This exaggerated illustration shows how the Moon induces tides on both the near and far sides of Earth. (a) The lunar gravitational force is greatest on the side nearest the Moon and smallest on the opposite side. (b) The differences between the lunar forces experienced at the locations shown in part (a) and the force exerted by the Moon on Earth's center. Closest to the Moon, the oceans tend to be pulled away from Earth; on the far side, Earth tends to be pulled away from the oceans. The result is a tidal bulge.

force. It is still the familiar force of gravity, just viewed in a different context—its effect on extended bodies such as planets and moons. We still use the word *tidal* in these other contexts, even though we are not discussing oceanic tides, and sometimes not even planets at all.

Notice in Figure 5.2 that the side of Earth opposite the Moon also experiences a tidal bulge. The different gravitational pulls—greatest on that part of Earth closest to the Moon, weaker at Earth's center, and weakest of all on Earth's opposite side—cause average tides on opposite sides of our planet to be approximately equal in height. On the side nearer the Moon, the ocean water is pulled slightly toward the Moon. On the opposite side, the ocean water is left behind as Earth is pulled closer to the Moon. Thus, high tide occurs *twice*, not once, every day.

Both the Moon and the Sun exert tidal forces on our planet. The tidal force drops off very rapidly with distance, but even though the Sun is around 375 times farther away from Earth than is the Moon, the Sun's mass is so much greater (by about a factor of 27 million) that its tidal influence is still significant—about half that of the Moon. Thus, instead of one tidal bulge being created on Earth, there are two: one pointing toward the Moon, the other toward the Sun. The interaction between them accounts for the changes in the height of the tides over the course of a month or a year. When Earth, Moon, and Sun are roughly lined up—at new or full Moon—the gravitational effects reinforce one another, and the highest tides occur (Figure 5.3a). These tides are known as *spring tides*. When the Earth–Moon line is perpendicular to the Earth–Sun line (at the first and third quarters), the daily tides are smallest (Figure 5.3b). These are termed *neap tides*.

Tidal Locking

The Moon rotates once on its axis in 27.3 days—exactly the same time as it takes to complete one revolution around Earth. ∞ *(Sec. 0.3)* As a result, the Moon presents the same face toward Earth at all times; that is, the Moon has a near side, which is always visible from Earth, and a far side, which never is (Figure 5.4a). This condition, in which the rotation period of a body is precisely equal to (or *synchronized* with) its orbital period around another body, is known as a **synchronous orbit.**

The fact that the Moon is in a synchronous orbit around Earth is no accident. It is an inevitable consequence of the tidal gravitational interaction between those two bodies. To see how this can occur, let's continue our discussion of tides on Earth. Because of Earth's rotation, the tidal bulge raised by the Moon does not point directly at the Moon, as indicated in Figure 5.2. Instead, through the effects of friction, both between the crust and the oceans and within Earth's interior, Earth's rotation tends to drag the tidal bulge around with it, causing the bulge to be displaced slightly ahead of the Earth–Moon line (Figure 5.4b). The Moon's asymmetrical gravitational pull on this slightly offset bulge acts to slow our planet's rotation.

According to fossil measurements, the rate of decrease is just 2 milliseconds every century—not much on the scale of a human lifetime, but it means that half a billion years ago Earth's day was just over 21 hours long and its year contained 410 days. At the same time, the Moon spirals slowly away from Earth, increasing its average distance from our planet by about 4 cm per century. This process will continue until (billions of years from now) Earth rotates on its axis at exactly the same rate as the Moon orbits Earth. Earth's rotation will become synchronized, or **tidally locked,** with the Moon's motion. At that time the Moon

▲ **FIGURE 5.3 Solar and Lunar Tides INTERACTIVE**   The combined effects of the Sun and the Moon produce variations in high and low tides. (a) When the Moon is either full or new, Earth, Moon, and Sun are approximately aligned, and the tidal bulges raised in Earth's oceans by the Moon and the Sun reinforce one another. (b) When the Moon is in either its first or its third quarter, the tidal effects of the Moon and the Sun partially cancel each other, and the tides are smallest.

will always be above the same point on Earth and will no longer lag behind the bulge it raises.

Basically the same process is responsible for the Moon's synchronous orbit. Earth and the Moon have evolved together. Just as the Moon raises tides on Earth, Earth also produces a tidal bulge in the Moon. Indeed, because Earth is so much more massive than the Moon, the lunar tidal bulge is considerably larger, and the synchronization process correspondingly faster. The Moon's rotation became tidally locked to Earth long ago. Most of the moons in the solar system are similarly locked by the tidal forces of their parent planets.

▶ **FIGURE 5.4 Tidal Locking INTERACTIVE**   (a) As the Moon orbits Earth, it keeps one face permanently pointed toward our planet. To the astronaut shown here, Earth is always directly overhead. Because of Earth's tides, the Moon is slightly elongated in shape, with its long axis pointing toward Earth. (b) The tidal bulge raised in Earth by the Moon does not point directly at the Moon. Instead, because of the effects of friction, the bulge points slightly ahead of the Moon, in the direction of Earth's rotation. (Both figures greatly exaggerated for clarity.)

CONCEPT CHECK

In what ways do tidal forces differ from the familiar inverse-square force of gravity?

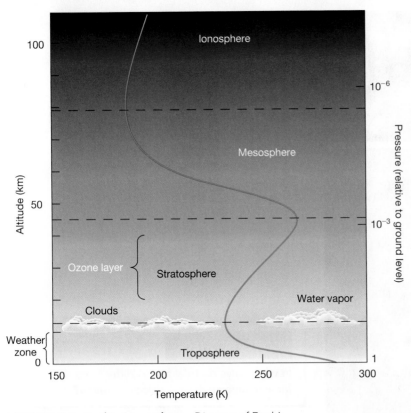

▲ **FIGURE 5.5 Earth's Atmosphere** Diagram of Earth's atmosphere, showing the changes in temperature (blue curve, bottom axis) and pressure (right-hand axis) from the planet's surface to the bottom of the ionosphere. Pressure decreases steadily with increasing altitude, but the temperature may fall or rise, depending on height above the ground.

5.3 Atmospheres

Earth's atmosphere is much more than just air for us to breathe or a blanket to keep us warm. As we will see, this thin layer of gas clinging to the edge of a planet or a moon also shields that body from potentially damaging solar and cosmic radiation and may radically alter conditions on the surface.

Earth's Atmosphere

Our planet's atmosphere is a mixture of gases, the most abundant of which are nitrogen (78 percent by volume), oxygen (21 percent), argon (0.9 percent), and carbon dioxide (0.03 percent). Water vapor is a variable part of the atmosphere, from 0.1 to 3 percent, depending on location and climate. The presence of a large amount of free oxygen makes our atmosphere unique in the solar system—Earth's oxygen is a direct consequence of the emergence of life on our planet.

Figure 5.5 shows a cross-section of Earth's atmosphere. Compared with Earth's overall dimensions, the extent of the atmosphere is tiny. Half of it lies within 5 km of the surface, and all but 1 percent of it is found below 30 km. The region below about 12 km is called the **troposphere.** Everything on Earth's surface lies within it—even Mount Everest, the tallest mountain on our planet, rises only 9 km above sea level. Above the troposphere, extending up to an altitude of 40–50 km, lies the **stratosphere.** Between 50 and 80 km from the surface lies the **mesosphere.** Above about 80 km, in the **ionosphere,** the atmosphere is kept partly ionized by solar ultraviolet radiation. These various atmospheric regions are distinguished from one another by the behavior of the temperature (decreasing or increasing with altitude) in each. Atmospheric pressure decreases steadily with increasing altitude.

The troposphere is the region of Earth's (or any other planet's) atmosphere where **convection** occurs. Convection is the constant upwelling of warm air and the concurrent downward flow of cooler air to take its place. In Figure 5.6, part of Earth's surface is heated by the Sun. The air immediately above the warmed surface is heated, expands a little, and becomes less dense. As a result, it becomes buoyant and starts to rise. At higher altitudes the opposite effect occurs: The air gradually cools, grows denser, and sinks back to the ground. The cool air at the surface rushes in to replace the hot, buoyant air that has risen, establishing a circulation pattern. These **convection cells** of rising and falling air contribute to atmospheric heating and are responsible for surface winds and all the weather we experience. Above the troposphere the atmosphere is stable and the air is calm.

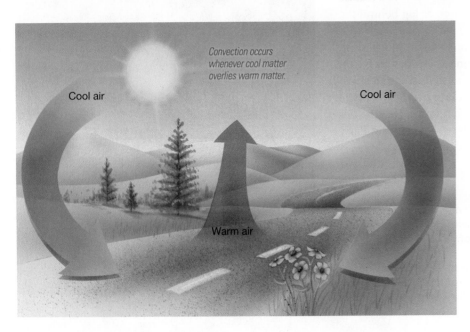

◀ **FIGURE 5.6 Convection** Convection occurs whenever cool matter overlies warm matter. The resulting circulation currents are familiar to us as the winds in Earth's atmosphere, caused by the solar-heated ground. Over and over, hot air rises, cools, and falls back to Earth. Eventually, steady circulation patterns are established and maintained, provided that the source of heat (the Sun, in the case of Earth) remains intact.

5.1 MORE PRECISELY

Why Air Sticks Around

Why does Earth have an atmosphere but the Moon does not? Why doesn't our atmosphere disperse into space? The answer is that *gravity* holds it down. However, gravity is not the only influence acting. If it were, all of Earth's air would have fallen to the surface long ago. *Heat* competes with gravity to keep the atmosphere buoyant. Our atmosphere stays in place because the buoyancy force balances the force of gravity.

Let's explore this competition between gravity and heat in a little more detail. All gas molecules are in constant random motion. The temperature of any gas is a direct measure of this motion—the hotter the gas, the faster the molecules are moving. ∞ *(More Precisely 2-1)* The rapid movement of heated molecules creates *pressure* that tends to oppose the force of gravity, preventing our atmosphere from collapsing under its own weight.

An important measure of the strength of a body's gravity is its *escape speed*—the speed needed for any object to escape forever from its surface. This speed increases with increased mass or decreased radius of the parent body. In fact, the escape speed is proportional to the speed of a circular orbit at the body's surface. In convenient units, it can be expressed as

$$\text{escape speed (in km/s)} = 11.2 \sqrt{\frac{\text{mass of body (in Earth masses)}}{\text{radius of body (in Earth radii)}}}.$$

In other words, you need high speed to escape the gravitational pull of a very massive or very small body, but you can escape from a less massive or larger body at lower speeds. If the *mass* of the parent body is quadrupled, the escape speed doubles. If the parent body's *radius* quadruples, then the escape speed is halved.

To determine whether or not a planet will retain an atmosphere, we must compare the planet's escape speed with the average speed of the gas particles making up the atmosphere. This speed depends not only on the temperature of the gas but also on the mass of the individual molecules—the hotter the gas or the smaller the molecular mass, the higher the average speed of the molecules:

$$\text{average molecular speed (in km/s)} = 0.157 \sqrt{\frac{\text{gas temperature (K)}}{\text{molecular mass (hydrogen atom masses)}}}.$$

Thus, increasing the absolute temperature of a sample of gas by a factor of four—from, for example, 100 K to 400 K—doubles the average speed of its constituent molecules. And at a given temperature, molecules of hydrogen in air move, on average, four times faster than molecules of oxygen, which are 16 times heavier.

At any instant, a tiny fraction of the molecules in any gas have speeds much greater than average, and some molecules are always moving fast enough to escape, even when the average molecular speed is much less than the escape speed. As a result, all planetary atmospheres slowly leak away into space. Don't be alarmed, however—the leakage is usually very gradual. As a rule of thumb, if the escape speed from a planet or moon exceeds the average speed of a given type of molecule by a factor of 6 or more, then molecules of that type will not have escaped from the planet's atmosphere in significant quantities since the solar system formed.

For air on Earth (with a temperature of about 300 K), the mean molecular speed of oxygen (mass = 32 hydrogen masses) and nitrogen (mass = 28) is about 0.6 km/s, comfortably below one-sixth of the escape speed (11.2 km/s). As a result, Earth is able to retain its atmosphere. However, if the Moon originally had an Earth-like atmosphere, the lunar escape speed of just 2.4 km/s means that any original atmosphere long ago dispersed into interplanetary space. The same reasoning can be used to understand atmospheric composition. For example, hydrogen molecules (mass = 2) move, on average, at about 2 km/s in Earth's atmosphere at sea level. Consequently, they have had plenty of time to escape since our planet formed, which is why we find very little hydrogen in Earth's atmosphere today.

Within the stratosphere lies the **ozone layer,** where atmospheric oxygen, ozone, and nitrogen absorb incoming solar ultraviolet radiation. Ozone is a form of oxygen. Most atmospheric oxygen consists of two oxygen atoms combined into a molecule; ozone consists of three oxygen atoms combined into each molecule. Ozone is formed in the atmosphere when solar ultraviolet radiation interacts with oxygen molecules. Its concentration is greatest at an altitude of around 25 km. The ozone layer is one of the insulating layers that protect life on Earth from the harsh environment of outer space. By absorbing potentially dangerous high-frequency radiation, it acts as a planetary umbrella. Without it, advanced life (at least on Earth's surface) would be at best impaired and at worst impossible. *Discovery 5-1* discusses how human activities may now be affecting this vital part of our atmosphere.

5.1 DISCOVERY

Earth's Growing Ozone "Hole"

During the last two centuries, human technology has begun to produce measurable—and possibly permanent—changes in our planet. You can probably think of many instances, mostly negative, of such changes, from the threat of nuclear war to the reality of air and water pollution worldwide. One example very much in the news and relevant to the discussion in this chapter is the depletion of Earth's ozone layer. (See *Discovery 5–2* for an unrelated, but even more serious, threat to our environment.)

A particularly undesirable by-product of human technology is a group of chemicals known as chloro-fluorocarbons (CFCs), relatively simple compounds once widely used for a variety of purposes—propellant in aerosol cans, dry-cleaning products, and in air conditioners and refrigerators. In the 1970s scientists discovered that, instead of quickly breaking down after use as had previously been thought, CFCs accumulate in the atmosphere and are carried high into the stratosphere by convection. There they are broken down by sunlight, releasing chlorine, which quickly reacts with ozone (O_3; see Section 5.3), turning it into oxygen (O_2). In chemical terms the chlorine is said to act as a *catalyst*—it is not consumed in the reaction, and so it survives to react with many more ozone molecules. A single chlorine atom can destroy up to 100,000 ozone molecules before being removed by other, less frequent chemical reactions. In this way, even a small quantity of CFCs turns out to be extraordinarily efficient at destroying atmospheric ozone.

As we saw in the text, ozone is part of the protective "blanket" of gases in our atmosphere that protect us from the harsh realities of outer space—in this case, solar ultraviolet radiation. Thus, the result of CFC emission is a substantial increase in ultraviolet radiation levels at Earth's surface, with detrimental effects to most living organisms. The accompanying figure shows the development of an ozone "hole" (colored magenta) over the Antarctic—a region where atmospheric circulation and low temperatures conspire each Antarctic spring to create a vast circumpolar cloud of ice crystals, that act to promote the ozone-destroying reactions, resulting in ozone levels about 50 percent below normal for the region.

Since the hole (really a relative lack of ozone) was discovered in the 1980s, its depth and area have grown each year. Its peak size is now larger than North America. Ozone depletion is not confined to the Antarctic, although the effect is greatest there. Smaller holes have been observed in the Arctic, and occasional ozone depletions of up to 20 percent have been reported at lower northern latitudes.

In the late 1980s, when the effect of CFCs on the atmosphere was realized, the world moved remarkably rapidly to curtail their production and use, with the goal of phasing them out entirely by 2030. Substantial cuts have already been made. Still, scientists think that even if all remaining CFC emissions were to stop today, it would still take several decades for CFCs to leave the atmosphere completely.

The Greenhouse Effect

The Sun emits most of its energy in the visible and near-infrared (that is, wavelengths only slightly longer than red light) regions of the electromagnetic spectrum. Because Earth's atmosphere is largely transparent to radiation of this type, almost all of the solar radiation not absorbed by or reflected from clouds in the upper atmosphere shines directly onto Earth's surface, heating it up. ∞ *(Sec. 2.3)*

Earth's surface reradiates the absorbed energy. As the temperature rises, the amount of radiated energy increases rapidly, according to Stefan's law, and eventually our planet radiates as much energy back into space as it receives from the Sun. In the absence of any complicating effects, this balance would be achieved at an average surface temperature of about 250 K (–23°C). At that temperature,

Wien's law tells us that most of the reemitted energy is in the form of far-infrared (roughly 10 μm) radiation. ∞ *(More Precisely 2-2)*

However, there is a complication. Long-wavelength infrared radiation is partially blocked by Earth's atmosphere Fig. 2.9, mainly because of carbon dioxide and water vapor, both of which absorb very efficiently in the infrared portion of the spectrum. Even though these two gases account for only a tiny fraction of our atmosphere, they manage to absorb a large fraction of all the infrared radiation emitted from the surface. Consequently, only some of that radiation escapes back into space. The rest is radiated back to the surface, causing the temperature to rise.

This partial trapping of solar radiation is known as the **greenhouse effect.** The name comes about because a similar process operates in a greenhouse—sunlight passes relatively unhindered through glass panes, but much of the infrared radiation reemitted by the plants is blocked by the glass and cannot get out.[1] Consequently, the interior of the greenhouse heats up, and flowers, and vegetables can grow even on cold winter days. The radiative processes that determine the temperature of Earth's atmosphere are illustrated in Figure 5.7. Earth's greenhouse effect makes our planet about 40 K hotter than would otherwise be the case—a very important difference, as it raises the average temperature *above* the freezing point of water.

The magnitude of the greenhouse effect is very sensitive to the concentration of *greenhouse gases* (that is, gases that absorb infrared radiation efficiently) in the atmosphere. As just mentioned, water vapor and carbon dioxide are both important greenhouse gases. Carbon dioxide is of particular concern today, because its concentration in Earth's atmosphere is increasing, largely as a result of the burning of fossil fuels (principally oil and coal) in the industrialized world. Carbon dioxide levels have increased by more than 20 percent in the last century, and they continue to rise at a present rate of 4 percent per decade. Many scientists think that this increase, if left unchecked, may result in global temperature increases of several kelvins over the next half-century, enough to melt much of the polar ice caps and cause dramatic, perhaps even catastrophic, changes in Earth's climate (see *Discovery 5-2*).

Lunar Air?

What about the Moon's atmosphere? That's easy—for all practical purposes, there is none! All of it escaped long ago. More massive objects have a better chance of retaining their atmospheres because the more massive an object, the greater the speed needed for atoms and molecules to escape (see *More Precisely 5-1*). The Moon's escape speed is only 2.4 km/s, compared with 11.2 km/s for Earth (Table 5.1). Simply put, the Moon has a lot less pulling power—any atmosphere it might once have had is gone forever.

Lacking the moderating influence of an atmosphere, the Moon experiences wide variations in surface temperature. Noontime temperatures can reach 400 K, well above the boiling point of water (373 K). ∞ *(More Precisely 2-1)* At night (which lasts nearly 14 Earth days) or in the shade, temperatures fall to about 100 K, well below water's freezing point (273 K). With such high daytime temperatures and no atmosphere to help retain it, any water that ever existed on the Moon's surface has most likely evaporated and escaped. Not only is there no lunar hydrosphere, but all the lunar samples returned by the U.S. and Soviet Moon programs were absolutely bone dry. Lunar rock doesn't even contain minerals with water molecules locked within their crystal structure. Terrestrial rocks, conversely, are almost always 1 or 2 percent water.

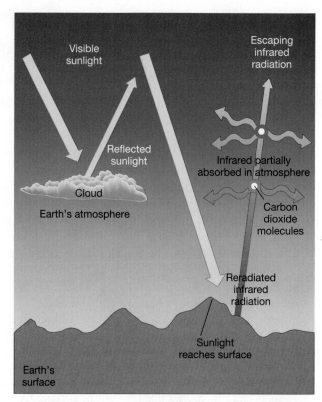

▲ **FIGURE 5.7 Greenhouse Effect INTERACTIVE** Sunlight not reflected by clouds reaches Earth's surface, warming it up. Infrared radiation reradiated from the surface is partially absorbed by carbon dioxide in the atmosphere, causing the overall surface temperature to rise.

(MA) **SELF-GUIDED TUTORIAL**
The Greenhouse Effect

(MA) **ANIMATION/VIDEO**
Seasonal Changes on the Earth

(MA) **SELF-GUIDED TUTORIAL**
Atmospheric Lifetimes

(MA) **ANIMATION/VIDEO**
Ozone Hole Over the Antarctic

CONCEPT CHECK

Why is the greenhouse effect important for life on Earth? What might be the consequences if the effect continues to strengthen?

---

[1]Actually, although this process does contribute to warming the interior of a greenhouse, it is not the most important effect. A greenhouse works mainly because its glass panes prevent convection from carrying heat up and away from the interior. Nevertheless, the name "greenhouse effect" to describe the heating effect due to Earth's atmosphere has stuck.

5.2 DISCOVERY

The Greenhouse Effect and Global Warming

Discovery 5-1 outlined the danger to Earth's ozone layer posed by CFCs—products of modern technology with unexpected global consequences. However, an older, and potentially far more serious, hazard comes from humanity's contribution to the *greenhouse effect* on our planet.

We saw the greenhouse effect in Section 5.3: so-called greenhouse gases in Earth's atmosphere, notably water vapor, carbon dioxide (CO_2), and methane (CH_4) tend to trap heat leaving the surface, raising our planet's temperature by several tens of degrees Celsius. The greenhouse effect is not a bad thing— it is the reason that water exists in the liquid state on Earth's surface, and thus it is crucial to the existence and survival of life on our planet (see Chapter 18). However, if atmospheric greenhouse gas levels rise unchecked, the consequences can be catastrophic (for a particularly extreme example, see Section 6.8).

Since the Industrial Revolution in the 18th century, and particularly over the past few decades, human activities on Earth have steadily raised the level of carbon dioxide in our atmosphere, as illustrated in the first figure. Fossil fuels (coal, oil, and gas), still the dominant energy source of modern industry, all release CO_2 when burned, and the present rate of increase of atmospheric CO_2 is consistent with our current consumption of fossil-fuel energy. At the same time, the extensive forests that once covered much of our planet are being systematically destroyed to make room for human expansion. Forests play an important role in this problem because vegetation absorbs carbon dioxide, thus providing a natural control mechanism for atmospheric CO_2. Deforestation therefore also tends to increase the level of greenhouse gases in Earth's atmosphere.

Global warming is the slow rise in Earth's surface temperature caused by the increased greenhouse effect resulting from higher levels of atmospheric carbon dioxide. As shown in the second figure, average global temperatures have risen by more than 1°C during the past century. This may not seem like much, but climate models predict that if CO_2 levels continue to rise, a further increase of as much as 5°C may be possible by the end of the 21st century. Such a rise would be enough to cause serious climatic change on a global scale.

We saw in *Discovery 5-1* how, once the environmental impact of CFCs was identified, rapid steps were taken to curb their use. Curiously, given the more serious nature of the potential damage, a concerted response to global warming has been much slower in coming. Most scientists see the human-enhanced greenhouse effect as a real threat to Earth's climate and urge prompt and deep reductions in CO_2 emissions, along with steps to slow and ultimately reverse deforestation. Some, however, particularly those connected with the industries most responsible for the production of greenhouse gases, argue that Earth's long-term response to increased greenhouse emissions is too complex for simple conclusions to be drawn and that immediate action is unnecessary. They suggest that the current temperature trend may be part of some much longer cycle or that natural environmental factors may in time stabilize, or even reduce, the level of CO_2 in the atmosphere without human intervention.

Given the stakes, it is perhaps not surprising that these arguments have become far more political than scientific in tone. The basic observations and much of the basic science are generally not seriously questioned, but the interpretation, long-term consequences, and proper response are all hotly debated. Separating the issues sometimes is not easy, but the outcome may be of vital importance to life on Earth.

The Moon is not entirely devoid of water, however. In November 1996, radar echoes from the U.S. *Clementine* spacecraft suggested the presence of water ice at the lunar poles. Since the Sun never rises more than a few degrees above the horizon, as seen from these regions, temperatures on the permanently shaded floors of craters near the poles never exceed about 100 K. Consequently, scientists theorized, ice could have remained permanently frozen there since being dumped on the Moon by comets during the very early days of the solar system, never melting or vaporizing and hence never escaping into space. ∞ *(Sec. 4.3)* In March 1998, NASA announced that the *Lunar Prospector* mission had confirmed *Clementine's* findings, detecting large amounts of water ice—possibly totaling trillions of tons—at both lunar poles, lying perhaps half a meter below the surface.

In 2009, in an attempt to gain more information about lunar ice, NASA placed the *Lunar Reconnaissance Orbiter (LRO)* spacecraft into a polar orbit just 50 km above the Moon's surface. During its 1-year mission, *LRO* collected detailed information on the lunar surface, with particular emphasis on the polar regions. In late 2009, mission scientists carried out a very direct experiment in their search for ice. The Centaur rocket that boosted both *LRO* and its sister spacecraft, *LCROSS* (the *Lunar CRater Observation and Sensing Satellite*), into lunar orbit was crashed into a deeply shadowed crater near the lunar south pole. *LCROSS* followed and watched from a few thousand kilometers behind, radioing detailed spectroscopic data back to Earth via *LRO* until it too impacted the Moon minutes later. Detailed analysis of the *LCROSS* data confirmed the presence of water molecules in the ejecta. The amount of water was not great—only about 1 part in 100,000, less than in desert sand on Earth—but was more than enough to corroborate the earlier reports.

5.4 Internal Structure of Earth and the Moon

Although we reside on Earth, we cannot easily explore our planet's interior, because drilling gear can penetrate rock only so far before breaking. No substance—not even diamond, the hardest known material—can withstand the conditions below a depth of about 10 km. That's rather shallow compared with Earth's 6400-km radius. Fortunately, geologists have developed techniques that can indirectly probe the deep recesses of our planet.

Seismology

A sudden dislocation of rocky material in Earth's crust—an **earthquake**—causes the entire planet to vibrate. It rings like a giant bell (but one tuned so low that human ears cannot detect the sound). These vibrations are not random. They are systematic waves, called **seismic waves** (after the Greek word for "earthquake"), that move outward from the site of the quake. Like all waves, they carry information. This information can be detected and recorded using sensitive equipment—a *seismograph*—designed to monitor Earth tremors.

Decades of earthquake research have demonstrated the existence of many kinds of seismic waves. Two are of particular importance to the study of Earth's internal structure (Figure 5.8). First to arrive at a monitoring site after a distant earthquake are the primary waves, or *P-waves*. These are *pressure* waves, a little like ordinary sound waves in air, that alternately expand and compress the medium (the core or mantle) through which they move. Seismic P-waves travel at 5–6 km/s and can travel through both liquids and solids. Some time later, secondary waves, or *S-waves*, arrive. These are *shear* waves causing side-to-side motion, like waves in a guitar string. S-waves travel through Earth's interior at 3–4 km/s; however, they cannot travel through liquid, which absorbs them. The speed of each type depends on the density and physical state of the matter through which it is traveling. By measuring the time taken for waves to move from the site of an earthquake to one or more monitoring stations on Earth's surface, geologists can infer the density of matter in the interior.

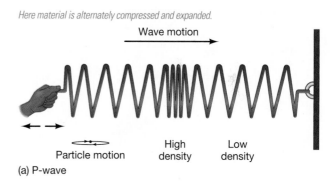

Here material is alternately compressed and expanded.

Wave motion →

Particle motion — High density — Low density

(a) P-wave

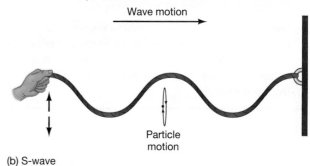

Here material moves up and down.

Wave motion →

Particle motion

(b) S-wave

▲ **FIGURE 5.8 P- and S-waves** (a) Pressure (P) waves traveling through Earth's interior cause material to vibrate in a direction parallel to the direction of motion of the wave. (b) Shear (S) waves produce motion perpendicular to the direction in which the wave travels, pushing material from side to side.

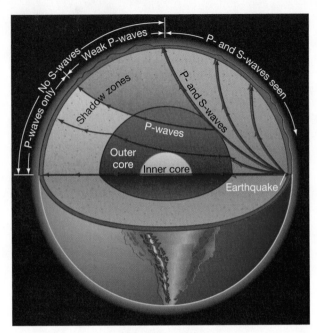

▲ FIGURE 5.9 Seismic Waves Earthquakes generate pressure (P, or primary) and shear (S, or secondary) waves that can be detected at seismographic stations around the world. The waves bend while moving through Earth's interior because of the variation in density and temperature within our planet. S-waves (red) are not detected by stations "shadowed" by the liquid core of Earth. P-waves (green) do reach the side of Earth opposite the earthquake, but their interaction with Earth's core produces another shadow zone, where almost no P-waves are registered.

PROCESS OF SCIENCE CHECK

Describe how scientists combine theory and observation to create models of Earth's interior. Give examples of some Earth properties that we actually observe and some that are derived purely from the models.

Modeling Earth's Interior

Because earthquakes occur often and at widespread places across the globe, geologists have accumulated a large amount of data about seismic-wave properties. They have used these data, along with direct knowledge of surface rocks, to model Earth's interior. Our knowledge of the deepest recesses of our planet is based almost entirely on modeling and indirect observation.

Figure 5.9 illustrates some paths followed by seismic waves from the site of an earthquake. Monitoring stations on the side of Earth opposite a quake never detect S-waves (colored red in the figure)—they are blocked by material within Earth's interior. Furthermore, while P-waves (green) always arrive at stations diametrically opposite the quake, other parts of Earth's surface receive almost none. Most geologists think that S-waves are absorbed by a liquid core at Earth's center and that P-waves are refracted at the core boundary, much as light is refracted by a lens. The result is the S- and P-wave "shadow zones" we observe. The fact that every earthquake exhibits these shadow zones is the best evidence that the core of our planet is molten. The radius of this **outer core,** as determined from seismic data, is about 3500 km. There is also evidence that the P-waves, which can pass through the liquid outer core, are reflected off the surface of a solid **inner core** of radius 1300 km.

Figure 5.10 presents a model that most scientists accept. Earth's outer core is surrounded by a thick mantle and topped with a thin crust. The mantle is about 3000 km thick and accounts for the bulk (80 percent) of our planet's volume. The crust has an average thickness of only 15 km—a little less (around 8 km) under the oceans and somewhat more (20–50 km) under the continents. Note just how thin the crust is relative to the size of our planet: 15 km, about twice the height of the tallest mountains on our planet, is a mere one quarter of 1 percent of Earth's radius. The average density of crust material is around 3000 kg/m^3. Density and temperature both increase with depth. From Earth's surface to its very center, the density increases from roughly 3000 kg/m^3 to a little more than 12,000 kg/m^3, while the temperature rises from just under 300 K to well over 5000 K. Much of the mantle has a density midway between the densities of the core and crust: about 5000 kg/m^3.

The high central density suggests to geologists that Earth's core must be rich in nickel and iron. Under the heavy pressure of the overlying layers, these metals (whose densities under surface conditions are around 8000 kg/m^3) are compressed to the high densities predicted by the model. The sharp density increase at the core–mantle boundary results from the difference in composition between these two regions. Unlike the dense metallic core, the mantle is composed of less dense, rocky material. Note that there is no comparable jump in density or temperature at the inner core boundary—the material there simply changes from liquid to solid.

Because geologists have been unable to drill deeper than about 10 km, no experiment has yet recovered a sample of the mantle. However, we are not entirely ignorant of the mantle's properties. In a **volcano,** molten rock upwells from below the crust, bringing a little of the mantle to us in the form of lava and providing some inkling of the composition of Earth's interior. The composition of the upper mantle is probably quite similar to the dark gray material known as *basalt* often found near volcanoes. With a density between 3000 and 3300 kg/m^3, it contrasts with the lighter *granite* (density 2700–3000 kg/m^3) that constitutes much of Earth's crust.

Differentiation

Earth, then, is not a homogeneous ball of rock. Instead, it has a layered structure, with a low-density rocky crust at the surface, intermediate-density rocky material in the mantle, and a high-density metallic core. (As we will see, the other terrestrial planets have similar internal structures.) This variation in density and composition is known as **differentiation.** Why isn't our planet just one big, rocky ball of uniform density? The answer is that at some time in the distant past, much of Earth was *molten,* allowing the higher-density matter to sink to the core, displacing lower-density material toward the surface. A remnant of this ancient heating exists today: Earth's central temperature is nearly equal to that of the Sun's surface. Two processes played important roles in heating Earth to the point where differentiation could occur. First, very early in its history, our planet experienced a violent bombardment by interplanetary debris, an essential part of the process by which Earth and the other planets formed and grew. ∞ *(Sec. 4.3)* This bombardment probably generated enough heat to melt much of our planet.

The second process that heated Earth after its formation and contributed to differentiation was **radioactivity**—the release of energy by certain unstable elements, such as uranium and thorium, that were present in the solar nebula when Earth formed. ∞ *(Sec. 4.3)* These elements emit energy as their complex, heavy nuclei break up into simpler, lighter ones. Rock is such a poor conductor of heat that the energy released through radioactivity took a very long time to reach the surface and leak away into space. As a result, the heat built up in the interior, adding to the energy left there by Earth's formation. Geologists think that enough radioactive elements were originally spread throughout the primitive planet, like raisins in a cake, that the entire planet—from crust to core—could have melted and remained molten, or at least semisolid, for about a billion years.

The Lunar Interior

The Moon's average density, about 3300 kg/m³, is quite similar to the density of the lunar surface rock obtained by U.S. and Soviet missions. This similarity all but eliminates any chance that the Moon has a large, massive, dense nickel–iron core like that within Earth. The low average lunar density suggests that the Moon contains substantially fewer heavy elements (such as iron) than does Earth.

Much of our detailed information on the Moon's interior comes from seismic data obtained from equipment left on the lunar surface by astronauts. These measurements indicate only very weak *moonquakes* deep within the lunar interior. Even if you stood directly above one of these quakes, you would not feel the vibrations. The average moonquake releases about as much energy as a firecracker, and no large quakes have ever been detected. This barely perceptible seismic activity confirms the idea that the Moon is geologically dead. Nevertheless, researchers can use these weak lunar vibrations to obtain information about the Moon's interior.

Combining all available data with mathematical models of the lunar interior indicates that most of the Moon is of almost uniform density, but chemically differentiated—that is, the chemical properties change from core to surface. As noted in Figure 5.1(b), the models suggest a central core about 330 km in radius, surrounded by a 400-km-thick inner mantle of semisolid rock having properties similar to those of Earth's upper mantle. Above these regions lies a 900-km-thick outer mantle of solid rock, topped by a 40-km-thick crust.

Data from the gravity experiment aboard the *Lunar Prospector* probe (Section 5.3), combined with magnetic measurements made as the Moon passed through Earth's magnetic "tail" (see Figure 5.23), imply that the lunar core is denser and more iron-rich than the rest of the Moon. Theoretical models predict a central temperature as low as 1500 K, too cool to melt rock or iron, but careful analysis of the *Apollo* seismic data suggests that parts of the core may be molten, indicating a higher temperature, perhaps due to heating by radioactive elements. As best we can tell, the Moon has a solid, mainly iron inner core roughly 240 km in radius. The rest of the core, as well as the innermost 150 km of the surrounding inner mantle, is liquid.

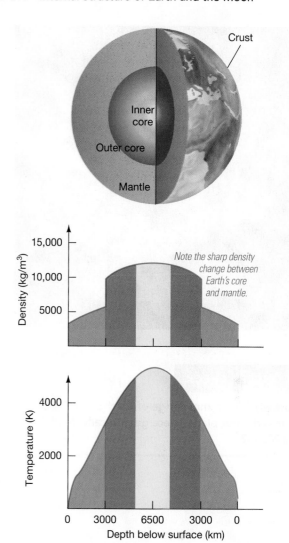

▲ **FIGURE 5.10 Earth's Interior** Computer models of Earth's interior imply that the density and temperature vary greatly through the mantle and the core.

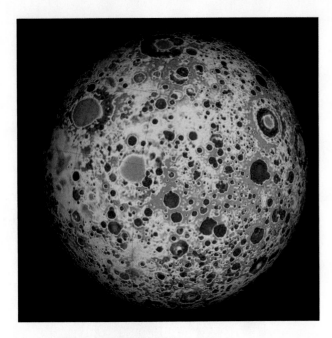

NASA's *Gravity Recovery and Interior Laboratory (GRAIL)* mission placed two small spacecraft (called *Ebb* and *Flow*) in coordinated orbits around the Moon for several months in 2012. They measured the lunar gravity field in unprecedented detail, giving scientists a new view of the Moon's subsurface structure. Figure 5.11 is a map based on *GRAIL* data, showing deviations of the Moon's gravity from the average value, allowing otherwise invisible mass concentrations to be detected. *GRAIL* found that the Moon's crust is fractured and shattered to a depth of several kilometers. In addition, while the crust averages about 30 km thick, its thickness is quite variable, ranging from as much as 60 km in some locations on the far side to almost zero under the larger near-side basins.

The *GRAIL* data also indicate that, on average, the crust on the lunar far side is some 10-15 km thicker than on the side facing Earth. The reason is most likely related to Earth's gravitational pull. Just as heavier material tries to sink to the center of Earth, the denser far-side lunar mantle tended to sink below the lighter far-side crust in the presence of Earth's gravitational field. In other words, while the Moon was cooling and solidifying, the far-side lunar mantle was pulled a little closer to Earth than the far-side crust. In this way, the crust and mantle became slightly off-center with respect to one another. The result was the thicker crust on the Moon's far side.

CONCEPT CHECK

How would our knowledge of Earth's interior be changed if our planet were geologically dead, like the Moon?

5.5 Surface Activity on Earth

Earth is geologically alive today. Its interior seethes, and its surface constantly changes. Many clear indicators of geological activity, in the form of earthquakes and volcanic eruptions, are scattered across our globe. Erosion by wind and water has obliterated much of the evidence from ancient times, but the sites of more recent activity are well documented.

Continental Drift

Figure 5.12 is a map of the currently active areas of our planet. Nearly all these sites have experienced surface activity during this century. The intriguing aspect of the figure is that the active sites are not spread evenly across our planet. Instead, they trace well-defined lines of activity, where crustal rocks shift (resulting in earthquakes) or mantle material upwells (in volcanoes).

In the mid-1960s, scientists realized that these lines are the outlines of gigantic *plates*, or slabs of Earth's surface and that these plates are slowly drifting around the surface of our planet. These plate motions, popularly known as *continental drift*, have created the mountains, oceanic trenches, and many other large-scale features across the face of planet Earth. The technical term for the study of plate movement and the reasons for it is **plate tectonics.** Some plates are made mostly of continental landmasses, some are made of a continent plus a large part of an ocean floor, and some contain no continental land at all and are made solely of ocean floor. For the most part, the continents are just passengers riding atop much larger plates.

The plates move at an extremely slow rate. Typical speeds amount to only a few centimeters per year—about the same speed as your fingernails grow. Nevertheless, over the course of Earth history, the plates have had plenty of time to

◄ **FIGURE 5.12 Global Plates**   The red dots represent active sites where major volcanoes or earthquakes occurred in the 20th century. Taken together, the sites outline vast "plates," indicated in dark blue, that drift around on the surface of our planet. The white arrows show the general directions and speeds of the plate motions.

move large distances. For example, a drift rate of 2 cm per year can cause two continents (for example, Europe and North America) to separate by 4000 km—the width of the Atlantic Ocean—in 200 million years. That's a long time by human standards, but it represents only about 5 percent of Earth's age.

As the plates drift around, we might expect collisions to be routine. Indeed, plates do collide, but they are driven by enormous forces and do not stop easily. Instead, they just keep crunching into one another. Figure 5.13 shows the result of a collision between ancient continents in what is now western America. As Earth's rocky crust crumbles and folds, mountains are formed—in this case, the Rocky Mountains that range from New Mexico to British Columbia, where the surface has been lifted nearly 5 km by tectonic forces.

Not all plates collide head-on. As indicated by the arrows on Figure 5.12, sometimes plates slide or shear past one another. A good example is the most famous active region in North America—the San Andreas Fault in California (Figure 5.14), which forms part of the boundary between the Pacific and North American plates. Along the fault, these two plates are not moving in quite the same direction, nor at quite the same speed. Like parts in a poorly oiled machine, their motion is neither steady nor smooth. Instead, they tend to stick, then suddenly lurch forward as surface rock gives way. The resulting violent, jerky movements have been the cause of many major earthquakes along the fault line.

At still other locations, such as the seafloor under the Atlantic Ocean, the plates are moving apart. As they recede, new mantle material wells up between them, forming midocean ridges. Today, hot mantle material is rising through a crack all along the Mid-Atlantic Ridge, which extends, like a seam on an enormous baseball, all the way from the North Atlantic to the southern tip of South America. Radioactive dating indicates that material has been upwelling along the ridge more or less steadily for the past 200 million years. The Atlantic seafloor is slowly growing as the North and South American plates move away from the Eurasian and African plates.

R I V U X G

▲ **FIGURE 5.13 Rocky Mountains**   Mountain building results mostly from plate collisions. Here, the folding and jostling of rock is visible high in the northern Rockies, a mountain range upthrust about 70 million years ago. For scale, note at bottom right the evergreen trees, which are about 10 m tall. *(M. Chaisson)*

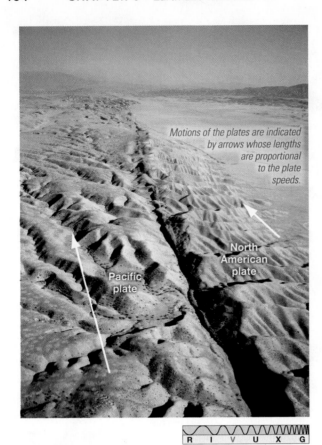

Motions of the plates are indicated by arrows whose lengths are proportional to the plate speeds.

North American plate

Pacific plate

R   I   V   U   X   G

▲ **FIGURE 5.14 Californian Fault**   A small portion of the San Andreas fault in California. The fault is the result of the North American and Pacific plates sliding past one another. The Pacific plate, which includes a large slice of the California coast, is drifting to the northwest relative to the North American plate. *(D. Parker/Science Photo Library)*

What Drives the Plates?

What is responsible for the enormous forces that drag plates apart in some locations and ram them together in others? The answer (Figure 5.15) is *convection*—the same process we encountered earlier in our study of Earth's atmosphere (Section 5.3). Each plate is made up of crust plus a small portion of upper mantle. Below the plates, at a depth of perhaps 50 km, the temperature is sufficiently high that the mantle at that depth is soft enough to flow, very slowly, although it is not molten.

This is a perfect setting for convection—warm matter underlying cool matter. The warm mantle rock rises, just as hot air rises in our atmosphere, and large circulation patterns become established. Riding atop these convection patterns are the plates. The circulation is extraordinarily sluggish. Semisolid rock takes millions of years to complete one convection cycle. Although the details are far from certain and remain controversial, many researchers suspect that it is large-scale convection patterns near plate boundaries that cause the plates to move.

Figure 5.16 illustrates how all the continents nearly fit together like pieces of a puzzle. Geologists think that sometime in the past a single gargantuan landmass dominated our planet. This ancestral supercontinent, known as *Pangaea* (meaning "all lands"), is shown in Figure 5.16(a). The rest of the planet was covered with water. The present locations of the continents, along with measurements of their current drift rates, suggest that Pangaea was the major land feature on Earth approximately 200 million years ago. Dinosaurs, which were then the dominant form of life, could have sauntered from Russia to Texas via Boston without getting their feet wet. The other frames in Figure 5.16 show how Pangaea split apart, its separate pieces drifting across Earth's surface, eventually becoming the familiar continents we know today. Very probably there has been a long series of "Pangaeas" stretching back in time over much of Earth's history, as tectonic forces have continually formed, destroyed, and re-formed our planet's landmasses. Probably there will be many more.

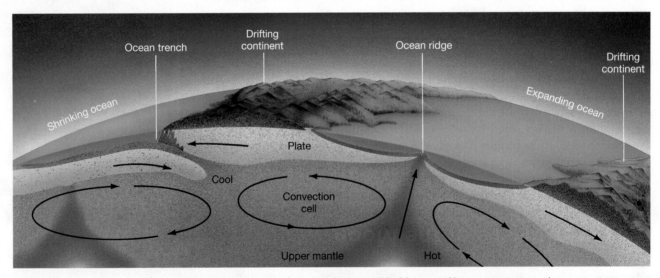

▲ **FIGURE 5.15 Plate Drift INTERACTIVE**   The motion of Earth's plates is probably caused by convection—in this case, giant circulation patterns in the upper mantle that drag the plates across the surface.

▲ FIGURE 5.16 **Pangaea** Given the current estimated drift rates and directions of the plates, we can trace their movements back into the past. About 200 million years ago, they would have been at the approximate positions shown in (a). The continents' current positions are shown in (d).

Plate Tectonics on the Moon

There is no evidence for plate tectonics on the Moon today—no obvious extensive fault lines, no significant seismic activity, no ongoing mountain building. Plate tectonics requires both a relatively thin outer rocky layer, which is easily fractured into continent-sized pieces, and a soft, convective region under it, to make the pieces move. On the Moon neither of these ingredients exists. The Moon's thick crust and solid upper mantle make it impossible for pieces of the surface to move relative to one another. There simply isn't enough energy left in the lunar interior for plate tectonics to work.

5.6 The Surface of the Moon

On Earth the combined actions of air, water, and geological activity erode our planet's surface and reshape its appearance almost daily. As a result, most of the ancient history of our planet's surface is lost to us. The Moon, though, has no air, no water, and no ongoing volcanic or other geological activity (although, as we will see, there is strong evidence that the young Moon underwent a period of intense geological activity before settling into its current inactive state). Consequently, features dating back almost to the formation of the Moon are still visible today. For this reason, studies of the lunar surface are of great importance to Earth geologists and have played a major role in shaping theories of the early development of our planet.

Large-Scale Features

The first observers to point their telescopes at the Moon noted large, roughly circular, dark areas that resemble (they thought) Earth's oceans. They called these regions **maria,** a Latin word meaning "seas" (singular: *mare*). The largest of them (Mare Imbrium, the Sea of Showers) is about 1100 km in diameter. Today we know that the maria are actually extensive flat plains that resulted from the spread of lava during an earlier, volcanic period of lunar evolution (Section 5.8). In a sense, then, the maria *are* oceans—ancient seas of molten lava, now solidified.

Early observers also saw light-colored areas that resembled Earth's continents. Originally dubbed *terrae*, from the Latin word for "land," these regions are now known to be elevated several kilometers above the maria. Accordingly, they are usually called the lunar **highlands.** Both types of region are visible in Figure 5.17, a photographic mosaic of the full Moon. These light and dark surface features are also evident to the naked eye, creating the face of the familiar "man-in-the-Moon."

Based on analyses of lunar rock brought back to Earth by *Apollo* astronauts and robot Soviet landers, geologists have found important differences in both

PROCESS OF SCIENCE CHECK

Go to a library or the Web and research the work of Alfred Wegener on the theory of continental drift—the forerunner of modern plate tectonics. What seemingly unrelated facts did the theory connect and explain? Why did the theory encounter such resistance in the scientific community?

CONCEPT CHECK

Describe the causes and some consequences of plate tectonics on Earth.

 ANIMATION/VIDEO
One Small Step

▲ FIGURE 5.17 **Full Moon, Near Side** A photographic mosaic of the full Moon, north pole at the top. Because the Moon emits no visible radiation of its own, we can see it only by the reflected light of the Sun. (*UC/Lick Observatory*)

R I V U X G

▲ **FIGURE 5.18 Full Moon, Far Side**   This image of the far side of the Moon, which we can never see from home, is actually a collection of more than 15,000 small photos taken by the *Lunar Reconnaissance Orbiter*. Only a few small maria exist on the heavily cratered far side. *(NASA)*

composition and *age* between the highlands and the maria. The highlands are made largely of rocks rich in aluminum, making them lighter in color and lower in density (2900 kg/m³). The maria's basaltic matter contains more iron, giving it a darker color and greater density (3300 kg/m³). Loosely speaking, the highlands represent the Moon's crust, while the maria are made of mantle material. Maria rock is quite similar to terrestrial basalt, and geologists think that it arose on the Moon much as basalt did on Earth, through the upwelling of molten material through the crust. Radioactive dating indicates ages of more than 4 billion years for highland rocks and from 3.2 to 3.9 billion years for those from the maria.[2]

Until spacecraft flew around the Moon, no one on Earth had any idea what the Moon's hidden half looked like. To the surprise of most astronomers, when the far side of the Moon was mapped, first by Soviet and later by U.S. spacecraft, no major maria were found. The lunar far side (Figure 5.18) is composed almost entirely of highlands.

Cratering

Because the smallest lunar features we can distinguish with the naked eye are about 200 km across, we see little more than the maria and highlands when we gaze at the Moon. Through a telescope, however (Figure 5.19), we find that the lunar surface is scarred by numerous bowl-shaped **craters** (after the Greek word for "bowl").

Most craters formed eons ago as the result of meteoritic impact. ∞ *(Sec. 4.2)* (Compare the ancient craters in Figures 5.19c and 5.21 with the much more recent

---

[2] Radioactive dating compares the rates at which different radioactive elements in a sample of rock decay into lighter elements. The age returned by this technique is the time since the rock solidified.

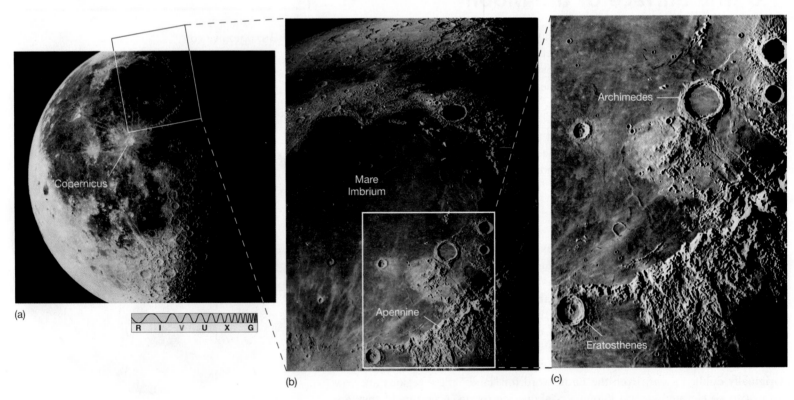

(a)

R I V U X G

(b)

(c)

ANIMATION/VIDEO (MA)
Full Rotation of the Moon

ANIMATION/VIDEO (MA)
Ranger Spacecraft Descent to Moon

▲ **FIGURE 5.19 Moon, Close-up**   (a) The Moon near third quarter makes visible surface features near the *terminator*, where sunlight strikes at a sharp angle and the lit lunar globe changes to dark. (b) Magnified view of a region near the terminator, as seen from Earth through a large telescope. (c) Enlargement of a portion of (b). The smallest craters visible here have diameters of about 2 km, about twice the size of the Barringer Crater shown in Figure 4.19. *(UC/Lick Observatory; Caltech)*

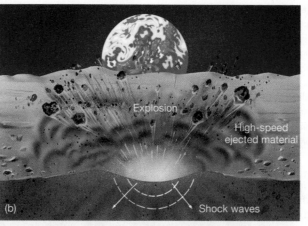

◀ FIGURE 5.20 **Meteoroid Impact INTERACTIVE** Stages in the formation of a crater by meteoritic impact. (a) The meteoroid strikes the surface, releasing a large amount of energy. The resulting explosion ejects material from the impact site and (c) sends shock waves through the underlying surface. (d) Eventually, a characteristic crater results, surrounded by a field of ejected material.

Barringer Crater in Figure 4.19.) Meteoroids generally strike the Moon at speeds of several kilometers per second. At these speeds even a small piece of matter carries an enormous amount of energy—for example, a 1-kg object hitting the Moon's surface at 10 km/s would release as much energy as the detonation of 10 kg of TNT. As illustrated in Figure 5.20, impact by a meteoroid causes sudden and tremendous pressures to build up on the lunar surface, heating the normally brittle rock and deforming the ground. The ensuing explosion pushes previously flat layers of rock up and out, forming a crater.

The material thrown out by the explosion surrounds the crater in a layer called an *ejecta blanket,* the ejected debris ranging in size from fine dust to large boulders. The larger pieces of ejected material may themselves create secondary craters. Many of the rock samples brought back by the *Apollo* astronauts show patterns of repeated shattering and melting—direct evidence of the violent shock waves and high temperatures produced in meteoritic impacts.

Lunar craters come in all sizes, reflecting the range in sizes of the impactors that create them. The largest craters are hundreds of kilometers in diameter, the smallest microscopic. Because the Moon has no protective atmosphere, even tiny interplanetary fragments can reach the lunar surface unimpeded. Figure 5.21(a) shows the result of a large meteoritic impact on the Moon, Figure 5.21(b) a crater formed by a micrometeoroid.

Craters are found everywhere on the Moon's surface, but the older highlands are much more heavily cratered than the younger maria. Knowing the ages of the highlands and maria, researchers can estimate the rate of cratering in the past. They conclude that the Moon, and presumably the entire inner solar system, experienced a sudden sharp drop in meteoritic bombardment rate about 3.9 billion years ago. The rate of cratering has been roughly constant since that time.

CONCEPT CHECK

Describe three important ways in which the lunar maria differ from the highlands.

▶ FIGURE 5.21 Lunar Craters (a) Two smaller craters called Reinhold and Eddington sit amid the secondary cratering resulting from the impact that created the 90-km-wide Copernicus Crater (near the horizon) about a billion years ago. (b) Craters of all sizes litter the lunar landscape. Some shown here, embedded in glassy beads retrieved by *Apollo* astronauts, measure only 0.01 mm across. (The scale at the top is in millimeters.) The beads themselves were formed during the explosion following a meteoroid impact, when surface rock was melted, ejected, and rapidly cooled. *(NASA)*

(a)

(b)

▲ FIGURE 5.22 Lunar Surface Despite the complete lack of wind and water, the lunar surface is slowly eroded by the constant "rain" of impacting meteoroids, especially micrometeoroids. Note the soft edges of the background hills of this image. The *Apollo* astronaut's boot prints in the foreground lunar dust are only a few centimeters deep, but they will survive for more than a million years. *(NASA)*

This time—3.9 billion years in the past—represents the end of the accretion process through which planetesimals became planets. ∞ *(Sec. 4.3)* The lunar highlands solidified and received most of their craters before that time. The great basins that formed the maria were created during the final stages of heavy meteoritic bombardment between about 4.1 and 3.9 billion years ago. The largest impacts were so violent that they cracked the lunar crust (see Figure 5.1b). Molten lava from the mantle welled up through these cracks to fill the large basins, then solidified as it cooled, resulting in the maria we see today.

Lunar Erosion

Meteoritic impact is the only important source of erosion on the Moon. Over billions of years, collisions with meteoroids, large and small, have scarred, cratered, and sculpted the lunar landscape. At the present average rates, one new 10-km-diameter lunar crater is formed every 10 million years, one new 1-m-diameter crater is created about once a month, and 1-cm-diameter craters appear every few minutes. In addition, a steady "rain" of micrometeoroids also eats away at the lunar surface (see Figure 5.22). The accumulated dust from countless impacts (called the lunar *regolith*) covers the lunar surface to an average depth of about 20 m, thinnest on the maria (about 10 m) and thickest on the highlands (more than 100 m in places).

Despite this barrage from space, the Moon's present-day erosion rate is still very low—about 10,000 times less than on Earth. For example, the Barringer Meteor Crater (Figure 4.19) in the Arizona desert, one of the largest meteor craters on Earth, is only 25,000 years old, but it is already decaying. It will probably disappear completely in a mere million years, quite a short time geologically. If a crater that size had formed on the Moon even a billion years ago, it would still be plainly visible today.

5.7 Magnetospheres

Simply put, the magnetosphere is the region around a planet that is influenced by that planet's magnetic field. It forms a buffer zone between the planet and the high-energy particles of the solar wind. It can also provide important insights into the planet's interior structure.

Earth's Magnetosphere

Shown in Figure 5.23, Earth's magnetic field extends far above the atmosphere, completely surrounding our planet. The *magnetic field lines*, which indicate the strength and direction of the field at any point in space, run from south to north, as indicated by the white arrowheads in the figure. The north and south *magnetic poles*, where the axis of an imaginary bar magnet within our planet (Figure 5.24) intersects Earth's surface, are very roughly aligned with Earth's spin axis.

Earth's inner magnetosphere contains two doughnut-shaped zones of high-energy charged particles, one located about 3000 km and the other 20,000 km above Earth's surface. These zones are named the **Van Allen belts,** after the American physicist whose instruments on board some early rocket flights during the late 1950s first detected them. We call them *belts* because they are most pronounced near Earth's equator and because they completely surround the planet. Figure 5.24 shows how these invisible regions envelop Earth except near the North and South Poles.

The particles that make up the Van Allen belts originate in the solar wind. Traveling through space, neutral particles and electromagnetic radiation are unaffected by Earth's magnetism, but electrically charged particles are strongly influenced. As illustrated in the inset to Figure 5.24, a magnetic field exerts a force on a moving charged particle, causing the particle to spiral around the magnetic field lines. In this way, charged particles—mainly electrons and protons—from the solar wind can become trapped by Earth's magnetism. Earth's magnetic field exerts electromagnetic control over these particles, herding them into the Van Allen belts. The outer belt contains mostly electrons; the much heavier protons accumulate in the inner belt.

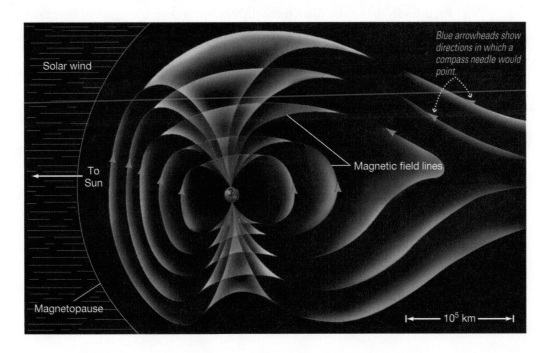

Solar wind

Blue arrowheads show directions in which a compass needle would point.

Magnetic field lines

To Sun

Magnetopause

|← 10⁵ km →|

◄ **FIGURE 5.23 Earth's Magnetosphere** The magnetosphere is the region surrounding a planet wherein particles from the solar wind are trapped by the planet's magnetic field. Far from Earth, the magnetosphere is greatly distorted by the solar wind, with a long "tail" extending from the nighttime side of Earth (here, at right) far into space. The magnetopause is the boundary of the magnetosphere in the sunward direction.

▶ **FIGURE 5.24 Van Allen Belts** Earth's magnetic field resembles somewhat the field of an enormous bar magnet buried inside our planet. High above Earth's atmosphere, the magnetosphere (light blue-green area) contains two doughnut-shaped regions (grayish areas) of magnetically trapped charged particles. These are the Van Allen belts.

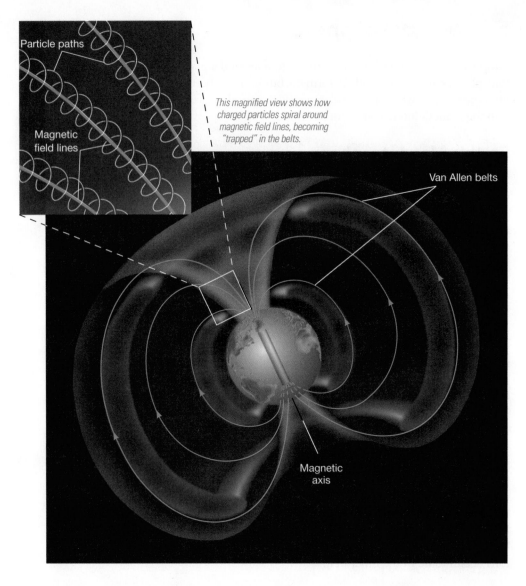

Particle paths

Magnetic field lines

This magnified view shows how charged particles spiral around magnetic field lines, becoming "trapped" in the belts.

Van Allen belts

Magnetic axis

Particles from the Van Allen belts often escape from the magnetosphere near Earth's north and south magnetic poles, where the field lines intersect the atmosphere. Their collisions with air molecules create a spectacular light show called an **aurora** (Figure 5.25). This colorful display results when atmospheric atoms, excited upon collision with the charged particles, fall back to their ground states and emit visible light. ∞ *(Sec. 2.6)* Aurorae are most brilliant at high latitudes, especially inside the Arctic and Antarctic circles. In the north the spectacle is called the *aurora borealis*, or *Northern Lights*. In the south it is called the *aurora australis*, or *Southern Lights*.

Just as Earth's magnetosphere influences the charged particles of the solar wind, the stream of incoming solar wind particles also affects our magnetosphere. As shown in Figure 5.23, the sunward (daytime) side is squeezed toward Earth's surface, while the opposite side has a long "tail" often extending hundreds of thousands of kilometers into space.

Earth's magnetosphere plays an important role in controlling many of the potentially destructive charged particles that venture near our planet. Without the magnetosphere, Earth's atmosphere—and perhaps the planet surface, too—would be bombarded by harmful particles, possibly damaging many forms of life. Some researchers have even suggested that had the magnetosphere not existed in the first place, life might never have arisen on planet Earth.

Earth's magnetic field is not an intrinsic part of our planet. Instead, it is continuously generated in Earth's core and exists only because the planet is rotating. As in the dynamos that generate electrical power in automobiles and power stations, Earth's magnetism is produced by the spinning, electrically conducting, liquid metal core deep within our planet. Indeed, the theory describing the creation of planetary magnetic fields is called **dynamo theory.** Both rapid rotation and a conducting liquid core are needed for this mechanism to work. This link between magnetism and internal structure is very important to studies of other planets, as we have few other probes of planetary interiors.

Lunar Magnetism

No Earth-based observation or spacecraft measurement has ever detected any lunar magnetic field. Based on our current understanding of how Earth's magnetic field is created, this is not surprising. As we have just seen, researchers think that planetary magnetism requires a rapidly rotating liquid metal core. Because the Moon rotates slowly and because its core is probably neither molten nor particularly rich in metals, the absence of a lunar magnetic field is exactly what we expect.

5.8 History of the Earth–Moon System

Given all the data, can we construct a consistent history of Earth and the Moon? The answer seems to be yes. Many specifics are still debated, but a consensus now exists.

Formation of the Moon

Sometime around 4.6 billion years ago Earth formed by accretion in the solar nebula. ∞ *(Sec. 4.3)* The formation of the Moon is less clear. The basic problem is that Earth and the Moon are too *dissimilar* in both density and composition for them simply to have formed together out of the same preplanetary matter. However, there are enough *similarities*, particularly between their mantles, to make it unlikely that Earth and the Moon formed independently and subsequently became bound to one another, presumably after a close encounter shortly after the solar system formed.

Most astronomers think that the Moon formed following a glancing collision between a large, Mars-sized object and a youthful, molten Earth. As we have seen, such collisions were probably quite frequent in the early solar system. ∞ *(Sec. 4.3)* Computer simulations of such a catastrophic event (Figure 5.26) show that most of the bits and pieces of splattered Earth could have recombined into a stable orbit, forming the Moon. If Earth had already formed an iron core by the time the collision occurred, the Moon could indeed have ended up with a composition similar to Earth's mantle. During the collision, any iron core in the impacting object would have been left behind, eventually to become part of Earth's core. Thus, both the Moon's overall similarity to Earth's mantle and its lack of a dense central core are explained. The condensation theory of solar system formation provides a natural context in which such a scenario could have occurred.

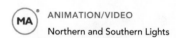

▲ **FIGURE 5.25 Aurorae** A colorful aurora rapidly flashes across the sky, resembling huge wind-blown curtains glowing in the dark. *(D. Vongprasert/Shutterstock)*

(MA) ANIMATION/VIDEO
Northern and Southern Lights

CONCEPT CHECK

What does the existence of a planetary magnetic field tell us about the planet's interior?

Red and blue represent rocky and metallic material.

The metals (blue) move toward Earth after the collision.

This is the newly formed Moon.

◄ FIGURE 5.26 Moon Formation   This sequence shows a simulated collision between Earth and an object the size of Mars. (The sequence proceeds from top to bottom and zooms out dramatically in the later frames.) Note how most of the impactor's metallic core becomes part of Earth, leaving the Moon composed primarily of rocky material. *(W. Benz)*

Lunar Evolution

The approximate age of the oldest rocks discovered in the lunar highlands is 4.4 billion years, so we know that at least part of the lunar crust must already have solidified by that time. The oldest known Earth rocks are of roughly the same age, although samples more than 4 billion years old are rare, suggesting that erosion on Earth has been quite effective at hiding the details of our planet's distant past. At formation, the Moon was already depleted in heavy metals relative to Earth.

Earth was at least partially molten during most of its first billion years. Denser matter sank toward the core while lighter material rose to the surface—Earth became differentiated. The intense meteoritic bombardment that helped melt Earth at early times subsided about 3.9 billion years ago. Radioactive heating in the interior continued even after Earth's surface cooled and solidified, but it, too, diminished with time. As our planet cooled, it did so from the outside in, because regions closest to the surface could most easily lose their excess heat to space. In this way, the surface developed a crust, and the differentiated interior attained the layered structure now implied by seismic studies. Today, radioactive heating continues throughout Earth, but there is probably not enough of it to melt any part of our planet. The high temperatures in the core are mainly the trapped remnant of a much hotter Earth that existed eons ago.

The lunar interior evolved quite differently from Earth's, in large part because of the Moon's smaller size. Small objects cool more quickly than large ones (basically because heat from the interior has less distance to go to reach the surface), and so the Moon rapidly lost its internal heat to space. During the earliest phases of the Moon's existence—roughly the first half-billion years—the meteoritic bombardment was violent enough to heat, and keep molten, most of the surface layers, perhaps to a depth of 400 km in places. However, the intense heat derived from these collisions probably did not penetrate into the deep lunar interior, because rock is a very poor conductor of heat. As on Earth, radioactivity probably heated the Moon, but (because the heat could escape more easily) not sufficiently to transform it from a warm, semisolid object to a completely liquid one. The Moon must have differentiated during this period. Its small iron core also formed at this time.

About 3.9 billion years ago, when the heaviest bombardment ceased, the Moon was left with a solid crust dented with numerous large basins (Figure 5.27a). The crust ultimately became the highlands, and the basins soon flooded with lava and became the maria. Between 3.9 and 3.2 billion years ago, lunar volcanism filled the basins with the basaltic material we see today. The age of the youngest maria—3.2 billion years—apparently indicates the time when this volcanic activity finally subsided (Figure 5.27b). The maria are the sites of the last extensive lava flows on the Moon. Their smoothness compared with the more rugged highlands disguises their great age. Not all these great craters became flooded with lava, however. One of the youngest is the Orientale Basin (Figure 5.28), which formed about 3.9 billion years ago. It did not undergo much subsequent volcanism, so we can recognize it as an impact crater rather than a mare. Similar "unflooded" basins can be seen on the lunar far side.

(a) 4 billion years ago   (b) 3 billion years ago   (c) Today

▲ **FIGURE 5.27 Lunar Evolution**   Paintings of the Moon (a) about 4 billion years ago, after much of the meteoritic bombardment had subsided and the surface had somewhat solidified; (b) about 3 billion years ago, after molten lava had made its way up through surface fissures to fill the low-lying impact basins and create the smooth maria; and (c) today, with much of the originally smooth maria now pitted with craters formed at various times within the past 3 billion years. *(U.S. Geological Survey)*

Because of Earth's gravitational pull, the lunar crust became thicker on the far side than on the near side. Therefore, lava from the interior had a shorter route through the crust to the surface on the Moon's Earth-facing side. As a result, relatively little volcanic activity occurred on the far side, and no large maria were created—the crust was simply too thick to allow that to occur.

As the Moon continued to cool, volcanic activity ended as the thickness of the solid surface layer increased. The crust is now far too thick for volcanism or plate tectonics to occur. With the exception of a few meters of surface erosion from eons of meteoritic bombardment (Figure 5.27c), the lunar landscape has remained more or less structurally frozen for the past 3 billion years. The Moon is dead now, and it has been dead for a long time.

CONCEPT CHECK

What single factor is most responsible for the evolutionary differences between Earth and the Moon?

R I V U X G

▲ **FIGURE 5.28 Large Lunar Crater**   A large lunar crater, the Orientale Basin. The meteoroid that produced this crater upthrust much surrounding matter, which can be seen as concentric rings of cliffs called the Cordillera Mountains. Notice the smaller, sharper, younger craters that have impacted this ancient basin more recently. *(NASA)*

THE BIG QUESTION

Will humans eventually colonize the Moon—perhaps "terraforming," or changing it to suit our needs? Fifty years ago, humankind appeared well on its way toward establishing permanent lunar habitats, but those early (and spectacular) exploratory programs soon stalled. Today, there seems little political will or economic means to mount human missions to even our closest neighbor in space. Should governments, or maybe the UN, lead the return to the Moon, or will this goal be best accomplished with private, entrepreneurial ventures like those that settled the Americas centuries ago?

CHAPTER REVIEW

SUMMARY

LO1 The six main regions of Earth are (from inside to outside) a central metallic **core** (p. 140), which is surrounded by a thick rocky **mantle** (p. 140) and topped with a thin **crust** (p. 140). Above the surface is the **atmosphere** (p. 140), composed primarily of nitrogen and oxygen. Surface winds and weather in the lower atmosphere, or **troposphere** (p. 144) are caused by **convection** (p. 144), whereby heat moves from one place to another by rising or sinking streams of air. Higher still lies the **magnetosphere** (p. 140), where charged particles from the Sun are trapped by Earth's magnetic field. The Moon has a partially differentiated interior consisting of a central core, a mantle, and a thick crust. It has no atmosphere, because its gravity is too weak to retain one, and no magnetosphere.

LO2 The daily **tides** (p. 141) in Earth's oceans are caused by the gravitational effects of the Moon and the Sun, which raise **tidal bulges** (p. 141) in the oceans. Their size depends on the orientations of the Sun and the Moon relative to Earth. A differential gravitational force is called a **tidal force** (p. 141), even when no oceans or even planets are involved. The tidal interaction between Earth and the Moon is causing Earth's spin to slow and is responsible for the Moon's **synchronous orbit** (p. 142), in which the same side of the Moon always faces our planet.

LO3 At high altitudes, in the **ionosphere** (p. 144), the atmosphere is kept ionized by the absorption of high-energy radiation and particles from the Sun. Between the ionosphere and the troposphere lies the **ozone layer** (p. 145), where incoming solar ultraviolet radiation is absorbed. Both these layers help protect us from dangerous radiation from space. The **greenhouse effect** (p. 147) is the absorption and trapping of infrared radiation emitted by Earth's surface by atmospheric gases (primarily carbon dioxide and water vapor). It makes our planet's surface some 40 K warmer than would otherwise be the case.

LO4 We study Earth's interior by observing how **seismic waves** (p. 149) produced by **earthquakes** (p. 149) travel through the mantle. Earth's iron core consists of a solid **inner core** (p. 150) surrounded by a liquid **outer core** (p. 150). The process by which dense material sinks to the center of a planet while lighter material rises to the surface is called **differentiation** (p. 150). The differentiation of Earth implies that our planet must have been at least partially molten in the past, because of bombardment by material from interplanetary space and heat released by **radioactivity** (p. 151) in Earth's interior.

LO5 Earth's surface is made up of enormous slabs, or plates. The slow movement of these plates across the surface is known as continental drift, or **plate tectonics** (p. 152). Earthquakes, volcanism, and mountain building are associated with plate boundaries, where plates may collide, move apart, or rub against one another. The "fit" of the continents to one another and the ages of rocks near oceanic ridges argue in favor of this theory. The motion of the plates is thought to be driven by convection in Earth's mantle. On the Moon, the crust is too thick and the mantle too cool for plate tectonics to occur.

LO6 The main surface features on the Moon are the dark **maria** (p. 155) and the lighter-colored **highlands** (p. 155). **Craters** (p. 156) of all sizes, caused by impacting meteoroids, are found everywhere on the lunar surface. The highlands are older than the maria and are much more heavily cratered. Meteoritic impacts are the main source of erosion on the lunar surface. There is no volcanic activity on the Moon because all volcanism was stifled by the Moon's cooling mantle shortly after extensive lava flows formed the maria more than 3 billion years ago. Astronomers can use the amount of cratering to deduce the ages of surfaces on the Moon and elsewhere in the solar system.

LO7 Charged particles from the solar wind are trapped by Earth's magnetic field lines to form the **Van Allen belts** (p. 159). When particles from the Van Allen belts hit Earth's atmosphere, they heat

and ionize the atoms there, causing the atoms to glow in an **aurora** (p. 160). Planetary magnetic fields are produced by the motion of rapidly rotating, electrically conducting fluid (such as molten iron) in a planet's core. The Moon rotates slowly and lacks a conducting liquid core, which accounts for the absence of a lunar magnetic field.

LO8 The most likely explanation for the formation of the Moon is that the newly formed Earth was struck by a Mars-sized object. The core of the impacting body remained behind as part of the core of our planet, and

debris splattered into space formed the Moon. Lunar volcanism was stifled by the Moon's cooling mantle shortly after the extensive lava flows formed the maria more than 3 billion years ago.

MasteringAstronomy® For instructor-assigned homework go to www.masteringastronomy.com

Problems labeled POS explore the process of science. VIS problems focus on reading and interpreting visual information. LO connects to the introduction's numbered Learning Outcomes.

REVIEW AND DISCUSSION

1. **LO2** Explain how the Moon produces tides in Earth's oceans.

2. What does it mean to say the Moon is in a synchronous orbit around Earth? How did the Moon come to be in such an orbit?

3. What is convection? What effect does it have on (a) Earth's atmosphere and (b) Earth's interior?

4. In contrast to Earth, the Moon undergoes extremes in temperature. Why?

5. Use the concept of escape speed to explain why the Moon has no atmosphere.

6. **LO3** Is the greenhouse effect operating in Earth's atmosphere helpful or harmful? Give examples. What are the consequences of an enhanced greenhouse effect?

7. **LO1** The density of water in Earth's hydrosphere and the density of rocks in the crust are both lower than the average density of the planet as a whole. What does this fact tell us about Earth's interior?

8. **LO4 POS** Give two reasons why geologists think that part of Earth's core is liquid.

9. **POS** What clue does Earth's differentiation provide to our planet's history?

10. **LO5** What process is responsible for the surface mountains, oceanic trenches, and other large-scale features on Earth's surface?

11. In what sense were the lunar maria once "seas"?

12. What is the primary source of erosion on the Moon? Why is the average rate of lunar erosion so much less than on Earth?

13. **LO6 POS** Name two pieces of evidence indicating that the lunar highlands are older than the maria.

14. **LO7** Give a brief description of Earth's magnetosphere. Why does the Moon have no magnetosphere?

15. **LO8 POS** Describe the theory of the Moon's origin currently favored by many astronomers.

CONCEPTUAL SELF-TEST: TRUE OR FALSE?/MULTIPLE CHOICE

1. Because of tidal forces, the Moon is in a synchronous orbit around Earth. (T/F)

2. Lunar maria are extensive lava-flow regions. (T/F)

3. Except for the layer of air closest to Earth's surface, the ozone layer is the warmest part of the atmosphere. (T/F)

4. Earth's magnetic field is the result of our planet's large, permanently magnetized iron core. (T/F)

5. Motion of the crustal plates is driven by convection in Earth's upper mantle. (T/F)

6. Volcanic activity continues today on the surface of the Moon. (T/F)

7. Like Earth, the Moon has a molten metal core. (T/F)

8. If you were making a scale model of Earth, representing our planet by a 12-inch basketball, the inner core would be about the size of (a) a half-inch ball bearing; (b) a 2-inch golf ball; (c) a 4-inch tangerine; (d) a 7-inch grapefruit.

9. Earth's average density is about the same as that of (a) a glass of water; (b) a heavy iron meteorite; (c) an ice cube; (d) a chunk of black volcanic rock.

10. Sunlight absorbed by Earth's surface is reemitted in the form of (a) microwave, (b) infrared, (c) visible, (d) ultraviolet radiation.

11. The deepest that geologists have drilled into Earth is about the same as (a) the height of the Statue of Liberty; (b) the altitude most commercial jet airplanes fly; (c) the distance between New York and Los Angeles; (d) the distance between the United States and China.

12. If Earth had no Moon, then tides would (a) not occur; (b) occur more often and with more intensity; (c) still occur, but not really be measurable; (d) occur with the same frequency, but would not be as strong.

13. The most likely theory of the formation of the Moon is that it (a) was formed by the gravitational capture of a large asteroid; (b) formed simultaneously with Earth's formation; (c) was created from a collision scooping out the Pacific Ocean; (d) formed from a collision of Earth with a Mars-sized object.

14. VIS According to Figure 5.5 (Earth's Atmosphere), commercial jet airplanes flying at 10 km are in (a) the troposphere; (b) the stratosphere; (c) the ozone; (d) the mesosphere.

15. VIS The first figure in *Discovery* 5-2 shows that CO_2 levels in Earth's atmosphere began to rise rapidly (a) in the middle ages; (b) in 1600; (c) in the mid-19th century; (d) in the late 20th century.

PROBLEMS

The number of squares preceding each problem indicates its approximate level of difficulty.

1. ■ What would Earth's surface gravity and escape speed be if the entire planet had a density equal to that of the crust (3000 kg/m³, say)?

2. ■ The Moon's mass is 1/80 that of Earth, and the lunar radius is 1/4 Earth's radius. Based on these figures, calculate the total weight on the Moon of a 100-kg astronaut with a 50-kg spacesuit and backpack, relative to his or her weight on Earth.

3. ■■ Most of Earth's ice is found in Antarctica, where permanent ice caps cover approximately 0.5 percent of Earth's surface area and are 3 km thick, on average. Earth's oceans cover roughly 71 percent of our planet to an average depth of 3.6 km. Assuming that water and ice have roughly the same density, estimate by how much sea level would rise if global warming were to cause the Antarctic ice caps to melt.

4. ■■ You are standing on Earth's surface during a total eclipse, and both the Moon and the Sun are directly overhead. What fraction is your weight decreased due to their combined tidal gravitational force?

5. ■ Based on the data presented in the text, estimate the fractions of Earth's volume represented by (a) the inner core, (b) the outer core, (c) the mantle, and (d) the crust.

6. ■■ Approximating Earth's atmosphere as a layer of gas 7.5 km thick, with a uniform density of 1.3 kg/m³, calculate the atmosphere's total mass of the atmosphere. Compare your result with Earth's mass.

7. ■■■ As discussed in the text, without the greenhouse effect, Earth's average surface temperature would be about 250 K. With the greenhouse effect, it is some 40 K higher. Use this information and Stefan's law to calculate the fraction of infrared radiation leaving Earth's surface that is absorbed by greenhouse gases in the atmosphere. ∞ *(Sec. 2.4)*

8. ■ Following an earthquake, how long would it take a P-wave moving in a straight line with a speed of 5 km/s to reach the opposite side of Earth?

9. ■■ Using the rate given in the text for the formation of 10-km craters on the Moon, estimate how long it would take to cover the Moon with new craters of this size. How much higher must the cratering rate have been in the past to cover the entire lunar surface with such craters in the 4.6 billion years since the Moon formed?

10. ■ The *Hubble Space Telescope* has a resolution of about 0.05 arc second. What is the smallest object it could see on the surface of the Moon? Give your answer in meters.

ACTIVITIES

Collaborative

1. Go online and read about global warming. How much carbon dioxide is produced each year by human activities? How does this compare with the total amount of carbon dioxide in Earth's atmosphere? Do all—or most—scientists agree that global warming is an inevitable consequence of carbon dioxide production? What political initiatives are currently under way to address the problem? As a group, which if any do you think are likely to succeed?

2. The estimated cost of transporting a gallon of water from Earth to the Moon is about $100,000. By determining how much water each group member uses in a single day, estimate the cost of taking a single day's supply of water for your group to the Moon.

Individual

1. Go to a sporting goods store and get a tide table; many stores near the ocean provide them free. Choose a month, and plot the height of one high and one low tide versus the day of the month. Now mark the dates when the primary phases of the Moon occur. How well does the phase of the Moon predict the tides?

2. Observe the Moon during an entire cycle of phases. When does the Moon rise, set, and appear highest in the sky at each major phase? What is the interval of time between each phase?

3. If you have binoculars, turn them on the Moon when it appears at twilight and when it appears high in the sky. Draw pictures of what you see. What differences do you notice in your two drawings? What color is the Moon when seen near the horizon? What color is the Moon when seen high in the sky? Why is there a difference?

4. Watch the Moon over a period of hours on a night when you can see one or more bright stars near it. Estimate how many Moon diameters it moves per hour, relative to the stars. Knowing the Moon is about 0.5° in diameter, how many degrees per hour does it move? Based on this, what is your estimate of the Moon's orbital period?

6

The Terrestrial Planets

A Study in Contrasts

With Earth and the Moon as our guides, we now expand our field of view to study the other terrestrial planets. As we explore these worlds and seek to understand the similarities and differences among them, we begin our comparative study of the only planetary system we know. Mercury in many ways is kin to Earth's Moon, and much can be learned by comparing Mercury with our own satellite. Venus and Mars both have properties more like Earth's, and we learn about these two terrestrial worlds by drawing parallels with our planet. It is possible that Venus, Earth, and Mars had many similarities when they formed, yet Earth today is vibrant, teeming with life, while Venus is an uninhabitable inferno and Mars is a dry, dead world. What were the factors leading to these present conditions? In answering this question, we will discover that a planet's environment, as well as its composition, can play a critical role in determining its future.

THE BIG PICTURE

Mars has been reconnoitered by humans more than any other cosmic object beyond Earth and the Moon. It intrigues us, frustrates us, and invites us to visit. Although Mars today is as dry as any desert on Earth, many scientists think it was much wetter billions of years ago, when its atmosphere was thicker and the climate was warmer. Mars remains high on most astronomers' list of likely sites for life in the solar system.

LEARNING OUTCOMES

Studying this chapter will enable you to:

LO1 Explain how Mercury's rotation has been influenced by its orbit around the Sun.

LO2 Describe how the atmospheres of Venus and Mars differ from one another and from Earth's.

LO3 Compare the surface of Mercury with that of the Moon, and describe how Mercury's surface features formed.

LO4 Compare the surfaces of Venus and Mars with that of Earth.

LO5 Explain why many scientists think that Mars once had running water and a thick atmosphere.

LO6 Identify the major similarities and differences in the internal structures and geological histories of the four terrestrial planets.

LO7 List the basic factors influencing atmospheric evolution, and explain why the atmospheres of Venus, Mars, and Earth are now so different from one another.

◄ This true-color mosaic of many images was made in 2013 by the *Curiosity* spacecraft, which soft-landed in Mars's equatorial region. Amid scattered rocks holding memories of the ancient events that formed them, this view shows wind-swept areas inside Gale crater, which once may have been filled with water, and where subsurface soil and ice have been scooped up and tested in *Curiosity*'s onboard chemistry lab. *(NASA/JPL)*

Mastering Astronomy®

Visit www.masteringastronomy.com for quizzes, animations, videos, interactive figures, and self-guided tutorials.

6.1 Orbital and Physical Properties

Mercury, the innermost planet, lies close to the Sun and is visible above the horizon for at most 2 hours before the Sun rises or after it sets. Orbiting somewhat farther from the Sun (but still within Earth's orbit), the next planet, Venus, is visible for a little longer—up to 3 hours, depending on the time of year (Figure 6.1). The interior orbits of these planets ensure that we never see them far from the Sun in the sky. Contrast this with the exterior orbit of the planet Mars. From our earthly viewpoint, Mars appears to traverse the entire sky, always keeping close to the ecliptic (and occasionally executing retrograde loops, as we saw in Chapter 1). ∞ *(Sec. 1.1)*

Venus is the third brightest object in the entire sky (only the Sun and the Moon are brighter). Like all the planets, it shines by reflected sunlight. It is so bright because almost all the sunlight reaching it is reflected from thick clouds that envelop the planet. You can even see Venus in the daytime if you know where to look. The much fainter Mercury is visible to the naked eye only when the Sun's light is blotted out—just before dawn, just after sunset, and during a total solar eclipse. Orange-red Mars is also quite easy to spot in the night sky. Because of its less reflective surface, smaller size, and greater distance from the Sun, Mars is not as bright as Venus, as seen from Earth. However, at its brightest—at closest approach, as marked on Figure 6.1(a)—Mars is still brighter than any star.

Table 6.1 expands Table 5.1 to include Mercury, Venus, and Mars and adds two additional properties: surface temperature and surface atmospheric pressure. In seeking to understand the other terrestrial planets, astronomers are guided by our much more detailed knowledge of Earth and the Moon. Mercury's high average density, for instance, tells us that the planet must have a large iron core, broadly consistent with the condensation theory's account of its formation. ∞ *(Sec. 4.3)* However, in many other respects Mercury is more similar to Earth's Moon, and we can use the Moon as a model for understanding Mercury's past. Venus and, to a lesser extent, Mars are more similar to Earth, so Earth provides the natural starting point for studies of those planets. For example, despite the lack of seismic data, we nevertheless assume that Venus has a metallic core and rocky mantle similar to Earth's. ∞ *(Sec. 5.1, 5.4)*. Even the widely different atmospheres of these two worlds can be explained in familiar earthly terms.

6.2 Rotation Rates

In principle, we can determine the rate at which a planet spins simply by observing the motion of some prominent surface feature across the planet's disk. Unfortunately, for Mercury and Venus surface markings are hard to see from Earth. Astronomers had to develop other techniques to probe the rotations of these bodies.

Mercury's Curious Spin

From Earth, even through a large telescope, we see Mercury only as a slightly pinkish, almost featureless disk. The largest ground-based telescopes can resolve features on the surface of Mercury about as well as we can perceive features on

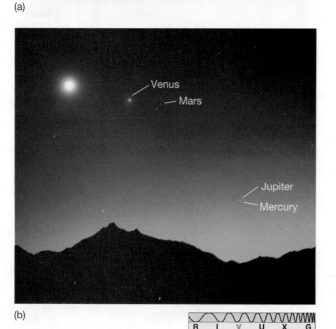

(a)

(b)

R I V U X G

◀ FIGURE 6.1 **Terrestrial Planet Orbits**   (a) The orbits of Mercury and Venus mean that neither planet is ever found far from the Sun, as seen from Earth. Mercury's greatest angular distance from the Sun is 28°, that of Venus 47°. Mars, on the other hand, lies outside Earth's orbit and can appear anywhere on the ecliptic plane, depending on the two planets' positions in their respective orbits. (b) Mercury, Venus, and Mars in the evening sky, along with Jupiter and the Moon. *(J. Sanford/Science Photo Library)*

TABLE 6.1 Some Properties of the Terrestrial Planets and Earth's Moon

	(kg)	(Earth = 1)	(km)	(Earth = 1)	Average Density (kg/m³)	Surface Gravity (Earth = 1)	Escape Speed (km/s)	Surface Rotation Period (solar days)	Surface Temperature (K)	Atmospheric Pressure (Earth = 1)
Mercury	3.3×10^{23}	0.055	2400	0.38	5400	0.38	4.2	59	100–700	—
Venus	4.9×10^{24}	0.82	6100	0.95	5300	0.91	10	−243[1]	730	90
Earth	6.0×10^{24}	1.00	6400	1.00	5500	1.00	11	1.00	290	1.0
Mars	6.4×10^{23}	0.11	3400	0.53	3900	0.38	5.0	1.03	180–270	0.007
Moon	7.3×10^{22}	0.012	1700	0.27	3300	0.17	2.4	27.3	100–400	—

[1]The minus sign indicates retrograde rotation.

our Moon with our unaided eyes. Figure 6.2 is one of the few photographs of Mercury taken from Earth that shows any indication of surface features.

In the mid-19th century, an Italian astronomer named Giovanni Schiaparelli attempted to measure Mercury's rotation rate by watching surface features move around the planet. He concluded that Mercury always keeps one side facing the Sun, much as our Moon always presents only one face to Earth. His explanation for this synchronous rotation was the same as for the Moon: The tidal bulge raised in Mercury by the Sun had modified the planet's rotation rate until the bulge always points directly at the Sun. ∞ *(Sec. 5.2)* Although the surface features could not be seen clearly, the combination of Schiaparelli's observations and a plausible physical explanation was enough to convince most astronomers.

In 1965 astronomers making radar observations of Mercury from the Arecibo radio telescope in Puerto Rico discovered that this long-held view was in error. ∞ *(Sec. 3.4)* The technique they used is illustrated in Figure 6.3, which shows a radar signal reflecting from the surface of a hypothetical planet. Because the planet is rotating, a pulse of outgoing radiation of a single frequency is broadened— "smeared out"—by the Doppler effect by an amount that depends on the planet's rotation speed. ∞ *(Sec. 2.7)* The radiation reflected from the side moving toward us returns at a slightly higher frequency than does the radiation reflected from the receding side. (Think of the two hemispheres as separate sources of radiation moving at slightly different velocities, one toward us and one away from us.) By measuring the broadening, we can determine the rate at which the planet rotates.

In this way, astronomers found that the rotation period of Mercury is not 88 days, as had previously been thought, but just under 59 days—in fact, exactly two-thirds of Mercury's year. This odd state of affairs surely did not occur by chance. In fact, 19th-century astronomers were correct in thinking that Mercury's rotation was governed by the tidal effect of the Sun. However, the combination of the Sun's gravity and Mercury's eccentric orbit has caused the planet's rotation to be more complicated than that of the Moon. Unable to come into a state of precisely synchronous rotation (because Mercury's orbital speed changes significantly from place to place in its orbit), Mercury did the next best thing. It presents the same face to the Sun not every time around, but every other time. Figure 6.4 illustrates the implications of this odd rotation for a hypothetical inhabitant of Mercury. The planet's solar day—the time from one noon to the next—is 2 Mercury years long!

R   I   V   U   X   G

▲ **FIGURE 6.2 Mercury** Photograph of Mercury taken from Earth with a large ground-based optical telescope. Only a few faint surface features are discernible. *(Palomar Observatory/Caltech)*

▶ **FIGURE 6.3 Planetary Radar** A radar beam (blue waves) reflected from a rotating planet yields information about both the planet's line-of-sight motion and its rotation rate.

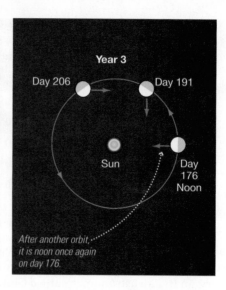

Pink arrows depict an observer standing on Mercury's surface.

Year 1

Day 30   Day 15

Rotation

Day 44

Sun   Day 0 Noon

Orbital motion

Day 59 (one full rotation completed)   Day 74

At day 0 at noon, the Sun is directly overhead.

Year 2

Day 118   Day 103

Day 132

Sun   Day 88 Midnight

Day 147   Day 162

By day 88—one full orbit later—Mercury has rotated 1.5 times, so that it is then midnight.

Year 3

Day 206   Day 191

Sun   Day 176 Noon

After another orbit, it is noon once again on day 176.

▲ **FIGURE 6.4 Mercury's Rotation** INTERACTIVE   Mercury's orbital and rotational motions combine to produce a solar day that is 2 Mercury years long.

ANIMATION/VIDEO   (MA)
Transit of Mercury

CONCEPT CHECK

How has the Sun's gravity influenced Mercury's rotation?

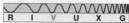
R   I   V   U   X   G

The Sun also influences the tilt of Mercury's spin axis. Because of the Sun's tides, Mercury's rotation axis is almost exactly perpendicular to its orbital plane. Thus, the noontime Sun is always directly overhead for someone standing on the equator and always on the horizon for someone standing at either pole.

Venus and Mars

The same clouds whose reflectivity make Venus so easy to see in the night sky also make it impossible for us to see any surface features, at least in visible light. Figure 6.5, one of the best photographs of Venus taken with an Earth-based telescope, shows an almost uniform yellow-white disk, with rare hints of clouds. Because of the cloud cover, astronomers did not know Venus's rotation period until the development of radar techniques in the 1960s, when Doppler broadening of returning radar echoes indicated an unexpectedly sluggish 243-day rotation period. Furthermore, Venus's spin was found to be *retrograde*—that is, opposite that of Earth and most other solar system objects and in the direction opposite Venus's orbital motion.

We have no "evolutionary" explanation for Venus's anomalous rotation. It is not the result of any known interaction with the Sun, Earth, or any other solar system body. At present, the best explanation astronomers can offer is that during the final stages of Venus's formation in the early solar system, the planet was struck by a large body, much like the one that may have hit Earth and formed the Moon. ∞ *(Secs. 4.3, 5.8)* That impact was sufficient to reduce the planet's spin almost to zero, leaving Venus rotating as we now observe.

Surface markings are easily seen on Mars (Figure 6.6), thus astronomers can track the planet's rotation. Mars rotates once on its axis every 24.6 hours—close to 1 Earth day. The planet's equator is inclined to the orbital plane at an angle of 24.0°, very similar to Earth's inclination of 23.5°. Thus, as Mars orbits the Sun, we find both daily and seasonal cycles, just as on Earth, although the Martian seasons are complicated somewhat by variations in solar heating due to the planet's eccentric orbit. Figure 6.7 summarizes the rotations and orbits of the four terrestrial planets.

◀ **FIGURE 6.5 Venus**   This photograph, taken from Earth, shows Venus with its creamy yellow mask of clouds. No surface detail can be seen because the clouds completely obscure our view of what lies beneath them. *(NOAO)*

▲ FIGURE 6.6 **Mars** This deep-red (800-nm) image of Mars was taken at Pic du Midi Observatory, an exceptionally clear site in the French Alps. One of the planet's polar caps appears at the top, and a few other surface markings are visible. *(CNRS and Université Paul Sabatier)*

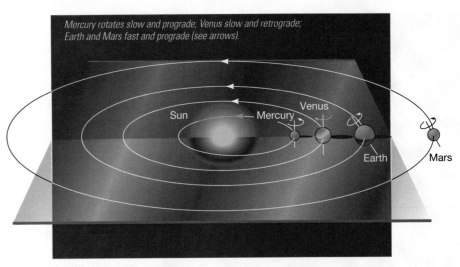

▲ FIGURE 6.7 **Terrestrial Planets' Spin** The inner planets of the solar system—Mercury, Venus, Earth, and Mars—display widely different rotational properties. All orbit the Sun in the same direction and in nearly the same plane, but Venus rotates clockwise as seen from above the plane of the ecliptic, whereas Mercury, Earth, and Mars all spin counterclockwise. This is a perspective view, roughly halfway between a flat edge-on view and a direct overhead view.

6.3 Atmospheres

In this section we outline what is known of the present-day atmospheres of Mercury, Venus, and Mars. We will return to this subject at the end of the chapter, when we discuss how the terrestrial atmospheres evolved into their present states.

ANIMATION/VIDEO
Transit of Venus

Mercury

To date, only two spacecraft have visited Mercury. NASA's *Mariner 10* made three flybys during 1974 and 1975. (A "flyby," in NASA parlance, is any space mission in which a probe passes relatively close to a planet—within a few planetary radii, say—but does not go into orbit around it.) Those close encounters were the source of virtually all the detailed information we had on Mercury until NASA's next mission, *Messenger*, made the first of three flybys in 2008, before going into orbit around the planet in 2011.

Mercury has no appreciable atmosphere. *Mariner 10* did find a trace of what was at first thought to be an atmosphere, but the gas is now known to be mostly temporarily trapped hydrogen and helium from the solar wind. Mercury holds this material for just a few weeks before it leaks away again into space. *Messenger* measured the composition of the gas and found that, while it is indeed composed largely of hydrogen and helium like the Sun, it also contains atoms that must have been blasted off the planet's surface by interactions with the solar wind.

The absence of any significant atmosphere on Mercury is readily explained by the planet's high surface temperature (up to 700 K at noon on the equator) and low mass (only 4.5 times the mass of the Moon). Any atmosphere Mercury might once have had escaped long ago. ∞ *(More Precisely 5-1)* With no atmosphere to retain heat, Mercury's surface temperature falls to about 100 K during the planet's long night. Mercury's 600 K temperature range is the largest of any planet or moon in the solar system. Near the poles, where the Sun's light arrives almost parallel to the surface, the temperature remains low at all times. Recent Earth-based radar studies suggest that Mercury's polar temperature could be as low as 125 K and that the poles may be permanently covered with extensive thin sheets of water ice.

(a)
R I V U X G

(b)
R I V U X G

▲ **FIGURE 6.8 Venus, Up Close** (a) Venus as photographed in the ultraviolet by the *Pioneer* spacecraft's cameras 200,000 km away from the planet. This image was made by capturing solar radiation reflected from the planet's clouds, which are composed mostly of sulfuric acid droplets, much like the corrosive acid in a car battery. (b) Infrared images of Venus's south pole, taken by *Venus Express* a few hours apart, allow us to see the polar vortex deeper in the planet's clouds. *(NASA/ESA)*

CONCEPT CHECK

Describe some important differences between the atmospheres of Venus and Earth.

Venus

In the 1930s, astronomers used spectroscopy to measure the temperature of Venus's upper atmosphere. They found it to be about 240 K, not much different from that of Earth's stratosphere. ∞ *(Sec. 5.3)* Taking into account the cloud cover and the planet's nearness to the Sun, and assuming that Venus had an atmosphere much like our own, researchers concluded that Venus might have an average surface temperature only a few degrees higher than Earth's. In the 1950s, however, when radio observations of the planet penetrated the cloud layer and gave the first indication of conditions near the surface, they revealed a temperature exceeding 600 K! Almost overnight, the popular conception of Venus changed from lush tropical jungle to arid, uninhabitable desert.

Since then, spacecraft data have revealed the full extent of the differences between the atmospheres of Venus and Earth. Venus's atmosphere is much more massive than our own, and it extends to a much greater height above the planet's surface. The surface pressure on Venus is about 90 times the pressure at sea level on Earth, equivalent to an (Earth) underwater depth of about 1 km. (Unprotected humans cannot dive much below 100 m.) The surface temperature is a sizzling 730 K.

The main component (96.5 percent) of Venus's atmosphere is carbon dioxide. Almost all of the remaining 3.5 percent is nitrogen. Given Venus's similarity to Earth in mass, radius, and location in the solar system, it is commonly assumed that Venus and Earth must have started off looking somewhat alike. However, there is no sign of the large amount of water vapor that would be present if a volume of water equivalent to Earth's oceans had once existed on Venus and later evaporated. If Venus started off with Earth-like composition, something happened to its water, for the planet is now an exceedingly dry place. Even the highly reflective clouds are composed not of water vapor, as on Earth, but of sulfuric acid droplets.

Venus's atmospheric patterns are much more evident when examined in the ultraviolet. Some of Venus's upper-level clouds absorb this high-frequency radiation, thereby increasing the contrast. Figure 6.8(a) is an ultraviolet image taken in 1979 by the U.S. *Pioneer Venus* spacecraft from a distance of 200,000 km (compare the optical image shown in Figure 6.5). The large, fast-moving cloud patterns lie between 50 and 70 km above the surface. Upper-level winds reach speeds of 400 km/h relative to the planet. Below the clouds, extending down to an altitude of 30 km, is a layer of haze. Below 30 km, the air is clear.

Figure 6.8(b) is a pair of infrared images taken in 2006 by the European *Venus Express* orbiter, whose cameras could partially penetrate the planet's thick haze. It shows a **polar vortex**—a relatively stable, long-lived wind flow circling the planet's south pole. Polar vortices are well known to atmospheric scientists and are expected in any rotating body (planet or moon) with an atmosphere. Earth's south polar vortex plays an important role in confining and concentrating the gases responsible for our planet's Antarctic ozone hole.(see *Discovery 5-1*).

Mars

Well before the arrival of spacecraft, astronomers knew from Earth-based spectroscopy that the Martian atmosphere is quite thin and composed primarily of carbon dioxide. Spacecraft measurements revealed that the atmospheric pressure is only about 1/150 the pressure of Earth's atmosphere at sea level. The Martian atmosphere is 95.3 percent carbon dioxide, 2.7 percent nitrogen, and 1.6 percent argon, plus small amounts of oxygen, carbon monoxide, and water vapor. While there is some superficial similarity in composition between the atmospheres of Mars and Venus, the two planets clearly must have had very different atmospheric histories. Average surface temperatures on Mars are about 70 K cooler than on Earth.

6.4 The Surface of Mercury

Figure 6.9 is a mosaic of images taken by *Messenger* as it orbited Mercury in 2011. Much of the planet's cratered surface bears a strong resemblance to the Moon's highlands. The crater walls are generally not as high as on the Moon, the craters are not as deep, and the ejected material landed closer to the impact site, exactly as we would expect given Mercury's greater surface gravity (which is a little more than twice that of the Moon).

Mercury's craters are less densely packed than their lunar counterparts, and there are extensive **intercrater plains** between the larger craters. These plains are the oldest visible parts of the planet's surface. They formed just under 4 billion years ago. In addition, Mercury has numerous **smooth plains**, akin to the lunar maria, where lava has filled depressions produced by large meteoritic impacts. The smooth plains are less obvious than the maria on the Moon because their color is comparable to the rest of Mercury's surface. They formed hundreds of millions of years after the intercrater plains.

As on the Moon, Mercury's craters are the result of meteoritic bombardment. The explanation for Mercury's relative lack of craters is that the older impact craters have been erased by volcanic activity, much as the Moon's maria filled in craters as they formed. Many geologists think that much of Mercury's crust may have formed through repeated volcanic eruptions. However, this *flood volcanism* was apparently not associated with impact basins, suggesting that Mercury's volcanic past may have been significantly different from the Moon's.

Figure 6.10 shows a **scarp**, or cliff, on Mercury that does not appear to be the result of volcanic or other familiar geological activity. It cuts across several craters, indicating that whatever produced it occurred after most of the meteoritic bombardment was over. Mercury shows no evidence of crustal motions, so the scarps did not form by tectonic processes such as those responsible for fault lines on Earth. ∞ *(Sec. 5.5)* Rather, they probably formed when the crust cooled, shrank, and split long ago, much as wrinkles form on the skin of an old, shrunken apple. One side of the crack moved upward relative to the other side, forming the scarp cliff face. If we can apply to Mercury the cratering age estimates we use for the Moon, the scarps appeared about 4 billion years ago. In 2010 *LRO* observed many much smaller scarps on the Moon, probably formed by similar processes operating on smaller scales.

A unique aspect of Mercury's landscape discovered by *Messenger* is a set of features known as *hollows*—small, shallow, irregularly shaped, rimless depressions that are often found in clusters near the centers of impact craters (Figure 6.11). Many have a bright, fresh appearance, suggesting that they may be actively forming today. They are not impact craters, but they do appear to be a secondary result of crater formation. Scientists theorize that meteoritic impacts can excavate material that becomes unstable when exposed to the harsh environment on Mercury's surface. The intense heat and solar wind may vaporize some light minerals, weakening the remaining rock and causing it to subside.

Figure 6.12 shows the result of what was probably the last great event in the geological history of Mercury—an immense bull's-eye crater called the Caloris Basin, formed eons ago by the impact of a large asteroid. ∞ *(Sec. 5.8)* Compare this basin with the Orientale Basin on the Moon (Figure 5.28). *Messenger* studies of this basin and the surrounding surface have been instrumental in demonstrating that volcanism was mainly responsible for the intercrater plains.

Notice the young, extensively rayed craters (arrows).

R I V U X G

▲ **FIGURE 6.9 Mercury Mosaic** Mercury is imaged here as a mosaic of photographs—a composite image constructed from many individual images—taken by the *Messenger* spacecraft in 2011 as it orbited the planet. Notice the young, extensively rayed craters, here imaged with a resolution of a few kilometers. *(NASA)*

CONCEPT CHECK

How do scarps on Mercury differ from geological faults on Earth?

This cliff extends about 400 km long and is 3 km high in places.

100 km

R I V U X G

▶ **FIGURE 6.10 Mercury's Surface** Scarps on Mercury's surface are cliffs that appear to have formed when the planet's crust cooled and shrank early in its history, causing a crease in the surface. This is Discovery Scarp, as imaged by *Messenger* in 2013. *(NASA)*

▲ FIGURE 6.11 Hollows on Mercury   *Messenger* also discovered numerous peculiar "hollows"—features not seen anywhere else in the solar system. They appear here as lighter-colored depressions on this crater's rim and floor. *(NASA)*

(a)

(b)

▲ FIGURE 6.12 Caloris Basin   Mercury's most prominent geological feature—the Caloris Basin—measures about 1400 km across and is ringed by concentric mountain ranges that reach more than 3 km high in places. This huge circular basin, shown here in orange in this false-colored visible image from *Messenger,* is similar in size to the Moon's Mare Imbrium and spans more than half of Mercury's radius. *(NASA)*

6.5 The Surface of Venus

Although the clouds of Venus are thick and the surface is totally shrouded, we are by no means ignorant of the planet's surface. Astronomers have bombarded Venus with radio signals from Earth and from orbit, and analysis of the radar echoes yields a map of surface features. Except for Figure 6.17, all the views of Venus in this section are "radargraphs" (not photographs) created in this way.

Large-Scale Topography

Figure 6.13(a) shows a relatively low-resolution map of Venus made by *Pioneer Venus* in 1979. Surface elevation above the average radius of the planet's surface is indicated by color, with white representing the highest elevations, blue the lowest. (Note that the blue has nothing to do with oceans!) For comparison, Figure 6.13(b) shows a map of Earth to the same scale and at the same spatial resolution. Figure 6.14 is a 1995 mosaic of *Magellan* images of Venus. The orange color is based on optical data returned from spacecraft that landed on the planet.

Venus's surface appears to be mostly smooth, resembling rolling plains with modest highlands and lowlands. Only two continent-sized features, called Ishtar Terra and Aphrodite Terra, adorn the landscape, and these contain mountains

◀ FIGURE 6.13 Venus Radar Map   (a) Radar map of the surface of Venus, based on *Pioneer Venus* data. Color represents elevation, with white the highest areas and blue the lowest. (b) A similar map of Earth, at the same spatial resolution. *(NASA)*

▶ **FIGURE 6.14 Venus Magellan Map**   A planetwide mosaic of Venus made from *Magellan* images, colored in roughly the same way as Figure 6.13 to represent altitude: blue is lowest, white highest. *(NASA)*

This large, dragon-shaped continent is called Aphrodite Terra. ....➤

comparable in height to those on Earth. The highest peaks rise some 14 km above the level of the deepest surface depressions. (The highest point on Earth, the summit of Mount Everest, lies about 20 km above the deepest section of the ocean floor.) The elevated "continents" occupy only 8 percent of Venus's total surface area, whereas continents on Earth make up about 25 percent of our planet's surface.

The larger continent-sized formation, Aphrodite Terra, is located on Venus's equator. It is comparable in size to Africa. Before *Magellan*'s arrival, some researchers had speculated that Aphrodite Terra might have been the site of something equivalent to seafloor spreading on Earth, where two tectonic plates moved apart and molten rock rose to the surface in the gap between them, forming an extended ridge. This is just what is happening today at Earth's Mid-Atlantic Ridge, which is clearly visible in Figure 6.13(b). However, the *Magellan* images seem to rule out any plate-tectonic activity on Venus, and the Aphrodite region shows no signs of spreading. The crust appears buckled and fractured, suggesting large compressive forces, and there seem to have been numerous periods when extensive lava flows occurred.

Volcanism and Cratering

Although erosion by the planet's atmosphere may play some part in obliterating surface features, the most important factor is volcanism (volcanic activity), which appears to resurface the planet every few hundred million years. Many areas of Venus have volcanic features. Figure 6.15(a) shows a *Magellan* image of seven pancake-shaped lava domes, each about 25 km across—a little bigger than Washington, D.C. They probably formed when lava oozed out of the surface, formed the dome, then withdrew, leaving the crust to crack and subside. Lava domes such as these are found in several locations on Venus. Figure 6.15(b) shows a computer-generated three-dimensional view of the domes.

The most common volcanoes on the planet are of the type known as **shield volcanoes.** Those on Earth are associated with lava welling up through a "hot spot" in the crust (like the Hawaiian Islands). They are built up over long periods of time by successive eruptions and lava flows. A characteristic of shield volcanoes is the formation of a *caldera,* or crater, at the summit when the underlying lava withdraws and the surface collapses. A large shield volcano, called Gula Mons, is shown (again as a computer-generated view) in Figure 6.15(c).

 ANIMATION/VIDEO
Topography of Venus

 ANIMATION/VIDEO
Flight Over Alpha Regio

▼ **FIGURE 6.15 Venus Surface Features**   (a) These dome-shaped structures on Venus resulted when molten rock bulged out of the ground and then retreated, leaving behind a thin, solid crust that later cracked and subsided. (b) A computer-generated three-dimensional representation of four of the lava domes. (c) A *Magellan* view of the large shield volcano known as Gula Mons, whose volcanic caldera at the summit is about 100 km across and 4 km high. *(NASA)*

(a)

(b)

(c)

▶ FIGURE 6.16 **Venus Corona** This corona, called Aine, lies in the plains south of Aphrodite Terra and is about 300 km across. Note the pancake-shaped lava domes at top, the many fractures in the crust around the corona, and the large impact craters with their surrounding white (rough) ejecta blankets that stud the region. *(NASA)*

The largest volcanic structures on Venus are huge, roughly circular regions known as **coronae.** A large corona, called Aine, can be seen in Figure 6.16. Coronae are unique to Venus. They appear to be the result of upwelling motions in the mantle that caused the surface to bulge outward, but never developed into full-fledged convection as on Earth. Coronae generally have volcanoes both in and around them, and their rims usually show evidence of extensive lava flows into the plains below.

Two pieces of indirect evidence suggest that volcanism on Venus continues today. First, the level of sulfur dioxide above Venus's clouds shows large and fairly frequent fluctuations, possibly the result of volcanic eruptions on the surface. Second, orbiting spacecraft have observed bursts of radio energy from the planet's surface similar to those produced by the lightning discharges that often occur in the plumes of erupting volcanoes on Earth. However, these pieces of evidence are circumstantial. No "smoking gun" (or erupting volcano) has yet been seen, so the case for active volcanism is not yet complete.

ANIMATION/VIDEO **MA**

Flight Over Sif Mons Volcano

▼ FIGURE 6.17 **Venus in Situ** One of the first true-color views of the surface of Venus, radioed back to Earth from the Russian *Venera 14* spacecraft, which made a soft landing in 1975. The amount of sunlight penetrating Venus's cloud cover is about the same as that reaching Earth's surface on a heavily overcast day. *(Russian Space Agency)*

A few Soviet *Venera* spacecraft have landed on Venus's surface. Each survived for about an hour before being destroyed by the intense heat, its electronic circuitry melting in this planetary oven. Figure 6.17 shows one of the first photographs of the surface of Venus radioed back to Earth. The flat rocks visible in this image show little evidence of erosion and are apparently quite young, supporting the idea of ongoing surface activity. Later Soviet landers performed simple chemical analyses of the surface. Some of the samples studied were found to be predominantly basaltic, again implying a volcanic past. Others resembled terrestrial granite.

Not all the craters on Venus are volcanic in origin. Some were formed by meteoritic impact. The largest

(b)

▲ **FIGURE 6.18 Impact Craters on Venus**   (a) Venus's largest crater, named after anthropologist Margaret Mead, has a double-ringed structure. (b) A *Magellan* image of a multiple-impact crater in Venus's southern hemisphere displays irregularly shaped light-colored ejected debris that was probably caused by a meteorite that fragmented just prior to impact; dark regions in the crater may be pools of solidified lava. *(NASA)*

(a)

R I V U X G

impact craters on Venus are generally circular, but those less than about 15 km in diameter can be quite asymmetric. Figure 6.18(b) shows a *Magellan* image of a relatively small impact crater, about 10 km across, in Venus's southern hemisphere. Geologists think that the light-colored region is an ejecta blanket—material thrown from the crater following the impact. Its irregular shape may be the result of a large meteorite that broke up just before impact, with the separate pieces hitting the surface near one another. This seems to be a fairly common fate for medium-sized bodies (1 km or so in diameter) that plow through Venus's dense atmosphere.

Figure 6.18(a) shows the largest-known impact feature on Venus, the 280-km-diameter crater Mead. Its double-ringed structure is in many ways similar to the Moon's Mare Orientale. ∞ *(Sec. 5.8)* Numerous impact craters, again identifiable by their ejecta blankets, can also be seen in Figure 6.16.

(MA)° SELF-GUIDED TUTORIAL
SuperSpaceship—Voyage to Venus

CONCEPT CHECK

Are the volcanoes on Venus mainly associated with the movement of tectonic plates, as on Earth?

6.6 The Surface of Mars

Earth-based observations of Mars at closest approach can distinguish surface features as small as 100 km across—about the same resolution as the unaided human eye can achieve when viewing the Moon. However, when Mars is closest to us and most easily observed, it is also full, so the angle of the Sun's rays does not permit us to see any topographical detail, such as craters or mountains. Even through a large telescope Mars appears only as a reddish disk with some light and dark patches and prominent polar caps (Figure 6.6).

Mars's surface features undergo slow seasonal changes over the course of a Martian year—a consequence of Mars's axial tilt and somewhat eccentric orbit. The polar caps grow or shrink according to the seasons, almost disappearing during the Martian summer. The dark features also vary in size and shape. To fanciful observers around the start of the 20th century, these changes suggested the annual growth of vegetation—and much more (see *Discovery 6-1*)—but, as with Venus, these speculations were not confirmed. The changing polar caps are mostly frozen carbon dioxide, not water ice as at Earth's North and South Poles (although

6.1 DISCOVERY

Martian Canals?

The year 1877 was an important one in the human study of the planet Mars. The Red Planet came unusually close to Earth, affording astronomers an especially good view. Of particular note was the discovery, by U.S. Naval Observatory astronomer Asaph Hall, of the two moons circling Mars. But most exciting was the report of the Italian astronomer Giovanni Schiaparelli on his observation of a network of linear markings that he termed *canali*. In Italian, *canali* can simply mean "grooves" or "channels," but it can also mean "canals." As far as we know, Schiaparelli did not intend to imply that the canali were anything other than natural, but the word was translated into English as *canals*, suggesting that the grooves had been constructed by intelligent beings. The world's press (especially in the United States) sensationalized these observations, and some astronomers began drawing elaborate maps of Mars, showing oases and lakes where canals met in desert areas.

Percival Lowell (see photo), a successful Boston businessman (and brother of the poet Amy Lowell and Harvard president Abbott Lawrence Lowell), became fascinated by these reports. He abandoned his business and purchased a clear-sky site at Flagstaff, Arizona, where he built a major observatory. He devoted his life to achieving a better understanding of the Martian "canals." In doing so, he championed the idea that Mars was drying out

and that an intelligent society had constructed the canals to transport water from the wet poles to the arid equatorial deserts.

Alas, the Martian valleys and channels photographed by robot spacecraft during the 1970s are far too small to be the canali that Schiaparelli, Lowell, and others thought they saw on Mars. The entire episode represents a classic case in the history of science—a case in which well-intentioned observers, perhaps obsessed with the notion of life on other worlds, let their personal opinions and prejudices seriously affect their interpretations of reasonable data. The accompanying figures of Mars show how surface features (which were probably genuinely observed by astronomers at the turn of the century) might have been imagined to be connected. The figure on the left is a photograph of how Mars actually looked in a telescope at the end of the 19th century. The sketch at right is an interpretation (done at the height of the canal hoopla) of the pictured view. The human eye, under physiological stress, tends to connect dimly observed yet distinctly separated features. Humans saw patterns and canals where none in fact existed.

The chronicle of the Martian canals illustrates how the scientific method requires scientists to acquire new data to sort out sense from nonsense, fact from fiction. Rather than simply believing the claims about the Martian canals, other scientists demanded further observations to test Lowell's hypothesis. Eventually, improved observations, climaxing in the *Mariner* and *Viking* exploratory missions to the Red Planet nearly a century after all the fuss began, totally disproved the existence of canals. It often takes time, but the scientific method does eventually lead to progress in understanding reality.

(Lowell Observatory)

(Lowell Observatory)

smaller "residual" caps of water ice, which persist even through the warmer summer months, also exist). The dark regions are just highly cratered and eroded areas on the surface. During summer in the Martian southern hemisphere, planetwide dust storms sweep up the dry dust and carry it aloft, sometimes for months at a time, eventually depositing it elsewhere on the planet. Repeated covering and uncovering of the Martian landscape gives the impression from a distance of surface variability, but it is only the thin dust cover that changes.

ANIMATION/VIDEO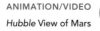

Hubble View of Mars

▶ **FIGURE 6.19 Mars Globe**   This highly detailed mosaic of Mars is based on images from a *Viking* spacecraft in orbit around the planet. Mars's Tharsis region, 5000 km across, bulges out from the equator, rising to a height of about 10 km. The two large volcanoes on the left mark the approximate peak of the Tharsis bulge. Dominating the center of the field of view is a vast "canyon" known as Valles Marineris—the Mariner Valley. *(NASA)*

Large-Scale Topography

There have been far more space missions to Mars than to any other planet, and many of these missions have played important roles in our understanding of the planet. Figure 6.19 shows a detailed mosaic of Mars, constructed from images made by one of the U.S. *Viking* spacecraft in orbit around the planet.

A striking feature of the terrain of Mars is the marked difference between the northern and southern hemispheres (Figure 6.20). The northern hemisphere is made up largely of rolling volcanic plains, somewhat like the lunar maria but much larger than any plains found on Earth or the Moon. They were apparently formed by eruptions involving enormous volumes of lava. The plains are strewn with blocks of volcanic rock, as well as with boulders blasted out of impact areas by infalling meteoroids. The southern hemisphere consists of heavily cratered highlands lying several kilometers above the level of the lowland north.

The northern plains are cratered much less than the southern highlands, suggesting that the northern surface is younger—perhaps 3 billion years old, compared with 4 billion in the south. In places, the boundary between the southern highlands and the northern plains is quite sharp. The surface level can drop by as much as 4 km in a distance of 100 km or so. Most scientists assume that the southern terrain is the original crust of the planet. How most of the northern hemisphere could have been lowered in elevation and flooded with lava remains a mystery.

▲ **FIGURE 6.20 Mars, Up Close**   (a) Mars's northern hemisphere consists of rolling volcanic plains. (b) The southern Martian highlands are heavily cratered. Both of these *Mars Express* photographs are in true color and show roughly the same scale, nearly 1000 km across. *(ESA)*

▲ FIGURE 6.21 **Mars Map** INTERACTIVE A computer-generated map of planet Mars, based on detailed measurements made by *Mars Global Surveyor.* Color represents height of the surface according to the scale at right; note the great difference in elevation between the northern and southern hemispheres. Some surface features are labeled, such as the Mariner Valley and the Hellas impact basin, as are the *Viking, Pathfinder, Exploration Rover,* and *Phoenix* robot landing sites. *(NASA)*

ANIMATION/VIDEO **MA**®
Flight Over Columbia Hills

ANIMATION/VIDEO **MA**®
Flight Over Tharsis

ANIMATION/VIDEO **MA**®
Flight Over Mariner Valley

In the late 1990s, NASA's *Mars Global Surveyor* satellite mapped the entire Martian surface to an accuracy of a few meters. Figure 6.21 shows a map of Mars based on these data, with some prominent surface features marked. The north–south asymmetry is very obvious in this map. The major geological feature on the planet is the *Tharsis bulge,* visible at the left of Figure 6.19 and marked in Figure 6.21. Roughly the size of North America, Tharsis lies on the Martian equator and rises some 10 km above the rest of the Martian surface. To the east and west of Tharsis lie wide depressions, hundreds of kilometers across and up to 3 km deep. If we wished to extend the idea of "continents" from Earth and Venus to Mars, we would say that Tharsis is the only continent on the Martian surface. However, as on Venus, there is no sign of plate tectonics—the continent of Tharsis is not drifting like its Earth counterparts. ∞ *(Sec. 5.5)* Tharsis appears to be even less heavily cratered than the northern hemisphere, making it the youngest region on the planet. It is estimated to be only 2–3 billion years old.

Associated with the Tharsis bulge is a great canyon known as Valles Marineris (the Mariner Valley). Shown in its entirety in Figure 6.21 and cutting across the center of Figure 6.19, this enormous crack in the Martian surface is not really a canyon in the terrestrial sense, because running water played no part in its formation. Geologists think it was formed by the same crustal forces that pushed the Tharsis region upward, causing the surface to split and crack. Cratering studies suggest that Valles Marineris is at least 2 billion years old. Similar (but smaller) cracks, formed in a similar way, have been found in the Aphrodite Terra region of Venus.

Valles Marineris runs for almost 4000 km along the Martian equator and extends about one-fifth of the way around the planet. At its widest, it is 120 km across, and it is as deep as 7 km in places. Earth's Grand Canyon in Arizona would easily fit into one of its side "tributary" cracks. It is so large that it can even be seen from Earth. This Martian geological feature is not a result of Martian plate tectonics, however. For some reason, the crustal forces that formed it never developed into full-fledged plate motion as on Earth.

Almost diametrically opposite Tharsis lies the Hellas Basin (Figure 6.21). Located in the southern highlands, it paradoxically contains the lowest point on the Martian surface. Some 3000 km across, the floor of the basin lies nearly 9 km below its rim and over 6 km below the average level of the planet's surface. Its shape and structure identify it as an impact feature. Its formation must have caused a major redistribution of the young Martian crust—maybe even enough to account for a substantial portion of the highlands, according to some researchers. Its heavily cratered floor indicates that the impact occurred very early in Martian history—perhaps 4 billion years ago.

The giant Borealis Basin around the Martian north pole—most of the blue region at the top of Figure 6.21 (see also Figure 6.26)—may be the result of one of the largest known impacts in the solar system. Recent research, based on computer simulations of the collision and detailed data from the *Mars Global Surveyor* and *Mars Reconnaissance Orbiter* spacecraft, suggests that the basin could have formed when a giant impactor some 2000 km across—twice the size of the largest asteroid, Ceres—struck the planet a grazing blow during the formative stages of the solar system. ∞ *(Sec. 4.3)* These ideas are controversial and still hotly debated by planetary scientists, but proponents claim that the resulting impact feature would be comparable in size to the observed basin, and the collision might explain why the northern hemisphere of Mars is so much lower than and differs so radically from the south.

Martian Volcanism

Mars contains the largest known volcanoes in the solar system. Four particularly large ones are found on the Tharsis bulge, two of them visible in Figure 6.19. Biggest of all is Olympus Mons (Figure 6.22), which lies on the northwestern slope of Tharsis, just over the left horizon of Figure 6.19, but marked on Figure 6.21. Olympus Mons is 700 km in diameter at its base—only slightly smaller than the state of Texas—and rises to a height of 25 km above the surrounding plains.

Like the volcanoes on Venus, those on Mars are not associated with plate motion, but instead are shield volcanoes, sitting atop hot spots in the Martian mantle. Spacecraft images of the Martian surface reveal many hundreds of volcanoes. Most of the largest are associated with the Tharsis bulge, but many smaller ones are also found in the northern plains. It is not known whether any of them are still active. However, from the extent of impact cratering on their slopes, it appears that some of them erupted as recently as 100 million years ago.

The great height of Martian volcanoes is a direct consequence of the planet's low surface gravity. ∞ *(Sec. 5.1)* As a shield volcano forms and lava flows and spreads, the new mountain's height depends on its ability to support its own weight. The lower the surface gravity, the less the lava weighs and the higher the mountain can be. Mars has a surface gravity only 40 percent that of Earth, and its volcanoes rise roughly 2.5 times as high.

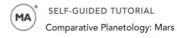

▲ **FIGURE 6.22 Martian Volcano** Olympus Mons, the largest volcano (now dormant) known on Mars or anywhere else in the solar system, is nearly three times taller than Mount Everest on Earth, measuring about 700 km across its base and rising 25 km high at its peak. By comparison, the largest volcano on Earth, Hawaii's Mauna Loa, measures a mere 120 km across and peaks just 9 km above the Pacific Ocean floor. *(NASA)*

(MA)° **SELF-GUIDED TUTORIAL**
Comparative Planetology: Mars

PROCESS OF SCIENCE CHECK

Use examples drawn from the study of Mercury, Venus, or Mars to illustrate how improvements in observational data can radically change the interpretation of a scientific theory.

▲ **FIGURE 6.24 Martian Outflow**   This outflow channel near the Martian equator bears witness to a catastrophic flood that occurred about 3 billion years ago. *(NASA)*

▲ **FIGURE 6.23 Martian Channel**   (a) This runoff channel on Mars is about 400 km long and up to 5 km wide in places. (b) The Red River on Earth runs from the Texas Panhandle to the Mississippi River. The two differ mainly in that there is currently no liquid water in this, or any other, Martian valley. *(ESA/NASA)*

Evidence for Past Water on Mars

Although the Valles Marineris was not formed by running water, photographic evidence reveals that liquid water once existed in great quantity on the surface of Mars. Two types of flow features were seen by the *Viking* orbiters: **runoff channels** and **outflow channels.**

The runoff channels (Figure 6.23) are found in the southern highlands. They are extensive systems—sometimes hundreds of kilometers in total length—of interconnecting, twisting channels that merge into larger, wider channels. They bear a strong resemblance to river systems on Earth, and geologists think that this is just what they are: the dried-up beds of long-gone rivers that once carried rainfall on Mars from the mountains down into the valleys. The systems of runoff channels are often referred to as *valley networks*. They speak of a time 4 billion years ago (the age of the Martian highlands) when the atmosphere was thicker, the surface warmer, and liquid water widespread.

The outflow channels (Figure 6.24) are probably relics of catastrophic flooding on Mars long ago. They appear only in equatorial regions and generally do not form extensive interconnected networks like the runoff channels. Instead, they are probably the paths taken by huge volumes of water draining from the southern highlands into the northern plains. Judging from the width and depth of the channels, the flow rates must have been truly enormous—perhaps as much as a hundred times greater than the 10^5 tons per second carried by the Amazon River, the largest river system on Earth. Flooding shaped the outflow channels about 3 billion years ago, about the same time as the northern volcanic plains formed.

Scientists speculate that Mars may have enjoyed an extended early period during which rivers, lakes, and perhaps even oceans adorned its surface. Figure 6.25 is

◀ **FIGURE 6.25 Martian River Delta**   Did this fan-shaped region of twisted streams form as a river flowed into a larger sea? If it did, the *Mars Global Surveyor* image supports the idea that Mars once had large bodies of liquid water on its surface. Not all scientists agree with this interpretation, however. *(NASA)*

a *Mars Global Surveyor* image showing what mission specialists think may be a delta—a fan-shaped network of channels and sediments where a river once flowed into a larger body of water, in this case a lake filling a crater in the southern highlands. Other researchers go even farther, suggesting that the data provide evidence for large open expanses of water on the early Martian surface. Figure 6.26(a) is a computer-generated view of the Martian north polar region, showing the extent of what may have been an ancient ocean covering much of the northern lowlands. The Hellas basin (Figure 6.21) is another candidate for an ancient Martian sea.

These ideas are controversial. Proponents point to features such as the terraced "beaches" shown in Figure 6.26(b), which might have been left behind as a lake or ocean evaporated and the shoreline receded. But detractors maintain that the terraces could also be due to geological activity, perhaps related to tectonic forces that depressed the northern hemisphere far below the level of the south, in which case they have nothing whatever to do with Martian water. Furthermore, *Mars Global Surveyor* data seem to indicate that the Martian surface contains too few *carbonate* rock layers—layers containing compounds of carbon and oxygen—that should have been formed in abundance in an ancient ocean. Their absence supports the picture of a cold, dry Mars that never experienced the extended mild period required to form lakes and oceans. Still, data from the most recent *NASA* landers (see below) imply that at least some parts of the planet did in fact experience long periods in the past during which liquid water existed on the surface. The debate is far from over.

Where Is the Water Today?

Astronomers have no direct evidence for liquid water anywhere on the Martian surface today, and the amount of water vapor in the Martian atmosphere is tiny. Yet the extent of the outflow channels and the other evidence just described indicates that a huge total volume of water existed on Mars in the past. Where did all that water go? The most likely answer is that much of Mars's original water is now locked in a layer of **permafrost,** a layer of water ice lying just below the planet's surface, much like that found in Earth's arctic regions, with more contained in the polar caps.

Figure 6.27 shows indirect evidence for the permafrost layer in the form of a fairly typical Martian impact crater named Yuty. Unlike the lunar craters discussed in Chapter 5, Yuty's ejecta blanket gives the distinct impression of a liquid that has splashed or flowed out of the crater. Most likely, the explosive impact heated and liquefied the permafrost, resulting in the fluid appearance of the ejecta. ∞ *(Sec. 5.4)* More direct evidence for subsurface ice came in 2002, when the *Mars Odyssey* orbiter detected extensive deposits of water ice crystals (actually, the hydrogen they contain) mixed with the Martian surface layers. In some locations ice appears to make up as much as 50 percent by volume of the planet's soil.

As noted earlier, Mars's polar caps actually consist of two distinct components. The *seasonal caps* vary in size as atmospheric carbon dioxide alternately freezes and evaporates during the winter and summer months. However, the permanently frozen *residual caps* (Figure 6.28) are now known to be composed of

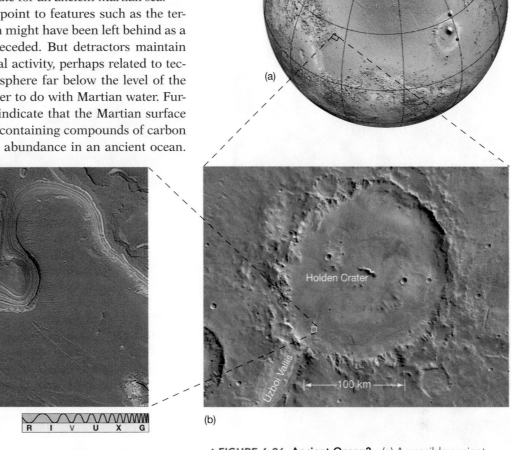

(a)

Holden Crater

Uzboi Vallis

|← 100 km →|

(b)

R   I   V   U   X   G

▲ **FIGURE 6.26 Ancient Ocean?**   (a) A possible ancient Martian ocean might once have spanned the polar regions. The blue areas in this computer-generated map indicate depth below the average radius of the planet, thus approximate possible ocean depth. (Color scale is the same as in Figure 6.21.) (b) This high-resolution image shows tentative evidence for erosion by standing water in the floor of Holden Crater, about 140 km across. *(NASA)*

The residual (permanent) polar caps on Mars may be major storehouses for water on the planet. This is the smaller southern cap, roughly 350 km in diameter. *(ESA)*

▲ **FIGURE 6.27 Martian Crater**   The ejecta from Mars's crater Yuty evidently was once liquid. This type of crater is sometimes called a "splosh" crater. *(NASA)*

(a)

1999   2005

(b)

water ice. The presence of ice in the larger northern cap has long been known from spectroscopic observations of water vapor above them as some small fraction of the cap evaporates in the Sun's heat during the summer months. The composition of the southern residual cap was finally established only in 2004 by spectroscopic imaging observations made by the European Space Agency's *Mars Express* orbiter. The thickness of the caps is uncertain, but it is quite possible that they represent a major storehouse for water on Mars.

Prior to the arrival of *Mars Global Surveyor*, astronomers thought that all the water on and below the Martian surface existed in the form of ice. However, in 2000, *Surveyor* mission scientists reported the discovery of numerous small-scale "gullies" in Martian cliffs and crater walls that apparently were carved by running water in the relatively recent past. Figure 6.29(a) shows one such gully, found in the inner rim of a Martian impact crater in the southern highlands. Its structure has many similarities to the channels carved by flash floods on Earth. The ages of these intriguing features are uncertain, and might be as great as a million years in some cases, but some *Surveyor* data suggest that some of them may still be active today, implying that liquid water might exist in some regions of Mars at depths of less than 500 meters. Some scientists have disputed this interpretation, however, arguing that the "fluid" responsible for the gullies could have been solid (granular) or even liquid carbon dioxide, expelled under pressure from the Martian crust.

Further data taken from orbit have deepened the mystery of the Martian gullies. Figure 6.29(b) shows two images of an unnamed impact crater in the southern highlands. The white streak in the second image is thought (by some) to be a frozen mudslide, where liquid water briefly flowed down the inside of the crater wall, carrying rocky debris with it, then froze on the chilly Martian surface. Its composition is uncertain, but one thing is clear—whatever it is, it formed recently, demonstrating that the production of these features is an ongoing process. In September 2015, NASA scientists announced that *Mars Reconnaissance Observer* had found spectroscopic

◄ **FIGURE 6.29 Running Water on Mars?**   (a) This high-resolution *Mars Global Surveyor* view of a crater wall near the Mariner Valley shows evidence of "gullies" apparently formed by running water in the relatively recent past. (b) Comparison of two images taken 6 years apart of another Martian impact crater shows that something—the white streak (lower right), possibly water—flowed across the surface. *(NASA)* evidence of hydrated minerals in similar gullies, strongly supporting the view that these features did indeed form from flowing salty water in the relatively recent past.

Scientists think that 4 billion years ago, as the Martian climate changed, the running water that formed the runoff channels began to freeze, forming the permafrost and drying out the river beds. Mars remained frozen for about a billion years, until volcanic (or some other) activity heated large regions of the surface, melting the permafrost and causing the flash floods that created the outflow channels. Subsequently, volcanic activity subsided, the water refroze, and Mars once again became a dry world. The present level of water vapor in the Martian atmosphere is the maximum possible given the atmosphere's present density and temperature. Estimates of the total amount of water stored as permafrost and in the polar caps are uncertain, but it is likely that if all the water on Mars were to become liquid, it would cover the surface to a depth of roughly 10 meters.

Exploration by Martian Landers

Seven U.S. spacecraft have successfully landed on the Martian surface. Their landing sites, marked on Figure 6.21, spanned a variety of Martian terrains. Their goals included detailed geological and chemical analyses of Martian surface rocks, the search for life, and the search for water (an essential requirement for life as we know it—see Chapter 18).

Both U.S. *Viking* missions dispatched landers to the Martian surface. Figure 6.30 shows the view from *Viking* 2, which touched down near the planet's equator just east of Tharsis. The photograph shows a windswept, desolate plain littered with rocks of all sizes, not unlike a high desert on Earth. This view may be quite typical of the northern Martian plains. The *Viking* landers performed numerous chemical analyses of the rock on Mars's surface and performed experiments designed specifically to test for life (see *Discovery* 6-2). One important finding of these studies was the high iron content of the planet's crust. Chemical reactions between the iron-rich surface soil and oxygen in the atmosphere are responsible for the iron oxide (rust) that gives Mars its characteristic red color.

The next successful mission to the Martian surface was *Mars Pathfinder* in 1997. The lander performed measurements of the Martian atmosphere and atmospheric dust while its robot rover *Sojourner* carried out chemical analyses of the soil and rocks within about 50 m of the parent craft. The soil chemistry in the vicinity of the landing site was similar to that found by the *Viking* landers. Interestingly, analyses of the nearby rocks revealed chemical makeups different from that of the Martian meteorites found on Earth (see *Discovery* 6-2). The landing site for the *Pathfinder* mission had been chosen to lie near the mouth of an outflow channel, and the properties of the many rocks and boulders surrounding the lander were indeed consistent with their having been deposited there by floodwater.

In 2004 the twin landers *Spirit* and *Opportunity* of the *Mars Exploration Rover* mission arrived at their targets on opposite sides of the planet (see Figure 6.21). The two robots were designed to operate for just 3 months, but each continued to send back vital data for several years. During their lifetime on Mars, the landers roamed the surface within several kilometers of their landing sites, making chemical and geological studies of rocks they encountered. The primary goal of the mission was to look for evidence of liquid water on Mars at any time in the past, and in that they succeeded. The rovers' findings

▲ FIGURE 6.30 Viking View   This is the view from the *Viking 2* spacecraft, now parked on Mars's northern Utopian plains. The fine-grained soil and the reddish rock-strewn terrain stretching toward the horizon contain substantial amounts of iron ore. The sky is pale pink, the result of airborne dust. The discarded canister in the foreground is about 20 cm long; the 0.5-m scars in the dirt were made by the robot's shovel. (NASA)

6.2 DISCOVERY

Life on Mars?

Even before the *Viking* missions reached Mars in 1976, astronomers had abandoned hope of finding life on the planet. Scientists knew there were no large-scale canal systems, no surface water, almost no oxygen in the atmosphere, and no seasonal vegetation changes. The present lack of liquid water on Mars especially dims the chances for life there now. However, running water and possibly a dense atmosphere in the past may have created conditions suitable for the emergence of life long ago. In the hope that some form of microbial life might have survived to the present day, the *Viking* landers carried out experiments designed to detect biological activity. The first pair of photographs show the robot arm of one of the landers digging a shallow trench.

All three *Viking* biological experiments assumed some basic similarity between hypothetical Martian bacteria and those found on Earth. A *gas-exchange* experiment offered a nutrient broth to any residents of a sample of Martian soil and looked for gases that would signal metabolic activity. A *labeled-release* experiment added compounds containing radioactive carbon to the soil, then waited for results signaling that Martian organisms had either eaten or inhaled this carbon. Finally, a *pyrolitic-release* experiment added radioactively tagged carbon dioxide to a sample of Martian soil and atmosphere, waited awhile, then removed the gas and tested the soil (by heating it) for signs that something had absorbed the tagged gas. In all cases, contamination by terrestrial bacteria was a major concern. Indeed, any release of Earth organisms would have invalidated these and all future such experiments on Martian soil. Both *Viking* landers were carefully sterilized prior to launch.

Initially, all three experiments appeared to be giving positive signals! However, subsequent studies showed that the results could all be explained by inorganic (that is, nonliving) chemical reactions. Thus, we have no evidence for even microbial life on the Martian surface. The *Viking* robots detected peculiar reactions that mimic in some ways the basic chemistry of living organisms, but they did not detect life itself.

One criticism of the *Viking* experiments is that they searched only for life now living. Today, Mars seems locked in an ice age—the kind of numbing cold that would prohibit sustained life as we know it. If bacterial life did arise on an Earth-like early Mars, however, then we might be able to find its fossilized remains preserved on or near the Martian surface. This is one reason that scientists are so eager to land more spacecraft on the planet, particularly at the poles, whose ice caps may offer the best environment for finding Martian life (or its remains).

Surprisingly, one alternative place to look for evidence of life on Mars is right here on Earth. The second figure shows ALH84001, a blackened, 2-kg meteorite about 17 cm across, found in 1984 in Antarctica. Chemical analysis strongly indicates that it originated on Mars. It was apparently blasted off that planet long ago by a meteoritic impact of some sort, thrown into space, and eventually captured by Earth's gravity. Based on the estimated cosmic-ray exposure it received before reaching Earth, it left Mars about 16 million years ago.

On the basis of all the data accumulated from studies of ALH84001, in 1996 a group of scientists argued that they had discovered fossilized evidence for life on Mars. They pointed to globular structures similar to those produced by bacteria on Earth (top left inset), the presence of chemical compounds sometimes associated with Earth biology, and curved, rodlike structures (right image) resembling Earth bacteria, which the researchers interpreted as fossils of primitive Martian organisms. In 1999 the team released an

(NASA)

R   I   V   U   X   G

(Before)   (After)

provided strong evidence for standing water on the ancient Martian surface and changed the minds of many skeptical scientists on the subject of water on Mars.

Spirit's landing site was rocky and similar in many ways to the terrains encountered by earlier landers, although closer study by the science team revealed that most of the rocks in the lander's vicinity appear to have been extensively altered by water long ago. Halfway around the planet, *Opportunity* appears analysis of a second meteorite also thought to have come from Mars, again reporting evidence for microbial life having similarities in size, shape, and arrangement to known nanobacteria on Earth.

These claims were and remain very controversial. Many experts categorically disagree that evidence of Martian life has been found—not even fossilized life. They maintain that, as in the case of the *Viking* experiments, the evidence could all be due to chemical reactions not requiring any kind of biology. The scale of the supposed fossil structures is also important. They are only about 0.5 μm across, 1/30 the size of ancient bacterial cells found fossilized on Earth. Furthermore, several key experiments have not yet been done, such as testing the suspected fossils for evidence of cell walls or of any internal cavities where body fluids would have resided. Nor has anyone yet found in either meteorite any amino acids, the basic building blocks of life as we know it (see Chapter 18). In addition, there is the huge problem of contamination—after all, ALH84001 was found on Earth and apparently sat in the Antarctic ice fields for 13,000 years before being picked up by meteorite hunters.

As things now stand, it's a matter of interpretation—at the frontiers of science, issues are rarely clear-cut. Only additional analysis and new data—perhaps in the form of samples returned directly from the Martian surface—will tell for sure if primitive Martian life existed long ago. Most workers in the field seem to have concluded that, taken as a whole, the results do not support the claim of ancient life on Mars. Still, even some skeptics concede that as much as 20 percent of the organic material in ALH84001 could have originated on the Martian surface—although that is a far cry from proving the existence of life there.

Whether or not ALH84001 actually turns out to contain evidence of ancient Martian life, the scientific uproar it has caused has forced scientists to think much more carefully about the kinds of life that might be able to survive in extreme environments—those with temperatures, pressures, or chemical makeup not normally regarded as hospitable to life as we know it on Earth. These studies have fueled the emerging field of *exobiology*—the search for and study of life on other planets, in our own solar system and beyond. As we will see in Chapters 8 and 18, the subsurface oceans of Jupiter's moon Europa and the hydrocarbon seas of Saturn's frigid moon Titan have joined Mars as prime candidates in the search for extraterrestrial life.

Should the claim of life on Mars hold up against the weight of healthy skepticism in the scientific community, these findings may go down in history as one of the greatest scientific discoveries of all time. We are—or at least were—not alone in the universe! Maybe . .

MA ANIMATION/VIDEO
Meteorites Ejected
From Mars

MA ANIMATION/VIDEO
Microscopic Martian
Fossils?

Carbonate globule

Structure inside globules

⊢— 5 cm —⊣

(NASA)

to have hit the jackpot in its quest, finding itself surrounded by rocks showing every chemical and geological indication of having been very wet—possibly immersed in salt water—in the distant past. The rocks near *Opportunity*'s landing site seem to have been alternately underwater and dry for extended periods of time, possibly as a shallow lake alternately filled and evaporated numerous times during the course of Martian history.

▶ **FIGURE 6.31 Phoenix** The *Phoenix* lander's robotic arm is shown here with a surface sample in its scoop, just before delivering it to an onboard chemistry laboratory. The circular hardware at bottom is one of its solar panels. The inset shows one of the first trenches (about the size of this book) dug by *Phoenix* to a depth of about 8 centimeters. The white material near the top is almost surely water ice, which melted soon after excavation. *(NASA)*

ANIMATION/VIDEO (**MA**)

Mars Exploration Rover Landing

ANIMATION/VIDEO (**MA**)

Flight Over Opportunity at Gustav Crater

CONCEPT CHECK

Why do astronomers think that the Martian climate was once quite different from today's?

NASA's 2008 *Phoenix* mission landed in the planet's north polar region (see Figure 6.21). Scientists had long been eager to send a spacecraft to the Martian polar regions because of the expectation of finding water ice there. The spacecraft was not a rover, but instead contained a sophisticated array of equipment to collect and analyze the surrounding soil. *Phoenix* confirmed the presence of subsurface water ice at the landing site (Figure 6.31) and found clay and carbonates in the soil, both indicators of a wet environment at some time in the past, although scientists do not know if the water resulted from seasonally melting ice or is a feature of the much more distant past. *Phoenix* touched down in late fall in the Martian northern hemisphere, and during its final weeks, its sensors reported the first snow as winter closed in. The mission ended when diminishing sunlight and extreme low temperatures shut down the lander's power supply.

The most recent arrival on Mars is NASA's *Curiosity* rover (Figure 6.32), which soft-landed on the planet in 2012. This car-sized robot is by far the most sophisticated lander NASA has ever placed on the Martian surface. Mission goals include studying the Martian climate and geology, probing the past and present roles of water, determining whether the landing site (inside Gale Crater) has ever been conducive to microbial life, and evaluating the suitability of the planet for future human exploration.

Curiosity landed in a region thought to be an ancient streambed that may have hosted, in the distant past, flowing water up to a meter deep. The search for carbonate rocks, which would strongly support the idea of ancient rivers and oceans on Mars, has been a key element of the mission. However, despite spectroscopic evidence of carbonates from orbit and some early hints in the data, no definitive detection of carbonates has yet been made. The Martian soil does contain significant quantities of water and carbon dioxide, a fact that may become important when humans finally arrive on the planet.

The *Curiosity* analysis did reveal complex chemistry within the Martian soil, including minerals containing what are believed to be the key ingredients for life—carbon, hydrogen, oxygen, phosphorus, sulfur, and chlorine. The origin—biological or nonbiological—of these minerals remains unclear. Analyzing the salinity of nearby soil, the rover found that the environment might well have been habitable billions of years ago. In 2014 *Curiosity* measured a spike in the level of atmospheric methane in its vicinity. Since living organisms, as we know them, can produce methane, its detection supports the idea that life might have once existed there, although the evidence is far from conclusive.

The inset to Figure 6.32 shows holes bored into a surface rock by an on-board drill, allowing the first sample of subsurface Martian rock, presumably unaffected by weathering processes that may have altered the chemistry of the rock's outer layers, to be analyzed. In 2014, samples taken from a rock named Cumberland provided the first definitive detection of organic molecules on the Martian surface.

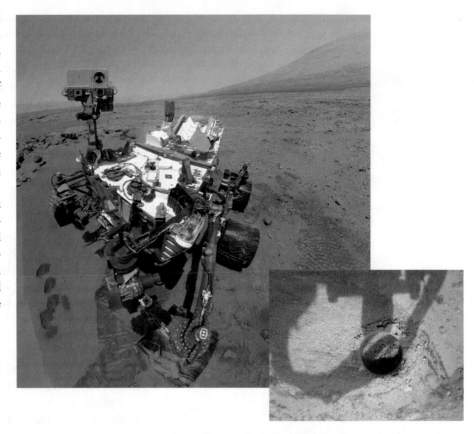

6.7 Internal Structure and Geological History

▲ **FIGURE 6.32 Curiosity**   This "selfie" of *Curiosity* on the floor of Gale crater on Mars was taken using a camera attached to the lander's sampling arm, which has subsequently been removed from the photo. The inset shows a 1.6-cm by 2-cm hole made in the Martian bedrock by *Curiosity*'s drill in search of an pristine rock. *(NASA/JPL)*

As with Earth and the Moon, scientists combine measurements of a planet's bulk properties with detailed observations of its gravity and magnetic field to build a model of the interior. ∞ *(Secs. 5.4, 5.7)* Our understanding of the internal structure and evolution of Mercury, Venus, and Mars has steadily grown as models have improved and spacecraft have radioed back more and more detailed data.

Mercury

Mercury's magnetic field, discovered by *Mariner 10*, is about 1/100 of Earth's field. The discovery that Mercury has a magnetic field came as a surprise to planetary scientists, who, having detected no magnetic field in the Moon, expected Mercury to have none either. In Chapter 5 we saw how a combination of liquid metal core and rapid rotation is necessary for the production of a planetary magnetic field. ∞ *(Sec. 5.7)* Mercury certainly does not rotate rapidly, and it may also lack a liquid metal core, yet a magnetic field undeniably surrounds it. Although weak, the field is strong enough to deflect the solar wind and create a small magnetosphere around the planet.

Before *Messenger*'s arrival, scientists thought it most likely that Mercury's magnetic field was a "fossil remnant" dating back to the distant past when the planet's core solidified. However, detailed *Messenger* observations now suggest that the field is in fact generated by dynamo action, as on Earth. Curiously, the field is significantly offset from the planet's center, making the field at the north pole much stronger than at the south. The details of how such a relatively strong and asymmetric field can be produced by a slowly rotating planet remain to be resolved.

Mercury's magnetic field and high average density (roughly (5400 kg/m^3) together imply that most of the planet's interior is dominated by a large, iron-rich core. *Messenger* measurements suggest a solid inner core with a radius of

Earth

Mars

Mercury

Moon

▲ FIGURE 6.33 **Terrestrial Planet Interiors**   The internal structures of Earth, the Moon, Mercury, and Mars, drawn to the same scale. Note how large a fraction of Mercury's interior is core. The interior structure of Venus is largely unknown, but is thought to be similar to that of Earth.

perhaps 1600 km, surrounded by a liquid outer core of radius 2100 km. A less-dense lunar-like mantle some 350 km thick lies above the core. The core accounts for some 80 percent of the planet's total mass. The ratio of core volume to total planet volume is greater for Mercury than for any other object in the solar system. Figure 6.33 illustrates the relative sizes and internal structures of Earth (and perhaps also Venus), the Moon, Mercury, and Mars. Mercury's large iron core seems to be a consequence of the planet's location in the hot inner regions of the early solar system when it formed some 4.6 billion years ago, as described in our discussion of the condensation theory in Chapter 4, although the core's size still poses problems to theorists. ∞ *(Sec. 4.3)*

Like the Moon, Mercury seems to have been geologically dead for roughly the past 4 billion years. Again as on the Moon, the lack of present-day geological activity on Mercury results from the mantle's being solid, preventing volcanism or tectonic motion. Scientists have pieced together the following outline of Mercury's early history.

When Mercury formed 4.6 billion years ago, its location in the hot inner region of the early solar system ensured a dense, largely metallic overall composition. ∞ *(Sec. 4.3)* During the next half-billion years, Mercury melted and differentiated, just as the other terrestrial worlds did. It suffered the same intense meteoritic bombardment as the Moon. Being more massive than the Moon, Mercury cooled more slowly, so its crust was thinner than the Moon's and volcanic activity more common. Lava erased more craters, leading to the intercrater plains found by *Mariner 10*.

As the planet's large iron core formed and then cooled, the planet began to shrink, causing the surface to contract. This compression produced the scarps seen on Mercury's surface and may have prematurely terminated volcanic activity by squeezing shut the cracks and fissures on the surface. Thus, Mercury did not experience the subsequent extensive volcanic outflows that formed the lunar maria. Despite its larger mass and greater internal temperature, Mercury has probably been geologically inactive for even longer than the Moon.

Venus

Venus has no detectable magnetic field or magnetosphere. Given that Venus's average density is similar to Earth's, it seems likely that Venus has an Earth-like overall composition and a partly molten iron-rich core. The lack of a magnetic field, then, is almost surely the result of the planet's extremely slow rotation. ∞ *(Sec. 5.7)*

Because none of the *Venera* landers carried seismic equipment, no direct measurements of the planet's interior have ever been made, and theoretical models of the interior have very little hard data to constrain them. The physical similarities between Venus and Earth suggest that the core/mantle structure of Venus resembles that of our own planet. However, to many geologists the surface of Venus resembles that of the young Earth, at an age of perhaps a billion years. At that time, volcanic activity had already begun on Earth, but the crust was still relatively thin and the convective processes in the mantle that drive plate tectonic motion were not yet established.

Why has Venus remained in that immature state and not developed plate tectonics as Earth did? That question is unanswered. Some geologists speculate that the high surface temperature on Venus has slowed the planet's cooling rate. Possibly the high surface temperature has made the crust too soft for Earth-style plates to develop. Or perhaps the high temperature and soft crust led to more volcanism, tapping energy that might otherwise have gone into convective motion.

Mars

In 1997 the orbiting *Mars Global Surveyor* succeeded in detecting a very weak Martian field, about 1/800 that of Earth. However, this is most likely a local anomaly, not a global field. Because Mars rotates rapidly, the weakness of the magnetic field is taken to mean that the planet's core is nonmetallic, or nonliquid, or both. ⚭ *(Sec. 5.7)*

Mars's small size means that any internal heat would have been able to escape more easily than in a larger planet like Earth or Venus. The evidence of ancient surface activity, especially volcanism, suggests that at least parts of the Martian interior must have melted at some time in the past, but the lack of current activity and the absence of any significant magnetic field indicate that the melting was never as extensive as on Earth. The latest data indicate that Mars's core has a diameter of about 2500 km, is composed largely of iron sulfide (a compound about twice as dense as surface rock), and is still at least partly molten.

Mars appears to be a planet where large-scale tectonic activity almost started but was stifled by the planet's rapidly cooling outer layers. On a larger, warmer planet, the upwelling that formed the Tharsis bulge might have developed into full-fledged plate tectonics, but the Martian mantle became too rigid, the crust too thick, for that to occur. Instead, although the upwelling fired ancient volcanic activity, geologically much of the planet died 2 billion years ago.

CONCEPT CHECK

Why do Venus and Mars lack magnetic fields?

6.8 Atmospheric Evolution on Earth, Venus, and Mars

Now that we have studied the structure and history of the three terrestrial worlds that still have atmospheres today—Venus, Earth, and Mars—let's return to the question of why their atmospheres are so different from one another.

MA ANIMATION/VIDEO
Terrestrial Planets II

The Runaway Greenhouse Effect on Venus

Why is Venus so hot? And if, as seems likely, Venus started off like Earth, then why is its atmosphere now so different from Earth's? The answer to the first question is easy. Given the present composition of its atmosphere, Venus is hot because of the *greenhouse effect*. ⚭ *(Sec. 5.3)* Venus's dense atmosphere is made up almost entirely of a prime greenhouse gas, carbon dioxide (Figure 6.34). This thick blanket absorbs about 99 percent of all the infrared radiation released from the surface of Venus and is the immediate cause of the planet's sweltering 730 K surface temperature. The answer to the second question is more complex, however, and requires us to consider in detail how the atmospheres of the terrestrial planets formed and evolved.

The terrestrial atmospheres did not come into being along with the planets themselves. Instead, they developed over many millions of years as **secondary atmospheres,** made up of gases released from the planets' interiors by volcanic activity—a process called *outgassing*. Volcanic gases are rich in water vapor, carbon dioxide, sulfur dioxide, and compounds containing nitrogen. On Earth, as the surface temperature fell and the water vapor condensed, oceans formed. Most of the carbon dioxide and sulfur dioxide dissolved in the oceans or combined with surface rocks. Solar ultraviolet radiation liberated nitrogen from its chemical bonds with other elements, and a nitrogen-rich atmosphere slowly appeared.

DATA POINTS

The Greenhouse Effect

Almost 40 percent of students encountered problems describing the greenhouse effect on the terrestrial planets. Some points to remember:

■ The greenhouse effect occurs when atmospheric gases (particularly carbon dioxide) prevent reemitted infrared radiation from escaping. The result is that the planet's temperature rises.

■ The strength of the greenhouse effect increases with the amount of carbon dioxide in the atmosphere. Mars has only a little, resulting in a weak effect. Earth has significantly more, raising our planet's temperature by some 40 K. Venus has a lot, resulting in very strong greenhouse heating.

The basic mechanism controlling the level of carbon dioxide in Earth's atmosphere is the competition between its production by volcanic (and human) activity and its absorption by the rocks and oceans that make up our planet's surface. ∞ *(Sec. 5.4)* Illustrated in Figure 6.35(a), this constant recycling of atmospheric carbon dioxide is known as the **carbon cycle.** The presence of liquid water accelerates the absorption process—carbon dioxide dissolves in water, eventually reacting with surface material to form carbonate rocks. At the same time, plate tectonics steadily releases carbon dioxide back into the air. The level of carbon dioxide in Earth's atmosphere today is the result of a balance between these opposing forces.

The initial stages of atmospheric development on Venus probably took place in much the same way as just described for our own planet. The real difference between Earth and Venus is that the greenhouse gases in Venus's secondary atmosphere never left the atmosphere as they did on Earth. Indeed, if all the dissolved or chemically combined carbon dioxide on Earth were released back into our present-day atmosphere, its new composition would be 98 percent carbon dioxide and 2 percent nitrogen, and it would have a pressure about 70 times its current value. In other words, apart from the presence of oxygen (which is a consequence of life on Earth) and water (whose absence on Venus will be explained in a moment), Earth's atmosphere would look a lot like that of Venus.

To understand what happened on Venus, imagine taking Earth from its present orbit and placing it in that of Venus. Being closer to the Sun, our planet would warm up. More water would evaporate from the oceans, leading to an increase in atmospheric water vapor. Because the ability of both the oceans and surface rocks to hold carbon dioxide diminishes with increasing temperature, more carbon dioxide would enter the atmosphere. The additional greenhouse heating would warm our planet still further, leading to a further increase in atmospheric greenhouse gases, and so on. This **runaway greenhouse effect** would eventually lead to the complete evaporation of the oceans, restoring all the original greenhouse gases to the atmosphere. Essentially the same thing must have happened on Venus long ago (Figure 6.35b), leading to the planetary inferno we see today.

The greenhouse effect on Venus was even more extreme in the past, when the atmosphere also contained water vapor. By adding to the blanketing effect of the carbon dioxide, the water vapor helped push the surface of Venus to temperatures perhaps twice as hot as at present. At those high temperatures, the water vapor was able to rise high into the planet's upper atmosphere—so high that solar ultraviolet radiation broke it up into its components, hydrogen and oxygen. The light hydrogen rapidly escaped, the reactive oxygen quickly combined with other atmospheric gases, and all water on Venus was lost forever.

Evolution of the Martian Atmosphere

The runaway greenhouse effect on Venus defines the "hot" edge of the habitable zone that features so prominently in discussions of Earth-like exoplanetary systems. ∞ *(Sec. 4.4)* The "cold" edge is more closely related to the history of Mars.

◀ **FIGURE 6.34 Venus's Atmosphere** Venus's atmosphere is much deeper and denser than Earth's, so a much smaller fraction of the infrared radiation leaving the planet's surface escapes into space. The result is a much stronger greenhouse effect and a correspondingly hotter planet.

(a)   (b)   (c)

▲ FIGURE 6.35 **Carbon Cycle**   (a) On Earth, competition between volcanic activity and absorption into surface rocks and water keeps atmospheric carbon dioxide at a modest level, leading to a small but beneficial greenhouse effect. (b) On Venus, the temperature is so high that relatively little carbon can be absorbed, leading to the runaway greenhouse effect. (c) On Mars, there is little volcanic activity, and so most of the carbon dioxide now resides in the surface rocks.

Presumably, Mars also had an outgassed secondary atmosphere early in its history. As we saw in Section 6.6, around 4 billion years ago Mars may have had a relatively dense atmosphere rich in carbon dioxide, perhaps even with blue skies and rain. Planetary scientists estimate that the greenhouse effect could have kept average surface temperatures above 0°C (Figure 6.36). Sometime during the next billion years, however, most of the Martian atmosphere disappeared. Possibly some of it escaped due to impacts with large bodies in the early solar system, and a large part may have leaked away into space because of the planet's weak gravity, but most of the remainder probably was lost in a kind of "reverse runaway greenhouse effect." ∞ *(More Precisely 5-1)*

As we have just seen, Mars cooled faster than Earth and apparently never developed large-scale plate tectonic motion. As a result, even taking into consideration the large volcanoes discussed earlier in this chapter, Mars has had on average far less volcanism than Earth, so the processes depleting carbon dioxide were much more effective than those replenishing it. The level of atmospheric carbon dioxide steadily declined, its greenhouse effect diminished, and the planet cooled, causing still more carbon dioxide to be absorbed in the surface layers and leave the atmosphere. The carbon cycle turned into a "one-way street" of ever lower temperatures and decreasing levels of carbon dioxide in the

▶ FIGURE 6.36 **Ancient Mars**   Artist's conception of Mars some 4 billion years ago, with a dwindling atmosphere and some lingering surface water.

atmosphere. Calculations suggest that much of the Martian atmospheric carbon dioxide could have been lost in this way in a relatively short period of time, perhaps as quickly as a few hundred million years.

As the temperature continued to fall, water froze out of the atmosphere, lowering still further the level of atmospheric greenhouse gases and accelerating the cooling. Eventually, even carbon dioxide began to freeze out, particularly at the poles, and Mars the frigid state we see today—a cold, dry planet with most of its original complement of atmospheric gases now residing in or under the barren surface.

CONCEPT CHECK

If Venus and Mars had formed at Earth's distance from the Sun, what might their climates be like today?

THE BIG QUESTION

For well over a century, people have wondered, is there life on Mars? After years of careful observation, this intriguing world has yielded no hint of living creatures, intelligent or otherwise—not even bugs hiding beneath the dusty rocks. Perhaps future robots will find archaeological evidence for life now extinct, or maybe they will find no organisms at all, dead or alive. Either way, the consequences for our place in the cosmos will be profound.

CHAPTER REVIEW

SUMMARY

LO1 Mercury's rotation rate is strongly influenced by the tidal effect of the Sun, which causes the planet to rotate exactly one and a half times for every one revolution around the Sun.

LO2 Mercury has no permanent atmosphere. The atmospheres of Venus and Mars are mainly carbon dioxide. Venus's atmosphere is extremely hot and 90 times denser than Earth's. The density of the cool Martian atmosphere is only 0.7 percent that of Earth's.

LO3 Mercury's surface is heavily cratered, much like the lunar highlands. Mercury lacks lunarlike maria but has extensive **intercrater plains** (p. 175) and cliff-forming cracks, or **scarps** (p. 175), in its crust. The plains were caused by lava flows early in Mercury's history. The scarps were apparently formed when the planet's core cooled and shrank, causing the surface to crack.

LO4 Because of its thick cloud cover, Venus's surface cannot be seen in visible light from Earth, although it has been thoroughly mapped by radar. Many lava domes and **shield volcanoes** (p. 177) have been found. No eruptions have been observed, although there is indirect evidence that Venus is still volcanically active today. The northern hemisphere of Mars consists mainly of rolling plains, while the south is rugged highlands. The reason for this asymmetry is unknown. Mars's major surface feature is the Tharsis bulge. Associated with the bulge is the largest-known volcano in the solar system and a huge crack, called the Valles Marineris, in the planet's surface.

LO5 There is clear evidence that water once existed in great quantity on Mars. The **runoff channels** (p. 184) are the remains of ancient Martian rivers. The **outflow channels** (p. 184) are the paths taken by floods that cascaded from the southern highlands into the northern plains. Today, a large amount of that water may be locked up in the polar caps and in a layer of **permafrost** (p. 185) lying under the Martian surface. Images from orbiting spacecraft seem to show what may be ancient coastlines; others suggest that there may still be liquid water below the surface in some places. The most recent craft to land on the surface have found strong geological evidence for standing water.

LO6 Mercury's weak magnetic field seems to be a fossil remnant from the time when the planet's iron core solidified. Geological activity on Mercury ceased long ago. Venus does not have a detectable magnetic field even though, like Earth, it probably has a molten core. This is due to the planet's slow rotation. The surface of Venus shows no sign of plate tectonics. Features called **coronae** (p. 178) are thought to have been caused by an upwelling of mantle material that never developed into full convective motion. Mars rotates rapidly, but has a very weak magnetic field, implying that its core is

nonmetallic, nonliquid, or both. Mars also appears to have had some form of tectonic activity long ago, but plate tectonics never developed because the planet cooled too rapidly.

LO7 Any initial atmospheres of light gases that Earth, Venus, and Mars had at formation rapidly escaped. All the terrestrial planets probably developed **secondary atmospheres** (p. 193), outgassed from volcanoes, early in their lives. On Mercury the atmosphere escaped, as on the Moon. On Earth much of the outgassed material became absorbed in surface rocks or dissolved in the oceans. On Venus the **runaway greenhouse effect** (p. 194) has resulted in all the planet's carbon dioxide being left in the atmosphere, leading to the extreme conditions we observe today. On Mars part of the secondary atmosphere escaped into space. Of the remainder, most of the carbon dioxide is now locked up in surface rock, and most of the water vapor is now stored in the permafrost and the polar caps.

MasteringAstronomy®   For instructor-assigned homework go to www.masteringastronomy.com

Problems labeled **POS** explore the process of science. **VIS** problems focus on reading and interpreting visual information. **LO** connects to the introduction's numbered Learning Outcomes.

REVIEW AND DISCUSSION

1. In contrast to Earth, Mercury undergoes extremes in temperature. Why?

2. What do Mercury's magnetic field and large average density imply about the planet's interior?

3. **LO3** How is Mercury's evolutionary history like that of the Moon? How is it different?

4. **LO1 POS** Mercury used to be called "the Moon of the Sun." Why do you think it had this name? What piece of scientific evidence proved the name inappropriate?

5. Why does Venus appear so bright to the naked eye?

6. **LO6** Venus probably has a molten iron-rich core like Earth. Why doesn't it also have a magnetic field?

7. **LO4 POS** How did radio observations of Venus made in the 1950s change our conception of that planet?

8. What are the main constituents of Venus's atmosphere? What are clouds in the upper atmosphere made of?

9. What is the runaway greenhouse effect, and how might it have altered the climate of Venus?

10. Why is Mars red?

11. **LO5 POS** What is the evidence that water once flowed on Mars? Is there water on Mars today?

12. Why were Martian volcanoes able to grow so large?

13. **LO2** Since Mars has an atmosphere and it is composed mostly of a greenhouse gas, why isn't there a significant greenhouse effect to warm its surface?

14. Do you think that sending humans to Mars in the near future is a reasonable goal? Why or why not?

15. **LO7** Compare and contrast the evolution of the atmospheres of Mars, Venus, and Earth. Include a discussion of the importance of volcanoes and water.

CONCEPTUAL SELF-TEST: TRUE OR FALSE/MULTIPLE CHOICE

1. Mercury's solar day is longer than its solar year. (T/F)

2. Numerous surface features on Venus can be seen in Earth-based images made in the ultraviolet part of the spectrum. (T/F)

3. Evidence of lava flows is common on the surface of Venus. (T/F)

4. There is strong circumstantial evidence that active volcanism continues on Venus. (T/F)

5. Mars has the largest volcanoes in the solar system. (T/F)

6. There are many indications of past plate tectonics on Mars. (T/F)

7. Valles Marineris is comparable in size to Earth's Grand Canyon. (T/F)

8. Mercury's large overall density suggests that the planet (a) has an interior structure similar to that of Earth's Moon; (b) has a dense metal core; (c) has a weaker magnetic field than the Moon; (d) is younger than the Moon.

9. Venus's surface is permanently obscured by clouds. As a result, the surface has been studied primarily by (a) robotic landers; (b) orbiting satellites using radar; (c) spectroscopy; (d) radar signals from Earth.

10. Compared with Earth, Venus has a level of plate tectonic activity that is (a) much more rapid; (b) somewhat less rapid; (c) about the same; (d) virtually nonexistent.

11. Venus's atmospheric temperature is (a) about the same as Earth's; (b) cooler than temperatures on Mercury; (c) hotter than temperatures on Mercury; (d) high due to the presence of sulfuric acid.

12. In terms of area, the extinct Martian volcano Olympus Mons is about the size of (a) Mt. Everest; (b) Colorado; (c) North America; (d) Earth's Moon.

13. In comparison to the atmosphere of Venus, the vastly different atmospheric character of Mars is likely due to a(n) (a) ineffective greenhouse effect; (b) a reverse greenhouse effect; (c) absence of greenhouse gases in the Martian atmosphere; (d) greater distance from the Sun.

14. VIS In Figure 6.3 (Planetary Radar), part (b), the highest-frequency reflected radiation is received from (a) the upper part of the planet; (b) the center of the planet; (c) the lower part of the planet.

15. VIS Of the following, the best evidence for the existence of liquid water on an ancient Mars is Figure (a) 6.19; (b) 6.22; (c) 6.23; (d) 6.30.

PROBLEMS

The number of squares preceding each problem indicates its approximate level of difficulty.

1. ■ How long does it take a radar signal to travel from Earth to Mercury and back when Mercury is at its closest point to Earth?

2. ■■ Assume that a planet will have lost its initial atmosphere by the present time if the average (daytime) molecular speed exceeds one-sixth of the escape speed (see *More Precisely 5-1*). What would Mercury's mass have to be in order for it still to have a nitrogen (molecular weight 28) atmosphere?

3. ■■ Approximating Venus's atmosphere as a layer of gas 50 km thick with a uniform density of 21 kg/m^3, calculate its total mass. Compare your answer with the mass of Earth's atmosphere (Chapter 5, Problem 6) and with the mass of Venus.

4. ■ *Pioneer Venus* observed high-level clouds moving around Venus's equator in 4 days. What was their speed in kilometers per hour? In miles per hour?

5. ■■ (a) What is the size of the smallest feature that can be distinguished on the surface of Venus (at closest approach) by the Arecibo radio telescope at an angular resolution of 1'? (b) Could an infrared telescope with an angular resolution of 0.1″ distinguish impact craters on the surface of Venus?

6. ■■■ Calculate the orbital period of the *Magellan* spacecraft, moving around Venus on an elliptical orbit with a minimum altitude of 294 km and a maximum altitude of 8543 km above the planet's surface. In 1993 the spacecraft's orbit was changed to have minimum and maximum altitudes of 180 km and 541 km, respectively. What was the new period?

7. ■ Verify that the surface gravity on Mars is 40 percent that of Earth. What would you weigh on Mars?

8. ■■ The outflow channel shown in Figure 6.24 is about 10 km across and 100 m deep. If it carried 10^7 metric tons (10^{10} kilograms) of water per second, as stated in the text, estimate the speed at which the water must have flowed.

9. ■■ Calculate the total mass of a uniform layer of water covering the entire Martian surface to a depth of 2 m (see Section 6.8). Compare it with the mass of Mars and with the mass of Venus's atmosphere (Problem 3).

10. ■■ Mars has two small, irregular moons, Phobos and Deimos, which orbit at distances of 9400 km and 23,500 km, respectively, from the planet. Their longest diameters are 28 km (Phobos) and 16 km (Deimos). Calculate their maximum angular diameters, as seen by an observer on the Martian surface directly under their orbits. Would the observer ever see a total solar eclipse?

ACTIVITIES

Collaborative

1. If Mars is visible in the night sky, observe it with as large a telescope as is available to you; binoculars will not be of much use. Use your almanac (or go online) to find out which Martian season is occurring at the time of your observation, which hemisphere is tilted in Earth's direction, and what longitude is pointing toward Earth. Sketch what you see. Look carefully and take your time. Repeat this observation several times over the course of a night. Take turns making the sketches. Afterward, try to identify the various features you have seen by referring to known objects on Mars (e.g., as shown in Figure 6.6). You should also be able to see the planet's rotation by watching the surface features move. Do the same the next night. Because Mars's rotation period is so similar to Earth's, you should see the same surface features again.

Individual

Consult an almanac to determine where Mercury, Venus, and Mars are in the sky this year.

1. Try to spot Mercury in the morning or evening twilight—not an easy task! From the Northern Hemisphere, the best evening sightings of the planet take place in the spring; the best morning sightings take place in the fall.

2. Find out when Venus will next pass between Earth and the Sun. How many days before and after this event can you glimpse the planet with the naked eye? Using binoculars or a small telescope, examine Venus as it goes through its phases. Note the phase and the relative size (you can compare its size to the field of view in a telescope; always use the same eyepiece). Observe it every few days or once a week. Make a table of its shape, size, and relative brightness. Can you see the correlations between these properties first recognized by Galileo?

3. Several months before opposition, Mars begins retrograde motion. Chart the planet's motion relative to the stars to determine when it stops moving eastward and begins moving toward the west. Notice the increase in Mars's brightness as it approaches opposition.

7

The Jovian Planets
Giants of The Solar System

Beyond the orbit of Mars, the solar system is a very different place. In sharp contrast to the small, rocky terrestrial bodies found near the Sun, the outer solar system presents us with a totally unfamiliar environment—huge gas balls, peculiar moons, planet-circling rings, and a wide variety of physical and chemical properties, many of which are only poorly understood even today. U.S. spacecraft visiting the outer planets over the past two decades have revealed these worlds in detail that was only dreamed of by centuries of Earth-bound astronomers. The jovian planets—Jupiter, Saturn, Uranus, and Neptune—differ from Earth and from each other in many ways, but they have much in common, too. As with the terrestrial planets, we learn from their differences as well as from their similarities.

THE BIG PICTURE

Engineering feats of today's space age have allowed us to explore the planets and moons of our rich and varied solar system. Such thrilling exploration calls to mind earlier voyages of discovery, launched from Europe toward the Americas by Columbus, Magellan, Cortes, de Champlain, and many other seafarers who ventured forth across an unknown Atlantic and eventually around our world. Now, in the 21st century, robot spacecraft controlled by humans back on Earth act as modern spacefarers, exploring new, unknown worlds—and what they are finding is absolutely amazing.

◄ NASA's *Cassini* probe captured this true-color view of Saturn and its rings in 2013 as the spacecraft orbited high above the planet's north pole. The Sun is to the right, casting a long shadow across the rings at the left. Since its arrival at Saturn in 2004, *Cassini* has explored the planet and its moon system, using each moon's gravity to propel it on to its next target. The probe has also monitored seasonal atmospheric changes on Saturn as the planet enters northern spring and the north pole emerges from 15 years of darkness. Note the hexagonal weather pattern around Saturn's north pole, apparently defined by a polar jet stream circling the planet. *(NASA/JPL)*

(a)

(b)

R I V U X G

▲ FIGURE 7.1 Jupiter (a) Photograph of Jupiter made with an Earth-based telescope, showing the planet and its four Galilean moons. (b) A 2014 *Hubble Space Telescope* image of Jupiter, in true color—that is, as the human eye would see it. (*AURA/NASA*)

7.1 Observations of Jupiter and Saturn

The View from Earth

Named after the most powerful god of the Roman pantheon, Jupiter is the third-brightest object in the night sky (after the Moon and Venus), making it easy to locate and study from Earth. Ancient astronomers could not have known the planet's true size, but their name choice was very apt—Jupiter is by far the largest planet in the solar system. Figure 7.1(a) is a photograph of Jupiter taken through a small telescope on Earth, showing the planet's overall reddish coloration and the alternating light and dark bands paralleling its equator. Figure 7.1(b) is a *Hubble Space Telescope* image of Jupiter, showing the colored bands in much more detail and displaying numerous oval structures within them (note in particular the large orange oval at lower right). These atmospheric features are quite unlike anything found on the terrestrial planets. In further contrast to the inner planets, Jupiter has many moons that vary greatly in size and other properties. The four large *Galilean moons* (named after Galileo Galilei, who discovered them) are visible in Figure 7.1(a). ∞ *(Sec. 1.2)* They can be seen from Earth with a small telescope (and, for a few people, with the naked eye).

Saturn (Figure 7.2), the next major body we encounter as we continue outward beyond Jupiter, was the most distant planet known to Greek astronomers. Named after the father of Jupiter in Greco-Roman mythology, Saturn orbits at almost twice Jupiter's distance from the Sun, making it considerably fainter (as seen from Earth) than either Jupiter or Mars. The planet's banded atmosphere is somewhat similar to Jupiter's, but Saturn's atmospheric bands are much less distinct. Overall, the planet has a rather uniform butterscotch hue. Again, like Jupiter and unlike the inner planets, Saturn has many moons orbiting it. Saturn's best-known feature, its spectacular *ring system*, is clearly visible in Figure 7.2. The moons and rings of the jovian planets are the subject of Chapter 8.

Spacecraft Exploration

Discovery 7-1 presents an overview of some key planetary missions over the past half century. Most of our detailed knowledge of the jovian planets has come from NASA spacecraft visiting those worlds. Of particular importance are the two *Voyager* probes and the more recent *Galileo* and *Cassini* missions.

The *Voyager* spacecraft left Earth in 1977, reaching Jupiter in March (*Voyager 1*) and July (*Voyager 2*) of 1979. Both craft subsequently used Jupiter's strong gravity to send them on to Saturn in a maneuver called a *gravity assist*, in which a spacecraft loops around a planet, using the planet's motion to increase its speed and the planet's gravitational pull to redirect its trajectory to the next target. Virtually all recent planetary missions have used multiple gravity assists to reach their eventual destinations.

Voyager 2 took advantage of a rare planetary configuration that enabled it to use Saturn's gravity to propel it to Uranus and then on to Neptune in a spectacularly successful "grand tour" of the outer planets. Each craft carried equipment to study planetary magnetic fields and magnetospheres, as well as radio, visible-light, and infrared sensors to analyze reflected and emitted radiation from the jovian planets and their moons. The data they returned revolutionized our knowledge of all the jovian worlds and remain our primary source of

R I V U X G

◀ FIGURE 7.2 Saturn Saturn as seen by the *Hubble Space Telescope* in 2003 when the planet was at its maximum tilt (27°) toward Earth in its 30-year orbit about the Sun. The colors in this photo are natural—that is, as we would see it if we had telescopic vision. (See also the full-page opening photo on page 200.) (*NASA*) detailed information on Uranus and Neptune. The two *Voyager* craft are now headed out of the solar system, still returning data as they race toward interstellar space.

Galileo was launched in 1989 and reached Jupiter in December 1995, after a circuitous route that took it three times through the inner solar system, receiving gravity assists from both Venus and Earth before finally reaching its target. The mission had two components: an atmospheric probe and an orbiter. The probe, slowed by a heat shield and a parachute, made measurements and chemical analyses as it descended into Jupiter's atmosphere. The orbiter executed a complex series of gravity-assisted maneuvers through Jupiter's moon system, returning to some moons studied by *Voyager* and visiting others for the first time. Many of *Galileo's* findings are described in Chapter 8.

The mission ended in 2003, when the spacecraft began to run out of the fuel needed to fine-tune its orbit and keep its antenna pointed toward Earth. Knowing that they would ultimately lose control of the probe, and wishing to avoid at all costs any possible contamination of Jupiter's moon Europa (see Section 8.1), the *Galileo* team elected to end the mission by steering the spacecraft into Jupiter's atmosphere in September 2003.

In 1997, NASA launched the *Cassini* mission to Saturn. The craft reached Saturn in 2004 after four planetary gravity assists—two from Venus and one each from Earth and Jupiter—en route. Figure 7.3 shows *Cassini's* close-up view of Jupiter during its encounter with that planet in 2001. In December 2004 the craft released a probe built by the European Space Agency into the atmosphere of Titan, Saturn's largest moon (see Section 8.2). *Cassini* is now orbiting among Saturn's moons, much as *Galileo* did at Jupiter, returning detailed data on the moons, the planet, and its famous ring system. The mission was originally scheduled to last for 4 years after arrival at Saturn, but the gravity-assisted maneuver that placed the probe in orbit around the planet was so accurate that almost no fuel had to be used to fine-tune the trajectory, extending *Cassini's* lifetime by several years. The current mission, now renamed *Cassini Solstice*, is scheduled to end in 2017.

7.2 The Discoveries of Uranus and Neptune

British astronomer William Herschel discovered Uranus in 1781. Herschel was mapping the faint stars in the sky when he came across an unusual object that he described as "a curious either nebulous star or perhaps a comet." Repeated observations showed that it was neither. The object appeared as a disk in Herschel's 6-inch telescope and moved relative to the stars, but it traveled too slowly to be a comet. Herschel soon realized that he had found the seventh planet in the solar system.

Because this was the first new planet discovered in recorded history, the event caused quite a stir. The story goes that Herschel's first instinct was to name the new planet *Georgium Sidus* (Latin for "George's star") after his king, George III of England. The world was saved from a planet named George by the wise advice of another astronomer, Johann Bode, who suggested instead that the tradition of using names from ancient mythology be continued and that the planet be named after Uranus, the father of Saturn.

Uranus is just barely visible to the naked eye, under the best possible observing conditions. It looks like a faint, undistinguished star. Even through a large optical telescope, Uranus appears hardly more than a tiny, pale, slightly greenish disk. Figure 7.4 shows a close-up visible-light image taken by *Voyager 2* in 1986. Uranus's apparently featureless atmosphere contrasts sharply with the bands and spots visible in the atmospheres of the other jovian planets.

ANIMATION/VIDEO
Galileo Mission to Jupiter

▲ **FIGURE 7.3 Jupiter from *Cassini*** This *Cassini* spacecraft image of Jupiter shows intricate atmospheric clouds of different heights, thicknesses, and chemical composition. The image is approximately true color. *(NASA)*

▶ **FIGURE 7.4 Uranus** A close-up view of Uranus sent back to Earth by *Voyager 2* while the spacecraft was whizzing past the planet at 10 times the speed of a rifle bullet. Color is natural, and the upper atmosphere is nearly featureless, except for a few wispy clouds in the northern hemisphere. *(NASA)*

7.1 DISCOVERY

Space Exploration of the Solar System

Since the 1960s, dozens of uncrewed space missions have traveled throughout the solar system. All of the planets have been probed at close range, and robot spacecraft have visited comets, asteroids, and dwarf planets. The impact of these missions on our understanding of our planetary system has been nothing short of revolutionary. The timeline stretching across these two pages shows all major missions since the dawn of the space age in the early 1960s. Together, they have redefined our view of our cosmic backyard. Here we highlight just a few of them.

Mercury has been visited by just two spacecraft to date. In the mid-1970s, Mariner 10 executed a series of flybys of the planet, snapping thousands of images revealing it to be almost as heavily cratered as Earth's Moon. More than 30 years later, NASA's *Messenger* probe went into orbit around Mercury in 2011, mapping the surface in detail and probing the planet's interior during its 4-year mission.

More spacecraft have explored Venus than any other planet. The Soviet Union took the lead in the 1960s when nearly a dozen *Venera* probes orbited (and some landed on) Venus—the photograph below shows one of the heavily armored landers. Since then, the United States (Pioneer and Magellan) and Europe (Venus Express) have sent several more craft to spy on this hellish world. The new data have taught us much about Venus's surface and atmosphere, and the results have helped us better understand weather here on planet Earth.

Mars has been the target of very active robotic exploration. The United States has sent more than a dozen probes to orbit the "red planet" and often to land on its surface; Russia and Europe have also aimed craft at

Mars, but most have missed the planet or crash-landed. Several U.S. *Mariner* craft paved the way in the 1960s, showing the planet to be surprisingly inhospitable yet geologically intriguing. The *Viking* program in the 1970s was one of NASA's finest missions, not only safely landing two spacecraft but also imaging the surface for the first time and searching for life (which it did not find).

Since the late 1990s, a series of robots has orbited Mars and landed on its surface, sampling the air and dirt, drilling into stones, and digging for ice. The true-color panoramic view on the facing page was taken in 2005 from the rover *Spirit*, whose mother ship is shown in the foreground. The latest generation of rovers has mainly been searching for water, or evidence of it in the past. "Follow the water" has become a popular mantra in the search for extraterrestrial life. So far, though, neither liquid water nor life of any kind has been found, although there seems to be ample evidence for water in Mars's past.

Two pairs of U.S. spacecraft launched in the 1970s—*Pioneer* and *Voyager*—rewrote our knowledge of the jovian planets. Pioneer 10 and 11 took many photographs and made numerous scientific discoveries. The *Voyager* spacecraft carried radio, visible-light, and infrared sensors as well as magnetometers to study planetary magnetic fields. *Pioneer 11*, and both *Voyager I* and *II*, used Jupiter's gravity to propel them onward to Saturn. *Voyager I* was programmed to visit Titan, Saturn's largest moon, and so did not come close

Russian engineers with *Venera 11* in 1978

Voyager 1/2 1979–1989

Science instrument boom
High gain antenna
Magnetometer boom
Radioisotope generators
Cameras and spectrometer
Cosmic ray detector
Star trackers
Low gain antenna

Viking 1/2 1976–1982

Antenna to Earth
Cameras
Weather Instrument
Power Supply
Fuel
Biology Inlets
Descent Engine
Surface Sampler

Craters on Mars discovered by *Mariner 4* in 1964

1960

1962 Venus
Mariner 2 (NASA)

1964 Mars
Mariner 4 (NASA)

1967 Venus
Venera 4 (USSR)

1969 Mars
Mariner 6, 7 (NASA)

1970 Venus *Venera 7* (USSR)

1970

1971 Mars
Mariner 9 (NASA)

1972 Venus
Venera 8 (USSR)

1973 Jupiter
Pioneer 10 (NASA)

1974 Jupiter
Pioneer 11 (NASA)

1974 Mercury
Mariner 10 (NASA)

1976 Mars
Viking 1, 2 (NASA)

1978 *Venera 11, 12* (USSR)

1978 Venus *Pioneer Venus* (NASA)

1979 Jupiter
Voyager 1, 2 (NASA)

1979 Saturn *Pioneer 11* (NASA)

1980

1981 Saturn
Voyager 1 (NASA)

1982 Saturn *Voyager 2* (NASA)

1983 Venus
Venera 15, 16 (USSR)

1986 Uranus
Voyager 2 (NASA)

enough to the planet to receive a gravity-assisted boost to Uranus. However, *Voyager II* went on to visit both Uranus and Neptune in a spectacularly successful "Grand Tour" of the outer planets.

During the past 2 decades, two multibillion-dollar missions have explored the outer solar system. The *Galileo* mission to Jupiter (arriving in 1995) toured Jupiter's moons for 8 years, transforming our understanding of their structure and history. The U.S./ European *Cassini* mission to Saturn arrived in 2004 and is still gliding among Saturn's moons, providing spectacular images of the Saturn system, as well as key new insights into its evolution.

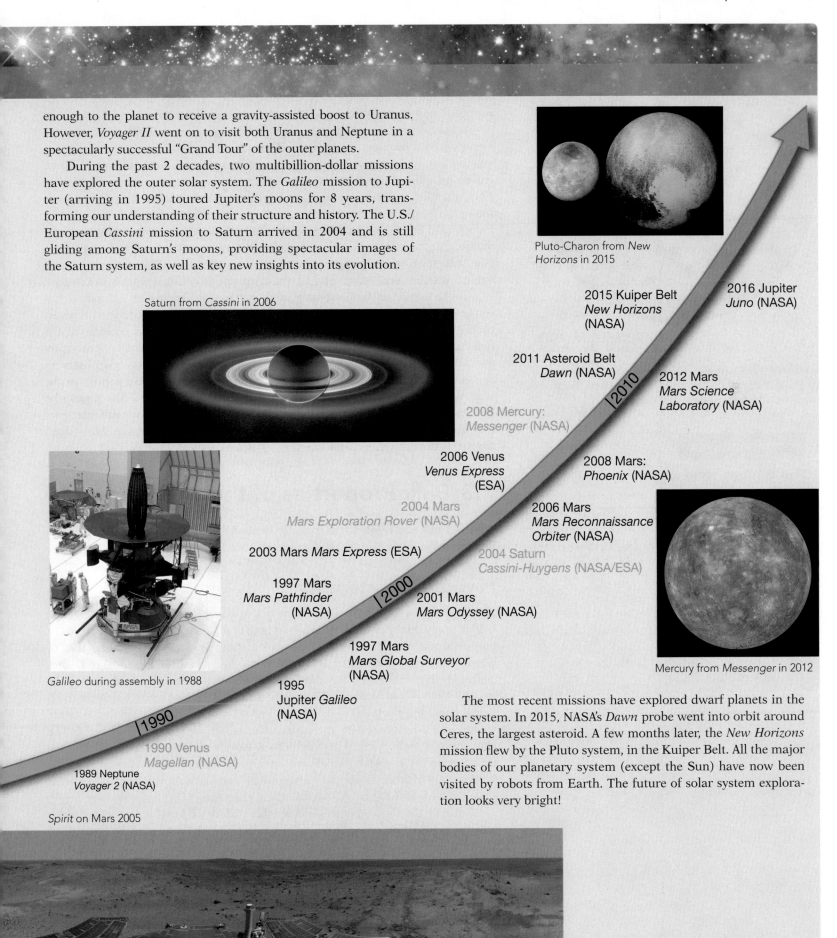

Pluto-Charon from *New Horizons* in 2015

2016 Jupiter *Juno* (NASA)

2015 Kuiper Belt *New Horizons* (NASA)

2011 Asteroid Belt *Dawn* (NASA)

2012 Mars *Mars Science Laboratory* (NASA)

|2010

2008 Mercury: *Messenger* (NASA)

2006 Venus *Venus Express* (ESA)

2008 Mars: *Phoenix* (NASA)

2004 Mars *Mars Exploration Rover* (NASA)

2006 Mars *Mars Reconnaissance Orbiter* (NASA)

2003 Mars *Mars Express* (ESA)

2004 Saturn *Cassini-Huygens* (NASA/ESA)

1997 Mars *Mars Pathfinder* (NASA)

|2000

2001 Mars *Mars Odyssey* (NASA)

1997 Mars *Mars Global Surveyor* (NASA)

1995 Jupiter *Galileo* (NASA)

Mercury from *Messenger* in 2012

Saturn from *Cassini* in 2006

Galileo during assembly in 1988

|1990

1990 Venus *Magellan* (NASA)

1989 Neptune *Voyager 2* (NASA)

Spirit on Mars 2005

The most recent missions have explored dwarf planets in the solar system. In 2015, NASA's *Dawn* probe went into orbit around Ceres, the largest asteroid. A few months later, the *New Horizons* mission flew by the Pluto system, in the Kuiper Belt. All the major bodies of our planetary system (except the Sun) have now been visited by robots from Earth. The future of solar system exploration looks very bright!

▲ **FIGURE 7.5 Neptune**   Neptune as seen in natural color by *Voyager* 2 from a distance of 1,000,000 km. Its cloud streaks range in width from 50 km to 200 km. *(NASA)*

CONCEPT CHECK

How did observations of Uranus lead to the discovery of Neptune?

Once Uranus was discovered, astronomers set about charting its orbit. They quickly discovered a discrepancy between the planet's predicted and observed positions. Try as they might, they could not find an elliptical orbit that fit the planet's path. By the early 19th century the discrepancy had grown to a quarter of an arc minute, far too large to be explained away as observational error. Uranus seemed to be violating—slightly—Kepler's laws of planetary motion. ∞ *(Sec. 1.3)*

Astronomers realized that although the Sun's gravitational pull dominates the planet's orbital motion, the small deviation meant that some unknown body was exerting a much weaker but still measurable gravitational force on Uranus. There had to be *another* planet in the solar system influencing Uranus. In September 1845, after almost 2 years of work, an English mathematician named John Adams solved the problem of determining the new planet's mass and orbit. In June 1846 a French mathematician, Urbain Leverrier, independently came up with essentially the same answer. Later that year the German astronomer Johann Galle found the new planet within one degree of the predicted position. The new planet was named Neptune, and Adams and Leverrier are jointly credited with its discovery.

Unlike Uranus, distant Neptune cannot be seen with the naked eye, although it can be seen through binoculars or a small telescope. In fact, according to his notes, Galileo may actually have seen Neptune, although he surely had no idea what it really was at the time. Through a large telescope on Earth, the planet appears only as a small, bluish disk, with few features visible. Even under the best observing conditions, only a few markings, suggestive of multicolored cloud bands, can be seen. With *Voyager 2*'s arrival, much more detail emerged (Figure 7.5)—in fact, the bulk of our knowledge of the two outermost jovian worlds has come from just this one spacecraft. Superficially, at least, Neptune resembles a blue-tinted Jupiter, with atmospheric bands and spots clearly evident.

7.3 Bulk Properties of the Jovian Planets

Table 7.1 extends Table 6.1 to include the four jovian planets. Figure 7.6 shows them to scale, along with Earth for comparison.

Physical Characteristics

The large masses and radii and relatively low average densities of these gigantic worlds imply that they differ radically in both composition and structure from the terrestrial planets. ∞ *(Sec. 4.1)* Saturn's density is actually less than that of water—the planet would float, if only we could find a bathtub big enough! Hydrogen and helium make up most of the mass of Jupiter and Saturn and about half the mass of Uranus and Neptune. These light gases have densities on Earth (at room temperature and sea level) of 0.08 kg/m³ and 0.16 kg/m³, respectively, but they are compressed enormously by the jovian planets' strong gravitational fields. The abundance of hydrogen and helium on these worlds is a consequence of the strong jovian gravity. The jovian planets are massive enough to have retained even the lightest gas—hydrogen—and very little of their original atmospheres have escaped since the birth of the solar system 4.6 billion years ago. ∞ *(More Precisely 5-1)*

None of the jovian planets has a solid surface of any kind. Their gaseous atmospheres just become hotter and denser with depth because of the pressure of the overlying layers, eventually becoming liquid in the interior. What we see from Earth is simply the outermost layer of atmospheric clouds.

At the heart of each jovian planet lies a dense, compact core many times more massive than Earth. As we will see in Section 7.6, three of the four (Jupiter, Saturn, and Neptune) have significant internal heating, which impacts the behavior and appearance of their atmospheres.

TABLE 7.1	Planetary Properties										
	Mass		Radius		Average Density (kg/m³)	Surface[1] Gravity (Earth = 1)	Escape Speed (km/s)	Rotation Period (solar days)	Axial Tilt (degrees)	Surface[1] Temperature (K)	Surface[1] Magnetic Field (Earth = 1)
	(kg)	(Earth = 1)	(km)	(Earth = 1)							
Mercury	3.3×10^{23}	0.550	2,400	0.38	5400	0.38	4.3	59	0	100–700	0.01
Venus	4.9×10^{24}	0.82	6,100	0.95	5300	0.91	10	-243^{2}	179	730	0.0
Earth	6.0×10^{24}	1.00	6,400	1.00	5500	1.00	11	1.00	23	290	1.0
Mars	6.4×10^{23}	0.11	3,400	0.53	3900	0.38	5.0	1.03	24	180–270	0.0
Jupiter	1.9×10^{27}	320	71,000	11	1300	2.5	60	0.41	3	120	14
Saturn	5.7×10^{26}	95	60,000	9.5	710	1.1	36	0.43	27	97	0.7
Uranus	8.7×10^{25}	15	26,000	4.0	1200	0.91	21	-0.69^{2}	98	58	0.7^{3}
Neptune	1.0×10^{26}	17	25,000	3.9	1700	1.1	24	0.72	30	59	0.4^{3}

[1]For the jovian planets, "surface" refers to the cloud tops.
[2]The minus sign indicates retrograde rotation.
[3]Average values. Because of the planet's asymmetric magnetic field geometry (see Section 7.6), the field strength varies substantially from one point on the surface to another.

Rotation Rates

Because the jovian planets have no solid surfaces to "tie down" the gas flow, different parts of their atmospheres can and do move at different speeds. This state of affairs—when the rotation rate is not constant from one location to another—is called **differential rotation.** It is not possible in solid objects like the terrestrial worlds but is normal for fluid bodies such as the jovian planets (and the Sun—see Chapter 9).

The amount of differential rotation on Jupiter is small—the equatorial regions rotate once every $9^{h} 50^{m}$, while the higher latitudes take about 6 minutes longer. On Saturn the difference between the equatorial and polar rotation rates is 26 minutes, again slower at the poles. On Uranus and Neptune the differences are greater still—more than 2 hours for Uranus and 6 for Neptune—with the poles rotating more rapidly in both cases. The differential rotation observed in the clouds reflects large-scale wind flows in the planets' atmospheres.

More meaningful measurements of overall planetary rotation rates are provided by observations of their magnetospheres. All four jovian worlds have strong magnetic fields and emit radiation at radio wavelengths. The strength of this radio emission varies with time and repeats periodically. Scientists assume that this period matches the rotation of the planet's deep interior, where (as on Earth) the magnetic field arises. ∞ *(Sec. 5.7)* The rotation periods listed in Table 7.1 were obtained in this way. There is no clear relationship between the interior and atmospheric rotation rates of the jovian planets. In Jupiter and Saturn the interior rotation rate matches the polar rotation. In Uranus the interior rotates more slowly than any part of the atmosphere. In Neptune the reverse is true.

▶ FIGURE 7.6 **Jovian Planets**   Jupiter, Saturn, Uranus, and Neptune, showing their relative scales compared to Earth. Note the very different sizes and atmospheric features of these five worlds. *(NASA)*

Jupiter

Saturn

Earth

Uranus

Neptune

Water in a very big bathtub illustrates two useful facts:

······Saturn is less dense than water and would therefore float.

······Water swirling down the drain rotates differentially—faster at the center than at the edges.

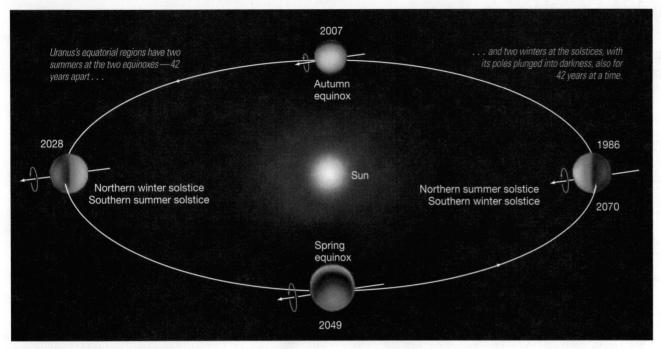

Uranus's equatorial regions have two summers at the two equinoxes—42 years apart . . .

2007

Autumn equinox

. . . and two winters at the solstices, with its poles plunged into darkness, also for 42 years at a time.

2028

1986

Sun

Northern winter solstice
Southern summer solstice

Northern summer solstice
Southern winter solstice

2070

Spring equinox

2049

▲ FIGURE 7.7 **Seasons on Uranus**   Because of Uranus's axial tilt of 98°, the planet experiences the most extreme seasons known in the solar system.

SELF-GUIDED TUTORIAL
Jupiter—Differential Rotation

Curiously, the *Cassini* measurement of Saturn's rotation period using the planet's magnetic field is about 6 minutes longer than the corresponding result obtained by *Voyager* more than 20 years earlier. Scientists don't think that Saturn's actual rotation rate has changed during this relatively short time. Rather, it seems that the planet's magnetic field may not be as good an indicator of interior rotation as was previously thought.

The observations used to measure a planet's rotation period also let us determine the orientation of its rotation axis. Jupiter's rotation axis is almost perpendicular to the orbit plane, the axial tilt being only 3°, quite small compared with Earth's tilt of 23°. The tilts of Saturn and Neptune (Table 7.1) are similar to those of Earth and Mars.

Each planet in our solar system seems to have some outstanding peculiarity. In the case of Uranus, it is its rotation axis. Unlike the other major planets, which have their spin axes (very) roughly perpendicular to the ecliptic plane, Uranus's rotation axis lies almost within that plane. We might say that Uranus is "tipped over" onto its side, for its axial tilt is 98°. (Because the north pole lies below the ecliptic plane, the planet's rotation, like that of Venus, is classified as retrograde.) As a result, the north pole[1] of Uranus, at some point in the planet's orbit, points almost directly toward the Sun. Half a Uranian year later the south pole faces the Sun, as illustrated in Figure 7.7. When *Voyager 2* encountered the planet in 1986, the north pole happened to be pointing nearly directly at the Sun, so it was midsummer in the northern hemisphere. At the same time, the south polar regions were halfway through a 42-year period of permanent darkness.

No one knows why Uranus is tilted in this way. Astronomers speculate that a catastrophic event, such as a grazing collision between the planet and another planet-size body, might have altered the planet's spin axis. ∞ *(Sec. 4.3)* However, there is no direct evidence for such an occurrence and no theory to tell us how we should seek to confirm it.

CONCEPT CHECK

Why do observations of a planet's magnetosphere allow astronomers to measure the rotation rate of the interior?

---

[1]We adopt here the convention that a planet's rotation is counterclockwise as seen from above the north pole (that is, planets always rotate from west to east).

7.4 Jupiter's Atmosphere

Just as Earth has been our yardstick for the terrestrial worlds, Jupiter will be our guide to the outer planets. We therefore begin our study of jovian atmospheres with Jupiter itself.

Overall Appearance and Composition

Visually, Jupiter is dominated by two atmospheric features (both clearly visible in Figure 7.1): a series of ever-changing atmospheric cloud bands arranged parallel to the equator and an oval atmospheric blob called the **Great Red Spot.** The cloud bands display many colors—pale yellows, light blues, deep browns, drab tans, and vivid reds, among others. The Red Spot is one of many features associated with Jupiter's weather. This egg-shaped region, twice the size of Earth, seems to be a hurricane that has persisted for hundreds of years.

What is the cause of Jupiter's colors? The most abundant gas in Jupiter's atmosphere is molecular hydrogen (roughly 86 percent of all molecules), followed by helium (nearly 14 percent). Small amounts of atmospheric methane, ammonia, and water vapor are also found. None of these gases can, by itself, account for Jupiter's observed coloration. For example, frozen ammonia and water vapor would produce white clouds, not the many colors we see. Scientists think that complex chemical processes occurring in Jupiter's turbulent atmosphere are responsible for these colors, although the details are not fully understood. The trace elements sulfur and phosphorus may play important roles in influencing the cloud colors—particularly the reds, browns, and yellows. The energy that powers the reactions comes in many forms: the planet's own internal heat, solar ultraviolet radiation, aurorae in the planet's magnetosphere, and lightning discharges within the clouds. ∞ *(Sec. 5.7)*

Astronomers describe the banded structure of Jupiter's atmosphere as consisting of a series of lighter-colored **zones** and darker **belts** crossing the planet. The zones and belts vary in both latitude and intensity during the year, but the general pattern is always present. These variations appear to be the result of convective motion in the planet's atmosphere. (Recall from Chapter 5 that convection can occur whenever cool gas overlies hotter material, as is the case here.) ∞ *(Sec. 5.3)* *Voyager* sensors indicated that the zones lie above upward-moving convective currents, while the belts are the downward part of the cycle, where material is generally sinking, as illustrated in Figure 7.8. Thus, because of the upwelling material below them, the zones are regions of high pressure; the belts, conversely, are low-pressure regions. However, observations made during the *Cassini* flyby have challenged this standard view, suggesting that upward convection is actually confined to the belts. For now, planetary scientists have no clear resolution of this contradiction between the *Voyager* and *Cassini* results.

The belts and zones are Jupiter's equivalents of the familiar high- and low-pressure systems that control the weather on Earth. A major difference from Earth is that Jupiter's rapid differential rotation causes these systems to wrap all the way around the planet, instead of forming localized circulating storms as on our own world. Because of the pressure difference between them, the belts and zones lie at slightly different heights in Jupiter's atmosphere. Cloud chemistry is very sensitive to temperature, and the temperature difference between the two levels is the main reason for the different colors we see.

Underlying the bands is an apparently stable pattern of east-west winds called Jupiter's **zonal flow.** The equatorial regions of Jupiter's atmosphere rotate faster than the planet as a whole, with an average flow speed of about 500 km/h in the easterly direction, somewhat similar to the jet stream on Earth. At higher

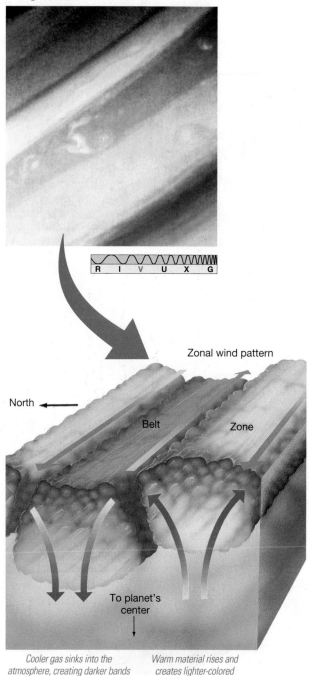

This is a real Voyager *photo of Jupiter's clouds, showing its actual banded structure.*

R I V U X G

Zonal wind pattern

North

Belt

Zone

To planet's center

Cooler gas sinks into the atmosphere, creating darker bands atop lower-pressure regions.

Warm material rises and creates lighter-colored zones.

▲ **FIGURE 7.8 Jupiter's Convection** The colored bands in Jupiter's atmosphere are associated with vertical convective motion. As on Earth, winds tend to blow from high-pressure regions to low-pressure regions. Jupiter's rapid rotation channels these winds into an east–west flow pattern, as indicated by the three yellow-red arrows drawn atop the belts and zones. *(NASA)*

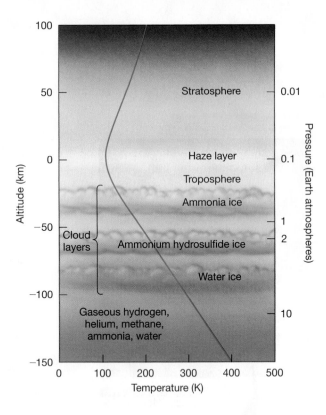

Models of the vertical structure of Jupiter's atmosphere indicate that the planet's clouds are arranged in three main layers, each with quite different colors and chemistry. The white regions are the tops of the upper ammonia clouds. The yellows, reds, and browns are associated with the second cloud layer, which is composed of ammonium hydrosulfide ice. The lowest (bluish) cloud layer is water ice. The blue curve shows how Jupiter's atmospheric temperature depends on altitude. (For comparison with Earth, see Figure 5.5.)

latitudes there are alternating regions of westward and eastward flow, the alternations corresponding to the pattern of belts and zones. The flow speed decreases toward the poles. Near the poles, where the zonal flow disappears, the band structure vanishes also.

Atmospheric Structure

Jupiter's clouds are arranged in several layers, with white ammonia clouds generally overlying the colored layers, whose composition we will discuss in a moment. Above the ammonia clouds lies a thin, faint layer of haze created by chemical reactions similar to those that cause smog on Earth. When we observe Jupiter's colors, we are looking down to many different depths in the planet's atmosphere.

Figure 7.9 is a diagram of Jupiter's atmosphere, based on observations and computer models. Because the planet lacks a solid surface to use as a reference level for measuring altitude, scientists conventionally take the top of the troposphere (the turbulent region containing the clouds we see) to lie at 0 km. With this level as the zero point, the troposphere's colored cloud layers all lie at negative altitudes in the diagram. As on other planets, weather on Jupiter is the result of convection in the troposphere. The haze layer lies at the upper edge of Jupiter's troposphere. The temperature at this level is about 110 K. Above the troposphere, as on Earth, the temperature rises as the atmosphere absorbs solar ultraviolet light.

Below the haze layer, at an altitude of –30 km, lie white, wispy clouds of ammonia ice. The temperature at this level is 125–150 K. A few tens of kilometers below the ammonia clouds, the temperature is a little warmer—above 200 K— and the clouds are made up mostly of droplets or crystals of ammonium hydrosulfide, produced by reactions between ammonia and hydrogen sulfide. At deeper levels the ammonium hydrosulfide clouds give way to clouds of water ice or water vapor. The top of this lowest cloud layer, which is not seen in visible-light images of Jupiter, lies some 80 km below the top of the troposphere.

The *Galileo* atmospheric probe arrived at Jupiter in December 1995. It survived for about an hour before being crushed by atmospheric pressure at an altitude of –150 km (that is, just at the bottom of Figure 7.9). The probe's entry location (Figure 7.10) was in Jupiter's equatorial zone and, as luck would have it, coincided with an atypical hole almost devoid of upper-level clouds and with abnormally low water content. However, once these factors were taken into account, *Galileo*'s findings on wind speed, temperature, and composition were in good agreement with the description presented above.

Weather on Jupiter

In addition to the large-scale zonal flow, Jupiter has many small-scale weather patterns. The Great Red Spot (Figure 7.11) is a prime example. Observational records indicate that it has existed continuously in one form or another for more than 300 years, and it may well be much older. *Voyager* and *Galileo* observations show the Spot to be a region of swirling, circulating winds, like a whirlpool or a terrestrial hurricane, a persistent and vast atmospheric storm. Scientists think that the Spot's color may be due to its enormous size and strength, which lifts its cloud tops high above the surrounding clouds where solar ultraviolet radiation

▲ FIGURE 7. 10 *Galileo's* Entry Site   The arrow on this image shows where the *Galileo* atmospheric probe plunged into Jupiter's cloud deck on December 7, 1995. Until its demise, the probe took numerous weather measurements, transmitting those signals to the orbiting mother ship, which then relayed them to Earth. *(NASA)*

causes chemical reactions that produce the colors we observe. The Spot varies in length, averaging about 25,000 km. It rotates around Jupiter at a rate similar to that of the planet's interior, suggesting that its roots lie far below the atmosphere. Astronomers think that the Spot is somehow sustained by Jupiter's large-scale atmospheric motion. The zonal motion north of the Spot is westward, while that to the south is eastward (Figure 7.11), supporting the idea that the Spot is confined and powered by the zonal flow. Turbulent eddies form and drift away from the spot's edge. The center, however, remains tranquil in appearance, like the eye of a hurricane on Earth.

The *Voyager* mission discovered many smaller spots that are also apparently circulating storm systems. Examples of these smaller systems are the **white ovals** seen in many images of Jupiter (see, for example, the one just south of the Great Red Spot in Figure 7.11). Their high cloud tops give these regions their white color. The white oval in Figure 7.11 is known to be at least 80 years old. The inset shows a **brown oval**—a long-lived hole in the overlying clouds that allows us to look down into the lower atmosphere. Many more oval storm systems—dark and light—can be seen in the *Cassini* view of Jupiter shown in Figure 7.3.

The simplified picture presented in Figure 7.8, with wind direction alternating between adjacent bands as Jupiter's rotation deflects the winds into eastward or westward streams, is really too crude to describe the many details observed in Jupiter's atmosphere. More likely, the interaction between convection in Jupiter's atmosphere and the planet's rapid differential rotation channels the largest eddies into the observed zonal pattern, but smaller eddies—like the Red Spot and white ovals—tend to cause localized irregularities in the flow.

In the late 1990s, astronomers observed the collision and merger of three relatively small white ovals in Jupiter's atmosphere. For several years the resultant larger oval remained white, but in early 2006 it changed from white to brown to red, becoming in a matter of months a smaller version of the Great Red Spot! Merger and growth may well be the mechanism by which large storms on the jovian planets form and strengthen. However, the "baby red spot" did not live long. Figure 7.12(a) is a time sequence of *Hubble* images showing how the smaller spot was shredded by the Great Red Spot's swirling winds over a period of a few months in 2008. Incidentally, the large white oval at the bottom of the frames in Figure 7.12(a) also formed by similar process in 1930s—but it was a little larger and farther from the Great Red Spot, so it is still around today.

These observations may help explain the longevity of storm systems on Jupiter. On Earth a large storm, such as a hurricane, forms over the ocean and may survive for many days, but it dies quickly once it encounters land because the land disrupts the energy supplied and flow patterns that sustain the storm. Jupiter has no continents, so once a storm is established and has reached a size at which other storm systems cannot destroy it, apparently little affects it. Still, encounters with other storms may take a toll. As illustrated in Figure 7.12(b), the Great Red Spot has been steadily shrinking since detailed measurements began, possibly as a result of these interactions.

A typical break ("brown oval") in Jupiter's atmosphere

5000 km

Note the complex turbulence to the left of both the Red Spot and the smaller white oval below it.

This is Earth, drawn to scale relative to Jupiter's clouds.

Direction of gas flow

10,000 km

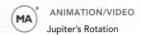

R I V U X G

▲ **FIGURE 7.11 Jupiter's Red Spot** **INTERACTIVE**   *Voyager 1* took this photograph of Jupiter's Great Red Spot (upper right) from a distance of about 100,000 km. The arrows indicate direction of gas flow above, below, and inside the Great Red Spot. *(NASA)*

MA ANIMATION/VIDEO
Jupiter's Rotation

▶ FIGURE 7.12 **Red Spot Evolution** (a) In mid-2008, the *Hubble* telescope recorded this sequence of images at monthly intervals (left to right), showing a "baby red spot" (arrows) approaching the Great Red Spot and being destroyed by it. (b) Perhaps because of such interactions, the Spot has steadily decreased in size since detailed observations began. It was estimated in the late 19th century to be 40,000 km long. In 1979 *Voyager 1* measured the long axis at 23,000 km. These *Hubble* observations indicate lengths of 21,000 km in 1995, 18,000 km in 2009, and 15,000 km in 2014. *(NASA)*

(a)

(b)

1995   2009   2014

▶ FIGURE 7.13 **Saturn, Up Close** (a) The banded structure of Saturn's atmosphere appears clearly in this *Voyager 2* image. (Three of Saturn's moons appear at bottom; a fourth, not visible here, casts a black shadow on the cloud tops.) (b) This 2004 *Cassini* image is actually a mosaic of many images taken in true color. *(NASA)*

(a)

(b)

7.5 Atmospheres of the Outer Jovian Worlds

Composition of Saturn's Atmosphere

Saturn is not as colorful as Jupiter. Figure 7.13(a) shows yellowish and tan cloud bands that parallel the equator, but these regions display less atmospheric structure than do the bands on Jupiter. No obvious large spots or ovals adorn Saturn's cloud decks. Storms do exist, but the color changes that distinguish them on Jupiter are largely absent on Saturn. Figure 7.13(b) is the highest-resolution full-global image ever made of Saturn—a mosaic of more than a hundred photographs taken in true color—showing clarity yet subtlety in the structure of its cloud deck.

Spectroscopic studies show that Saturn's atmosphere consists of molecular hydrogen (92.4 percent) and helium (7.4 percent), with traces of methane and ammonia. As on Jupiter, hydrogen and helium dominate. However, the fraction of helium on Saturn is substantially less than on Jupiter. It is unlikely that the processes that created the outer planets preferentially stripped Saturn of nearly half its helium or that the missing helium somehow escaped from the planet while the hydrogen remained behind. Instead, astronomers theorize that at some time in Saturn's past the helium liquefied and then sank toward the center of the planet, reducing the abundance of that gas in the outer layers. The reason for this gas-to-liquid transformation is discussed in Section 7.6.

Figure 7.14 illustrates the vertical structure of Saturn's atmosphere. In many respects it is similar to Jupiter's (Figure 7.9), except that Saturn's overall atmospheric temperature is a little lower because of its greater distance from the Sun. As on Jupiter, the troposphere contains clouds arranged in three distinct layers, composed (in order of increasing depth) of ammonia ice, ammonium hydrosulfide ice, and water ice. Above the clouds lies a layer of haze.

The total thickness of the three cloud layers in Saturn's atmosphere is roughly 250 km—about three times the cloud thickness on Jupiter—and each layer is thicker than its counterpart on Jupiter. The reason for this difference is Saturn's weaker gravity. At the haze level, Jupiter's gravitational field is nearly 2.5 times stronger than Saturn's, so Jupiter's atmosphere is pulled much more powerfully toward the center of the planet and the cloud layers are squeezed more closely together.

The colors of Saturn's cloud layers, as well as the planet's overall yellowish coloration, are likely due to the same basic photochemistry as on Jupiter. However, because Saturn's clouds are thicker, there are fewer holes and gaps in the top layer, with the result that we rarely glimpse the more colorful levels below. Instead, we see only the topmost layer, which accounts for Saturn's less-varied appearance.

Weather on Saturn

Computer-enhanced images of Saturn clearly show bands, oval storm systems, and turbulent flow patterns, many looking very much like those on Jupiter. In addition, there is an apparently quite stable east–west zonal flow, although the wind speed is considerably greater than on Jupiter and shows fewer east–west alternations. The equatorial eastward jet stream reaches a brisk 1500 km/h. Astronomers do not fully understand the reasons for the differences between Jupiter's and Saturn's flow patterns.

Figure 7.15(a) shows a sequence of *Cassini* images of a storm system on Saturn in 2011. The storm appeared as a small white spot (marked in the second frame) and rapidly developed into a band of clouds completely encircling the planet. Astronomers think that the white coloration arose from crystals of ammonia ice that formed when an upwelling plume of warm gas penetrated the cool upper cloud layers. Because the crystals were freshly formed, they had not yet been affected by

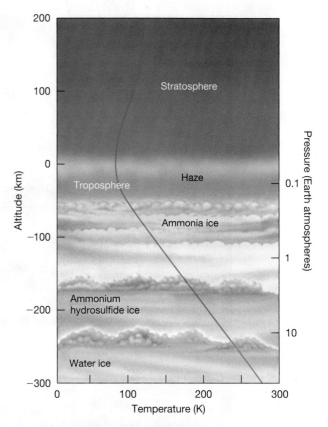

▲ **FIGURE 7.14 Saturn's Atmosphere**   The vertical structure of Saturn's atmosphere contains several cloud layers, like Jupiter, but Saturn's weaker gravity results in thicker clouds and a more uniform appearance.

ANIMATION/VIDEO
Saturn Cloud Rotation

(a)

(b)

▲ **FIGURE 7.15 Storm on Saturn**   (a) This huge storm was observed by the *Cassini* spacecraft in 2011. While churning its way through the northern hemisphere, it left behind a "tail" wrapping around the planet. (b) After the storm faded from view in visible light, infrared observations continued to show its ongoing activity. *(NASA)*

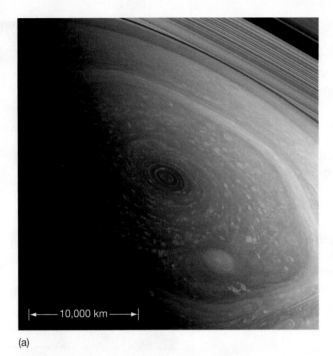

|←——— 10,000 km ———→|

(a)

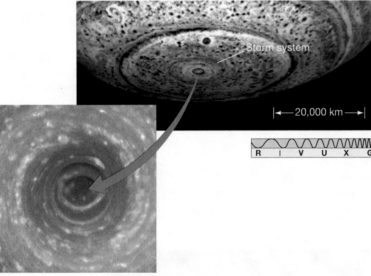

Storm system

|←— 20,000 km —→|

R I V U X G

This is a close-up of the polar vortex at Saturn's south pole.

(b)

▲ **FIGURE 7.16 Saturn's Polar Vortices** These *Cassini* images show giant swirling vortices at Saturn's poles, apparently the result of jet streams circling the planet. (a) The south polar vortex, in the infrared. (b) The north polar vortex, in enhanced color, with the rings in the background (blue). For a true color view, see the chapter-opening image on p. 200. *(NASA)*

CONCEPT CHECK

List some similarities and differences between Jupiter's belts and zones and weather systems on Earth.

the chemical reactions that color the planet's other clouds. The turbulent flow patterns seen in and around the spot had many similarities to the flow around Jupiter's Great Red Spot. Even after the storm faded from view at visible wavelengths, its ferocity was still evident in the infrared (Figure 7.15b). Routine observations of temporary atmospheric phenomena on the outer worlds are giving scientists greater insight into the dynamics of planetary atmospheres.

These large, complex storms may be Saturn's equivalent of thunderstorms on Earth (although on scales comparable to our entire planet). *Cassini*'s detectors have measured strong bursts of radio waves associated with them, most likely produced by intense lightning discharges deep below the cloud tops. The lightning is powered by convection and precipitation (water and ammonia "rain"), as on Earth, but the bursts are millions of times stronger than anything ever witnessed here at home. Astronomers suspect that these storms may actually be long-lived phenomena rooted deep in Saturn's atmosphere—perhaps a little like Jupiter's Great Red Spot—but normally hidden below the upper cloud layers. Occasionally they flare up, producing bright plumes visible from outside.

Figure 7.16 shows *Cassini*'s view of two other storm systems of great interest to planetary scientists—the planet's *polar vortices*. The orbiter's close-up imagery has allowed scientists to study these systems with unprecedented clarity. The north polar vortex (Figure 7.16b) is surrounded by smaller storms, all confined within a peculiar hexagonal jet stream, still not fully understood. The Earth-sized southern vortex (Figure 7.16a) is a hurricane-like storm with an eye wall at its center some 50 km in height. As mentioned in Chapter 6, *Venus Express* has mapped the south polar vortex on Venus in great detail. ∞ *(Sec. 6.3)* The opportunity to compare polar vortices in planets with such different atmospheres, orbits, and rotational properties as Venus, Earth, and Saturn offers astronomers unique opportunities for comparative planetology.

Atmospheric Composition of Uranus and Neptune

The atmospheres of Uranus and Neptune are very similar in composition to that of Jupiter—the most abundant gas is molecular hydrogen (84 percent), followed by helium (about 14 percent) and methane, which is more abundant on Neptune (about 3 percent) than on Uranus (2 percent). Ammonia is not observed in any significant quantity in the atmosphere of either planet, most likely because, at the low atmospheric temperatures of those planets, the ammonia exists in the form of ice crystals, which are much harder to detect spectroscopically than gaseous ammonia.

The blue color of the outer jovian planets is the result of their relatively high percentages of methane. Because methane absorbs long-wavelength red light quite efficiently, sunlight reflected from these planets' atmospheres is deficient in red and yellow photons and appears blue-green or blue. The greater the concentration of methane, the bluer the reflected light. Therefore Uranus, having less methane, looks blue-green, while Neptune, having more methane, looks distinctly blue.

Weather on Uranus and Neptune

Voyager 2 detected only a few atmospheric features on Uranus, and even they became visible only after extensive computer enhancement. Few clouds exist in Uranus's cold upper atmosphere. Clouds of the kind found on Jupiter and Saturn are found only at lower, warmer atmospheric levels, so we must look deep into the planet's atmosphere to see any structure, and the bands and spots that characterize flow patterns on the other jovian worlds are largely washed out on Uranus by intervening stratospheric haze. Figure 7.17(a) and (b) show some of the best images of atmospheric features on Uranus, including circulating clouds, massive weather events, and unusual convective features, as well as an Earth-sized dark spot. Figures 7.17(c) and (d) shown atmospheric clouds and flow patterns moving around the planet in the same direction as the planet's rotation, with wind speeds ranging from 200 to 500 km/h.

Although Neptune lies farther from the Sun, its internal heat (see Section 7.6) means that its upper atmosphere is actually slightly warmer than that of Uranus. This is probably the reason why Neptune's atmospheric features are more easily visible than those of Uranus (Figure 7.18a). The extra warmth causes Neptune's stratospheric haze to be thinner than the haze on Uranus and also makes the cloud layers less dense than they would otherwise be, with the result that they lie at higher levels in the atmosphere.

Voyager 2 detected numerous white methane clouds on Neptune (some of them visible in Figure 7.5) lying about 50 km above the main cloud tops. Neptune's equatorial winds blow from east to west with speeds of over 2000 km/h relative to the planet's interior. Why such a cold planet should have such rapid winds, and why they should be retrograde relative to the west-to-east rotation of the interior, remains a mystery.

Neptune sports several storm systems similar in appearance to those seen on Jupiter, and probably produced and sustained by the same basic processes. The largest storm, the **Great Dark Spot** (Figure 7.18b), was discovered by *Voyager 2* in 1989. Comparable in size to Earth, the Spot was located some 20° south of the planet's equator—quite similar to the Great Red Spot on Jupiter. The Dark Spot also exhibited other characteristics in common with Jupiter's Red Spot. The flow around it was counterclockwise, and there appeared to be turbulence where the winds associated with the Dark Spot interacted with the zonal flow to its north and south. Astronomers did not have long to study the Dark Spot's properties, however. By the time detailed *Hubble Space Telescope* observations of the planet became available in the mid-1990s, the Spot had disappeared (see Figure 7.18a). At around the same time, a new Dark Spot of comparable size became visible in Neptune's northern hemisphere. The origins and lifetimes of large storm systems on the jovian planets are currently not well understood.

(a)

(b)

(c)

(d)

▲ **FIGURE 7.17 Uranus's Atmosphere** (a) This infrared image of Uranus was captured in 2004 by an adaptive optics system on an exceptionally clear night at the Keck Observatory in Hawaii. It shows unprecedented atmospheric detail, as well as a very thin ring. (b) The *Hubble Space Telescope* took this image of a rare dark spot on Uranus in 2006. (c, d) These computer-enhanced *HST* images, taken roughly 4 hours apart, show the motion of a pair of bright clouds (labeled A and B) in Uranus's southern hemisphere, allowing astronomers to track the planet's rotation. (*L. Sromovsky and P. Fry; NASA*)

CONCEPT CHECK

Why are atmospheric features much less evident on Uranus than on the other jovian planets?

(a)

(b)

R   I   V   U   X   G

▲ FIGURE 7.18 **Neptune's Dark Spot**   (a) These 2011 *Hubble Space Telescope* views of Neptune were taken at roughly 4-hour intervals, offering a full view of the planet. The cloud features (mostly methane ice crystals) are tinted pink here because they were imaged in the infrared, but they are white in visible light. (b) Close-up views, taken by *Voyager 2* of the Great Dark Spot of Neptune, show a large storm system in the planet's atmosphere, possibly similar in structure to Jupiter's Great Red Spot. The entire dark spot is roughly the size of Earth. *(NASA)*

Molecular hydrogen

Depth 100 km
Temperature 300 K
Pressure 10 atm

Depth 20,000 km
Temperature 11,000 K
Pressure 3×10^6 atm

Depth 60,000 km
Temperature 18,000 K
Pressure 4×10^7 atm

Icy, rocky core
Depth 70,000 km
Temperature 25,000 K
Pressure 6×10^7 atm

Metallic hydrogen

▲ FIGURE 7.19 **Jupiter's Interior**   Jupiter's internal structure, as deduced from spacecraft measurements and theoretical modeling. Pressure and temperature increase with depth, and the atmosphere gradually liquefies over the outermost few thousand kilometers. Below 20,000 km, the hydrogen behaves like a liquid metal. At the center of the planet lies a rocky core (red region), itself larger than any of the terrestrial planets.

7.6 Jovian Interiors

The visible cloud layers of the jovian planets are all less than a few hundred kilometers thick. How, then, do we determine the conditions beneath the clouds, in the unseen interiors of these distant worlds? We have no seismographic information on these regions (but see *Discovery 7-2*), nor do we happen to live on a more or less similar planet that we can use for comparison. Instead, we must use all available bulk data on each planet, along with our knowledge of physics and chemistry, to construct a model of its interior that agrees with observations. ∞ *(Sec. 0.5)* Our statements about the interiors of the jovian worlds, then, are really statements about the models that best fit the facts. As we will see throughout this book, astronomers rely heavily on model building, often aided by powerful computers, as a tool to interpret data about objects too far away for direct exploration.

Internal Structure

As illustrated in Figure 7.19, both the temperature and the pressure of Jupiter's atmosphere increase with depth. At a depth of a few thousand kilometers, the gas makes a gradual transition into the liquid state. By a depth of about 20,000 km, the pressure is about 3 million times greater than atmospheric pressure on Earth.

Under these conditions the hot liquid hydrogen is compressed so much that it undergoes another transition, this time to a metallic state, with properties in many ways similar to some liquid metals on Earth. Of particular importance for Jupiter's magnetic field, this metallic hydrogen is an excellent conductor of electricity.

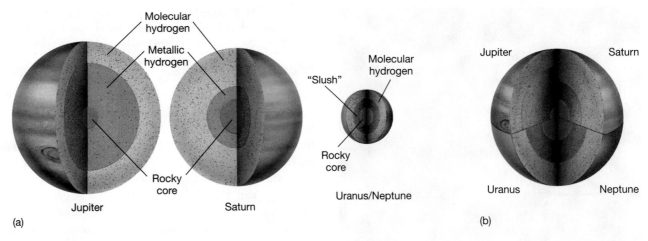

(a)

(b)

▲ FIGURE 7.20 Jovian Interiors INTERACTIVE A comparison of the interior structures of the four jovian planets. (a) The planets drawn to scale. (b) The relative proportions of the various internal zones.

Jupiter's rotation produces a strong outward push that causes a pronounced bulge at the planet's equator—Jupiter's equatorial radius (Table 7.1) exceeds its polar radius by nearly 7 percent. However, careful calculations indicate that if Jupiter were composed of hydrogen and helium alone, it should be *more* flattened than is observed. The actual degree of flattening implies a dense *core* having a mass perhaps 10–20 times the mass of Earth. The core's precise composition is unknown, but many scientists think it consists of materials similar to those found in the terrestrial worlds: molten, perhaps semisolid, rock. Because of Jupiter's enormous central pressure—approximately 50 million times that on Earth's surface, or 10 times that at Earth's center—the core must be compressed to very high densities (perhaps twice the core density of Earth). It is probably not much more than 20,000 km in diameter.

Saturn has the same basic internal parts as Jupiter, but their relative proportions are different: Its metallic hydrogen layer is thinner, although the central dense core (again inferred from studies of the planet's polar flattening) is similar—around 15 Earth masses. Because of its lower mass, Saturn has a less extreme core temperature, density, and pressure than Jupiter. The central pressure is around one-tenth of Jupiter's—not too different from the pressure at the center of Earth.

The interior pressures of Uranus and Neptune are sufficiently low that hydrogen stays in its molecular form all the way into the planets' cores. Currently, astronomers theorize that deep below the cloud layers, Uranus and Neptune may have high-density, "slushy" interiors containing thick layers of highly compressed water clouds. It is also possible that much of the planets' ammonia could be dissolved in the water, producing a thick, electrically conducting layer that could conceivably explain the planets' magnetic fields. At the present time, however, we don't know enough about the interiors to assess the correctness of this picture.

The cores of Uranus and Neptune are inferred mainly from considerations of those planets' densities. Because of their lower masses, we would not expect the hydrogen and helium in the interiors of Uranus and Neptune to be compressed as much as in the two larger jovian worlds, yet the average densities of Uranus and Neptune are actually *greater* than the density of Saturn and similar to that of Jupiter. Astronomers interpret this to mean that Uranus and Neptune each have Earth-sized cores some 10 times more massive than our planet, with compositions similar to the cores of Jupiter and Saturn.

Our current knowledge of the internal structures of the four jovian worlds is summarized in Figure 7.20. The next space mission to significantly expand our knowledge of jovian interiors will be NASA's *Juno* probe, launched in August 2011 and due to arrive at Jupiter in 2016. From its polar orbit, *Juno* will map Jupiter's gravity and magnetism to study the planet's internal structure. It will also observe the composition and circulation of the planet's atmosphere.

MA° ANIMATION/VIDEO
The Gas Giants Part II

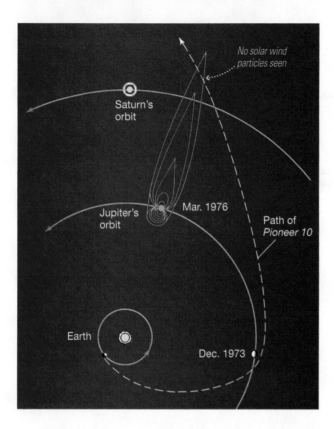

◀ FIGURE 7.21 **Pioneer** Mission   The *Pioneer 10* spacecraft (a forerunner to the *Voyager* missions) did not detect any solar particles while moving far behind Jupiter in March 1976, indicating that the tail of Jupiter's magnetosphere (drawn in red) extends beyond the orbit of Saturn.

Magnetospheres

The combination of rapid overall rotation and an extensive region of highly conductive fluid in its interior gives Jupiter by far the strongest planetary magnetic field in the solar system. ∞ *(Sec. 5.7)* The field at the cloud tops is some 14 times that of Earth, but taking the planet's radius into account implies an intrinsic field strength some 20,000 times Earth's. Jupiter is surrounded by a vast sea of energetic charged particles (mostly electrons and protons), somewhat similar to Earth's Van Allen belts but very much larger. Direct spacecraft measurements show Jupiter's magnetosphere to be almost 30 million km across (north to south), roughly a million times more voluminous than Earth's magnetosphere and far larger than the entire Sun. All four Galilean moons lie within it.

On the sunward side the boundary of Jupiter's magnetic influence on the solar wind lies about 3 million km from the planet. On the opposite side, however, the magnetosphere has a long "tail" extending away from the Sun at least as far as Saturn's orbit (some 4 AU, or 600 million kilometers), as sketched in Figure 7.21. Near the planet the magnetic field channels particles from the magnetosphere into the upper atmosphere, forming aurorae vastly larger and more energetic than those observed on Earth (Figure 7.22). ∞ *(Sec. 5.7)*

Saturn's electrically conducting interior and rapid rotation also produce a strong magnetic field and an extensive magnetosphere. However, because of the considerably smaller mass of Saturn's metallic hydrogen zone, the magnetic field strength at the planet's cloud tops is only about 1/20 that of Jupiter. Saturn's magnetosphere extends about 1 million kilometers toward the Sun and is large enough to contain the planet's ring system and most of its moons.

Voyager 2 found that both Uranus and Neptune have fairly strong magnetic fields, comparable to those of Saturn and Earth. Taking the planets' radii into account, we find intrinsic field strengths a few percent that of Saturn, or 30–40 times Earth's. Both planets have substantial magnetospheres, populated largely by electrons and protons either captured from the solar wind or created from hydrogen gas escaping from the planets below. Figure 7.23 compares the magnetic field structures of Earth and the four jovian planets. The locations and orientations of the bar magnets represent the observed planetary fields, and the sizes of the bars indicate overall magnetic field strength. Note that the fields of Uranus and Neptune are not aligned with the planets' rotation axes and are significantly offset from the planets' centers. Astronomers do not yet understand why this should be so.

Internal Heating

On the basis of Jupiter's distance from the Sun, astronomers expected to find the temperature of the cloud tops to be around 105 K. At that temperature, they reasoned, Jupiter would radiate back into space the same amount of energy as it receives from the Sun. However, radio and infrared observations of the planet revealed a black-body spectrum corresponding to a temperature of 125 K, implying (from Stefan's law) that Jupiter emits about twice—$(125/105)^4$—as much energy as reaches the planet in the form of sunlight. ∞ *(Sec. 2.4)*

Astronomers theorize that the source of Jupiter's excess energy is heat left over from the planet's formation. ∞ *(Sec. 4.3)* In the distant past, Jupiter's interior was probably very much hotter than it is today (although not nearly hot enough for Jupiter ever to have become a star like the Sun). That early heat is slowly leaking out through the planet's heavy atmospheric blanket, resulting in the excess emission we observe today.

Infrared measurements indicate that Saturn, too, has an internal energy source. Saturn's cloud-top temperature is 97 K, substantially higher than the

◀ FIGURE 7.22 **Aurorae on Jupiter**   The main (underlying) image was taken by the *Hubble* telescope in visible (true-color) light, but the two insets at the poles were taken in the ultraviolet part of the spectrum. The oval-shaped aurorae, extending hundreds of kilometers above Jupiter's surface, result from charged particles escaping the jovian magnetosphere and colliding with the atmosphere, causing the gas to glow. (See Figure 5.24.) *(NASA)*

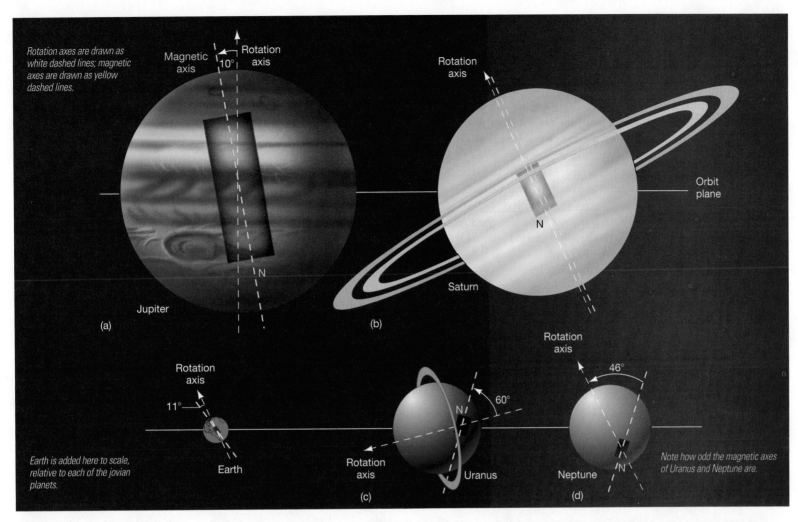

Rotation axes are drawn as white dashed lines; magnetic axes are drawn as yellow dashed lines.

Magnetic axis   10°   Rotation axis

N

Jupiter

(a)

Rotation axis

Orbit plane

N

Saturn

(b)

Earth is added here to scale, relative to each of the jovian planets.

Rotation axis

11°

Earth

(c)

Rotation axis

60°

N

Rotation axis

Uranus

Rotation axis

46°

N

Neptune

(d)

Note how odd the magnetic axes of Uranus and Neptune are.

▲ **FIGURE 7.23 Jovian Magnetic Fields**   Comparison of magnetic field strengths, orientations, and offsets in the four jovian planets. The magnetic fields are depicted here as simple bar magnets. The size and location of each magnet represent the strength and orientation of the planetary field.

temperature at which Saturn would reradiate all the energy it receives from the Sun. In fact, Saturn radiates away almost *three times* more energy than it absorbs. However, the explanation for Jupiter's excess emission doesn't work for Saturn. Saturn is smaller than Jupiter and so must have cooled more rapidly—fast enough, in fact, that its original supply of heat should have been used up long ago. What then is happening inside Saturn to produce this extra heat?

The explanation for this strange state of affairs also explains Saturn's atmospheric helium deficit. At the temperatures and high pressures found in Jupiter's interior, liquid helium *dissolves* in liquid hydrogen. In Saturn, where the internal temperature is lower, the helium doesn't dissolve so easily and tends to form droplets instead. (The basic phenomenon is familiar to cooks who know that it is generally much easier to dissolve ingredients in hot liquids than in cold ones.) Saturn probably started out with a fairly uniform mix of hydrogen and helium, but the helium tended to condense out of the surrounding hydrogen, much as water vapor condenses out of Earth's atmosphere to form a mist. The amount of helium condensation was greatest in the planet's cool outer layers, where the mist turned to rain about 2 billion years ago. A light shower of liquid helium has been falling through Saturn's interior ever since. This helium precipitation is responsible for depleting the outer layers of their helium content.

As the helium sinks toward the center, the planet's gravitational field compresses it and heats it up. The energy thus released is the source of Saturn's internal heating. In the distant future the helium rain will stop and Saturn will cool until its outermost layers radiate only as much energy as they receive from the Sun. When that happens, the temperature at Saturn's cloud tops will be 74 K.

PROCESS OF SCIENCE CHECK

Why is mathematical modeling essential to understanding the interior structures of the jovian planets?

7.2 DISCOVERY

A Cometary Impact

In 1994, astronomers were granted a rare alternative means of studying Jupiter's atmosphere and interior when a comet, called Shoemaker-Levy 9, collided with the planet. Discovered in 1993, the comet had an unusual flattened nucleus made up of several large pieces, as shown in the first image (a) below. The largest piece was no larger than 1 km across. Tracing the orbit backward in time, researchers found that the comet had narrowly missed Jupiter in 1992. They realized that the objects shown in the figure were fragments produced when a previously "normal" comet was captured by Jupiter and torn apart by the planet's strong tidal forces. ∞ *(Sec. 5.2)* The data also revealed an even more remarkable fact—the comet would collide with Jupiter on its next approach!

Between July 16 and July 22, 1994, as most of the telescopes on (and in orbit around) Earth watched, the comet fragments struck Jupiter's upper atmosphere at speeds of over 60 km/s, causing a series of enormous explosions. Each impact created, for a period of a few minutes, a brilliant fireball hundreds of kilometers across and having a temperature of many thousands of kelvins. The energy released in each explosion was comparable to a billion terrestrial nuclear detonations. One of the largest pieces of the comet, fragment G, produced the spectacular fireball shown in the second image (b). The effects on the planet's atmosphere and the vibrations produced throughout Jupiter's interior were observable for days after the impact.

As far as we can determine, none of the cometary fragments breached the jovian clouds. Only *Galileo* had a direct view of the impacts on the back side of Jupiter, and in every case the explosions seemed to occur high in the atmosphere, above the uppermost cloud layer. Spectral lines from water vapor, silicon, magnesium, and iron were observed, but all apparently came from the comet itself—which really did resemble a loosely packed snowball. ∞ *(Sec. 4.2)* Debris from the impacts, like the "black eye" shown in image (c), spread slowly around Jupiter and eventually encircled the planet. It took years for all the cometary matter to settle into the interior.

The Shoemaker-Levy 9 impact has scientific significance far beyond its direct effect on Jupiter. Since the 1960s, planetary scientists have become increasingly aware that collisions between small bodies—asteroids, comets, and meteoroids—and the major planets are relatively commonplace events in the solar system. As we have seen, the consequences for life can be catastrophic. ∞ *(Discovery 4-1)* The fact that we witnessed such an event during the few decades since it became technologically feasible for us to do so underscores just how frequent these encounters probably are and has convinced many scientists of the importance of planetary impacts to solar system evolution.

The final figure (d) shows a 2009 comet collision with Jupiter, comparable in violence to the 1994 event, discovered by pure chance on the 15th anniversary of the Shoemaker-Levy 9 impact! Another collision was observed in 2012. Such collisions are becoming important sources of information about planetary interiors.

(a) S G D

Toward Sun

R I V U X G

Toward Jupiter

(b) *(Palomar)* R I V U X G

(c) R I V U X G

(d) *(NASA)* R I V U X G

Uranus apparently has no internal energy source, radiating into space as much energy as reaches it from the Sun. It never experienced helium precipitation, and its initial heat was lost to space long ago. Surprisingly, given its other similarities to Uranus, Neptune *does* have an internal heat source—it emits 2.7 times more energy than it receives from the Sun. The reason is still uncertain. Some scientists think that Neptune's relatively high concentration of methane insulates the planet, helping to maintain its initially high internal temperature.

CONCEPT CHECK

Why do all the jovian planets have strong magnetic fields?

THE BIG QUESTION

Knowing the structure and evolution of the jovian planets is key to understanding the formation of our solar system, but these distant worlds guard their secrets well. Will the next generation of planetary probes dig deep enough below the surfaces of these worlds to let us discriminate among competing theories? Or will the breakthrough instead come from studies of exoplanets, using the statistics of thousands of remote planetary systems to shed light on our own?

CHAPTER REVIEW

SUMMARY

LO1 Jupiter and Saturn were the outermost planets known to ancient astronomers. Uranus was discovered in the 18th century, by chance. Neptune was discovered after mathematical calculations of Uranus's slightly non-Keplerian orbit revealed the presence of an eighth planet.

LO2 The four jovian planets are all much more massive than the terrestrial worlds and are of much lower density. They are composed primarily of hydrogen and helium. However, all have large, dense cores, each roughly 10 times Earth's mass and of chemical composition similar to that of the terrestrial planets. The jovian planets all rotate more rapidly than Earth. Having no solid surfaces, they display **differential rotation** (p. 207)—the rotation rate varies with latitude and with depth below the cloud tops. For unknown reasons, Uranus's spin axis lies nearly in the ecliptic plane, leading to extreme seasonal variations in solar heating on the planet as it orbits the Sun.

LO3 The cloud layers on all the jovian worlds are arranged into bands of light-colored zones and darker belts crossing the

planets parallel to their equators. The bands are the result of convection in the planets' interiors and the planets' rapid rotation. The lighter zones are the tops of upwelling, warm currents, and the darker bands are cooler regions, where gas is sinking. Underlying them is a stable pattern of eastward or westward wind flow called a **zonal flow.** The wind direction alternates as we move north or south away from the equator. Jupiter's atmosphere consists of three main cloud layers. The colors we see are the result of chemical reactions at varying depths below the cloud tops.

LO4 A major weather pattern on Jupiter is the **Great Red Spot** (p. 209), a huge hurricane that apparently has been raging for at least three centuries. Other, smaller, atmospheric features— **white ovals** (p. 211) and **brown ovals** (p. 211)—are also observed. Similar systems are found on the other jovian planets, although they are less distinct on Saturn and Uranus. Long-lived storms on Saturn may lie hidden below the planet's clouds. The **Great Dark Spot** (p. 215) on Neptune had many similarities to Jupiter's Red Spot. It disappeared within a few years of its discovery by *Voyager 2*. Storms in the jovian atmospheres are long-lived because there is no solid surface to dissipate their energy.

LO5 The atmospheres of the jovian planets become hotter and denser with depth, eventually becoming liquid. In Jupiter and Saturn the interior pressures are so high that the hydrogen near the center exists in a liquid-metallic form rather than as molecular hydrogen. In Uranus and Neptune this change from molecular hydrogen to metallic hydrogen apparently does not occur. Radio waves emitted from the planets' magnetospheres provide a measure of their interior rotation rates. The combination of rapid rotation and conductive interiors means that all four jovian planets have strong magnetic fields and extensive magnetospheres.

LO6 Three jovian planets radiate more energy than they receive from the Sun. On Jupiter and Neptune the source of this energy is most likely heat left over from the planet's formation. On Saturn the heating is the result of helium precipitation in the interior, where helium liquefies and forms droplets that then fall toward the center of the planet. This process is also responsible for reducing the amount of helium observed in Saturn's outer layers.

MasteringAstronomy®   For instructor-assigned homework go to www.masteringastronomy.com

Problems labeled **POS** explore the process of science. **VIS** problems focus on reading and interpreting visual information. **LO** connects to the introduction's numbered Learning Outcomes.

REVIEW AND DISCUSSION

1. How was Uranus discovered?

2. **LO1 POS** Why did astronomers suspect an eighth planet beyond Uranus?

3. Why has Jupiter retained most of its original atmosphere?

4. What is differential rotation, and how is it observed on Jupiter?

5. Why does Saturn have a less varied appearance than Jupiter?

6. **LO2** How (and why) do the thicknesses of Saturn's various layers (clouds, molecular and metallic hydrogen, core) compare to those in Jupiter?

7. **LO4** What is the Great Red Spot? What is the source of its energy?

8. Describe the cause of the colors in the jovian planets' atmospheres.

9. **LO3** Compare and contrast the atmospheres of the jovian planets, describing how the differences affect the appearance of each.

10. Explain the theory that accounts for Jupiter's internal heat source.

11. **LO6** Describe the theoretical model that accounts for Saturn's missing helium and the surprising amount of heat the planet radiates.

12. What is unusual about the rotation of Uranus?

13. What is responsible for Jupiter's enormous magnetic field?

14. **LO5** How are the interiors of Uranus and Neptune thought to differ from those of Jupiter and Saturn?

15. **POS** How much do you think we might know about the outer solar system without the *Voyager, Galileo,* and *Cassini* missions?

CONCEPTUAL SELF-TEST: TRUE OR FALSE?/MULTIPLE CHOICE

1. The solid surface of Jupiter lies just below the cloud layers that are visible from Earth. (T/F)

2. Storms in the jovian planets' atmospheres are generally much longer lived than storms on Earth. (T/F)

3. The element helium plays an important role in producing Saturn's atmospheric colors. (T/F)

4. Jupiter emits more energy than it receives from the Sun. (T/F)

5. Although often referred to as a gaseous planet, Jupiter is mostly liquid in its interior. (T/F)

6. Neptune probably does not have a rocky core. (T/F)

7. Around northern midsummer on Uranus, an observer near the north pole would observe the Sun high and almost stationary in the sky. (T/F)

8. The magnetic fields of Uranus and Neptune are highly tilted relative to their rotation axes and offset from the planets' centers. (T/F)

9. The main constituent of Jupiter's atmosphere is (a) hydrogen; (b) helium; (c) ammonia; (d) carbon dioxide.

10. Saturn's cloud layers are much thicker than those of Jupiter because Saturn has (a) more mass; (b) lower density; (c) a weaker magnetic field; (d) weaker gravity.

11. Compared to Earth, Saturn's core is roughly (a) half the mass; (b) the same mass; (c) twice as massive; (d) 10 times more massive.

12. Uranus was discovered about the same time as (a) Columbus reached North America; (b) the U.S. Declaration of Independence; (c) the American Civil War; (d) the Depression.

13. Compared with Uranus, the planet Neptune is (a) much smaller; (b) much larger; (c) roughly the same size; (d) closer to the Sun.

14. VIS According to Figure 7.9 (Jupiter's Atmosphere), if ammonia and ammonium hydrosulfide ice were transparent to visible light, Jupiter would appear (a) bluish; (b) red; (c) tawny brown; (d) exactly as it does now.

15. VIS According to Figure 7.21, Jupiter's magnetosphere extends about (a) 1 AU; (b) 5 AU; (c) 10 AU; (d) 20 AU beyond the planet.

PROBLEMS

The number of squares preceding each problem indicates its approximate level of difficulty.

1. ■ The *Hubble Space Telescope* has an angular resolution of 0.05″. What is the size of the smallest feature visible on (a) Jupiter, at 4.2 AU, and (b) Neptune, at 29 AU?

2. ■ What is the gravitational force on Uranus from Neptune, at closest approach? Compare with the Sun's gravitational force on Uranus.

3. ■ What is the round-trip travel time of light from Earth to Jupiter (at the minimum Earth–Jupiter distance of 4.2 AU)? How far would a spacecraft orbiting Jupiter at a speed of 20 km/s travel in that time?

4. ■■ What would be the mass of Jupiter if it were composed entirely of hydrogen at a density of 0.08 kg/m^3 (the density of hydrogen at sea level on Earth)?

5. ■ Verify the statement made in the text that the force of gravity at Jupiter's cloud tops is nearly 2.5 times that at Saturn's cloud tops.

6. ■ Given Jupiter's current atmospheric temperature, what is the smallest possible mass the planet could have and still have retained its hydrogen atmosphere? ∞ *(More Precisely 5-1)*

7. ■ How long does it take for Saturn's equatorial flow, moving at 1500 km/h, to encircle the planet?

8. ■■ If the temperature at Saturn's cloud tops is 97 K and the planet radiates three times more energy than it receives from the Sun, use Stefan's law to calculate what the temperature would be in the absence of any internal heat source. ∞ *(More Precisely 2-2)*

9. ■■ If the core of Uranus has a radius twice that of planet Earth and an average density of 8000 kg/m^3, calculate the mass of Uranus outside the core. What fraction of the planet's total mass is its core?

10. ■ From Wien's law, at what wavelength does Neptune's thermal emission peak? In what part of the electromagnetic spectrum does this wavelength lie? ∞ *(More Precisely 2-2, Fig. 2.9)*

ACTIVITIES

Collaborative

1. Jupiter's rapid (10-hour) rotation and ever-changing atmospheric features make it a highly dynamic telescopic object. Even a small telescope should reveal several of the planet's cloud belts, as well as its four brightest moons. Can you see the Great Red Spot? Make a series of sketches of the planet. You'll have to work fast because of the planet's rotation! Try to complete each sketch within 20–30 minutes, and prepare in advance a series of sheets of paper with a circle to represent the planet. Note the date and time of each sketch. By carefully timing some easily recognizable surface feature, such as the Great Red Spot, measure Jupiter's rotation period. (You may have to come back on another night to be able to conveniently observe the chosen feature cross the entire planetary disk.) Repeat your observations a week or two later. Have any of the surface features, or their relation to one another, changed?

Individual

1. Consult a sky chart and find Jupiter in the night sky. It's an easy-to-find naked-eye object. Do any stars look as bright as Jupiter? What other differences do you notice between Jupiter and the stars?

2. Use a telescope to look at Saturn. Does the planet appear flattened? Can you see any atmospheric features?

3. Locate Uranus in the night sky. It may be barely visible to the naked eye, but binoculars will make the search much easier. (Hint: Uranus shines more steadily than the background stars.) Can you detect the planet's color?

4. The search for Neptune requires a much more determined effort! A telescope is best, but high-powered binoculars mounted on a steady support will also do. Comparing Uranus and Neptune, which planet appears bluer? Through a telescope, does either planet appear as a disk, or do they look more like points of light?

8

Moons, Rings, and Plutoids

Small Worlds Among Giants

All four jovian planets have systems of moons and rings that exhibit fascinating variety and complexity. The six largest jovian moons have many planetary features, providing us with insight into the terrestrial worlds. Two of these moons, Jupiter's Europa and Saturn's Titan, have become prime candidates for study in the search for life elsewhere in the solar system. Like the moons, the rings of the jovian planets also differ greatly. The rings of Saturn—the best-known ring system among the jovian planets—are one of the most spectacular displays in the sky. Plutoids are dwarf planets outside the orbit of Neptune—small icy bodies having much in common with the large jovian moons and little similarity to the eight major planets. For these reasons we study them here, along with the moons and rings of their giant neighbors.

THE BIG PICTURE

Our knowledge of the moons and other small bodies of the solar system has steadily increased over the past half century as bigger telescopes and more sophisticated space probes have enabled astronomers to better explore our cosmic neighborhood. But the more closely we examine these alien worlds, the more complex and bizarre they seem to be, and in many cases their properties depend critically on their environment, so that planetary moon and ring systems seem much more like giant ecosystems than collections of individual objects.

◄ Jupiter's innermost large moon Io is a world unlike any other in our planetary system. This 1999 color-enhanced mosaic from the *Galileo* space probe spans slightly more than the visible spectrum, and shows a landscape of volcanoes, lava flows and sulfurous deposits formed as Io's superheated interior squirts through the surface. This ongoing volcanism continually resurfaces the moon and blasts plumes of ionized gas out into the larger jovian environment. *(NASA/JPL/University of Arizona)*

LEARNING OUTCOMES

Studying this chapter will enable you to:

LO1 Describe how the Galilean moons form a miniature solar system around Jupiter and exhibit a wide range of properties.

LO2 Describe the composition and likely origin of the atmosphere on Titan, Saturn's largest moon.

LO3 Explain why astronomers think Neptune's moon Triton was captured by that planet.

LO4 Outline the properties of the medium-sized moons of the outer planets.

LO5 Describe the nature and detailed structure of Saturn's rings.

LO6 Explain how planetary rings form and evolve.

LO7 List the major differences between the rings of Saturn and those of the other jovian planets.

LO8 Describe the composition of the solar system beyond Neptune, and explain why astronomers no longer regard Pluto as a planet.

MasteringAstronomy®

Visit www.masteringastronomy.com for quizzes, animations, videos, interactive figures, and self-guided tutorials.

8.1 The Galilean Moons of Jupiter

All four jovian planets have extensive moon systems—Jupiter has at least 67 natural satellites orbiting it, Saturn has 62, Uranus 27, and Neptune 14. These moons range greatly in size and other physical properties, but they fall into three natural groups.

The first group consists of the 6 "large" jovian moons, each comparable in size to Earth's Moon. They are listed in Table 8.1, along with our own Moon for comparison. All have diameters greater than 2500 km and are large enough to have their own distinctive geologies and histories. Next are the 12 "medium-sized" moons—spherical bodies with diameters ranging from about 400 to 1500 km. Their heavily cratered surfaces offer tantalizing clues to the past and present states of the environment in the outer solar system. Finally, there are the many "small" moons—irregularly shaped chunks of ice, all less than 300 km across—that exhibit a bewildering variety of complex and fascinating motions. Most are probably little more than captured interplanetary debris. ∞ *(Sec. 4.3)* The number of known small moons has increased rapidly in recent years (and will probably continue to rise) thanks to steadily improving ground and space-based technology.

In this chapter we focus our attention on the large and midsize jovian moons, beginning with the Galilean satellites of Jupiter. ∞ *(Sec. 1.2)*

A "Miniature Solar System"

Jupiter's four large moons travel on nearly circular, prograde orbits in Jupiter's equatorial plane. Moving outward from the planet, they are named Io, Europa, Ganymede, and Callisto, after the mythical attendants of the Roman god Jupiter. The two *Voyager* spacecraft and, more recently, the *Galileo* mission returned many remarkable images of these small worlds, allowing scientists to discern much fine surface detail on each. Figure 8.1 is a *Voyager 1* image of Io and Europa, with Jupiter providing a spectacular backdrop. Figure 8.2 shows the four Galilean moons to scale, along with their relative orbits.

The Galilean moons range in size from slightly smaller than Earth's Moon (Europa) to slightly larger than Mercury (Ganymede). Their densities decrease with increasing distance from Jupiter in a manner very reminiscent of the falloff in density of the terrestrial planets with increasing distance from the Sun. ∞ *(Sec. 4.1)* Indeed, many astronomers think that the formation of Jupiter and its Galilean satellites may have mimicked on a small scale the formation of the Sun and the inner planets, with generally denser moons forming closer to the intense heat of the young Jupiter. (Sec. 4.3)

Detailed gravity measurements made by the *Galileo* orbiter support this picture of the origin of the Jupiter system. Io and Europa have large, iron-rich cores

ANIMATION/VIDEO MA®
Galilean Moons Transit Jupiter

| R | I | V | U | X | G |

▲ **FIGURE 8.1 Jupiter, Up Close** *Voyager 1* took this photograph of Jupiter with ruddy Io on the left and pearl-like Europa toward the right. Note the scale of objects here: Io and Europa are both comparable in size to our Moon; the Great Red Spot is roughly twice as big as Earth. *(NASA)*

TABLE 8.1   The Large Moons of the Solar System

Name	Parent Planet	Distance from (km)	Parent Planet (planet radii)	Orbit Period (Earth days)	Radius (km)	Mass (Earth's Moon = 1)	Density (kg/m³)
Moon	Earth	384,000	60.2	27.3	1740	1.00	3300
Io	Jupiter	422,000	5.90	1.77	1820	1.22	3500
Europa	Jupiter	671,000	9.38	3.55	1570	0.65	3000
Ganymede	Jupiter	1,070,000	15.0	7.15	2630	2.02	1900
Callisto	Jupiter	1,880,000	26.3	16.7	2400	1.46	1800
Titan	Saturn	1,220,000	20.3	16.0	2580	1.83	1900
Triton	Neptune	355,000	14.3	−5.88[1]	1350	0.29	2100

[1]The minus sign indicates a retrograde orbit.

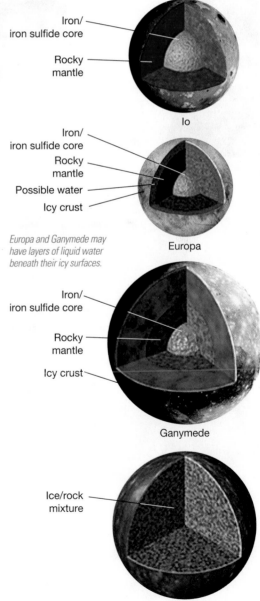

surrounded by thick mantles of generally rocky composition, perhaps similar to the crusts of the terrestrial planets. Europa has a water/ice outer shell between 100 and 200 km thick. Ganymede and Callisto are less dense overall; low-density materials, such as water ice, may account for as much as half their total mass. Ganymede has a relatively small metallic core topped by a rocky mantle and a thick icy outer shell. Callisto seems to be a largely undifferentiated mixture of rock and ice. Figure 8.3 summarizes our knowledge of the interiors of these bodies.

Io: The Most Active Moon

Io is the most geologically active object in the entire solar system. In mass and radius it is similar to Earth's Moon, but there the resemblance ends. Shown in Figure 8.4, Io's uncratered surface is a collage of oranges, yellows, and blackish browns.

As *Voyager 1* sped past Io, it made an outstanding discovery: Io has active volcanoes! The spacecraft photographed eight erupting volcanoes. Six of them were still erupting when *Voyager 2* passed by 4 months later. Inset (a) in Figure 8.4 shows one volcano, called Prometheus, ejecting matter at speeds of up to 2 km/s to an altitude of about 150 km. The orange color immediately surrounding the volcanoes most likely results from sulfur compounds in the ejected material. Io's exceptionally smooth surface is apparently the result of molten matter constantly filling in any "dents and cracks." Io has a thin atmosphere composed

▶ FIGURE 8.3 **Galilean Moon Interiors** Cutaway diagrams of the interior structure of the four Galilean satellites. Moving outward from Io to Callisto, densities steadily decrease as the composition shifts from rocky mantles and metallic cores in Io and Europa, to a thick icy crust and smaller core in Ganymede, to an almost uniform rock and ice mix in Callisto.

(a)

Volcanic plume

Surface

Volcano   Plume

(b)

◄ FIGURE 8.4 Io

INTERACTIVE Jupiter's innermost moon, Io, differs from the other three Galilean satellites. Its surface is kept smooth and brightly colored by constant volcanism, shown here as dark, circular features. The two inserts show Io's umbrella-like eruptions as *Galileo* flew past in 1997; both volcanoes are about 300 km across, and the plume at the left is about 150 km high. *(NASA)*

R I V U X G

ANIMATION/VIDEO **MA**
Jupiter Icy Moons Orbiter Mission

ANIMATION/VIDEO **MA**
Io Cutaway

ANIMATION/VIDEO **MA**
Galileo's View of Europa

mainly of sulfur dioxide, produced by volcanic activity and temporarily retained by the moon's gravity. ∞ *(More Precisely 5-1)*

By the time *Galileo* arrived, some of the volcanoes observed by *Voyager 2* had subsided, but many new ones were seen—in fact, *Galileo* found that Io's surface features can change significantly in as little as a few weeks. In all, more than 80 active volcanoes have been identified on Io. The largest, known as Loki (located on the "back" side of Io in Figure 8.4), is larger than the state of Maryland and emits more energy than all of Earth's volcanoes combined. According to *Galileo's* instruments, lava temperatures on Io generally range from 650 to 900 K, although temperatures as high as 2000 K—far hotter than any earthly volcano—have been measured at some locations. Mission scientists speculate that these superhot volcanoes may be similar to those that occurred on Earth more than 3 billion years ago.

Io is far too small to have geological activity similar to that on Earth. It should be long dead, like our Moon, its internal heat lost to space billions of years ago. Instead, the source of Io's energy is *external*—Jupiter's gravity. Because Io orbits very close to Jupiter, the planet's huge gravitational field produces strong tidal forces on the moon, resulting in a large (100 m) tidal bulge. ∞ *(Sec. 5.2)* If Io were Jupiter's only satellite, it would long ago have come into a state of synchronous rotation with the planet. In that case, Io would move in a perfectly circular orbit with one face permanently turned toward Jupiter. The tidal bulge would be stationary with respect to the moon, and there would be no internal stresses and no volcanism.

But Io is not alone. As it orbits, it is constantly tugged by the gravity of its nearest large neighbor, Europa. These tugs are small and not enough to cause any great tidal effect in and of themselves, but they are sufficient to make Io's orbit slightly noncircular, preventing the moon from settling into a precisely synchronous state. The moon's orbital speed varies from place to place as it revolves around the planet, but its rate of rotation on its axis remains constant. Thus, it cannot keep one face permanently turned toward Jupiter. Instead, Io rocks or wobbles slightly from side to side on its axis as it moves. The tidal bulge, however, always points directly toward Jupiter, so it moves back and forth across Io's surface as the moon wobbles. The conflicting forces acting on Io result in enormous tidal stresses that continually flex and squeeze its interior.

Just as repeated back-and-forth bending of a wire can produce heat through friction, this ever-changing distortion continually energizes Io. Researchers estimate that the total amount of heat generated within Io as a result of tidal flexing is about 100 million megawatts—five times the total power consumption of all the nations on Earth. The large amount of heat generated within Io causes huge jets of gas and molten rock to squirt out of the surface. It is likely that much of Io's interior is soft or molten, with only a relatively thin solid crust overlying it.

Europa: Liquid Water Locked in Ice

Europa (Figure 8.5) is a world very different from Io. Europa has relatively few craters on its surface, again suggesting that recent activity has erased the scars of ancient meteoric impacts. However, the surface displays a vast network of lines crisscrossing bright, clear fields of water ice. Many of these linear bands extend halfway around the satellite.

Based on the *Voyager* images, planetary scientists theorized that Europa is covered by an ocean of liquid water whose surface layers are frozen. Later images from *Galileo* seem to support this idea. Figure 8.5(b) is a high-resolution image of this weird moon, showing ice-filled surface cracks similar to those seen in Earth's polar ice floes. Figure 8.5(c) shows what appear to be icebergs—flat chunks of ice that have been broken apart and reassembled, perhaps by the action of water below. Mission scientists speculate that Europa's ice may be several kilometers thick and that the ocean below it may be 100 km deep. As on Io, the ultimate source of the energy that keeps the ocean liquid and also drives the internal motion leading to the cracks in the icy surface is Jupiter's gravity. The heating mechanism is essentially similar to that on Io, but the effects are less extreme on Europa because of its greater distance from the planet.

Other detailed images of the surface support this hypothesis. Figure 8.5(d) shows a region where Europa's icy crust appears to have been pulled apart and new material has filled in the gaps between the separating ice sheets. Elsewhere on the surface, *Galileo* found what appeared to be the icy equivalent of lava flows on Earth—regions where water apparently erupted through the surface and flowed for many kilometers before solidifying. The scarcity of impact craters on Europa implies that the processes responsible for these features did not stop long ago. Rather, they must be ongoing.

Further evidence comes from studies of Europa's magnetic field. *Galileo* found that Europa has a weak magnetic field that constantly changes strength and direction. This finding is consistent with the idea that the field is generated by the action of Jupiter's magnetism on a shell of electrically conducting fluid about 100 km below Europa's surface—in other words, the salty liquid water layer suggested by the surface observations. These results convinced quite a few skeptical scientists of the reality of Europa's deep ocean.

The likelihood that Europa has an extensive layer of liquid water below its surface ice opens up many interesting avenues of speculation about the possible development of life there. In the rest of the solar system, only Earth has liquid water on or near its surface, and most scientists agree that water played a key role in the appearance of life here (see Chapter 18). Europa may well contain more liquid water than exists on our entire planet! Of course, the existence of water does not *necessarily* imply the emergence of life. Europa, even with its liquid ocean, is still a hostile environment compared to Earth. Nevertheless, the possibility, even a remote one, of life on Europa was an important motivating factor in the decision to extend the *Galileo* mission for 6 additional years. ∞ *(Sec. 7.1)*.

▲ **FIGURE 8.5 Europa** (a) A *Voyager 1* mosaic of Europa. (b) Its icy surface is only lightly cratered, indicating that some ongoing process must be obliterating impact craters soon after they form. (c) This image from the *Galileo* spacecraft shows a smooth yet tangled surface, called Conamara Chaos, resembling the huge ice floes that cover Earth's polar regions. (d) This detailed *Galileo* image shows "pulled apart" terrain that suggests liquid water upwelling from the interior and freezing, filling in the gaps between separating surface ice sheets. *(NASA)*

(a)

(b)

|← 1000 km →|

(c)

R I V U X **G**

(d)

R I V U X **G**

(MA) ANIMATION/VIDEO
Galileo's View of Ganymede

◄ **FIGURE 8.6 Ganymede**   (a) and (b) *Voyager 2* images of Ganymede. The dark regions are the oldest parts of the moon's surface and may represent its original icy crust. Galileo Regio spans some 3200 km. The lighter, younger regions are the result of flooding and freezing that occurred within a billion years or so of Ganymede's formation. (c) Grooved terrain on Ganymede may have been caused by a process similar to plate tectonics on Earth. (d) This 2015 *Hubble* ultraviolet image shows an aurora in Ganymede's magnetosphere, here superimposed on a *Galileo* image of the moon itself. By studying Ganymede's aurorae, scientists can estimate the extent of liquid water in the moon's interior. *(NASA)*

Ganymede and Callisto: Fraternal Twins

Ganymede (Figure 8.6) is the largest moon in the solar system, exceeding both Earth's Moon and the planet Mercury in size. It has many impact craters on its surface, and light and dark patterns reminiscent of features on the Moon. Ganymede's history has many lunar parallels (with water ice replacing lunar rock).

As with the inner planets and Earth's Moon, we can estimate the ages of Ganymede's surface features by counting craters. The darker regions are the oldest parts of Ganymede's surface. They are the original icy surface, just as the ancient highlands on our Moon are its original crust. Ganymede's surface has darkened with age as micrometeorite dust has slowly covered it. The light-colored regions are much less heavily cratered, and so must be younger. They are Ganymede's "maria" and probably formed similarly to the maria on the Moon. ∞ *(Sec. 5.6)* Intense meteoritic bombardment caused liquid water—Ganymede's counterpart to the Moon's molten lava—to upwell from the interior, flooding the impacting regions before solidifying. Not all of Ganymede's surface features follow the lunar analogy, however. Ganymede has a system of grooves and ridges (Figure 8.6b, c) that may have resulted from tectonic motion, like mountain building and faulting at plate boundaries on Earth. ∞ *(Sec. 5.5)* The process stopped about 3 billion years ago when the cooling crust became too thick.

Callisto (Figure 8.7) is in many ways similar in appearance to Ganymede, although Callisto is more heavily cratered and has few fault lines. Callisto's most obvious feature is a huge series of concentric ridges surrounding two large basins, one clearly visible in Figure 8.7. The ridges resemble the ripples made as a stone hits water, but on Callisto they probably resulted from a cataclysmic impact with a meteorite. The upthrust ice was partially melted, but it resolidified quickly, before the ripples could subside. Today the crust is frigid ice and shows no obvious signs of geological activity. Apparently, Callisto froze before plate tectonic or other activity could start. The density of impact craters indicates that the impact basins formed long ago, perhaps 4 billion years in the past.

Ganymede's internal differentiation indicates that the moon was largely molten at some time in the past, yet Callisto is undifferentiated and hence apparently never melted (Figure 8.3). ∞ *(Sec. 5.4)* Researchers are unsure why two such similar bodies evolved so differently. The *Galileo* orbiter also found that Ganymede has a weak magnetic field, about 1 percent that of Earth, suggesting that this moon, too, may have liquid or "slushy" water below its surface. ∞ *(Sec. 5.7)* Figure 8.6(d) shows an aurora on Ganymede, the result of the moon's magnetic field interacting with its environment, strongly supporting this idea. The presence of an extensive saltwater ocean means that Ganymede's heating and differentiation must have happened relatively recently—less than a billion years ago, based on estimates of how rapidly the moon's heat escapes into space. Scientists have no clear explanation for how this could have occurred, although some speculate that interactions among the inner moons may have caused Ganymede's orbit to change significantly about 1 billion years ago and that tidal heating by Jupiter melted the moon's interior.

(a)

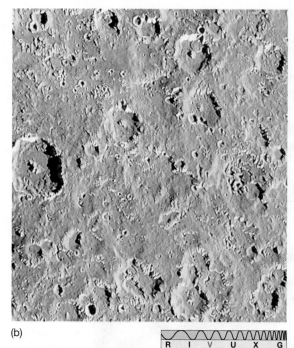

(b)

◄ **FIGURE 8.7 Callisto** (a) Callisto resembles Ganymede in overall composition but is more heavily cratered. Extending nearly 1500 km from the Valhalla basin center, its concentric ridges formed when "ripples" from a large meteoritic impact froze before they could disperse completely. (b) This higher-resolution *Galileo* image of Callisto's equatorial region, about 200 km across, displays more clearly its heavy cratering. *(NASA)*

8.2 The Large Moons of Saturn and Neptune

CONCEPT CHECK

What is the source of energy for all the activity observed on Jupiter's Galilean moons?

The other two jovian moons, Titan and Triton, are very different worlds. Titan's thick atmosphere continues to evolve even today, while distant Triton, the smallest of the six "large" moons, may well be a recent acquisition from the Kuiper belt. Both pose new riddles and offer unique insights into the formation and evolution of our planetary system.

Titan: A Moon with an Atmosphere

Titan is the largest and most intriguing of Saturn's moons. From Earth, it is visible only as a barely resolved orange disk. Long before the *Voyager* missions, astronomers knew from spectroscopic observations that the moon's ruddy coloration is caused by something quite special—an atmosphere. So anxious were mission planners to obtain a closer look, they programmed *Voyager 1* to pass very close to Titan, even though that meant the spacecraft could not then use Saturn's gravity to continue on to Uranus and Neptune.

Unfortunately, despite this close encounter, the moon's surface remained a mystery. The spacecraft's view was completely obscured by a thick, uniform haze layer, in some ways similar to the smog found over many cities on Earth, that envelops the moon (Figure 8.8). More detailed, close-up information on Titan is one of the principal goals of the ongoing *Cassini* mission. ∞ *(Discovery 7-1)* The spacecraft's carefully planned trajectory looping among Saturn's moons has included some 130 visits to Titan so far, with another 30 planned by the time the *Cassini Solstice* mission ends in 2017.

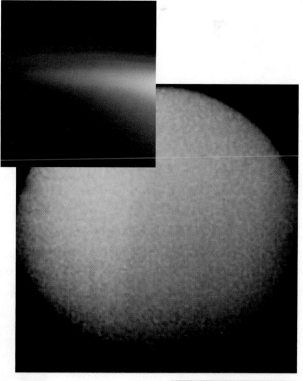

▶ **FIGURE 8.8 Titan** Larger than Mercury and roughly half the size of Earth, Titan was photographed from only 4000 km away as the *Voyager 1* spacecraft passed by in 1980. The inset shows a contrast-enhanced true color image of the haze layers in Titan's upper atmosphere, taken by the *Cassini* spacecraft in 2005. *(NASA)*

◄ **FIGURE 8.9 Titan Revealed**   *Cassini*'s telescopes revealed this infrared, false-color view of Titan's surface in late 2004. The inset shows a circular surface feature thought to be an icy volcano. *(NASA/ESA)*

Figure 8.9 shows how *Cassini*'s infrared instruments penetrated the moon's foggy atmosphere, revealing details of the surface. The image shows light and dark regions near the center of the field of view, thought to be icy plateaus, apparently smeared with hydrocarbon tar. Ridges and cracks on the moon's surface suggest that geological activity may be common. The rather blurred boundaries between the light and dark regions, the peculiar surface markings in the light-colored region, and the absence of extensive cratering suggest that some sort of erosion is occurring, perhaps as a result of wind or volcanic activity. Radar imaging has revealed few large (10–100 km diameter) craters on the moon's surface, but fewer small craters than expected given Titan's location in Saturn's congested ring plane (see Section 8.4). The inset in Figure 8.9 shows what appears to be an icy volcano, supporting the view that the moon's surface is geologically active.

Titan's atmosphere is thicker and denser than Earth's and is certainly more substantial than that of any other moon. Radio and infrared observations made by *Voyager 1* and *Cassini* show that the atmosphere is composed mostly of nitrogen (more than 98 percent), with methane and trace amounts of other gases making up the rest. The surface temperature is a frigid 94 K, roughly what we would expect on the basis of Titan's distance from the Sun. Under these conditions, as on Ganymede and Callisto, water ice plays the role of rock on Earth, and liquid water the role of lava. The presence of an atmosphere allows us to extend this analogy further—at the temperatures typical of the lower atmosphere, methane and ethane behave like water on Earth, resulting in methane rain, snow, and fog and ethane rivers and oceans!

On January 14, 2005, the *Huygens* probe, transported to Saturn by *Cassini* and released 3 weeks earlier, arrived at Titan and parachuted through the thick atmosphere to the moon's surface. Figure 8.10(a) shows an intriguing image radioed back from *Huygens* during its descent. It appears to show a network of drainage channels leading to a shoreline, but this interpretation is uncertain. The probe landed on solid ground, and for the next hour transmitted images and instrument readings to *Cassini* as it passed overhead. The hazy view from the surface (Figure 8.10b) reveals an icy landscape. The rocks in the foreground are a few centimeters across and show evidence of erosion by liquid of some sort.

▶ **FIGURE 8.10 Titan's Surface**   (a) This photograph of the surface, rendered in approximately true color, was taken from an altitude of about 15 km as the *Huygens* probe descended. (b) This enlargement of a portion of part (a) shows a network of dark channels reminiscent of streams (at center) or rivers draining from the light-shaded uplifted terrain into darker, low-lying regions (at bottom). (c) *Huygens*'s view of its landing site, again in approximately true color. The foreground icy "rocks" are about 10 centimeters across. *(NASA/ESA)*

(a)   (b)   (c)

Figure 8.11 shows the probable structure of Titan's atmosphere, based on *Voyager 1* and *Huygens* measurements. Despite Titan's low mass (a little less than twice that of Earth's Moon) and hence low surface gravity (one-seventh of Earth's), the atmospheric pressure at ground level is 60 percent greater than that on Earth. Titan's atmosphere contains about 10 times more gas than Earth's atmosphere and, because of Titan's weaker gravitational pull, extends 10 times farther into space than does our own. The top of the main haze layer lies about 200 km above the surface (Figure 8.8 inset), although there are additional layers, seen primarily through their absorption of ultraviolet radiation, at 300 km and 400 km.

Below the haze the atmosphere is clear, although rather gloomy because so little sunlight gets through. *Cassini* detected low-lying methane clouds at roughly the altitude (20 km) predicted by the theoretical models, but the clouds were less common than scientists had expected, with the result that the rain supposedly responsible for the channels in Figure 8.10(a) is not as widespread as was previously thought. Current models suggest that the methane rain may be a seasonal phenomenon, falling onto the poles in winter and evaporating again in summer. *Huygens* reached Titan almost at the end of winter at its landing site.

In 2003, radio astronomers using the Arecibo telescope reported the detection of liquid hydrocarbon lakes on Titan's surface. ∞ *(Sec. 3.4)* In 2007, *Cassini* mission specialists confirmed this finding, presenting radar images showing numerous lakes, some many tens of kilometers in length, near Titan's north polar regions (Figure 8.12). As with the *Magellan* radar images of Venus, the darkest regions are extremely smooth, in Titan's case implying that they are composed of liquid; the shapes of the regions also strongly suggest liquid bodies. ∞ *(Sec. 6.5)*

Subsequently, *Cassini* confirmed the presence of liquid ethane in the lakes, although their exact composition remains uncertain. Methane is surely present; the methane rain feeds the lakes, but it is much more volatile than ethane under the conditions found on Titan, so it probably evaporates rapidly, leaving heavier hydrocarbons behind. Computer models indicate that the lakes are mostly ethane (75 percent), with methane (10 percent) and propane (7 percent) making up much of the remainder. No waves have been detected on the lakes, hinting that they may also contain heavier tar-like hydrocarbons, increasing their viscosity. The number of lakes observed near Titan's south pole is significantly smaller than the number seen in the north, suggesting that the lakes grow during the rainy winter season. Scientists hope that repeated observations on future *Cassini* passes will reveal changes in the size and structure of the lakes as spring on Titan gives way to summer.

Titan's atmosphere seems to act like a gigantic chemical factory. Powered by the energy of sunlight, it is undergoing a complex series of chemical reactions that ultimately result in the observed composition. The upper atmosphere is thick with smoggy haze, and the surface may be covered with hydrocarbon sediment that has settled from the clouds. Spectrometers aboard *Cassini* have detected organic molecules below the haze layers. Exploration of the lakes and other surface features by *Cassini* and future missions may afford scientists the opportunity to study the kind of chemistry thought to have occurred billions of years ago on Earth—the prebiotic chemical reactions that eventually led to life on our own planet (see Chapter 18).

Why does Titan have a thick atmosphere when Ganymede and Callisto do not? The answer lies largely in Titan's 94 K surface temperature, which is considerably lower than that of Jupiter's moons, making it easier for the moon to retain an atmosphere. *(More Precisely 5-1)* Also, gases may have become trapped in the surface ice when the moon formed. As Titan's internal radioactivity warmed it, the ice released the trapped gases, forming a thick methane–ammonia atmosphere.

FIGURE 8.11 Titan's Atmosphere The structure of Titan's atmosphere, as deduced from *Voyager 1* observations. The blue curve represents temperature at different altitudes.

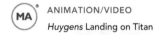 ANIMATION/VIDEO
Huygens Landing on Titan

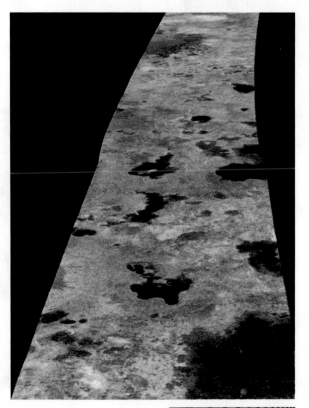

▶ **FIGURE 8.12 Titan's Lakes** Radar aboard *Cassini* detected many smooth regions (colored dark blue in this false-colored radio image) thought to be lakes of liquid methane, near Titan's north pole. This scene spans about 200 km vertically, with the largest features much smaller than any of the Great Lakes on Earth. *(NASA/ESA)*

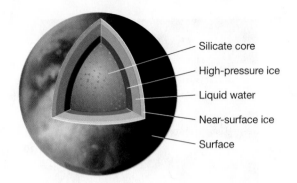

▲ **FIGURE 8.13 Titan's Interior** Based on measurements of Titan's gravitational field during numerous flybys, Titan's interior seems to be largely a rocky silicate mix. Most intriguing is the subsurface layer of liquid water, similar to that hypothesized on Jupiter's Europa and Ganymede.

Sunlight split the ammonia into hydrogen, which escaped into space, and nitrogen, which remained in the atmosphere. The methane, which was less easily broken apart, survived intact. Detailed studies by *Cassini* scientists of the composition of Titan's atmosphere suggest that the atmosphere is steadily escaping, in large part due to the constant buffeting by Saturn's harsh magnetosphere, and may have been much thicker—perhaps 5–10 times denser—in the distant past.

Each pass of *Cassini* allows scientists to probe the gravity of Titan, and repeated passes, coupled with knowledge of the properties of Titan's likely constituents, have allowed construction of a remarkably detailed model of the moon's interior (Figure 8.13). The model indicates that Titan has a rocky core surrounded by a thick mantle of water ice, but intriguingly, also predicts the presence of a thick layer of liquid water a few tens of kilometers below the surface. Thus, Titan joins Europa, Ganymede, Earth, and (as we will see) Enceladus on the list of solar system objects containing large bodies of liquid water, with all that implies for the prospects of life developing there.

Triton: Captured from the Kuiper Belt?

Neptune's large moon, Triton, is the smallest of the seven large moons in the solar system, with about half the mass of the next smallest, Europa. Lying 4.5 billion kilometers from the Sun and possessing an icy surface that reflects much of the solar radiation reaching it, Triton has a temperature of just 37 K. *Voyager 2* found that this moon has an extremely thin nitrogen atmosphere and a solid, frozen surface probably composed primarily of water ice.

A *Voyager 2* mosaic of Triton's south polar region is shown in Figure 8.14. The moon's low temperatures produce a layer of nitrogen frost that forms, evaporates, and reforms seasonally over the polar caps. The frost is visible as the pinkish region at the bottom right of Figure 8.14. Overall, there is a marked lack of cratering on Triton, presumably indicating that surface activity has obliterated the evidence of most impacts. There are many other signs of an active past. Triton's face is scarred by large fissures similar to those seen on Ganymede, and the moon's odd cantaloupe-like terrain may indicate repeated faulting and deformation over the moon's lifetime. In addition, Triton has numerous frozen lakes of water ice, thought to be volcanic in origin.

Triton's surface activity is not just a thing of the past. As *Voyager 2* passed the moon, its cameras detected great jets of nitrogen gas erupting several kilometers into the sky. These geysers may result when liquid nitrogen below Triton's surface is heated and vaporized by some internal energy source, or perhaps even by the Sun's feeble light. Scientists conjecture that nitrogen geysers may be very common on Triton and are perhaps responsible for much of the moon's thin atmosphere.

Triton is unique among the large moons of the solar system in that its orbit around Neptune is *retrograde*. Also, with an orbital inclination of about 20°, Triton is the only large jovian moon not to orbit in its parent planet's equatorial plane. The event(s) that placed Triton on a tilted, retrograde orbit are the subject of considerable speculation. Triton's peculiar orbit and surface features suggest to many astronomers that the moon did not form as part of the Neptune system but instead

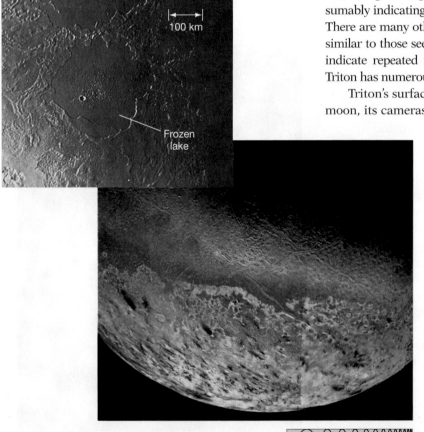

◄ **FIGURE 8.14 Triton** The south polar region of Triton, showing a variety of terrains ranging from deep ridges and gashes to what appear to be frozen water lakes, all indicative of past surface activity. The pinkish region at the lower right is nitrogen frost, forming the moon's polar cap. The roughly circular lakelike feature shown in the inset may have been caused by the eruption of an ice volcano. *(NASA)*

was captured from the Kuiper belt, probably not too long ago. The surface deformations on Triton certainly suggest violent and relatively recent events in the moon's past, but the details remain uncertain. The absence of a "normal" midsize moon system around Neptune (see Section 8.3) is also often cited as evidence for some catastrophic event in the planet's history.

Whatever Triton's past, its future seems clear. Because of its retrograde orbit, the tidal bulge that Triton raises on Neptune tends to make the moon spiral *toward* the planet rather than away from it (as our Moon moves away from Earth). ∞ *(Sec. 5.2)* Calculations indicate that Triton is doomed to be torn apart by Neptune's tidal gravitational field, probably in no more than about 100 million years.

8.3 The Medium-Sized Jovian Moons

Table 8.2 lists the 12 medium-sized satellites of the solar system, conventionally defined as bodies having radii between about 200 and 800 km. The 6 that orbit Saturn are shown in Figure 8.15. The remainder (5 orbiting Uranus, 1 orbiting Neptune) are shown in Figure 8.16 at the same scale as Figure 8.15. Their densities suggest that all 12 are composed largely of rock and water ice. All move on nearly circular paths and are tidally locked by their parent planet's gravity, and so have permanently "leading" and "trailing" faces. ∞ *(Sec. 5.2)*

Most of these moons have old, heavily cratered surfaces, with no evidence for extensive geological activity. Prior to *Cassini*'s arrival, scientists thought that the light-colored *wispy terrain* on Saturn's Rhea and Dione (see Figure 8.15) was the result of events in the distant past, during which water was released from the moons' interiors and condensed on their surfaces. However, *Cassini* images reveal that the markings are

CONCEPT CHECK

Why was *Voyager 1* unable to photograph the surface of Titan? Why do we now have a much better understanding of the moon's surface?

ANIMATION/VIDEO
Saturn Satellite Transit

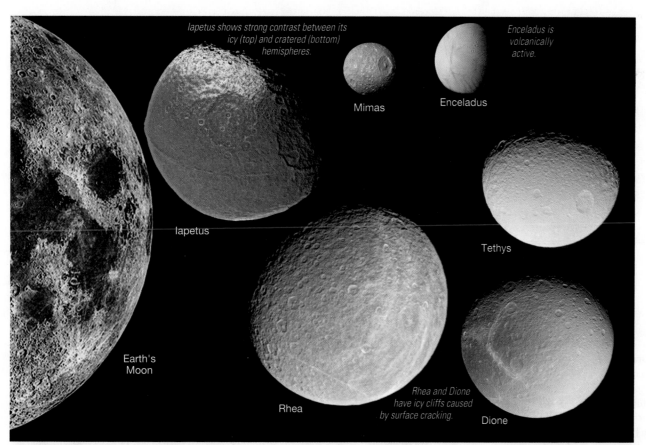

◀ **FIGURE 8.15 Saturnian Moons** Saturn's six medium-sized satellites, to scale, as imaged by the *Cassini* spacecraft. All are heavily cratered and all are shown here in natural color. For scale, part of Earth's Moon is shown at left. *(UC/Lick Observatory; NASA)*

TABLE 8.2   The Medium-Sized Moons of the Solar System

Name	Parent Planet	Distance from (km)	Parent Planet (planet radii)	Orbit Period (Earth days)	Radius (km)	Mass (Moon = 1)	Density (kg/m3)
Mimas	Saturn	186,000	3.10	0.94	200	0.00051	1100
Enceladus	Saturn	238,000	3.97	1.37	250	0.00099	1000
Tethys	Saturn	295,000	4.92	1.89	530	0.0085	1000
Dione	Saturn	377,000	6.28	2.74	560	0.014	1400
Rhea	Saturn	527,000	8.78	4.52	760	0.032	1200
Iapetus	Saturn	3,560,000	59.3	79.3	720	0.022	1000
Miranda	Uranus	130,000	5.09	1.41	240	0.00090	1200
Ariel	Uranus	191,000	7.48	2.52	580	0.018	1700
Umbriel	Uranus	266,000	10.4	4.14	590	0.016	1400
Titania	Uranus	436,000	17.1	8.71	790	0.048	1700
Oberon	Uranus	583,000	22.8	13.5	760	0.041	1600
Proteus	Neptune	118,000	4.76	1.12	210	—	—

in fact bright ice cliffs created by *tectonic fractures*, where stresses in the moons' icy interiors as they cooled and contracted caused the surface layers to crack and buckle. Dione also has "maria" of sorts, where flooding from ice volcanism appears to have obliterated the older craters. Both Tethys in the Saturn system and Ariel orbiting Uranus have extensive cracks on their surfaces, which may also be tectonic fractures, or possibly the results of violent ancient impacts.

The Uranian moons Titania, Oberon, and Umbriel are heavily cratered and show few signs of geological activity—they are comparable to Rhea in appearance (and probable history), except they lack Rhea's wispy streaks. All the moons of Uranus are darker than those of Saturn, indicating much less reflective surfaces. The most likely explanation is **radiation darkening**, where solar radiation and high-energy particles break up surface molecules, leading to chemical reactions that slowly form a layer of dark, organic material. However, our knowledge of these distant moons is quite limited—our only source of detailed information is the 1988 *Voyager 2* flyby. ∞ *(Discovery 7-1)*

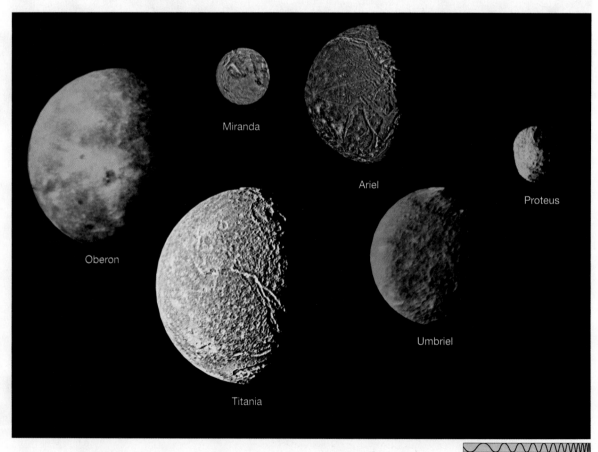

▶ **FIGURE 8.16 Moons of Uranus and Neptune** The five medium-sized moons of Uranus and the single medium-sized moon of Neptune, Proteus, shown at the same scale as Figure 8.15. All six moons are heavily cratered, and most show no clear evidence of any present or past geological activity. *(NASA)*

R I V U X G

▶ **FIGURE 8.17 Enceladus** (a) Saturn's tiny ice-covered moon Enceladus shows evidence of youthful terrain in the south where craters are mostly absent. The long blue "tiger stripe" streaks (about 1 km wide) are fractures in the ice through which gas escapes to form a thin but real atmosphere. (b) Some of the jets emerging from cracks near the south pole. The inset to part (a) shows a model of the moon's interior, featuring an extensive salt-water ocean below the south pole. *(JPL)*

Saturn's moon Iapetus has a distinctly "two-faced" appearance. Its leading hemisphere is very dark, while the other is very bright. For years astronomers puzzled over how the dark markings could adorn only the leading face, as there seemed to be no nearby material for the moon to sweep up as it orbited the planet. However, in 2009 the *Spitzer Space Telescope* discovered a new, huge, diffuse ring more than 6 million kilometers from Saturn. ∞ *(Sec. 3.5)* Iapetus lies at the inner edge of this ring, and the steady accumulation of ring particles over billions of years, coupled with the effects of solar heating and radiation darkening, could naturally account for the moon's asymmetric appearance.

Iapetus's other prominent surface feature is a giant 20-km-high, 1400-km-long ridge spanning half of the moon's circumference. Discovered by *Cassini* in 2005, it is clearly visible cutting across the bottom third of the moon in Figure 8.15. It is unique in the solar system, and so far defies explanation.

One surprising *Cassini* finding has been that Saturn's moon Enceladus (Figure 8.17a) has ongoing geological activity. Its surface is bright and shiny—reflecting virtually 100 percent of the sunlight falling on it-because it is completely coated with fine crystals of pure ice, the icy "ash" of water volcanoes. Impact craters on the surface have been erased by "lava flows" of liquid water emerging from the moon's interior and freezing on the surface.

Cassini detected icy jets emerging from cracks near Enceladus's south pole (Figure 8.17b) and a transient water vapor atmosphere surrounding the moon. The jets steadily replenish the atmosphere as it escapes from the moon's weak gravity to form Saturn's E ring (see Section 8.4). Models of the jets suggest that an extensive liquid ocean lies just below the south polar crust, as indicated in the inset to Figure 8.17(a). Interestingly, some of the jets contain organic material, fueling speculation about the possibility of life in the moon's interior.

Why is there so much activity on such a small moon? The best explanation seems to be that the internal heating is the result of tidal stresses, much like those operating in Io and Europa (Section 8.1). The orbit of Enceladus is slightly noncircular due to the gravity of the other moons, especially Dione. Saturn's tidal force on Enceladus is only one-quarter of the force exerted by Jupiter on Io, but this may still be enough to partially liquefy the interior and cause the activity observed.

Why is the jet activity concentrated at the south pole? Scientists speculate that upwelling material at an interior warm spot may have caused the entire moon to "roll over" to place the low-density warm region on the rotation axis—at the pole—in much the same way as a spinning bowling ball will tend to rotate so as to place the (low-density) holes on its spin axis.

Most of the midsized moons show heavy cratering, indicating the cluttered and violent planetary environment that existed just after the outer planets formed, as the remaining planetesimals were accreted by larger bodies or ejected from the vicinity of the jovian planets to form the Oort cloud. ∞ *(Sec. 4.3)* The innermost midsize moons, Mimas (Saturn) and Miranda (Uranus), show perhaps the clearest evidence of violent meteoritic impacts. The impact that caused the large crater visible on the image of Mimas in Figure 8.15 must have come very close to shattering the moon. Miranda (Figure 8.18) displays a wide range of surface terrains, including ridges, valleys, faults, and many other geological features. To explain why Miranda seems to combine so many types of terrain, some researchers have hypothesized that it has been catastrophically disrupted several times, with the pieces falling back together in a chaotic, jumbled way. However, it will be a long time before we can obtain more detailed information to test this theory.

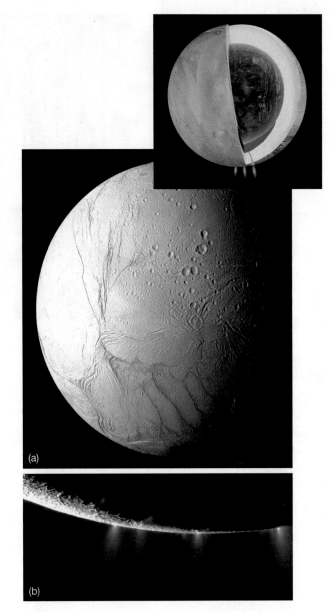

(a)

(b)

CONCEPT CHECK

What surface property of the medium-sized moons tells us that the outer solar system was once a very violent place?

▶ FIGURE 8.18 Miranda The asteroid-sized innermost moon of Uranus, photographed by *Voyager 2*. Miranda has a fractured surface suggestive of a violent past, but the cause of the grooves and cracks is unknown. *(NASA)*

DATA POINTS

Planetary Rings

More than half of all students had difficulty describing the properties of Saturn's rings. Some points to remember:

- The disruptive tidal force of a planet on a moon increases rapidly as the moon approaches the planet.

- The Roche limit is the distance at which the planet's tidal force overcomes the moon's gravity, tearing the moon apart.

- Generally speaking, we expect to find rings and small moons inside the Roche limit and larger moons outside.

ANIMATION/VIDEO

Saturn Ring Plane Crossing

TABLE 8.3   The Rings of Saturn

Ring	Inner Radius (km)	(planet radii)	Outer Radius (km)	(planet radii)	Width (km)
D	60,000	1.00	74,000	1.23	14,000
C	74,000	1.23	92,000	1.53	18,000
B	92,000	1.53	117,600	1.96	25,600
A	122,200	2.04	136,800	2.28	14,600
F	140,500	2.34	140,600	2.34	100
E	210,000	3.50	300,000	5.00	90,000

8.4 Planetary Rings

All four jovian worlds have planetary ring systems girdling their equators. The properties and appearance of each planet's rings are intimately connected with the planet's small and medium-sized moons, with many of the inner jovian moons orbiting close to (or even within) the parent planet's rings. Saturn's rings are by far the best known and best observed, and we study them here in greatest detail. However, the rings of all the jovian planets have much to tell us about the environment in the outer solar system.

Saturn's Spectacular Ring System

Figure 8.19 shows the main features of Saturn's rings. From Earth, astronomers identified three rings, which they simply labeled A, B, and C. The **A ring** lies farthest from the planet and is separated from the inner **B** and **C rings** by a dark gap called the **Cassini Division**, named for the 17th century French–Italian astronomer Giovanni Domenico Cassini. A smaller gap, known as the **Encke Gap**, is found in the outer part of the A ring. Its width is 270 km. No finer ring details are visible from Earth. Of the three main rings, the B ring is brightest, followed by the somewhat fainter A ring, and then by the almost translucent C ring. The additional rings listed in Table 8.3 cannot be seen in Figure 8.19.

Because Saturn's rings lie in the planet's equatorial plane and the planet's rotation axis is tilted with respect to the ecliptic, the appearance of the rings (as seen from Earth) changes seasonally, as shown in Figure 8.20. As Saturn orbits the Sun, the angle at which the rings are illuminated varies. When the planet's north or south pole is tipped toward the Sun (which means it is either summer or winter on Saturn), the highly reflective rings are at their brightest. In Saturn's spring and fall, the rings are close to being edge-on both to the Sun and to observers on Earth, so they seem to disappear. One important deduction we can make from this simple observation is that the rings are very *thin*. In fact, we now know that their thickness is less than a few hundred meters, even though they are more than 200,000 km in diameter.

Saturn's rings are not solid objects. In 1857 Scottish physicist James Clerk Maxwell realized that they must be composed of a great number of small solid *particles,* all independently orbiting Saturn like so many tiny moons. Measurements of the Doppler shift of sunlight reflected from the rings reveal that the orbital speeds decrease with increasing distance from the planet in precise accordance with Kepler's laws and Newton's law of gravity. ∞ *(Sec. 1.4)* The highly reflective nature of the rings suggests that the particles are made of ice, and infrared observations in the 1970s confirmed that water ice is in fact a prime ring constituent. *Cassini* also detected substantial amounts of "mud"—small rocky particles and dust—mixed in. Radar observations and later *Voyager* and *Cassini* studies of scattered sunlight showed that the sizes of the icy ring particles range from fractions of a millimeter to tens of meters in diameter, with most particles being about the size (and composition) of large, dirty snowballs. The rings are thin because collisions between ring particles tend to keep them moving in circular orbits in a single plane. Any particle straying from this orderly configuration soon runs into other ring particles and is eventually forced back into the fold.

The Roche Limit

To understand why rings form, consider the fate of a small moon orbiting close to a massive planet such as Saturn (Figure 8.21). As we bring our hypothetical moon closer to the planet, the tidal force exerted on it by the planet increases, stretching it along the direction to the planet. ∞ *(Sec. 5.2)* The tidal force increases rapidly as the distance to the planet decreases. Eventually, a point is reached where the planet's tidal force becomes greater than the internal forces holding the moon together, and the moon is torn apart by the planet's gravity. The pieces of the satellite move on their own individual orbits around that planet, eventually spreading all the way around it in the form of a ring.

For any given planet and any given moon, this critical distance inside which the moon is destroyed is known as the **Roche limit,** after the 19th-century French mathematician Edouard Roche, who first calculated it. If our hypothetical moon is held together by its own gravity, and if its average density is similar to that of the parent planet (both reasonably good assumptions for Saturn's larger moons), then the Roche limit is approximately 2.4 times the radius of the planet. For example, in the case of Saturn (of radius 60,000 km), the Roche limit lies roughly 144,000 km from the planet's center—just outside the outer edge of the A ring. The rings of Saturn occupy the region inside Saturn's Roche limit. These considerations apply equally well to the other jovian worlds. As indicated in Figure 8.22, all four ring systems in our solar system are found within (or close to) the Roche limit of their parent planet.

Note that the calculation of the Roche limit applies only to moons massive enough for their own gravity to be the dominant binding force. Space probes and sufficiently small moons can survive within the Roche limit because they are held together not by gravity but rather by interatomic (electromagnetic) forces, so the above arguments do not apply.

▲ **FIGURE 8.19 Saturn's Rings** Saturn as observed by the *Cassini* spacecraft in 2005. The main ring features are marked and are shown here in false color in order to enhance contrast. *(NASA/ESA)*

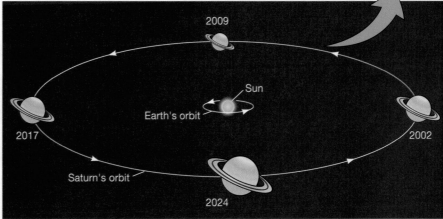

▶ **FIGURE 8.20 Saturn's Tilt** (a) Over time Saturn's rings change their appearance to terrestrial observers as the tilted ring plane orbits the Sun. The roughly true-color images (inset) span a period of several years from 2003 (bottom) to nearly the present (top) and show how the rings change from our perspective on Earth, from almost edge-on to more nearly face-on. *(NASA)*

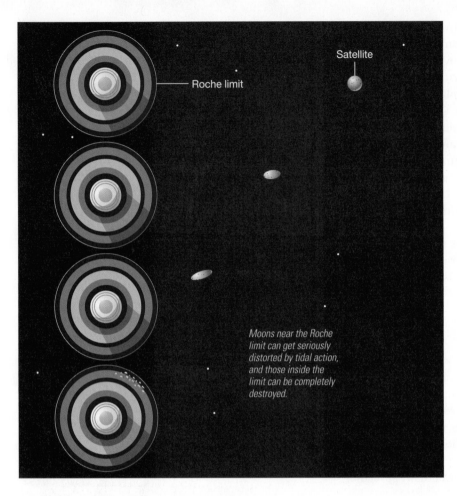

ANIMATION/VIDEO (MA)

Voyager Ring Spokes Animation

◀ **FIGURE 8.21 Roche Limit** The tidal field of a planet first distorts (top) and then destroys (bottom) a moon that strays too close.

Fine Structure in Saturn's Rings

In 1979 the *Voyager* probes found that Saturn's main rings are actually composed of tens of thousands of narrow **ringlets.** A *Cassini* view of this fine ring structure is shown in Figure 8.23. This intricate structure is not fixed, but instead varies both with time and with location in the rings. The ringlets form when the self-gravity of the ring particles, possibly in combination with the effects of Saturn's inner moons, creates waves of matter—regions of high and low density—that move in the plane of the rings like ripples on the surface of a pond. Ring particles bunch up in some places and move apart in others, creating the ringlets we see.

The narrower gaps in the rings—about 20 in all—are the result of small *moonlets* embedded in them. Because these moonlets are much larger (perhaps 10 or 20 km across) than the largest ring particles, they sweep up ring material as they go. Prior to *Cassini*'s arrival, only one of these moonlets had been found, despite many careful searches of the *Voyager* images. *Cassini* has now identified numerous small moonlets (a few kilometers across) in the gaps between the main rings.

The largest gap—the Cassini Division—has a different origin. It owes its existence to the gravitational influence of Saturn's innermost medium-sized moon, Mimas. Over time, particles orbiting within the Division have been deflected by Mimas's gravity into eccentric orbits that cause them to collide with other ring particles, effectively moving them into new orbits at different radii. The net result is that the number of ring particles in the Cassini Division is greatly reduced.

Outside the A ring lies perhaps the strangest ring of all: the faint, narrow **F ring** (Figure 8.24). It was discovered by *Pioneer 11* in 1979, but its full complexity

▶ **FIGURE 8.22 Jovian Ring Systems** All of the distances to the rings of Jupiter, Saturn, Uranus, and Neptune are expressed in planetary radii. The red line represents the Roche limit, and all the rings lie within (or very close to) this limit of the parent planet.

▲ **FIGURE 8.23 Saturn's Rings, Up Close INTERACTIVE** *Cassini* took this true-color mosaic of Saturn's dazzling ring structure just before flying through the tenuous outer rings. Earth is superposed, to scale, for a size comparison. *(NASA/ESA)*

became evident only when the *Voyager* spacecraft took a closer look. Its oddest feature is that it looks as though it is made up of several strands braided together. The ring's complex structure, as well as its thinness, arises from the influence of two small moons, known as **shepherd satellites**, that orbit about 1000 km on either side of it. Their gravitational effect gently guides any particle straying too far out of the F ring back into it. *Cassini* has discovered several more braided rings within the main ring system. The braids are the result of waves created in the rings by the moons. However, the details of how the braids are produced, and why the shepherd moons are there at all, remain subjects of active research.

    Voyager 2 also found a faint series of rings, now known collectively as the **D ring**, inside the inner edge of the C ring, stretching down almost to Saturn's cloud tops. The D ring contains relatively few particles and is so dark that it is completely invisible from Earth. The faint **E ring,** also a *Voyager 2* discovery, lies well outside the main ring structure. It is thought to be the result of volcanism on the moon Enceladus, discussed earlier. Figure 8.25 is a *Cassini* image showing the rings from a perspective never before seen—behind the planet looking back toward the eclipsed Sun. Just as diffuse airborne dust is most easily seen against the light streaming in through a sunlit window, the planet's faint rings show up clearly in this remarkable back-lit view. In addition, this image reveals several more faint rings, some of them also associated with the orbits of various small moons.

The Rings of Jupiter, Uranus, and Neptune

One of the many remarkable findings of the 1979 *Voyager* missions was the discovery of a faint ring of matter encircling Jupiter in the plane of the planet's equator (Figure 8.26). This ring lies roughly 50,000 km above the top cloud layer, inside the orbit of the innermost moon and is probably composed of dark fragments of rock and dust chipped off the inner moons by meteorites. The outer edge of the ring is quite sharply defined, although a thin sheet of material may extend all the way down to Jupiter's cloud tops. In the direction perpendicular to the equatorial plane, the ring is only a few tens of kilometers thick.

    The ring system surrounding Uranus was discovered in 1977, when astronomers observed the planet passing in front of a bright star, momentarily dimming the star's light. Such a **stellar occultation** happens a few times per decade and allows astronomers to measure planetary structures that are too small and faint to be detected directly. The ground-based observations revealed 9 thin rings encircling Uranus. Two more were discovered by *Voyager 2*. The locations of all 11 rings are indicated in Figure 8.22.

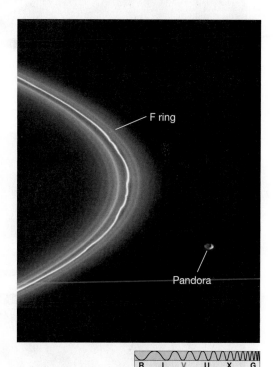

▶ **FIGURE 8.24 Shepherd Moon** Saturn's narrow F ring appears to contain kinks and braids, making it unlike any of Saturn's other rings. Its thinness is caused by two shepherd satellites that orbit the ring, one a few hundred kilometers inside the ring and the other a similar distance outside it. One of the potato-shaped shepherding satellites, called Pandora and roughly 100 km across, can be seen at right. *(NASA)*

Similar to the way the combined efforts of shepherd and sheepdogs contain sheep within a somewhat wavy line, shepherd moons confine loose material within the braided strands of Saturn's rings.

▶ **FIGURE 8.25 Back-Lit Rings** *Cassini* took this spectacular image of Saturn's rings as it passed through Saturn's shadow. The normally hard to see F, G, and E outer rings are clearly visible in this contrast-enhanced image. The inset shows the moon Enceladus orbiting within the E ring; its eruptions give rise to the ring's icy particles. *(NASA)*

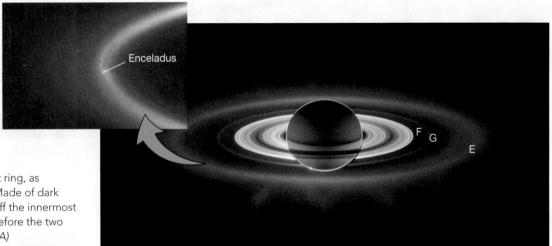

▼ **FIGURE 8.26 Jupiter's Ring**   Jupiter's faint ring, as photographed nearly edge-on by *Voyager 2*. Made of dark fragments of rock and dust possibly chipped off the innermost moons by meteorites, the ring was unknown before the two *Voyager* spacecraft arrived at the planet. *(NASA)*

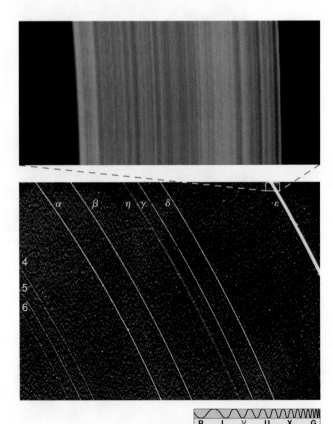

The nine rings known from Earth-based observation are shown in Figure 8.27, a *Voyager 2* image that makes it readily apparent that Uranus's rings are quite different from those of Saturn. While Saturn's rings are bright and wide, with relatively narrow, dark gaps between them, those of Uranus are dark, narrow, and widely spaced. However, like Saturn's rings, the rings of Uranus are all less than a few tens of meters thick (measured in the direction perpendicular to the ring plane). Like the F ring of Saturn, Uranus's narrow rings require shepherd satellites to keep them in place. During its 1986 flyby, *Voyager 2* detected the shepherds of the Epsilon ring (Figure 8.28). Many other, undetected shepherd satellites must also exist.

As shown in Figure 8.29, Neptune is surrounded by four dark rings. Three are quite narrow, like the rings of Uranus, and one is quite broad and diffuse, more like Jupiter's ring. The outermost ring is noticeably clumped in places. From Earth we see not a complete ring, but only partial arcs; the unseen parts of the ring are simply too thin (unclumped) to be detected. The connection between the rings and the planet's small inner satellites has not yet been firmly established, although many astronomers now think the clumping is caused by shepherd satellites.

The Formation of Planetary Rings

The dynamic behavior—waves, collisions, and interactions with moons—observed in the rings of Saturn and the other jovian planets suggests to many researchers that the rings are quite young—perhaps no more than 50 million years old, just 1 percent of the age of the solar system. If the rings are indeed so young, then either they are replenished from time to time, perhaps by fragments of moons chipped off by meteoritic impact, or they are the result of a relatively recent possibly catastrophic event in the planet's system—a moon torn apart by tidal forces, or destroyed in an impact with a large comet or even with another moon.

Astronomers estimate that the total mass of Saturn's rings is enough to make a satellite about 250 km in diameter. If such a satellite strayed inside Saturn's Roche limit or was destroyed (perhaps by a collision) near that radius, a ring system could have resulted. While we do not know if Saturn's rings originated in this way, this is the likely fate of Neptune's large moon Triton. As mentioned earlier (in Section 8.2), Triton will be torn apart by Neptune's tidal gravitational field within 100 million years or so. By that time it is quite possible that Saturn's ring system will have disappeared, and Neptune will then be the planet in the solar system with spectacular rings!

◀ **FIGURE 8.27 Uranus's Rings**   The main rings of Uranus, as imaged by *Voyager 2*. All nine of the rings known before the spacecraft's arrival can be seen in this photo. The inset at top shows a closeup of the Epsilon ring, revealing some internal structure. *(NASA)*

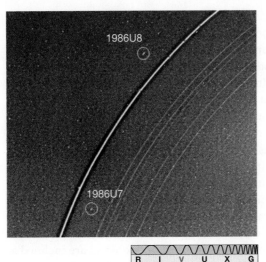

◀ **FIGURE 8.28 More Shepherd Moons** These two small moons, now named Cordelia and Ophelia, were discovered by *Voyager 2* in 1986. They shepherd Uranus's Epsilon ring, keeping it from diffusing away. *(NASA)*

▲ **FIGURE 8.29 Neptune's Faint Rings** In this long-exposure image, Neptune (center) is heavily overexposed and has been artificially obscured (by an instrument) to make the rings easier to see. One of the two fainter rings lies between the inner bright ring and the planet. The others lie between the two bright rings. *(NASA)*

8.5 Beyond Neptune

The inventory of known Kuiper belt objects has expanded greatly since the early 1990s. Ground-based telescopes have led the way in the painstaking effort to capture the meager amounts of sunlight reflected from these dark inhabitants of the outer solar system. However, one Kuiper belt resident has been known for decades—Pluto.

The Discovery of Pluto

In Chapter 7, we saw how irregularities in Uranus's orbit led to the discovery of Neptune in the mid-19th century. ∞ *(Sec. 7.2)* By the end of that century, observations of the orbits of Uranus and Neptune suggested that Neptune's influence was not sufficient to account for all of the observed irregularities in Uranus's motion. Furthermore, it seemed that Neptune itself was affected by some other unknown body. Astronomers hoped to pinpoint the location of this new planet using techniques similar to those that had guided them to the discovery of Neptune.

One of the most ardent planet hunters was Percival Lowell, one of the best-known astronomers of his day. ∞ *(Discovery 6-1)* Basing his investigation primarily on the motion of Uranus (Neptune's orbit was still quite poorly known at the time), Lowell calculated where the supposed ninth planet should be. He searched for it, without success, until his death in 1916. In 1930 the American astronomer Clyde Tombaugh, working with improved equipment and photographic techniques at the Lowell Observatory, finally succeeded in finding Lowell's ninth planet just 6° away from Lowell's predicted position. The new object was named Pluto for the Roman god of the dead, who reigns over eternal darkness (and also because its first two letters and its astronomical symbol ♇ are Lowell's initials).

On the face of it, the discovery of Pluto looks like another spectacular success for Newtonian mechanics. However, the supposed irregularities in the motions of Uranus and Neptune did not in fact exist, and the mass of Pluto, not measured accurately until the 1980s, is far too small to have caused them anyway. The discovery of Pluto owed much more to simple luck than to complex mathematics!

The Pluto–Charon System

At nearly 40 AU from the Sun, Pluto is hard to distinguish from background stars. It is never visible to the naked eye. In 1978 astronomers at the U.S. Naval Observatory discovered that Pluto has a satellite. It is now named Charon, after the mythical boatman who ferried the dead across the river Styx into Hades, Pluto's domain. Figure 8.30 is a 2012 *Hubble* view that clearly resolves Pluto and Charon and shows four additional small satellites orbiting the Pluto–Charon system. The two largest

CONCEPT CHECK

What is the Roche limit, and what is its relevance to planetary rights?

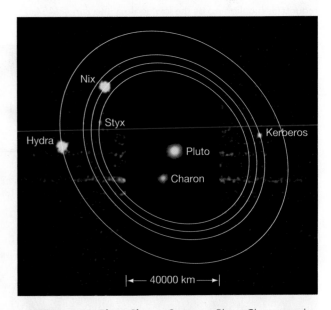

▲ **FIGURE 8.30 Pluto Charon System** Pluto, Charon, and four smaller moons can be seen in this 2012 *Hubble Space Telescope* image. The vertical black stripe in the middle of the frame is there because Pluto and Charon were imaged with a much shorter exposure than the other much fainter moons. The orbits of the outer moons are marked. *(NASA)*

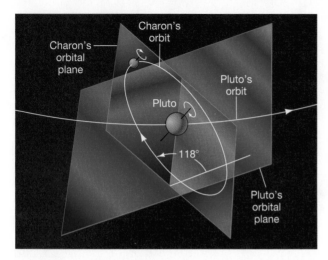

▲ **FIGURE 8.31 Pluto–Charon Orbit** Charon's path around Pluto is circular, synchronous, and inclined at 118° to the orbit plane of the Pluto–Charon system about the Sun. The Pluto–Charon orbit plane is itself inclined at 17.2° to the plane of the ecliptic.

▼ **FIGURE 8.32 Pluto, In Detail** This enhanced-color *New Horizons* image of Pluto shows a remarkable variety of terrains on this small world, ranging from smooth plains to towering water-ice mountains. The cream-colored basin at center contains ices of nitrogen and methane, while the bright white upland region at right may be coated in nitrogen snow. The inset shows "snakeskin terrain" — oddly textured mountains highlighted by the low angle of the Sun. *(NASA/New Horizons)*

are called Nix (for the goddess of darkness and mother of Charon) and Hydra (a mythical nine headed monster). They orbit Pluto at roughly twice the distance of Charon. Despite their overexposed appearance in Figure 8.30, both are very small—roughly 50 km in diameter. The other two moons, Kerberos and Styx, are even smaller (10 km); they were spotted in 2011 and 2012, respectively.

By pure chance, Charon's orbit between 1985 and 1991 produced a series of eclipses as Pluto and Charon repeatedly passed in front of one another, as seen from Earth, allowing astronomers to calculate the masses and radii of both bodies to great accuracy. Pluto weighs in at just 0.0022 Earth masses—about 0.17 times the mass of Earth's Moon. Its radius is 1190 km, about one-fifth that of Earth. The mass of Charon is about 0.12 that of Pluto; Charon is about 600 km in radius and orbits 19,600 km from Pluto. By studying sunlight reflected from their surfaces, astronomers also found that Pluto and Charon are tidally locked as they orbit each other. Their orbital period, and the rotation period of each, is 6.4 days. The Pluto-Charon orbit and the spin axes of both objects are inclined at an angle of 118° to the plane of the ecliptic (Figure 8.31).

The masses and radii of Pluto and Charon imply average densities of 1900 kg/m^3—just what we would expect for bodies of that size made up mostly of water ice, like several of the large moons of the outer planets. In fact, Pluto is very similar in both mass and radius to Neptune's large moon, Triton—which, as we have just seen, is thought to be a captured Kuiper belt object.

Pluto and Charon, Up Close

For decades following Pluto's discovery, every new fact about this remote world was gleaned from painstaking observations from Earth or Earth-orbiting instruments. This situation changed radically in July 2015 when NASA's *New Horizons* mission flew by Pluto and Charon after a 7-year journey from Earth. *New Horizons* spent just a few hours in the Pluto-Charon system, and is now speeding toward a 2019 rendezvous with its next target (a Kuiper belt object called 2014 MU69, a billion kilometers farther away), but those few hours provided a wealth of information.

Figure 8.32 is a color-enhanced view of Pluto from New Horizons, taken from a distance of about 14,000 km. It reveals unexpectedly complex surface features, including icy mountains and frozen plains. Figure 8.33 shows the spacecraft's view of Charon. Both Pluto and Charon are remarkably free of impact craters, suggesting relatively young surfaces—as little as 100 million years old—that have been shaped by recent geologic activity. Some of the canyons on Charon's surface are up to 8 km deep. Scientists don't yet understand what processes can power enough activity on these small, icy worlds to obliterate impact craters and produce the observed mountain ranges and other surface features.

As on Triton, water ice on Pluto plays the structural role of rock on Earth while nitrogen behaves somewhat like water on our planet. The central plain in Figure 8.32 (unofficially named Sputnik Planum) is covered with ices of nitrogen, methane, and carbon monoxide, which apparently collect on Pluto's higher elevations (white areas at right) and then

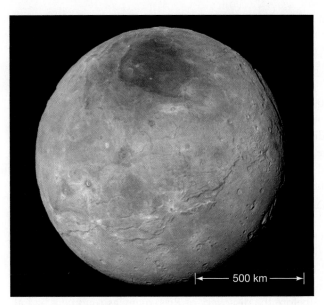

◄ FIGURE 8.33 **Charon, In Detail** Pluto's large moon Charon also displays a surprisingly complex geological history, including smooth plains, surface fracturing, tall mountains, and heavy cratering. (*NASA/New Horizons*).

▲ FIGURE 8.34 **Pluto's Atmosphere** As *New Horizons* sped away from Pluto, it captured this backlit image of the dwarf planet, showing a 40-km deep nitrogen atmosphere containing more than a dozen layers of haze. (*NASA/New Horizons*)

flow as glaciers into lower-lying regions. *New Horizons* also detected nitrogen fog in Pluto's valleys and, as shown in the backlit view shown in Figure 8.34, multiple layers of nitrogen haze in the dwarf planet's thin but surprisingly extended atmosphere.

The New Horizons data are being slowly downloaded to Earth as the spacecraft continues on its mission. The last Pluto data won't arrive until late 2016, and we can be sure that they will keep scientists on their toes long after the last bit is received.

Plutoids and the Kuiper Belt

Most Kuiper belt objects are not nearly as well observed as Pluto, which happens to be the largest known member of the class and orbits near the inner edge of the belt, making it appear relatively bright as seen from Earth. Objects beyond Neptune— even the large ones—are very faint. Nevertheless, as observational techniques have improved, the number of objects detected beyond Neptune has risen rapidly, and astronomers have pieced together a better understanding of their sizes and orbits. For example, we now know that about 25 percent of all Kuiper belt objects have exactly the same orbital period as Pluto. This is not a coincidence, of course—it is caused by gravitational interactions between these bodies and the planet Neptune, which maintains all their orbital periods at precisely 1.5 times that of Neptune.

Almost 2000 "trans-Neptunian" objects are currently known. Most orbit in the Kuiper belt, conventionally defined to lie between 40 and 50 AU from the Sun. Because they are so small and distant, researchers reason that only a tiny fraction of the total has so far been observed. The total number of Kuiper belt objects larger than 100 km across is estimated at more than 100,000. If so, the combined mass of all the debris beyond Neptune could be hundreds of times larger than the mass of the inner asteroid belt (although still less than the mass of Earth).

As the details were filled in and the number of known objects increased, it became increasingly clear to astronomers that Pluto was not distinctly different from the other small bodies in the outer solar system, as had once been supposed. The Kuiper belt object Quaoar (pronounced "Kwah-o-ar," and named for a native American creation god), discovered in 2002, is roughly 1200 km across—larger than the largest asteroid, Ceres, and more than half the size of Pluto. The Kuiper belt objects

CONCEPT CHECK

What do Pluto and Triton have in common?

---

[1]Technically, because its eccentric orbit extends well beyond the Kuiper belt, Eris is not classified as a Kuiper belt object. However, like the members of the Kuiper belt, it most likely formed as an icy planetesimal in the outer solar system and was subsequently kicked out by Neptune, so in terms of composition and history, it probably differs little from "true" Kuiper belt objects. ∞ (*Sec. 4.3*)

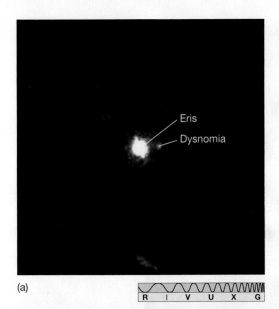

(a)

◀ **FIGURE 8.35 Kuiper Belt Object** (a) The trans-Neptunian object Eris and its small moon Dysnomia (named after the Greek goddess of discord and her daughter of chaos and lawlessness) were imaged in the infrared at the Keck Observatory in Hawaii. (b) An artist's conception of the Eris system, with Dysnomia at the top and the distant Sun at right. *(Keck)*

(b)

Haumea, discovered in 2003, and Makemake, in 2005, are larger still—both some 1500–2000 km across. (Both names are drawn from Hawaiian mythology.) But the final blow came with the discovery in 2005 of the (non-Kuiper belt)[1] object Eris (appropriately named for the Greek goddess of discord, and shown in Figure 8.35), whose diameter, measured in 2006 by *HST,* is 2330 km—virtually identical to Pluto.

Figure 8.36 shows the sizes of some of the largest bodies orbiting beyond Neptune. It also includes yet another intriguing object called Sedna, discovered in 2003 and thought to have a diameter of about 1500 km. It is the farthest known object in the solar system. Its highly elliptical orbit takes it out to almost 1000 AU from the Sun—almost to the (theoretical) inner edge of the Oort cloud. There may well be more Pluto-sized (or larger) objects still out there waiting to be discovered—the possibility has not been conclusively ruled out.

Even before the discovery of Eris, many astronomers had already concluded that Pluto was simply a large Kuiper belt object, playing much the same role in the Kuiper belt as Ceres does among the asteroids. Once it became clear that Eris was larger than Pluto, pressure mounted to find a classification that reflected astronomers' new understanding of the outer solar system. As mentioned in Chapter 4, the result was the invention of a new category of solar system object—*dwarf planet*—containing Pluto, Eris, Ceres, and now also Haumea and Makemake. ∞ *(Sec. 4.1)*

Some astronomers are unhappy at Pluto's demotion from the solar system "A list." They argue (probably correctly) that the new dwarf planet category was concocted largely to exclude Pluto and Eris from planetary status. Others applaud the redefinition as long-overdue recognition of Pluto's true identity. The arguments over terminology are not over, but it seems unlikely that future changes will restore Pluto to its former status. Few astronomers doubt that if Pluto were discovered today, its classification as a member (the biggest yet, the headlines would say!) of the Kuiper belt would be assured. As if for consolation, in 2008 the *International Astronomical Union* decided that the icy dwarf planets beyond Neptune would collectively be known as **plutoids.**

▶ **FIGURE 8.36**
Trans-Neptunian Objects
Some large trans-Neptunian objects, including Pluto and the largest known, called Eris, with part of Earth and the Moon added for scale. Most diameters are approximate, as they are estimated from each object's observed brightness. *(NASA; Caltech)*

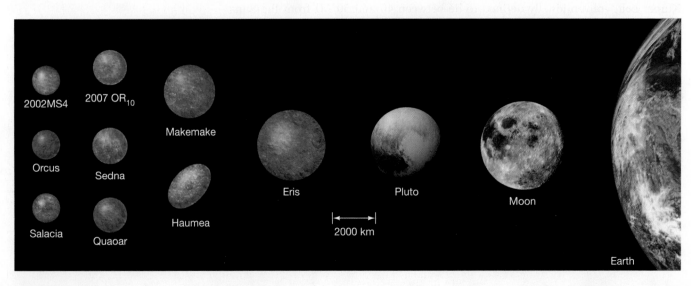

Pluto's reclassification illustrates the way in which science evolves. Our conception of the cosmos has undergone many changes—some radical, others more gradual—since the time of Copernicus. ∞ *(Sec. 1.2)* As our understanding grows, our terminology and classifications change. The situation with Pluto has a close parallel in the discovery of the first asteroids in the early 19th century. They, too, were initially classified as planets; indeed, in the 1840s, leading astronomy texts listed no fewer than 11 planets in our solar system, including as numbers 5 through 8 the asteroids Vesta, Juno, Ceres, and Pallas. Within a couple of decades, however, the discovery of several dozen more asteroids had made it clear that these small bodies represented a new class of solar system objects, separate from the planets, and the number of planets fell to eight (including the newly discovered Neptune).

Much of the observational work on the Kuiper belt began as a search for a tenth planet. It is ironic that the end result of these efforts has been a reduction in the number of "true" (that is, major) solar system planets back to eight.

> **PROCESS OF SCIENCE CHECK**
>
> Why is Pluto no larger regarded as a major planet?

THE BIG QUESTION

In studies of exoplanetary systems, attention often focuses on the planetary habitable zone surrounding the parent star, where temperatures are suitable for liquid water to exit. But in our own solar system there are many examples of jovian planetary moons, far outside the usual habitable zone, where subsurface liquid water has apparently been found. If extrasolar life exists, might these unearthly environments be the ones more likely to host it?

CHAPTER REVIEW

SUMMARY

LO1 Six large moons in the outer solar system are comparable to or larger than Earth's Moon in size and mass. The four Galilean moons of Jupiter have densities that decrease with increasing distance from the planet. Io has active volcanoes powered by the constant flexing of the moon by Jupiter's tidal forces. Europa's cracked, icy surface may well conceal an ocean of liquid water. Ganymede and Callisto have heavily cratered surfaces. Ganymede's magnetic field implies relatively recent geological activity, but Callisto shows none.

LO2 Saturn's large moon Titan has a thick atmosphere that obscures the moon's surface and may be the site of complex cloud and surface chemistry. The moon's surface is so cold that water behaves like rock and liquid methane and ethane flow like water. Sensors aboard *Cassini* have mapped the moon's surface, revealing evidence for ongoing erosion and volcanic activity. The *Huygens* probe landed on the icy surface and photographed what may be channels carved by flowing methane.

LO3 Neptune's moon Triton has a fractured surface of water ice and a thin atmosphere of nitrogen, probably produced by nitrogen "geysers" on its surface. Triton is the only large moon in the solar system to have a retrograde orbit around its parent planet. This orbit is unstable and will eventually cause Triton to be torn apart by Neptune's gravity.

LO4 The medium-sized moons of Saturn and Uranus are made up predominantly of rock and water ice. Many are heavily cratered, and some must have come close to being destroyed by the impacts whose craters we now see. Some show evidence of recent geological activity; in most cases the cause of the activity is not known.

LO5 From Earth, the main visible features of Saturn's rings are the **A, B,** and **C rings,** the **Cassini Division,** and the **Encke Gap** (p. 238). The Cassini Division is a dark region between the A and B rings. The Encke Gap lies near the outer edge of the A ring. The rings are made up of trillions of individual particles, ranging in size from dust grains to boulders. Their total mass is comparable to that of a small moon. Interactions between the ring particles and the planet's inner moons are responsible for tens of thousands of narrow **ringlets** (p. 240). The tenuous **E ring** (p. 241) is associated with volcanism on the moon Enceladus. Saturn's narrow **F ring** (p. 240) lies just outside the A ring. Its kinked, braided structure is due to two small **shepherd satellites** (p. 241) that orbit close to it and prevent it from breaking up.

LO6 The **Roche limit** (p. 239) of a planet is the distance within which the planet's tidal field overwhelms the internal gravity of an orbiting moon, tearing the moon apart and forming a ring. All the rings of the four jovian planets lie inside (or close to) their parent planets' Roche limits.

LO7 Jupiter has a faint, dark ring extending down to the planet's cloud tops. Uranus has a series of dark, narrow rings first

detected from Earth by **stellar occultation** (p. 241), which occurs when a body passing between Earth and a distant star obscures the light from that star. Shepherd satellites keep Uranus's rings from breaking apart. Neptune has three narrow rings like Uranus's and one broad ring, like Jupiter's.

LO8 Pluto was discovered in 1930 after a laborious search for a planet that was supposedly affecting Uranus's orbital motion. We now know that Pluto is far too small to have had any detectable influence on Uranus's path. Pluto has a large moon, Charon, and two smaller moons. Studies of Charon's orbit around Pluto have allowed the masses and radii of both bodies to be accurately determined. Several bodies beyond Neptune's orbit have masses comparable to or greater than that of Pluto; Eris is the most massive. Eris, Pluto, and two other Kuiper belt objects are examples of **plutoids** (p. 246).

MasteringAstronomy®   For instructor-assigned homework go to www.masteringastronomy.com

Problems labeled **POS** explore the process of science. **VIS** problems focus on reading and interpreting visual information. **LO** connects to the introduction's numbered Learning Outcomes.

REVIEW AND DISCUSSION

1. **LO1** How do the density and composition of the Galilean moons vary with increasing distance from Jupiter?

2. What is unusual about Jupiter's moon Io?

3. **POS** Why is there speculation that the Galilean moon Europa might be an abode for life?

4. **POS** Why do scientists think that Ganymede's interior may have been heated as recently as a billion years ago?

5. **LO2** What properties of Saturn's largest moon, Titan, make it of particular interest to astronomers?

6. **LO3** What is the predicted fate of Triton?

7. **LO4** What is the evidence for geological activity on Saturn's midsized moons?

8. **LO5** Seen from Earth, Saturn's rings sometimes appear broad and bright but at other times seem to disappear. Why?

9. **LO6 POS** Why do many astronomers think Saturn's rings formed quite recently?

10. What effect does Mimas have on the rings of Saturn?

11. Describe the behavior of shepherd satellites.

12. **LO7** How do the rings of Neptune differ from those of Uranus and Saturn?

13. **LO8 POS** Why don't astronomers regard Pluto as a major planet?

14. In what respect is Pluto more like a moon than a planet?

15. How was the mass of Pluto determined?

CONCEPTUAL SELF-TEST: TRUE OR FALSE?/MULTIPLE CHOICE

1. Scientists speculate that Europa may have liquid water below its frozen surface. (T/F)

2. Saturn's ring particles are made up mainly of water ice. (T/F)

3. Saturn's rings lie within the planet's Roche limit. (T/F)

4. Two small shepherd satellites are responsible for the unusually complex form of Saturn's F ring. (T/F)

5. Titan's surface is obscured by thick clouds of ammonia ice. (T/F)

6. Uranus has no large moons. (T/F)

7. Triton's orbit is unusual because it is retrograde. (T/F)

8. Pluto was discovered via its gravitational effect on Neptune. (T/F)

9. Io's surface appears very smooth because it (a) is continually resurfaced by volcanic activity; (b) is covered with ice; (c) has been shielded by Jupiter from meteorite impacts; (d) is liquid.

10. The Galilean moons of Jupiter are sometimes described as a miniature inner solar system because (a) there are the same number of Galilean moons as there are terrestrial planets; (b) the moons have generally "terrestrial" composition; (c) the moons' densities decrease with increasing distance from Jupiter; (d) the moons all move on circular, synchronous orbits.

11. Compared to Earth's surface, the pressure at the surface of Titan is (a) less than; (b) about the same as; (c) about 1.5 times greater than that of Earth.

12. The rings of Uranus are (a) broad and bright; (b) narrow and dark; (c) narrow and bright; (d) undetectable from Earth.

13. The solar system object most similar to Pluto is (a) Mercury; (b) the Moon; (c) Titan; (d) Triton.

14. **VIS** In Figure 8.11 (Titan's Atmosphere), the atmospheric temperature at the top of the main haze layer is approximately (a) 50 K; (b) 100 K; (c) 150 K; (d) 200 K.

15. **VIS** From Figure 8.20 (Saturn's Tilt), the next time Saturn's rings will appear roughly edge-on as seen from Earth will be around (a) 2025; (b) 2015; (c) 2020; (d) 2035.

PROBLEMS

The number of squares preceding each problem indicates its approximate level of difficulty.

1. ■ What is the orbital speed, in kilometers per second, of ring particles at the inner edge of Saturn's B ring? Compare the particles speed with that of a satellite in low Earth orbit (500 km altitude, say). Why are these speeds so different?

2. ■■ Io orbits Jupiter at a distance of six planetary radii from the planet's center once every 42 hours. If Jupiter completes one rotation every 10 hours, use Kepler's third law to calculate how far from the center a satellite must orbit in order to appear stationary above the planet. ∞ *(Sec. 1.3)*

3. ■■■ Calculate the difference between the acceleration due to Jupiter's gravity at Io's center and the acceleration on Io's surface at the point closest to the planet. This is the tidal acceleration exerted by Jupiter on the moon. ∞ *(Sec. 5.2)* Compare this with Io's surface gravity. Repeat the comparison for Earth and Earth's Moon.

4. ■ Show that Titan's surface gravity is about one-seventh that of Earth's. What is Titan's escape speed? ∞ *(More Precisely 5-1)*

5. ■■ Compare the apparent sizes of the Galilean moons, as seen from Jupiter's cloud tops, with the angular diameter of the Sun at Jupiter's distance. Would you expect ever to see a total solar eclipse from Jupiter's cloud tops? ∞ *(More Precisely 4-1)*

6. ■■ Assuming a spherical shape and a uniform density of 2000 kg/m^3, calculate how small an icy moon of one of the outer planets would have to be before a 40 m/s (about 90 mph) fastball could escape.

7. ■ The total mass of material in Saturn's rings is about 10^5 tons (10^8 kg). Suppose the average ring particle is 6 cm in radius (the size of a large snowball) and has a density of 1000 kg/m^3. How many ring particles are there?

8. ■ How close is Charon to Pluto's Roche limit?

9. ■ What would be your weight on Pluto? On Charon?

10. ■■■ What is the round-trip travel time of light from Earth to Pluto (at a distance of 40 AU)? How far would a spacecraft moving in a circular orbit 500 km above Pluto's surface travel during that time? To how many orbits does this correspond?

ACTIVITIES

Collaborative

1. Observe the orbits of the Galilean moons of Jupiter. You'll need a telescope and a tripod or other mount to keep the image steady. Center the planet in the field of view. You should be able to see its red-and-tan cloud bands, as well as three or four points of light along a line from Jupiter's center. These are the Galilean moons. Carefully sketch what you see, or if you can attach a camera, make an image of the field of view. In your sketch, show Jupiter, the nearby moons, any bright background stars, and the full field of view of the telescope. Keep track of the time at which the observation was made. Repeat this procedure roughly every hour over the next 3–4 nights— work in shifts so no one has to lose too much sleep! The inner two moons will complete at least one orbit during this time. Your sketches should allow you to visualize the orbits of the moons. Estimate their orbital periods and radii relative to Jupiter's radius and compare your results with the numbers in Table 8.1.

Individual

1. Use binoculars to look at Jupiter. Can you see any of the four large moons? If you come back the following evening, the moons' positions will have changed. Have some changed more than others?

2. Through a telescope, you should have a much better view of Jupiter's moons. Before observing, look up the positions of the Galilean moons in a current magazine such as *Astronomy* or *Sky & Telescope*. Identify each of the moons. Watch Io over a period of an hour. Can you see its motion? Do the same for Europa.

3. Use a telescope to look at Saturn's rings. Can you see a dark line in them? This is the Cassini Division. Can you see the shadow of the rings on the planet's surface? Can you see any moons? They line up with the rings; Titan is often the farthest out and is always the brightest. Use an almanac to identify each moon you find.

PART 3

The Stars

All stars evolve—they are born, mature, and die. They form in huge clouds of gas and dust, their nuclear fires ignite much as in the Sun, and eventually they end their lives by returning to space some of the heavy elements created within their cores. The poetic notion that we are made of stardust is a romantic one—but it also happens to be true!

Part 3 describes many diverse stellar and interstellar regions that resemble a "galactic ecosystem"—an evolutionary balance nearly as intricate and delicate as life in a tide pool or a tropical forest. The theory of stellar evolution gives astronomers a powerful, unifying theme with which to study all stars.

The images here illustrate a sampling of the stars and stellar systems of Part 3, all residents of an even larger Milky Way Galaxy.

Red Giant ~ 10^{11} m

Dusty Disk ~ 10^{14} m

Nebula ~ 10^{16} m

Galaxy ~ 10^{21} m

Star Cluster ~ 10^{18} m

Nearly a trillion trillion stars inhabit our universe; the Sun is but one of these stars.

Mar., 2011

Jan., 2012

Sept., 2012

May, 201

May, 2010

Mar., 2014

9

The Sun

Our Parent Star

Living in the solar system, we have the chance to study, at close range, perhaps the most common type of cosmic object—a star. Our Sun is a star, and a fairly average star at that, but with one unique feature: It is very close to us—some 300,000 times closer than the next nearest star, Alpha Centauri. Whereas Alpha Centauri is 4.3 light-years away, the Sun is only 8 light-minutes from us. Consequently, astronomers know far more about the properties of the Sun than about any of the other distant points of light in the universe. Indeed, a good fraction of all our astronomical knowledge is based on modern studies of the Sun. Just as we studied our parent planet, Earth, to set the stage for our exploration of the solar system, we now study our parent star, the Sun, as the next step in our exploration of the universe.

THE BIG PICTURE

The Sun is our star—the main source of energy that powers weather, climate, and life on Earth. Imagine our planet without the Sun—no light, no heat, no comforting "parent" in the sky. Although we take it for granted each and every day, the Sun is vitally important to us in the cosmic scheme of things. Simply put, without the Sun, we would not exist.

◄ These images of the Sun, taken in the extreme ultraviolet part of the spectrum, show the increase of solar activity since the last solar minimum in 2010 to the maximum in 2014. Active regions appear as brighter areas. (The changing size of the Sun is artistic license, not physics.) These images were captured by the *Solar Dynamics Observatory*—a 2-ton robot that orbits Earth but stares at the Sun unblinkingly 24 hours a day, eavesdropping on this gas ball's surface, atmosphere, and interior. *(SDO/NASA)*

SELF-GUIDED TUTORIAL
SuperSpaceship–Voyage to the Sun

R   I   V   U   X   G

▲ **FIGURE 9.1 The Sun**   The inner part of this composite, filtered image of the Sun shows a sharp edge, although our star, like all stars, is composed of a gradually thinning gas. The edge appears sharp because the solar photosphere is so thin. The outer portion of the image is the solar corona, normally too faint to be seen, but visible during an eclipse, when the light from the solar disk is blotted out. (Note the blemishes; they are sunspots.) *(NOAO)*

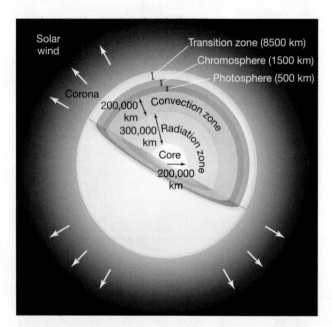

▲ **FIGURE 9.2 Solar Structure**   The main regions of the Sun, not drawn to scale, with some physical dimensions labeled.

9.1 The Sun in Bulk

The Sun is the sole source of light and heat for the maintenance of life on Earth. It is a **star**, a glowing ball of gas held together by its own gravity and powered by nuclear fusion (a process in which two or more atomic nuclei combine and release energy—see Section 9.5) at its center. In its physical and chemical properties, the Sun is very similar to most other stars, regardless of when and where they formed. Far from detracting from our interest in the Sun, this very mediocrity is one of the main reasons we study it—most of what we know about the Sun also applies to many other stars in the sky. The Sun is the best-studied star in the universe, and it is no exaggeration to say that our knowledge of the cosmos rests squarely on our understanding of solar physics.

Overall Structure

Table 9.1 lists some basic solar data. With a radius larger than 100 Earth radii, a mass greater than 300,000 Earth masses, and a surface temperature well above the melting point of any known material, the Sun is clearly very different from any other body we have encountered so far.

The Sun has a surface of sorts—not a solid surface (the Sun contains no solid material), but rather that part of the brilliant gas ball we perceive with our eyes or view through a heavily filtered telescope. The part of the Sun that emits the radiation we see is called the **photosphere** (after the Greek word *photos*, meaning "light"). Its radius (the radius of the Sun in Table 9.1) is about 700,000 km. However, as we will see, its thickness is probably no more than 500 km—less than 0.1 percent of the radius—which is why we perceive the Sun as having a well-defined, sharp edge (Figure 9.1).

The main regions of the Sun are illustrated in Figure 9.2. Just above the photosphere is the Sun's lower atmosphere, called the **chromosphere**, about 1500 km thick. Above that lies a region called the **transition zone**, where the temperature rises dramatically. Above 10,000 km, and stretching far beyond, is a thin, hot upper atmosphere, the solar **corona**. At still greater distances the corona turns into the **solar wind**, which flows away from the Sun and permeates the entire solar system. ∞ *(Sec. 4.2)*

Below the photosphere, extending down some 200,000 km, is the **convection zone**, a region where the material of the Sun is in constant convective motion. ∞ *(Sec. 5.3)* Below the convection zone lies the **radiation zone**, where solar energy is transported toward the surface by radiation rather than by convection. The term *solar interior* is often used to mean both the radiation and convection zones. The central **core**, roughly 200,000 km in radius, is the site of powerful nuclear fusion reactions that generate the Sun's enormous energy output.

Solar rotation can be measured by timing sunspots and other surface features as they cross the solar disk. ∞ *(Sec. 1.2)* These observations indicate that the Sun completes one rotation in about a month, but it does not do so as a solid body. Instead, it spins differentially—faster at the equator and slower at the poles, like Jupiter and Saturn ∞ *(Sec. 7.3)*. The equatorial rotation period is about 25 days. Sunspots are never seen above latitude 60° (north or south), but at that latitude they indicate a 31-day period. Other measurement techniques, such as those described in Section 9.2, reveal that the rotation period continues to increase as we approach the poles. The polar period may be as long as 36 days.

Luminosity

The Sun radiates an enormous amount of energy into space. We can measure the Sun's total energy output in two stages, as follows. First, by holding a light-sensitive device—a solar cell, perhaps—above Earth's atmosphere perpendicular to the Sun's rays we can measure how much solar energy is received per square meter of area every second at Earth's distance from the Sun. This quantity, known as the **solar constant,** is approximately 1400 watts per square meter (W/m^2). About 50–70 percent of this energy reaches Earth's surface; the rest is intercepted by the atmosphere (30 percent) or reflected away by clouds (0–20 percent). Thus, for example, a sunbather's body having a total surface area of about 0.5 m^2 receives solar energy at a rate of roughly 500 watts, equivalent to the output of a small electric room heater or five 100-W lightbulbs.

Second, knowing the solar constant, we can calculate the total amount of energy radiated in all directions from the Sun. Imagine a three-dimensional sphere that is centered on the Sun and is just large enough that its surface intersects Earth's center (Figure 9.3). The sphere's radius is 1 AU, and its surface area is therefore $4\pi \times (1\ AU)^2$, or approximately $2.8 \times 10^{23}\ m^2$. Assuming that the Sun radiates energy equally in all directions, we can determine the total rate at which energy leaves the Sun's surface simply by multiplying the solar constant by the total surface area of our imaginary sphere. The result is just under 4×10^{26} W. This quantity is called the **luminosity** of the Sun. As we will see, it is quite typical of the luminosities of many of the stars in the sky.

Take a moment to consider the magnitude of the solar luminosity. The Sun is an enormously powerful source of energy. *Every second* it produces an amount of energy equivalent to the detonation of about 100 billion 1-megaton nuclear bombs. There is simply nothing on Earth comparable to a star!

TABLE 9.1 Some Solar Properties	
Radius	696,000 km
Mass	1.99 × 1030 kg
Average density	1410 kg/m³
Rotation period	25.1 days (equator)
	30.8 days (60° latitude)
	36 days (poles)
	26.9 days (interior)
Surface temperature	5780 K
Luminosity	3.86 × 10²⁶ W

CONCEPT CHECK

Why must we assume that the Sun radiates equally in all directions when computing the solar luminosity from the solar constant?

9.2 The Solar Interior

How have we learned about the inside of the Sun? After all, the sunlight we see comes from the photosphere; none of the radiation from the interior reaches us directly. In this section we discuss in a little more detail how we know what we know about conditions below the solar surface.

Lacking any direct measurements of the Sun's interior, researchers must use indirect means to probe in more detail the inner workings of our parent star. To accomplish this, they construct *mathematical models* of the Sun, combining all available observations with theoretical insight into solar physics. ∞ *(Sec. 0.5)* The resulting **standard solar model** has gained widespread acceptance among astronomers. We can construct similar models of other stars, but because the Sun is so close and well studied, the standard solar model is by far the best tested.

Modeling the Structure of the Sun

The Sun's bulk properties—mass, radius, temperature, and luminosity—do not vary much from day to day or from year to year. We will see in Chapter 12 that stars like the Sun do change significantly over periods of *billions* of years, but for our purposes here we can ignore this slow evolution. On "human" time scales, the Sun may reasonably be thought of as unchanging.

Accordingly, as illustrated in Figure 9.4, theoretical models generally begin by assuming that the Sun is in a state of **hydrostatic equilibrium**, in which pressure's outward push exactly counteracts gravity's inward pull. This stable balance between opposing forces is the basic reason that the Sun neither collapses under

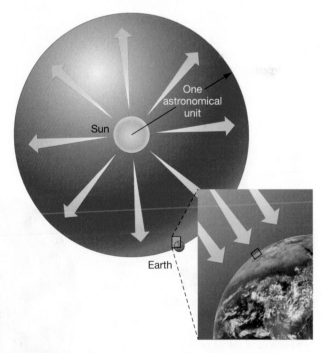

▲ FIGURE 9.3 **Solar Luminosity** If we draw an imaginary sphere around the Sun so that the sphere's surface passes through Earth's center, then the radius of this imaginary sphere is 1 AU. The "solar constant" is the amount of power striking a 1-m² detector at Earth's distance, as suggested in the inset. The Sun's luminosity is then determined by multiplying the sphere's surface area by the solar constant. *(NASA)*

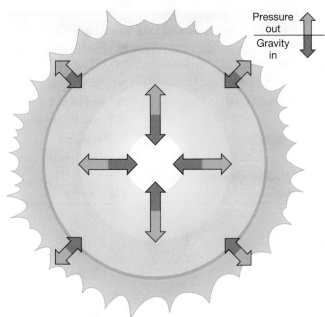

Pressure out
Gravity in

▲ **FIGURE 9.4 Stellar Balance**   In the interior of a star such as the Sun, the outward pressure of hot gas balances the inward pull of gravity. This is true at every point within the star, guaranteeing its stability.

its own weight nor explodes into interstellar space. ∞ *(More Precisely 5-1)* It also has an important consequence for the solar interior. Because the Sun is very massive, its gravitational pull is very strong, so very high internal pressure is needed to maintain hydrostatic equilibrium. This high pressure in turn requires a very high central temperature, a fact crucial to our understanding of solar energy generation (Section 9.5). In this way the assumption of hydrostatic equilibrium lets us determine the density and temperature inside the Sun. This information, in turn, allows the model to make predictions about other observable solar properties—luminosity, radius, spectrum, and so on. Scientists fine-tune the internal details of the model until the predictions agree with observations. This is the scientific method at work; the standard solar model is the result. ∞ *(Sec. 0.5)*

To test and refine the standard solar model, astronomers need information about the solar interior. In the 1960s, measurements of the Doppler shifts of solar spectral lines revealed that the surface of the Sun oscillates, or vibrates, like a complex set of bells. ∞ *(Sec. 2.7)* These vibrations, illustrated in Figure 9.5, are the result of internal pressure ("sound") waves that reflect off the photosphere and repeatedly cross the solar interior. These waves penetrate deep inside the Sun, and analysis of their surface patterns allows scientists to study conditions far below the Sun's surface. This process is similar to the way in which seismologists study Earth's interior by observing seismic waves produced by earthquakes. ∞ *(Sec. 5.4)* For this reason, study of solar surface patterns is usually called **helioseismology**, even though solar pressure waves have nothing whatever to do with solar seismic activity (which doesn't exist, as the Sun is entirely gaseous).

The most extensive study of solar vibrations is the ongoing GONG (short for Global Oscillations Network Group) project. By making continuous observations of the Sun from many clear sites around Earth, solar astronomers can obtain uninterrupted high-quality solar data spanning weeks at a time. Space-based observatories (see *Discovery 9-1*) have provided continuous monitoring of the Sun's surface and atmosphere since 1995. Analysis of these datasets provides detailed information about the temperature, density, rotation, and convective state of the solar interior, permitting direct comparison with theory over a large portion of the Sun's volume. The agreement between the standard solar model and observations is spectacular—the frequencies and wavelengths of observed solar oscillations are within 0.1 percent of the model predictions.

▶ FIGURE 9.5 **Solar Oscillations**   (a) By observing the motion of the solar surface, scientists can determine the wavelengths and frequencies of the individual waves and deduce information about the Sun's complex vibrations. (b) Waves contributing to the observed oscillations can travel deep inside the Sun, providing vital information about the solar interior. *(National Solar Observatory)*

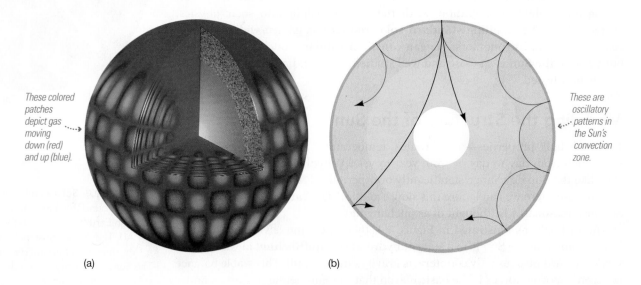

These colored patches depict gas moving down (red) and up (blue).

These are oscillatory patterns in the Sun's convection zone.

(a)   (b)

9.1 DISCOVERY

Eavesdropping on the Sun

Throughout the few decades of the Space Age, various nations, led by the United States, have sent spacecraft to most of the major bodies in the solar system. One of the as-yet-unexplored bodies is Pluto, the most notable member of the Kuiper belt, which has never been visited by a robot orbiter or even a flyby craft—although this may change soon. ∞ *(Sec. 8.5)* The other unexplored body is the Sun. Currently, the next best thing to a dedicated reconnoitering spacecraft is the *Solar and Heliospheric Observatory (SOHO)*, launched in 1995 and (as of 2015) still operating, and the *Solar Dynamics Observatory (SDO)*, launched in 2010. Both spacecraft have radioed back to Earth volumes of new data—and a few new puzzles—about our parent star.

SOHO is a billion-dollar mission operated primarily by the European Space Agency. The 2-ton robot is now on-station about 1.5 million kilometers sunward of Earth—about 1 percent of the distance from Earth to the Sun. This is the so-called L_1 Lagrangian point, where the gravitational pull of the Sun and Earth are precisely equal—a good place to park a monitoring platform. By contrast, the American *SDO* spacecraft travels around Earth in a high orbit. Both automated vehicles stare at the Sun 24 hours a day, and both carry several instruments capable of measuring almost everything from the Sun's corona and magnetic field to its solar wind and internal vibrations. The accompanying figure is a false-color *SDO* image of the Sun's lower corona in the ultraviolet part of the spectrum, showing widespread magnetic activity and (at bottom) a large *coronal hole* (Section 9.4), where the solar wind streams into interplanetary space.

Both spacecraft are positioned just beyond Earth's magnetosphere, so their instruments can study cleanly the high-speed charged particles of the solar wind. Coordinating these on-site measurements with *SOHO* and *SDO* images of the Sun itself, astronomers now think they can follow solar magnetic field loops expanding and breaking as the Sun prepares itself for mass ejections several days before they actually occur. Given that such coronal storms can endanger pilots and astronauts and play havoc with communications, power grids, satellite electronics, and other human activity, the ability to make accurate forecasts of disruptive solar events is of global importance.

In fact, there do seem to be some correlations between the 22-year solar cycle (two sunspot cycles, with oppositely directed magnetic fields) and periods of climatic dryness here on Earth. For example, near the start of the past eight solar cycles, there have been droughts in North America—at least within the middle and western plains from South Dakota to New Mexico. The most recent of these droughts, which typically last 3 to 6 years, came in the late 1950s. The one expected in the early 1980s, however, did not occur as clearly as anticipated.

Other possible Sun–Earth connections include a link between solar activity and increased atmospheric circulation on our planet. As circulation increases, terrestrial storm systems deepen, extend over wider ranges of latitude, and carry more moisture. The relationship is complex, and the subject controversial, because no one has yet shown any physical mechanism (other than the Sun's heat, which does not vary much during the solar cycle) that would allow solar activity to stir our terrestrial atmosphere. Without a better understanding of the physical mechanism involved, none of these effects can be incorporated into our weather-forecasting models.

The data also allow scientists to monitor global circulation patterns in the solar interior, including two gigantic "conveyor belts" that transport subsurface material from the equator to the poles, then return it to the equator at a depth of some 300,000 km, far below the convection zone. These circulation patterns, which move at 10–15 m/s and take roughly 40 years to complete a single loop, are thought to be important in regulating the Sunspot cycle (see Section 9.4).

(a)

(b)

(c)

Figure 9.6 shows the solar density and temperature according to the standard solar model, plotted as functions of distance from the Sun's center. Notice how the density drops sharply at first, then more slowly near the photosphere. The variation in density is large, ranging from a core value of about 150,000 kg/m³, 20 times the density of iron, to an extremely small photospheric value of 2×10^{-4} kg/m³, about 10,000 times less dense than air. The *average* density of the Sun (Table 9.1) is 1400 kg/m³, about the same as the density of Jupiter. As shown in Figure 9.6(c), the solar temperature also decreases with increasing radius, but not as rapidly as the density. The temperature is about 15 million K at the center and decreases steadily, reaching the observed value of 5800 K at the photosphere.

Energy Transport

The very hot solar interior ensures violent and frequent collisions among gas particles. In and near the core, the extremely high temperatures guarantee that the gas is completely ionized. ∞ *(Sec. 2.6)* With no electrons left on atoms to capture photons and move into more excited states, the deep solar interior is quite transparent to radiation. Only occasionally does a photon interact with a free electron. The energy produced by nuclear reactions in the core travels outward toward the surface with relative ease in the form of radiation.

As we move outward from the core, the temperature falls, and eventually some electrons can remain bound to nuclei. With more and more atoms retaining electrons that can absorb the outgoing radiation, the gas in the interior changes from being relatively transparent to almost totally opaque. By the outer edge of the radiation zone, 500,000 km from the center, *all* of the photons produced in the Sun's core have been absorbed. Not one of them reaches the surface. But what happens to the energy they carry?

The photon's energy must travel beyond the solar interior—the fact that we see sunlight proves that the energy escapes. The energy is carried to the solar surface by *convection*—the same basic physical process we saw in our study of Earth's atmosphere, although operating in a very different environment in the Sun. ∞ *(Sec. 5.3)* Convection can occur whenever cooler material overlies warmer material, and this is just what happens in the outer part of the Sun's interior. Hot solar gas moves outward, while cooler gas above it sinks, creating a characteristic pattern of convection cells. All through the convection zone, energy is transported to the surface by physical motion of the solar gas. (Note that this actually represents a departure from hydrostatic equilibrium, as defined above, but it can still be handled within the standard solar model.) Remember that there is no physical movement of material when radiation is the energy-transport mechanism; convection and radiation are *fundamentally different* ways in which energy can be transported from one place to another.

Figure 9.7 is a schematic diagram of the solar convection zone. There is a hierarchy of convection cells, organized in tiers at different depths. The deepest tier, about 200,000 km below the photosphere, is thought to contain cells tens of thousands of kilometers in diameter. Energy is carried upward through

◄ FIGURE 9.6 **Solar Interior** Density and temperature distributions in the interior of the Sun. Parts (b) and (c) show the large variations in solar density and temperature, relative to a cutaway diagram of the Sun's interior (a).

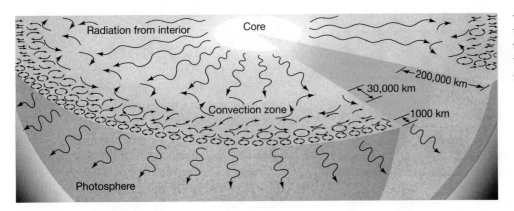

◀ FIGURE 9.7 Solar Convection Energy is physically transported in the Sun's convection zone, which is here visualized as a boiling, seething sea of gas. As drawn, the convective cell sizes become progressively larger at greater depths.

a series of progressively smaller cells, stacked one upon another until, at a depth of about 1000 km below the photosphere, the individual cells are about 1000 km across. The top of this uppermost tier of convection lies just below the solar photosphere.

Convection does not proceed into the solar atmosphere. In and above the photosphere the density is so low that the gas is transparent and radiation once again becomes the mechanism of energy transport. Photons reaching the photosphere escape freely into space. The solar density falls off so rapidly in this region of the Sun that the gas makes the transition from completely opaque to completely transparent over a very small distance—just a few hundred kilometers—accounting for the thinness of the photosphere and the observed sharp "edge" of the Sun.

Evidence for Solar Convection

In part, our knowledge of solar convection is derived indirectly, from computer models of the solar interior. However, astronomers also have some direct evidence of conditions in the convection zone. Figure 9.8 is a high-resolution photograph of the solar surface, filtered to accentuate brightness variations across the field of view. The visible surface is highly mottled with regions of bright and dark gas known as *granules*. This **granulation** of the solar surface is a direct reflection of motion in the convection zone. Each bright granule measures about 1000 km across—comparable to one of the larger states in the United States—and has a lifetime of between 5 and 10 minutes.

Each granule is the topmost part of a solar convection cell. Spectral lines in the bright granules are slightly blueshifted, indicating Doppler-shifted matter approaching us—that is, moving upward—at about 1 km/s. ∞ *(Sec. 2.7)* Similarly, material in the dark granules is sinking into the solar interior, exactly as we would expect for the topmost tier of convection in Figure 9.7. The variation in brightness is due to differences in temperature. The upwelling gas is hotter and therefore (by Stefan's law) emits more radiation than the cooler downwelling gas. ∞ *(Sec. 2.4)* The bright and dark regions appear to contrast considerably, but in reality their temperature difference is less than about 500 K.

Careful measurements also reveal a much larger-scale flow beneath the solar surface. **Supergranulation** is a flow pattern quite similar to granulation except that supergranulation cells measure some 30,000 km across. As with granulation, material upwells at the center of the cells, flows across the surface, then sinks down again at the edges. Scientists think that supergranules are the imprint on the photosphere of the deepest tier of large convective cells depicted in Figure 9.7.

CONCEPT CHECK

Describe the two distinct ways in which energy moves outward from the solar core to the photosphere.

▲ FIGURE 9.8 Solar Granulation This photograph of the granulated solar photosphere, taken with the 1-m Swedish Solar Telescope, shows solar granules comparable in size to Earth's continents. The bright portions of the image are regions where hot material is upwelling from below. The dark regions are cooler gas that is sinking back down into the interior. *(SST/Royal Swedish Academy of Sciences)*

▲ FIGURE 9.9 **Solar Spectrum** A detailed visible spectrum of our Sun shows thousands of dark absorption lines, indicating the presence of 67 different elements in various stages of excitation and ionization in the lower solar atmosphere. The numbers give wavelengths in nanometers. *(Palomar Observatory/Caltech)*

9.3 The Solar Atmosphere

Astronomers can obtain an enormous amount of information about the Sun (or any star) by analyzing the absorption lines in its spectrum. ∞ *(Sec. 2.5)* Figure 9.9 (see also Figure 2.14) is a detailed visible spectrum of the Sun, spanning the range of wavelengths from 360 to 690 nm. Table 9.2 lists the 10 most common elements in the Sun based on spectral analysis. This distribution is very similar to what we saw on the jovian planets, and it is what we will find for the universe as a whole. Hydrogen is by far the most abundant element, followed by helium.

Strictly speaking, analysis of spectral lines allows us to draw conclusions only about the part of the Sun where the lines form—the photosphere and chromosphere. However, the data in Table 9.2 are thought to be representative of the entire Sun (with the exception of the solar core, where nuclear reactions are steadily changing the composition—see Section 9.5).

The Chromosphere

Density continues to decrease rapidly as we move outward through the solar atmosphere. The low density of the chromosphere means that it emits very little light of its own and cannot be observed visually under normal conditions. The photosphere is just too bright, dominating the chromosphere's radiation. Still, astronomers have long been aware of its existence. Figure 9.10 shows the Sun

TABLE 9.2	The Composition of the Sun	
Element	Percentage of Total Number of Atoms	Percentage of Total Mass
Hydrogen	91.2	71.0
Helium	8.7	27.1
Oxygen	0.078	0.97
Carbon	0.043	0.40
Nitrogen	0.0088	0.096
Silicon	0.0045	0.099
Magnesium	0.0038	0.076
Neon	0.0035	0.058
Iron	0.0030	0.14
Sulfur	0.0015	0.040

◄ FIGURE 9.10 **Solar Chromosphere** This photograph of a total solar eclipse shows the solar chromosphere a few thousand kilometers above the Sun's surface. Note the prominence at left. *(G. Schneider)*

▲ **FIGURE 9.11 Solar Spicules** Short-lived, narrow jets of gas that typically last mere minutes can be seen sprouting up from the chromosphere in this ultraviolet image of the Sun. These so-called spicules are the thin spikelike regions where gas leaves the surface at speeds of about 100 km/s. *(NASA)*

 ANIMATION/VIDEO
Solar Granulation

 ANIMATION/VIDEO
Solar Chromosphere

during an eclipse in which the photosphere was obscured by the Moon, but the chromosphere was not. The chromosphere's characteristic pinkish hue, the result of the red H-alpha emission line of hydrogen, is plainly visible. ∞ *(Sec. 2.6)*

The chromosphere is far from tranquil. Every few minutes, small solar storms erupt there, expelling spikes of hot matter known as *spicules* into the Sun's upper atmosphere (Figure 9.11). These long, thin jets leave the Sun's surface at typical speeds of about 100 km/s, reaching thousands of kilometers above the photosphere. Spicules tend to accumulate around the edges of supergranules, and the Sun's magnetic field is also somewhat stronger than average in those regions. Scientists think that the downwelling material there amplifies the solar magnetic field and that spicules are the result of magnetic disturbances in the Sun's churning outer layers.

The Transition Zone and Corona

During the brief moments of a solar eclipse, if the Moon's angular size is large enough that both the photosphere and the chromosphere are blocked, the ghostly solar corona can be seen, as in Figure 9.12. With the photospheric light removed, the pattern of spectral lines changes dramatically. The spectrum shifts from absorption to emission, and an entirely new set of spectral lines suddenly appears.

The shift from absorption to emission is entirely in accordance with Kirchhoff's laws, because we see the corona against the blackness of space, not against the bright continuous spectrum from the photosphere below. ∞ *(Sec. 2.5)* The new lines arise because atoms in the corona are much more highly ionized than atoms in the photosphere or chromosphere. ∞ *(Sec. 2.6)* Their internal electronic structure and hence their spectra are therefore quite different from those of atoms at lower atmospheric levels. The cause of this extensive electron stripping is the high coronal temperature. The degree of ionization inferred from spectra observed during solar eclipses tells us that the upper chromosphere is hotter than the photosphere, and the corona, where even more ionization is seen, is hotter still.

CONCEPT CHECK

Describe two ways in which the spectrum of the solar corona differs from that of the photosphere.

▲ **FIGURE 9.12 Solar Corona** When both the photosphere and the chromosphere are obscured by the Moon during a solar eclipse, the faint solar corona becomes visible. This 1979 photograph shows the emission of radiation from an active corona. *(NCAR High Altitude Observatory)*

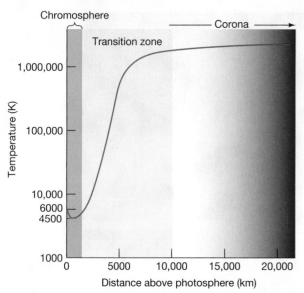

▲ FIGURE 9.13 **Solar Atmospheric Temperature**   The change of gas temperature in the lower solar atmosphere is dramatic. The temperature, indicated by the blue line, reaches a minimum of 4500 K in the chromosphere and then rises sharply in the transition zone, finally leveling off at around 3 million K in the corona.

ANIMATION/VIDEO (MA)®

Sunspot

DATA POINTS

The Solar Cycle

About one-third of all students had difficulty describing the solar cycle. Remember:

■ The sunspot cycle is a somewhat irregular process taking roughly 11 years to complete. Sunspots increase in number and tend to be found progressively closer to the solar equator as the cycle proceeds.

■ The peak of the sunspot cycle is associated with more and stronger solar activities—flares, coronal mass ejections, etc.

■ The Sun's entire magnetic field reverses direction from one sunspot cycle to the next, so the full solar cycle actually takes 22 years to complete.

Figure 9.13 shows how the temperature varies with altitude above the photosphere. The temperature decreases to a minimum of about 4500 K some 500 km above the photosphere, after which it rises steadily. About 1500 km above the photosphere, in the transition zone, the gas temperature increases sharply, reaching more than 1 million K at an altitude of 10,000 km. Thereafter, in the corona, the temperature remains roughly constant at around 3 million K, although orbiting instruments have detected coronal "hot spots" having temperatures many times higher than this average value. The rapid temperature rise in the transition zone is not fully understood. Moving away from a heat source, we would normally expect the heat to diminish, but this is not the case in the lower atmosphere of the Sun. The corona must have another energy source. Astronomers think that magnetic disturbances in the solar photosphere (see Section 9.4) are ultimately responsible for heating the corona.

The Solar Wind

Electromagnetic radiation and fast-moving particles—mostly protons and electrons—escape from the Sun all the time. The radiation moves away from the photosphere at the speed of light, taking 8 minutes to reach Earth. The particles travel much more slowly, although at the still considerable speed of about 500 km/s, reaching Earth in a few days. This constant stream of escaping solar particles is the solar wind.

The solar wind is a direct consequence of the high coronal temperature. About 10 million km above the photosphere, the coronal gas is hot enough to escape the Sun's gravity, and it begins to flow outward into space. At the same time, the solar atmosphere is continuously replenished from below. If that were not the case, the corona would disappear in about a day. The Sun is, in effect, "evaporating"—constantly shedding mass through the solar wind. The wind is an extremely thin medium, however. Although it carries away about a million tons of solar matter each second, less than 0.1 percent of the Sun's mass has been lost since the solar system formed 4.6 billion years ago.

9.4 The Active Sun

Most of the Sun's luminosity results from continuous emission from the photosphere. However, superimposed on this steady, predictable aspect of our star's energy output is a much more irregular component, characterized by explosive and unpredictable surface activity. This **solar activity** contributes little to the Sun's total luminosity and has little effect on the evolution of the Sun as a star, but it does affect us here on Earth. The strength of the solar wind is strongly influenced by the level of solar activity, and this, in turn, directly affects Earth's magnetosphere.

Sunspots

Figure 9.14 is an optical photograph of the entire Sun showing numerous dark **sunspots** on the surface. ∞ *(Sec. 1.2)* They typically measure about 10,000 km across—about the size of Earth. At any given instant, the Sun may have hundreds of sunspots, or it may have none at all.

Detailed observations of sunspots show an *umbra*, or dark center, surrounded by a grayish *penumbra*. The close-up view of a pair of sunspots in Figure 9.15(a) shows both of these dark areas against the bright background of the undisturbed

Sunspots appear dark because they are slightly cooler than the surrounding gas.

(a) |← 50,000 km →|

Penumbra

Umbra

This spot is about the size of Earth.

(b) |←10,000 km→|

R I V U X G

▲ **FIGURE 9.14 Sunspots** This photograph of the entire Sun, taken during a period of maximum solar activity, shows several groups of sunspots. The largest spots in this image are more than 20,000 km across, nearly twice the diameter of Earth. *(Palomar Observatory/Caltech)*

▲ **FIGURE 9.15 Sunspots, Up Close** (a) An enlarged photo of the largest pair of sunspots in Figure 9.14 shows how each spot consists of a cool, dark umbra surrounded by a warmer, brighter penumbra. (b) A high-resolution image of a single typical sunspot shows details of its structure as well as the granules surrounding it. *(Palomar Observatory/Caltech; SST/ Royal Swedish Academy of Sciences)*

photosphere. This gradation in darkness indicates a gradual change in photospheric temperature. Sunspots are simply cooler regions of the photospheric gas. The temperature of the umbra is about 4500 K, that of the penumbra 5500 K. The spots, then, are certainly composed of hot gases. They seem dark only because they appear against an even brighter background (the 5800 K photosphere).

As mentioned in Sec. 9.1, studies of sunspots indicate that the Sun does not rotate as a solid body (see Table 9.1). Instead, it spins *differentially*—faster at the equator and more slowly at the poles, like Jupiter and Saturn. The combination of differential rotation and convection radically affects the character of the Sun's magnetic field, which in turn plays a major role in determining the numbers and location of sunspots.

Solar Magnetism

Spectroscopic studies indicate that the magnetic field in a typical sunspot is about 1000 times greater than the field in the surrounding, undisturbed photosphere (which is itself several times stronger than Earth's magnetic field). Furthermore, the field lines are not randomly oriented, but instead are directed roughly perpendicular to (out of or into) the Sun's surface. Scientists think that sunspots are cooler than their surroundings because these abnormally strong fields interfere with the normal convective flow of hot gas toward the solar surface.

(a)

(b)

▶ **FIGURE 9.16 Sunspot Magnetism**
(a) The Sun's magnetic field lines emerge from the surface through one member of a sunspot pair and reenter the Sun through the other member. If the magnetic field lines are directed into the Sun in one leading spot, they are inwardly directed in all other leading spots in that hemisphere. The opposite is the case in the southern hemisphere. (b) A far-ultraviolet image taken by NASA's *Transition Region and Coronal Explorer (TRACE)* satellite, showing magnetic field lines arching between two sunspot groups. (c) A wider view from *NASA's Solar Dynamics Observatory* shows the complex large-scale solar field structure. *(NASA)*

(c)

The **polarity** of a sunspot indicates which way its magnetic field is directed. We conventionally label spots as S where the field lines emerge from the interior and as N where the lines dive below the photosphere (so that field lines above the surface run from S to N, as on Earth). Sunspots almost always come in pairs whose members lie at roughly the same latitude and have opposite magnetic polarities. As illustrated in Figure 9.16(a), magnetic field lines emerge from the solar interior through one member (S) of a sunspot pair, loop through the solar atmosphere, then reenter the photosphere through the other member (N). As in Earth's magnetosphere, charged particles tend to follow the solar magnetic field lines. ∞ *(Sec. 5.7)* Figure 9.16(b) shows an actual image of solar magnetic loops; the light we see here is emitted by charged particles trapped by the solar field.

Despite the irregular appearance of the sunspots themselves, there is a great deal of order in the underlying solar field. *All* the sunspot pairs in the same solar hemisphere (north or south) at any instant have the *same* magnetic configuration. That is, if the leading spot (measured in the direction of the Sun's rotation) has N polarity, as shown in the figure, then all leading spots in that hemisphere have the same polarity. What's more, in the other hemisphere at the same time, all sunspot pairs have the *opposite* magnetic configuration (S polarity leading). To understand these regularities in sunspot polarities, we must look in more detail at the Sun's magnetic field.

As illustrated in Figures 9.17 (a–c), the Sun's differential rotation greatly distorts the solar magnetic field, "wrapping" it around the solar equator and eventually causing any north–south magnetic field to reorient itself in an east–west direction. At the same time, convection causes the magnetized gas to upwell toward the surface, twisting and tangling the magnetic field pattern. In some places, the field lines become kinked like a knot in a garden hose, causing the field strength to increase. Occasionally, the field becomes so strong that it overwhelms the Sun's gravity, and a "tube" of field lines bursts out of the surface and loops through the lower atmosphere, forming a sunspot pair (Figure 9.17c). The general east–west organization of the underlying solar field accounts for the observed polarities of the resulting sunspot pairs in each hemisphere.

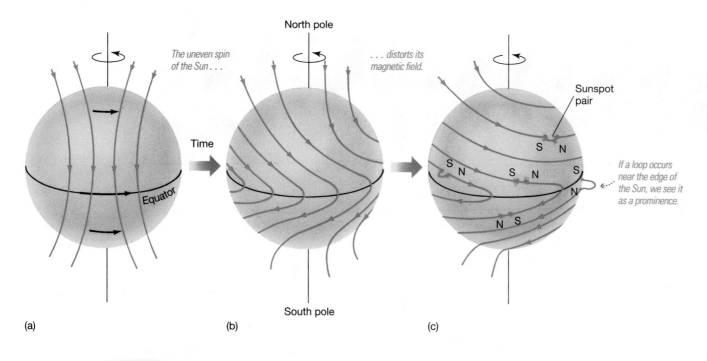

North pole

*The uneven spin
of the Sun . . .*

*. . . distorts its
magnetic field.*

Time

Equator

South pole

Sunspot
pair

S N

S S N S

N N

*If a loop occurs
near the edge of
the Sun, we see it
as a prominence.*

N S

(a)   (b)   (c)

A good analogy to a
tangled solar magnetic
field is a garden hose
with loops and kinks.

▲ **FIGURE 9.17 Solar Rotation**   (a, b) The Sun's differential
rotation wraps and distorts the solar magnetic field.
(c) Occasionally, the field lines burst out of the surface and
loop through the lower atmosphere, thereby creating a
sunspot pair. The underlying pattern of the solar field lines
explains the observed pattern of sunspot polarities.

The Solar Cycle

Sunspots are not steady. Most change their size and shape, and all come and go.
Individual spots may last anywhere from 1 to 100 days. A large group of spots
typically lasts 50 days. However, centuries of observations have established a clear
sunspot cycle. Figure 9.18(a) shows the number of sunspots observed monthly
during the 20th century. The average number of spots reaches a maximum every
11 or so years, then falls off almost to zero before the cycle begins afresh. The latitudes at which sunspots appear vary as the sunspot cycle progresses. Individual
sunspots do not move up or down in latitude, but new spots appear closer to the
equator as older ones at higher latitudes fade away. Figure 9.18(b) is a plot of observed sunspot latitudes as a function of time.

In fact, the 11-year sunspot cycle is only half of a 22-year **solar cycle**—the
period needed for both the average number of spots *and* the Sun's overall magnetic polarity to repeat themselves. For the first 11 years of the solar cycle, the
leading spots of all sunspot pairs in the same solar hemisphere have one polarity,
and spots in the other hemisphere have the opposite polarity (Figure 9.16). These
polarities then reverse for the next 11 years.

Astronomers think that the Sun's differential rotation and convection cause
constant stretching, twisting, and folding of magnetic field lines. This ongoing
process both generates and amplifies the solar magnetic field. The theory is similar to the dynamo theory that accounts for the magnetic fields of Earth and the
jovian planets, except that the solar dynamo operates much faster and on a much
larger scale. ∞ *(Sec. 5.7)* One important prediction of this theory is that the
Sun's magnetic field should rise to a maximum, then fall to zero and reverse itself

▲ **FIGURE 9.18 Sunspot Cycle** (a) The monthly number of sunspots during the 20th century clearly displays the (roughly) 11-year solar cycle. At the time of minimum solar activity, hardly any sunspots are seen. About 4 years later, at maximum solar activity, about 100 spots are observed per month. (b) Sunspots cluster at high latitudes when solar activity is at a minimum. They appear at lower and lower latitudes as the number of sunspots peaks. They are again prominent near the Sun's equator as solar minimum is again approached.

▲ **FIGURE 9.19 Maunder Minimum** Peaks and troughs of the sunspot cycle over the past four centuries. Note the near total absence of spots during the late 17th century.

in a roughly periodic way, just as observed. Solar surface activity, such as the sunspot cycle, simply follows the changes in the underlying field.

Figure 9.19 plots all sunspot data recorded since the invention of the telescope. As can be seen, the 11-year "periodicity" of the solar sunspot cycle is far from exact. Not only does the period vary from 7 to 15 years, but also the sunspot cycle has disappeared entirely in the relatively recent past. The lengthy period of solar inactivity from 1645 to 1715 is called the *Maunder minimum*, after the British astronomer who drew attention to these historical records. (Interestingly, the period also seems to correspond fairly well to the coldest years of the so-called Little Ice Age that chilled northern Europe during the late 17th century, suggesting a link between solar activity and climate on Earth.) Absent a complete understanding of the solar cycle, the cause of the Sun's century-long variations remains a mystery.

In fact, the most recent sunspot minimum (in 2009) resulted in the least active Sun in almost a century—no sunspots at all were seen on almost 80 percent of the days during that year, and the solar wind was anomalously weak. The current cycle, which peaked in late 2013, has been the least active in the last 100 years. Scientists attribute this reduced activity to changes in the subsurface "conveyor belt" flow described in Section 9.2, which is thought to directly affect sunspot behavior. For unknown reasons, the flow sped up during the 1990s by a few meters per second, greatly suppressing the numbers of spots during the next cycle (and quite possibly the one after that, now underway). It has since slowed, but for now no one is sure how long the effects will last.

Active Regions

Sunspots are relatively gentle aspects of solar activity. However, the photosphere surrounding them occasionally erupts violently, spewing forth into the corona large quantities of energetic particles. The sites of these explosive events are known simply as **active regions.** Most pairs or groups of sunspots have active regions associated with them. Like all other aspects of solar activity, these phenomena tend to follow the solar cycle and are most frequent and violent around the time of sunspot maximum.

▲ FIGURE 9.20 Solar Prominences   (a) This particularly large solar prominence was observed by ultraviolet detectors aboard the *SOHO* spacecraft in 2002. (b) Like a phoenix rising from the solar surface, this filament of hot gas measures more than 100,000 km in length. Dark regions in this *TRACE* image have temperatures less than 20,000 K; the brightest regions are about 1 million kelvins. Most of the gas will subsequently cool and fall back into the photosphere. *(NASA)*

Figure 9.20 shows two solar **prominences**—loops or sheets of glowing gas ejected from an active region on the solar surface. Prominences move through the inner parts of the corona under the influence of the Sun's magnetic field. Magnetic instabilities in the strong fields found in and near sunspot groups may cause the prominences, although the details are still not completely understood. A typical solar prominence measures some 100,000 km in extent, nearly 10 times the diameter of planet Earth. Some may persist for days or even weeks. Prominences as large as the one shown in Figure 9.20(b), which spanned almost half a million kilometers of the solar surface (half the Sun's diameter), are less common and usually appear only at times of greatest solar activity.

Flares are another type of solar activity observed near active regions (Figure 9.21). Also the result of magnetic instabilities, flares are much more violent (and even less well understood) than prominences. They flash across a region of the Sun in minutes, releasing enormous amounts of energy as they go. Temperatures in the extremely compact hearts of flares can reach 100 million K (roughly six times hotter than the solar core). So energetic are these cataclysmic explosions that some researchers have likened them to bombs exploding in the lower regions of the Sun's atmosphere. Unlike the trapped gas that makes up the characteristic loop of a prominence, the particles produced by a flare are so energetic that the Sun's magnetic field is unable to hold them and shepherd them back to the surface. Instead, the particles are simply blasted into space by the violence of the explosion. Flares are thought to be responsible for the internal pressure waves that give rise to solar surface oscillations.

Sometimes (but not always) a flare or a prominence is associated with a **coronal mass ejection** (Figure 9.22a). This giant magnetic "bubble" of ionized gas separates from the rest of the solar atmosphere and escapes into interplanetary

(MA) ANIMATION/VIDEO
Solar Flare

▲ FIGURE 9.21 Solar Flare   Much more violent than a prominence, a solar flare is an explosion on the Sun's surface that sweeps across an active region in a matter of minutes, accelerating solar material to high speeds and blasting it into space. *(NASA)*

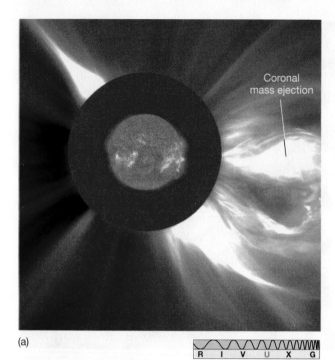

(a)

R I V U X G

▲ FIGURE 9.22 Coronal Mass Ejection   (a) A few times per week, on average, a giant magnetized "bubble" of solar material detaches itself from the Sun and rapidly escapes into space, as shown in this *SOHO* image taken in 2002. The circles are artifacts of an imaging system designed to block out the light from the Sun itself and exaggerate faint features at larger radii. (b) Should such a coronal mass ejection encounter Earth with its magnetic field oriented opposite to our own, as illustrated, the field lines can join together as in part (c), allowing high-energy particles to enter and possibly severely disrupt our planet's magnetosphere. By contrast, if the fields are in the same direction, the coronal mass ejection can slide by Earth with little effect. *(NASA/ESA)*

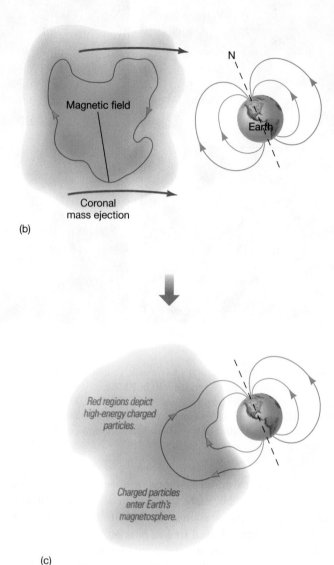

space. Such ejections occur about once per week at times of sunspot minimum, but up to two or three times per day at solar maximum. Carrying an enormous amount of energy, they can—if their fields are properly oriented—connect with Earth's magnetic field via a process known as *reconnection,* dumping some of their energy into the magnetosphere and potentially causing widespread communications and power disruptions on our planet (Figures 9.22b and c).

The Sun in X-Rays

Unlike the 5800 K photosphere, which emits most strongly in the visible part of the electromagnetic spectrum, the million-kelvin coronal gas radiates at much higher frequencies—primarily in X-rays. ∞ *(Sec. 2.4)* For this reason, X-ray telescopes have become important tools in studying the solar corona. Figure 9.23 shows a sequence of X-ray images of the Sun. The full corona extends well beyond the regions shown, but the density of coronal particles emitting the radiation diminishes rapidly with distance from the Sun. The intensity of X-ray radiation farther out is too dim to be seen here.

In the mid-1970s instruments aboard NASA's *Skylab* space station revealed that the solar wind escapes mostly through solar "windows" called **coronal holes.** The

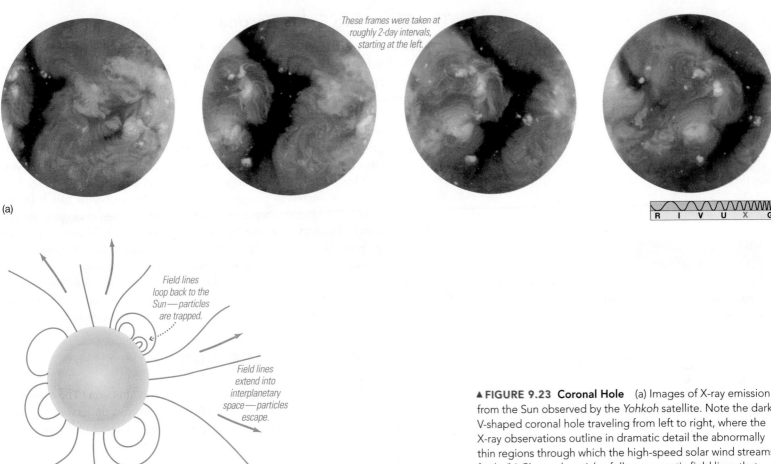

These frames were taken at roughly 2-day intervals, starting at the left.

(a)

R I V U X G

(b)

Field lines loop back to the Sun—particles are trapped.

Field lines extend into interplanetary space—particles escape.

▲ **FIGURE 9.23 Coronal Hole** (a) Images of X-ray emission from the Sun observed by the *Yohkoh* satellite. Note the dark, V-shaped coronal hole traveling from left to right, where the X-ray observations outline in dramatic detail the abnormally thin regions through which the high-speed solar wind streams forth. (b) Charged particles follow magnetic field lines that compete with gravity. When the field is trapped and loops back toward the photosphere, the particles are also trapped; otherwise, they can escape as part of the solar wind. *(ISAS/ Lockheed Martin)*

dark area moving from left to right in Figure 9.23(a), which shows more recent data from the Japanese *Yohkoh* X-ray solar observatory, is an example. Not really holes, such structures are simply deficient in matter—vast regions of the Sun's atmosphere where the density is about 10 times lower than the already tenuous, normal corona. The underlying solar photosphere looks black in these images because it is too cool to emit X-rays in any significant quantity. The largest coronal holes can be hundreds of thousands of kilometers across. Structures of this size are seen only a few times each decade. Smaller holes—perhaps only a few tens of thousands of kilometers in size—are much more common, appearing every few hours.

Coronal holes are lacking in matter because the gas there is able to stream freely into space at high speeds, driven by disturbances in the Sun's atmosphere and magnetic field. Figure 9.23(b) illustrates how, in coronal holes, the solar magnetic field lines extend from the surface far out into interplanetary space. Because charged particles tend to follow the field lines, they can escape, particularly from the Sun's polar regions, according to findings from NASA's *Ulysses* spacecraft, which flew high above the ecliptic plane to explore the Sun's polar regions. (Note, however, that the coronal hole shown in Figure 9.23a spans the Sun's equator.) The speed of the solar wind can reach 800 km/s over some coronal holes. In other regions of the corona, the solar magnetic field lines stay close to the Sun, keeping charged particles near the surface and inhibiting the outward flow of the solar wind (just as Earth's magnetic field tends to prevent the incoming solar wind from striking Earth), so the density remains relatively high. ∞ *(Sec. 5.7)*

ANIMATION/VIDEO
Coronal Mass Ejections

The Changing Solar Corona

The solar corona also varies in step with the sunspot cycle. The photograph of the corona in Figure 9.10 shows the Sun at sunspot minimum. The corona is fairly regular in appearance and surrounds the Sun more or less uniformly. Compare this image with Figure 9.12, which was taken in 1994, close to a peak of the sunspot cycle. The active corona here is much more irregular than the inactive one in Figure 9.10 and extends farther from the solar surface. The streamers of coronal material pointing away from the Sun are characteristic of this phase.

Astronomers think that the corona is heated primarily by solar surface activity, which can inject large amounts of energy into the upper solar atmosphere. *SOHO* observations have implicated a "magnetic carpet" as the source of much of the heating, where small magnetic loops continually sprout up, then disappear, dumping vast amounts of energy into the lower solar atmosphere. These small-scale magnetic loops probably provide most of the energy needed to heat the corona. In addition, more extensive disturbances often move through the corona above an active site in the photosphere, distributing the energy throughout the coronal gas. Given this connection, it is not surprising that both the appearance of the corona and the strength of the solar wind are closely correlated with the solar cycle.

CONCEPT CHECK

What do observations of sunspot polarities tell us about the solar magnetic field?

9.5 The Heart of the Sun

How does the Sun shine, day after day, year after year, eon after eon? The answer to this question is central to all of astronomy. Without it we can understand neither the evolution of stars and galaxies in the universe nor the existence of life on Earth.

Nuclear Fusion

Only one known energy-generation mechanism can conceivably account for the Sun's enormous energy output. That process is **nuclear fusion**—the combining of light nuclei into heavier ones. Fusion may be contrasted with nuclear *fission*, which powers nuclear reactors on Earth. In this process, a heavy nucleus (such as uranium or plutonium) splits into lighter nuclei, releasing energy in the process. We can represent a typical fusion reaction as

$$\text{nucleus 1} + \text{nucleus 2} \rightarrow \text{nucleus 3} + \text{energy}.$$

For powering the Sun and other stars, the most important piece of this equation is the energy produced.

The crucial point here is that, during a fusion reaction, the total mass *decreases*—the mass of nucleus 3 is less than the combined masses of nuclei 1 and 2. Where does the mass go? It is converted to energy according to Einstein's famous equation

$$E = mc^2,$$

or

$$\text{energy} = \text{mass} \times (\text{speed of light})^2.$$

This equation says that to determine the amount of energy corresponding to a given mass, simply multiply the mass by the square of the speed of light (c in the equation). The speed of light is so large that even a small amount of mass translates into an enormous amount of energy.

The production of energy by a nuclear fusion reaction is an example of the **law of conservation of mass and energy,** which states that the sum of mass and energy (properly converted to the same units, using Einstein's equation) must always remain constant in any physical process. There are no known exceptions. The fact that we see light coming from the Sun means that the Sun's mass must be slowly, but steadily, decreasing with time.

The Proton–Proton Chain

The lightest and most common element in the universe is hydrogen, and it is the fusion of hydrogen nuclei (protons) to form nuclei of helium, the next lightest element, that powers the Sun. But all atomic nuclei are positively charged, so they repel one another. Furthermore, by the inverse-square law, the closer two nuclei come to one another, the greater is the repulsive force between them (Figure 9.24a). ∞ *(Sec. 2.2)* How then do nuclei—two protons, say—ever manage to fuse into anything heavier? The answer is that if they collide at high enough speeds one proton can momentarily plow deep into the other, eventually coming within the exceedingly short range of the **strong nuclear force,** which binds nuclei together (see *More Precisely 9-1*). At distances less than about 10^{-15} m, the attraction of the nuclear force overwhelms the electromagnetic repulsion, and fusion can occur (Figure 9.24b). Speeds in excess of a few hundred kilometers per second are needed to slam protons together fast enough to initiate fusion. Such high speeds are associated with extremely high temperatures—at least 10 million kelvins. ∞ *(More Precisely 5-1)* Such conditions are found in the core of the Sun and at the centers of all stars. The radius at which the temperature drops below this value defines the outer edge of the solar core (Figure 9.6).

In the reaction shown in Figure 9.24(b), two protons combine to form a **deuteron**—a nucleus of *deuterium*, or "heavy hydrogen," containing a proton and an extra neutron. Energy is released in the form of two new particles: a *positron* and a *neutrino*. The **positron** is the positively charged *antiparticle* of an electron. Its properties are identical to those of a normal, negatively charged electron, except for the positive charge. Particles and antiparticles annihilate (destroy) one another when they meet, producing pure energy in the form of gamma-ray photons. The **neutrino** is a chargeless and virtually massless elementary particle. (The name derives from the Italian for "little neutral one.") Neutrinos move at nearly the speed of light and interact with hardly anything. They can penetrate, without stopping, several light-years of lead. Their interactions with matter are governed by the **weak nuclear force**, described in *More Precisely 9-1*. Yet despite their elusiveness, neutrinos can be detected with carefully constructed instruments. In the final section we discuss some rudimentary neutrino "telescopes" and the important contribution they have made to solar astronomy.

The basic set of nuclear reactions powering the Sun (and the vast majority of all stars) is sketched in Figure 9.25. In fact, it is not a single reaction, but rather a sequence of reactions called the **proton–proton chain:**

(I) Two protons combine, as shown in Figure 9.24(b), forming a deuteron and releasing a positron and a neutrino.

(II) The deuteron then combines with another proton to create a type of helium called helium-3, which contains only one neutron, while the positron annihilates with an electron, producing gamma rays. The neutrino escapes into space. Figure 9.25 shows two of each of these sets of reactions.

(III) Finally, two helium-3 nuclei combine to form helium-4 and two protons.

Setting aside the temporary intermediate nuclei produced, the net effect is that four hydrogen nuclei (protons) combine to create one nucleus of the next lightest element, helium-4 (containing 2 protons and 2 neutrons, for a total mass of 4), creating two neutrinos and releasing energy in the form of gamma-ray radiation: ∞ *(Sec. 2.6*

$$4 \text{ protons} \rightarrow \text{helium-4} + 2 \text{ neutrinos} + \text{energy}.$$

Other reaction sequences, involving the more massive nuclei of carbon, nitrogen, and oxygen, can produce the same end result—four protons fuse into helium-4, and energy is produced in the form of gamma rays. However, the sequence shown in Figure 9.25 is the simplest and is responsible for almost 90 percent of the Sun's

(a)

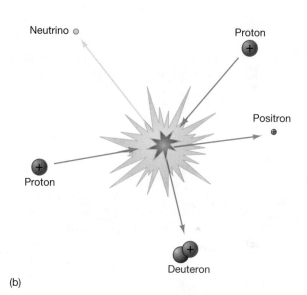

(b)

▲ **FIGURE 9.24 Proton Interactions** (a) Since like charges repel, two low-speed protons veer away from one another, never coming close enough for fusion to occur. (b) Higher-speed protons may succeed in overcoming their mutual repulsion, approaching close enough for the strong force to bind them together—in which case they collide violently, triggering nuclear fusion that ultimately powers the Sun.

9.1 MORE PRECISELY

The Strong and Weak Nuclear Forces

As far as we can tell, the behavior of all matter in the universe, from elementary particles to clusters of galaxies, is ruled by just four (or fewer) basic forces, which are fundamental to everything in the universe. We have already encountered two of them in this text—the *gravitational force*, which binds planets, stars, and solar systems together, and the much stronger *electromagnetic force*, which governs the properties of atoms and molecules, as well as magnetic phenomena on planets and stars. ∞ *(Secs. 1.4, 2.2)*

Both of these forces are governed by inverse-square laws and are long range. In contrast, the other two fundamental forces, known as the *weak nuclear force* and the *strong nuclear force* are of extremely short range—about 10^{-18} m and 10^{-15} m, respectively, less than the size of an atomic nucleus.

The weak force governs the emission of radiation during some radioactive decays, as well as interactions between neutrinos and other particles. In fact, according to current theory, the electromagnetic force and the weak force are not really distinct entities, but are actually just different aspects of a single "electroweak" force.

The strong force holds atomic nuclei together. It binds particles with enormous strength, but only when they approach one another very closely. Two protons must come within about 10^{-15} m of one another before the attractive strong force can overcome their electromagnetic repulsion.

Very loosely speaking, we can say that the strong force is about 100 times stronger than electromagnetism, 100,000 times stronger than the weak force, and 10^{39} times stronger than gravity. But not all particles are subject to all types of force. All particles interact through gravity because all have mass (or equivalently, energy). However, only charged particles interact electromagnetically. Protons and neutrons are affected by the strong force, but electrons are not. Under the right circumstances, the weak force can affect any type of subatomic particle, regardless of its charge.

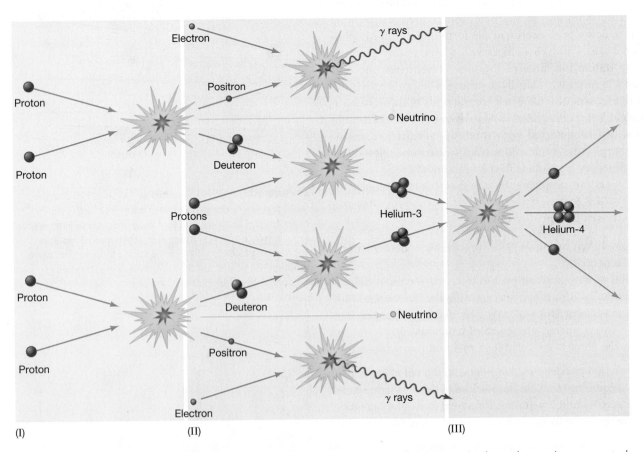

▲ FIGURE 9.25 Solar Fusion INTERACTIVE   In the proton–proton chain, a total of six protons (and two electrons) are converted to two protons, one helium-4 nucleus, and two neutrinos. The two leftover protons are available as fuel for new proton–proton reactions, so the net effect is that four protons are fused to form one helium-4 nucleus. Energy, in the form of gamma rays, is produced at each stage. The three stages indicated here correspond to reactions (I), (II), and (III) described in the text.

luminosity. Notice how, at each stage of the chain, more complex nuclei are created from simpler ones. We will see later that not all stars are powered by hydrogen fusion; nevertheless, this basic idea of simpler elements combining to form more complex ones, releasing energy in the process, continues to apply.

Gargantuan quantities of protons are fused into helium by the proton–proton chain in the core of the Sun every second. As discussed in more detail in *More Precisely 9-2*, to fuel the Sun's present energy output, hydrogen must be fused into helium in the core at a rate of 600 million tons per second—a lot of mass, but only a tiny fraction of the total amount available. As we will see in Chapter 12, the Sun will be able to sustain this rate of core burning for about another 5 billion years.

The Sun's nuclear energy is produced in the core in the form of gamma rays. However, as the energy passes through the cooler overlying layers of the solar interior and photons are repeatedly absorbed and reemitted, the radiation's blackbody spectrum steadily shifts toward lower and lower temperatures and, by Wien's law, the characteristic wavelength of the radiation increases. ∞ *(Sec. 2.4)* The energy eventually leaves the solar photosphere mainly in the form of visible and infrared radiation. A comparable amount of energy is carried off by the neutrinos, which escape unhindered into space at almost the speed of light.

CONCEPT CHECK

Why does the fact that we see sunlight imply that the Sun's mass is slowly decreasing?

9.2 MORE PRECISELY

Energy Generation in the Proton–Proton Chain

Let's look in a little more detail at the energy produced by fusion in the solar core and compare it with the energy needed to account for the Sun's luminosity. As discussed in the text and illustrated below, the net effect of the fusion process is that four protons combine to produce a nucleus of helium-4, in the process creating two neutrinos and two positrons (which are quickly converted into energy by annihilation with electrons).

We can calculate the total amount of energy released by accounting carefully for the total masses of the nuclei involved and applying Einstein's famous formula $E = mc^2$. This allows us to relate the Sun's total luminosity to the consumption of hydrogen fuel in the core. Careful laboratory experiments have determined the masses of all the particles involved in the above reaction: The total mass of the protons is 6.6943×10^{-27} kg, the mass of the helium-4 nucleus is 6.6466×10^{-27} kg, and the neutrinos are virtually massless. We omit the positrons here—their masses will end up being counted as part of the total energy released. The difference between the total mass of the four protons and that of the final helium-4 nucleus, 0.0477×10^{-27} kg, is not great—only about 0.71 percent of the original mass—but it is easily measurable.

Multiplying the vanished mass by the square of the speed of light yields 0.0477×10^{-27} kg $\times (3.00 \times 10^8$ m/s$)^2 = 4.28 \times 10^{-12}$ J. This is the energy produced in the form of radiation when 6.69×10^{-27} kg (the rounded-off mass of the four protons) of hydrogen fuses to helium. It follows that fusion of 1 kg of hydrogen generates $4.28 \times 10^{-12}/6.69 \times 10^{-27} = 6.40 \times 10^{14}$ J.

Thus we have established a direct connection between the Sun's energy output and the consumption of hydrogen in the core. The Sun's luminosity of 3.86×10^{26} W (see Table 9.1), or 3.86×10^{26} J/s (joules per second), implies a mass consumption rate of 3.86×10^{26} J/s $\div 6.40 \times 10^{14}$ J/kg $= 6.03 \times 10^{11}$ kg/s— roughly 600 million tons of hydrogen every second. That sounds like a lot—the mass of a small mountain and 600 times the mass loss rate in the solar wind—but it represents only a few million million millionths of the total mass of the Sun. Our parent star will be able to sustain this loss for a very long time—this is the subject of Chapter 12.

Observations of Solar Neutrinos

Because the gamma-ray energy created in the proton–proton chain is transformed into visible and infrared radiation by the time it emerges from the Sun, astronomers have no direct electromagnetic evidence of the core nuclear reactions. Instead, the *neutrinos* created in the proton–proton chain are our best bet for learning about conditions in the solar core. They travel cleanly out of the Sun, interacting with virtually nothing, and escape into space a few seconds after being created.

Of course, the fact that they can pass through the entire Sun also makes neutrinos very difficult to detect on Earth! Nevertheless, with knowledge of neutrino physics it is possible to construct neutrino detectors. During the last four decades of the 20th century, several experiments (two of them shown in Figure 9.26) were constructed to detect solar neutrinos reaching Earth's surface. The experiments' designs differed greatly, and they were sensitive to neutrinos of very different energies, but they all agreed on one very important point: The number of solar neutrinos reaching Earth is substantially less (by 50 to 70 percent) than the prediction of the standard solar model. This discrepancy became known as the **solar neutrino problem.**

Theorists argued that it was very unlikely that the resolution of the solar neutrino problem would lie in the physics of the Sun's interior. The nuclear reactions just described are too well understood, and the agreement between the standard solar model and helioseismological observations (Section 9.2) is far too close for conditions in the core to deviate much from the standard solar model. But the experimental results were clear. How could we explain this stark disagreement between theory and observation? As it turned out, the answer involved the properties of the neutrinos themselves and caused scientists to rethink some very fundamental concepts in particle physics.

At the time, it was thought that neutrinos were massless. But if neutrinos have even a minute amount of mass, theory indicates that they can change their properties—even transform into other particles—during their 8-minute flight from the solar core to Earth, through a process known as **neutrino oscillations**. In that case, neutrinos could be produced in the Sun at the rate required by the standard solar model, but some could then turn into something else—actually, other types of neutrinos—on their way to Earth, and hence go undetected in the experiments just described. (In the jargon of the field, the neutrinos are said to "oscillate" into other particles.)

In 1998 the Japanese group operating the Super Kamiokande detector shown in Figure 9.26(a) reported the first experimental evidence of neutrino oscillations (and hence of nonzero neutrino masses), although the observed oscillations did not involve neutrinos of the type produced in the Sun. In 2001, measurements made at the Sudbury Neutrino Observatory (SNO) in Ontario, Canada (Figure 9.26b) observed the "other" neutrinos into which the Sun's neutrinos had transformed. The total numbers of neutrinos observed were completely consistent with the standard solar model. The solar neutrino problem was solved—and neutrino astronomy claimed its first major triumph!

> **PROCESS OF SCIENCE CHECK**
>
> Using the solar neutrino problem as an example, discuss how scientific theory and observation evolve and adapt when they come into conflict.

(a)

(b)

◄ **FIGURE 9.26 Neutrino Telescope**   (a) The Super Kamiokande neutrino detector is buried beneath a mountain near Tokyo, Japan. It is filled with 50,000 tons of purified water and contains 13,000 individual light detectors to sense the telltale signature—a brief burst of light—of a neutrino passing through the apparatus. (b) The Sudbury Neutrino Observatory, some 2 km underground in Ontario, Canada, is similar in design to the Kamiokande facility, but is also sensitive to other neutrino types. It contains 10,000 light-sensitive detectors arranged on the inside of the large sphere shown here. *(ICRR; SNO)*

THE BIG QUESTION

How does the Sun's corona get so hot? What causes the 11-year solar cycle of activity? Why do sunspots exist, and why do they look so messy? There is not just one big question regarding the Sun, but many smaller, nagging ones that have perplexed scientists for decades. Although we understand the basic physics of how our star shines, we still have much to learn about its unpredictable behavior, which can sometimes affect life here on Earth.

CHAPTER REVIEW

SUMMARY

LO1 Our Sun is a **star** (p. 254), a glowing ball of gas held together by its own gravity and powered by nuclear fusion at its center. The **photosphere** (p. 254) is the region at the Sun's surface from which virtually all the visible light is emitted. The main interior regions of the Sun are the **core** (p. 254), where nuclear reactions generate energy; the **radiation zone** (p. 254), where the energy travels outward in the form of electromagnetic radiation; and the **convection zone** (p. 254), where the Sun's matter is in constant convective motion.

LO2 The amount of solar energy reaching a square meter area at the top of Earth's atmosphere each second is the **solar constant** (p. 255). The Sun's **luminosity** (p. 255) is the total amount of energy radiated from the solar surface per second.

LO3 Much of our knowledge of the solar interior comes from mathematical models. The model that best fits the observed properties of the Sun is the **standard solar model** (p. 255). **Helioseismology** (p. 256)—the study of vibrations of the solar surface caused by pressure waves in the interior—provides further insight into the Sun's structure. The effect of the solar convection zone can be seen on the surface in the form of **granulation** (p. 259) of the photosphere. Lower levels in the convection zone also leave their mark on the photosphere in the form of larger transient patterns called **supergranulation** (p. 259).

LO4 Above the photosphere lies the **chromosphere** (p. 254), the Sun's lower atmosphere. In the **transition zone** (p. 254) above the chromosphere, the temperature increases from a few thousand kelvins to around a million kelvins. Above the transition zone is the Sun's thin, hot upper atmosphere, the solar **corona** (p. 256). At a distance of about 15 solar radii, the gas in the corona is hot enough to escape the Sun's gravity, and the corona begins to flow outward as the **solar wind** (p. 254).

LO5 **Sunspots** (p. 262) are Earth-sized regions of intense magnetism on the solar surface that are a little cooler than the surrounding photosphere. The numbers and locations of sunspots vary in a roughly 11-year **sunspot cycle** (p. 265) as the Sun's magnetic field rises and falls. The overall direction of the field reverses from one sunspot cycle to the next. The 22-year cycle that results when the direction of the field is taken into account is called the **solar cycle** (p. 265). Solar activity tends to be concentrated in **active regions** (p. 266) associated with sunspot groups; this activity heats the corona and drives the solar wind, which flows along open field lines from low-density regions called **coronal holes** (p. 268).

LO6 The Sun generates energy by converting hydrogen to helium in its core by the process of **nuclear fusion** (p. 270). When four protons are converted to a helium nucleus in the **proton–proton chain** (p. 271), some mass is lost. The **law of conservation of mass and energy** (p. 270) requires that this mass appear as energy, eventually resulting in the light we see.

LO7 Neutrinos (p. 274) are nearly massless particles that are produced in the proton–proton chain and escape from the Sun. It is possible to detect a small fraction of the neutrinos streaming from the Sun. Observations over several

decades led to the **solar neutrino problem** (p. 274)—substantially fewer neutrinos are observed than were predicted by theory. The explanation, supported by observations, is that **neutrino oscillations** (p. 274) convert some neutrinos to other (undetected) particles en route from the Sun to Earth.

MasteringAstronomy®   For instructor-assigned homework go to www.masteringastronomy.com

Problems labeled **POS** explore the process of science. **VIS** problems focus on reading and interpreting visual information. **LO** connects to the introduction's numbered Learning Outcomes.

REVIEW AND DISCUSSION

1. **LO1** Name and briefly describe the main regions of the Sun.

2. How massive is the Sun, compared with Earth?

3. How hot is the solar surface? The solar core?

4. **LO2** How is the Sun's luminosity measured?

5. **LO3 POS** How is helioseismology used to model the Sun?

6. Describe how energy from the solar core eventually reaches Earth.

7. Why does the Sun appear to have a sharp edge?

8. What evidence do we have for solar convection?

9. **LO4** What is the solar wind, and how is it related to the Sun's outer atmosphere?

10. **LO5** What is the cause of sunspots, flares, and prominences?

11. What fuels the Sun's enormous energy output?

12. **POS** What is the law of conservation of mass and energy? How is it relevant to nuclear fusion in the Sun?

13. **LO6** What are the ingredients and the end result of the proton–proton chain? Why is energy released in the process?

14. **LO7 POS** Why are observations of solar neutrinos important to both astronomers and physicists?

15. What would we observe on Earth if the Sun's internal energy suddenly shut off? How long do you think it would take for the Sun's light to begin to fade? Answer the same question for solar neutrinos.

CONCEPTUAL SELF-TEST: TRUE OR FALSE?/MULTIPLE CHOICE

1. The most abundant element in the Sun is hydrogen. (T/F)

2. There are as many absorption lines in the solar spectrum as there are elements present in the Sun. (T/F)

3. The density and temperature in the solar corona are much higher than in the photosphere. (T/F)

4. Sunspots appear dark because they are hotter than the surrounding gas of the photosphere. (T/F)

5. Flares are caused by magnetic disturbances in the lower atmosphere of the Sun. (T/F)

6. The proton–proton chain releases energy because mass is created in the process. (T/F)

7. Neutrinos have never been detected experimentally. (T/F)

8. The Sun is roughly (a) the same size; (b) 10 times larger; (c) 100 times larger; (d) 1 million times larger than Earth.

9. The Sun spins on its axis roughly once each (a) hour; (b) day; (c) month; (d) year.

10. Astronomers on Venus would measure a solar constant (a) larger than; (b) smaller than; (c) the same as on Earth.

11. A typical solar granule is about the size of (a) a U.S. city; (b) a large U.S. state; (c) the Moon; (d) Earth.

12. The time between successive sunspot maxima is about (a) a month; (b) a year; (c) a decade; (d) a century.

13. Solar energy is produced by (a) fusion of light nuclei into heavier ones; (b) fission of heavy nuclei into lighter ones; (c) the release of heat left over from the Sun's formation; (d) solar magnetism.

14. **VIS** In the standard solar model (Figure 9.6, Solar Interior), as the distance from the center increases, (a) the density and temperature decrease at about the same rate; (b) the density decreases faster than the temperature; (c) the density decreases more slowly than the temperature; (d) the density decreases but the temperature increases.

15. **VIS** Comparing Figures 9.6 (Solar Interior) and 9.13 (Solar Atmospheric Temperature), the temperature in the solar corona is about the same as in (a) the core; (b) the bottom of the radiation zone; (c) the bottom of the convection zone; (d) the photosphere.

PROBLEMS

The number of squares preceding each problem indicates its approximate level of difficulty.

1. ■ Use the reasoning presented in Section 9.1 to calculate the value of the "solar constant" (a) on Mercury, at perihelion, and (b) on Jupiter.

2. ■■ The largest-amplitude solar pressure waves have periods of about 5 minutes and move at about 10 km/s. (a) How far does the wave move during one wave period? (b) Roughly how many wavelengths are needed to circle the Sun? (c) Compare the wave period with the orbital period in the solar photosphere.

3. ■ Given that the solar spectrum corresponds to a temperature of 5800 K and peaks at a wavelength of 500 nm, use Wien's law to determine the wavelength corresponding to the peak of the black-body curve (a) in the core of the Sun, where the temperature is 10^7 K, (b) in the solar convection zone (10^5 K), and (c) just below the solar photosphere (10^4 K). ∞ *(More Precisely 2-2)* What form (visible, infrared, X-ray, etc.) does the radiation take in each case?

4. ■ If convected solar material moves at 1 km/s, how long does it take to cross the 1000-km expanse of a typical granule? Compare this time with the 10-minute lifetimes observed for most solar granules.

5. ■■ Use Stefan's law to calculate how much less energy is emitted per unit area of a 4500-K sunspot than is emitted per unit area of the surrounding 5800-K photosphere. ∞ *(More Precisely 2-2)*

6. ■■ The solar wind carries mass away from the Sun at a rate of about 2 million ton/s (where 1 ton = 1000 kg). (a) Compare this rate with the rate at which the Sun loses mass in the form of radiation. (b) At this rate, how long would it take for all of the Sun's mass to escape?

7. ■■ The Sun's differential rotation is responsible for wrapping the solar magnetic field around the Sun (Figure 9.17). Using the figures presented in Table 9.1, estimate how long it takes for material at the solar equator to "lap" the material near the poles—that is, to complete one extra trip around the Sun's rotation axis.

8. ■■ How long does it take the Sun to convert one Earth mass of hydrogen to helium?

9. ■■ Assuming that (1) the solar luminosity has been constant since the Sun formed, and (2) the Sun was initially of uniform composition throughout, as described by Table 9.2, estimate how long it would take the Sun to convert all of its original hydrogen into helium.

10. ■■■ The entire reaction sequence shown in Figure 9.25 generates 4.3×10^{-12} joules of electromagnetic energy and releases two neutrinos. If all of the Sun's energy comes from this sequence and neutrino oscillations transform half of the neutrinos produced into other particles by the time they reach Earth, estimate the total number of solar neutrinos passing through Earth each second.

ACTIVITIES

Never Look Directly At The Sun Without A Filter!

Collaborative

The safest way to observe the Sun is to project its image onto a screen. Here are two ways to do it.

1. To build a "pinhole camera," you'll need two sheets of stiff white paper and a pin. Use the pin to punch a hole in the center of one of your pieces of paper. Go outside, hold the paper up and aim the hole at the Sun. ***Don't look directly at the Sun!*** Now find the image of the Sun coming through the hole. Move the other piece of paper back and forth until the image looks best. What happens to the image as you vary the size of the pinhole?

2. Alternatively, you can project an image of the Sun using a pair of binoculars or a small telescope. You'll need a sheet of stiff white paper and a tripod. Mount the binoculars or telescope firmly on the tripod and point it at the Sun. ***Don't attempt to view the Sun directly!*** Hold the paper about 10 inches behind the eyepiece to act as a screen. You should see a bright blurred circle of light. Focus the instrument until the circle is sharp. This is the disk of the Sun. Experiment with moving the card closer and farther away from the telescope. What effect does this have on the image?

Use one of these projectors to study some sunspots, as in Project 1 below. Have half your group construct a pinhole camera and the other half build a telescopic projector, and compare results. What are the pros and cons of each?

Individual

1. A filtered telescope will easily show you sunspots. Count the number of sunspots you see on the Sun's surface. Notice that sunspots often come in pairs or groups. Look again a few days later and you'll see that the Sun's rotation has caused the spots to move, and the spots themselves have changed. If a sufficiently large sunspot (or sunspot group) is seen, continue to watch it as the Sun rotates. It will be out of sight for about 2 weeks.

2. View some solar prominences and flares. Hydrogen-alpha (Hα) filters are available for small telescopes. You can often see prominences and flares even during times of sunspot minimum. You are actually viewing the chromosphere rather than the photosphere, so the Sun looks quite different from normal.

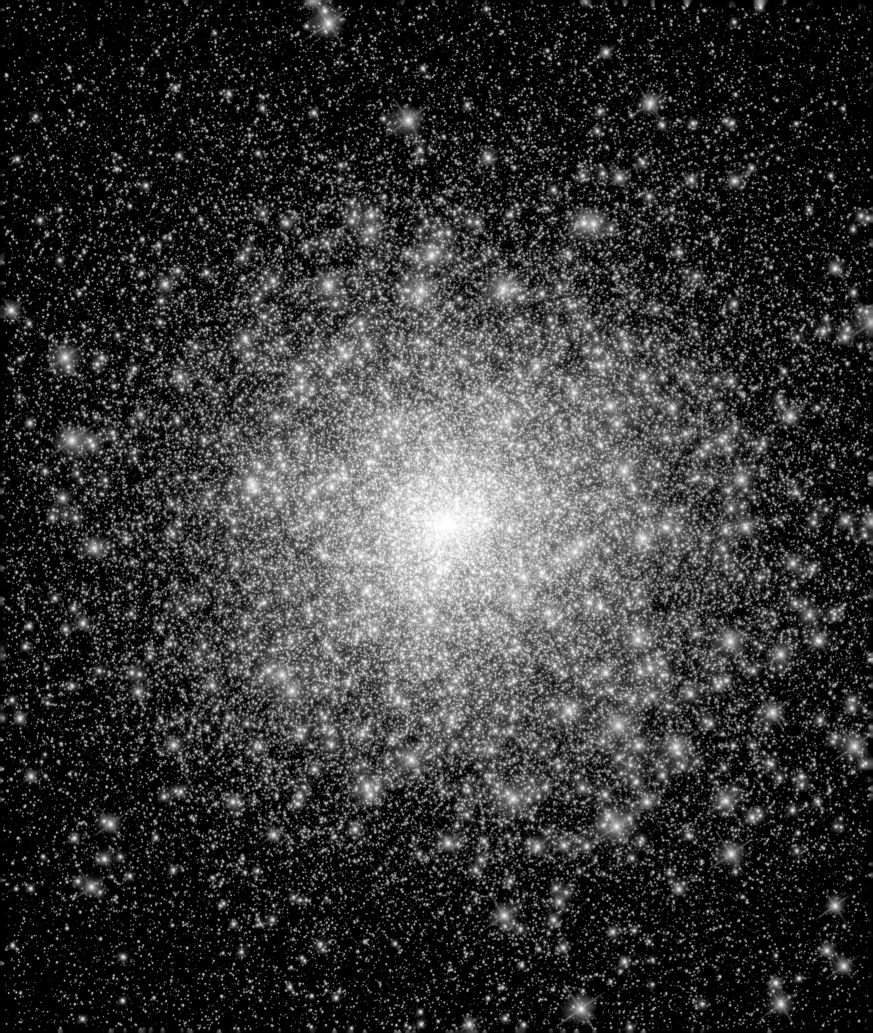

10

Measuring the Stars

Giants, Dwarfs, and the Main Sequence

We have now studied Earth, the Moon, the solar system, and the Sun. To continue our inventory of the contents of the universe, we must move away from our local environment into the depths of space. In this chapter we take a great leap in distance and consider stars in general. Our primary goal is to comprehend the nature of the stars that make up the constellations, as well as the myriad more distant stars we cannot perceive with unaided eyes. Rather than studying their individual peculiarities, however, we will concentrate on determining the physical and chemical properties they share. There is order in the legions of stars scattered across the sky. Like comparative planetology in the solar system, comparing and cataloging the stars play vital roles in furthering our understanding of the galaxy and the universe we inhabit.

LEARNING OUTCOMES

Studying this chapter will enable you to:

LO1 Explain how stellar distances are determined.

LO2 Distinguish between luminosity and apparent brightness, and explain how stellar luminosity is determined.

LO3 Describe how stars are classified according to their colors, surface temperatures, and spectral characteristics.

LO4 Explain how physical laws can be used to estimate stellar sizes.

LO5 Describe how a Hertzsprung–Russell diagram is constructed and used to identify stellar properties.

LO6 Explain how knowledge of a star's spectroscopic properties can lead to an estimate of its distance.

LO7 Explain how stellar masses are measured and how mass is related to other stellar properties.

THE BIG PICTURE

Stars are everywhere in the night sky. The naked eye can distinguish about 6000 of them, spread across 88 constellations. With binoculars or even a small telescope, millions more become visible. The total number of stars, even in our local cosmic neighborhood, is virtually beyond our ability to count. By analyzing the light from millions of distant stars, astronomers have learned a great deal about stellar properties—their masses and radii, their luminosities, even their ages and destinies. Stars tell us more about the fundamentals of astronomy than any other class of objects in the universe.

◄ In 2013 the *Hubble Space Telescope* captured this spectacular image of the globular star cluster M15, a rich group of roughly 1 million stars spread across some 100 light-years of space. It lies 33,000 light-years away in the direction of the constellation Pegasus, and is estimated to be 12 billion years old—more than twice the age of the Sun. Its central regions represent one of the densest stellar environments known in our Galaxy, with 100,000 stars packed into a volume a few light-years across—the distance from our Sun to its nearest neighbor in space. *(ESA/NASA)*

MasteringAstronomy®

Visit www.masteringastronomy.com for quizzes, animations, videos, interactive figures, and self-guided tutorials.

10.1 The Solar Neighborhood

Looking deep into space, astronomers can observe and study millions of individual stars. By observing distant galaxies, we can infer statistically the properties of trillions more. All told, the observable universe probably contains several tens of sextillions (1 sextillion = 10^{21}) stars. Yet, despite their incredible numbers, the essential properties of stars—their appearance in the sky, their births, lives, deaths, even their interactions with their environment—can be understood in terms of just a few basic physical stellar quantities: luminosity (brightness), temperature (color), chemical composition, size, and mass. In this chapter we examine these basic stellar properties, describe how we measure them, and show how they lead to a simple but powerful means of classifying stars of all kinds. With this foundation we can then discuss the life cycles of stars in the next three chapters.

As with the planets, knowing the distances to the stars is essential to understanding many of their other properties. We therefore begin our study of stellar astronomy by reviewing the use of simple geometry in determining the distances to our neighbors.

Stellar Parallax

The distances to the nearest stars can be measured using *parallax*—the apparent shift of an object relative to some distant background as the observer's point of view changes. ∞ *(Sec. 0.4)* However, even the closest stars are so far away that no baseline on Earth is adequate to measure their parallaxes. But, by comparing observations made of a star *at different times of the year*, as shown in Figure 10.1 (compare Figure 0.20), we can effectively extend the baseline to the diameter of Earth's orbit around the Sun, 2 AU. The parallax is determined by comparing photographs made from the two ends of the baseline. As indicated on the figure, a star's parallactic angle—or, more commonly, just its "parallax"—is conventionally defined to be *half* of its apparent shift relative to the background as we move from one side of Earth's orbit to the other.

Because the stars are so far away, stellar parallaxes are always very small, and astronomers find it convenient to measure them in arc seconds rather than in degrees. ∞ *(More Precisely 0-1)* If we ask at what distance from the Sun a star must lie in order for its observed parallax to be exactly 1″, we get an answer of 206,265 AU, or 3.1×10^{16} m. Astronomers call this distance one **parsec** (1 pc), from "*par*allax in arc *sec*onds." Because parallax decreases as distance increases, we can relate a star's parallax to its distance from the Sun by the following simple formula:

$$\text{distance (in parsecs)} = \frac{1}{\text{parallax (in arc seconds)}}.$$

Thus, a star with a measured parallax of 1″ lies at a distance of 1 pc from the Sun. The parsec is defined so as to make the conversion between parallactic angle and distance easy. An object with a parallax of 0.5″ lies at a distance of 1/0.5 = 2 pc, an object with a parallax of 0.1″ lies at 1/0.1 = 10 pc, and so on. One parsec is approximately equal to 3.3 light-years.

Background stars

Star

Parallactic angle

Earth's orbit

January   1 AU   1 AU   July

Sun

Baseline

(a)

This is the view as seen in January . . .   . . . and in July, when the star shifts.

(b)

◄ FIGURE 10.1 **Stellar Parallax INTERACTIVE**   (a) For observations made 6 months apart, the baseline is twice the Earth–Sun distance, or 2 AU. (b) The parallactic shift (exaggerated here, and indicated by the white arrow) is usually measured photographically, as illustrated by the red star. Images of the same region of the sky made at different times of the year determine a star's apparent movement relative to background stars.

Our Nearest Neighbors

The closest star to the Sun is called Proxima Centauri. It is part of a triple-star system (three stars orbiting one another, bound by gravity) known as the Alpha Centauri complex. Proxima Centauri has the largest known stellar parallax, 0.77″, meaning that it is about 1/0.77 = 1.3 pc away—about 270,000 AU, or 4.3 light-years. That's the *nearest* star to the Sun—at almost 300,000 times the Earth–Sun distance! This is a fairly typical interstellar distance in the Milky Way Galaxy.

Vast distances can sometimes be grasped by means of analogies. Imagine Earth as a grain of sand orbiting a golfball-sized Sun at a distance of 1 m. The nearest star, also a golfball-sized object, is then 270 *kilo*meters away. Except for the planets in our solar system, ranging in size from grains of sand to small marbles within 50 m of the "Sun," and the Kuiper belt and Oort cloud, trillions of tiny dust grains scattered over a volume 100 km across, nothing of consequence exists in the 270 km separating the Sun and the other star.

The next nearest neighbor to the Sun beyond the Alpha Centauri system is called Barnard's Star. Its parallax is 0.55″, so it lies at a distance of 1.8 pc, or 6.0 light-years—370 km in our model. All told, fewer than 100 stars lie within 5 pc (1000 km in our model) of the Sun. Such is the void of interstellar space. Figure 10.2 is a map of our nearest galactic neighbors—the 30 or so stars lying within 4 pc of Earth. *Discovery 10-1* (p. 285) describes how some of the stars in this figure got their peculiar names.

Ground-based images of stars are generally smeared out into a seeing disk of radius 1″ or so by turbulence in Earth's atmosphere. ∞ *(Sec. 3.3)* However, astronomers use special techniques to routinely measure stellar parallaxes of 0.03″ or less, corresponding to stars within about 30 pc (100 light-years) of Earth. Several thousand stars lie within this range, most of them of much lower luminosity than the Sun and invisible to the naked eye. Adaptive optics systems allow more accurate measurements of stellar positions, extending the parallax range to over 100 pc. ∞ *(Sec. 3.3)* In the 1990s, the European *Hipparcos* satellite expanded this range to well over 200 pc, encompassing nearly a million stars. Even so, almost all of the stars in our Galaxy are far more distant. Upcoming next-generation space missions will extend out to 25,000 pc, spanning our entire Galaxy and likely revolutionizing our knowledge of its structure.

Stellar Motion

In addition to the apparent motion caused by parallax, stars have real spatial motion, too. A star's *radial* velocity—along the line of sight—can be measured using the Doppler effect. ∞ *(Sec. 2.7)* For many nearby stars the *transverse* velocity—perpendicular to our line of sight—can also be determined by careful monitoring of the star's position on the sky.

Figure 10.3(a) shows two photographs of the sky around Barnard's star, taken on the same day of the year, but 22 years apart. Note that the star (marked by the arrow) has moved during the 22-year interval shown. Because Earth was at the same point in its orbit when the photographs were taken, the observed displacement is *not* the result of parallax. Instead, it indicates *real* space motion of Barnard's star relative to the Sun. This annual

CONCEPT CHECK

Why can't astronomers use simultaneous observations from different parts of Earth's surface to determine stellar distances?

▼ FIGURE 10.2 **Sun's Neighborhood**   A plot of the 30 closest stars to the Sun, projected so as to reveal their three-dimensional relationships. All lie within 4 pc (about 13 light-years) of Earth. The gridlines represent distances in the Galactic plane, and the vertical lines denote distances perpendicular to that plane.

(a)

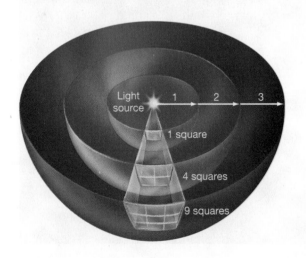

(b)

movement of a star across the sky, as seen from Earth and corrected for parallax, is called **proper motion.** Barnard's star moved 228″ in 22 years. Its proper motion is therefore 228″/22 years, or 10.3″/yr.

A star's transverse velocity is easily calculated once its proper motion and its distance are known. At the distance of Barnard's star (1.8 pc), an angle of 10.3″ corresponds to a distance of 0.00009 pc, or about 2.8 billion kilometers. Barnard's star takes a year to travel this distance, so its transverse velocity is 2.8×10^9 km 3.2×10^7 s, or 88 km/s. Even though stars' transverse velocities are often quite large—tens or even hundreds of kilometers per second—their great distances from the Sun mean that it usually takes many years for us to discern their movement across the sky. In fact, Barnard's star has the largest-known proper motion of any star. Only a few hundred stars have proper motions greater than 1″/yr.

A star's total space velocity is the combination of its radial and transverse components. Figure 10.3(b) illustrates the relationship among these quantities for another near neighbor, Alpha Centauri. Spectral lines from Alpha Centauri are blueshifted by a tiny amount—about 0.0067 percent—allowing astronomers to measure the star system's radial velocity (relative to the Sun) as 300,000 km/s × 6.7 $\times 10^{-5}$ = 20 km/s toward us. ∞ *(Sec. 2.7)* Alpha Centauri's proper motion has been measured to be 3.7″/yr. At Alpha Centauri's distance of 1.35 pc (corresponding to a parallax of 0.74″), that measurement implies a transverse velocity of 24 km/s. We can combine the transverse (24 km/s) and radial (20 km/s) velocities according to the Pythagorean theorem, as indicated in the figure. The total velocity is $\sqrt{24^2 + 20^2}$ or about 31 km/s, in the direction shown by the horizontal red arrow.

10.2 Luminosity and Apparent Brightness

Luminosity—the amount of radiation leaving a star per unit time—is an *intrinsic* stellar property. ∞ *(Sec 9.1)* It does not depend in any way on the location or motion of the observer. It is sometimes referred to as the star's *absolute brightness.* However, when we observe a star, we see not its luminosity but rather its **apparent brightness**—the amount of energy striking unit area of some light-sensitive surface or device, such as a human eye or a CCD chip, per unit time. (Notice that this is precisely the same definition as given in Chapter 9 for the solar constant—in our new terminology, the solar constant is just the apparent brightness of the Sun.) In this section we discuss how these important quantities are related to one another.

Another Inverse-Square Law

Figure 10.4 shows light leaving a star and traveling through space. Moving outward, the radiation passes through imaginary spheres of increasing radius surrounding the source. The farther the light travels from the source, the less energy passes through each unit of area. Think of the energy as being spread out over an ever larger area and therefore spread more thinly, or "diluted," as it expands into space. Because the area of a sphere grows as the square of the radius, the energy per unit area—the star's apparent brightness—is inversely proportional to the square of the distance from the star. Doubling the distance from a star makes it appear 2^2, or 4, times fainter. Tripling the distance reduces the apparent brightness by a factor of 3^2, or 9, and so on.

Of course, the star's luminosity also affects its apparent brightness. Doubling the luminosity doubles the energy crossing any spherical shell surrounding the star and hence doubles the apparent brightness. The apparent brightness of a star is therefore directly proportional to the star's luminosity and inversely proportional to the square of its distance:

$$\text{apparent brightness} \propto \frac{\text{luminosity}}{\text{distance}^2}.$$

(Reminder: $\propto$ is the symbol representing proportionality.) Thus, two identical stars can have the same apparent brightness if (and only if) they lie at the same distance from Earth. However, as illustrated in Figure 10.5, two nonidentical stars can also have the same apparent brightness if the more luminous one lies farther away. A bright star, having large apparent brightness, is a powerful emitter of radiation (high luminosity), is near Earth, or both. A faint star (small apparent brightness) is a weak emitter (low luminosity), is far from Earth, or both.

This is the view seen by the observer looking out into space.

▲ FIGURE 10.5 Luminosity Two stars A and B of different luminosities can appear equally bright to an observer on Earth if the brighter star B is more distant than the fainter star A.

10.1 DISCOVERY

Naming the Stars

Looking at Figure 10.2, you might well wonder how those stars got their names. In fact, there is no coherent convention for naming stars—or rather, there are many, often leading to many names for any given object! Here is a brief (and roughly chronological) description of some common naming schemes:

1. The brightest stars in the sky usually have ancient names. Many of the names are of Arabic origin, dating back to around the 10th century A.D., when Muslim astronomy flourished, preserving the scientific knowledge of Greek and other early cultures while European science floundered in the Dark Ages. Examples are Betelgeuse (*Yad al Jauzah*, Hand of the Central One, later corrupted to Bed Elgueze), Rigel (*Rijl Jauzah al Yusra*, Left Foot of the Central One), Aldebaran (*Al Dabaran*, the Follower), and Deneb (*Al Dhanab al Dulfin*, the Dolphin's Tail). The names Procyon and Sirius (see Figure 10.2) are Greek.

2. A more systematic scheme was introduced in 1603 by German lawyer Johann Bayer. He ranked the stars in a given constellation (Orion, say) by brightness, labeling them with Greek letters: Alpha Orionis (Betelgeuse), Beta Orionis (Rigel), or Epsilon Eridani (Figure 10.2). ∞ *(Sec. 0.1)*

3. There are only 24 Greek letters, so eventually the Bayer scheme broke down. In the early 18th century, British Astronomer Royal John Flamsteed suggested simply numbering the stars from west to east within a constellation. In

Flamsteeds scheme, Betelgeuse is 58 Ori, Rigel is 19 Ori, and so on (see also 61 Cygni in Figure 10.2). Sometimes the Bayer and Flamsteed designations are combined, as in "58 α Ori."

4. As telescopes improved and more and more stars were discovered, astronomers started compiling their own catalogs. Some catalogs listed stars by celestial coordinates—for example, star BD + 5°1668 in Figure 10.2, which appears in the 19th-century German Bonner Durchmusterung (BD) catalog and lies just over 5° above the celestial equator. ∞ *(Sec. 0.1)* Other catalogs simply named stars by the order in which they were added to the list—the star Lalande 21185 in Figure 10.2 is the 21,185th entry in a catalog compiled by 18th-century French astronomer Joseph de Lalande. Unfortunately, the catalogs often overlap, so stars can appear in many different catalogs with very different names. For example, Lalande 21185 is also known as HD 95735 (in the Henry Draper catalog), while Betelgeuse is also known as HD 39801 and SAO 113271 (in the Smithsonian Astrophysical Observatory).

5. Today, newly discovered stars are defined simply by their coordinates in the sky and not given any special name. If you like, you can (for a fee, of course) register your own star name, but don't expect any astronomer ever to cite that name, or to take the registry seriously—only "official" names approved by *International Astronomical Union* are used in astronomy.

DATA POINTS

Luminosity and Apparent Brightness

Forty-six percent of students had difficulty putting the inverse-square law into practice when discussing the apparent brightnesses of stars. Some points to remember:

- The apparent brightness of a star is directly proportional to its luminosity (which measures its true energy emission). Doubling the luminosity doubles the apparent brightness.

- The apparent brightness of a star is inversely proportional to the square of its distance. Doubling the distance decreases the apparent brightness by a factor of four.

- If two stars appear equally bright, the more luminous one is the more distant.

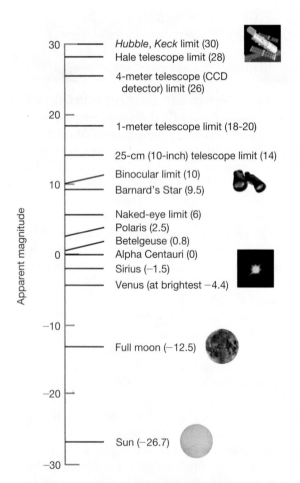

▲ **FIGURE 10.6 Apparent Magnitude** This graph illustrates the apparent magnitudes of some astronomical objects and the limiting magnitudes (that is, the faintest magnitudes attainable) of some telescopes used to observe them.

Determining a star's luminosity is a twofold task. First, the astronomer must determine the star's apparent brightness by measuring the amount of energy detected through a telescope in a given amount of time. Second, the star's distance must be measured—by parallax for nearby stars and by other means (to be discussed later) for more distant stars. The luminosity can then be found using the inverse-square law. This is the same reasoning used earlier in our discussion of how astronomers measure the luminosity of the Sun. ∞ *(Sec. 9.1)*

The Magnitude Scale

Instead of measuring apparent brightness in SI units (for example, watts per square meter, W/m^2, the unit in which we expressed the solar constant in Section 9.1), optical astronomers find it more convenient to work in terms of a construct called the **magnitude scale**—a system of ranking stars by their apparent brightness. This scale dates back to the second century B.C., when the Greek astronomer Hipparchus ranked the naked-eye stars into six groups. He categorized the brightest stars as first magnitude, labeled the next brightest stars as second magnitude, and so on, down to the faintest stars visible to the naked eye, which he classified as sixth magnitude. The range 1 (brightest) through 6 (faintest) spanned all the stars known to the ancients. Notice that a *large* magnitude means a *faint* star.

When astronomers began using telescopes with sophisticated detectors to measure the light received from stars, they quickly discovered two important facts about the magnitude scale. First, the 1 through 6 magnitude range defined by Hipparchus spans about a factor of 100 in apparent brightness—a first-magnitude star is approximately 100 times brighter than a sixth-magnitude star. Second, the characteristics of the human eye are such that a change of one magnitude corresponds to a factor of about 2.5 in apparent brightness. In other words, to the human eye a first-magnitude star is roughly 2.5 times brighter than a second-magnitude star, which in turn is roughly 2.5 times brighter than a third-magnitude star, and so on. By combining factors of 2.5, we confirm that a first-magnitude star is $(2.5)^5$, or roughly 100 times brighter than a sixth-magnitude star.

Because we are really talking about apparent (rather than absolute) brightnesses, astronomers refer to the numbers in Hipparchus's ranking system as **apparent magnitudes.** In the modern version of the magnitude scale, we *define* a change of 5 in the magnitude of an object to correspond to *exactly* a factor of 100 in apparent brightness. In addition, the scale is no longer limited to whole numbers and has expanded to include magnitudes outside the original range of 1 to 6. Very bright objects can have apparent magnitudes less than 1, and very faint objects can have apparent magnitudes much greater than 6. Figure 10.6 illustrates the apparent magnitudes of some astronomical objects, ranging from the Sun at −26.8 to the faintest object detectable by the *Hubble* or *Keck* telescopes, at an apparent magnitude of +30—about as faint as a firefly seen from a distance equal to Earth's diameter.

Apparent magnitude measures a star's apparent brightness when seen at the star's actual distance from us. To compare intrinsic, or absolute, properties of stars, however, astronomers imagine looking at all stars from a standard distance of 10 pc. (There is no particular reason to use 10 pc—it is simply convenient.) A star's **absolute magnitude** is its apparent magnitude when viewed from a distance of 10 pc. Because distance is fixed in this definition, absolute magnitude is a measure of a star's absolute brightness, or luminosity. Despite the Sun's large negative (that is, very bright) apparent magnitude, our star's absolute magnitude is 4.8. In other words, if the Sun were moved to a distance of 10 pc from Earth, it would be only a little brighter than the faintest stars visible in the night sky. As discussed further in *More Precisely 10-1*, the numerical difference between a star's absolute and apparent magnitudes is a measure of the distance to the star.

10.1 MORE PRECISELY

More on the Magnitude Scale

Let's restate our discussion of two important topics—stellar luminosity and the inverse-square law—in terms of magnitudes.

Absolute magnitude is equivalent to luminosity—an intrinsic property of a star. Given that the Sun's absolute magnitude is 4.8, we can construct a conversion chart (shown at right) relating these two quantities. Since an increase in brightness by a factor of 100 corresponds to a decrease in magnitude by 5 units, it follows that a star with luminosity 100 times that of the Sun has absolute magnitude $4.8 - 5 = -0.2$, while a 0.01-solar luminosity star has absolute magnitude $4.8 + 5 = 9.8$. We can fill in the gaps by noting that 1 magnitude corresponds to a factor of $100^{1/5} \approx 2.512$, 2 magnitudes to $100^{2/5} \approx 6.310$, and so on. A factor of 10 in brightness corresponds to 2.5 magnitudes. You can use this chart to convert between solar luminosities and absolute magnitudes in many of the figures in this and later chapters.

To cast the *inverse-square law* in these terms, recall that increasing the distance to a star by a factor of 10 decreases its apparent brightness by a factor of 100 (by the inverse-square law) and hence increases its apparent magnitude by 5 units. Increasing the distance by a factor of 100 increases the apparent magnitude by 10, and so on. Every increase in distance by a factor of 10 increases the apparent magnitude by 5. Since absolute magnitude is simply apparent magnitude at a distance of 10 pc, we can write:

apparent magnitude − absolute magnitude

$$= 5 \log_{10}\left(\frac{\text{distance}}{10 \text{ pc}}\right).$$

(The *logarithm* function—the LOG key on your calculator—is defined by the property that if $a = \log_{10}(b)$, then $b = 10a$.) Even though it doesn't look much like it, this equation is precisely equivalent to the inverse-square law presented in the text!

For stars more than 10 pc from Earth, the apparent magnitude is greater than the absolute magnitude, while the reverse is true for stars closer than 10 pc. Thus, for example, the Sun, with an absolute magnitude of 4.8, seen from a distance of 100 pc would have an apparent magnitude of $4.8 + 5 \log_{10}(100) = 14.8$, since $\log_{10}(100) = 2$. This is well below the threshold of visibility for binoculars or even a large amateur telescope (see Figure 10.6). We can also turn this around to illustrate how knowledge of a star's absolute and apparent magnitudes tells us its distance. The star Rigel Kentaurus (also known as Alpha Centauri) has absolute magnitude + 4.34 and is observed to have apparent magnitude −0.01. Its magnitude difference is −4.35, and its distance must therefore be 10 pc $\times$ $10^{-4.34/5} = 1.35$ pc, in agreement with the result (obtained by parallax) given in the text.

10.3 Stellar Temperatures

Color and the Blackbody Curve

Looking at the night sky, you can tell at a glance which stars are hot and which are cool. In Figure 10.7(a), which shows the constellation Orion as it appears through a small telescope, the colors of the cool red star Betelgeuse (α) and the hot blue star Rigel (β) are clearly evident. ∞ *(Sec. 0.2)* Astronomers can determine a star's surface temperature by measuring its apparent brightness (radiation intensity) at several frequencies, then matching the observations to the appropriate blackbody curve. ∞ *(Sec. 2.4)* In the case of the Sun, the theoretical curve that best fits the emission describes a 5800-K emitter. The same technique works for any star, regardless of its distance from Earth.

Because the basic shape of the blackbody curve is so well understood, astronomers can estimate a star's temperature using as few as *two* measurements at selected wavelengths. This is accomplished by using telescope filters that block out all radiation except that within specific wavelength ranges. For example, a B

CONCEPT CHECK

Two stars are observed to have the same apparent magnitude. Based on this information, what, if anything, can be said about their luminosities?

(blue) filter rejects all radiation except for a certain range of violet to blue light. Similarly, a V (visual) filter passes only radiation in the green to yellow range (the part of the spectrum to which human eyes happen to be particularly sensitive).

Figure 10.8 shows how these filters admit different amounts of light for objects of different temperatures. In curve (a), corresponding to a very hot 30,000-K emitter, considerably more radiation is received through the B filter than through the V filter. The temperature of the emitter in curve (b) is 10,000 K, and the B and V intensities are about the same. The cool 3000-K emitter represented in curve (c) produces far more energy in the V range than in the B range. In each case it is possible to reconstruct the entire blackbody curve on the basis of only those two measurements because no other blackbody curve can be drawn through both measured points.

To the extent that a star's spectrum is well approximated as a blackbody, measurements of the B and V intensities are enough to specify the star's blackbody curve and thus yield its surface temperature. Astronomers often refer to the ratio of a star's B to V intensities (or, equivalently, the difference between the star's apparent magnitudes measured through the B and V filters) as the star's *color index* (or just its "color"). Table 10.1 lists surface temperatures and colors for a few well-known stars.

Stellar Spectra

Color is a useful way to describe a star, but astronomers often use a more detailed scheme to classify stellar properties, incorporating additional knowledge of stellar physics obtained through spectroscopy. ∞ *(Sec. 2.5)* Figure 10.9 compares the spectra of seven stars, arranged in order of decreasing surface temperature. All the spectra extend from 400 to 650 nm. Like the solar spectrum, each shows a series of dark absorption lines superimposed on a background of continuous color. ∞ *(Sec. 9.3)* However, the line patterns differ greatly from one star to another. Some stars display strong (prominent) lines in the long-wavelength part of the spectrum, while others have their strongest lines at short wavelengths. Still others show strong absorption lines spread across the whole visible spectrum. What do these differences tell us?

Although spectral lines of many elements are present at widely varying strengths, the differences among the spectra in Figure 10.9 are not due to differences in chemical composition—all seven stars are, in fact, more or less solar in their makeup. Instead, the differences are due almost entirely to the stars' temperatures. The spectrum at the top of Figure 10.9 is exactly what we expect from a star having solar composition and a surface temperature of about 30,000 K, the second from a 20,000-K star, and so on, down to the 3000-K star at the bottom.

The main differences among the spectra in Figure 10.9 are as follows:

- Spectra of stars having surface temperatures exceeding 25,000 K usually show *strong* absorption lines of singly ionized helium (i.e., helium atoms that have lost one orbiting electron) and multiply ionized heavier elements, such as oxygen, nitrogen, and silicon (the latter lines are not shown in the figure). These strong lines are not seen in the spectra of cooler stars because only very hot stars can excite and ionize such tightly bound atoms.
- In contrast, the hydrogen absorption lines in the spectra of very hot stars are relatively *weak*. The reason is not a lack of hydrogen, which is by far the most abundant element in all stars. Rather, at these high temperatures, much of the hydrogen is ionized, so there are few intact hydrogen atoms to produce strong spectral lines.
- Hydrogen lines are strongest in stars having intermediate surface temperatures of around 10,000 K. This temperature is just right for electrons to move frequently between hydrogen's second and higher orbitals, producing the characteristic visible hydrogen spectrum. ∞ *(Sec. 2.6)*

(a)

(b) R I V U X G

▲ FIGURE 10.7 Star Colors (a) The different colors of the stars composing the constellation Orion are easily distinguished in this photograph. The bright red star at the upper left (α) is Betelgeuse; the bright blue-white star at the lower right (β) is Rigel. The scale of this photograph is about 20° across. (b) An incredibly rich field of colorful stars, this time in the direction of the center of the Milky Way. Here, the field of view is just 2 arc minutes across—much smaller than in (a). (P. Sanz/Alamy; NASA)

TABLE 10.1	Stellar Colors and Temperatures	
Surface Temperature (K)	Color	Familiar Examples
30,000	electric blue	Mintaka (δ Orionis)
20,000	blue	Rigel
10,000	white	Vega, Sirius
7,000	yellow-white	Canopus
6,000	yellow	Sun, Alpha Centauri
4,000	orange	Arcturus, Aldebaran
3,000	red	Betelgeuse, Barnard's Star

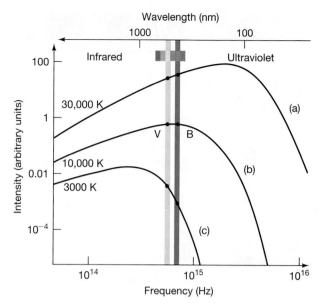

▲ FIGURE10.8 **Blackbody Curves Star** (a) is very hot—30,000 K—so its B (blue) intensity is greater than its V (visual) intensity. Star (b) has roughly equal B and V readings and so appears white. Its temperature is about 10,000 K. Star (c) is red; its V intensity greatly exceeds the B value, and its temperature is 3000 K.

■ Hydrogen lines are again weak in stars with surface temperatures below about 4000 K because the temperature is too low to boost many electrons out of the ground state. ∞ *(Sec. 2.6)* The most intense spectral lines in these stars are due to weakly excited heavy atoms and from some molecules, as noted at the bottom of Figure 10.9.

Stellar spectra are the source of all the detailed information we have on stellar composition, and they do reveal significant composition differences among stars, particularly in the abundances of carbon, nitrogen, oxygen, and heavier elements. However, as we have just seen, these differences are *not* the primary reason for the different spectra observed. Instead, the main determinant of a star's spectral appearance is its temperature, and stellar spectroscopy is a powerful and precise tool for measuring this important stellar property.

Spectral Classification

Astronomers obtained stellar spectra like those shown in Figure 10.9 for numerous stars around the start of the 20th century, as observatories around the world amassed stellar data from both hemispheres of the sky. Between 1880 and 1920, researchers correctly identified many of the observed spectral lines by comparing them with lines obtained in the laboratory. However, modern atomic theory had not yet been developed, so the correct interpretation of the line strengths, as just described, was impossible at the time. Lacking full understanding of how atoms produce spectra, astronomers organized their data by classifying stars according to their most prominent spectral features—especially their hydrogen-line intensities. They adopted an alphabetic scheme in which A stars, with the strongest hydrogen lines, were thought to have more hydrogen than did B stars, and so on. The classification extended as far as the letter P.

In the 1920s scientists began to understand the intricacies of atomic structure and the causes of spectral lines. Astronomers realized that stars could be more meaningfully classified according to surface temperature. But instead of adopting an entirely new scheme, they chose to shuffle the existing alphabetical categories—based on the strength of hydrogen lines—into a new sequence based on temperature. In order of decreasing temperature, the letters now run O, B, A, F, G, K, M. (The other letter classes have been dropped.) These stellar designations are called **spectral classes** (or *spectral types*). Use the mnemonic "**O**h, **B**e **A** **F**ine **G**uy/**G**irl, **K**iss **M**e" to remember them in the correct order.

Astronomers further divide each lettered spectral class into 10 subdivisions, numbered 0 to 9. The lower the number, the hotter the star. For example, our Sun is classified as a G2 star (a little cooler than G1 and a little hotter than G3), Vega is A0, Barnard's star is M5, Betelgeuse is M2, and so on. Table 10.2 lists the main properties of each stellar spectral class for the stars presented in Table 10.1.

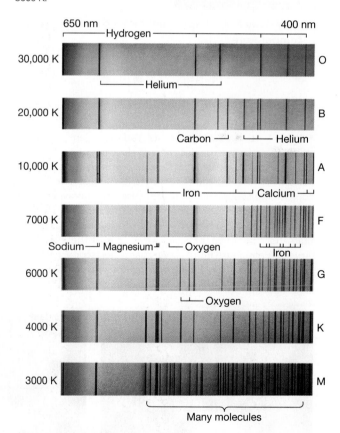

▲ FIGURE10.9 **Stellar Spectra** Comparison of spectra observed for seven different stars having a range of surface temperatures. The spectra of the hottest stars, at the top, show lines of helium and multiply ionized heavy elements. In the coolest stars, at the bottom, helium lines are absent, but lines of neutral atoms and molecules are plentiful. At intermediate temperatures, hydrogen lines are strongest.

TABLE 10.2	Spectral Classes			
Spectral Class	**Temperature (K)**	**Prominent Absorption Lines**		**Familiar Examples**
O	30,000	Ionized helium strong; multiply ionized heavy elements; hydrogen faint		Mintaka (O9)
B	20,000	Neutral helium moderate; singly ionized heavy elements; hydrogen moderate		Rigel (B8)
A	10,000	Neutral helium very faint; singly ionized heavy elements; hydrogen strong		Vega (A0), Sirius (A1)
F	7000	Singly ionized heavy elements; neutral metals; hydrogen moderate		Canopus (F0)
G	6000	Singly ionized heavy elements; neutral metals; hydrogen relatively faint		Sun (G2), Alpha Centauri (G2)
K	4000	Singly ionized heavy elements; neutral metals strong; hydrogen faint		Arcturus (K2), Aldebaran (K5)
M	3000	Neutral atoms strong; molecules moderate; hydrogen very faint		Betelgeuse (M2)
				Barnard's Star (M5)

PROCESS OF SCIENCE CHECK

The spectral classification scheme developed by astronomers around the start of the 20th century was based on assumptions now known to be incorrect. Why then did it prove to be so useful?

CONCEPT CHECK

Why does a star's spectrum depend on its temperature?

We should not underestimate the importance of the early work in classifying stellar spectra. Even though the original classification was based on erroneous assumptions, the painstaking accumulation of large quantities of accurate and well-categorized data paved the way for rapid improvements in understanding once a theory came along that explained the observations.

10.4 Stellar Sizes

Direct and Indirect Measurements

Virtually all stars are unresolvable points of light in the sky, even when viewed through the largest telescopes. Still, a few are big enough, bright enough, and close enough to allow us to measure their sizes directly. In some cases, the results are even detailed enough to allow us to distinguish a few surface features (Figure 10.10). By measuring a star's angular size and knowing its distance from the Sun, astronomers can determine its radius using simple geometry. ∞ *(More Precisely 4-1)* The sizes of a few dozen stars have been measured in this way.

Most stars are too distant or too small for such direct measurements to be possible. Their sizes must be inferred by more indirect means, using the radiation laws. ∞ *(Sec. 2.4)* According to the Stefan-Boltzmann law, the rate at which a star emits energy into space—the star's luminosity—is proportional to the fourth power of the star's surface temperature. ∞ *(More Precisely 2-2)* However, the luminosity also depends on the star's surface area, because large bodies radiate more energy than do small bodies having the same surface temperature. Because the surface area of a star is proportional to the square of its radius, we can say

$$\text{luminosity} \propto \text{radius}^2 \times \text{temperature}^4.$$

▼ **FIGURE 10.10 Betelgeuse** (a) The supergiant star Betelgeuse is large enough and close enough for astronomers to resolve its size directly. Betelgeuse is roughly 600 times the size of the Sun, making its photosphere comparable to Jupiter's orbit. This false-color ultraviolet view was taken by the *Hubble Space Telescope*. Hints of surface features, thought to be storms similar to (but much larger than) those found on the Sun can be seen. (b) These infrared views of Betelgeuse from the Very Large Telescope in Chile show the swollen star puffing out huge plumes of gas into its surroundings. *(NASA/ESA/ESO)*

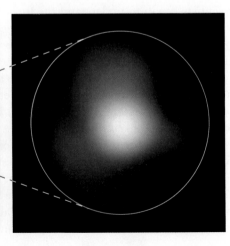

(a) 2 AU

(b) 100 AU

▶ **FIGURE 10.11 Stellar Sizes**   Shown here are the greatly varying sizes of several well-known stars. Only part of the red-giant star Antares can be shown on this scale, and the supergiant Betelgeuse would fill the entire page. (Here and in other figures, the symbol "⊙" stands for the Sun, so the symbol "$R_\odot$" here means "solar radius.")

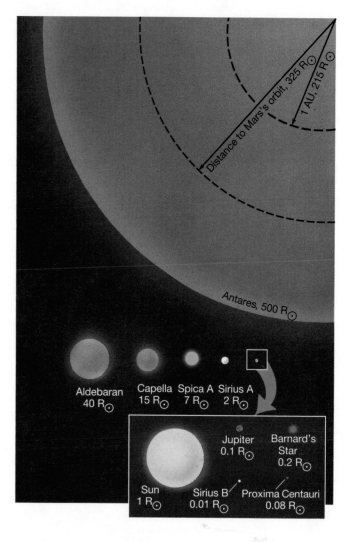

This **radius–luminosity–temperature relationship** is important because it demonstrates that knowledge of a star's luminosity and temperature can yield an estimate of its radius—an *indirect* determination of stellar size.

Giants and Dwarfs

Let's consider some examples to clarify these ideas. *More Precisely 10-2* (p. 294) presents more detail on the calculations involved. The star known as Aldebaran (the orange-red "eye of the bull" in the constellation Taurus) has a surface temperature of about 4000 K and a luminosity of 1.3×10^{29} W. Thus, its surface temperature is 0.7 times, and its luminosity about 330 times, the corresponding quantities for our Sun. The radius–luminosity–temperature relationship then implies that the star's radius is almost 40 times the solar value. If our Sun were that large, its photosphere would extend halfway to the orbit of Mercury and, seen from Earth, would cover more than 20° on the sky.

A star as large as Aldebaran is known as a *giant*. More precisely, **giants** are stars having radii between 10 and 100 times that of the Sun. Since any 4000-K object is reddish in color, Aldebaran is known as a **red giant.** Even larger stars, ranging up to 1000 solar radii in size, are known as **supergiants.** Betelgeuse is a prime example of a **red supergiant.**

Now consider Sirius B—a faint companion to Sirius A, the brightest star in the night sky. Sirius B's surface temperature is roughly 24,000 K, four times that of the Sun. Its luminosity is 10^{25} W, about 0.025 times the solar value. The radius–luminosity–temperature relationship then gives us a radius of 0.01 times the solar radius—roughly the size of Earth. Sirius B is much hotter but smaller and fainter than our Sun. Such a star is known as a *dwarf*. In astronomical parlance, the term **dwarf** refers to any star of radius comparable to or smaller than the radius of the Sun (including the Sun itself). Because any 24,000 K object glows bluish-white, Sirius B is an example of a **white dwarf.**

Stellar radii range from less than 0.01 to more than 100 times the radius of the Sun. Figure 10.11 illustrates the sizes of a few well-known stars.

CONCEPT CHECK

Can we measure the radius of a star without knowing its distance?

10.5 The Hertzsprung–Russell Diagram

Figure 10.12 plots luminosity versus temperature for a few well-known stars. A figure of this sort is called a *Hertzsprung–Russell diagram*, or **H–R diagram,** after Danish astronomer Ejnar Hertzsprung and U.S. astronomer Henry Norris Russell, who independently pioneered the use of such plots in the 1920s. The vertical luminosity scale, expressed in units of solar luminosity (3.9×10^{26} W), extends over a large range, from 10^{-4} to 10^4. The Sun appears right in the middle of the luminosity range, at a luminosity of 1. Surface temperature is plotted on the horizontal axis, although in the unconventional sense of temperature increasing to the *left*, so that the spectral sequence O, B, A, . . . reads left to right.

 SELF-GUIDED TUTORIAL
Hertzsprung–Russell Diagram

Spectral classification

◄ **FIGURE 10.12 H–R Diagram of Well-Known Stars** A plot of luminosity against surface temperature (or spectral class) is a useful way to compare stars. Plotted here are the data for some stars mentioned earlier in the text, including the Sun, which has a luminosity of 1 solar unit and a temperature of 5800 K—a G-type star. The B-type star Rigel, at top left, has a temperature of about 11,000 K and a luminosity more than 10,000 times that of the Sun. The M-type star Proxima Centauri, at bottom right, has a temperature of about 3000 K and a luminosity less than 1/10,000 that of the Sun.

The Main Sequence

The few stars plotted in Figure 10.12 give little indication of any particular connection between stellar properties. However, as Hertzsprung and Russell plotted more and more stellar temperatures and luminosities, they found that a relationship does in fact exist. Stars are *not* uniformly scattered across the H–R diagram. Instead, most are confined to a fairly well-defined band stretching diagonally from top left (high-temperature, high-luminosity) to bottom right (low-temperature, low-luminosity). In other words, cool stars tend to be faint (less luminous) and hot stars tend to be bright (more luminous). This band of stars spanning the H–R diagram is known as the **main sequence.** Figure 10.13 is an H–R diagram for stars lying within 5 pc of the Sun. Note that most stars in the solar neighborhood lie on the main sequence.

A useful analogy involves human beings. Our height and weight are also correlated along a "main sequence." That is, as shown near Figure 10.13, the majority of the human population falls within a diagonal area (bounded by the shaded area). This is true since tall people usually weigh more than shorter ones. Some exceptions—including dwarfs and basketball players—lie outside this well-defined range.

The surface temperatures of main-sequence stars range from about 3000 K (spectral class M) to more than 30,000 K (spectral class O). This temperature range is relatively small—only a factor of 10. In contrast, the observed range in luminosities is very large, covering eight orders of magnitude (that is, a factor of 100 million), from 10^{-4} to 10^4 the luminosity of the Sun.

Using the radius–luminosity–temperature relationship (Section 10.4), astronomers find that stellar radii also vary along the main sequence. The faint, red M-type stars in the bottom right of the H–R diagram are only about 1/10 the size of the Sun, whereas the bright, blue O-type stars in the upper left are about 10 times larger than the Sun. The dashed lines in Figure 10.13 represent constant stellar radius, meaning that any star lying on a given line has the same radius, regardless of its luminosity or temperature. Along a constant-radius line, the radius–luminosity–temperature relationship implies

$$\text{luminosity} \propto \text{temperature}^4.$$

By including such lines on our H–R diagrams, we can indicate stellar temperatures, luminosities, and radii on a single plot.

At the top end of the main sequence, the stars are large, hot, and very luminous. Because of their size and color, they are referred to as **blue giants.** The very largest are called **blue supergiants.** At the other end, stars are small, cool, and faint. They are known as **red dwarfs.** Our Sun lies right in the middle.

Figure 10.14 shows an H–R diagram for a different group of stars—the 100 stars of known distance having the greatest apparent brightness, as seen from

Analogous to an H–R diagram, people's height and weight fall along a "main sequence."

◄ **FIGURE 10.13 H–R Diagram of Nearby Stars INTERACTIVE** Most stars have properties within the long, thin shaded region of the H–R diagram known as the main sequence. The points plotted here are for stars lying within about 5 pc of the Sun. Each dashed diagonal line corresponds to a constant stellar radius. (Recall that the symbol "$R_\odot$" means "solar radius.")

Earth. Notice that here there are many more stars at the upper end of the main sequence than at the lower end. The reason for this excess of blue giants is simple—we can see very luminous stars a long way off. The stars shown in Figure 10.14 are scattered through a much greater volume of space than those shown in Figure 10.13, and the Figure 10.14 sample is heavily biased toward the most luminous objects. In fact, of the 20 brightest stars in the sky, only 6 lie within 10 pc of us. The rest are visible, despite their great distances, because of their high luminosities.

If very luminous blue giants are overrepresented in Figure 10.14, low-luminosity red dwarfs are surely underrepresented. In fact, no dwarfs appear on this diagram. This absence is not surprising, because low-luminosity stars are difficult to observe from Earth. In the 1970s, astronomers began to realize that they had greatly underestimated the number of red dwarfs in our Galaxy. As hinted at by the H–R diagram in Figure 10.13, which shows an unbiased sample of stars in the solar neighborhood, red dwarfs are actually the most common type of star in the sky. They probably account for upward of 80 percent of all stars in the universe. In contrast, O- and B-type supergiants are extremely rare—only about one star in 10,000 falls into this category.

The White-Dwarf and Red-Giant Regions

Some of the points plotted in Figures 10.12–10.14 clearly do not lie on the main sequence. One such point in Figure 10.12 represents Sirius B, the white dwarf we met in Section 10.4 (see *More Precisely 10-2*). Its surface temperature (24,000 K) is about four times greater than that of the Sun, but its luminosity is hundreds of times less than the solar value. A few more such faint bluish-white stars can be seen in the bottom left-hand corner of Figure 10.13, in the **white-dwarf region**.

Also shown in Figure 10.12 is Aldebaran, with a surface temperature (4000 K) about two-thirds that of the Sun, but with a luminosity more than 300 times greater than the Sun's. Another point represents Betelgeuse, the ninth brightest star in the sky, a little cooler than Aldebaran, but nearly 100 times more luminous. The upper right-hand corner of the H–R diagram (marked on Figure 10.14), where these stars are found, is called the **red-giant region.** No red giants are found within 5 pc of the Sun (Figure 10.13), but many of the brightest stars seen in the sky are in fact red giants (Figure 10.14). Red giants are relatively rare, but they are so bright that they are visible at very great distances. They form a third distinct class of stars in the H–R diagram, very different in their properties from both main-sequence stars and white dwarfs.

The *Hipparcos* mission (Section 10.1), in addition to determining hundreds of thousands of stellar parallaxes with unprecedented accuracy, also measured the colors and luminosities of more than 2 million stars. Figure 10.15 shows an H–R diagram based on a tiny portion of the enormous *Hipparcos* dataset. The main sequence and red giant regions are clearly evident. Few white dwarfs appear, however, because the telescope was limited to observations of relatively bright objects—brighter than apparent magnitude 12. Almost no white dwarfs lie close enough to Earth that their magnitudes fall below this limit.

About 90 percent of all stars in our solar neighborhood, and presumably a similar percentage elsewhere in the universe, are main-sequence stars. About 9 percent of stars are white dwarfs, and 1 percent are red giants.

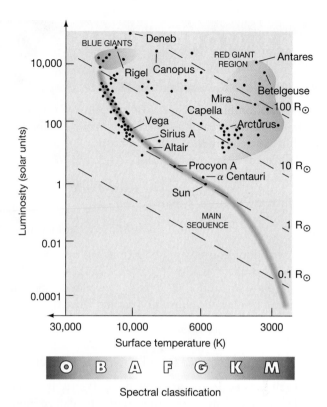

▲ **FIGURE 10.14 H–R Diagram of Brightest Stars**
INTERACTIVE An H–R diagram for the 100 brightest stars in the sky is biased in favor of the most luminous stars—which appear toward the upper left—because we can see them more easily than we can the faintest stars. (Compare with Figure 10.13, which shows only the closest stars.)

▶ **FIGURE10.15** *Hipparcos* **H–R Diagram** This simplified version of the most complete H–R diagram ever compiled represents more than 20,000 data points, as measured by the European *Hipparcos* spacecraft for stars within a few hundred parsecs of the Sun.

10.2 MORE PRECISELY

Estimating Stellar Radii

We can combine the Stefan-Boltzmann law $F = \sigma T^4$ with the formula for the area of a sphere, $A = 4\pi R^2$, to obtain the relationship between radius (R), luminosity (L), and temperature (T) described in the text ⚬⚬ *(More Precisely 2-2):*

$$L = FA = 4\pi\sigma R^2 T^4$$

or

$$\text{luminosity} \propto \text{radius}^2 \times \text{temperature}^4.$$

If we adopt convenient solar units, in which L is measured in solar luminosities (3.9×10^{26} W), R is measured in solar radii (696,000 km), and T in units of the solar temperature (5800 K), we can remove the constant $4\pi\sigma$ and write this equation as

L(in solar luminosities)

$$= R^2 \text{ (in solar radii)} \times T^4 \text{(in units of 5800 }K).$$

As illustrated in the accompanying figure, both the radius and the temperature are important in determining the star's luminosity.

To compute the radius of a star from its luminosity and temperature, we rearrange this equation to read (in the same units)

$$R = \frac{\sqrt{L}}{T^2}.$$

This simple application of the radiation laws is the basis for almost every estimate of stellar size made in this book.

Let's take as examples the two stars discussed in the text. The red giant star Aldebaran has luminosity $L = 330$ units and temperature $T = 4000/5800 = 0.69$ unit. According to the foregoing equation for R, its radius is therefore $\sqrt{330}/0.69^2 = 18/0.48 = 39$ times the radius of the Sun. At the opposite extreme, the white dwarf Sirius B has $L = 0.025$ and $T = 4.7$, so its radius is $\sqrt{10.025}/4.7^2 = 0.16/22 \approx 0.007$ is solar radii.

Blue giant	Red giant	Red dwarf
$10^4 \, L_\odot$	$80 \, L_\odot$	$0.05 \, L_\odot$
$20 \, R_\odot$	$20 \, R_\odot$	$0.5 \, R_\odot$
13,000 K	4000 K	4000 K

Only a tiny fraction of all stars are giants. Why, then, do giants account for so many of the stars visible to the naked eye in the night sky?

ANIMATION/VIDEO   (MA)

White Dwarfs in Globular Cluster

10.6 Extending the Cosmic Distance Scale

We have already discussed the connections between luminosity, apparent brightness, and distance. Knowledge of a star's apparent brightness and its distance allows us to determine its luminosity using the inverse-square law. But we can also turn the problem around. If we somehow knew a star's luminosity and then measured its apparent brightness, the inverse-square law would tell us its distance from the Sun.

Spectroscopic Parallax

Most of us have a rough idea of the approximate intrinsic brightness (that is, the luminosity) of a typical traffic signal. Suppose you are driving down an unfamiliar street and see a red traffic light in the distance. Your knowledge of the light's luminosity enables you immediately to make a mental estimate of its distance. A light that appears relatively dim (low apparent brightness) must be quite distant (assuming it's not just dirty). A bright one must be relatively close. Thus, *measurement of the apparent brightness of a light source, combined with some knowledge of its luminosity, can yield an estimate of its distance.*

For a star the trick is to find an independent measure of luminosity without knowing the distance. The H–R diagram can provide just that. For example, suppose we observe a star and determine its apparent magnitude to be 10. By itself, that doesn't tell us much—the star could either be faint and close, or bright and distant (Figure 10.5). But suppose we have some additional information: The star lies on the main sequence and has spectral type A0. Then we can read the star's luminosity off a graph such as Figure 10.13 or Figure 10.14. A main-sequence A0 star has a luminosity of approximately 100 solar units. According to *More Precisely 10-1*, this corresponds to an absolute magnitude of 0 and hence to a distance of 1000 pc.

This process of using stellar spectra to infer distances is called **spectroscopic parallax.** The key steps are as follows:

1. measure the star's apparent brightness and spectral type *without* knowing how far away it is;
2. use the spectral type to estimate the star's luminosity, assuming that it lies on the main sequence; and
3. finally, apply the inverse-square law to determine the distance to the star.

Note that despite its name, the method has nothing in common with stellar (geometric) parallax other than its use as a means of determining stellar distances. The main sequence allows us to make a connection between an easily measured quantity (spectral type) and the star's luminosity, which would otherwise be unknown. As we will see in upcoming chapters, this essential logic (with a variety of different techniques replacing step 2) is used again and again as a means of distance measurement in astronomy. In practice, the "fuzziness" of the main sequence translates into a small (10 − 20 percent) uncertainty in the distance, but the basic idea remains valid.

In Chapter 1 we introduced the first rung on a ladder of distance-measurement techniques that will ultimately carry us to the edge of the observable universe. That rung is radar ranging on the inner planets. ∞ *(Sec.1.3)* It establishes the scale of the solar system and defines the astronomical unit. At the beginning of this chapter we discussed a second rung on the cosmic distance ladder—stellar parallax—which is based on the first, since Earth's orbit is the baseline. Now, having used the first two rungs to determine the distances and other physical properties of many nearby stars, we use that knowledge to construct the third rung—spectroscopic parallax. Figure 10.16 illustrates schematically how this third rung builds on the first two and expands our cosmic field of view still deeper into space.

Spectroscopic parallax can be used to determine stellar distances out to several thousand parsecs. Beyond that, spectra and colors of individual stars are difficult to obtain. Note that, in using this method, we are assuming (without proof) that distant stars are basically similar to those nearby—in particular, that main-sequence stars *fall on the same main sequence*. Only by making this assumption can we use spectroscopic parallax to expand the boundaries of our distance-measurement techniques.

Each rung in the distance ladder is calibrated using data from the lower rungs, so changes made at any level will affect measurements made on *all* larger scales. As a result, the impact of the enormous dataset of high-quality observations returned by the *Hipparcos* mission (Section 10.1) extends far beyond the volume of space actually surveyed by the satellite. By recalibrating the local foundations of the cosmic distance scale, *Hipparcos* has revised estimates of distances on *all* scales—up to and including the scale of the universe itself. All distances quoted throughout this text reflect updated values based on *Hipparcos* data.

A useful analogy to light's dependence on distance is a traffic light that is distant and dim . . .

. . . compared to one that is nearby and bright.

▲ **FIGURE 10.16 Stellar Distances INTERACTIVE**   Knowledge of a star's luminosity and apparent brightness can yield an estimate of its distance. Astronomers use this third rung on the distance ladder, called spectroscopic parallax, to measure distances as far out as individual stars can be clearly discerned—several thousand parsecs.

CONCEPT CHECK

Suppose astronomers discover that due to a calibration error all distances measured by geometric parallax are 10 percent larger than currently thought. What effect would this have on the "standard" main sequence used in spectroscopic parallax?

(a)

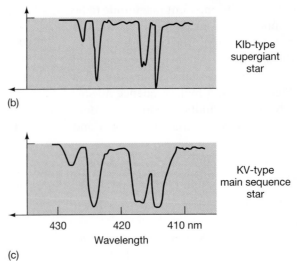

(b)

(c)

◀ **FIGURE 10.17 Luminosity Classes** (a) Approximate locations of the standard stellar luminosity classes in the H–R diagram. The widths of absorption lines also provide information on the density of a star's atmosphere. The denser atmosphere of a main-sequence K-type star has broader lines (c) than a giant star of the same spectral class (b).

Luminosity Class

What if the star in question happens to be a red giant or a white dwarf and does not lie on the main sequence? As mentioned in Chapter 2, detailed analysis of spectral *line widths* can provide information on the pressure, and hence the density, of the gas where the line formed. ∞ *(Sec. 2.6)* The atmosphere of a red giant is much less dense than that of a main-sequence star, and this in turn is much less dense than the atmo-sphere of a white dwarf. Figures 10.17(a) and (b) illustrate the differences be-tween the spectra of a main-sequence star and a red giant of the same spectral type.

Over the years, astronomers have developed a system for classifying stars ac-cording to the widths of their spectral lines. Because line width depends on pres-sure in the stellar photosphere, and because this pressure in turn is well correlated with luminosity, this stellar property is known as **luminosity class.** The standard luminosity classes are listed in Table 10.3 and shown on the H–R diagram in Fig-ure 10.17(c). By determining a star's luminosity class, astronomers can usually tell with a high degree of confidence what sort of object it is and the full specification of a star's spectral properties includes its luminosity class. For example, the Sun, a G2 main-sequence star, is of class G2V, the B8 blue supergiant Rigel is B8Ia, the red dwarf Barnard's star is M5V, the red supergiant Betelgeuse is M2Ia, and so on.

Consider, for example, a K2-type star (Table 10.4) with a surface temperature of around 4500 K. If the widths of the star's spectral lines tell us that it lies on the main sequence (i.e., it is a K2V star), then its luminosity is about 0.3 times the solar value. If the lines are narrower than those normally found in main-sequence stars, the star may be recognized as a K2III giant, with a luminosity 100 times that of the Sun (Figure 10.17c). If the lines are very narrow, the star might instead be classified as a K2Ib supergiant, brighter by a further factor of 40, at 4000 solar luminosities. In each case, knowledge of luminosity classes allows astronomers to identify the ob-ject and make a useful estimate of its luminosity and hence its distance.

PROCESS OF SCIENCE CHECK

What assumptions are we making when we apply spectroscopic parallax to measure the distance to a star?

TABLE 10.3   Stellar Luminosity Classes

Class	Description
Ia	bright supergiants
Ib	supergiants
II	bright giants
III	giants
IV	subgiants
V	main-sequence stars/dwarfs

TABLE 10.4   Variation in Stellar Properties within a Spectral Class

Surface Temperature (K)	Luminosity (solar luminosities)	Radius (solar radii)	Object	Example
4900	0.3	0.8	K2V main-sequence star	ε Eridani
4500	110	21	K2III red giant	Arcturus
4300	4000	140	K2Ib red supergiant	ε Pegasi

10.7 Stellar Masses

As with all other astronomical objects, we measure a star's mass by observing its gravitational influence on some nearby body—another star, perhaps, or a planet. If we know the distance between the two bodies, we can use Newton's laws to calculate their masses. ∞ *(Sec. 1.4)*

MA SELF-GUIDED TUTORIAL
Binary Stars–Radial Velocity Curves

MA SELF-GUIDED TUTORIAL
Eclipsing Binary Stars–Light Curves

Binary Stars

Most stars are members of *multiple-star systems*—groups of two or more stars in orbit around one another. The majority form **binary-star systems** (or simply *binaries*), which consist of two stars in orbit about their common center of mass, held together by their mutual gravitational attraction. (The Sun is not part of a multiple star system; if it has anything at all uncommon about it, it is this lack of stellar companions.)

Astronomers classify binaries according to their appearance from Earth and the ease with which they can be observed. **Visual binaries** have widely separated components bright enough to be observed and monitored separately (Figure 10.18a). The more common **spectroscopic binaries** are too distant from us to be resolved into two distinct stars, but they can be indirectly perceived by monitoring the back-and-forth Doppler shifts of their spectral lines as the stars orbit one another and their line-of-sight velocities vary periodically. ∞ *(Sec. 2.7)*

In a *double-line* spectroscopic binary, two distinct sets of spectral lines—one for each component star—shift back and forth as the stars move. The periodic shifts of the lines tell us that the objects emitting them are in orbit. In the more common *single-line* systems (Figure 10.18b), one star is too faint for its spectrum to be distinguished, so we see only one set of lines shifting back and forth. This shifting means that the detected star must be in orbit around another star, even though the companion cannot be observed directly. (If this idea sounds familiar, it should—all the extrasolar planetary systems discovered to date are extreme examples of single-line spectroscopic binaries.) ∞ *(Sec. 4.4)*

In the much rarer **eclipsing binaries,** the orbital plane of the pair of stars is almost edge-on to our line of sight. In this situation, we observe a periodic decrease of starlight intensity as one member of the binary passes in front of the other (Figure 10.18c). By studying the variation of the light from an eclipsing binary system—called the binary's **light curve**—astronomers can derive detailed information not only about the stars' orbits and masses, but also about their radii.

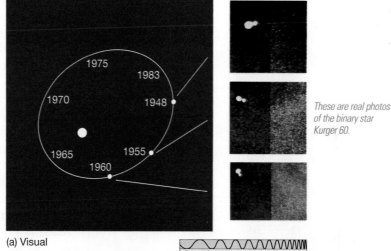

(a) Visual

These are real photos of the binary star Kurger 60.

R I V U X G

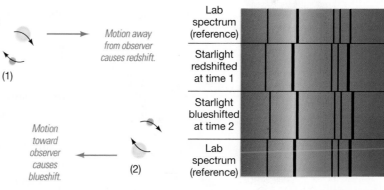

Motion away from observer causes redshift.

(1)

Motion toward observer causes blueshift.

(2)

Lab spectrum (reference)

Starlight redshifted at time 1

Starlight blueshifted at time 2

Lab spectrum (reference)

(b) Spectroscopic

#1 #2 #3 #4 #5 #6

Intensity

Time

(c) Eclipsing

▶ **FIGURE 10.18 Binary Stars INTERACTIVE** (a) The period and separation of a double-star system can be observed directly if each star is clearly seen. (b) Binary properties can be determined indirectly by measuring the periodic Doppler shift of one star relative to the other while moving in their orbits. This is a single-line system, in which only one spectrum (from the brighter component) is visible. (c) If two stars eclipse one another, additional information can be obtained by observing the periodic decrease in starlight as one star passes in front of the other. *(Harvard College Observatory)*

INTERACTIVE

Exploring the Light Curve of an Eclipsing Binary Star System

Note that these categories are not mutually exclusive. For example, an eclipsing binary may also be (and in fact often is) a spectroscopic binary system.

Mass Determination

The period of a binary can be measured by observing the orbits of the component stars, the back-and-forth motion of the stellar spectral lines, or the dips in the light curve—whatever information is available. Observed binary periods span a broad range—from hours to centuries. How much additional information can be extracted depends on the type of binary involved.

If the distance to a *visual binary* is known, then the orbits of each component can be individually tracked and the masses of the components can be determined. ∞ *(Sec.1.4)*

For *spectroscopic binaries*, Doppler-shift measurements give us information only on the radial velocities of the component stars, and this limits the information we can obtain. Simply put, we cannot distinguish between a slow-moving binary seen edge-on and a fast-moving binary seen almost face-on (so that only a small component of the orbital motion is along the line of sight). For a double-line system, only lower limits on the individual masses can be obtained. For single-line systems, even less information is available, and only a rather complicated relation between the component masses (known as the mass function) can be derived.

If a spectroscopic binary also happens to be an eclipsing system, then the uncertainty in the orbital inclination is removed, as the binary is known to be edge-on (or very nearly so). In that case, both masses can be determined for a double-line binary. For a single-line system, the mass function is simplified to the point where the mass of the unseen component is known if the mass of the brighter component can be obtained by other means (for example, if it is recognized as a main-sequence star of a certain spectral class—see Figure 10.19).

Despite all these qualifications, individual component masses have been obtained for many nearby binary systems. A simple example is presented in *More Precisely 10-3*. Virtually all we know about the masses of stars is based on such observations.

Mass and Other Stellar Properties

Table 10.5 summarizes the various observational and theoretical techniques used to measure the stars, listing the quantities that are assumed known (usually as a result of the application of other techniques listed in the table), those that are measured, and the theory that is applied to turn the observations into the desired result. We end our introduction to the stars with a brief look at how mass is correlated with the other stellar properties discussed in this chapter.

Figure 10.19 is a schematic H–R diagram showing how stellar mass varies along the main sequence. There is a clear progression from low-mass red dwarfs to high-mass blue giants. With few exceptions, main-sequence stars range in mass from about 0.1 to 20 times the mass of the Sun. The hot O- and B-type stars are generally about 10 to 20 times more massive than our Sun. The coolest K- and M-type stars contain only a few tenths of a solar mass. The mass of a star at the time of its formation determines its location on the main sequence. Based on observations of stars within a few hundred light-years of the Sun, Figure 10.20 illustrates how the masses of main-sequence stars are distributed. Notice the huge fraction of low-mass stars, as well as the tiny fraction contributed by stars of more than a few solar masses.

Figure 10.21 illustrates how the radii and luminosities of main-sequence stars depend on mass. The two plots are called the *mass–radius* (part a) and *mass–luminosity* (part b) relationships. Along the main sequence, both radius and luminosity increase with mass. The radius rises slightly more slowly than the

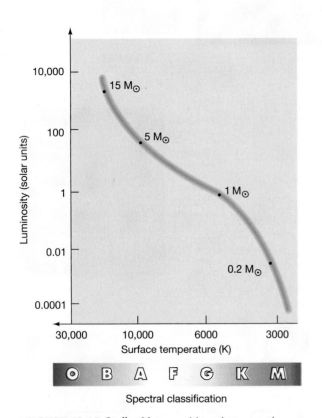

▲ **FIGURE 10.19 Stellar Masses** More than any other stellar property, mass determines a star's position on the main sequence. Low-mass stars are cool and faint; they lie at the bottom of the main sequence. Very massive stars are hot and bright; they lie at the top of the main sequence. (The symbol "$M_\odot$" means "solar mass.")

10.3 MORE PRECISELY

Measuring Stellar Masses in Binary Stars

As discussed in the text, most stars are members of binary systems—where two stars orbit one another, bound together by gravity. Here we describe how—in an idealized case where the relevant orbital parameters are known—we can use the observed orbital data, together with our knowledge of basic physics, to determine the masses of the component stars.

Consider the nearby visual binary system made up of the bright star Sirius A and its faint companion Sirius B, sketched in the accompanying figure. The binary's orbital period can be measured simply by watching the stars orbit one another, or alternatively by following the back-and-forth velocity wobbles of Sirius A due to its faint companion. It is almost exactly 50 years. The orbital semimajor axis can also be obtained by direct observation of the orbit, although in this case we must use some additional knowledge of Kepler's laws to correct for the binary's 46° inclination to the line of sight. ∞ *(Sec. 1.3)* It is 20 AU—an angular size of at a distance of 2.7 pc. ∞ *(Sec. 1.3)* Once we know these two key orbital parameters, we can use the modified version of Kepler's third law to calculate the sum of the masses of the two stars. The result is $20^3/50^2 = 3.2$ times the mass of the Sun. ∞ *(Sec. 1.4)*

Further study of the orbit allows us to determine the individual stellar masses. Doppler observations show that Sirius A moves at approximately half the speed of its companion relative to their center of mass. ∞ *(Secs. 1.4, 2.7)* This implies that Sirius A must have twice the mass of Sirius B. It then follows that the masses of Sirius A and Sirius B are 2.1 and 1.1 solar masses, respectively. Often the calculation of the masses of binary components is complicated by the fact that only partial information is available—we might only be able to see one star, or perhaps only spectroscopic velocity information is available (see Section 10.7). Nevertheless, this technique of combining elementary physical principles with detailed observations is how virtually every stellar mass quoted in this text has been determined.

mass, whereas luminosity increases much faster—more like the *fourth power* of the mass (indicated by the straight line in Figure 10.21b). For example, a 2-solar mass main-sequence star has a radius roughly twice that of the Sun and a luminosity of 16 (2^4) solar luminosities. A 0.2-solar mass main-sequence star has a radius of about 0.2 solar radii and a luminosity of 0.0016 (0.2^4) solar luminosities.

TABLE 10.5   Measuring the Stars

Stellar Property	Measurement Technique	"Known" Quantity	Measured Quantity	Theory Applied	Section
Distance	stellar parallax spectroscopic parallax	astronomical unit main sequence	parallactic angle spectral type apparent magnitude	elementary geometry inverse-square law	10.1 10.6
Radial velocity		speed of light atomic spectra	spectral lines	Doppler effect	10.1
Transverse velocity	astrometry	distance	proper motion	elementary geometry	10.1
Luminosity		distance main sequence	apparent magnitude spectral type color	inverse-square law	10.2 10.6
Temperature	photometry spectroscopy		spectral type	blackbody law atomic physics	10.3 10.3
Radius	direct indirect	distance	angular size luminosity temperature	elementary geometry radius–luminosity–temperature relationship	10.4 10.4
Composition	spectroscopy		spectrum	atomic physics	10.3
Mass	observations of binary stars	distance	binary period binary orbit orbital velocity	Newtonian gravity and dynamics	10.7

Star	Spectral Type/ Luminosity Class[1]	Mass (solar masses)	Central Temperature (10^6 K)	Luminosity (solar luminosities)	Estimated Lifetime (millions of years)
Spica B[*]	B2V	6.8	25	800	90
Vega	A0V	2.6	21	50	500
Sirius	A1V	2.1	20	22	1000
Alpha Centauri	G2V	1.1	17	1.6	7000
Sun	G2V	1.0	15	1.0	10,000
Proxima Centauri	M5V	0.1	0.6	0.00006	16,000,000

TABLE 10.6   Key Properties of Some Well-Known Main-Sequence Stars

[*]The "star" Spica is, in fact, a binary system comprising a B1III giant primary (Spica A) and a B2V main-sequence secondary (Spica B).

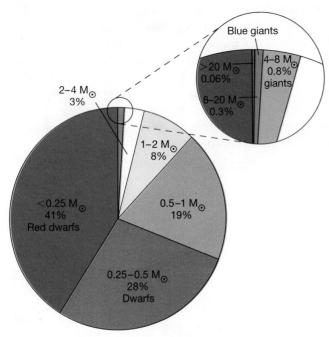

▲ **FIGURE 10.20 Stellar Mass Distribution** The distribution of masses of main-sequence stars, as determined from careful measurement of stars in the solar neighborhood.

Table 10.6 compares key properties of some well-known main-sequence stars, arranged in order of decreasing mass. Notice that the central temperature (obtained from mathematical models similar to those discussed in Chapter 9) differs relatively little from star to star, compared to the large spread in stellar luminosities. The final column in the table presents an estimate of each star's *lifetime*, obtained by dividing the amount of fuel available (that is, the star's mass) by the rate at which the fuel is consumed (the luminosity),

$$\text{stellar lifetime} \propto \frac{\text{stellar mass}}{\text{stellar luminosity}},$$

and noting that the lifetime of the Sun (see Chapter 12) is about 10 billion years.

Because luminosity increases so rapidly with mass, the most massive stars are by far the shortest lived. For example, according to the mass–luminosity relationship, the lifetime of a 10-solar mass O-type star is roughly $10/10^4 = 1/1000$ that of the Sun, or about 10 million years. We can be sure that all the O- and B-type stars we now observe are quite young—less than a few tens of millions of years old. Their nuclear reactions proceed so rapidly that their fuel is quickly depleted despite their large masses. At the opposite end of the main sequence, the low core density and temperature of an 0.1-solar mass M-type star

(a)

(b)

◄ **FIGURE 10.21 Stellar Radii and Luminosities** (a) Actual measurements of main-sequence stars show that radius increases almost in proportion to mass over much of the range (as indicated by the straight line drawn through the data). (b) Stellar luminosity increases roughly as the fourth power of the mass (indicated again by the straight line).

mean that its proton–proton reactions churn away much more sluggishly than in the Sun's core, leading to a very low luminosity and a correspondingly long lifetime. (*Sec. 9.5*) Many of the K- and M-type stars now visible in the sky will shine on for at least another trillion years. The evolution of stars—large and small—is the subject of the next two chapters.

THE BIG QUESTION

The *Hipparcos* satellite measured the detailed properties of hundreds of thousands of stars in the neighborhood of the Sun, greatly expanding our understanding of stellar properties and orbits in our Galaxy. In December 2013, the European Space Agency launched the GAIA mission, designed to extend the coverage of *Hipparcos* to roughly 1 billion objects, spanning a large portion of the entire Milky Way Galaxy. What new insights will GAIA yield? The impact on stellar and galactic astrophysics may be immeasurable.

CHAPTER REVIEW

SUMMARY

LO1 The distances to the nearest stars can be measured by stellar parallax. A star with a parallax of 1 arc second is **1 parsec** (p. 280)—about 3.3 light-years—away from the Sun. A star's **proper motion** (p. 282), which is its true motion across the sky, is a measure of the star's velocity perpendicular to our line of sight.

LO2 The **apparent brightness** (p. 282) of a star is the rate at which energy from the star reaches a detector. Apparent brightness is proportional to luminosity and falls off as the inverse square of the distance. Optical astronomers use the **magnitude scale** (p. 284) to express and compare stellar brightnesses. The greater the magnitude, the fainter the star; a difference of five magnitudes corresponds to a factor of 100 in brightness. **Apparent magnitude** (p. 284) is a measure of apparent brightness. The **absolute magnitude** (p. 284) of a star is the apparent magnitude it would have if placed at a standard distance of 10 pc from the viewer. It is a measure of the star's luminosity.

LO3 Astronomers often determine the temperatures of stars by measuring their brightnesses through two or more optical filters, then fitting a blackbody curve to the results. Spectroscopic observations of stars provide an accurate means of determining both stellar temperatures and stellar composition.

Astronomers classify stars according to the absorption lines in their spectra. The standard stellar **spectral classes** (p. 287), in order of decreasing temperature, are O B A F G K M.

LO4 Only a few stars are large enough and close enough to Earth that their radii can be measured directly. The sizes of most stars are estimated indirectly through the **radius–luminosity–temperature relationship** (p. 289). Stars categorized as **dwarfs** (p. 289) are comparable in size to, or smaller than, the Sun; **giants** (p. 289) are up to 100 times larger than the Sun; and **super-giants** (p. 289) are more than 100 times larger than the Sun. In addition to "normal" stars such as the Sun, two other important classes of star are **red giants** (p. 289), which are large, cool, and luminous, and **white dwarfs** (p. 289), which are small, hot, and faint.

LO5 A plot of stellar luminosity versus stellar spectral class (or temperature) is called an **H–R diagram** (p. 289). About 90 percent of all stars plotted on an H–R diagram lie on the **main sequence** (p. 290), which stretches from hot, bright **blue supergiants** (p. 290) and **blue giants** (p. 290), through intermediate stars such as the Sun, to cool, faint **red dwarfs** (p. 290). Most main-sequence stars are red dwarfs; blue giants are quite rare. About 9 percent of stars lie in the **white-dwarf region** (p. 291) and the remaining 1 percent lie in the **red-giant region** (p. 291).

LO6 If a star is known to lie on the main sequence, measurement of its spectral type allows its luminosity to be estimated and its distance to be measured. This method of distance determination, which is valid for stars up to several thousand parsecs from Earth, is called **spectroscopic parallax** (p. 293). A star's **luminosity class** (p. 294) allows astronomers to distinguish main-sequence stars from giants and supergiants of the same spectral type, making distance estimates much more reliable.

LO7 Most stars are not isolated in space but instead orbit other stars in **binary-star systems** (p. 295). In a **visual binary** (p. 295), both stars can be seen and their orbits charted. In a **spectroscopic binary** (p. 295), the stars cannot be resolved, but their orbital motion can be detected spectroscopically. In an **eclipsing binary** (p. 295), the orbits are oriented in such a way that one star periodically passes in front of the other as seen from Earth and dims the light we receive. Studies of binary stars often allow stellar masses to be measured. The mass of a star determines its size, temperature, and brightness. Hot blue giants are much more massive than the Sun; cool red dwarfs are much less massive. High-mass stars burn their fuel rapidly and have much shorter lifetimes than the Sun. Low-mass stars consume their fuel slowly and may remain on the main sequence for trillions of years.

MasteringAstronomy®   For instructor-assigned homework go to www.masteringastronomy.com

Problems labeled **POS** explore the process of science. **VIS** problems focus on reading and interpreting visual information. **LO** connects to the introduction's numbered Learning Outcomes.

REVIEW AND DISCUSSION

1. **LO1 POS** How is stellar parallax used to measure the distances to stars?

2. What is a parsec? Compare it to the astronomical unit.

3. Explain how a star's real motion through space translates into motion observable from Earth.

4. **LO4** Describe some characteristics of red giants and white dwarfs.

5. **LO2** What is the difference between absolute and apparent brightness?

6. **POS** How do astronomers measure star temperatures?

7. **LO3** Briefly describe how stars are classified according to their spectral characteristics.

8. **LO5** What information is needed to plot a star on the H–R diagram?

9. What is the main sequence? What basic property of a star determines where it lies on the main sequence?

10. **LO6** How are distances determined using spectroscopic parallax?

11. Which stars are most common in the Galaxy? Why don't we see many of them in H–R diagrams?

12. **LO7 POS** How can stellar masses be determined by observing binary-star systems?

13. High-mass stars start off with much more fuel than low-mass stars. Why don't high-mass stars live longer?

14. In general, is it possible to determine the life span of an individual star simply by noting its position on an H–R diagram?

15. Why are visual binaries and eclipsing binaries relatively rare compared to spectroscopic binaries?

CONCEPTUAL SELF-TEST: TRUE OR FALSE?/MULTIPLE CHOICE

1. Star A appears brighter than star B, as seen from Earth. Therefore, star A must be closer to Earth than star B. (T/F)

2. A star of apparent magnitude +5 looks brighter than one of apparent magnitude + 2. (T/F)

3. Red giants are very bright because they are extremely hot. (T/F)

4. The radius of a star can be determined if the star's distance and luminosity are known. (T/F)

5. Astronomers can distinguish between main-sequence and giant stars by purely spectroscopic means. (T/F)

6. In a spectroscopic binary, the orbital motion of the component stars appears as variations in the overall apparent brightness of the system. (T/F)

7. There are no billion-year-old main-sequence O- or B-type stars. (T/F)

8. From a distance of 1 parsec, the angular size of Earth's orbit would be (a) 1 degree; (b) 2 degrees; (c) 1 arc minute; (d) 2 arc seconds.

9. Compared with a star of absolute magnitude –2 at a distance of 100 pc, a star of absolute magnitude 5 at a distance of 10 pc will appear (a) brighter; (b) fainter; (c) to have the same brightness; (d) bluer.

10. Stars of spectral class M do not show strong lines of hydrogen in their spectra because (a) they contain very little hydrogen; (b) their surfaces are so cool that most hydrogen is in the ground state; (c) their surfaces are so hot that most hydrogen is ionized; (d) the hydrogen lines are swamped by even stronger lines of other elements.

11. Cool stars can be very luminous if they are very (a) small; (b) hot; (c) large; (d) close to our solar system.

12. The mass of a star may be determined (a) by measuring its luminosity; (b) by determining its composition; (c) by measuring its Doppler shift; (d) by studying its orbit around a binary companion.

13. VIS Pluto's apparent magnitude is approximately 14. According to Figure 10.6 (Apparent Magnitude), Pluto can be seen (a) with the naked eye on a dark night; (b) using binoculars; (c) using a 1-meter telescope; (d) only with the *Hubble Space Telescope*.

14. VIS According to Figure 10.12 (H–R Diagram of Well-Known Stars), Barnard's star must be (a) hotter; (b) larger; (c) closer to us; (d) bluer than Proxima Centauri.

15. VIS According to Figures 10.19 (Stellar Masses) and 10.20 (Stellar Mass Distribution), compared to the Sun, most stars are (a) much hotter; (b) much cooler; (c) much larger; (d) much brighter.

PROBLEMS

The number of squares preceding each problem indicates its approximate level of difficulty.

1. ■ How far away is the star Spica, whose parallax is 0.013″? What would Spica's parallax be if it were measured from an observatory on Neptune's moon Triton as Neptune orbited the Sun?

2. ■■ A star lying 20 pc from the Sun has proper motion of 0.5″ /yr. What is its transverse velocity? If the star's spectral lines are observed to be redshifted by 0.01 percent, calculate the magnitude of its three-dimensional velocity relative to the Sun.

3. ■■ (a) What is the luminosity of a star having three times the radius of the Sun and a surface temperature of 10,000 K? (b) A certain star has a temperature twice that of the Sun and a luminosity 64 times greater than the solar value. What is its radius, in solar units?

4. ■■ Two stars—A and B, of luminosities 0.5 and 4.5 times the luminosity of the Sun, respectively—are observed to have the same apparent brightness. Which star is more distant, and how much farther away is it than the other?

5. ■■ Two stars—A and B with absolute magnitudes 3 and 8, respectively—are observed to have the same apparent magnitude. Which star is more distant, and how much farther away is it than the other?

6. ■■ Calculate the amount of energy received per unit area per unit time from the sun, as seen from a distance of 10 pc. Compare this with the solar constant at Earth. ∞ *(Sec. 9.1)*

7. ■ Astronomical objects visible to the naked eye range in apparent brightness from faint sixth-magnitude stars to the Sun, with magnitude –27. What is the range in energy flux corresponding to this magnitude range?

8. ■ A star has apparent magnitude 10.0 and absolute magnitude 2.5. How far away is it?

9. ■■■ Two stars in an eclipsing spectroscopic binary system are observed to have an orbital period of 25 days. Further observations reveal that the orbit is circular, with a separation of 0.3 AU, and that one star is 1.5 times the mass of the other. What are the masses of the stars?

10. ■■ Given that t he Sun's lifetime is about 10 billion years, estimate the life expectancy of (a) a 0.2-solar mass, 0.01-solar luminosity red dwarf; (b) a 3.0-solar mass, 30-solar luminosity star; (c) a 10-solar mass, 1000-solar luminosity blue giant.

ACTIVITIES

Collaborative

1. Estimate the total number of stars visible in the night sky. Each member of your group should be equipped with identical cardboard tubes—the tube at the center of a roll of paper towels or toilet paper is perfect for the task. On a clear, moonless night, hold your tube up to your eye and count the total number of stars you can see. Do this several times, randomly choosing different areas of the sky and avoiding obvious obstacles such as clouds and trees. Try to sample all directions roughly equally. Be sure to allow time for your eyes to adapt to the dark—10 to 15 minutes at least—before taking any measurements. Add up all your measurements and divide by the total number of observations to calculate the average number of stars observed—call it n. You can convert this number into an estimate of the total number N of visible stars by multiplying by the square of the ratio of the length L of the tube to its diameter D, that is: $N = (L/D)^2 \times n$. (Can you figure out where this formula came from?) Repeat your measurement at a variety of sites—a city, the suburbs, and a dark rural location. Can you understand why astronomers are so concerned about light pollution?

Individual

1. The Winter Circle is an asterism—or pattern of stars—made up of six bright stars in five different constellations: Sirius, Rigel, Betelgeuse, Aldebaran, Capella, and Procyon. These stars span nearly the entire range of colors (and therefore temperatures) of normal stars. Rigel is of type B; Sirius, A; Procyon, F; Capella, G; Aldebaran, K; and Betelgeuse, M. The color differences are easy to see. Why do you suppose there is no O-type star in the Winter Circle?

2. In the winter sky, find the red supergiant Betelgeuse in the constellation Orion. It's easy to see because it's one of the brightest stars in the entire night sky. Betelgeuse is a variable star with a period of about 6.5 years. Its brightness changes as it expands and contracts. At maximum size, it fills a volume of space that would extend from the Sun to beyond the orbit of Jupiter. Betelgeuse is thought to be about 10 to 15 times more massive than our Sun and probably between 4 and 10 million years old. A similar star can be found shining prominently in midsummer. This is the red supergiant Antares in the constellation Scorpius.

11

The Interstellar Medium

Star Formation in the Milky Way

Stars and planets are not the only inhabitants of our Galaxy. The space around us harbors invisible matter throughout the dark voids between the stars. The density of this interstellar matter is extremely low—approximately a trillion trillion times less dense than matter in either stars or planets, far thinner than the best vacuum attainable on Earth. Only because the volume of interstellar space is so vast does the mass of interstellar matter amount to anything at all. Why bother to study this near-perfect vacuum? We do so for three important reasons. First, there is as much mass in the "voids" between the stars as there is in the stars themselves. Second, interstellar space is the region out of which new stars are born. Third, it is the region into which old stars expel their matter when they die. It is one of the most significant crossroads through which matter passes in the history of our universe.

LEARNING OUTCOMES

Studying this chapter will enable you to:

LO1 Summarize the composition and properties of interstellar matter.

LO2 Describe the characteristics of emission nebulae, and explain their connection with star formation.

LO3 Outline the properties of dark interstellar dust clouds and molecular clouds.

LO4 Specify some of the radio techniques used to probe the nature of interstellar gas.

LO5 Summarize the sequence of events leading from a cold, dense interstellar cloud to the formation of a star like our Sun.

LO6 Describe some of the observational evidence supporting the modern theory of star formation.

LO7 Explain how the formation of a star depends on its mass.

LO8 Describe the properties of star clusters, and discuss how they form.

◄ This *Hubble Space Telescope* image of the young star cluster Westerlund 2 combines visible-light and near-infrared exposures of the cluster itself (right center), with visible-light observations of the surrounding gas out of which it formed. The cluster is about 3000 times the mass of the Sun, but the interstellar gas in its vicinity is much more massive. The intense radiation from young stars in the cluster is sculpting and heating the nearby gas, dispersing it into space and possibly triggering new episodes of star formation.

MasteringAstronomy®

Visit www.masteringastronomy.com for quizzes, animations, videos, interactive figures, and self-guided tutorials.

Interstellar Gas

Sixty percent of students had difficulty describing and distinguishing the properties of interstellar gas clouds. Some key points to remember:

- Dark dust clouds are cool regions of gas that emit no visible light and absorb visible light from stars behind them. They emit radiation primarily in the radio and infrared parts of the electromagnetic spectrum.

- Emission nebulae are hot regions of gas surrounding newborn stars. They emit visible light.

- Emission nebulae are often found embedded in dark dust clouds in which stars are forming.

▼ **FIGURE 11.1 Milky Way Mosaic**   The Milky Way photographed panoramically, across 360° of the entire southern and northern celestial sphere. This band, which constitutes the central plane of our Galaxy, contains high concentrations of stars, as well as interstellar gas and dust. The gray box outlines the field of view of Figure 11.4. *(ESO/S. Brunier)*

11.1 Interstellar Matter

Figure 11.1 is a mosaic of photographs covering a much greater expanse of universal "real estate" than anything we have studied thus far. From our vantage point on Earth, the panoramic view shown here stretches all the way across the sky. On a clear night it is visible to the naked eye as the *Milky Way.* In Chapter 14 we will come to recognize this band as the flattened disk, or *plane,* of our own Galaxy.

The bright regions in this image are made up of innumerable unresolved distant stars, merging together into a continuous blur at the resolution of the telescope. However, the dark areas are not simply "holes" in the stellar distribution. They are regions of space where *interstellar matter* obscures the light from stars beyond, blocking from our view what would otherwise be a smooth distribution of bright starlight. In this chapter we focus not on the stars themselves but also on the vast reaches of interstellar space in which they form.

Gas and Dust

The matter between the stars is collectively termed the **interstellar medium.** It is made up of two components—*gas* and *dust*—intermixed throughout all of space. The gas is made up mainly of individual atoms and small molecules. The dust consists of clumps of atoms and molecules—not unlike the microscopic particles that make up smoke or soot.

Apart from numerous narrow atomic and molecular absorption lines, the gas alone does not block electromagnetic radiation to any great extent. The patchy obscuration evident in Figure 11.1 is caused by dust. Light from distant stars cannot penetrate the densest accumulations of interstellar dust any more than a car's headlights can illuminate the road ahead in a thick fog. As a rule of thumb, a beam of light can be absorbed or scattered only by particles having diameters comparable to or larger than the wavelength of the radiation involved. The amount of obscuration (absorption or scattering) produced by particles of a given size increases with decreasing wavelength. The typical diameter of an interstellar dust

▶ **FIGURE 11.2 Reddening** (a) Starlight passing through a dusty region of space is both dimmed and reddened, but spectral lines are still recognizable in the light that reaches Earth. (b) This dusty interstellar cloud, called Barnard 68, is opaque to visible light, except near the edges. The cloud spans about 0.2 pc and lies about 160 pc away. Frame (c) illustrates (in false color) how infrared radiation can penetrate Barnard 68. *(ESO)*

particle—or **dust grain**—is about 10^{-7} m comparable in size to the wavelength of visible light. Consequently, dusty regions of interstellar space are transparent to long-wavelength radio and infrared radiation but opaque to shorter-wavelength optical, ultraviolet, and X-ray radiation.

Because the opacity of the interstellar medium increases with decreasing wavelength, light from distant stars is preferentially robbed of its higher-frequency ("blue") components. Hence, in addition to being generally diminished in overall brightness, stars also appear redder than they really are. Illustrated in Figure 11.2, this effect, known as **reddening**, is similar to the process that produces spectacular red sunsets here on Earth (see Figure 11.3). Reddening can be seen very clearly in Figure 11.2(b), which shows a type of compact, dusty interstellar cloud called a *globule*. (We will discuss such dark interstellar clouds in more detail in Section 11.3.) The center of this cloud, called Barnard 68, is opaque to all optical wavelengths, so starlight cannot pass through it. However, near the edges, where there is less intervening cloud matter, some light does make it through; in the infrared image of Figure 11.2(c), even more radiation gets through. Notice how stars seen through the cloud are both dimmed and reddened relative to those seen directly.

As indicated in Figure 11.2(a), absorption lines in a star's spectrum are still recognizable in the radiation reaching Earth, allowing the star's spectral class, and hence its true luminosity and color, to be determined. ∞ *(Sec. 10.6)* Astronomers can then measure the degree to which the star's light has been dimmed and reddened en route to Earth, allowing them to estimate the amount of dust along the line of sight.

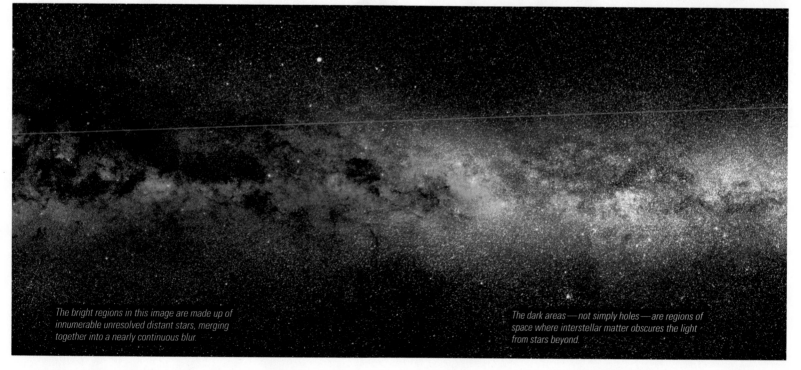

The bright regions in this image are made up of innumerable unresolved distant stars, merging together into a nearly continuous blur.

The dark areas—not simply holes—are regions of space where interstellar matter obscures the light from stars beyond.

R I V U X G

◀ **FIGURE 11.3 Reddening in Earth's Atmosphere**   Sunlight is reddened as it passes through Earth's atmosphere at the end of a hot summer day. Airborne dust particles and water molecules scatter the Sun's blue light, leaving only its dimmed, red light—a spectacular sunset. (*Joyce Photographics*)

Density and Composition of the Interstellar Medium

By measuring its effect on the light from many different stars, astronomers have built up a picture of the distribution and chemical properties of interstellar matter in the solar neighborhood.

Gas and dust are found everywhere in interstellar space. No part of our Galaxy is truly devoid of matter, although the density of the interstellar medium is extremely low. Overall, the gas averages roughly 10^6 atoms per cubic meter—just one atom per cubic centimeter. (For comparison, the best vacuum presently attainable in laboratories on Earth contains about 10^9 atoms per cubic meter.) Interstellar dust is even rarer—about one dust particle for every trillion or so atoms. The space between the stars is populated with matter so thin that harvesting all

11.1 DISCOVERY

Ultraviolet Astronomy and the "Local Bubble"

The ultraviolet (UV) region is that part of the spectrum where we expect to witness events and see objects involving temperatures in the hundreds of thousands, even millions, of kelvins—hot regions like the seething atmospheres or eruptive flares of stars; violent events, such as massive stars caught in the act of exploding; and active galaxies whose spinning hearts may harbor black holes. Yet ultraviolet astronomy has also contributed greatly to the study of the interstellar medium, by allowing a unique mapping of our local cosmic neighborhood.

Only 30 years ago astronomers had presumed that the gases filling the spaces among the stars would absorb virtually all short-wavelength UV radiation before it had a chance to reach Earth. But in 1975, during a historic linkup of the *Apollo* (U.S.) and *Soyuz* (USSR) space capsules, the onboard astronauts and cosmonauts performed a key experiment—they used a small telescope to detect extreme ultraviolet radiation from a few nearby, very hot stars. Shortly thereafter, theoretical ideas changed as astronomers began to realize that interstellar gas is distributed very unevenly in space, in cool dense clumps interspersed by irregularly shaped regions of hot, low-density gas.

Since then, space-based ultraviolet observations have shown that some regions of interstellar space are much thinner (5000 atoms/m^3) and hotter (500,000 K) than previously expected. Part of the space among the dust clouds and the emission nebulae seems to contain extremely dilute yet seething plasma, probably the result of the concussion and expanding debris from stars that exploded long ago. These superheated interstellar "bubbles," or *intercloud medium*, may extend far into interstellar space beyond our local neighborhood and conceivably into the even vaster spaces among the galaxies.

The Sun seems to reside in one such low-density region—a huge cavity called the *Local Bubble*, sketched in the accompanying

figure. Only because we live within this vast, low-density bubble can we detect so many stars in the far UV; the hot, tenuous interstellar gas is virtually transparent to this radiation. The Local Bubble contains about 200,000 stars and extends for hundreds of trillions of kilometers, or nearly 100 pc. It was probably carved out by a supernova explosion that occurred in these parts roughly 300,000 years ago (see Section 12.4). Our distant ancestors must have seen it—a stellar catastrophe as bright as the full Moon.

(Rice University/NASA)

the gas and dust in an interstellar region the size of Earth would yield barely enough matter to make a pair of dice.

Despite such low densities, over sufficiently large distances interstellar matter accumulates slowly but surely to the point where it can block visible light and other short-wavelength radiation from distant sources. All told, space in the vicinity of the Sun contains about as much mass in the form of interstellar gas and dust as exists there in the form of stars.

Interstellar matter is distributed very unevenly (see *Discovery 11-1*). In some directions it is largely absent, allowing astronomers to study objects literally billions of parsecs from the Sun. In other directions there are small amounts of interstellar matter, so the obscuration is moderate, preventing us from seeing objects more than a few thousand parsecs away, but allowing us to study nearby stars. Still other regions are so heavily obscured that starlight from even relatively nearby stars is completely absorbed before reaching Earth. Even in regions of dense obscuration, however, there are often "windows" through which we can see to great distances—several can be seen in Figure 11.1.

The composition of interstellar gas is reasonably well known from spectroscopic studies of interstellar absorption lines. ∞ *(Sec. 2.5)* It mirrors the composition of other astronomical objects, such as the Sun, the stars, and the Jovian planets: Most of the gas—about 90 percent—is atomic or molecular hydrogen, 9 percent is helium, and the remaining 1 percent consists of heavier elements. The gas is deficient in some heavy elements, such as carbon, oxygen, silicon, magnesium, and iron, most likely because these elements have gone to form the interstellar dust.

In contrast, the composition of the dust is not well known, although there is some infrared evidence for silicates, carbon, and iron, supporting the theory that interstellar dust forms out of interstellar gas. The dust probably also contains some "dirty ice," a frozen mixture of water ice contaminated with trace amounts of ammonia, methane, and other compounds, much like cometary nuclei in our own solar system. ∞ *(Sec. 4.2)*

CONCEPT CHECK

If space is a near-perfect vacuum, how can there be enough dust in it to block starlight?

11.2 Star-Forming Regions

Figure 11.4 shows a magnified view of the central part of Figure 11.1, in the general direction of the constellation Sagittarius. The field of view is mottled with innumerable stars and dark foreground interstellar matter. Historically, astronomers have used the term **nebula** (plural: nebulae) to refer to any "fuzzy" patch (dark or bright) on the sky—any region of space that is clearly distinguishable through a telescope, but not sharply defined, unlike a star or a planet. We now know that many (although by no means all) nebulae are clouds of interstellar dust and gas. If such a cloud happens to obscure stars lying behind it, then we see it as a dark patch on a bright background, as in Figure 11.2(b)—a *dark nebula*. But if something within the cloud—a group of hot young stars, for example—causes it to glow, we see something very different.

▶ **FIGURE 11.4 Star Field in Sagittarius** This enlargement of the central portion of Figure 11.1, shows regions of brightness (vast fields of stars) as well as regions of darkness (where interstellar matter obscures the light from more distant stars). The field of view is about 10° across. *(ESO/S. Guisard)*

◄ **FIGURE 11.5 Galactic Plane** This 35° wide swath of the sky near the center of the Milky Way corresponding to the gray rectangle on Figure 11.1 shows stars, gas, and dust, as well as several distinct fuzzy patches of light known as emission nebulae. *(Harvard College Observatory)*

Reflection nebula

Dust lanes

Emission nebula

(a)

(b)

▲ **FIGURE 11.6 M20–M8 Region** A true-color enlargement of the bottom of Figure 11.5, showing M20 (top) and M8 (bottom) more clearly. *(R. Gendler)*

▲ **FIGURE 11.7 Trifid Nebula** (a) Further enlargement of the top of Figure 11.6, showing only M20 and its interstellar environment. Called the Trifid Nebula because of the dust lanes (in black) that trisect its midsection, the nebula itself (in red) is about 20 light-years across. (b) A false-color infrared image taken by the *Spitzer Space Telescope* reveals bright regions of star-forming activity mostly in those lanes of dust. *(NASA)*

Figure 11.5 expands the field of view of Figure 11.4. The approximate location of this 30 degree swath relative to the Milky Way is indicated by the gray rectangle in Figure 11.1. In addition to stars and dark nebulae, two fuzzy patches of light are visible near the top of the figure. Labeled M8 and M20, they correspond to the 8th and 20th entries in a catalog compiled by Charles Messier, an 18th-century French astronomer. They are called **emission nebulae**—glowing clouds of hot interstellar gas surrounding groups of young, bright stars. Two more emission nebulae (with catalog names NGC 6334 and NGC 5357) can be seen at bottom center. The method of spectroscopic parallax applied to stars visible within these nebulae indicates that their distances from Earth range from 1200 pc (M8) to 2400 pc (NGC 5357) ∞ *(Sec. 10.6)*.

We can gain a better appreciation of these nebulae by examining progressively smaller fields of view. Figure 11.6 is an enlargement of the region near the bottom of Figure 11.5, showing M20 at the top and M8 at the bottom. Figure 11.7 further enlarges the top of Figure 11.6, presenting a close-up of M20 and its immediate environment in both the visible and infrared parts of the spectrum. The total area displayed measures some 12 pc across. Emission nebulae are among the most spectacular objects in the universe, yet they appear only as small, undistinguished patches of light when viewed in the larger context of the Milky Way. Perspective is crucial in astronomy.

The nebulae shown in Figures 11.4–11.7 are regions of glowing, ionized gas. At or near the center of each is at least one newly formed hot O- or B-type star producing copious amounts of ultraviolet light. As ultraviolet photons travel outward from the star, they heat and ionize the surrounding gas. As electrons recombine with nuclei, they emit visible radiation, causing the gas to glow. The predominant red coloration is the result of hydrogen atoms emitting Hα light in the red part of the visible spectrum. ∞ *(Sec. 2.6)*

Woven through the glowing nebular gas, and plainly visible in Figures 11.5–11.7, are dark **dust lanes** obscuring the nebular light. These dust lanes are part of the nebulae and are not just unrelated dust clouds that happen to lie along our line of sight. The bluish region visible in Figure 11.7 immediately above M20 is another type of nebula unrelated to the red emission nebula itself. Called a *reflection nebula*, it is caused by starlight reflecting off intervening dust particles along the line of sight. The blue coloration occurs because short-wavelength blue light is more easily scattered by interstellar matter back toward Earth and into our detectors. Figure 11.8 sketches some of the key features of emission nebulae, illustrating the connection between the central stars, the nebula itself, and the surrounding interstellar medium.

Figure 11.9 shows two other nebulae (located just outside the top-left corner of Figure 11.5). Notice again the hot bright stars embedded within the glowing nebular gas and the overall red hue of the reemitted radiation in parts (a) and (c). (The colors in the *Hubble* images parts [b] and [d] accentuate observations made at different wavelengths.) The relationship between the nebula and the dust lanes

INTERACTIVE

Multiwavelength Cloud with Embedded Stars

▼ **FIGURE 11.8 Nebular Structure** An emission nebula results when ultraviolet radiation from one or more hot stars ionizes part of an interstellar cloud. If starlight happens to encounter another dusty cloud, some of the radiation, particularly at the shorter wavelength blue end of the spectrum, may be scattered back toward Earth, forming a reflection nebula.

(a)

(b)

(c)

R I V U X G

(d)

ANIMATION / VIDEO (MA)

Gaseous Pillars of Star Birth

▲ **FIGURE 11.9 Emission Nebulae** (a) M16, the Eagle Nebula and (b) a close-up of its huge pillars of cold gas and dust within, showing delicate sculptures created by the action of stellar ultraviolet radiation on the original cloud. (c) M8, the Lagoon Nebula and (d) a high-resolution view of its core, a region known as the Hourglass. The varied colors of the insets result from observations at different wavelengths. Both these nebulae lie in the Milky Way plane, just to the upper left of the field of view of Figure 11.5. *(NASA/AURA/ESO)*

is again evident in Figure 11.9 (b) and (d), where regions of gas and dust are simultaneously silhouetted against background nebular emission and illuminated by foreground nebular stars. The interaction between stars and gas is particularly striking in Figure 11.9(b). The three dark "pillars" visible in this spectacular image are part of the interstellar cloud from which the stars formed. The rest of the cloud in the vicinity of the new stars has already been dispersed by their radiation. The fuzz around the edges of the pillars, especially at top right and center, is the result of this ongoing process, where intense stellar radiation continues to "eat away" the cloud, heating and dispersing the dense molecular gas. The process removes the less dense material first, leaving behind delicate sculptures composed of the denser parts of the original cloud, much as wind and water create spectacular

TABLE 11.1 Nebular Properties					
Object	Approx Distance (pc)	Average Diameter (pc)	Density (10^6 particles/m^3)	Mass (Solar Masses)	Temperature (K)
M8	1200	14	80	2600	7500
M16	1800	8	90	600	8000
M17	1500	7	120	500	8700
M20	1600	6	100	250	8200

CONCEPT CHECK

If emission nebulae are powered by ultraviolet light from hot stars, why do they appear red?

▼ FIGURE 11.10 **Nebular Spectrum** The visible spectrum of the hot gases in a nearby star-forming region known as the Omega Nebula (M17). Shining by the light of several very hot stars, the nebula produces a complex spectrum of bright and dark lines (bottom), also shown here as an intensity trace from red to blue (center). *(ESO)*

structures in Earth's deserts and shores by eroding away the softest rock. The pillars will eventually be destroyed, but probably not for another hundred thousand or so years.

Table 11.1 lists some vital statistics for some of the nebulae shown in this section. Unlike stars, nebulae are large enough for their sizes to be measurable by simple geometry. ∞ *(More Precisely 4-1)* Coupling this size information with estimates of the amount of matter along our line of sight (as revealed by absorption of the nebula's light), we can find a nebula's density. Emission nebulae typically contain a few hundred particles per cubic centimeter (10^8 per cubic meter), mostly protons and electrons. Figure 11.10 shows the emission nebula M17, along with its spectrum at visible and near-ultraviolet wavelengths. ∞ *(Sec. 2.5)*. Numerous emission lines can be seen, providing further information on the nebula. Analyses of such spectra reveal nebular compositions similar to the Sun and other stars, as well as elsewhere in the interstellar medium. Spectral line widths imply temperatures of around 8000 K. ∞ *(Sec. 2.8)*

11.3 Dark Dust Clouds

Emission nebulae are only one small component of interstellar space. Most of space—in fact, more than 99 percent of it—is devoid of such regions and contains no stars. It is simply dark. The average temperature of a typical dark region of interstellar space is about 100 K. Compare this with 273 K, at which water freezes, and 0 K, at which atomic and molecular motions cease. ∞ *(More Precisely 2-1)*

Within the dark voids among the nebulae and the stars lurks another distinct type of astronomical object, the **dark dust cloud.** These clouds are cooler than their surroundings, with temperatures as low as a few tens of kelvins, and thousands or even millions of times denser. In some regions, densities exceeding 10^9 atoms/m^3 (1000 atoms/cm^3) are found. Researchers often refer to dark dust clouds as *dense* interstellar clouds, but recognize that even these densest interstellar regions are barely denser than the best vacuum achievable in terrestrial laboratories. Still, it is because their density is much larger than the average value of 10^6 atoms/m^3 in interstellar space that we can distinguish these clouds from the surrounding expanse of the interstellar medium.

Obscuration of Visible Light

Dark dust clouds bear little resemblance to terrestrial clouds. Most are bigger than our solar system, and some are many parsecs across, although they still make up no more than a few percent of the entire volume of interstellar space. Despite their name, these clouds are composed primarily of gas, just like the rest

(a)

(b)

▲ FIGURE 11.11 Obscuration and Emission   (a) At optical wavelengths, this dark dust cloud (known as L977) can be seen only by its obscuration of background stars. (b) At radio wavelengths, it emits strongly in the CO molecular line, with the most intense radiation coming from the densest part of the cloud. *(C. and E. Lada)*

of the interstellar medium. However, their absorption of starlight is due almost entirely to the dust they contain.

Figure 11.2(b) is a good example of a dark dust cloud. Figure 11.11(a), a region called L997 in the constellation Cygnus, is another. Some early (18th-century) observers thought that these dark patches on the sky were simply empty regions of space that happened to contain no bright stars. However, by the late 19th century, astronomers had discounted this idea. They realized that seeing clear spaces among the stars would be like seeing clear tunnels between the trees in a forest, and it was statistically impossible that so many tunnels would lead directly away from Earth.

Before the advent of radio astronomy, astronomers had no direct means of studying clouds like L977. Emitting no visible light, they are generally undetectable to the eye except by the degree to which they dim starlight. However, as shown in Figure 11.11(b), the cloud's radio emission—in this case from carbon monoxide (CO) molecules contained within its volume—outlines it clearly at radio wavelengths, providing an indispensable tool for the study of such objects.

Figure 11.12 is a spectacular wide-field image of another dark dust cloud. Taking its name from a neighboring star system, Rho Ophiuchus, this dust cloud resides relatively nearby, making it one of the most intensely studied regions of star formation in the Milky Way. Pockets of heavy blackness mark regions where the dust and gas are especially concentrated and the light from the background stars is completely obscured. Measuring several parsecs across, the Ophiuchus cloud is also visible on the right side of Figure 11.5. Note that this cloud, like most interstellar clouds, is very irregularly shaped. Some of the features labeled in the image are part of the cloud itself, whereas others are newly formed stars near the edge of the cloud. Still other objects have no connection to the cloud and just happen to lie along the line of sight.

▶ FIGURE 11.12 Dark Dust Cloud   The Ophiuchus dark dust cloud resides only 170 pc away, surrounded by colorful stars and nebulae that are actually small illuminated parts of a much bigger and invisible molecular cloud engulfing much of the region shown. *(R. Gendler/J. Misti/S. Mazlin)*

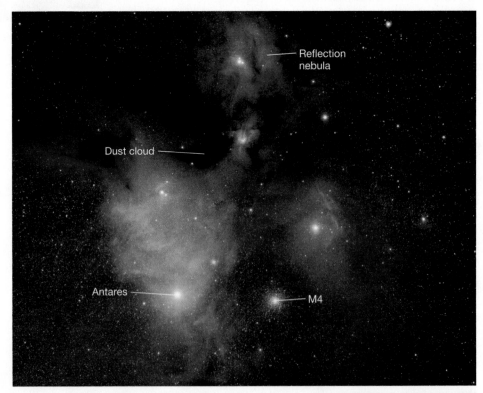

Reflection nebula

Dust cloud

Antares

M4

Figure 11.13 shows a particularly striking and well-known example of a dark dust cloud—the Horsehead Nebula in Orion. This curiously shaped finger of gas and dust projects out from the much larger dark cloud in the bottom half of the image and stands out clearly against the red glow of a background emission nebula.

21-Centimeter Radiation

Often, dark dust clouds absorb so much light from background stars that they cannot easily be studied by the optical techniques described in Section 11.1. Fortunately, astronomers have an important alternative means of probing their structure that relies on low-energy radio emission from the interstellar gas itself.

Much of the gas in interstellar space is atomic hydrogen. Recall that a hydrogen atom consists of one electron orbiting a single-proton nucleus. ∞ *(Sec. 2.6)* Besides its orbital motion around the central proton, the electron also has some rotational motion—that is, *spin*—about its own axis. The proton also spins. This model parallels a planetary system, in which, in addition to the orbital motion of a planet about a central star, both the planet (electron) and the star (proton) rotate about their axes.

The laws of physics dictate that there are just two possible spin configurations for a hydrogen atom in its ground state. As illustrated in Figure 11.14, the electron and proton can either rotate in the *same* direction, with their spin axes parallel, or they can rotate in *opposite* directions, with their axes parallel but oppositely oriented. The former configuration has slightly higher energy than the latter. When a slightly excited hydrogen atom having the electron and proton spinning in the same direction drops down to the less-energetic, opposite-spin state, the transition releases a photon with energy equal to the energy difference between the two states.

Because the energy difference between the two states is very small, the energy of the emitted photon is very low. ∞ *(Sec. 2.6)* Consequently, the wavelength of the radiation is long—about 21 cm. That wavelength lies in the radio portion of the electromagnetic spectrum. Researchers refer to the spectral emission line that results from this hydrogen–spin-flip process as the **21-centimeter line.** It provides a vital probe into any region of the universe containing atomic hydrogen gas. Needing no visible starlight to help calibrate their signals, radio astronomers can observe any interstellar region that contains enough hydrogen to produce a detectable signal. Even the low-density regions between dark dust clouds can be studied.

Of great importance is the fact that the wavelength of 21-cm radiation is much larger than the typical size of interstellar dust particles. Accordingly, this radio radiation reaches Earth completely unaffected by interstellar debris. The opportunity to observe interstellar space well beyond a few thousand parsecs, and in directions lacking detectable background stars, makes 21-cm observations among the most important and useful tools in all of astronomy. As we will see in Chapters 10 and 15, it has been indispensable in allowing astronomers to map out the large-scale structure of our own galaxy and many others.

(a)   (b)

R I V U X G

▲ **FIGURE 11.13 Horsehead Nebula** (a) The Horsehead Nebula is a striking example of a dark dust cloud, silhouetted against the bright background of an emission nebula. (b) A stunning image of the Horsehead, taken at highest resolution by the Very Large Telescope (VLT) in Chile. *(Sec. 3.2)* The "neck" of the horse is about 0.25 pc across. This nebular region lies roughly 1500 pc from Earth, in the constellation Orion. *(Royal Observatory of Belgium; ESO)*

MA ANIMATION / VIDEO
Pillars Behind the Dust

MA ANIMATION / VIDEO
The Tarantula Nebula

MA ANIMATION / VIDEO
Horsehead Nebula

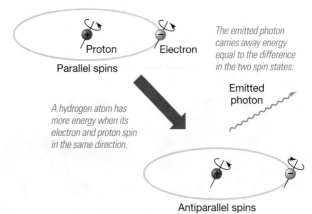

Proton   Electron
Parallel spins

The emitted photon carries away energy equal to the difference in the two spin states.

A hydrogen atom has more energy when its electron and proton spin in the same direction.

Emitted photon

Antiparallel spins

▲ **FIGURE 11.14 Hydrogen 21-cm Emission** A ground-state hydrogen atom switches from a higher-energy state (top) to a lower-energy state (bottom), emitting a radio photon in the process.

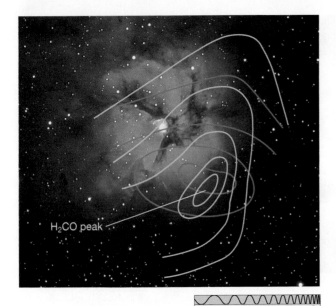

R I V U X G

▲ **FIGURE 11.15 Molecules Near M20** Contour map of formaldehyde near the M20 nebula, showing how the molecule is more abundant in the darkest interstellar regions. The maximum density of formaldehyde lies just to the bottom right of the visible nebula. The colors show the intensity of formaldehyde spectral lines at two different frequencies. (*AURA*)

PROCESS OF SCIENCE CHECK

In mapping molecular clouds, why do astronomers use observations of "minority" molecules such as carbon monoxide and formaldehyde when these molecules constitute only a tiny fraction of the total in interstellar space?

▼ **FIGURE 11.16 Molecular Cloud Complexes** This false-color radio map shows the outer portion of the Milky Way as it appears in CO emission. The bright regions are molecular cloud complexes, dense regions of interstellar space where molecules abound and, apparently, stars are forming. The map is huge, extending over about a quarter of the sky. It was made from 1,696,800 observations of CO spectra. (*Five College Radio Astronomy Observatory*)

Molecular Gas

In certain interstellar regions of cold (typically 10–20 K) neutral gas, densities can reach as high as 10^{12} particles/m^3, and most gas particles are not atoms, but *molecules*. These regions are known as **molecular clouds**. Only long-wavelength radio radiation can escape from these dense, dusty parts of interstellar space.

Molecular hydrogen (H_2) is by far the most common constituent of molecular clouds; but it does not emit or absorb radio radiation, so it cannot easily be used as a probe of cloud structure. Nor are 21-cm observations helpful—they are sensitive only to *atomic* hydrogen, not to the molecular form of the gas. Instead, astronomers use radio observations of "tracer" molecules, such as carbon monoxide (CO), water (H_2O), and formaldehyde (H_2CO), to study the dark interiors of these dusty regions. These molecules are produced by chemical reactions within the clouds and emit energy in the radio part of the spectrum. They are rare—perhaps one per billion hydrogen molecules—but when we observe them we can be confident that large amounts of molecular hydrogen and other important constituents are also present. Figure 11.15 shows a contour map of the distribution of formaldehyde molecules around M20. Notice that the emission peaks well away from the visible nebula.

Molecular observations reveal that the spectacular star-forming regions we see are actually parts of much larger molecular clouds. The bright emission nebula "bursts out" of its dark parent cloud when the radiation from a group of new-born stars heats and ionizes the cold molecular gas. And molecular clouds do not exist as distinct objects in space. Rather, they form huge **molecular cloud complexes** tens of parsecs across, some containing enough gas to make millions of stars like our Sun. About 1000 such complexes are known in our Galaxy. Figure 11.16 is a radio map of a large region of the sky, made using spectral lines of the CO molecule, showing molecular cloud complexes stretching across the entire field of view.

Why are molecules found only in the densest interstellar clouds? One reason is that the dust protects the fragile molecules from the harsh interstellar environment. The same absorption that prevents high-frequency radiation from getting out to our detectors also prevents it from getting in to destroy the molecules. The dust may also help form the molecules. Dust grains provide both a place where atoms can stick and react and a means of dissipating any heat associated with the reaction, which might otherwise destroy the newly formed molecules.

In recent years, astronomers have come to realize that the interstellar medium is a dynamic environment in which energy released by new-born stars (Section 11.5) and supernovae (see Chapter 12) drives large-scale, turbulent motion in the gas. The cold molecular clouds we see may simply be regions of dense gas temporarily compressed by the large-scale flow—transient islands in a sea of surrounding chaos.

R I V U X G

11.4 Formation of Stars Like the Sun

Let us now turn our attention to the connection between the interstellar medium and the stars in our Galaxy. How do stars form? What factors determine their masses, luminosities, and spatial distribution? In short, what basic processes are responsible for the appearance of our night sky?

Gravity and Heat

Simply stated, star formation begins when part of the interstellar medium—one of the cold, dark clouds discussed in the previous section—collapses under its own weight. An interstellar cloud is maintained in hydrostatic equilibrium by a balance between two basic opposing influences: gravity (which is always directed inward) and heat (in the form of outwardly directed pressure). ∞ *(Sec. 9.2)* If gravity somehow begins to dominate over heat, then the cloud can lose its equilibrium and start to contract.

Consider a small portion of a large interstellar cloud. Concentrate first on just a few atoms, as shown in Figure 11.17(a). Even though the cloud's temperature is very low, each atom has some random motion. ∞ *(More Precisely 2-1)* Each atom is also influenced by the gravitational attraction of all its neighbors. The gravitational force is not large, however, because the mass of each atom is so small. When a few atoms accidentally cluster for an instant, as sketched in Figure 11.17(b), their combined gravity is insufficient to bind them into a lasting, distinct clump of matter. This accidental cluster disperses as quickly as it formed (Figure 11.17c). The effect of heat—the random motion of the atoms—is much stronger than the effect of gravity.

As the number of atoms increases, their gravitational attraction increases too, and eventually the collective gravity of the clump is strong enough to prevent it from dispersing back into interstellar space. How many atoms are required for this to be the case? The answer, for a typical cool (100 K) cloud, is about 10^{57}—in other words, the clump must have mass comparable to that of the Sun. If our hypothetical clump is more massive than that, it will not disperse. Instead, its gravity will cause it to contract, ultimately to form a star.

Of course, 10^{57} atoms don't just clump together by random chance. Rather, star formation is *triggered* when a sufficiently massive pocket of gas is squeezed by some external event. Perhaps it is compressed when two blobs of gas collide, or by the shock wave produced when a nearby group of O- or B-type stars form and heat their surroundings, creating an emission nebula, or when a neighboring star explodes as a supernova (see Section 12.4). Or possibly part of an interstellar cloud simply becomes too cold for its internal pressure to support it against its own gravity. Whatever the cause, theory suggests that once the collapse begins, star formation inevitably follows.

Table 11.2 lists seven evolutionary stages that an interstellar cloud goes through before becoming a main-sequence star like our Sun. These stages are characterized by different central temperatures, surface temperatures, central densities, and radii

(a)   (b)   (c)

If more than a few atoms interact, the group would come together, start to slide by, but then pause, and finally contract into a clump.

◄ **FIGURE 11.17 Atomic Motions** The motions of a few atoms within an interstellar cloud are influenced by gravity so slightly that the atoms' paths are hardly changed (a) before, (b) during, and (c) after an accidental, random encounter.

TABLE 11.2   Prestellar Evolution of a Sun-like Star

Stage	Approximate Time to Next Stage (yr)	Central Temperature (K)	Surface Temperature (K)	Central Density (particles/m³)	Diameter[1] (km)	Object
1	2×10^6	10	10	10^9	10^{14}	Interstellar cloud
2	3×10^4	100	10	10^{12}	10^{12}	Cloud fragment
3	10^{15}	10,000	100	10^{18}	10^{10}	Cloud fragment/protostar
4	10^6	1,000,000	3000	10^{24}	10^8	Protostar
5	10^7	5,000,000	4000	10^{28}	10^7	Protostar
6	2×10^6	10,000,000	4500	10^{31}	2×10^6	Star
7	10^{10}	15,000,000	6000	10^{32}	1.5×10^6	Main-sequence star

[1]For comparison, recall that the diameter of the Sun is 1.4×10^6 km and that of the solar system roughly 1.5×10^{10} km.

of the prestellar object. They trace its progress from a quiescent interstellar cloud to a genuine star. The numbers given in Table 11.2 and the following discussion are valid only for stars of approximately the same mass as the Sun. In the next section we will relax this restriction and consider the formation of stars of other masses.

Stage 1—An Interstellar Cloud

The first stage in the star-formation process is a dense interstellar cloud—the core of a dark dust cloud or perhaps part of a molecular cloud. These clouds are truly vast, sometimes spanning tens of parsecs (10^{14} to 10^{15} km) across. Typical temperatures are about 10 K throughout, with a density of perhaps 10^9 particles/m³. Stage-1 clouds contain thousands of times the mass of the Sun, mainly in the form of cold atomic and molecular gas. (The dust they contain is important for cooling the cloud as it contracts and also plays a crucial role in planet formation, but it constitutes a negligible fraction of the total mass.) ∞ *(Sec. 4.3)*

The dark region outlined by the red and green radio contours in Figure 11.15 (*not* the emission nebula itself, where stars have already formed) probably represents a stage-1 cloud just starting to contract. Doppler shifts of the observed formaldehyde lines indicate that it is probably infalling. Less than a light year across, this region has a total mass over 1000 times the mass of the Sun—considerably greater than the mass of M20 itself.

Once the cloud is compressed past the point where gravity overcomes gas pressure, theory indicates that it will naturally fragment into smaller and smaller clumps of matter due to further gravitational instabilities in the gas. As illustrated in Figure 11.18, a typical cloud may form tens, hundreds, even thousands of fragments, each destined to form a star or group of stars. The whole process, from initial unstable cloud to many collapsing fragments, takes perhaps a few hundred thousand years. Depending on the precise conditions in the cloud, it may form either a few dozen stars, each much larger than our Sun, or a collection of hundreds or thousands of stars comparable to or smaller than our Sun. There is little evidence for stars born in isolation, one star from one cloud. Most stars—perhaps

▶ **FIGURE 11.18 Cloud Fragmentation**   As an interstellar cloud contracts, gravitational instabilities cause it to fragment into smaller pieces. The pieces themselves continue to fragment, eventually forming many tens or hundreds of individual stars.

all—appear to originate as members of multiple systems. Figure 11.19 presents recent observational evidence that directly supports the scenario just described. It shows a dense, dark molecular cloud near the center of our Galaxy, containing within it several dozen even denser prestellar fragments, having masses ranging from a few tens to a few hundred times the mass of the Sun.

The process of continuing fragmentation is eventually stopped by the increasing density within the shrinking cloud. As fragments continue to contract, they eventually become so dense that radiation cannot get out easily. The trapped radiation causes the temperature to rise, the pressure to increase, and the fragmentation to stop. However, the contraction continues.

Stages 2 and 3—A Contracting Cloud Fragment

As it enters stage 2, a fragment destined to form a star like the Sun—the end product of the process sketched in Figure 11.18—contains between one and two solar masses of material. Estimated to span a few hundredths of a parsec across, this fuzzy, gaseous blob is still about 100 times the size of our solar system. Its central density is about 10^{12} particles/m^3.

Even though the fragment has shrunk substantially, its average temperature is not much different from that of its parent cloud. The reason is that the gas constantly radiates large amounts of energy into space. The material of the fragment is so thin that photons produced anywhere within it easily escape without being reabsorbed, so virtually all the energy released in the contraction is radiated away and does not raise the temperature. Only at the center, where the radiation must traverse the greatest amount of material in order to escape, is there any appreciable temperature increase. The gas there might be as warm as 100 K by this stage. For the most part, however, the fragment stays cold as it shrinks.

Several tens of thousands of years after it first began contracting, the stage-2 fragment has shrunk to a gaseous sphere with a diameter roughly the size of our solar system (still 10,000 times the size of our Sun). Now, at the start of stage 3, the inner regions have become opaque to their own radiation and have started to heat up considerably, as noted in Table 11.2. The central temperature has reached about 10,000 K—hotter than the hottest steel furnace on Earth. However, the gas near the edge is still able to radiate its energy into space and so remains cool. The central density by this time is approximately 10^{18} particles/m^3 (still only 10^{-9} kg/m^3 or so).

For the first time, our fragment is beginning to resemble a star. The dense, opaque region at the center is called a **protostar.** Its mass increases as more and more material rains down on it from outside, although its radius continues to shrink because its pressure is still unable to overcome the relentless pull of gravity. By the end of stage 3, we can distinguish a "surface" on the protostar—its *photosphere.* Inside the photosphere, the protostellar material is opaque to the radiation it emits. (Note that this is the same operational definition of *surface* we used for the Sun.) ∞ *(Sec. 9.1)* From here on, the surface temperatures listed in Table 11.2 refer to the photosphere and not to the edge of the collapsing fragment, where the temperature remains low.

Figure 11.20 shows a star-forming region in Orion. Lit from within by several O-type stars, the bright Orion Nebula is partly surrounded by a vast molecular cloud that extends well beyond the roughly 1-parsec-square region bounded by the photograph in Figure 11.20(c). The Orion complex harbors several smaller sites of intense radio emission from molecules deep within its core. Shown in Figure 11.20(d), they measure about 10^{10} km, about the diameter of our solar system. Their density is about 10^{15} particles/m^3, much higher than the density of the surrounding cloud. Although the temperatures of these regions cannot be estimated reliably, many researchers regard the regions as objects between stages 2 and 3, on the threshold of becoming protostars. (See Figures 11.25 and 11.31 for visible and infrared images of part of the nebula showing other evidence for protostars.)

R I V U X G

▲ **FIGURE 11.19 Prestellar Fragments**   This remarkable image combines infrared, millimeter, and radio observations of small molecular cloud called G0.253+0.016 near the Galactic center. The cyan image is an infrared view from *Spitzer*, while the white contours trace radio emission from molecules in the cloud, outlining its full extent. The cloud is so dense that it absorbs the infrared background, and so appears dark. The red blobs are high-resolution *ALMA* observations of dense prestellar fragments within the cloud. ∞ *(Sec. 5.4)* These fragments are well on their way to forming stars; their total mass is a few thousand times the mass of the Sun. *(J. Rathborne/Astrophysical Journal)*

Stage 4—A Protostar

As the protostar evolves, it shrinks, its density increases and its temperature rises, both in the core and at the photosphere. Some 100,000 years after the fragment formed, it reaches stage 4, where its center seethes at about 1,000,000 K. The electrons and protons ripped from atoms whiz around at hundreds of kilometers per second, but the temperature is still short of the 10^7 K needed to ignite the proton–proton nuclear reactions that fuse hydrogen into helium. Still much larger than the Sun, our gassy heap is now about the size of Mercury's orbit. Its surface temperature has risen to a few thousand kelvins.

Knowing the protostar's radius and surface temperature, we can calculate its luminosity using the radius–luminosity–temperature relationship. ∞ *(Sec. 10.4)* Remarkably, it turns out to be around 1000 times the luminosity of the Sun. Because nuclear reactions have not yet begun in the protostar's core, this luminosity is due entirely to the release of gravitational energy as the protostar continues to shrink and material from the surrounding fragment (which we called the solar nebula back in Chapter 4) rains down on its surface. ∞ *(Sec. 4.3)*

In the hunt for examples of more advanced stages of star formation, radio techniques become less useful because objects in stages 4, 5, and 6 have increasingly higher temperatures. By Wien's law, their emission shifts toward shorter wavelengths, and so they shine most strongly in the infrared. ∞ *(Sec. 2.4)* One particularly bright infrared emitter, known as the Becklin–Neugebauer object,

▼ **FIGURE 11.20 Orion Nebula, Up Close** INTERACTIVE
(a) The constellation Orion, with the region around its famous emission nebula marked by a rectangle (see Figure 0.2).
(b) Enlargement of the framed region in part (a), but here shown in the infrared, revealing how the nebula is partly surrounded by a vast molecular cloud. Various parts of this cloud are probably fragmenting and contracting, with even smaller sites forming protostars. The three frames at the right show some of the evidence for those protostars: (c) real-color visible image of embedded nebular "knots" within the Orion Nebula, (d) false-color radio image of some intensely emitting molecular sites, and (e) high-resolution image of one of several young stars surrounded by disks of gas and dust where planets might ultimately form. *(P. Sanz/Alamy; SST; CfA; NASA)*

(a) ⟨— 100 arc min —⟩ R I V U X G

(b) R I V U X G

(c) ⟨— 0.2 pc —⟩ R I V U X G

(d) ⟨— 0.03 pc —⟩ R I V U X G

(e) ⟨— 0.03 pc —⟩ R I V U X G

was detected in the core of the Orion molecular cloud in the 1970s. Its luminosity is around 1000 times the luminosity of the Sun. Most astronomers agree that this warm, dense blob is a high-mass protostar, probably around stage 4.

By the time stage 4 is reached, our protostar's physical properties can be plotted on a Hertzsprung–Russell (H–R) diagram. ∞ (Sec. 10.5) At each phase of a star's evolution, its surface temperature and luminosity can be represented by a single point on the diagram. The motion of that point around the diagram as the star evolves is known as the star's **evolutionary track.** It is a graphical representation of a star's life. Note that it has nothing to do with any real spatial motion of the star. The red track on Figure 11.21 depicts the approximate path followed by our interstellar cloud fragment since it became a protostar at the end of stage 3 (which itself lies off the right-hand edge of the figure). Figure 11.22 is an artist's sketch of an interstellar gas cloud proceeding along the evolutionary path outlined so far.

Stage 5—Protostellar Evolution

Our protostar is still not in equilibrium. Even though its temperature is now so high that outward-directed pressure has become a powerful countervailing influence against gravity's inward pull, the balance is not yet perfect. The protostar's internal heat gradually diffuses out from the hot center to the cooler surface, where it is radiated away into space. As a result, the contraction slows, but it does not stop completely.

After stage 4, the protostar moves downward on the H–R diagram (toward lower luminosity) and slightly to the left (toward higher temperature), as shown in Figure 11.23. By stage 5, the protostar has shrunk to about 10 times the size of the Sun, its surface temperature is about 4000 K, and its luminosity has fallen to about 10 times the solar value. The central temperature has risen to about 5,000,000 K. The gas is completely ionized by now, but the protons still do not have enough thermal energy for nuclear fusion to begin.

Protostars often exhibit violent surface activity during this phase of their evolution, resulting in extremely strong protostellar winds, much denser than that of our Sun. This portion of the evolutionary track is often called the **T-Tauri phase,** after T-Tauri, the first "star" (actually protostar) to be observed in this stage of prestellar development. The interaction between the strong winds and the nebular disk out of which the protostar (and possibly planets) is still forming often results in a *bipolar flow*—expelling two "jets" of matter perpendicular to the disk. Figure 11.24(a) shows a particularly striking example.

These outflows can be very energetic. Figure 11.24(b) shows a portion of the Orion molecular cloud, south of the Orion Nebula, where a protostar is seen still surrounded by a bright nebula, its turbulent wind spreading out into the interstellar medium. Below it (enlarged in the inset) are twin jets known as HH1 and HH2. (HH stands for Herbig–Haro, after the investigators who first cataloged such objects.) Formed in another (unseen) protostellar disk, these jets have traveled outward for almost half a light-year before colliding with interstellar matter. More Herbig–Haro objects can be seen in the upper right portion of the figure.

Events proceed more slowly as the protostar approaches the main sequence. The initial contraction and fragmentation of the interstellar cloud occurred quite rapidly, but by

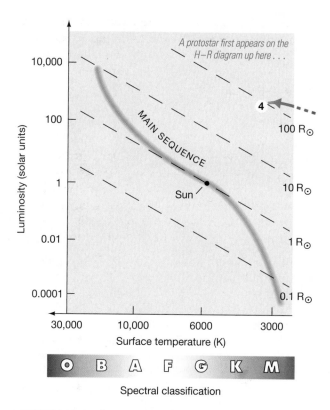

▲ FIGURE 11.21 **Protostar on the H–R Diagram** The red arrow indicates the approximate evolutionary track followed by an interstellar cloud fragment before becoming a stage-4 protostar. The boldface numbers on this and subsequent H–R plots refer to the prestellar evolutionary stages listed in Table 11.2.

▼ FIGURE 11.22 **Interstellar Cloud Evolution** Artist's conception of the changes in an interstellar cloud during the early evolutionary stages outlined in Table 11.2. (Not drawn to scale.) The duration of each stage, in years, is indicated.

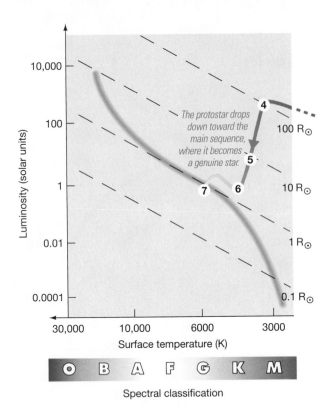

Luminosity (solar units) (vertical axis): 10,000 · 100 · 1 · 0.01 · 0.0001

The protostar drops down toward the main sequence, where it becomes a genuine star.

100 R☉ · 10 R☉ · 1 R☉ · 0.1 R☉

Surface temperature (K): 30,000 · 10,000 · 6000 · 3000

O B A F G K M

Spectral classification

◄ **FIGURE 11.23 Newborn Star on the H–R Diagram INTERACTIVE** The changes in a protostar's observed properties are shown by the path of decreasing luminosity, from stage 4 to stage 6. At stage 7, the newborn star has arrived on the main sequence.

▼ **FIGURE 11.24 Protostellar Outflow** (a) This remarkable image (at right) shows two jets emanating from the young star system HH30, the result of matter accreting onto an embryonic star near the center. (b) The view below of the Orion molecular cloud shows the outflow from a newborn star, still surrounded by nebular gas. The inset shows a pair of jets called HH1/HH2, formed when matter falling onto another protostar (still obscured by the dusty cloud fragment from which it formed) creates a pair of high-speed jets of gas perpendicular to the flattened protostellar disk. The jets are nearly 1 light-year across. Several more Herbig–Haro objects can be seen at top right—one of them resembling a "waterfall." *(AURA; NASA)*

Jet

Disk

(a)

R I V U X G

MA ANIMATION / VIDEO
Bi-Polar Outflow

R I V U X G

(b)

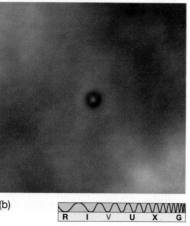

(a)

R I V U X G

(b)

R I V U X G

◄ **FIGURE 11.25 Protostars**   (a) An infrared, edge-on image of a planetary system–size dusty disk in the Orion region, showing heat and light emerging from its center. This unnamed source seems to be a low-mass protostar around stage 5 in the H–R diagram. (b) An optical, face-on image of a slightly more advanced circumstellar disk surrounding an embedded protostar in Orion. *(NASA)*

stage 5, as the protostar nears the status of a full-fledged star, its evolution slows. Its contraction is governed largely by the rate at which it can radiate its internal energy into space. As the luminosity decreases, so, too, does the contraction rate.

Until the *Infrared Astronomy Satellite (IRAS)* was launched in the early 1980s, astronomers were aware only of very massive stars forming in clouds far away. But *IRAS* showed that stars are forming much closer to home, and some of these protostars have masses comparable to that of our Sun. Figure 11.25 shows two examples of low-mass protostars—both spotted by the *Hubble Space Telescope* in a rich star-forming region in Orion. Their infrared heat signatures are those expected of a stage-5 object.

Stages 6 and 7—A Newborn Star

Some 10 million years after reaching stage 4, the protostar finally becomes a true star. By stage 6, when our roughly one-solar-mass object has shrunk to a radius of about 1,000,000 km, the contraction has raised the central temperature to 10,000,000 K—enough to ignite nuclear burning. Protons begin fusing into helium nuclei in the core, and a star is born. As shown in Figure 11.23, the star's surface temperature at this point is about 4500 K, still a little cooler than the Sun. Even though the radius of the newly formed star is slightly larger than that of the Sun, its lower temperature means that its luminosity is slightly less than (actually, about two-thirds of) the present solar value.

Indirect evidence for stage-6 stars comes from infrared observations of objects that seem to be luminous hot stars hidden from optical view by surrounding dark clouds. Their radiation is mostly absorbed by a "cocoon" of dust, then reemitted by the dust as infrared radiation. Two considerations support the idea that the hot stars responsible for heating the clouds have only recently ignited: (1) dust cocoons are predicted to disperse quite rapidly once their central stars form, and (2) these objects are invariably found in the dense cores of molecular clouds, consistent with the star-formation sequence just outlined.

Over the next 30 million years or so, the stage-6 star contracts a little more. Its central density rises to about 10^{32} particles/m^3 (more conveniently expressed as 10^5 kg/m^3), the central temperature increases to 15,000,000 K, and the surface temperature reaches 6000 K. By stage 7 the star finally reaches the main sequence just about where our Sun now resides. Pressure and gravity are finally balanced in the stellar interior, and the rate at which nuclear energy is generated in the core exactly matches the rate at which energy is radiated from the surface.

The journey from interstellar cloud to star occurs over the course of 40–50 million years. Although this is a long time by human standards, it is still less than 1 percent of the Sun's lifetime on the main sequence. Once an object begins fusing hydrogen and establishes a "gravity-in/pressure-out" hydrostatic equilibrium,

CONCEPT CHECK

What distinguishes a collapsing cloud from a protostar, and a protostar from a star?

it burns steadily for a very long time. The star's location on the H–R diagram—that is, its surface temperature and luminosity—will remain almost unchanged for the next 10 billion years.

11.5 Stars of Other Masses

The numerical values and evolutionary track just described are valid only for 1-solar-mass stars. The temperatures, densities, and radii of prestellar objects of other masses exhibit similar trends, but the details differ, in some cases quite considerably. Perhaps not surprisingly, the most massive fragments formed within interstellar clouds tend to produce the most massive protostars and eventually the most massive stars. Similarly, low-mass fragments give rise to low-mass stars.

The Zero-Age Main Sequence

Figure 11.26 compares the theoretical pre-main-sequence evolutionary track taken by our Sun with the corresponding tracks of a 0.3-solar-mass star and a 3.0-solar-mass star. All three tracks traverse the H–R diagram in the same general manner, but cloud fragments that eventually form stars more massive than the Sun approach the main sequence along a higher track on the diagram, while those destined to form less massive stars take a lower track. The *time* required for an interstellar cloud to become a main-sequence star also depends strongly on its mass. ∞ *(Sec. 10.7)* The most massive fragments contract into O-type stars in a mere million years, roughly 1/50 the time taken by the Sun. The opposite is the case for prestellar objects having masses much less than our Sun. A typical M-type star, for example, requires nearly a billion years to form.

Whatever the mass, the endpoint of the prestellar evolutionary track is the main sequence. A star is considered to have reached the main sequence when hydrogen burning begins in its core and the star's properties settle down to stable values. The main-sequence band predicted by theory is usually called the **zero-age main sequence** (ZAMS). It agrees quite well with the main sequences observed for stars in the vicinity of the Sun and those observed in more distant stellar systems.

It is important to realize that *the main sequence is not an evolutionary track—stars do not evolve along it.* Rather, it is a "waystation" on the H–R diagram where stars stop and spend most of their lives—low-mass stars at the bottom, high-mass stars at the top. Once on the main sequence, a star stays in roughly the same location in the H–R diagram during its whole time as a stage-7 object. In other words, a star that arrives on the main sequence as, say, a G-type star can never "work its way up" to become a B- or an O-type main-sequence star, nor move down to become an M-type red dwarf. As we will see in Chapter 12, the next stage of stellar evolution occurs when a star leaves the main sequence. When this occurs, the star will have pretty much the same surface temperature and luminosity it had when it arrived on the main sequence millions (or billions) of years earlier.

"Failed" Stars

Some cloud fragments are too small ever to become stars. Pressure and gravity come into equilibrium before the central temperature becomes hot enough to fuse hydrogen, so they never evolve beyond the protostar stage. Rather than turning into stars, such low-mass fragments continue to cool, eventually becoming compact, dark "clinkers"—cold fragments of unburned matter—in interstellar space. Small, faint, and cool (and growing ever colder), these objects are known collectively as **brown dwarfs.** On the basis of theoretical modeling, astronomers calculate that the minimum mass of gas needed to generate core temperatures high enough to begin nuclear fusion is about 0.08 solar masses—80 times the mass of Jupiter.

CONCEPT CHECK

Do stars evolve along the main sequence?

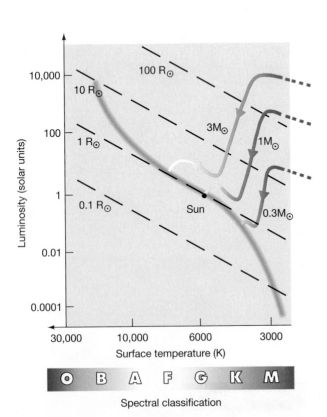

▲ **FIGURE 11.26 Prestellar Evolutionary Tracks**   Prestellar evolutionary paths for stars more massive and less massive than our Sun.

(a)

(b)

R I V U X G

(c)

R I V U X G

Vast numbers of brown dwarfs may well be scattered throughout the universe—fragments frozen in time somewhere in the cloud-contraction phase. With our current technology, we have great difficulty in detecting them, whether they orbit other stars or float alone in space. We can telescopically detect stars and spectroscopically infer atoms and molecules, but low-mass astronomical objects far beyond our solar system are very hard to see. However, recent advances in observational hardware and image-processing techniques have identified many likely brown dwarf candidates, often using techniques similar to those employed in the search for extrasolar planets. ∞ *(Sec. 4.4)* (Recall that most stars are found in binaries; the same may very well be true of brown dwarfs. ∞ *(Sec. 10.7)*)

Figure 11.27 shows Gliese 623, a binary system containing a brown dwarf candidate, originally identified by radial velocity measurements and Gliese 229, first identified as a possible brown dwarf by ground-based infrared observations. More recent infrared images from space show many more brown dwarf candidates. Figure 11.27(c) is a *Spitzer* image of a small star cluster showing not only many bright stars, but also numerous much fainter objects thought to be brown dwarfs. Based on current observations, it seems that up to 100 billion cold, dark, brown dwarfs may lurk in the depths of interstellar space—comparable to the total number of "real" stars in our Galaxy. Figure 11.27(d) compares brown dwarfs, notably Gliese 229 of part (b), to some other well-known objects.

(d)

▲ **FIGURE 11.27 Brown Dwarfs** (a) This image shows Gliese 623, a binary system that may contain a brown dwarf (marked by an arrow). (b) This image of the binary-star system Gliese 229 shows two objects only 7″ apart; the fainter "star" (arrow) has a luminosity only a few millionths that of the Sun and an estimated mass about 50 times that of Jupiter. (The rings in part a and the spike in part b are instrumental artifacts.) (c) An infrared image of a star cluster just north of the Orion Nebula, showing bright objects that are stars and many faint specks that are brown-dwarf candidates. (d) This rendering compares the sizes of some stars, brown dwarfs, and planets. *(NASA)*

11.6 Star Clusters

The end result of the contraction and fragmentation of an interstellar gas cloud is a group of stars, all formed from the same parent cloud and lying in the same region of space. Such a collection of stars is called a **star cluster.** Figure 11.28 shows a spectacular view of a newborn star cluster and (part of) the interstellar cloud from which it came. Because all the stars formed at the same time out of the same cloud and under the same environmental conditions, clusters in many cases are near-ideal "laboratories" for stellar studies—not in the sense that astronomers can perform experiments on them, but because the properties of the stars are very tightly constrained. The only factor distinguishing one star from another in the same cluster is mass, so theoretical models of star formation and evolution can be compared with reality without the complications introduced by broad spreads in age, chemical composition, and place of origin.

Clusters and Associations

Figure 11.29(a) shows a small star cluster called the Pleiades, or Seven Sisters, a well-known naked-eye object in the constellation Taurus, lying about 120 pc from Earth. This type of loose, irregular cluster, found mainly in the plane of the Milky

▲ **FIGURE 11.28 Newborn Cluster** The star cluster NGC 3603 and part of the larger molecular cloud in which it formed. The cluster contains about 2000 bright stars and lies some 6000 pc from Earth. The field of view shown here spans about 20 light-years. Radiation from the cluster has cleared a cavity in the cloud several light-years across. The inset shows the central area more clearly, revealing many small stars less massive than the Sun. *(ESO; NASA)*

Way, is called an **open cluster.** Open clusters typically contain from a few hundred to a few tens of thousands of stars and are a few parsecs across. Figure 11.29(b) shows the H–R diagram for stars in the Pleiades. The cluster contains stars in all parts of the main sequence. The blue stars must be relatively young, for, as we saw in Chapter 10, they burn their fuel very rapidly. ∞ *(Sec. 10.7)* Thus, even though we have no direct evidence of the cluster's birth, we can estimate its age at less than 20 million years, the lifetime of an O-type star. The wisps of leftover gas evident in the photograph are further evidence of the cluster's youth.

Less massive, but more extended, clusters are known as **associations.** These typically contain no more than a few hundred stars, but may span many tens of parsecs. Associations tend to be rich in very young stars and are very loosely bound—if they are bound at all. Many associations appear to be expanding freely into space and dissolving following their formation.

Figure 11.30(a) shows a very different type of star cluster, called a **globular cluster.** All globular clusters are roughly spherical (which accounts for their name), are generally found away from the Milky Way plane, and contain hundreds of thousands, and sometimes millions, of stars spread out over about 50 pc. The H–R diagram for this cluster (called Omega Centauri) is shown in Figure 11.30(b).

The most outstanding feature of globular clusters is their lack of upper main-sequence stars. In fact, globular clusters contain no main-sequence stars with masses greater than about 0.8 times the mass of the Sun. (The A-type stars in this plot are stars at a much later evolutionary stage that happen to be passing through the location of the upper main sequence.) Their more massive O- through F-type stars have long since exhausted their nuclear fuel and disappeared from the main sequence. On the basis of these and other observations, astronomers estimate that all globular clusters are at least 10 billion years old. They contain the oldest known stars in our Galaxy. Astronomers speculate that the 150 or so globular clusters observed today are just the survivors of a much larger population of clusters that formed long ago.

Clusters and Nebulae

How many stars form in a cluster, and of what type are they? What does the collapsed cloud look like once star formation has run its course? At present, although the main stages (3–7) in the formation of individual stars are fairly well established, the answers to these broader questions (involving stages 1 and 2) remain sketchy. Low-mass stars are much more common than high-mass ones, but an explanation of exactly why this is so still awaits a more thorough understanding of the star-formation process. ∞ *(Sec. 10.7)*

At present, researchers are divided on whether more massive stars simply form from denser or more massive clumps of interstellar gas or whether all stars form as relatively lightweight objects (with masses much less than the mass of the Sun). Stars subsequently grow by accretion from their surroundings, in a manner reminiscent of the growth of planetesimals in the early solar system. ∞ *(Sec. 4.3)* In the latter view, the most massive stars are the ones that win the competition for resources.

In either case, computer simulations of star-forming clouds (see Figure 11.31) suggest that the sequence of events leading to main-sequence stars may be strongly influenced by physical interactions—close encounters and even collisions—between protostars within the cluster. The simulations indicate that the strong gravitational fields of the most massive protostars give them a competitive advantage over their smaller rivals in attracting gas from the surrounding nebula, causing giant protostars to grow even faster. Close encounters with massive protostars also tend to disrupt smaller protostellar disks, terminating the growth of their central protostars and ejecting planets and low-mass brown dwarfs from the disk into intracluster space. In dense clusters these interactions may even lead to mergers and further growth of massive objects.

ANIMATION / VIDEO ⓂⒶ
Carina Nebula

Thus, even though many details remain to be filled in, we can see several ways in which the cluster environment affects the kinds of stars that form there. The most massive stars form fastest and in the process tend to prevent the formation of more high-mass stars, both by stealing their raw material and by helping destroy their disks. Eventually, the intense radiation from newborn O and B stars disrupts the environment in which lower-mass stars are forming, "freezing in" their masses. This also helps explain the existence of brown dwarfs by providing a mechanism whereby star formation can stop before nuclear fusion begins in a low-mass stellar core.

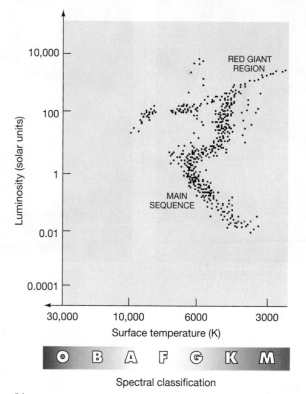

▲ **FIGURE 11.29 Open Cluster** (a) The Pleiades cluster (also known as the Seven Sisters because only six or seven of its stars can be seen with the naked eye) lies about 120 pc from the Sun. (b) An H–R diagram for all the stars of this well-known open cluster. *(NOAO)*

▲ **FIGURE 11.30 Globular Cluster** (a) The globular cluster Omega Centauri is approximately 5000 pc from Earth and some 40 pc in diameter. (b) An H–R diagram for some of its stars. *(P. Seitzer)*

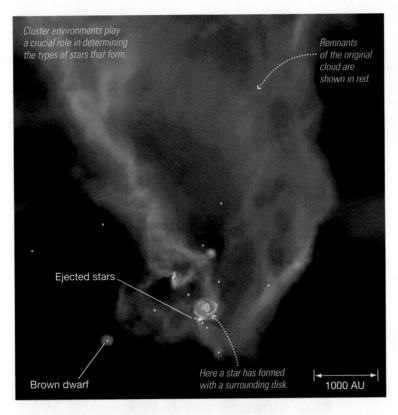

Cluster environments play a crucial role in determining the types of stars that form.

Remnants of the original cloud are shown in red.

Ejected stars

Brown dwarf

Here a star has formed with a surrounding disk.

1000 AU

◄ **FIGURE 11.31 Protostellar Collisions INTERACTIVE** In the congested environment of a young cluster, star formation is a competitive and violent process. Large protostars may grow by "stealing" gas from smaller ones, and the extended disks surrounding most protostars can lead to collisions and even mergers. This frame from a supercomputer simulation shows a small star cluster emerging from an interstellar cloud that originally contained some 50 solar masses of material spread over a volume 1 light-year across. (*M. Bate, I. Bonnell, and V. Bromm*)

Young star clusters are often shrouded in gas and dust, making them hard to see in visible light, but infrared observations clearly demonstrate that clusters really are found within star-forming regions. Figure 11.32 compares optical (a) and infrared (b) views of the central regions of the Orion Nebula. The optical image shows the Trapezium, the group of four bright stars responsible for ionizing the nebula; the infrared image reveals an extensive cluster of stars within and behind the visible nebula and shows many stages of star formation, including nearly 1000 new stars forming and interacting with the surrounding cloud.

Eventually, star clusters dissolve into individual stars. In some cases, the irregular star formation process illustrated in Figure 11.31 simply leaves the newborn cluster gravitationally unbound. In others, the ejection of leftover gas reduces the cluster's mass so much that it becomes unbound and quickly disperses. The tidal gravitational field of the Galaxy slowly destroys the remainder. ∞ *(Sec. 5.2)* The most massive systems—the globular clusters—may survive for billions of years, but most clusters surrender their stars to the Galactic population in less than a few hundred million years.

Take another look at the sky one clear, dark evening. Ponder all of the cosmic activity you have learned about as you peer upward at the stars. After studying this chapter, you may find that you have to modify your view of the night sky. Even the seemingly quiet nighttime darkness is dominated by continual change.

0.5 pc

(a)   R I V U X G   (b)   R I V U X G

◄ **FIGURE 11.32 Young Stars in Orion INTERACTIVE** (a) Visible *Hubble* and (b) infrared *Spitzer* views of the central part of the Orion Nebula. The visible image is dominated by the emission nebula and shows few stars. However, the infrared image shows an extensive star cluster containing stars of many masses, possibly including many brown dwarfs (see also Figures 3.24c and d). (*NASA*)

CONCEPT CHECK

If stars in a cluster all start to form at the same time, how can some influence the formation of others?

THE BIG QUESTION

When did the first stars form? We observe stars forming today throughout the Milky Way and myriad other galaxies, and studies of distant stars imply that they formed even more efficiently billions of years ago. Astronomers are pushing back the veil of ancient star formation, trying to understand how and when conditions in the early universe first allowed gas without walls—stars—to ignite as brilliant balls of fire.

CHAPTER REVIEW

SUMMARY

LO1 The **interstellar medium** (p. 304) occupies the space between stars. It is composed of cold (less than 100 K) gas, mostly atomic or molecular hydrogen and helium, and **dust grains** (p. 305) containing carbon, silicates, and iron. Interstellar dust is very effective at blocking our view of distant stars, even though the density of the interstellar medium is very low. The spatial distribution of interstellar matter is very patchy. The dust preferentially absorbs short-wavelength radiation, leading to **reddening** (p. 305) of light passing through interstellar clouds.

LO2 **Emission nebulae** (p. 308) are extended clouds of hot, glowing interstellar matter. Those associated with star formation are caused by hot O- and B-type stars heating and ionizing their surroundings. They are often crossed by dark **dust lanes** (p. 309)—part of the larger molecular cloud from which they formed.

LO3 **Dark dust clouds** (p. 311) are cold, irregularly shaped regions in the interstellar medium that diminish or completely obscure the light from background stars. The interstellar medium also contains many cold, dark **molecular clouds** (p. 314). Dust within these clouds probably both protects the molecules and acts as a catalyst to help them form. Molecular clouds are likely sites of future star formation. Often, several molecular clouds are found close to one another, forming a **molecular cloud complex** (p. 314) millions of times more massive than the Sun.

LO4 Astronomers can study dark interstellar clouds by observing their effect on the light from more distant stars. Another way to observe these regions of interstellar space is through spectral analysis of the **21-centimeter line** (p. 313) produced whenever the electron in a hydrogen atom reverses its spin, changing its energy very slightly in the process. Astronomers usually study these clouds through observations of other molecules that are less common than hydrogen but much easier to detect.

LO5 Stars form when an interstellar cloud collapses under its own gravity and breaks up into smaller pieces. The evolution of the contracting cloud can be represented as an **evolutionary track** (p. 319) on the Hertzsprung–Russell diagram. A cold interstellar cloud (stage 1) containing a few thousand solar masses of gas can fragment into tens or hundreds of smaller clumps of matter, from which stars eventually form. As a collapsing prestellar fragment heats up and becomes denser (stages 2 and 3), it eventually becomes a **protostar** (stage 4; p. 317)—a warm, very luminous object that emits radiation mainly in the infrared portion of the electromagnetic spectrum. The protostar contracts as it radiates its internal energy into space (stage 5). Eventually, the central temperature becomes high enough for hydrogen fusion to begin (stage 6), and the protostar becomes a star (stage 7).

LO6 Dark dust clouds and globules are examples of stage-1 clouds and stage-2 collapsing fragments. Examples of stage-3 objects have been observed in some star-forming regions, such as Orion. Stars in the **T-Tauri phase** (p. 319) are examples of stage-4/5 protostars. Low-mass stage-6 stars on their final approach to the main sequence are found in star-forming regions.

LO7 Stars of different masses go through similar formation stages, but end up at different locations along the main sequence, high-mass stars at the top, low-mass stars at the bottom. The **zero-age main sequence** (p. 322) is the main sequence predicted by stellar evolutionary theory. It agrees very well with observed main sequences. The most massive stars have the shortest formation times and the shortest main-sequence lifetimes. At the other extreme, some low-mass fragments never reach the point of nuclear ignition. Objects not massive enough to fuse hydrogen to helium are called **brown dwarfs** (p. 322). They may be very common in the universe.

LO8 A single contracting and fragmenting cloud can give rise to hundreds or thousands of stars—a **star cluster** (p. 323). **Open clusters** (p. 324), which are loose, irregular clusters that typically contain from a few tens to a few thousands of stars, are found mostly in the Milky Way plane. They typically contain many bright blue stars, indicating that they formed relatively recently. **Globular clusters** (p. 324) are roughly spherical and may contain millions of stars. They include no main-sequence stars more massive than the Sun, indicating that they formed long ago. Infrared observations have revealed young star clusters within several emission nebulae. Eventually, star clusters break up into individual stars, although the process may take billions of years to complete.

MasteringAstronomy®   For instructor-assigned homework go to www.masteringastronomy.com

Problems labeled POS explore the process of science. VIS problems focus on reading and interpreting visual information. LO connects to the introduction's numbered Learning Outcomes.

REVIEW AND DISCUSSION

1. LO1 What is the composition of interstellar gas? Of interstellar dust?

2. LO3 How is interstellar matter distributed through space?

3. What are some methods that astronomers use to study interstellar dust?

4. LO2 What is an emission nebula?

5. LO4 What is 21-centimeter radiation, and why is it useful to astronomers?

6. If our Sun were surrounded by a cloud of gas, would this cloud be an emission nebula? Why or why not?

7. LO5 Briefly describe the basic chain of events leading to the formation of a star like the Sun.

8. What is an evolutionary track?

9. Why do stars tend to form in groups?

10. What critical event must occur in order for a protostar to become a star?

11. LO7 What are brown dwarfs?

12. POS Stars live much longer than we do, so how do astronomers test the accuracy of theories of star formation?

13. LO6 At what evolutionary stages must astronomers use radio and infrared radiation to study prestellar objects? Why can't they use visible light?

14. POS Explain the usefulness of the H–R diagram in studying the evolution of stars. Why can't evolutionary stages 1–3 be plotted on the diagram?

15. LO8 Compare and contrast the properties of open and globular star clusters.

CONCEPTUAL SELF-TEST: TRUE OR FALSE? / MULTIPLE CHOICE

1. Interstellar matter is quite evenly distributed throughout the Milky Way Galaxy. (T/F)

2. Emission nebulae radiate mainly in the ultraviolet part of the electromagnetic spectrum. (T/F)

3. Twenty-one-centimeter radiation can be used to probe the interiors of molecular clouds. (T/F)

4. More massive stars form more rapidly. (T/F)

5. Brown dwarfs take a long time to form, but will eventually become visible as stars on the lower main sequence. (T/F)

6. The formation of the first high-mass stars in a collapsing cloud tends to inhibit further star formation within that cloud. (T/F)

7. Most stars form as members of groups or clusters. (T/F)

8. VIS The dark interstellar globule shown in Figure 11.2(b) is about the same size as (a) a cloud in Earth's atmosphere; (b) the entire planet Earth; (c) a star like the Sun; the Oort cloud.

9. The red glow of an emission nebula (a) is emitted by warm gas falling onto the stars at the center; (b) is produced by hydrogen gas heated to high temperatures by massive stars within the nebula; (c) is a reflection of light from stars near the nebula; (d) is the unresolved light of many faint red stars within the nebula.

10. VIS The Rho Ophiuchus cloud, shown in Figure 11.12(a) (Dark Dust Cloud), is dark because (a) there are no stars in this region; (b) the stars in this region are young and faint; (c) starlight from behind the cloud does not penetrate the cloud; (d) the region is too cold to sustain stellar fusion.

11. Of the following telescopes, the one best suited to observing dark dust clouds is (a) an X-ray telescope; (b) a large visible-light telescope; (c) an orbiting ultraviolet telescope; (d) a radio telescope.

12. A protostar that will eventually turn into a star like the Sun is significantly (a) smaller; (b) more luminous; (c) fainter; (d) less massive than the Sun.

13. VIS The main difference between the Pleiades cluster (Figure 11.29a) and Omega Centauri (Figure 11.30a) is that the Pleiades cluster is much (a) larger; (b) younger; (c) farther away; (d) denser.

14. VIS In Figure 11.2 (Reddening), the effect of interstellar absorption on spectral lines is (a) to shift them to longer wavelengths; (b) to shift them to shorter wavelengths; (c) to leave their wavelengths unchanged; (d) to erase them.

15. VIS In Figure 11.23 (Newborn Star on the H–R Diagram), as the star moves from stage 6 to stage 7 it becomes (a) cooler and fainter; (b) hotter and smaller; (c) redder and brighter; (d) larger and cooler.

PROBLEMS

The number of squares preceding each problem indicates its approximate level of difficulty.

1. ■ The average density of interstellar gas within the Local Bubble is much lower than the value mentioned in the text—in fact, it is roughly 10^3 hydrogen atoms/m^3. Given that the mass of a hydrogen atom is 1.7×10^{-7} kg, calculate the total mass of interstellar matter contained within a bubble volume equal in size to planet Earth.

2. ■■ Calculate the frequency of 21-cm radiation. If interstellar clouds along our line of sight have radial velocities in the range 75 km/s (receding) to 50 km/s (approaching), calculate the range of frequencies and wavelengths over which the 21-cm line will be observed. ∞ *(Sec. 2.7)*

3. ■ Calculate the radius of a spherical molecular cloud whose total mass equals the mass of the Sun. Assume a cloud density of 10^{12} hydrogen molecules per cubic meter.

4. ■■ A beam of light shining through a dense molecular cloud is diminished in intensity by a factor of 2 for every 5 pc it travels. By how many magnitudes is the light from a background star dimmed if the total thickness of the cloud is 60 pc?

5. ■■ Estimate the escape speeds near the edges of the four emission nebulae listed in Table 11.1, and compare them with the average speeds of hydrogen nuclei in those nebulae. ∞ *(More Precisely 5-1)* Is it likely that the nebulae are held together by their own gravity?

6. ■■ For an interstellar gas cloud to contract, the average speed of its constituent particles must be less than half the cloud's escape speed. Will a (spherical) molecular hydrogen cloud of mass of 1000 solar masses, a radius 10 pc, and temperature of 10 K begin to collapse? Why or why not? ∞ *(More Precisely 5-1)*

7. ■ A protostar evolves from a temperature $T = 3500$ K and a luminosity $L = 5000$ times that of the Sun to $T = 5000$ K and $L = 3$ solar units. What is its radius (a) at the start, and (b) at the end of the evolution? ∞ *(Sec. 10.4)*

8. ■ Through roughly how many magnitudes does a three-solar-mass protostar decrease in brightness between stage 4 and stage 6?

9. ■■ What is the luminosity, in solar units, of a brown dwarf whose radius is 0.1 solar radii and whose surface temperature is 600 K (0.1 times that of the Sun)? ∞ *(Sec. 10.4)*

10. ■■ Approximating the gravitational field of our Galaxy as a mass of 10^{11} solar masses at a distance of 8000 pc (see Chapter 14), estimate the "tidal radius" of a 20,000-solar-mass star cluster—that is, the distance from the cluster center outside of which the tidal force due to the Galaxy overwhelms the cluster's gravity. ∞ *(Sec. 5.2)*

ACTIVITIES

Collaborative

1. The Messier catalog is a list of nebulae and other "fuzzy" objects in the night sky, compiled by 18th century astronomer Charles Messier to avoid confusion with his main interest, comets. Today, the catalog is a treasure trove of fascinating deep-sky objects. Many Messier objects are star-forming regions or star clusters. Star-forming regions mentioned in this chapter include M8 (the Lagoon Nebula), M16 (Eagle Nebula), M17 (Omega Nebula), M20 (Trifid Nebula), M42 (Orion Nebula). Interesting star clusters and associations include M6, M7, M11, M35, M37, M44, M45, M52, M67, and M103. Not all of these objects are easily observable on any given night, so consult a sky chart or do some research online, and make a list of which are visible. A small telescope will give the best results in most cases, and you may want to observe in shifts over the course of the night. For each Messier object on your list, find out what sort of object it is (emission nebula, association, young or old open cluster), carefully follow the finding instructions to locate it, and sketch it (or photograph it, if you have the equipment). Compare your sketch or photograph to images found in this book or online.

Individual

2. Observe the Milky Way on a dark, clear night away from city lights. Is it a continuous band of light across the sky or is it mottled? The parts of the Milky Way that appear missing are actually dark dust clouds lying relatively near the Sun. Identify the constellations in which you see these clouds. Make a sketch and compare with a star atlas. Find other small clouds in the atlas and try to find them with your eye or with binoculars.

3. The constellation Orion the Hunter is prominent in the winter sky. Its most noticeable feature is a short, straight row of three medium-bright stars: the famous belt of Orion. A line of stars extends from the easternmost star of the belt, toward the south. This is Orion's sword. Toward the bottom of the sword is the sky's most famous emission nebula, the Orion Nebula (M42). Observe the Orion Nebula with your eye, with binoculars, and with a telescope. What is its color? How can you account for this? With the telescope, try to find the Trapezium, a grouping of four stars in the center of the nebula. These are hot, young stars; their energy causes the Orion Nebula to glow.

4. The Trifid Nebula (M20) is another place where new stars are forming. It has been called a "dark night revelation, even in modest apertures." An 8- to 10-inch telescope is needed to see the nebula's triple-lobed structure. Ordinary binoculars reveal the Trifid as a hazy patch located in the constellation Sagittarius, set against the richest part of the Milky Way. What are the dark lanes? Why are other parts of the nebula bright? There have been reports of large-scale changes occurring in this nebula over the last 150 years, based on old drawings showing the nebula looking slightly different from its appearance today. Do you think it possible for an interstellar cloud to undergo a change in appearance on a time scale of years, decades, or centuries?

12

Stellar Evolution
The Lives and Deaths of Stars

Once nuclear fusion begins, a newborn star changes little in outward appearance for more than 90 percent of its lifetime. However, at the end of this period, as the star begins to run out of fuel and die, its properties once again change greatly. Aging stars travel along evolutionary tracks that take them far from the main sequence. The fate of a star depends primarily on its mass. Low-mass stars are destined to end their lives quiescently, their outer layers eventually escaping into interstellar space, while very massive stars die violently, in spectacular explosions of almost unimaginable fury. In either case, the deaths of stars enrich the Galaxy with newly created heavy elements. By continually comparing theoretical calculations with detailed observations of stars of all types, astronomers have refined the theory of stellar evolution into a precise and powerful tool for understanding the universe.

LEARNING OUTCOMES

Studying this chapter will enable you to:

LO1 Explain why stars evolve off the main sequence.

LO2 Summarize the evolutionary stages a Sun-like star follows after it leaves the main sequence, and describe the resulting remnant.

LO3 Explain how white dwarfs in binary systems can become explosively active.

LO4 Summarize the sequence of events leading to the violent death of a massive star.

LO5 Describe the two types of supernova, and explain how each is produced.

LO6 Describe how observations of star clusters support the theory of stellar evolution.

LO7 Explain the origin of elements heavier than helium, and identify the significance of these elements for the study of stellar evolution.

THE BIG PICTURE

Understanding the story of the birth, development, and death of stars is one of the greatest accomplishments of 20th-century science. Yet no one has ever seen even a single star pass through all of its many varied evolutionary stages. Like archaeologists who examine bones and artifacts from long ago to learn more about the evolution of human culture, astronomers observe stars of many different ages to construct a consistent model of how stars evolve over billions of years.

◄ Resembling a cosmic hourglass or a celestial butterfly, this striking image captures hot gas released by a dying star about 3800 light-years away. Known as NGC 6302, or more informally as the Bug Nebula, this complex object is a planetary nebula—an old star shedding its outer layers over light-year dimensions as it ends its life. Its peculiar shape results from a belt of dust (dark lane at center) that obscures the dying star and partially blocks the rolling cauldrons of outwardly expelled gas. *(STScI)*

MasteringAstronomy®

Visit www.masteringastronomy.com for quizzes, animations, videos, interactive figures, and self-guided tutorials.

12.1 Leaving the Main Sequence

Most stars spend most of their lives on the main sequence of the H–R diagram. A star like the Sun, for example, after spending a few tens of millions of years in formation (stages 1–6 in Chapter 11), will reside on or near the main sequence (stage 7), with roughly constant temperature and luminosity, for some 10 billion years before evolving into something else. ∞ *(Sec. 11.4)* That "something else" is the topic of this chapter.

Stars and the Scientific Method

No one has ever witnessed the complete evolution of any star from birth to death. Stars take a very long time—millions, billions, even trillions of years—to evolve. ∞ *(Sec. 10.7)* Yet in less than a century, astronomers have developed a comprehensive theory of stellar evolution that is one of the best-tested in all of astronomy. How can we can talk so confidently about what took place billions of years in the past, and what will happen billions of years in the future? The answer is that we can observe billions of stars in the universe, enough to see examples of every stage of stellar development, allowing us to test and refine our theoretical ideas. Just as we can piece together a picture of the human life cycle by studying a snapshot of all the residents of a large city, so we can construct a picture of stellar evolution by studying the myriad stars we see in the night sky.

The modern theory of the lives and deaths of stars is another example of the scientific method in action. ∞ *(Sec. 0.5)* Faced with a huge volume of observational data, with little theory to organize or explain it, astronomers in the late 19th and early 20th centuries painstakingly classified and categorized the properties of the stars they observed. ∞ *(Sec. 10.5)* During the first half of the 20th century, as quantum mechanics yielded detailed insight into the behavior of light and matter on subatomic scales, theoretical understanding of key stellar properties fell into place. ∞ *(Sec. 2.6)* Since the 1950s, a truly comprehensive theory has emerged, tying together the basic disciplines of atomic and nuclear physics, electromagnetism, thermodynamics, and gravitation into a coherent whole. Today, theory and observation proceed hand in hand, each refining and validating the details of the other as astronomers continue to hone their understanding of stellar evolution.

Note that astronomers always use the term *evolution* in this context to mean change "during the lifetime of an individual star." Contrast this usage with the meaning of the term in biology, where evolution refers to changes in the characteristics of a *population* of plants or animals over many generations. In fact, populations of stars also evolve in the "biological" sense, as the overall composition of the interstellar medium (and hence of each new stellar generation) changes slowly over time due to nuclear fusion in stars. However, in astronomical parlance, *stellar evolution* refers to changes that occur during a single stellar lifetime.

Structural Change

On the main sequence, a star slowly fuses hydrogen into helium in its core. This process is usually called **core hydrogen burning,** but realize that here is another instance where astronomers use a familiar term in an unfamiliar way: To astronomers, "burning" always means nuclear fusion in a star's core, not a chemical reaction, such as the combustion of wood or gasoline in air, that we would normally mean in everyday speech. Chemical burning does not directly affect atomic nuclei.

As illustrated schematically in Figure 12.1, a main-sequence star is in a state of hydrostatic equilibrium, in which pressure's outward push exactly counteracts gravity's inward pull. ∞ *(Sec. 9.2)* This is a stable balance between

Pressure out

Gravity in

Normally, a star is in equilibrium, with gravity's inward pull exactly balanced by pressure's outward push.

If the star's interior gets hotter, pressure temporarily wins the battle against gravity and the star expands.

Eventually, as the star expands and cools, equilibrium is once again established.

▲ **FIGURE 12.1 Hydrostatic Equilibrium** In a steadily burning star on the main sequence, the outward pressure exerted by hot gas balances the inward pull of gravity.

gravity and pressure, in which a small change in one always results in a small compensating change in the other. For example, as shown in the figure, a small increase in the star's central temperature leads to an increase in pressure, causing the star to expand and cool, thus recovering its equilibrium. Conversely, a small temperature decrease leads to a slight pressure decrease, and gravity causes the star to contract and heat up; again, equilibrium is restored. You should keep Figure 12.1 in mind as you study the various stages of stellar evolution described next. Much of a star's complex behavior can be understood in these simple terms.

As the hydrogen in the star's core is consumed, the balance between these opposing forces starts to shift. Both the star's internal structure and its outward appearance begin to change. The star leaves the main sequence. The subsequent stages of stellar evolution—the end of the star's life—depend critically on the star's mass. As a rule of thumb, we can say that low-mass stars die gently, while high-mass stars die catastrophically. The dividing line between these two very different outcomes lies around eight times the mass of the Sun, and in this chapter we will refer to stars of more than 8 solar masses as "high-mass" stars. However, within both the "high-mass" and the "low-mass" (i.e., less than 8 solar masses) categories, there are substantial variations, some of which we will point out as we proceed.

Rather than dwelling on the many details, we will concentrate on a few representative evolutionary sequences. We begin by considering the evolution of a fairly low-mass star like the Sun. Later, we will broaden our discussion to include all stars, large and small.

12.2 Evolution of a Sun-like Star

Figure 12.2 illustrates how the interior composition of a main-sequence star changes as the star ages. The star's helium content increases fastest at the center, where temperatures are highest and the burning (creating helium from hydrogen) is fastest. ∞ *(Sec. 9.5)* The helium content also increases near the edge of the core, but more slowly because the burning rate is less rapid there. The inner, helium-rich region becomes larger and more hydrogen deficient as the star continues to shine.

Stages 8 and 9—Subgiant to Red Giant

Recall from Chapter 9 that a temperature of 10^7 K is needed to fuse hydrogen into helium. Only above that temperature do colliding hydrogen nuclei (that is, protons) have enough speed to overwhelm their mutual electromagnetic repulsion. ∞ *(Sec. 9.5)* Because helium nuclei (with two protons each, compared to one for hydrogen) carry a greater positive charge, their electromagnetic repulsion is greater, and so higher temperatures are needed to cause them to fuse. The core temperature at this stage is too low for helium fusion to begin. Eventually, hydrogen becomes depleted at the center, the nuclear fires there subside, and the location of principal burning moves to higher layers in the core. An inner core of nonburning pure helium starts to grow.

Without nuclear burning to maintain it, the outward-pushing gas pressure weakens in the helium inner core; however, the inward pull of gravity does not. Once the pressure is relaxed—even a little—structural changes in the star become inevitable. As soon as hydrogen becomes substantially depleted, about 10 billion years after the star arrived on the main sequence, the helium core begins to contract.

CONCEPT CHECK

In what sense is the Sun stable?

At birth, a star's helium abundance is about 10 percent.

After 5 billion years, only a small amount of hydrogen has changed into more helium.

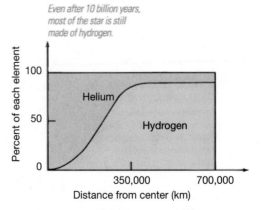

Even after 10 billion years, most of the star is still made of hydrogen.

▲ **FIGURE 12.2 Solar Composition Change**   Theoretical estimates of the changes in a Sun-like star's composition show how hydrogen (yellow) and helium (blue) abundance vary within the star from birth to death (top to bottom).

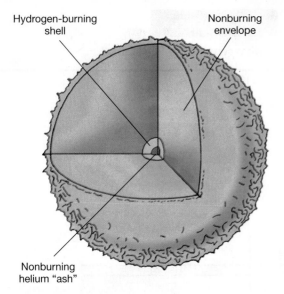

Hydrogen-burning shell

Nonburning envelope

Nonburning helium "ash"

▲ FIGURE 12.3 Hydrogen Shell Burning   As a star's core converts more and more of its hydrogen into helium, the hydrogen in the shell surrounding the nonburning helium "ash" burns ever more violently. By the time the star has reached the bottom of the giant branch, its core has shrunk to a few tens of thousands of kilometers in diameter, and its surface is 10 times the star's original size.

The shrinking helium core releases gravitational energy, driving up the central temperature, heating the overlying layers of the core, and causing the hydrogen there to fuse even more rapidly than before. Figure 12.3 depicts this **hydrogen shell-burning** stage, in which hydrogen burns at a furious rate in a relatively thin layer surrounding the nonburning inner core of helium "ash." The hydrogen shell generates energy faster than did the original main-sequence star's hydrogen-burning core, and the shell's energy production continues to increase as the helium core contracts. Strange as it may seem, the star's response to the disappearance of the nuclear fire at its center is to get brighter!

The pressure exerted by this enhanced hydrogen burning causes the star's nonburning outer layers to increase in radius. Not even gravity can stop them. Even as the core shrinks and heats up, the overlying layers expand and cool. The star, aged and unbalanced, is on its way to becoming a red giant. The change from main-sequence star to red giant takes about 100 million years.

We can trace these large-scale changes on an H–R diagram. Figure 12.4 shows the star's path away from the main sequence, labeled as stage 7. ∞ *(Sec. 11.3)* The star evolves to the right on the diagram, its surface temperature dropping while its luminosity increases only slightly. The star's roughly horizontal track from the main sequence (stage 7) to stage 8 on the figure is called the **subgiant branch.** By stage 8, the star's radius has increased to about three times the radius of the Sun. The nearly vertical (constant temperature) path followed by the star between stages 8 and 9 is known as the **red giant branch** of the H–R diagram. (Figure 12.3 corresponds to a point just after stage 8, as the star begins to ascend the giant branch.)

The red giant is huge. By stage 9 its luminosity is many hundreds of times the solar value. Its radius is around 100 solar radii—about the size of Mercury's orbit. In contrast, its helium core is surprisingly small, less than 1/1000 the size of the entire star, or just a few times larger than Earth. Continued shrinkage of the red giant's core has compacted its helium gas to a density of approximately 10^8 kg/m^3—10,000 times denser than any material found on Earth. About 25 percent of the mass of the entire star is packed into its planet-sized core.

A familiar example of a low-mass star in the red giant phase is the KIII giant Arcturus ∞ *(Figure 10.14)*, one of the brightest stars in our night sky. Its mass is about 1.5 times that of the Sun. Currently in the hydrogen shell-burning stage and ascending the red giant branch, its radius is some 21 times the solar value. ∞ *(Table 10.4)* It emits more than 160 times more energy than the Sun, much of it in the infrared part of the spectrum.

Stage 10—Helium Fusion

This simultaneous contraction of the red giant's core and expansion of its envelope—the nonburning layers surrounding the core—does not continue indefinitely. A few hundred million years after a solar-mass star leaves the main sequence, the central temperature reaches the 10^8 K needed for helium to fuse into carbon, and the nuclear fires reignite.

For stars comparable in mass to the Sun, the high densities and pressures found in the stage-9 core mean that the onset of helium fusion is a very violent event. Once the burning starts, the core cannot respond quickly enough to the rapidly changing conditions within it, and its temperature rises sharply in a

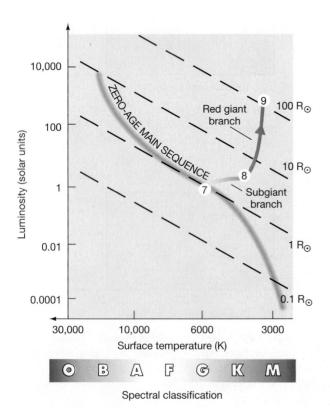

◀ FIGURE 12.4 Red Giant on the H–R Diagram   As its helium core shrinks and its outer envelope expands, the star leaves the main sequence (stage 7). At stage 8, the star is well on its way to becoming a red giant. The star continues to brighten and grow as it ascends the red giant branch to stage 9.

runaway explosion called the **helium flash.** For a few hours, the helium burns ferociously, like an uncontrolled bomb. Eventually, the star's structure "catches up" with the flood of energy dumped into it by helium burning. The core expands, its density drops, and equilibrium is restored as the inward pull of gravity and the outward push of gas pressure come back into balance. The core, now stable, begins to burn helium into carbon at temperatures well above 10^8 K.

The helium flash terminates the star's ascent of the red giant branch of the H–R diagram. Yet despite the explosive detonation of helium in the core, the flash does *not* increase the star's luminosity. Rather, the helium flash produces an expansion and cooling of the core that ultimately results in a *reduction* in the energy output as the star jumps from stage 9 to stage 10. As indicated in Figure 12.5, the surface temperature is now higher than it was on the red giant branch, although the luminosity is less than at the helium flash. This adjustment in the star's properties occurs quite quickly—in about 100,000 years.

At stage 10 our star is now stably burning helium in its inner core and fusing hydrogen in a shell surrounding that core. It resides in a well-defined region of the H–R diagram known as the **horizontal branch,** where core helium-burning stars remain for a time before resuming their journey around the H–R diagram. The star's specific position within this region is determined mostly by its mass—not its original mass, but whatever mass remains after the red giant phase. The two masses differ because strong stellar winds can eject large amounts of matter—as much as 20–30 percent of the initial mass—from a red giant's surface. The more massive stars have thicker envelopes and lower surface temperatures at this stage, but all stars have roughly the same luminosity after the helium flash. As a result, stage-10 stars tend to lie along a horizontal line on the H–R diagram, with more massive stars to the right and less massive ones to the left.

Stage 11—A Red Giant Again

Nuclear reactions in stars proceed at rates that increase very rapidly with temperature. At the extremely high temperatures found in the core of a horizontal-branch star, the helium fuel doesn't last long—no more than a few tens of millions of years after the initial flash.

As helium fuses to carbon, a new carbon-rich inner core begins to form and phenomena similar to the earlier buildup of helium occur. Helium becomes depleted at the center, and eventually fusion ceases there. The nonburning carbon core shrinks in size—even as its mass increases due to helium fusion—and heats up as gravity pulls it inward, causing the hydrogen- and helium-burning rates in the overlying layers to increase. The star now contains a contracting carbon-ash inner core surrounded by a helium-burning shell, which is in turn surrounded by a hydrogen-burning shell (Figure 12.6). The outer envelope of the star expands, much as it did earlier in the first red giant stage. By the time it reaches stage 11 in Figure 12.7, the star has become a swollen red giant for a second time.

The burning rates in the shells surrounding the inner core are much fiercer during the star's second trip into the red giant region, and the radius and luminosity increase to values even greater than those reached during the first visit (stage 9). Its carbon core continues to shrink, driving the hydrogen- and helium-burning shells to higher and higher temperatures and luminosities. To distinguish this second visit to the red giant region from the first, the star's evolutionary track during this stage is often referred to as the *asymptotic giant branch.*[1]

---

[1]The term is borrowed from mathematics. An asymptote to a curve is a second curve that approaches ever closer to the first as the two are extended to infinity. Theoretically, if the star remained intact, the asymptotic-giant branch would approach the red giant branch from the left as the luminosity increased and would effectively merge with the red giant branch near the top of Figure 12.10. However, a Sun-like star will not live long enough for that to occur.

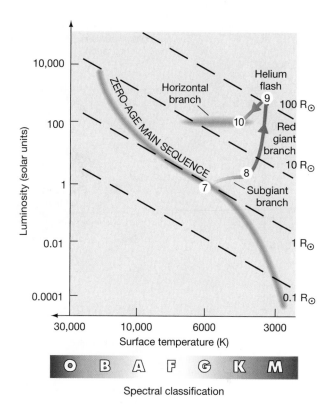

▲ **FIGURE 12.5 Horizontal Branch**   A large increase in luminosity occurs as a star ascends the red giant branch, ending in the helium flash. The star then settles down into another equilibrium state at stage 10, on the horizontal branch.

CONCEPT CHECK

Why does a star get brighter as it runs out of fuel in its core?

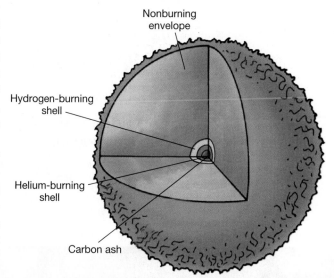

▲ **FIGURE 12.6 Helium Shell Burning**   Within a few million years after the onset of helium burning (stage 9), carbon ash accumulates in the star's inner core (not to scale). Above this core, hydrogen and helium are still burning in concentric shells.

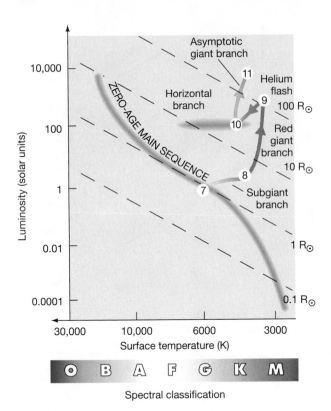

▲ FIGURE 12.7 **Reascending the Giant Branch** A carbon-core star reenters the giant region of the H–R diagram (stage 11) for the same reason it evolved there the first time around: Lack of nuclear fusion at the center causes the core to contract and the overlying layers to expand.

Table 12.1 summarizes the key stages through which our solar-mass star evolves. It is a continuation of Table 11.2, except that the density units are now the more convenient kilograms per cubic meter and we now express radii in units of the solar radius. As in Chapter 11, The numbers in the "Stage" column refer to the evolutionary stages noted in the figures and discussed in the text. Figure 12.8 illustrates the stages through which a G-type star like our Sun will evolve over the course of its lifetime.

12.3 The Death of a Low-Mass Star

As our star moves from stage 10 to stage 11, its envelope swells while its inner carbon core, too cool for further nuclear burning, continues to contract. If the central temperature could become high enough for carbon fusion to occur, still heavier elements could be synthesized, and the newly generated energy might again support the star, restoring for a time the equilibrium between gravity and pressure. For solar-mass stars, however, this does not occur—a lot more mass is needed, as we will see in the next section. The temperature never reaches the 600 million kelvins needed for new nuclear reactions to occur. The red giant is now very close to the end of its nuclear-burning lifetime.

Stage 12—A Planetary Nebula

Our aging star is now in quite a predicament. Its carbon core is, for all practical purposes, dead, while the outer hydrogen- and helium-burning shells consume fuel at a furious and increasing rate. As it expands, cools, and reascends the giant branch, the star begins to fall apart. Driven by the intense radiation from within, its outer layers begin to drift away into interstellar space. Slowly at first, then more rapidly as the core luminosity increases, the star loses virtually its entire envelope in less than a million years. The former red giant now consists of two distinct pieces: the exposed core, very hot and still very luminous, surrounded by a cloud of dust and cool gas—the escaping envelope—expanding away at a typical speed of a few tens of kilometers per second.

As the core exhausts its last remaining fuel, it contracts and heats up, moving to the left in the H–R diagram. Eventually, it becomes so hot that its ultraviolet radiation ionizes the inner parts of the surrounding cloud, producing a spectacular display called a **planetary nebula** (see Figure 12.9). The terminology is misleading, for these objects have nothing to do with planets. It originated in the

ANIMATION/VIDEO
Death of the Sun Part I

ANIMATION/VIDEO
Death of the Sun Part II

▼ FIGURE 12.8 **G-type Star Evolution** INTERACTIVE Artist's conception of the relative sizes and changing colors of a normal G-type star (such as our Sun) during its formative stages, on the main sequence, and while passing through the red giant and white dwarf stages. At maximum swelling, the red giant is approximately 70 times the size of its main-sequence parent; the core of the giant is about $1/15$ the main-sequence size and would be barely discernible if this figure were drawn to scale. The duration of time spent in the various stages—protostar, main-sequence star, red giant, and white dwarf—is roughly proportional to the lengths shown in this imaginary trek through space.

Protostar

Main-sequence G-type star

Stage 4

Stage 7

TABLE 12.1   Evolution of a Sun-like Star

Stage	Approx. Time to Next Stage (yr)	Central Temperature (K)	Surface Temperature (K)	Central Density (kg/m³)	Radius (km)	Radius (solar radii)	Object
7	10^{10}	1.5×10^7	6,000	10^5	7×10^5	1	Main-sequence star
8	10^8	5×10^7	4,000	10^7	2×10^6	3	Subgiant
9	10^5	10^8	4,000	108	7×10^7	100	Red giant/helium flash
10	5×10^7	2×10^8	5,000	10^7	7×10^6	10	Horizontal branch
11	10^4	2.5×10^8	4,000	10^8	4×10^8	500	Red giant (AGB)
	10^5	3×10^8	100,000	10^{10}	10^4	0.01	Carbon core
12	—	—	3,000	10^{-17}	7×10^8	1,000	Planetary nebula*
13	—	10^8	50,000	10^{10}	10^4	0.01	White dwarf
14	—	Close to 0	Close to 0	10^{10}	10^4	0.01	Black dwarf

*Values in columns 2–7 refer to the envelope.

18th century when, viewed at poor resolution through small telescopes, these shells of gas looked to some astronomers like the circular disks of planets in our solar system.

The mechanism by which planetary nebulae shine is very similar to that powering the emission nebulae we studied earlier—ionizing radiation from a hot star embedded in a cool gas cloud. ∞ *(Sec. 11.2)* However, these two classes of object have very different origins and represent completely separate phases of stellar evolution. The emission nebulae in Chapter 11 were the signposts of recent stellar birth. Planetary nebulae indicate impending stellar death.

Astronomers once thought that the escaping giant envelope would be more or less spherical in shape, completely surrounding the core in three dimensions, just as it had while still part of the star. Figure 12.9(a) shows an example in which this may well be the case. The nebula looks brighter near the edges because there is more emitting gas along the line of sight there, creating the illusion of a bright ring. However, such cases now seem to be in the minority. There is growing evidence that the final stages of red giant mass loss are often decidedly *non*spherical. The Ring Nebula shown in Figure 12.9(b) may very well *be* a ring and not just our view of a glowing shell of gas as was once thought. As illustrated in Figure 12.9(c) and (d), some planetary nebulae exhibit much more complex structures, suggesting that the star's environment—including the existence of a binary companion—can play an important role in determining the nebula's shape and appearance.

The central star fades and eventually cools, and the expanding gas cloud becomes more and more diffuse, gradually dispersing into interstellar space. In doing so it plays a vital role in the evolution of our Galaxy. During the final stages of the red giant's life, nuclear reactions between carbon and unburned helium in the core create oxygen, and in some cases even heavier elements, such as neon and magnesium. Some of these reactions also release neutrons

ANIMATION/VIDEO
Helix Nebula

ANIMATION/VIDEO
Helix Nebula Animation

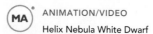
ANIMATION/VIDEO
Helix Nebula White Dwarf

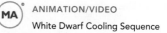
ANIMATION/VIDEO
White Dwarf Cooling Sequence

ANIMATION/VIDEO
Bi-Polar Planetary Nebula

Arrows in parts a and b mark the hot, yet dead, central stars of these planetary nebulae.

(a)

0.5 pc

R I V U X G

(b)

0.25 pc

R I V U X G

(c)

0.05 pc

R I V U X G

(d)

0.1 pc

R I V U X G

that, carrying no electrical charge, have no electrostatic repulsion to overcome and hence can interact with existing nuclei to form still heavier elements. ∞ *(Secs. 2.6, 9.5)* All of these elements—helium, carbon, oxygen, and heavier—are "dredged up" from the depths of the core into the envelope by convection during the star's final years and enrich the interstellar medium when the giant envelope escapes. ∞ *(Sec. 9.2)* The evolution of low-mass stars is the source of virtually all of the carbon-rich dust observed throughout the plane of our Galaxy. ∞ *(Sec. 11.1)*

Dense Matter

Before the shrinking carbon core can become hot enough for carbon to fuse, the rising pressure stops the contraction and stabilizes the temperature. However, this pressure is not the "normal" thermal pressure of high-speed particles in a very hot gas. ∞ *(More Precisely 5-1)* Instead, as we now discuss, the core enters a state in which the free electrons in the core—a vast sea of charged particles stripped from their parent nuclei by the ferocious heat in the stellar interior—play a critical role in determining the star's future.

The core density at this stage (stage 12 in Table 12.1) is enormous—about 10^{10} kg/m^3. This density is much higher than anything we have seen so far in our study of the cosmos. A single cubic centimeter of core matter would weigh 1000 kg on Earth—a ton of matter compressed into a volume about the size of a grape! Under these extreme conditions, a law of quantum physics known as the *Pauli exclusion principle* comes into play, preventing the electrons in the core from being crushed any closer together. (In fact, the core was in a similar high-density state at the moment of helium ignition, causing the helium flash.)

In essence, the electrons behave like tiny rigid spheres that can be squeezed relatively easily up to the point of contact, but become virtually incompressible thereafter. By stage 12 the pressure in the carbon core resisting the force of gravity is supplied almost entirely by tightly packed electrons, and this pressure brings the core back into equilibrium at a central temperature of about 300 million K—too cool to fuse carbon into heavier elements. This stage represents the maximum compression that the star can achieve—there is simply not enough matter in the overlying layers to bear down any harder. Supported by the resistance of its electrons to further compression, the contraction of the core stops.

Stage 13—A White Dwarf

Formerly concealed by the atmosphere of the red giant, the carbon core becomes visible as a *white dwarf* as the envelope recedes. ∞ *(Sec. 10.4)* The core is small, about the size of Earth, with a mass about half that of the Sun. Shining only by stored heat, not by nuclear reactions, this object has a white-hot surface when it first becomes visible, although it appears dim because of its small size. This is stage 13 of Table 12.1. The approximate path followed by the star on the H–R

◀ **FIGURE 12.9 Planetary Nebulae** The ejected envelopes of dying stars can take on a bewildering variety of forms. (a) Abell 39, some 2100 pc away, is a classic planetary nebula shedding a spherical shell of gas about 1.5 pc across; (b) the Ring Nebula, perhaps the most famous of all planetaries at 1500 pc distant and 0.5 pc across, is probably a genuine ring, yet too small and dim to be seen with the naked eye; (c) the Cat's Eye Nebula, about 1000 pc away and 0.1 pc across, is an example of a much more complex planetary nebula, possibly produced by a pair of binary stars (unresolved at center) that have both shed envelopes (red seen in visible light, blue in X-rays); (d) M2-9, some 600 pc away, shows surprising twin jets of glowing gas racing outward from the central star at speeds of about 300 km/s. *(AURA; NASA)*

► FIGURE 12.10 **White Dwarf on H–R Diagram** INTERACTIVE A star's passage from the horizontal branch (stage 10) to the white dwarf stage (stage 13) creates an evolutionary path that cuts across the H–R diagram.

diagram as it evolves from a stage 11 red giant to a stage 13 white dwarf is shown in Figure 12.10.

Not all white dwarfs are seen as the cores of planetary nebulae. Several hundred have been discovered "naked," their envelopes expelled to invisibility long ago. Figure 12.11 shows an example of a white dwarf, Sirius B, that happens to lie particularly close to Earth; it is the faint binary companion of the much brighter Sirius A. ∞ *(Sec. 10.7)* Some of its properties are listed in Table 12.2. With more than a solar mass of material packed into a volume smaller than Earth, Sirius B is about a million times denser than anything known in the solar system. Sirius B has an unusually high mass for a white dwarf. It is thought to be the descendent of a star roughly four times the mass of the Sun.

Hubble Space Telescope observations of nearby globular clusters have revealed the white dwarf sequences long predicted by theory but previously too faint to detect at such large distances. Figure 12.12(a) shows a ground-based view of the globular cluster M4, lying 2100 pc from Earth. Part (b) of the figure shows a *Hubble* close-up of a small portion of the cluster, revealing dozens of white dwarfs (some marked) among the cluster's much brighter main-sequence, red giant, and horizontal-branch stars. When plotted on an H–R diagram (see Figure 12.26), the white dwarfs fall nicely along the line indicated on Figure 12.10.

The white dwarf continues to cool and dim with time, following the white–yellow–red line near the bottom of Figure 12.10, eventually becoming a **black dwarf**—a cold, dense, burned-out cinder in space. This is stage 14 of Table 12.1, the graveyard of stars. The cooling dwarf does not shrink much as it fades away, however. Even though its heat is leaking away into space, gravity does not compress it further. Its tightly packed electrons will support the star even as its temperature drops (after trillions of years) almost to absolute zero. As the dwarf cools, it remains about the size of Earth.

Most white dwarfs are composed of carbon and oxygen, but theory predicts that very low-mass stars (less than about one-quarter the mass of the Sun) will never reach the point of helium fusion. Instead, their cores will become supported by the resistance of its electrons to further compression before the central temperature reaches the 100 million kelvins needed for helium to fuse. The interiors of such stars are completely convective, ensuring that fresh hydrogen continually mixes from the envelope into the core. As a result, unlike the case of the Sun illustrated in Figure 12.2, a nonburning helium inner core never appears, and eventually *all* of the star's hydrogen is converted to helium, forming a *helium white dwarf.*

The time needed for this kind of transformation to occur is very long—hundreds of billions of years—so no helium white dwarfs have ever actually formed in this way. ∞ *(Sec. 10.7)* However, if a solar-mass star is a member of a *binary* star system, it is possible for its envelope to be stripped away during the red giant stage by the gravitational pull of its companion, exposing the helium core

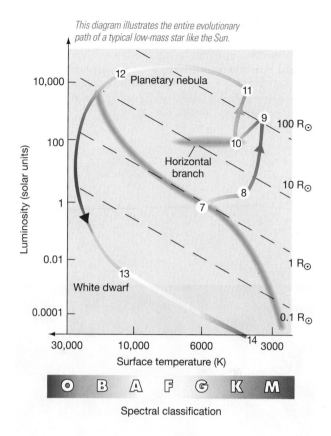

This diagram illustrates the entire evolutionary path of a typical low-mass star like the Sun.

TABLE 12.2	Sirius B—A Nearby White Dwarf
Mass	1.1 Solar masses
Radius	0.008 Solar radii (5500 km)
Luminosity (total)	0.04 Solar luminosities
Surface temperature	24,000 K
Average density	3×10^9 kg/m³

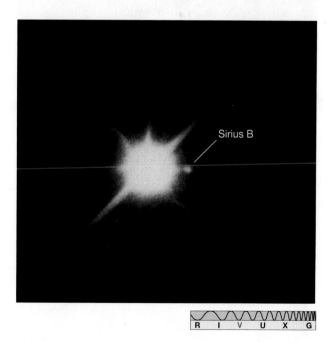

▲ FIGURE 12.11 **Sirius Binary System** Sirius B (the speck of light to the right of the much larger and brighter Sirius A) is a white dwarf star, a companion to Sirius A. The "spikes" on the image of Sirius A are not real; they are caused by the support struts of the telescope. *(Palomar Observatory)*

▶ **FIGURE 12.12 Distant White Dwarfs** (a) The globular cluster M4, as seen through a large ground-based telescope at Kitt Peak National Observatory in Arizona, is the closest globular cluster to us, at 1700 pc away; it spans some 16 pc. (b) A peek at M4's "suburbs" by the *Hubble Space Telescope* shows nearly a hundred white dwarfs within a small 0.2 square-parsec region. *(AURA; NASA)*

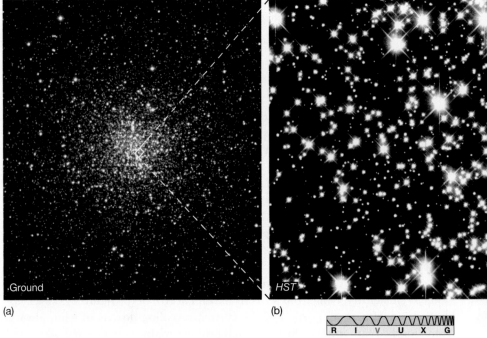

(a)

(b)

R I V U X G

(a)

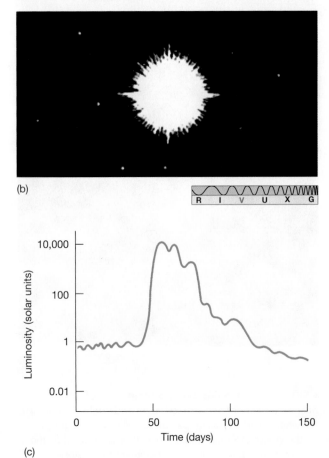

(b)

R I V U X G

and terminating the star's evolution before helium fusion can begin. Several such low-mass helium white dwarfs have in fact been detected in binary systems in our Galaxy.

In stars much more massive than the Sun (close to the 8-solar mass limit on "low-mass" stars at the time the carbon core forms—see Section 12.4), temperatures in the core may become high enough that oxygen can combine with helium to form neon, eventually leading to the formation of a rare *neon–oxygen white dwarf* when fusion finally stops.

Novae

In some cases, the white dwarf stage does not represent the end of the road for a Sun-like star. Given the right circumstances, it is possible for a white dwarf to become explosively active, in the form of a highly luminous **nova** (plural: novae). The word *nova* means "new" in Latin, and to early observers, these stars did indeed seem new as they suddenly appeared in the night sky. Astronomers now recognize that a nova is what we see when a white dwarf undergoes a violent explosion on its surface, resulting in a rapid, temporary increase in luminosity. Figures 12.13(a) and (b) illustrate the brightening of a typical nova over a period of 3 days. Figure 12.13(c) shows the nova's light curve, demonstrating how the luminosity rises dramatically in a matter of days, then fades slowly back to normal over the course of several months. On average, two or three novae are observed each year.

What could cause such an explosion on a faint, dead star? The energy involved is far too great to be explained by flares or other surface activity, and as we have just seen, there is no nuclear activity in the dwarf's interior. The answer to this question lies in the white dwarf's surroundings. If the white dwarf is isolated,

◀ **FIGURE 12.13 Nova** Shown here is Nova Herculis 1934 in (a) March 1935 and (b) May 1935, after brightening by a factor of 60,000. (c) The light curve of a typical nova displays a rapid rise followed by a slow decline in the light received from the star, in good agreement with the explanation of the nova as a nuclear flash on a white dwarf's surface. *(UC/Lick Observatory)*

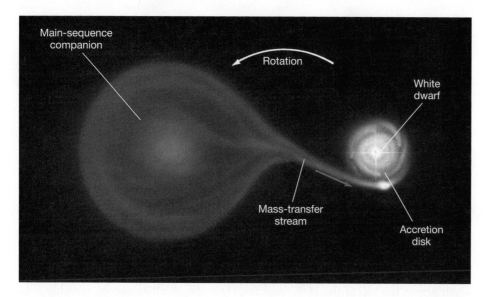

Main-sequence
companion

Rotation

White
dwarf

Mass-transfer
stream

Accretion
disk

◄ **FIGURE 12.14 Close Binary System**   If a white dwarf in a binary system is close enough to its companion, its gravitational field can tear matter from the companion's surface. Notice that the matter does not fall directly onto the white dwarf's surface. Instead, it forms an accretion disk of gas spiraling down onto the dwarf.

then it will indeed cool and ultimately become a black dwarf, as just described. However, should the white dwarf be part of a binary system in which the other star is either a main-sequence star or a giant, an important new possibility exists.

If the distance between the dwarf and the other star is small enough, the dwarf's gravitational field can pull matter—primarily hydrogen and helium—away from the surface of the companion. Because of the binary's rotation, material leaving the companion does not fall directly onto the white dwarf. Instead, it "misses" the dwarf, loops around behind it, and goes into orbit around it, forming a swirling, flattened disk of matter called an **accretion disk,** as shown in Figure 12.14. As it builds up on the white dwarf's surface, the stolen gas becomes hotter and denser. Eventually, its temperature exceeds 10^7 K and the hydrogen ignites, fusing into helium at a furious rate. This surface-burning stage is as brief as it is violent. The star suddenly flares up in luminosity, then fades away as some of the fuel is exhausted and the rest is blown off into space (Figure 12.15). If the event happens to be visible from Earth, we see a nova.

Once the nova explosion is over and the binary has returned to normal, the mass-transfer process can begin again. Astronomers know of many *recurrent novae*—stars that have been observed to "go nova" several times over the course of a few decades. Such systems can, in principle, repeat their violent outbursts many dozens, if not hundreds, of times.

DATA POINTS

Stellar Evolution

Forty percent of students had difficulty describing the basic evolutionary processes occurring once a star leaves the main sequence. Some key points:

■ Stars begin their lives on the Main Sequence, the most massive at the top left, the least massive at the lower right.

■ Stars do not evolve along the Main Sequence; they evolve away from it as they age.

■ Stars generally evolve toward the upper right of the Hertzsprung-Russell once they leave the main sequence. That is, they become cooler and brighter, as red giants, as their non-burning cores shrink and heat up.

■ High-mass stars explode as supernovae while still in the upper part of the Main Sequence; low-mass stars generally end up as white dwarfs, at the lower left.

CONCEPT CHECK

Will the Sun ever become a nova?

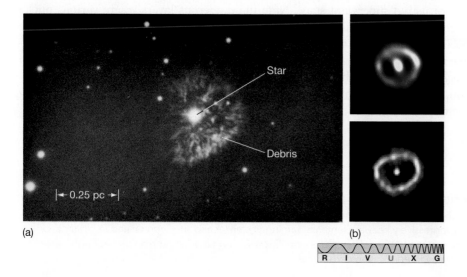

Star

Debris

|← 0.25 pc →|

(a)

(b)

R I V U X G

MA ANIMATION/VIDEO
Recurrent Nova

◄ **FIGURE 12.15 Nova Matter Ejection**   (a) The ejection of material from the star's surface can clearly be seen in this image of Nova Persei, taken some 50 years after it suddenly brightened by a factor of 40,000 in 1901. (b) Nova Cygni, imaged here with a European camera on the *Hubble Space Telescope*, erupted in 1992. At left, more than a year after the blast, a rapidly expanding bubble is seen; at right, 7 months after that, the shell continued to expand and distort. (*Palomar Observatory; ESA*)

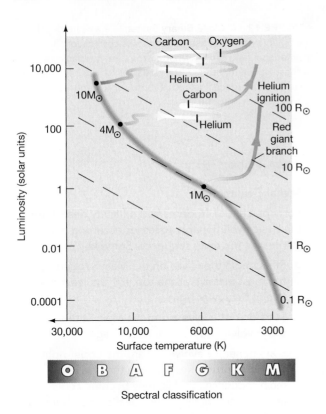

▲ FIGURE 12.16 High-Mass Evolutionary Tracks
INTERACTIVE Evolutionary tracks for stars of 1, 4, and 10 solar masses. Stars with masses comparable to the mass of the Sun ascend the giant branch almost vertically, whereas higher-mass stars move roughly horizontally across the H–R diagram from the main sequence into the red giant region. Some points are labeled with the element that has just started to fuse in the inner core.

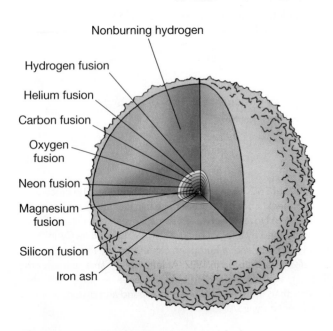

12.4 Evolution of Stars More Massive than the Sun

All stars leaving the main sequence on their way toward the red giant region of the H–R diagram have similar internal structure. Thereafter, however, their evolutionary tracks diverge.

Formation of Heavy Elements

Figure 12.16 compares the post-main-sequence evolution of three stars having masses 1, 4, and 10 times the mass of the Sun. Note the qualitative differences between the solar-mass and higher-mass evolutionary tracks. The solar-mass star ascends the red giant branch almost vertically after leaving the main sequence, its luminosity and radius increasing dramatically while its temperature changes only a little. The higher-mass stars tend to do just the opposite. They loop back and forth across the top of the H–R diagram, their luminosities staying roughly constant as their radii alternately increase and decrease and their surface temperatures fall and rise. The 4- and 10-solar-mass stars do not undergo helium flashes when helium fusion begins, and they do reach core temperatures high enough for helium and carbon to fuse into oxygen.

In the 4-solar-mass star, nuclear burning ceases after the carbon-fusion stage, eventually leading to a carbon–oxygen white dwarf, much as described previously. The fate of the 10-solar-mass star is a little less certain. Depending on how much mass it loses as it evolves (see below), it may also end up as a white dwarf, or it might retain enough mass for fusion to proceed to more massive nuclei. Even higher-mass stars can fuse not just hydrogen, helium, and carbon, but also oxygen, neon, magnesium, silicon, and even heavier elements as their inner cores continue to contract and their central temperatures continue to rise.

Evolution proceeds so rapidly for stars bigger than 10 solar masses that they don't even reach the red giant region before helium fusion begins. Such stars achieve central temperatures of 10^8 K while still quite close to the main sequence. As each element is burned to depletion at the center, the core contracts, heats up, and fusion starts again. A new inner core forms, contracts again, heats again, and so on. The star's evolutionary track continues smoothly across the H–R diagram, seemingly unaffected by each new phase of burning. The star's luminosity stays roughly constant as its radius increases and its surface temperature drops. It swells to become a **red supergiant.**

Figure 12.17 is a cutaway diagram of an evolved high-mass star nearing the end of its red supergiant stage. Note the numerous layers where various nuclei burn. As the temperature increases with depth, the product of each burning stage becomes the fuel for the next. At the relatively cool periphery of the core, hydrogen fuses into helium. In the intermediate layers, shells of helium, carbon, and oxygen burn to form heavier nuclei. Deeper down reside neon, magnesium, silicon, and other heavy nuclei, all produced by nuclear fusion in the layers overlying the nonburning inner core. The inner core itself is composed of iron.

◀ FIGURE 12.17 Heavy-Element Fusion Cutaway diagram of the interior of a highly evolved high-mass star shows how the interior resembles the layers of an onion, with shells of progressively heavier elements burning at smaller and smaller radii and at higher and higher temperatures. The core is actually only a few times larger than planet Earth, while the star is hundreds of times larger than the Sun.

Observations of Supergiants

A good example of a post-main-sequence blue supergiant is the bright star Rigel in the constellation Orion. With a radius some 70 times that of the Sun and some 60,000 solar luminosities, Rigel is thought to have had an original mass about 17 times that of the Sun. A strong stellar wind has probably carried away a significant fraction of the star's mass since it formed. Although still near the main sequence, Rigel is probably already fusing helium into carbon in its core.

Astronomers have discovered that, while stars of all spectral types have stellar winds, the highly luminous supergiants have by far the strongest. ∞ *(Sec. 9.3)* Satellite and rocket observations have shown that the winds of some blue supergiants may reach speeds of 3000 km/s and carry off more than 10^{-6} solar masses of material per year. Over the relatively short span of a few million years, these stars can blow a tenth or more of their total mass—a Sun's worth of matter—off into space. During violent outbursts near the end of the star's life, the mass loss rate may exceed 10^{-3} solar masses per year. Figure 12.18 shows a luminous blue variable star called Eta Carinae that is currently displaying this type of behavior. The glowing nebula surrounding the star itself contains up to 3 solar masses of gas shed during this and previous outbursts over the past few hundred years.

Perhaps the best-known red supergiant is Betelgeuse (shown in Figures 10.7 and 10.10), also in Orion and Rigel's rival for the title of brightest star in the constellation. Its luminosity is 10,000 times that of the Sun in visible light and perhaps four times that in the infrared. Astronomers think that Betelgeuse is currently fusing helium into carbon and oxygen in its core, but its eventual fate is uncertain. As best we can tell, the star's mass at formation was between 12 and 17 times the mass of the Sun. However, like Rigel and many other supergiants, Betelgeuse has a strong stellar wind and is known to be surrounded by a huge shell of dust of its own making. This suggests that Betelgeuse has lost a lot of mass since it formed, but just how much remains uncertain.

The End of the Road

Through each period of stability and instability and each new burning stage, the star's central temperature increases, the nuclear reactions speed up, and the newly released energy supports the star for ever shorter periods of time. For example, in round numbers, a star 20 times more massive than the Sun burns hydrogen for 10 million years, helium for 1 million years, carbon for 1000 years, oxygen for 1 year, and silicon for a week. Its iron core grows for less than a day.

Once the inner core begins to change to iron, our high-mass star is in trouble. Nuclear fusion involving iron does not produce energy, because iron nuclei are so tightly bound that energy cannot be extracted by combining them into heavier elements. In effect, iron plays the role of a fire extinguisher, damping the inferno in the stellar core. With the appearance of substantial quantities of iron, the central fires cease for the last time, and the star's internal support begins to dwindle. The star's foundation is destroyed, and its equilibrium is gone forever. Even though the temperature in the iron core has reached several billion kelvins by this stage, the enormous inward gravitational pull of matter ensures catastrophe in the very near future. Gravity overwhelms the pressure of the hot gas, and the star implodes, falling in on itself.

The core temperature rises to nearly 10 billion kelvins. At these temperatures individual photons are energetic enough to split iron into lighter nuclei and then break those lighter nuclei apart until only protons and neutrons remain. This

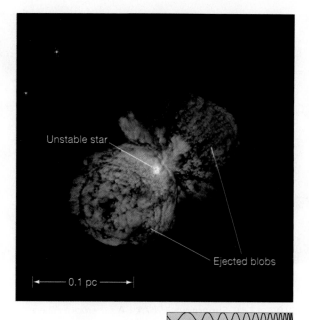

Unstable star

Ejected blobs

|← 0.1 pc →|

R I V U X G

▲ **FIGURE 12.18 Mass Loss from Supergiants** With an estimated mass of around 100 times the mass of the Sun and a luminosity of 5 million times the solar value, Eta Carinae is one of the most massive and most luminous stars known. This *HST* image shows blobs of ejected material racing away from the star at hundreds of kilometers per second, following a series of explosive outbursts that ejected 2 or 3 solar masses of material into space. *(NASA)*

MA ANIMATION/VIDEO
Light Echo

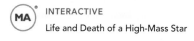

MA INTERACTIVE
Life and Death of a High-Mass Star

▲ **FIGURE 12.19 Supernova 1987A** A supernova called SN1987A (arrow) was exploding near this nebula (called 30 Doradus) at the moment the photograph at right was taken. The photograph at left is the normal appearance of the star field. (See *Discovery 12-1*.) *(AURA)*

CONCEPT CHECK

Why does the iron core of a high-mass star collapse?

TABLE 12.3	End Points of Evolution for Stars of Different Masses
Initial Mass (Solar Masses)	**Final State**
Less than 0.08	(Hydrogen) brown dwarf
0.08–0.25	Helium white dwarf
0.25–8	Carbon–oxygen white dwarf
8–12 (approx.)*	Neon–oxygen white dwarf
Greater than 12*	Supernova

*Precise numbers depend on the (poorly known) amount of mass lost while the star is on, and after it leaves, the main sequence.

process is known as *photodisintegration*. In less than a second, the collapsing core undoes all the effects of nuclear fusion that occurred during the previous 10 million years. But to split iron and lighter nuclei into smaller pieces requires a lot of energy. After all, this splitting is the opposite of the fusion reactions that generated the star's energy during earlier times. The process thus *absorbs* some of the core's heat energy, reducing the pressure and accelerating the collapse.

Now the core consists entirely of electrons, protons, neutrons, and photons at enormously high densities, and it is still shrinking. As the density continues to rise, the protons and electrons are crushed together, combining to form more neutrons and releasing neutrinos. Even though the central density by this time may exceed, 10^{12} kg/m^3, most of the neutrinos pass through the core as if it weren't there. ∞ *(Sec. 9.5)* They escape into space, carrying away still more energy as they go.

There is nothing to prevent the collapse from continuing all the way to the point at which the neutrons themselves come into contact with each other, at the incredible density of about 10^{15} kg/m^3. At this density, the neutrons in the shrinking core play a role similar in many ways to that of the electrons in a white dwarf. When far apart, they offer little resistance to compression, but when brought into contact, they produce enormous pressures that strongly oppose further contraction. The collapse finally begins to slow. By the time it is actually halted, however, the core has overshot its point of equilibrium and may reach a density as high as 10^{18} kg/m^3 before beginning to reexpand. Like a fast-moving ball hitting a brick wall, the core becomes compressed, stops, then rebounds—with a vengeance!

The events just described do not take long. Only about a second elapses from the start of the collapse to the "bounce" at nuclear densities. Driven by the rebounding core, an enormously energetic shock wave then sweeps outward through the star at high speed, blasting all the overlying layers—including the heavy elements outside the iron inner core—into space. Although the details of how the shock reaches the surface and destroys the star are still uncertain, the end result is not: The star explodes in one of the most energetic events known in the universe (Figure 12.19). This spectacular death rattle of a high-mass star is known as a **core-collapse supernova.**

Our earlier dividing line of 8 solar masses between "low mass" and "high mass" really refers to the mass at the time the carbon core forms. Since very luminous stars often have strong stellar winds, it is possible that main-sequence stars as massive as 10 to 12 times the mass of the Sun may still manage to avoid going supernova. Unfortunately, we do not know exactly how much mass either Rigel or Betelgeuse has lost, so we cannot yet tell whether they are above or below the supernova threshold. Either might explode or instead become a (neon–oxygen) white dwarf, but for now we can't say which. We may just have to wait and see—in a few million years we will know for sure!

Table 12.3 summarizes the possible outcomes of stellar evolution for stars of different masses. For completeness, brown dwarfs—the end product of low-mass protostars unable even to fuse hydrogen in their cores—are included in the list. ∞ *(Sec. 11.4)*

12.5 Supernova Explosions

Novae and Supernovae

Observationally, a *supernova*, like a nova, is a "star" that suddenly increases dramatically in brightness, then slowly dims again, eventually fading from view. However, despite some similarities in their light curves, novae and supernovae are very different phenomena. As we saw in Section 12.3, a nova is a violent explosion on the surface of a white dwarf in a binary system.[2] Supernovae are much more energetic—about a million times brighter than novae—and are driven by very different underlying physical processes. A supernova produces a burst of light billions of times brighter than the Sun, reaching that brightness within just a few hours of the start of the outburst, sometimes rivaling its parent galaxy in brightness (Figure 12.20a). The total amount of electromagnetic energy radiated by a supernova during the few months it takes to brighten and fade away is roughly the same as the Sun will radiate during its *entire* 10^{10}-year lifetime!

Astronomers divide supernovae into two classes. **Type I supernovae** contain very little hydrogen, according to their spectra, and have light curves (see Figure 12.20b) somewhat similar in shape to those of typical novae—a sharp rise in intensity followed by steady, gradual decline. **Type II supernovae** are hydrogen-rich and usually have a characteristic "plateau" in the light curve a few months after the maximum. Observed supernovae are divided roughly evenly between these two categories.

Type I and Type II Supernovae Explained

How do we explain the two types of supernova just described? In fact, we already have half of the answer—the characteristics of Type II supernovae are exactly consistent with the core-collapse supernovae discussed in the previous section. The Type II light curve is in good agreement with computer simulations of a stellar envelope expanding and cooling as it is blown into space by a shock wave sweeping up from below. And since the expanding material consists mainly of unburned hydrogen and helium, it is not surprising that those elements dominate the supernova's spectrum.

What of Type I supernovae? Is there more than one way for a supernova explosion to occur? The answer is yes. To understand the alternative supernova mechanism, we must reconsider the long-term consequences of the accretion–explosion cycle that causes a nova. A nova explosion ejects matter from a white dwarf's surface, but it does not necessarily expel or burn all the material that has accumulated since the last outburst. In other words, there is a tendency for the dwarf's mass to increase slowly with each new nova cycle. As its mass grows and the internal pressure required to support its weight rises, the white dwarf can enter into a new period of instability—with disastrous consequences.

Recall that a white dwarf is held up by the pressure of electrons that have been squeezed so close together that they have effectively come into contact with one another. However, there is a limit to the mass of a white dwarf, beyond which the electrons cannot support the star against its own gravity. Detailed calculations show that the maximum mass of a white dwarf is about 1.4 solar masses, a mass often called the *Chandrasekhar mass*, after the Indian-American astronomer Subramanyan Chandrasekhar, whose work in theoretical astrophysics earned him a Nobel Prize in Physics in 1983. More recently, in 1999, NASA's latest X-ray

This supernova exploded in 1994 in the disk of a distant galaxy.

|← 10,000 pc →|

(a)

(b)

▲ FIGURE 12.20 **Supernova Light Curves**   (a) The (Type I) supernova SN1994D in the galaxy NGC 4526 was for a time comparable in brightness to the entire galaxy in which it occurred. (b) The light curves of typical Type I and Type II supernovae both show that their maximum luminosities can sometimes reach that of a billion suns, but there are characteristic differences in the decline of luminosity after the initial peak. Type I light curves resemble those of novae (see Figure 12.13c), but the total energy released is much larger. Type II curves have a characteristic plateau during the declining phase.

---

[2]When discussing novae and supernovae, astronomers tend to blur the distinction between the observed event (the sudden appearance and brightening of an object in the sky) and the process responsible for it (a violent explosion in or on a star). The terms can have either meaning, depending on context.

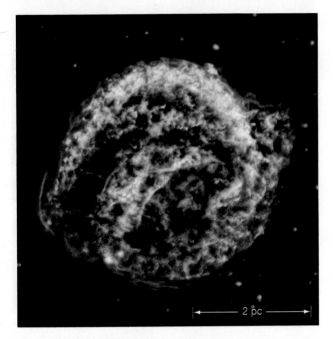

▲ FIGURE 12.21 Kepler's Supernova This X-ray image, captured by the orbiting *Chandra* telescope and released in 2013, shows the aftermath of a titanic stellar explosion that was observed by many people on Earth in 1604. At its peak it was the brightest object in the night sky; it was visible even during the daytime for several weeks. It lies about 6000 pc from Earth. ∞ *(Sec. 1.2)* (CXC)

telescope was named *Chandra* in his honor. ∞ *(Sec. 3.5)* Figure 12.21 shows a *Chandra* image of a (relatively) recent Type I supernova.

If an accreting white dwarf exceeds the Chandrasekhar mass, it starts to collapse. Its internal temperature rapidly rises to the point at which carbon can (at last) fuse into heavier elements. Carbon fusion begins everywhere throughout the white dwarf almost simultaneously, and the entire star explodes in a **carbon-detonation supernova**—an event comparable in violence to the core-collapse supernova associated with the death of a high-mass star. In an alternative and (many astronomers think) possibly more common scenario, two white dwarfs in a binary system may collide and merge to form a massive, unstable star. The end result is the same.

The detonation of a carbon white dwarf, the descendant of a *low-mass* star, is a Type I supernova.[3] Because the explosion occurs in a system containing virtually no hydrogen, we can readily see why the supernova spectrum shows little evidence of that element. The light curve is almost entirely due to the radioactive decay of unstable heavy elements produced during the explosion. These explosions have one other important property: Because they all originate in similar systems—white dwarfs whose masses have just exceeded the Chandrasekhar limit—Type I supernovae show remarkably little variation in their properties from one supernova to the next, no matter where (or when) they occur. As we will see in Chapters 15 and 17, this fact, coupled with the huge luminosities involved, make Type I supernovae invaluable tools for mapping out the structure and evolution of the universe on the largest scales.

Figure 12.22 summarizes the processes responsible for the two types of supernovae. Recognize that, despite the similarity in their peak luminosities, Type I

---

[3]In the jargon of the field, these explosions are technically known as Type Ia supernovae, to distinguish them from the rarer Type Ib and Ic explosions, which are core-collapse events in stars that have lost their hydrogen envelopes.

(a) Type I Supernova

Time

(b) Type II Supernova

▲ FIGURE 12.22 Two Types of Supernova Type I and Type II supernovae have different causes. These sequences depict the evolutionary history of each type. (a) A Type I supernova usually results when a carbon-rich white dwarf pulls matter onto itself from a nearby red giant or main-sequence companion. (b) A Type II supernova occurs when the core of a high-mass star collapses and then rebounds in a catastrophic explosion.

and Type II supernovae are unrelated to one another. They occur in stars of very different types, under very different circumstances. All high-mass stars become Type II (core-collapse) supernovae, but only a tiny fraction of low-mass stars evolve into white dwarfs that ultimately explode as Type I (carbon-detonation) supernovae. However, there are far more low-mass stars than high-mass stars, resulting in the remarkable coincidence that the two types of supernova occur at roughly the same rate.

Supernova Remnants

We have plenty of evidence that supernovae have occurred in our Galaxy. Occasionally, the supernova explosions are visible from Earth. In other cases, we can detect their glowing remains, or **supernova remnants.** One of the best-studied supernova remnants is the Crab Nebula (Figure 12.23a). Its brightness has

CONCEPT CHECK

How did astronomers know, even before the mechanisms were understood, that there were at least two distinct physical processes at work in creating supernovae?

 ANIMATION/VIDEO
Supernova Explosions

 INTERACTIVE
Supernova Remnant Cassiopeia A

(a)

── 0.5 pc ──

(b)

R I V U X G

◄ **FIGURE 12.23 Supernova Remnants INTERACTIVE** (a) This remnant of an ancient Type II supernova is called the Crab Nebula. It resides about 1800 pc from Earth, and its debris is scattered over a region about 2 pc across. The main image was taken with the Very Large Telescope of the European Southern Observatory in Chile, the inset by the *Hubble* telescope in orbit. (b) The glowing gases of the Vela supernova remnant are spread across 6° of the sky. *(ESO; NASA; ESO)*

12.1 DISCOVERY

Supernova 1987A

Although there have been no observable supernovae in our own Galaxy since the invention of the telescope, astronomers were treated to the next best thing in 1987 when a 15-solar mass B-type supergiant exploded in the Large Magellanic Cloud (LMC), a small satellite galaxy orbiting our own (see Section 15.1). For a few weeks the Type II supernova SN1987A, shown in Figure 12.19, outshone all the other stars in the LMC combined. Because the LMC is relatively close to Earth and the explosion was detected soon after it occurred, SN1987A provided astronomers with a wealth of detailed information on supernovae, allowing them to make key comparisons between theoretical models and observational reality. By and large, the theory of stellar evolution has held up very well. Still, SN1987A did hold a few surprises.

The light curve of SN1987A, shown at right, differed somewhat from the "standard" Type II shape (Figure 12.20). The peak brightness was only about 1/10 the standard value and occurred much later than expected. These differences stemmed from the fact that the parent star's envelope was deficient in heavy elements, significantly altering its evolutionary track. As a result, the star was a (relatively small) blue supergiant, at the top right of the H-R diagram with a surface temperature of around 20,000 K, when the rapid chain of events leading to the supernova occurred.

Because the parent star was small and quite tightly bound by gravity, a lot of the energy produced in the explosion was used in expanding SN1987A's stellar envelope, so far less was left over to be radiated into space. Thus the luminosity during the first few months was lower than expected, and the early peak evident in Figure 12.20 did not occur. The peak in the SN1987A light curve at about 80 days actually corresponds to the "plateau" in the Type II light curve in Figure 12.20.

About 20 hours before the supernova was detected optically, underground detectors in Japan and the United States recorded a brief (13-second) burst of neutrinos. The neutrinos preceded the light because they escaped during the collapse, whereas the first light of the explosion was emitted only after the supernova shock had plowed through the body of the star to the surface. Theoretical models, consistent with these observations, suggest that many tens of thousands of times more energy was emitted in the form of neutrinos than in any other form. Detection of this neutrino pulse was a brilliant confirmation of theory and may well herald a new age of astronomy. For the first time, astronomers received information from beyond the solar system by means outside the electromagnetic spectrum.

The accompanying photographs show the barely resolved remnant of SN1987A (at the center of each image) surrounded by a much larger shell of glowing gas (in yellow). The progenitor star expelled this shell during its red giant phase, some 40,000 years before the explosion. The image we see results from the flash of ultraviolet light from the supernova hitting the ring and causing it to glow brightly. The insets at the bottom show supernova debris moving outward toward the ring at nearly 3000 km/s. The fastest-moving ejecta have already reached the ring, forming the glowing regions that now surround the supernova remnant. The main image also revealed two additional faint rings that might be caused by radiation sweeping across an hourglass-shaped bubble of gas, itself perhaps the result of a nonspherical "bipolar" stellar wind from the progenitor star before the supernova occurred.

Buoyed by the success of stellar-evolution theory astronomers eagerly await further developments in the story of this remarkable object.

(NASA)

| 1994 | 1998 | 2001 | 2004 | 2007 | 2011 |

MA · ANIMATION/VIDEO
Shockwaves Hit the Ring
Around Supernova 1987A

R I V U X G

greatly dimmed now, but the original explosion in the year A.D. 1054 was so brilliant that it is prominently recorded in the manuscripts of Chinese and Middle Eastern astronomers. For nearly a month, this exploded star reportedly could be seen in broad daylight. Even today, the knots and filaments give a strong indication of past violence. The nebula—the envelope of the high-mass star that exploded to create this Type II supernova—is still expanding into space at a speed of several thousand kilometers per second. Figure 12.24 illustrates this motion by combining a positive image of the Crab taken in 1960 with a negative image taken in 1974. If the gas were not moving, the positive and negative images would overlap perfectly, but they do not. The gas moved outward in the intervening 14 years. Tracing the motion backward in time, astronomers have found that the explosion must have occurred about nine centuries ago, consistent with the Chinese and Middle Eastern observations.

Figure 12.23b shows another example (also of Type II). The expansion velocity of the Vela supernova remnant implies that its central star exploded around 9000 B.C. This remnant lies only 500 pc from Earth. Given its proximity, it may have been as bright as the Moon for several months. See also Figure 3.29 for an X-ray image of a nearby supernova remnant.

Although hundreds of supernovae have been observed in other galaxies during the 20th century, no one using modern equipment has ever observed one in our own Galaxy. (See *Discovery 12-1* for a discussion of a recent supernova in a galaxy very close to our own.) A viewable Milky Way star has not exploded since Galileo first turned his telescope to the heavens almost four centuries ago. Based on stellar evolutionary theory, astronomers estimate that an observable supernova ought to occur in our Galaxy every 100 years or so. Our local neighborhood seems long overdue for one. Unless stars explode much less frequently than predicted by theory, we should be treated to a (relatively) nearby version of nature's most spectacular cosmic event any day now.

Formation of the Heavy Elements

From the point of view of life on Earth, probably the most important aspect of stellar evolution is its role in creating, and then dispersing, the heavy elements out of which both our planet and our bodies are made. Since the 1950s, astronomers have come to realize that all of the hydrogen and most of the helium in the universe are *primordial*—that is, they date back to the very earliest times, long before the first stars formed (see Section 17.6). All other elements (and, in particular, virtually everything we see around us on Earth) formed later, in stars.

We have already seen how heavy elements are created from light ones by nuclear fusion. This is the basic process that powers all stars. Hydrogen fuses to helium, then helium to carbon and oxygen. Subsequently, in high-mass stars, carbon and oxygen can fuse to form still heavier elements. Neon, magnesium, sulfur, silicon—in fact, elements up to and including iron—are created in turn by fusion reactions in the cores of the most massive stars.

However, fusion stops at iron. The fact that iron nuclei will not fuse to release energy and create more massive nuclei is the basic underlying cause of Type II supernovae. How then were even heavier elements, such as copper, lead, gold, and uranium, formed? As mentioned in Section 12.3, some of these elements were formed during the late red giant stages of low-mass stars via

This superposition of two images taken 14 years apart shows how the Crab Nebula expanded during that time.

Blast center

R I V U X G

▲ FIGURE 12.24 **The Crab in Motion**   Positive and negative photographs of the Crab Nebula taken 14 years apart do not superimpose exactly, indicating that the gaseous filaments are still moving away from the site of the explosion. The positive image in white was taken first, and the black (negative) filaments were overlaid later—hence the black (but still glowing) outlying debris is farther from the center of the blast. The scale is roughly the same as in Figure 12.23(a). *(Harvard College Observatory)*

DATA POINTS

Novae and Supernovae

Two-thirds of students had difficulty distinguishing novae and the different kinds of supernovae. Some points to remember:

- Supernovae are millions of times brighter than novae.

- Novae are thermonuclear detonations on the surfaces of white dwarf stars. They are violent, but do not destroy the star on which they occur.

- A Type I supernova is a detonation involving an entire white dwarf that has exceeded its maximum stable mass. The explosion likely leaves no remnant behind.

- A Type II supernova results from the runaway collapse of the core of a massive star; it probably leaves a neutron star or black hole remnant behind.

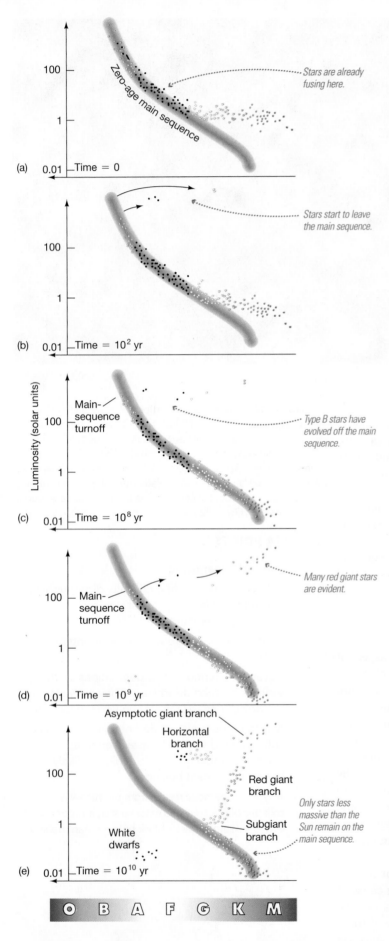

reactions involving neutrons. Similar reactions occur in supernovae too, where neutrons and protons, produced when some nuclei are ripped apart by the almost unimaginable violence of the blast, are crammed into other nuclei, creating heavy elements that cannot have formed by any other means. Many of the heaviest elements were formed *after* their parent stars had already died, even as the debris from the explosion signaling the stars' deaths was hurled into interstellar space. In this way stellar evolution steadily enriches the composition of each new generation of stars.

Although no one has ever observed directly the formation of heavy nuclei in stars, astronomers are confident that the chain of events just described really occurs. When nuclear reaction rates (determined from laboratory experiments) are incorporated into detailed computer models of stars and supernovae, the results agree very well, point by point, with the composition of the universe inferred from analysis of meteorites and spectroscopic studies of planets, stars, and the interstellar medium. The reasoning is indirect, but the agreement between theory and observation is so striking that most astronomers regard it as strong evidence supporting the entire theory of stellar evolution.

12.6 Observing Stellar Evolution in Star Clusters

Star clusters provide excellent test sites for the theory of stellar evolution. Every star in a given cluster formed at nearly the same time, from the same interstellar cloud, with virtually the same composition. ∞ *(Sec. 11.6)* Only the mass varies from one star to another. This uniformity allows us to check the accuracy of our theoretical models in a very straightforward way. Having studied in some detail the evolutionary tracks of individual stars, let us now consider how their collective appearance changes in time.

We begin our study shortly after the cluster's formation, with the high-mass stars already fully formed and burning steadily on the upper main sequence and lower-mass stars just beginning to arrive on the main sequence (Figure 12.25a). The appearance of the cluster at this early stage is dominated by its most massive stars—the bright blue supergiants.

Figure 12.25(b) shows our cluster's H–R diagram after 10 million years. The most massive O-type stars have evolved off the main sequence. Most have already exploded and vanished, but one or two may still be visible as supergiants traversing the top of the diagram. The remaining cluster stars are largely unchanged in appearance.

◀ FIGURE 12.25 Cluster Evolution on the H–R Diagram   (a) Initially, stars on the upper main sequence are already burning steadily while the lower main sequence is still forming. (b) At 10^7 years, O-type stars have already left the main sequence, and a few red giants are visible. (c) By 10^8 years, more red giants are visible, and the lower main sequence is almost fully formed. (d) At 10^9 years, the subgiant and red giant branches are just becoming evident, and the formation of the lower main sequence is complete. (e) At 10^{10} years, the cluster's subgiant, red giant, horizontal, and asymptotic giant branches are all discernible, and many white dwarfs have now formed.

The cluster's H–R diagram has a slightly shortened main sequence. Figure 12.26 shows the twin open clusters h and (the Greek letter chi) Persei, along with their observed H–R diagram. Comparing Figure 12.26(b) with Figure 12.25(b), astronomers estimate the age of this double cluster to be about 10 million years.

After 100 million years (Figure 12.25c), stars brighter than type B5 or so (about 4 to 5 solar masses) have left the main sequence, and a few more supergiants are visible. By this time most of the cluster's low-mass stars have finally arrived on the main sequence, although the dimmest M-type stars may still be in their contraction phase. The appearance of the cluster is now dominated by bright B-type stars and brighter supergiants.

At any time during the evolution, the cluster's original main sequence is intact up to some well-defined stellar mass, corresponding to the stars that are just leaving the main sequence at that instant. We can imagine the main sequence being "peeled away" from the top down, with fainter and fainter stars turning off and heading for the giant branch as time goes on. Astronomers refer to the high-luminosity end of the observed main sequence as the **main-sequence turnoff.** The mass of the star that is just evolving off the main sequence at any moment is known as the *turnoff mass*. Knowing the turnoff mass is equivalent to knowing the age of the cluster—stars less massive than the turnoff are still on the main sequence, while more massive stars have already evolved into something else.

At 1 billion years (Figure 12.25d) the main-sequence turnoff mass is around 2 solar masses, corresponding roughly to spectral type A2. The subgiant and giant branches associated with the evolution of low-mass stars are just becoming apparent, and the formation of the lower main sequence is now complete. In addition, the first white dwarfs have just appeared, although they are often too faint to be observed at the distances of most clusters. Figure 12.27 shows the Hyades open cluster and its H–R diagram. The H–R diagram appears to lie between Figures 12.25 (c) and (d). More careful measurements yield a cluster age of about 600 million years.

At 10 billion years, the turnoff point has reached solar-mass stars, of spectral type G2. The subgiant and giant branches are now clearly discernible in the H–R diagram (Figure 12.25e), and the horizontal branch appears as a distinct region. Many white dwarfs are also present in the cluster. Figure 12.28 shows a composite H–R diagram for several globular clusters in our Galaxy. The various evolutionary stages predicted by theory are all clearly visible in this figure, although the individual points are shifted somewhat to the left relative to our previous H–R diagrams because of differences in composition between stars like the Sun and stars in globular clusters. (Globular cluster stars tend to be hotter than solar-type stars of the same mass.) The clusters represented in Figure 12.28(b) are quite deficient in heavy elements compared with the Sun. Figure 12.28(a) shows a *Hubble* image of the globular cluster M80, which has a similarly low concentration of heavy elements—only about 2 percent the solar value—and whose H–R diagram looks qualitatively very similar to Figure 12.28.

By carefully adjusting their theoretical models until the main sequence, subgiant, red giant, and horizontal branches are all well matched, astronomers find that this diagram corresponds to a cluster age of roughly 12 billion years, a little older than the hypothetical cluster in Figure 12.25(e). The deficiency of heavy elements is consistent with these clusters being among the first objects to form in our Galaxy (see Section 12.7). In fact, globular cluster ages determined this way show a remarkably small spread. Most of the globular clusters in our Galaxy formed between about 10 and 12 billion years ago.

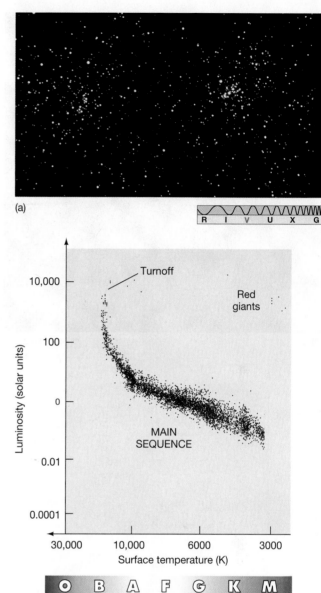

(a)

(b)

▲ **FIGURE 12.26 Newborn Cluster H–R Diagram** (a) The "double cluster" h and χ Persei, two open clusters that apparently formed at the same time. (b) The H–R diagram of the pair indicates that the stars are very young—probably only 10–15 million years old. Even so, the most massive stars have already left the main sequence. *(NOAO; data from T. Currie)*

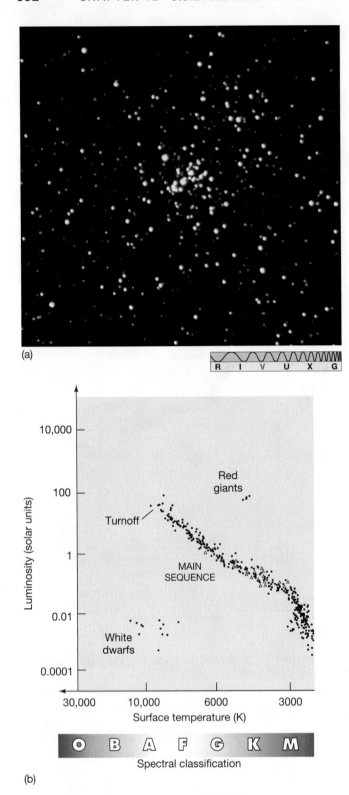

(a)

(b)

▲ FIGURE 12.27 Young Cluster H–R Diagram   (a) The Hyades cluster, a relatively young group of stars visible to the naked eye, is found 46 pc away in the constellation Taurus. (b) The H–R diagram for this cluster is cut off at about spectral type A, implying an age of about 600 million years. A few massive stars have already become white dwarfs. *(NOAO)*

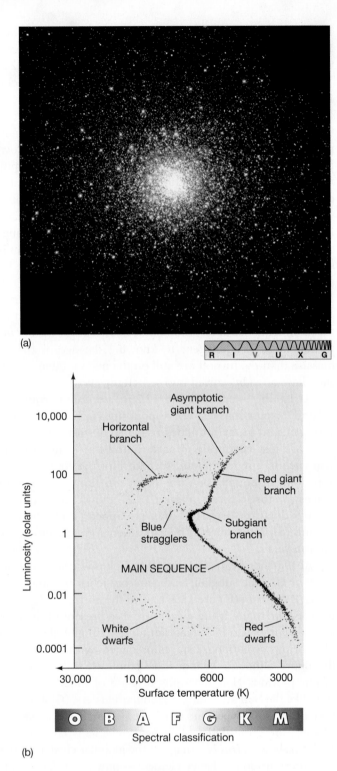

(a)

(b)

▲ FIGURE 12.28 Old Cluster H–R Diagram   (a) The globular cluster M80, which lies some 8 kpc from Earth. (b) Combined H–R diagram, based on ground- and space-based observations, for several globular clusters similar in overall composition to M80. Fitting the main-sequence turnoff and the giant and horizontal branches to theoretical models implies an age of about 12 billion years, making these clusters among the oldest known objects in the Milky Way Galaxy. *(NASA; data courtesy of William E. Harris)*

▶ **FIGURE 12.29 Multiple Stellar Generations**   The ground-based H–R diagram (b) of the globular cluster NGC 2808 (a) shows an apparently normal main sequence, but more precise observations from *HST* (c) reveal that the main sequence is actually made up of three distinct sequences (as noted), increasing in helium content from right to left. These observations imply multiple generations of star formation shortly after the cluster formed, but no theory can yet explain just how this occurred. *(NASA)*

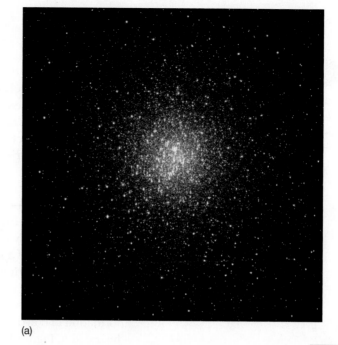

(a)

The objects labeled as *blue stragglers* in Figure 12.28(b) appear at first sight to contradict the theory just described. They lie on the main sequence, but above the turnoff, suggesting that they should have evolved into white dwarfs long ago, given the cluster's age of 12 billion years. Blue stragglers are observed in many star clusters. They are main-sequence stars, but they did not form when the cluster did. Instead, they formed much more recently, through *mergers* of lower-mass stars—so recently, in fact, that they have not yet had time to evolve into giants. In some cases, the mergers were probably the result of stellar evolution in binary systems, as the component stars evolved, grew, and came into contact. In others, the mergers are thought to be the result of actual *collisions* between stars. The dense central cores of globular clusters—with a million or more times as many stars per unit volume as are found in the neighborhood of the Sun—are among the few places in the entire universe where stellar collisions are likely to occur.

High-precision observations from *HST* have revealed a new, and as yet unresolved, mystery about globular clusters that may yet force astronomers to change significantly their views on how massive star clusters form. Figure 12.29 presents the H–R diagram for the cluster NGC 2808, showing (in the inset) what appear to be *three* distinct main sequences, undetected in earlier ground-based observations. The stars in the three sequences contain different amounts of helium, carbon, and nitrogen and are thought to be the result of multiple episodes of star formation occurring over the course of about 100 million years. Models suggest that the two more helium-

(c) *Three separate main sequences are evident here.*

(b)

rich generations formed from gas enriched by stellar evolution in the first-generation stars, but astronomers do not know how this could have occurred in the time available. Whatever happened, it seems to have been a common phenomenon, as high-resolution studies of many globular clusters now reveal similar multiple populations. Indeed, some observers would go so far as to claim that multiple stellar populations are the norm in the Galactic globular cluster population.

12.7 The Cycle of Stellar Evolution

The theory of stellar evolution outlined in this chapter can naturally account for the observed differences in the abundances of heavy elements between the old globular-cluster stars and stars now forming in our Galaxy. Even though an evolved star continuously creates new heavy elements in its interior, changes in the star's composition are confined largely to the core, and the star's spectrum may give little indication of the events within. Only at the end of the star's life are its newly created elements released and scattered into space.

Thus the spectra of the *youngest* stars show the *most* heavy elements because each new generation of stars increases the concentration of these elements in the interstellar clouds from which the next generation forms. Accordingly, the photosphere of a recently formed star contains a much greater abundance of heavy elements than that of a star formed long ago. Knowledge of stellar evolution allows astronomers to estimate the ages of stars from purely spectroscopic studies, even when the stars are isolated and are not members of any cluster.

We have now seen all the ingredients that make up the complete cycle of star formation and evolution. Let us now briefly summarize that process, which is illustrated in Figure 12.30.

1. Stars form when part of an interstellar cloud is compressed beyond the point at which it can support itself against its own gravity. The cloud collapses and fragments, forming a cluster of stars. The hottest stars heat and ionize the surrounding gas, sending shock waves through the surrounding cloud, modifying the formation of lower-mass stars, and possibly triggering new rounds of star formation. ∞ *(Secs. 11.2, 11.3)*

2. Within the cluster, stars evolve. The most massive stars evolve fastest, creating heavy elements in their cores and spewing them forth into the interstellar medium in supernova explosions. Low-mass stars evolve more slowly, but they too create heavy elements and contribute significantly to the "seeding" of interstellar space when they shed their envelopes as planetary nebulae. Roughly speaking, low-mass stars created most of the carbon, nitrogen, and oxygen that make life on Earth possible. High-mass stars produced the iron and silicon that make up Earth itself, as well as the heavier elements on which much of our technology is based.

3. The creation and explosive dispersal of new heavy elements are accompanied by further shock waves. The passage of these shock waves through the interstellar medium simultaneously enriches the medium and compresses it into further star formation. Each generation of stars increases the concentration of heavy elements in the interstellar clouds from which the next generation forms. As a result, recently formed stars contain a much greater abundance of heavy elements than stars that formed long ago.

In this way, although some material is used up in each cycle—turned into energy or locked up in stars less massive than the Sun (most of which have not yet left the main sequence)—the Galaxy continuously recycles its matter. Each new round of formation creates stars containing more heavy elements than the preceding generation had. From the old globular clusters, which are observed to be deficient in heavy elements relative to the Sun, to the young open clusters, which contain much larger amounts of these elements, we observe this enrichment process in action. Our Sun is the product of many such cycles. We ourselves are another. Without the heavy elements synthesized in the hearts of stars, life on Earth would not exist.

. . . throwing out heavy-element debris, such as N49 here, that then enrich interstellar space.

Supernova + heavy elements

Stellar recycling begins with dark interstellar clouds, such as this one called Barnard 68 . . .

Interstellar medium

Stellar evolution

. . . after which stars evolve, some eventually exploding, like NGC 7635 shown brightly here . . .

Star formation

. . . stars form when such clouds collapse under their own gravity, as here in the RCW 38 star-forming region . . .

▲ **FIGURE 12.30 Stellar Recycling INTERACTIVE**   The cycle of star formation and evolution continuously replenishes the Galaxy with new heavy elements and provides the driving force for the creation of new generations of stars. Clockwise from the top are an interstellar cloud (Barnard 68), a star-forming region in our Galaxy (RCW 38), a massive star ejecting a "bubble" and about to explode (NGC 7635), and a supernova remnant and its heavy-element debris (N49). *(ESO; NASA)*

Stellar evolution is one of the great success stories of astrophysics. Like all good scientific theories, it makes definite, testable predictions about the universe, at the same time remaining flexible enough to incorporate new discoveries as they occur. Theory and observation have advanced hand in hand. At the start of the 20th century, many scientists despaired of ever knowing the compositions of the stars, let alone why they shine and how they change. Today, the theory of stellar evolution is a cornerstone of modern astronomy.

CONCEPT CHECK

Why is stellar evolution important to life on Earth?

THE BIG QUESTION

If stars like the Sun end their lives quiescently and similarly, why do their scattered remnants look so different on the sky? Planetary nebulae display all sorts of weird shapes and sizes, some with rings and spheres, others with loops and jets. What causes these dissimilar structures? Are they somehow intrinsic to the stars themselves, or are they due to the complex environments through which dying stars expel their contents into interstellar space?

CHAPTER REVIEW

SUMMARY

LO1 Stars spend most of their lives on the main sequence, in the **core hydrogen-burning** (stage 7; p. 332) phase of stellar evolution. Stars leave the main sequence when the hydrogen in their cores is exhausted. With no internal energy source, the star's helium core is unable to support itself against its own gravity and begins to shrink. The star at this stage is in the **hydrogen shell-burning** (p. 334) phase, with nonburning helium at the center surrounded by a layer of burning hydrogen. The energy released by the contracting helium core heats the hydrogen-burning shell, greatly increasing the nuclear reaction rates there. As a result, the star becomes much brighter, while the envelope expands and cools. A solar-mass star moves off the main sequence on the H–R diagram first along the **subgiant branch** (stage 8; p. 334), then almost vertically up the **red giant branch** (stage 9; p. 334).

LO2 As the helium core contracts, it heats up. Eventually, helium begins to fuse into carbon. In a star like the Sun, helium burning begins explosively, in the **helium flash** (p. 335). The flash expands the core and reduces the star's luminosity, sending it onto the **horizontal branch** (stage 10; p. 335) of the H–R diagram. The star now has a core of burning helium surrounded by a shell of burning hydrogen. An inner core of nonburning carbon forms, shrinks, and heats the overlying burning layers, and the star once again becomes a red giant, even more luminous than before (stage 11). The core of a solar-mass star

never becomes hot enough to fuse carbon. Such a star continues to brighten and expand until its envelope is ejected into space, forming a **planetary nebula** (stage 12; p. 336). At that point the core becomes visible as a hot, faint, and extremely dense white dwarf (stage 13). The white dwarf cools and fades, eventually becoming a cold **black dwarf** (stage 14; p. 339).

LO3 Although most white dwarfs simply cool and dim with time, some may become explosively active in the form of a **nova** (p. 340). A nova is a star that suddenly increases greatly in brightness, then slowly fades back to its normal appearance over a period of months. It is the result of a white dwarf in a binary system drawing hydrogen-rich material from its companion. The gas spirals inward in an **accretion disk** (p. 341) and builds up on the white dwarf's surface, eventually becoming hot and dense enough for the hydrogen to burn explosively, temporarily causing a large increase in the dwarf's luminosity.

LO4 Stars more massive than about 8 solar masses form heavier and heavier elements in their cores, at a more and more rapid pace. As they evolve into **red supergiants** (p. 342), their cores form a layered structure consisting of burning shells of successively heavier elements. The process stops at iron, whose nuclei can neither be fused together nor split to produce energy. As a star's iron core grows in mass, it eventually becomes unable to support itself against gravity and begins to collapse. At the high temperatures produced during the collapse, iron nuclei are broken down into protons and neutrons. The protons combine with electrons to form more neutrons. Eventually, when the core becomes so dense that the neutrons are effectively brought into physical contact with one another, their resistance to further squeezing stops the collapse and the core rebounds, sending a violent shock wave out through the rest of the star. The star explodes in a **core-collapse supernova** (p. 344).

LO5 **Type I supernovae** (p. 345) are hydrogen poor and have light curves similar in shape to those of novae. **Type II supernovae** (p. 345) are hydrogen rich and have a characteristic bump in the light curve a few months after maximum. A Type II supernova is a core-collapse supernova. A Type I supernova is a **carbon-detonation supernova** (p. 346), which occurs when a white dwarf in a binary system collapses and then explodes as its carbon ignites. We can see evidence for a past supernova in the form of a **supernova remnant** (p. 347), a shell of exploded debris surrounding the site of the explosion and expanding into space at a speed of thousands of kilometers per second.

LO6 The theory of stellar evolution can be tested by observing star clusters. At any instant, stars with masses above the cluster's **main-sequence turnoff** (p. 351) have evolved off the main sequence. By comparing a cluster's main-sequence turnoff mass with theoretical predictions, astronomers can determine the cluster's age. Many globular clusters show evidence of multiple stellar populations, whose origin is not yet understood.

LO7 All elements heavier than helium formed in evolved stars or in supernova explosions. With each new generation of stars, the fraction of heavy elements in the universe increases. Comparisons between theoretical predictions of element production and observations of element abundances in the Galaxy provide strong support for the theory of stellar evolution.

MasteringAstronomy®   For instructor-assigned homework go to www.masteringastronomy.com

Problems labeled **POS** explore the process of science. **VIS** problems focus on reading and interpreting visual information. **LO** connects to the introduction's numbered Learning Outcomes.

REVIEW AND DISCUSSION

1. How long can a star like the Sun keep burning hydrogen in its core?

2. **LO1** Why is the depletion of hydrogen in the core of a star such an important event?

3. What makes an ordinary star become a red giant?

4. Roughly how big (in AU) will the Sun become when it enters the red giant phase?

5. **LO2** How long does it take for a star like the Sun to evolve from the main sequence to the top of the red giant branch?

6. What is the helium flash?

7. How do stars of low mass die? How do stars of high mass die?

8. What is a planetary nebula? With what stage of stellar evolution is it associated?

9. What are white dwarfs? What is their ultimate fate?

10. **LO3** Under what circumstances will a binary star produce a nova?

11. **LO4** What occurs in a massive star to cause it to explode?

12. What are the observational differences between Type I and Type II supernovae?

13. **LO5** How do the mechanisms that cause Type I and Type II supernovae explain their observed differences?

14. **LO6 POS** What evidence do we have that many supernovae have occurred in our Galaxy?

15. **LO7 POS** What do stars in clusters tell us about the various stages of stellar evolution?

CONCEPTUAL SELF-TEST: TRUE OR FALSE?/MULTIPLE CHOICE

1. All of the single red-dwarf stars that ever formed are still on the main sequence today. (T/F)

2. The Sun will get brighter as it begins to run out of fuel in its core. (T/F)

3. A planetary nebula is the disk of matter around a star that will eventually form a planetary system. (T/F)

4. Nuclear fusion in the core of a massive star cannot create elements much heavier than iron. (T/F)

5. A nova is a sudden outburst of light coming from an old main-sequence star. (T/F)

6. It takes less and less time to fuse heavier and heavier elements inside a high-mass star. (T/F)

7. In a core-collapse supernova, the outer part of the core rebounds from the inner, high-density core, destroying the entire outer part of the star. (T/F)

8. Because of stellar nucleosynthesis, the spectra of old stars show more heavy elements than those of young stars. (T/F)

9. A white dwarf is supported by the pressure of tightly packed (a) electrons; (b) protons; (c) neutrons; (d) photons.

10. A star like the Sun will end up as a (a) blue giant; (b) white dwarf; (c) binary star; (d) red dwarf.

11. A white dwarf can dramatically increase in brightness only if it (a) has another star nearby; (b) can avoid nuclear fusion in its core; (c) is spinning very rapidly; (d) is descended from a very massive star.

12. Nuclear fusion in the Sun will (a) never create elements heavier than helium; (b) create elements up to and including oxygen; (c) create all elements up to and including iron; (d) create some elements heavier than iron.

13. Most of the carbon in our bodies originated in (a) the core of the Sun; (b) the core of a red giant star; (c) a supernova; (d) a nearby galaxy.

14. **VIS** If the evolutionary track in *Overlay 3*, showing a Sun-like star, were instead illustrating a significantly more massive star, its starting point (stage 7) would be (a) up and to the right; (b) down and to the left; (c) up and to the left; (d) down and to the right.

15. **VIS** Figure 12.20 (Supernova Light Curves) indicates that a supernova whose luminosity declines steadily in time is most likely associated with a star that is (a) without a binary companion; (b) more than eight times the mass of the Sun; (c) on the main sequence; (d) comparable in mass to the Sun.

PROBLEMS

The number of squares preceding each problem indicates its approximate level of difficulty.

1. ■ Calculate the average density of a red giant core of mass 0.25 solar masses and radius 15,000 km. Compare this with the average density of the giant's envelope, if the mass of the envelope is 0.5 solar masses and its radius is 0.5 AU. Compare each with the central density of the Sun. ∞ *(Sec. 9.2)*

2. ■■■ A main sequence star at a distance of 20 pc is barely visible through a certain telescope. The star subsequently ascends the giant branch, during which time its temperature drops by a factor of three and its radius increases 100-fold. What is the new maximum distance at which the star would still be visible using the same telescope?

3. ■■ The Sun will reside on the main sequence for years. If the luminosity of a main-sequence star is proportional to the fourth power of the star's mass, what is the mass of a star that is just now leaving the main sequence in a cluster that formed (a) 400 million years ago? (b) 2 billion years ago?

4. ■ How long will it take the Sun's planetary nebula, expanding at a speed of 20 km/s, to reach the orbit of Neptune? How long to reach the nearest star?

5. ■ What are the escape speed (in km/s) and surface gravity (relative to Earth's gravity) of Sirius B (Table 12.2)? ∞ *(More Precisely 5-1)*

6. ■■■ A certain telescope could just detect the Sun at a distance of 10,000 pc. What is the apparent magnitude of the Sun at this distance? (For convenience, take the Sun's absolute magnitude to be 5.) What is the maximum distance at which the telescope could detect a nova having a peak luminosity of 10^5 solar luminosities? Repeat the calculation for a supernova having a peak luminosity 10^{10} times that of the Sun.

7. ■■ At what distance would the supernova in the previous question look as bright as the Sun (apparent magnitude −27)? Would you expect a supernova to occur that close to us?

8. ■■ A (hypothetical) supernova at a distance of 150 pc has an absolute magnitude of −20. Compare its apparent magnitude with that of (a) the full Moon; (b) Venus at its brightest (see Figure 10.6). Would you expect a supernova to occur so close to us?

9. ■ The Crab Nebula is now about 1 pc in radius. If it was observed to explode in A.D. 1054, roughly how fast is it expanding? (Assume constant expansion velocity. Is that a reasonable assumption?)

10. ■■ A supernova's energy is often compared to the total energy output of the Sun over its lifetime. Using the Sun's current luminosity, calculate the total solar energy output, assuming a 10^{10} year main-sequence lifetime. Using Einstein's formula $E = mc^2$ calculate the equivalent amount of mass, in Earth masses. ∞ *(Sec. 9.5)*

ACTIVITIES

Collaborative

1. Open clusters are generally found in the plane of the Galaxy. If you can see the hazy band of the Milky Way arcing across your night sky—in other words, if you are far from city lights and looking at an appropriate time of night and year—you can simply sweep with your binoculars along the Milky Way. Numerous "clumps" of stars will pop into view. Many will turn out to be open star clusters. The most easily visible clusters are those in the Messier catalog, but there are many others besides those. How many clusters can you find that are *not* on Messier's list?

2. Globular star clusters are harder to find. They are intrinsically larger, but they are also much farther away and therefore appear smaller in the sky. The most famous globular cluster visible from the Northern Hemisphere is M13 in the constellation Hercules, visible on spring and summer evenings. It contains half million or so of the Galaxy's most ancient stars. It may be glimpsed through binoculars as a little ball of light, located about one-third of the way from the star Eta to the star Zeta in the Keystone asterism of the constellation Hercules. Telescopes reveal this cluster as a magnificent, symmetrical grouping of stars. Can you find the following well-known globular clusters: M3, M4, M5, M13, M15? Look online or in a star chart for details on how to locate them.

Individual

1. Can you find the Hyades cluster? It lies about 46 pc away in the constellation Taurus, making up the "face" of the bull. It appears to surround the very bright star Aldebaran, the bull's eye, which makes it easy to locate in the sky. Aldebaran is a low-mass red giant, about twice the mass of the Sun, probably on the asymptotic giant branch of its evolution. Despite appearances, it is not part of the Hyades cluster. In fact, it lies only about half as far away—about 20 pc from Earth

2. In 1758, Charles Messier discovered the sky's most legendary supernova remnant, now called M1, or the Crab Nebula. It is located northwest of Zeta Tauri, the star that marks the southern tip of the horns of Taurus the Bull. Try to find it. An 8-inch telescope reveals the Crab's oval shape, but it will appear faint; a 10-inch or larger telescope reveals some of its famous filamentary structure.

3. The Ring Nebula (M57) is perhaps the most famous planetary nebula. At magnitude 9, it is faint, but a 6-inch or larger telescope should show its structure. To locate it, find Beta and Gamma Lyrae, the second and third brightest stars in the constellation Lyra. The ring lies between them, about one-third the way from Beta to Gamma. Don't expect the Ring to look as colorful as the *Hubble* images you may have seen! Can you see any color in the ring?

13

Neutron Stars and Black Holes

Strange States of Matter

Our study of stellar evolution has led us to some very unusual and unexpected objects. Red giants, white dwarfs, and supernova explosions surely represent extreme states of matter completely unfamiliar to us here on Earth. Yet, stellar evolution can have even more bizarre consequences. The strangest states of all result from the catastrophic implosion–explosion of stars much more massive than our Sun.

Neutron stars and black holes are among the most exotic objects in the universe. They are the end of the road for massive stars, and their bizarre properties boggle the imagination. Yet theory and observation seem to agree that, fantastic or not, they really do exist in space.

THE BIG PICTURE

The almost unimaginable violence of supernova explosions may create objects so extreme in their behavior that they require us to reconsider some of our most cherished laws of physics. They open up a science fiction writer's dream of fantastic phenomena that border on reality. They may even one day force scientists to construct a whole new theory of the universe.

Neutron stars and black holes are among the most exotic members of the vast population of stars throughout the universe. These objects represent the end states of stellar systems, yet despite their bizarre nature, they do seem to agree quite well with models of stellar evolution. This stunning image combines X-ray observations from the *Chandra* satellite, with red, green, and blue representing low, medium, and high X-ray energies. This debris field spread across 10 light-years is known as Cassiopeia A, the remnant of a supernova whose radiation first reached Earth about 300 years ago. The white dot at the center may be a neutron star created at the precise center of the blast. *(CXC)*

LEARNING OUTCOMES

Studying this chapter will enable you to:

LO1 List the key properties of neutron stars, and outline how these strange objects are formed.

LO2 Describe the nature and origin of pulsars, and explain how pulsars provide evidence for the existence of neutron stars.

LO3 List and explain some of the observable properties of neutron star binary systems.

LO4 Outline the basic characteristics of gamma-ray bursts and the leading theoretical attempts to explain them.

LO5 Describe Einstein's theories of relativity, and explain how they relate to neutron stars and black holes.

LO6 Explain how black holes are formed, and describe their effects on matter and radiation in their vicinity.

LO7 Relate the phenomena that occur near black holes to the warping of space around them.

LO8 Describe the difficulties in observing black holes, and list some ways in which black holes can be detected.

Mastering Astronomy®

Visit www.masteringastronomy.com for quizzes, animations, videos, interactive figures, and self-guided tutorials.

13.1 Neutron Stars

What remains after a supernova explosion? Is the original star blown to bits and dispersed into interstellar space, or does some portion of it survive? For a Type I (carbon-detonation) supernova, most astronomers regard it as unlikely that anything is left behind after the explosion. ∞ *(Sec. 12.5)* The entire star is shattered by the blast. However, for a Type II (core-collapse) supernova, theoretical calculations indicate that part of the star should survive the explosion.

Recall that in a Type II supernova, the iron core of a massive star collapses until its neutrons effectively come into contact with one another. At that point the central portion of the core rebounds, creating a powerful shock wave that races outward through the star, violently expelling matter into space. ∞ *(Sec. 12.4)* The key point here is that the shock wave does not start at the very center of the collapsing core. The innermost part of the core—the region that rebounds—remains intact as the shock wave it produces destroys the rest of the star. After the violence of the supernova has subsided, this ultracompressed ball of neutrons is all that is left. Researchers call this core remnant a **neutron star.**

Neutron stars are extremely small and very massive. Composed purely of neutrons packed together in a tight ball about 20 km across, a typical neutron star is not much bigger than a small asteroid or a terrestrial city (Figure 13.1), yet its mass is greater than that of the Sun. With so much mass squeezed into such a small volume, neutron stars are incredibly dense—some 10^{17} or even 10^{18} kg/m^3, nearly a billion times denser than a white dwarf. ∞ *(Sec. 12.3)* A thimbleful of neutron star material would weigh 100 million tons—about as much as a terrestrial mountain.

Neutron stars are solid objects. Provided that a sufficiently cool one could be found, you might even imagine standing on it. However, this would not be easy, as its gravity is immensely strong. A 70-kg human would weigh the Earth equivalent of about 1 billion kilograms (1 million tons). The severe pull of a neutron star's gravity would flatten you much thinner a piece of paper.

In addition to large mass and small size, newly formed neutron stars have two other very important properties. First, they *rotate* extremely rapidly, with periods measured in fractions of a second. This is a direct result of the law of conservation of angular momentum (Chapter 4), which tells us that any rotating body must spin faster as it shrinks, and the core of the parent star almost certainly had some rotation before it began to collapse. ∞ *(More Precisely 4-2)* Second, they have very strong *magnetic fields*. The original field of the parent star is amplified as the collapsing core squeezes the magnetic field lines closer together, creating a magnetic field trillions of times stronger than Earth's.

In time, theory indicates, a neutron star will spin more and more slowly as it radiates its energy into space, and its magnetic field will diminish. However, for a few million years after its birth, these two properties combine to provide the primary means by which this strange object can be detected and studied, as we now discuss.

13.2 Pulsars

The first observation of a neutron star occurred in 1967 when Jocelyn Bell, a graduate student at Cambridge University, observed an astronomical object emitting radio radiation in the form of rapid *pulses* (Figure 13.2). Each pulse consisted of a roughly 0.01-s burst of radiation, separated by precisely 1.34 s from the next.

More than 1500 of these pulsating objects are now known. They are called **pulsars.** Each has its own characteristic pulse shape and period. Observed pulsar periods are generally quite short, ranging from about a few milliseconds to about a second, corresponding to flashing rates of between one and several

Are all supernovae expected to lead to neutron stars?

▲ **FIGURE 13.1 Neutron Star** Neutron stars are not much larger than many of Earth's major cities. In this fanciful comparison, a typical neutron star sits alongside Manhattan Island. *(NASA)*

◄ **FIGURE 13.2 Pulsar Radiation** This recording shows the regular change in the intensity of the radio radiation emitted by the first pulsar discovered, known as CP1919. Some of the object's pulses are marked.

hundred times per second. In some cases, the periods are extremely stable—in fact, constant to within a few seconds in a million years—making pulsars the most accurate natural clocks known in the universe, more accurate even than the best atomic clocks on Earth.

A Pulsar Model

When Bell made her discovery in 1967, she did not know what she was looking at. Indeed, no one at the time knew what a pulsar was. The explanation won Bell's thesis advisor, Anthony Hewish, the 1974 Nobel Prize in physics. Hewish reasoned that the only physical mechanism consistent with such precisely timed pulses is a small, rotating source of radiation. Only rotation can cause the high degree of regularity of the observed pulses, and only a small object can account for the sharpness of each pulse (see Section 13.4). The best current model describes a pulsar as a compact, spinning neutron star that periodically flashes radiation toward Earth.

Figure 13.3 outlines the main features of this pulsar model. Two "hot spots," either on the surface of a neutron star or in the magnetosphere just above it, continuously emit radiation in a narrow beam. These spots are most likely localized regions near the neutron star's magnetic poles, where charged particles, accelerated to extremely high energies by the star's rotating magnetic field, emit radiation along the star's magnetic axis. The hot spots radiate more

▶ **FIGURE 13.3 Pulsar Model** The "lighthouse model" of neutron star emission explains many of the observed properties of pulsars. Charged particles, accelerated by the magnetism of the neutron star, flow along the magnetic field lines, producing radio radiation that beams outward. At greater distances from the star, the field lines channel these particles into a high-speed outflow in the star's equatorial plane, forming a pulsar wind. The beam sweeps across the sky as the neutron star rotates. If it happens to intersect Earth, we see a pulsar.

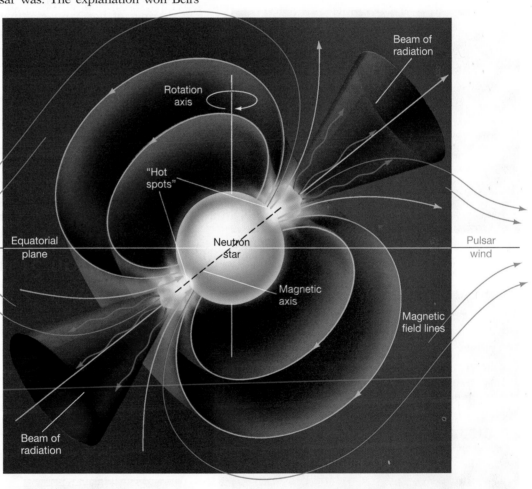

A lighthouse beacon is a good analogy for a rotating pulsar.

or less steadily, and the resulting beams sweep through space like a revolving lighthouse beacon as the neutron star rotates. Indeed, this pulsar model is often known as the **lighthouse model.** If the neutron star happens to be oriented such that one of the rotating beams sweeps across Earth, then we see the pulses. The period of the pulses is the star's rotation period.

A few pulsars are definitely associated with supernova remnants, clearly establishing those pulsars' explosive origin, although not all such remnants have a detectable pulsar within them. Figure 13.4(c) shows optical images of the Crab pulsar at the center of the Crab supernova remnant (Figure 13.4a). ∞ *(Sec. 12.5)* By observing the speed and direction of the Crab's ejected matter, astronomers can work backward to pinpoint the location in space at which the explosion must have occurred. It corresponds to the location of the pulsar. This is all that remains of the once massive star whose supernova was observed in 1054.

As indicated in Figure 13.3, the neutron star's strong magnetic field and rapid rotation channel high-energy particles from near the star's surface into the surrounding nebula. The result is an energetic *pulsar wind* that flows outward at almost the speed of light, primarily in the star's equatorial plane. Figure 13.4(b) shows this process in action in the Crab—this combined *Hubble/Chandra* image reveals rings of hot X-ray-emitting gas moving rapidly away from the pulsar. Eventually, the energy from the pulsar wind is deposited in the surrounding Crab Nebula, heating the gas to very high temperatures and powering the spectacular display we see from Earth.

Most pulsars emit pulses in the form of radio radiation. However, some also pulse in the visible, X-ray, and gamma-ray parts of the spectrum. Figure 13.5 shows the Crab and nearby Geminga pulsars in gamma rays. Geminga is unusual in that, while it pulsates strongly in gamma rays, it is barely detectable in visible light and not at all at radio wavelengths. Whatever types of radiation are produced, pulsar flashes at different frequencies all occur at regular, repeated time intervals—as we would expect since they arise from the same object—although the pulses may not all occur at the same instant.

Most known pulsars are observed (usually by Doppler measurements) to have high speeds—much greater than the typical speeds of stars in the Galaxy. The most likely explanation for these anomalously high speeds is that a neutron star can receive a substantial "kick" due to asymmetries in the supernova in which it forms. If the supernova's enormous energy is channeled even slightly in one direction, the newborn neutron star/pulsar can recoil in the opposite

CONCEPT CHECK

Why don't we see pulsars at the centers of all supernova remnants?

(a)

R I V U X G

Disk

Jet

Pulsar

Jet

|— 0.1 pc —|

(b)

R I V U X G

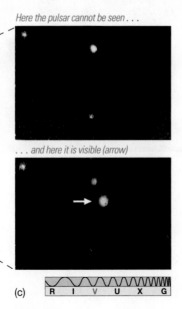

Here the pulsar cannot be seen . . .

. . . and here it is visible (arrow)

(c)

R I V U X G

◀ **FIGURE 13.4 Crab Pulsar**   In the core of the Crab Nebula (a), the Crab pulsar (c) blinks on and off about 30 times each second. In the top frame, the pulsar is off; in the bottom frame, it is on (arrow). (b) This *Chandra* X-ray image of the Crab, superimposed on a *Hubble* optical image, shows the central pulsar, as well as rings of hot X-ray-emitting gas in the equatorial plane, driven outward by the pulsar wind. Also visible in the image is a jet of hot gas escaping perpendicular to the equatorial plane. *(ESO; NASA; UC/Lick Observatory)*

(b) |←——————— 0.24 second ———————→|

direction with a speed of many tens or even hundreds of kilometers per second. Thus, observations of pulsar velocities give theorists additional insight into the detailed physics of supernovae.

Neutron Stars and Pulsars

All pulsars are neutron stars, but not all neutron stars are pulsars, for two reasons. First, the two ingredients that make the neutron star pulse—rapid rotation and strong magnetic field—both diminish with time, so the pulses gradually weaken and become less frequent. Theory indicates that within a few tens of millions of years, the pulses have all but stopped. Second, even a young, bright pulsar is not necessarily visible from Earth. The pulsar beam depicted in Figure 13.3 is relatively narrow—perhaps as little as a few degrees across. Only if the neutron star happens to be oriented in just the right way do we see pulses. When we see the pulses from Earth, we call the object a pulsar. Note that we are using the term "pulsar" here to mean the pulsing object we observe if the beam crosses Earth. However, many astronomers use the word more generically to mean *any* young neutron star producing beams of radiation as in Figure 13.3. Such an object will be a pulsar as seen from some directions—just not necessarily ours!

Figure 13.6 shows *Hubble* images of a neutron star that is not a pulsar. First identified by X-ray observations, this object is just 30 km across and has a surface temperature of about 700,000 K. Despite this enormous temperature, its small size means that it is very faint—just 25th magnitude in visible light. ∞ *(Sec. 10.4)* The star is thought to be about 1 million years old, lies about 60 pc away, and is moving at a speed of 110 km/s across our line of sight. Such faint objects are very difficult to detect, however, and a detection of a "bare" neutron star like this is a rare event. Most cataloged neutron stars have been detected either as pulsars or by their interaction with a "normal" stellar companion in a binary system (see Section 13.3).

Given our current knowledge of star formation, stellar evolution, and neutron stars, observations of pulsars are consistent with the ideas that (1) *every* high-mass star dies in a supernova explosion, (2) most supernovae leave a neutron star behind (a few result in black holes, as discussed in a moment), and (3) *all* young neutron stars emit beams of radiation, just like the pulsars we actually detect. On the basis of estimates of the rate at which massive stars have formed over the lifetime of the Milky Way, astronomers reason that, for every pulsar we know of, there must be several hundred thousand more neutron stars moving unseen somewhere in the Galaxy. Some formed relatively recently—less than a few million years ago—and simply happen not to be beaming their energy toward Earth. However, the vast majority are old, their youthful pulsar phase long past.

Neutron stars (and black holes too) were suggested by theory long before they were actually observed, although their extreme properties made many scientists doubt that they would ever be found in nature. The fact that we now have strong observational evidence, not just for their existence, but also for the vitally important roles they play in many areas of high-energy astrophysics, is yet another testament to the fundamental soundness of the theory of stellar evolution.

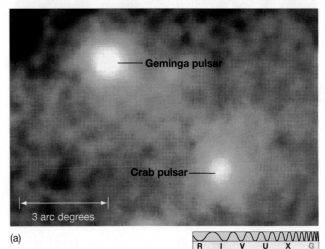

(a)

R I V U X G

▲ FIGURE 13.5 **Gamma-Ray Pulsars** (a) The Crab and Geminga pulsars lie fairly close to one another in the sky. Unlike the Crab, Geminga is barely visible at optical wavelengths, and undetectable in the radio region of the spectrum. (b) Sequence of *Compton Gamma-Ray Observatory* images showing Geminga's 0.24-s pulse period. *(NASA)*

R I V U X G

▲ FIGURE 13.6 **Isolated Neutron Star** This lone neutron star was first detected by its X-ray emission and subsequently imaged by *Hubble*. This 700,000-K object lies about 60 pc from Earth and is thought to be about 1 million years old. This triple exposure shows the star streaking across the sky at more than 100 km/s. *(NASA)*

(a)

BEFORE   DURING

(b)

▲ **FIGURE 13.7 X-Ray Burster**   An X-ray burster produces a sudden, intense flash of X-rays, followed by a period of relative inactivity lasting as long as several hours. Then another burst occurs. (a) An optical photograph of the globular star cluster Terzan 2, showing a 2″ dot at the center where the X-ray bursts originate. (b) X-ray images taken before and during the outburst. The most intense X-rays correspond to the position of the black dot shown in (a). *(SAO)*

▲ **FIGURE 13.8 Millisecond Pulsar**   As infalling matter strikes the star, it moves almost parallel to the surface, so it tends to make the star spin faster. Eventually, this process can result in a millisecond pulsar—a neutron star spinning at the incredible rate of hundreds of revolutions per second.

13.3 Neutron Star Binaries

We noted in Chapter 10 that most stars are members of binary systems. ∞ *(Sec. 10.7)* Although many pulsars are known to be isolated (that is, not part of any binary), at least some do have binary companions, and the same is true of neutron stars in general (that is, even the ones not seen as pulsars). One important consequence of this is that some neutron star masses have been determined very accurately. Most of the measured masses are fairly close to 1.4 times the mass of the Sun—the Chandrasekhar mass of the stellar core that collapsed to form the neutron star remnant—although a neutron star with a mass twice that of the Sun has recently been reported.

X-Ray Sources

With the launch of the first orbiting X-ray telescopes in the 1970s, numerous X-ray sources were found near the central regions of our Galaxy and also near the centers of a few globular star clusters. ∞ *(Sec. 11.6)* Some of these sources, known as **X-ray bursters,** emit much of their energy in violent eruptions, each thousands of times more luminous than our Sun but lasting only seconds. A typical burst is shown in Figure 13.7.

The X-ray emission is thought to arise on or near neutron stars that are members of binary systems. Matter torn from the surface of the companion by the neutron star's strong gravitational pull accumulates on the neutron star's surface. (Figure 12.14 shows the white dwarf equivalent.) ∞ *(Sec. 12.3)* The gas forms an accretion disk around the neutron star, then slowly spirals inward. The inner portions of the disk become extremely hot, releasing a steady stream of X-rays.

As gas builds up on the neutron star's surface, its temperature rises due to the pressure of overlying material. Eventually, it becomes hot enough to fuse hydrogen. The result is a sudden period of rapid nuclear burning that releases a huge amount of energy in a brief but intense X-ray burst. After several hours of renewed accumulation, a fresh layer of matter produces the next burst. The mechanism is similar to that responsible for a nova explosion on a white dwarf, except that it occurs on a far more violent scale because of the neutron star's much stronger surface gravity.

Millisecond Pulsars

In the mid-1980s astronomers discovered an important new category of pulsars—a class of very rapidly rotating objects called **millisecond pulsars.** These objects spin hundreds of times per second (that is, their pulse periods are a few milliseconds, 0.001 s). This speed is about as fast as a typical neutron star can spin without flying apart. In some cases the star's equator is moving at more than 20 percent of the speed of light. This suggests a phenomenon bordering on the incredible—a cosmic object just 20 km across, more massive than our Sun, spinning almost at breakup speed, making nearly 1000 complete revolutions *every second.* Yet the observations and their interpretation leave little room for doubt. More than 200 millisecond pulsars are now known.

The story of these remarkable objects is further complicated by the fact that many of them—about two-thirds—are found in globular clusters. This is odd because globular clusters are at least 10 billion years old, yet Type II supernovae (the kind that create neutron stars) are associated with massive stars that explode within a few tens of *millions* of years after their formation, and no stars have formed in any globular cluster since the cluster came into being. ∞ *(Secs. 11.6, 12.4)* Thus, no new neutron star has been produced in a globular cluster in a very long time. Furthermore, as mentioned earlier, the pulsar produced in a supernova explosion is

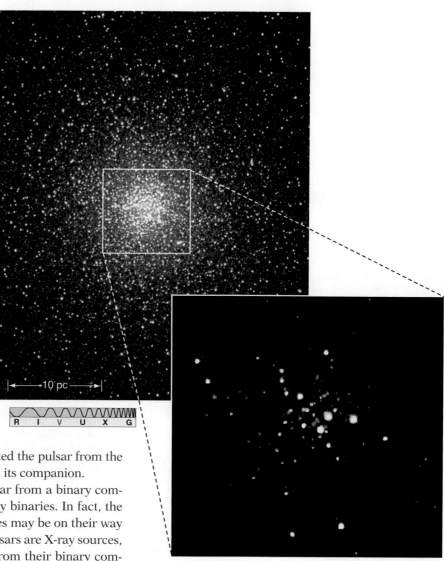

► FIGURE 13.9 **Cluster X-Ray Binaries** The dense core of the old globular cluster 47 Tucanae harbors more than 100 separate X-ray sources (shown in the *Chandra* image at bottom right). More than half of these are thought to be binary millisecond pulsars, still accreting small amounts of gas from their companions after an earlier period of mass transfer spun them up to millisecond speeds. *(ESO; NASA)*

expected to slow down and fade away in only a few tens of millions of years—after 10 billion years, its rotation rate should be very slow. The rapid rotation of the pulsars found in globular clusters therefore cannot be a relic of their birth. These pulsars must have been "spun up"—that is, had their rotation rates increased—by some other, much more recent, mechanism.

The most likely explanation for the high rotation rate of millisecond pulsars is that they have been spun up by drawing in matter from a companion star. As matter spirals down onto the neutron star's surface in an accretion disk, it provides the "push" needed to make the neutron star spin faster (Figure 13.8). Theoretical calculations indicate that this process can spin the star up to breakup speed in about 100 million years. This picture is supported by the finding that, of the 140 or so millisecond pulsars observed in globular clusters, roughly half are known currently to be members of binary systems. The remaining solo millisecond pulsars were probably formed when an encounter with another star ejected the pulsar from the binary or when the pulsar's own intense radiation destroyed its companion.

Notice that the scenario of accretion onto a neutron star from a binary companion is the same as the one we just used to explain X-ray binaries. In fact, the two phenomena are very closely linked. Many X-ray binaries may be on their way to becoming millisecond pulsars, and many millisecond pulsars are X-ray sources, powered by the trickle of material still falling onto them from their binary companions. Figure 13.9 shows the globular cluster 47 Tucanae, together with a remarkable *Chandra* image of its core showing more than 100 X-ray sources—about 10 times the number that had been known in the cluster prior to *Chandra's* launch. Roughly half of these sources are millisecond pulsars; the cluster also contains two or three "conventional" neutron star binaries. Most of the remaining sources are white dwarf binaries, similar to those discussed in Chapter 12. ∞ *(Sec. 12.3)*

Pulsar Planets

Radio astronomers can capitalize on the precision with which pulsar signals repeat themselves to make extremely accurate measurements of pulsar motion. In 1992 radio astronomers at the Arecibo Observatory found that the pulse period of a millisecond pulsar lying some 500 pc from Earth exhibits tiny but regular fluctuations (at the level of about 1 part in 10^7) on two distinct time scales—67 and 98 days.

These fluctuations are caused by the Doppler effect as the pulsar wobbles back and forth in space under the combined gravitational pulls of two small bodies, each about three times the mass of Earth ∞ *(Sec. 4.4)*. One orbits the pulsar at a distance of 0.4 AU, the other at a distance of 0.5 AU, with orbital periods of 67 and 98 days, respectively. Further observations not only confirmed these findings, but also revealed the presence of a third body, with a mass comparable to that of Earth's Moon, orbiting just 0.2 AU from the pulsar.

Some gamma ray bursts last just a fraction of a second . . .

. . . while others persist for minutes or even hours.

(a)

(b)

These remarkable findings were the first definite evidence of planet-sized bodies outside our solar system. A few other millisecond pulsars have since been found also to have planets. However, it is unlikely that any of these planets formed in the same way as our own. Any planetary system orbiting the pulsar's parent star was almost certainly destroyed in the supernova explosion that created the pulsar. Scientists are uncertain about how these planets came into being. One possibility involves the binary companion that provided the matter necessary to spin the pulsar up to millisecond speeds. Possibly, the pulsar's intense radiation and strong gravity destroyed the companion, then spread its matter out into a disk (a little like the solar nebula) in whose cool outer regions the planets might have condensed.

Astronomers searched for decades for planets orbiting main-sequence stars like our Sun on the assumption that planets are a natural by-product of star formation. ∞ *(Sec. 4.3)* These searches have now identified many extrasolar planets, although only a few planets comparable in mass to Earth have so far been detected. ∞ *(Sec. 4.4)* It is ironic that the first Earth-sized planets found outside the solar system orbit a dead star and have little or nothing to do with our own world.

13.4 Gamma-Ray Bursts

Discovered serendipitously in the late 1960s by military satellites looking for violations of the Nuclear Test Ban Treaty, and first made public in the 1970s, **gamma-ray bursts** consist of bright, irregular flashes of gamma rays typically lasting only a few seconds (Figure 13.10a). Until the 1990s it was thought that gamma-ray bursts were basically "scaled-up" versions of X-ray bursters, in which even more violent nuclear events released more energetic gamma rays. However, that is not the case.

Distances and Luminosities

Figure 13.10(b) shows an all-sky plot of the positions of 2704 bursts detected by the *Compton Gamma-Ray Observatory (CGRO)* during its 9-year operational lifetime. ∞ *(Sec. 3.5)* On average, *CGRO* detected gamma-ray bursts at the rate of about one per day. Note that the bursts are distributed uniformly across the sky, rather than being confined to the relatively narrow band of the Milky Way (compare Figure 3.30). This widespread distribution convinced most astronomers that the bursts do not originate within our own Galaxy, as had previously been assumed, but instead are produced at much greater distances.

Measuring the distance to a gamma-ray burst is no easy task. The gamma-ray observations do not provide enough information in and of themselves to tell us how far away the burst is, so astronomers must instead associate the burst with some other object in the sky—called the burst

◀ **FIGURE 13.10 Gamma-Ray Bursts** (a) Plots of intensity versus time for three gamma-ray bursts show that some of them are irregular and spiky, whereas others are much more smoothly varying. (b) Positions on the sky of all the gamma-ray bursts detected by the *Compton Gamma-Ray Observatory* during its 9 years of operation in orbit. The bursts appear to be distributed uniformly across the entire sky, which is mapped here with the plane of the Milky Way running horizontally across its center. *(NASA)* *counterpart*—whose distance can be measured by other means. The techniques for studying counterparts generally involve observations in the optical or X-ray parts of the electromagnetic spectrum. The problem is that the resolution of a gamma-ray telescope is quite poor, so burst positions are uncertain by up to 1 degree, and a relatively large region of the sky must be scanned in search of a counterpart. ∞ *(Sec. 3.5)* In addition, the "afterglow" of a burst at X-ray or optical wavelengths fades quite rapidly, severely limiting the time available to complete the search (see Figures 13.11a and b).

The most successful searches for burst counterparts have been carried out by satellites combining gamma-ray detectors with X-ray and/or optical telescopes. For example, NASA's *Swift* mission, launched in 2004 and still (as of 2015) operational, combines a wide-angle gamma-ray detector (to monitor as much of the sky as possible) with two telescopes: one X-ray and one optical/ultraviolet instrument. The gamma-ray detector system pinpoints the direction of the burst to an accuracy of about 4 arc minutes, and within seconds the onboard computer automatically repositions the satellite to point the X-ray and optical telescopes in that direction. At the same time, the craft relays the burst position to other instruments in space and on the ground. *Swift* detects burst counterparts at the rate of about one per week and has played a pivotal role in advancing our understanding of these violent phenomena. Figures 13.11(c) and (d) show *Swift* X-ray and optical images of GRB 080319B, one of the brightest bursts to date. Automated observations at many wavelengths began within seconds of *Swift's* detection, making this burst one of the most intensively studied on record.

The first direct measurement of the distance to a gamma-ray burst was made in 1997 when astronomers detected the visible afterglow created as the burst faded and cooled. Using spectroscopic distance-measurement techniques (to be discussed in more detail in Section 16.1, they found that the event occurred more than 2 billion parsecs from Earth. Figure 13.11 shows an optical image of another gamma-ray burst, this one almost 5 billion parsecs distant, along with what may be its host galaxy; this is one of the first images to associate a burst with a possible galactic host. To date, the distances to hundreds of gamma-ray bursts have been measured from their afterglows. All are very large, implying that the bursts must be extremely energetic, otherwise they wouldn't be detectable by our equipment.

By the late 1990s, detection techniques had improved to the point that astronomers could locate and study both the rapidly fading afterglows and the faint host galaxies of many gamma-ray bursts. They were then able to measure the distances to these violent events. Distances to several hundred gamma-ray bursts are currently known. As suggested by the bursts' distribution on the sky, *all* of the measured distances are very large—typically billions of parsecs. The two bursts shown in Figure 13.11 lie, respectively, 3.6 and 2.3 billion parsecs away.

These large distances imply that the bursts must be very energetic, since otherwise we wouldn't be able to detect them. But when astronomers tried to estimate the total energy emitted using the inverse-square law, they ran into a

(a)   (b)   (c)   (d)

▲ **FIGURE 13.11 Gamma-Ray Burst Counterpart** (a) Optical image of the gamma-ray burst GRB 971214 taken by the Keck telescope, shows the bright visible afterglow of the gamma-ray source (arrow). (b) By the time this *Hubble* image was taken 2 months later, the afterglow had faded, but a faint image of a host galaxy remains. (c and d) Another long-duration gamma-ray burst, GRB 080319B, was one of the brightest yet observed. Its light was emitted 7.5 billion years ago, yet for a few seconds the flash would have been visible to the unaided eye—if anyone had been looking in just the right spot! Only moments after detonation, it was observed in X-rays (c) and in visible light (d). *(Keck; NASA; ESO)* problem. ∞ *(Sec. 10.2)* If they assumed that the gamma rays were emitted equally in all directions, they found that each burst would generate more energy—in some cases hundreds of times more energy—than a typical supernova explosion, all in a matter of seconds! Such an immense amount of power seemed to defy the laws of physics. However, if the energy is emitted as a *jet*, as most experts now think, then the overall luminosity of the burst is reduced to much more manageable levels, as the energy we see is representative of only a small fraction of the sky, making the energetics of the brightest bursts easier to understand.

As an analogy, consider a handheld laser pointer of the sort commonly used in talks and lectures. It radiates only a few milliwatts of power, much less than a household lightbulb, but it appears enormously bright if you happen to look directly into the beam. (*Don't* do this, by the way!) The laser beam is bright because all of its energy is concentrated in almost a single direction instead of being radiated in all directions into space.

ANIMATION/VIDEO (MA)
Colliding Binary Neutron Stars

What Causes the Bursts?

Not only are gamma-ray burst sources extremely energetic, they are also very *small*. The millisecond flickering in the bursting gamma rays detected by [*CGRO*] implies that whatever their origin, all of their energy must come from a volume no larger than a few hundred kilometers across. The reasoning is as follows: If the emitting region were, say, 300,000 km—1 light-second—across, even an instantaneous change in intensity at the source would be smeared out over a time interval of 1 s as seen from Earth, because light from the far side of the object would take 1 s longer to reach us than light from the near side. For the gamma-ray variation not to be blurred by the light-travel time, the source cannot be more than 1 light-millisecond, or 300 km, in diameter.

▼ **FIGURE 13.12 Gamma-Ray Burst Models**   Two models have been proposed to explain gamma-ray bursts. Part (a) depicts the merger of two neutron stars; part (b) shows the collapse of a single star. Both models predict (at right) a relativistic fireball, perhaps releasing energy in the form of jets, as shown.

Neutron star binary system

Coalescence and merger

High-temperature accretion disk

(a) Merging stars

Time

Collapsing star

Supernova stalls and black hole forms

Accretion disk restarts supernova

(b) Hypernova

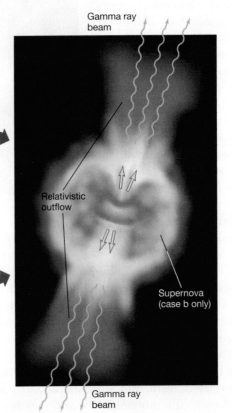

Gamma ray beam

Relativistic outflow

Supernova (case b only)

Gamma ray beam

Theoretical models of gamma-ray bursts describe the burst as a *relativistic fireball*—an expanding jet of superhot gas radiating furiously in the gamma-ray part of the spectrum. (The term "relativistic" here means that the gas particles are moving at speeds comparable to the speed of light, and Einstein's theory of relativity is needed to describe them—see Section 13.6.) The complex burst structure and afterglows we see are produced as the fireball expands, cools, and interacts with its surroundings.

Two leading models for the energy source have emerged. The first (Figure 13.12a) is the "true" end point of a binary star system. Suppose that both members of the binary become neutron stars. As the system continues to evolve, gravitational radiation

13.1 DISCOVERY

Gravity Waves—A New Window on the Universe

Electromagnetic waves are common everyday phenomena. They involve periodic changes in the strengths of electric and magnetic fields. ∞ *(Fig. 2.6)* They move through space and transport energy. Any accelerating charged particle, such as an electron in a broadcasting antenna or on the surface of a star, generates electromagnetic waves.

The modern theory of gravity—Einstein's theory of relativity—also predicts waves that move through space. A *gravity wave* is the gravitational counterpart of an electromagnetic wave. *Gravitational radiation* results from changes in the strength of a gravitational field. In theory, any time an object having any mass accelerates, a gravity wave is emitted at the speed of light. The passage of a gravity wave should produce small distortions in the space through which it passes. Gravity is an exceedingly weak force compared with electromagnetism, so these distortions are expected to be very small—in fact, much smaller than the diameter of an atomic nucleus for waves that might be produced by any known astrophysical source. Yet many researchers think that these tiny distortions should be measurable. No experiment has yet succeeded in detecting gravity waves, but their detection would provide such strong support for the theory of relativity that scientists are eager to find them.

The leading candidates for systems likely to produce gravity waves detectable on Earth include (1) close binary systems containing black holes, neutron stars, or white dwarfs and (2) the collapse of a star into a black hole. Each of these possibilities involves the acceleration of huge masses, resulting in rapidly changing gravitational fields. The first probably presents the best chance to detect gravity waves. Binary star systems should emit gravitational radiation as the component stars orbit one another. As energy escapes in the form of gravity waves, the two stars spiral toward one another, orbiting more rapidly and emitting even more gravitational radiation. For sufficiently close systems, the stars merge in a few tens or hundreds of millions of years (although most of the radiation is emitted during the last few seconds). As we saw in Section 13.4, neutron star mergers may well also be the origin of some gamma-ray bursts, so gravitational radiation might provide an alternative means of studying these violent and mysterious phenomena.

Such a slow but steady decay in the orbit of a binary system has in fact been detected. In 1974, radio astronomer Joseph Taylor and graduate student Russell Hulse at the University of Massachusetts discovered a very unusual binary system. Both components are neutron stars, and one is observable from Earth as a pulsar. This system has become known as the *binary pulsar*. Measurements of the periodic Doppler shift of the pulsar's radiation show that its orbit is slowly shrinking at exactly the rate predicted by relativity theory if the energy were being carried off by gravity waves. Even though the gravity waves themselves have not been detected, the binary pulsar is regarded by most astronomers as very strong evidence supporting general relativity. Taylor and Hulse received the 1993 Nobel Prize in Physics for their discovery.

In 2004, radio astronomers announced the discovery of a *double-pulsar* binary system with an even shorter period than the binary pulsar, implying stronger relativistic effects and a shorter merger time—less than 100 million years. Because both components are pulsars and the system, by pure luck, also happens to be seen almost exactly edge on by observers on Earth, leading to eclipses, this system has provided a wealth of information on both neutron stars and gravitational physics.

The figure below shows part of an ambitious gravity-wave observatory called LIGO—short for Laser Interferometric Gravity-Wave Observatory—which became operational in 2003. Twin detectors, one (shown here) in Hanford, Washington, the other in Livingston, Louisiana, use laser beams to measure the extremely small (less than one-thousandth the diameter of an atomic nucleus) distortions of space produced in the lengths of the 4-km-long arms should a gravity wave pass by. The instruments are in theory capable of detecting waves from many Galactic and extragalactic sources, although, so far, no gravity waves have actually been detected, despite upgrades in 2010 and 2015 that boosted the detector's sensitivity by more than a factor of 20.

If these experiments are successful, the discovery of gravity waves could herald a new age in astronomy, in much the same way that invisible electromagnetic waves, virtually unexplored a century ago, revolutionized classical astronomy and led to the field of modern astrophysics.

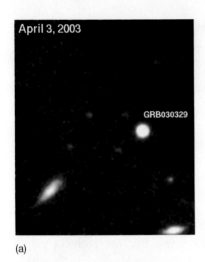

April 3, 2003

GRB030329

(a)

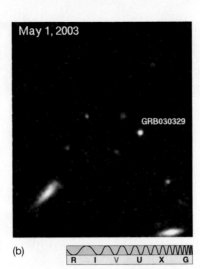

May 1, 2003

GRB030329

(b)

R   I   V   U   X   G

◀ **FIGURE 13.13 Hypernova?** The gamma-ray burst GRB 030329 may prove crucial to theorists' understanding of the physical processes underlying these violent phenomena. First detected by the *High Energy Transient Explorer 2* satellite and subsequently observed at radio, optical, and X-ray wavelengths, the burst counterpart has all the hallmarks of a high-mass supernova, lending strong support to the hypernova model. Here, the counterpart is shown (a) near the moment of the burst and (b) fading a month later. (c) This graph shows details of the emitted radiation from another, but similar, gamma-ray burst. *(ESO)*

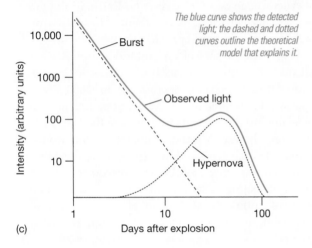

The blue curve shows the detected light; the dashed and dotted curves outline the theoretical model that explains it.

Intensity (arbitrary units)

10,000 — Burst

1000

Observed light

100

10

Hypernova

1   10   100

Days after explosion

(c)

(see *Discovery 13-1*) is released, and the two ultradense stars spiral in toward each other. Once they are within a few kilometers of one another, coalescence is inevitable. Such a merger will likely produce a violent explosion comparable in energy to a supernova, and energetic enough to explain the flashes of gamma rays we observe. The overall rotation of the binary system may channel the energy into a high-speed, high-temperature jet.

The second model (Figure 13.12b), sometimes called a *hypernova*, is a "failed" supernova—but what a failure! In this picture a very massive star undergoes core collapse just as described earlier for a Type II supernova, but instead of forming a neutron star, the core collapses in on itself, forming a black hole (see Section 13.5). ∞ *(Sec. 12.4)* Instead of being blown to pieces, the star implodes onto the black hole, forming an accretion disk and again generating a relativistic jet. The jet punches its way out of the star, producing a gamma-ray burst as it slams into the surrounding shells of gas expelled from the star during the final stages of its nuclear-burning lifetime. At the same time, intense radiation from the accretion disk may restart the stalled supernova, blasting what remains of the star into space.

The idea of a relativistic fireball has become widely accepted among workers in this branch of astrophysics. Most researchers regard it as likely that the energy is released as a jet and that the most intense gamma-ray bursts occur when the jet is directed toward us ("looking into the laser"). Which of the two models just described is correct? Experts in the field would say that the answer is probably *both*. The hypernova model predicts bursts of relatively long duration and is the leading explanation of the "long" gamma-ray bursts lasting more than about 2 s. Detailed calculations indicate that the neutron star merger model can naturally account for the shorter bursts. Thus, both of the scenarios depicted in Figure 13.12 contribute to the total.

Thanks to the "rapid response network" provided by *Swift* and other instruments, astronomers now have detailed observations of many afterglows of both types of burst. Figures 13.13(a) and (b) show the afterglow of the long burst GRB 030329, as seen by the 8.2-m VLT in Chile. ∞ *(Sec. 3.2)* Both the spectrum and the light curve were consistent with what astronomers expected from the supernova of a very massive (roughly 25-solar mass) star. ∞ *(Sec. 12.4)* Figure 13.13(c) shows a simplified light curve of another burst, illustrating how the "burst" and "hypernova" components can be distinguished.

The rapidly fading X-ray afterglows from the short bursts are consistent with the predictions of the neutron star merger scenario. Very recent observations also reveal that a few of these bursts may actually involve the theoretically predicted, but rare, merger between a neutron star and a black hole, which should have its own characteristic light signature.

13.5 Black Holes

Neutron stars are supported by the resistance of tightly packed neutrons to further compression. Squeezed together, these elementary particles form a hard ball of ultradense matter that not even gravity can compress further. Or do they? Is it possible that, given enough matter packed into a small enough volume, the collective pull of gravity can eventually crush even a neutron star? Can gravity continue to compress a massive star into an object the size of a planet, a city, a pinhead—even smaller? The answer, apparently, is yes.

The Final Stage of Stellar Evolution

Although the precise figure is uncertain, mainly because the behavior of matter at very high densities is not well understood, most researchers concur that the mass of a neutron star cannot exceed about three times the mass of the Sun. This is the neutron star equivalent of the white dwarf mass limit discussed in the previous chapter. ∞ *(Sec. 12.5)* Above this mass, not even tightly packed neutrons can withstand the star's gravitational pull. In fact, we know of *no* force that can counteract gravity beyond this point. Thus, if enough material is left behind after a supernova or hypernova explosion that the central core exceeds this limit, or if enough matter falls back onto the neutron star after the supernova has occurred, gravity wins the battle with pressure once and for all, and the central core collapses forever. Stellar evolution theory indicates that this is the fate of any star whose main-sequence mass exceeds about 25 times the mass of the Sun.

As the core shrinks, the gravitational pull in its vicinity eventually becomes so great that nothing—not even light—can escape. The resultant object therefore emits no light, no other form of radiation, no information whatsoever. Astronomers call this bizarre end point of stellar evolution, in which the core of a very-high-mass star collapses in on itself and vanishes forever, a **black hole.**

Black holes are much rarer than neutron stars, simply because very few stars have enough mass to form black holes. Nevertheless, astronomers think they have evidence for at least one relatively recent hypernova in our Galaxy that produced a black hole. Figure 13.14 shows W49B, a Galactic supernova remnant whose asymmetric shape and anomalous heavy element composition are consistent with the predictions of hypernova models of massive stars, and careful observations reveal no evidence of a neutron star. If these observations hold up, then W49B may contain the youngest known black hole in our Galaxy, formed just 1000 years ago.

Escape Speed

Newtonian mechanics—until now our indispensable guide to the universe—cannot adequately describe conditions in or near black holes. ∞ *(Sec. 1.4)* To comprehend these collapsed objects, we must turn instead to the modern theory of gravity, Einstein's **general theory of relativity,** discussed in Section 13.6. Still, we can usefully describe some aspects of these strange bodies in more or less Newtonian terms. Let's reconsider the familiar Newtonian concept of escape speed—the speed needed for one object to escape the gravitational pull of another—supplemented by two key facts from relativity: (1) nothing can travel faster than light and (2) all things, *including light,* are attracted by gravity.

A body's escape speed is proportional to the square root of the body's mass divided by the square root of its radius (see *More Precisely 5-1*). Earth's radius is 6400 km and the escape speed from Earth's surface is just over 11 km/s. Now consider a hypothetical experiment in which Earth is squeezed on all sides by a gigantic vise. As our planet shrinks under the pressure, the escape speed increases because the radius is decreasing. Suppose Earth were compressed to one-fourth

This possible hypernova remnant may contain the youngest known black hole in our Galaxy.

|← 5 pc →|

▲ **FIGURE 13.14 Hypernova Remnant?** This is a composite X-ray, infrared, and optical image of W49B, a supernova remnant lying some 8000 pc away. Its distorted, asymmetric shape and unusual composition suggest to many astronomers that this is actually the remnant of a hypernova—a gamma-ray burst in our own Galaxy—perhaps 1000 years old and containing a black hole. *(NASA)* its present size. The escape speed would double (because $1/\sqrt{1/4} = 2$). To escape from this compressed Earth, an object would need a speed of at least 22 km/s.

Imagine compressing Earth by an additional factor of 1000, making its radius hardly more than a kilometer. Now a speed of about 700 km/s—more than the escape speed of the Sun—would be needed to escape. Compress Earth still further, and the escape speed continues to rise. If our hypothetical vise were to squeeze Earth hard enough to crush its radius to about a centimeter, the speed needed to escape its surface would reach 300,000 km/s. But this is no ordinary speed. It is the speed of light, c, the fastest speed allowed by the laws of physics as we currently know them.

Thus, if by some fantastic means the entire planet Earth could be compressed to less than the size of a grape, the escape speed would exceed the speed of light. And because nothing can exceed that speed, the compelling conclusion is that nothing—absolutely nothing—could escape from the surface of such a compressed body. Even radiation—radio waves, visible light, X-rays, photons of all wavelengths—would be unable to escape its intense gravity. Our planet would be invisible and uncommunicative—no signal of any sort could be sent to the universe outside. The origin of the term *black hole* becomes clear. For all practical purposes, Earth could be said to have disappeared from the universe. Only its gravitational field would remain behind, betraying the presence of its mass, now shrunk to a point.

The Event Horizon

Astronomers have a special name for the critical radius at which the escape speed from an object would equal the speed of light and within which the object could no longer be seen. It is the **Schwarzschild radius**, after Karl Schwarzschild, the German scientist who first studied its properties. The Schwarzschild radius of any object is proportional to its mass. For Earth, it is 1 cm. For Jupiter, at about 300 Earth masses, it is about 3 m. For the Sun, at 300,000 Earth masses, it is 3 km. For a 3-solar-mass stellar core, the Schwarzschild radius is about 9 km. As a convenient rule of thumb, the Schwarzschild radius of an object is simply 3 km multiplied by the object's mass measured in solar masses.

Every object has a Schwarzschild radius—it is the radius to which the object would have to be compressed for it to become a black hole. Put another way, a black hole is an object that happens to lie within its own Schwarzschild radius. The surface of an imaginary sphere having a radius equal to the Schwarzschild radius and centered on a collapsing star is called the **event horizon.** No event occurring within this region can ever be seen, heard, or known by anyone outside. Even though there is no matter of any sort associated with it, we can think of the event horizon as the "surface" of a black hole.

Theory indicates that, if more than 3 solar masses of material remain behind after a supernova explosion, the remnant core will collapse catastrophically, diving below the event horizon in much less than a second. The event horizon is not a physical boundary of any kind—just a communications barrier. The core would shrink right past the Schwarzschild radius to ever-diminishing size on its way to being crushed to a point. It simply "winks out," disappearing and becoming a small dark region from which nothing can escape—a literal black hole in space. This is the predicted fate of stars having more than about 20–25 times the mass of the Sun.

SELF-GUIDED TUTORIAL (MA)
Escape Speed and Black Hole Event Horizons

CONCEPT CHECK

What happens if you compress an object within its own Schwarzschild radius?

13.6 Einstein's Theories of Relativity

By the latter part of the 19th century, physicists were well aware of the special status of the speed of light, c. It was, they knew, the speed at which all electromagnetic waves traveled, and, as best they could tell, it represented an upper limit on the speeds of *all* known particles. Scientists struggled without success to construct a theory of mechanics and radiation in which c was a natural speed limit.

Special Relativity

In 1887, a fundamental experiment carried out by American physicists A. A. Michelson and E. W. Morley compounded theorists' problems further by demonstrating another important aspect of light: The measured speed of a beam of light is *independent* of the motion of either the observer or the source.

Michelson and Morley attempted to determine Earth's motion relative to the "absolute" space in which light supposedly moved by measuring the speed of light at different times of the day (so that the orientation of their equipment would change as Earth rotated) and on different days of the year (so that Earth's velocity would vary as our planet orbited the Sun). As illustrated in Figure 13.15, they expected to measure a faster-moving beam of light when it was traveling opposite to Earth's motion (to the left in the figure) and a slower beam when Earth was "catching up" on it (to the right). But in fact they measured precisely the *same* velocity for any orientation of their apparatus. Far from establishing the properties of absolute space, the Michelson–Morley experiment ultimately demolished the entire concept—and with it, the 19th-century view of the universe.

All subsequent experiments confirm that no matter what our motion may be relative to the source of the radiation, we always measure precisely the same value for *c*: 299,792.458 km/s. A moment's thought tells us that this is a decidedly nonintuitive statement. For example, if we were traveling in a car moving at 100 km/h and we fired a bullet forward with a speed of 1000 km/h relative to the car, an observer standing at the side of the road would see the bullet pass by at 100 + 1000 = 1100 km/h, as illustrated in Figure 13.16(a). However, the Michelson–Morley experiment tells us that if we were traveling in a rocket ship at one-tenth the speed of light, 0.1*c*, and we shone a searchlight beam ahead of us (Figure 13.16b), an outside observer would measure the speed of the beam not as 1.1*c*, as the example of the bullet would suggest, but as *c*. The rules that apply to particles moving at or near the speed of light are different from those we are used to in everyday life.

The *special theory of relativity*, or just *special relativity*, was proposed by Einstein in 1905 to deal with the preferred status of the speed of light. The theory is the mathematical framework that allows us to extend the familiar laws of physics from low speeds (i.e., speeds much less than *c*, which are often referred to as *nonrelativistic*) to very high (or *relativistic*) speeds, comparable to *c*.

The essential features of special relativity are:

1. The speed of light, *c*, is the maximum possible speed in the universe, and all observers measure the *same* value for *c*, regardless of their motion. Einstein broadened this statement into the *principle of relativity*: The basic laws of physics are *the same* to all observers.
3. There is *no* absolute frame of reference in the universe; that is, there is no "preferred" observer relative to whom all other velocities can be measured. Instead, only relative velocities between observers matter (hence the term "relativity").
3. Neither space nor time can be considered independently of one another. Rather, they are each components of a single entity: *spacetime*. There is no absolute, universal time—observers' clocks tick at different rates, depending on the observers' motions relative to one another.

Special relativity is equivalent to Newtonian mechanics in describing objects that move much more slowly than the speed of light, but it differs greatly in its

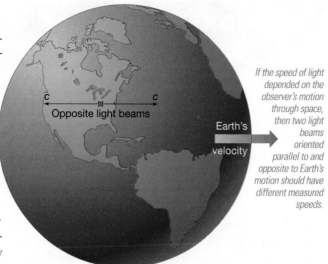

If the speed of light depended on the observer's motion through space, then two light beams oriented parallel to and opposite to Earth's motion should have different measured speeds.

▲ **FIGURE 13.15 Michelson–Morley Experiment** If light propagates at a fixed speed through space, then common sense suggests that the speed of a beam of light should depend on Earth's velocity. The Michelson–Morley experiment tried but failed to measure this dependence and in the process ushered in the era of modern physics.

MA INTERACTIVE
Exploring the Effects of Relative Motion

▲ **FIGURE 13.16 Speed of Light** (a) A bullet fired from a speeding car is measured by an outside observer to have a speed equal to the sum of the speeds of the car and of the bullet. (b) A beam of light shining forward from a high-speed spacecraft is still observed to have speed *c*, regardless of the speed of the spacecraft. The speed of light is thus independent of the speed of the source or of the observer.

13.1 MORE PRECISELY

Special Relativity

Einstein's theory of special relativity was constructed in large part to solve the puzzle of the 1887 Michelson–Morley experiment (see Section 13.6), which demonstrated that the observed speed of light is *independent* of the observer's motion through space. Einstein elevated the speed of light to the status of a constant of nature, rewrote the laws of mechanics to reflect that new fact, and opened the door to a flood of new physics and a much deeper understanding of the universe. But many common-sense ideas had to be abandoned in the process and replaced with some decidedly less-intuitive concepts. We describe here some of those odd consequences of Einstein's theory.

Imagine that you are an observer watching a rocket ship fly past at relative velocity *v* and that the craft is close enough for you to make detailed observations of the inside of its cabin. If *v* is much less than the speed of light, *c*, you would see nothing out of the ordinary—special relativity is consistent with familiar Newtonian mechanics at low velocities. However, if *v* is comparable to *c*, unexpected things start to happen.

As the ship's velocity increases, you begin to notice that the ship appears to *contract* in the direction in which it is moving. A meterstick on board, identical at launch to the one you keep in your laboratory, is now shorter than its twin. This is called *Lorentz contraction* (or sometimes Lorentz–Fitzgerald contraction). The first figure shows the stick's measured length aboard the moving ship: At low velocities (bottom) the meterstick measures 1 m, but at high velocities (top), the stick is shortened considerably. Because of Lorentz contraction, a meterstick moving at 90 percent of the velocity of light would shrink to a little less than half a meter. (This is not an optical illusion.)

At the same time, the ship's clock, synchronized prior to launch with your own, now ticks *more slowly* (second figure). This phenomenon, known as *time dilation,* has been observed many times in laboratory experiments in which fast-moving radioactive particles are observed to decay more slowly than if they were at rest in the lab. Their internal clocks are slowed by their rapid motion. Although no material particle can actually reach the speed of light, Einstein's theory implies that, as *v* approaches *c*, the measured length of the meterstick will shrink to nearly zero and the clock will slow to a virtual stop.

Of course, from the point of view of an astronaut on board the spaceship, *you* are the one moving rapidly. Hence, as seen from the ship, *you* appear to be compressed in the direction of motion, and *your* clock runs slowly! How can both observations be correct? After all, you both know that nothing has happened to your metersticks or clocks. You feel no force squeezing your bodies. What is going on?

When measuring the length of the moving meterstick, you do so by noting the positions of the two ends *at the same time, according to your clock.* But those two events—the two measurements you make—do *not* occur at the same time as seen by the astronaut on the spaceship. In relativity, time is relative, and simultaneity (the idea that two events happen "at the same time") is no longer a well-defined concept. From the astronaut's viewpoint, your measurement of the leading end of the meterstick occurs *before* your measurement of the trailing end. This time discrepancy ultimately results in the Lorentz contraction you observe. A similar argument applies to measurements of time, such as the period between two clock ticks. Time dilation occurs because the measurements occur at the same location and different times in one frame, but at *different places and times* in the other.

Einstein's ideas were revolutionary at the time, requiring physicists to abandon some long-held, cherished, and "obvious" facts about the universe. Reading the above discussion, you can perhaps begin to understand why special relativity initially encountered such resistance among some scientists (and why it still confuses nonscientists today!). Nevertheless, the gain in scientific understanding ultimately overcame the price in unfamiliarity. Within just a few years, special relativity had become almost universally accepted (and Einstein was well on his way to becoming the best-known scientist on the planet).

Speed of light

Speed (percent light speed)

Measured length (meters)

If the relative velocity of a spaceship actually reached light speed, time would appear to halt completely. From our point of view, its occupants would seem to stop aging.

Measured time (seconds)

predictions at relativistic velocities (see *More Precisely 13-1*). Yet, despite their often nonintuitive nature, all of the theory's predictions have been repeatedly verified to a high degree of accuracy. Today, special relativity lies at the heart of modern science. No scientist seriously doubts its validity.

General Relativity

In constructing his special theory of relativity, Einstein rewrote the laws of motion expounded by Newton more than two centuries previously. ∞ *(Sec. 1.4)* Fitting Newton's other great legacy—the theory of gravitation—into the framework of relativity was a much more complex mathematical problem, which took Einstein another decade to complete. The result once again overturned scientists' conception of the universe.

In 1915, Einstein illustrated the connection between special relativity and gravity with the following famous "thought experiment." Imagine that you are enclosed in an elevator with no windows, so that you cannot directly observe the outside world, and the elevator is floating in space. You are weightless. Now suppose that you begin to feel the floor press up against your feet. Weight has apparently returned. There are two possible explanations for this, as shown in Figure 13.17. A large mass could have come nearby, and you are feeling its downward gravitational attraction (Figure 13.17a), *or* the elevator has begun to accelerate upward, and the force you feel is that exerted by the elevator as it accelerates you at the same rate (Figure 13.17b). The crux of Einstein's argument is this: There is *no* experiment that you can perform within the elevator (without looking outside) that will let you distinguish between these two possibilities.

Thus, Einstein reasoned, there is no way to tell the difference between a gravitational field and an accelerated frame of reference (such as the rising elevator in the thought experiment). This statement is known more formally as the *equivalence principle.* Using it, Einstein succeeded in incorporating gravity into special relativity as a general acceleration of all particles. The resulting theory is called *general relativity.* However, Einstein found that another huge conceptual leap was needed. To include the effects of gravity, the mathematics forced Einstein to the unavoidable conclusion that spacetime—the single entity combining space and time, central to special relativity—has to be *curved*.

The central concept of general relativity is this: Matter—all matter—tends to "warp" or curve space in its vicinity. Objects such as planets and stars react to this warping by changing their paths. In the Newtonian view of gravity, particles move on curved trajectories because they are acted upon by a gravitational force. ∞ *(Sec. 1.4)* In Einsteinian relativity, those particles move on curved trajectories because they are falling freely through space, following the curvature of spacetime produced by some nearby massive object. The more the mass, the greater the warping. In general relativity, there is no such thing as a "gravitational force" in the Newtonian sense. Objects move as they do because they follow the curvature of spacetime, which is determined by the amount of matter present. Stated more loosely, in the words of renowned relativist John Archibald Wheeler, "Spacetime tells matter how to move, and matter tells spacetime how to curve."

Some props may help you visualize these ideas. Bear in mind, however, that these props are not real, but only tools to help you grasp some exceedingly strange concepts. Imagine a pool table with the tabletop made of a thin rubber sheet rather than the usual hard felt. As Figure 13.18 suggests, such a rubber sheet becomes distorted when a heavy weight (such as a rock) is placed on it. The heavier the rock (Figure 13.18b), the larger is the distortion.

Trying to play pool on this table, you would quickly find that balls passing near the rock are deflected by the curvature of the tabletop (Figure 13.18b). The

A person inside a windowless elevator could not distinguish between these two cases.

▲ **FIGURE 13.17 Einstein's Elevator** Einstein reasoned that no experiment conducted entirely within an elevator floating in space can tell a passenger whether the force he feels is (a) due to the gravity of a nearby massive object or (b) caused by the acceleration of the elevator itself.

DATA POINTS

Special Relativity

More than 35 percent of students had difficulty describing how special relativity affects the measured properties of fast-moving objects. Some points to remember:

■ As you observe a fast-moving spacecraft, it will appear to shrink in the direction of motion, its clock will tick more slowly, and its mass will increase.

■ An observer on board the spacecraft will say exactly the same about your size, clock, and mass!

■ The resolution of this paradox is that simultaneous measurements in one frame are not simultaneous in another, so that two observers can each see the other's clock apparently running slow without contradiction.

The amount of mass determines the amount of curvature and hence the amount of deflection.

(a)

Deflection of pool ball

(b)

◀ FIGURE 13.18 **Curved Space** INTERACTIVE (a) A pool table made of a thin rubber sheet sags when a weight is placed on it. Likewise, space is bent, or warped, in the vicinity of any massive object. (b) A ball shown rolling across the table is deflected by the curvature of the surface, in much the same way that a planet's curved orbit is determined by the curvature of spacetime produced by the Sun.

pool balls are not attracted to the rock in any way; rather, they respond to the curvature of the sheet produced by the rock's presence. In much the same way, anything that moves through space—matter *or radiation*—is deflected by the curvature of spacetime near a star. (See *More Precisely 13-2* for a solar system measurement, and Sections 14.6 and 16.3 for examples on much larger scales.) Earth's orbital path is the trajectory that results as our planet falls freely in the relatively gentle curvature of space created by our Sun. When the curvature is small (i.e., gravity is weak), both Einstein and Newton predict the same orbit—the one we observe. However, as the gravitating mass increases, the two theories begin to diverge.

Curved Space and Black Holes

Modern notions about black holes rest squarely on the general theory of relativity. Although white dwarfs and (to a lesser extent) neutron stars can be adequately described by the classical Newtonian theory of gravity, only the modern Einsteinian theory of relativity can properly account for the bizarre physical properties of black holes.

Let's consider another analogy. Imagine a large extended family of people living on a huge rubber sheet—a sort of gigantic trampoline. Deciding to hold a reunion, they converge on a given place at a given time. As shown in Figure 13.19, one person remains behind, not wishing to attend. She keeps in touch with her relatives by means of "message balls" rolled out to her (and back from her) along the surface of the sheet. These message balls are the analog of radiation carrying information through space.

As the people converge, the rubber sheet sags more and more. Their accumulating mass creates an increasing amount of space curvature. The message balls can still reach the lone person far away in nearly flat space, but they arrive less frequently as the sheet becomes more and more warped and stretched—as shown in Figures 13.19(b) and (c)—and the balls have to climb out of a deeper and deeper well. Finally, when enough people have arrived at the appointed spot, the mass becomes too great for the rubber to support them. As illustrated in Figure 13.19(d), the sheet pinches off into a "bubble," compressing the people into oblivion and severing their communications with the lone survivor outside. This final stage represents the formation of an event horizon around the reunion party.

Right up to the end—the pinching off of the bubble—two-way communication is possible. Message balls can reach the outside from within (but at a slower and slower rate as the rubber stretches), and messages from outside can get in without difficulty. However, once the event horizon (the bubble) forms, balls from the outside can still fall in, but they can no longer be sent back out to the person left behind, no

▼ FIGURE 13.19 **Space Warping** Mass causes a rubber sheet (or space) to be curved. As people assemble at a fixed spot on the sheet (marked by an "×"), the curvature grows larger, as shown in (a)–(c). The blue arrows represent some directions in which information can be sent from place to place. (d) The people are eventually sealed inside the bubble, forever trapped and cut off from the outside world.

This analogy again shows how more mass . . .

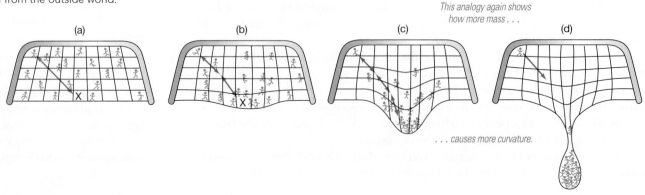

(a)   (b)   (c)   (d)

. . . causes more curvature.

13.2 MORE PRECISELY

Tests of General Relativity

Special relativity is the most thoroughly tested and most accurately verified theory in the history of science. General relativity, however, is on somewhat less firm experimental ground. The problem with verifying general relativity is that its effects on Earth and in the solar system—the places where we can most easily perform tests—are very small. Just as special relativity produces major departures from Newtonian mechanics only when speeds approach the speed of light, general relativity predicts large departures from Newtonian gravity only when extremely strong gravitational fields are involved.

Here we consider just two "classical" tests of the theory—solar system experiments that helped ensure acceptance of Einstein's theory. Bear in mind, however, that there are no known tests of general relativity in the "strong-field" regime—that part of the theory that predicts black holes, for example—so the full theory has never been experimentally tested.

At the heart of general relativity is the premise that everything, including light, is affected by gravity because of the curvature of spacetime. Shortly after he published his theory in 1915, Einstein noted that light from a star should be deflected by a measurable amount as it passes the Sun. The closer to the Sun the light comes, the more it is deflected. Thus, the maximum deflection should occur for a ray that just grazes the solar surface. Einstein calculated that the deflection angle should be 1.75, a small but detectable amount.

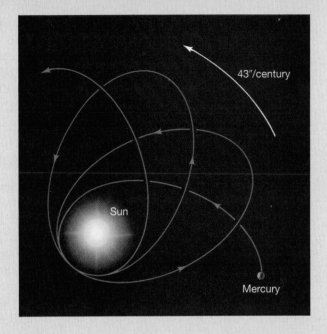

Of course, it is normally impossible to see stars so close to the Sun. During a solar eclipse, however, when the Moon blocks the Sun's light, the observation does become possible, as illustrated in the figure at lower left. In 1919 a team of observers led by British astronomer Sir Arthur Eddington succeeded in measuring the deflection of starlight during an eclipse. The results were in excellent agreement with the prediction of general relativity.

Another prediction of general relativity is that planetary orbits should deviate slightly from the perfect ellipses of Kepler's laws. Again, the effect is greatest where gravity is strongest—that is, closest to the Sun. Thus, the largest relativistic effects are found in the orbit of Mercury. Relativity predicts that Mercury's orbit is not exactly a closed ellipse. Instead, its orbit should rotate slowly, as shown in the second (highly exaggerated) diagram above. The amount of rotation is very small—only per century—but Mercury's orbit is so well charted that even this tiny effect is measurable.

matter how fast they are rolled. They cannot make it past the "lip" of the bubble in Figure 13.19(d). This analogy (very) roughly depicts how a black hole warps space completely around on itself, isolating its interior from the rest of the universe.

The essential ideas—the slowing down and eventual cessation of outward-going signals and the one-way nature of the event horizon once it forms—all have parallels in the case of stellar black holes. In Einsteinian terms, a *black hole* is a region of space where the gravitational field becomes overwhelming and the curvature of space extreme. At the event horizon itself, the curvature is so great that space "folds over" on itself, causing objects within to become trapped and disappear.

PROCESS OF SCIENCE CHECK

How do Newton's and Einstein's theories differ in their descriptions of gravity? Why were scientists initially resistant to Einstein's theories of relativity?

13.7 Space Travel Near Black Holes

Black holes are *not* cosmic vacuum cleaners. They don't cruise around interstellar space, sucking up everything in sight. The orbit of an object near a black hole is essentially the same as its orbit near a "normal" star of the same mass as the black hole. Only if the object happened to pass within a few Schwarzschild radii (perhaps 50–100 km for a typical 5- to 10-solar-mass black hole formed in a supernova explosion) of the event horizon would there be any significant difference between its actual orbit and the one predicted by Newtonian gravity and described by Kepler's laws. Of course, if some matter did fall into the black hole—that is, if its orbit happened to take it too close to the event horizon—it would be unable to get out. Black holes are like turnstiles, permitting matter to flow in only one direction—inward.

Tidal Forces

Matter flowing into a black hole is subject to enormous gravitational stress. An unfortunate explorer falling feet first into a solar mass black hole would find that the pull of gravity is much stronger at her feet (which are closer to the hole) than at her head. This force difference is a *tidal force*—precisely the same basic phenomenon that is responsible for ocean tides on Earth and the spectacular volcanoes on Io—except that the tidal forces at work in and near a black hole are far more powerful than any we are ever likely to find in the solar system. ∞ *(Secs. 5.2, 8.1)* The tidal force due to the black hole is so great that our explorer would be stretched enormously in height and squeezed unmercifully laterally. She would be torn apart long before she reached the event horizon.

The net result of all this stretching and squeezing is numerous and violent collisions among the resulting debris, causing a great deal of frictional heating of any infalling matter. As illustrated (with some artistic license) in Figure 13.20, material is simultaneously torn apart and heated to high temperatures as it plunges into the hole. So efficient is this heating that prior to reaching the hole's event horizon, matter falling into the hole emits radiation of its own accord. For a black hole of roughly solar mass, the energy is expected to be emitted in the form of X-rays. Thus, contrary to what we might expect from an object whose defining property is that nothing can escape from it, the region surrounding a black hole is expected to be a *source* of energy. Of course, once the hot matter falls below the event horizon, its radiation is no longer detectable—it can never leave the hole.

Approaching the Event Horizon

One safe way to study a black hole would be to go into orbit around it, safely beyond the disruptive influence of the hole's strong tidal forces. After all, Earth and the other planets of our solar system all orbit the Sun without falling into it and being torn apart. The gravity field around a black hole is basically no different. Let's imagine sending a less fragile (and more expendable) observer—a robot probe—toward the center of the hole, as illustrated in Figure 13.21. If our robot is small enough (to minimize the tidal force) and of sufficiently strong construction, it might conceivably survive to reach the event horizon, even if a human could not. Watching from a safe distance in our orbiting spacecraft, we can then examine the nature of space and time, at least down to the event horizon. After that boundary is crossed, there is no way for the probe to return any information about its findings.

Suppose our robot has an accurate clock and a light source of known frequency mounted on it. From our safe vantage point far outside the event horizon, we could use telescopes to read the clock and measure the frequency of the light we receive. What might we discover? We would find that the light from the robot would become more and more redshifted—shifted toward longer wavelengths—as the robot neared the event horizon (Figure 13.22). Even if the robot used rocket engines to

ANIMATION/VIDEO
Energy Released from a Black Hole?

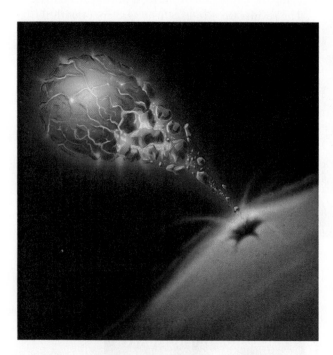

▲ **FIGURE 13.20 Black Hole Heating**   Any matter falling into the clutches of a black hole will become severely distorted and heated. This sketch shows an imaginary planet being pulled apart by a black hole's gravitational tides.

remain motionless, the redshift would still be detected. The redshift is not caused by motion of the light source—it is not the result of the Doppler effect. ∞ *(Sec. 2.7)* Rather, it is a result of the black hole's gravitational field, clearly predicted by Einstein's general theory of relativity and known as **gravitational redshift.**

We can explain the gravitational redshift as follows. Photons are attracted by gravity. As a result, in order to escape from a source of gravity, photons must expend some energy—they have to work to get out of the gravitational field. They don't slow down (photons always move at the speed of light); they just lose energy. Because a photon's energy is proportional to the frequency of its radiation, light that loses energy must have its frequency reduced (or, conversely, its wavelength increased).

As photons traveled from the robot's light source to our orbiting spacecraft, they would become gravitationally redshifted. From our standpoint, a yellow light would appear orange, then red as the robot astronaut approached the black hole. As the robot neared the event horizon, the light would become undetectable with an optical telescope. First infrared and then radio telescopes would be needed to detect it. Theoretically, light emitted *from the event horizon itself* would be gravitationally redshifted to infinitely long wavelengths.

What about the robot's clock? What time does it tell? Is there any observable change in the rate at which the clock ticks while moving deeper into the hole's gravitational field? We would find, from our orbiting spacecraft, that any clock close to the hole would appear to tick more *slowly* than an equivalent clock on board our ship. The closer the clock came to the hole, the slower it would appear to run. Upon reaching the event horizon, the clock would seem to stop altogether. All action would become frozen in time. Consequently, we would never actually witness the robot cross the event horizon. The process would appear to take forever.

This apparent slowing down of the robot's clock is known as **time dilation** and is closely related to gravitational redshift. To see why this is so, think of the robot's light source as a clock, with the passage of a wave crest (say) constituting a "tick," so the clock ticks at the frequency of the radiation. As the wave is redshifted, the frequency drops, and fewer wave crests pass us each second—the clock appears to slow down. This thought experiment demonstrates that the redshift of the radiation and the slowing of the clock are one and the same thing.

From the point of view of the robot, relativity theory predicts no strange effects at all. The light source hasn't reddened and the clock keeps perfect time. In the robot's frame of reference, everything is normal. Nothing prohibits it from approaching within the Schwarzschild radius of the hole (so long as the probe is strong enough to avoid being torn apart by tidal forces)—no law of physics constrains an object from passing through an event horizon. There is no barrier at the event horizon and no sudden lurch as it is crossed. It is just a boundary in space, but one that can be crossed in only one direction.

Perhaps counterintuitively, the tidal force at the event horizon is *smaller* for more massive black holes (see problems 8 and 9 at the end of this chapter). Although, as we saw, an intrepid explorer would be shredded long before reaching the event horizon of a 1-solar-mass black hole, travelers passing through the event horizon of a multimillion solar mass hole (such as might lurk in the heart of our own Galaxy, as we will see in the next chapter) might not feel any particular discomfort. They would remain able to contemplate their surroundings—and the all-important fact that they couldn't reverse course—for a few more seconds before the rapidly rising tidal forces in the black hole's interior finally finished them off.

▲ FIGURE 13.21 **Robot Astronaut** A hypothetical spacecraft launches a robotic probe that travels toward a black hole to perform experiments that humans, farther away in the craft, could monitor to learn about the nature of space near the event horizon.

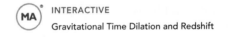

INTERACTIVE
Gravitational Time Dilation and Redshift

▶ FIGURE 13.22 **Gravitational Redshift** Photons escaping from the strong gravitational field close to a black hole must expend energy to overcome the hole's gravity. As a result, the photons change wavelength; their color changes, and their frequency lessens. This figure shows the effect on two beams of radiation, one of visible light and one of X-rays, emitted from a space probe as it nears the event horizon of a 1-solar-mass black hole.

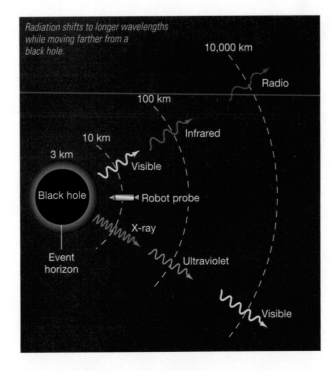

Deep Down Inside

No doubt, you are wondering what lies within the event horizon of a black hole. The answer is simple: No one knows. However, the question raises some very fundamental issues that lie at the forefront of modern physics.

General relativity predicts that without some agent to compete with gravity, the core remnant of a high-mass star will collapse all the way to a point at which both its density and its gravitational field become infinite—a so-called **singularity.** The same applies to any infalling matter. However, we should not take this prediction of infinite density too literally. Singularities always signal the breakdown of the theory producing them. The present laws of physics are simply inadequate to describe the final moments of a star's collapse.

As it stands today, the theory of gravity is incomplete because it does not incorporate a proper description of matter on very small scales. As our collapsing stellar core shrinks to smaller and smaller radii, we eventually lose our ability even to describe, let alone predict, its behavior. Perhaps matter trapped in a black hole never actually reaches a singularity. Perhaps it just approaches this bizarre state in a manner that we will someday understand as the subject of *quantum gravity*—the merger of general relativity (the theory of the universe on the largest scales) with quantum mechanics (the theory of matter on atomic and subatomic scales)—develops.

Having said that, we can at least estimate how small the core can get before current theory fails. It turns out that by the time that stage is reached, the core is already much smaller than any elementary particle. Thus, although a complete description of the end point of stellar collapse may well require a major overhaul of the laws of physics, for all practical purposes the prediction of collapse to a point is valid. Even if a new theory somehow succeeds in doing away with the central singularity, it is very unlikely that the external appearance of the hole, or the existence of its event horizon, will change. Any modifications to general relativity are expected to occur only on submicroscopic scales, not on the macroscopic (kilometer-sized) scale of the Schwarzschild radius.

Singularities are places where the rules break down, and some very strange things may occur near them. Many possibilities have been suggested—gateways to other universes, time travel, or the creation of new states of matter, for instance—but none has been proved, and certainly none has been observed. Because these regions are places where science fails, their presence causes serious problems for many of our cherished laws of physics, including causality (the idea that cause should precede effect, which runs into immediate problems if time travel is possible) and energy conservation (which is violated if material can hop from one universe to another through a black hole). It is unclear whether the removal of the singularity by some future, all-encompassing theory would necessarily also eliminate all of these problematic side effects.

CONCEPT CHECK

Why would you never actually witness an infalling object crossing the event horizon of a black hole?

ANIMATION/VIDEO (MA)
X-Ray Binary Star

13.8 Observational Evidence for Black Holes

Theoretical ideas aside, is there any observational evidence for black holes? What proof do we have that these strange invisible objects really do exist?

Black Holes in Binary Systems

Perhaps the most promising way to find black holes is to look for their effects on other astronomical objects. Our Galaxy harbors many binary star systems in which only one object can be seen, but we need observe the motion of only one star to infer the existence of an unseen companion and measure some of its properties. ∞ *(Sec. 10.7)* In the majority of cases, the invisible companion is simply small and dim, nothing more than an M-type star hidden in the glare of an O- or B-type partner, or perhaps shrouded by dust or other debris, making it invisible to even the best available equipment. In either case, the invisible object is not a black hole.

A few close binary systems, however, have peculiarities suggesting that one of their members may be a black hole. Figure 13.23(a) shows the area of the sky in the constellation Cygnus, where the evidence is particularly strong. The system of interest lies some 2000 pc from Earth. The black hole candidate is an X-ray source called Cygnus X-1, discovered in the early 1970s. Its visible companion is a blue B-type giant. Assuming that it lies on the main sequence, its mass must be around 25 times the mass of the Sun. Spectroscopic observations indicate that the binary system has an orbital period of 5.6 days. Combining this information with spectroscopic measurements of the visible component's orbital speed, astronomers estimate the total mass of the binary system to be around 35 solar masses, implying that Cygnus X-1 has a mass of about 10 times that of the Sun.

X-ray radiation emitted from the immediate neighborhood of Cygnus X-1 indicates the presence of high-temperature gas, perhaps as hot as several million kelvins. Furthermore, the X-ray emission has been observed to vary in intensity on time scales as short as a millisecond. As discussed earlier in the context of gamma-ray bursts, for this variation not to be blurred by the travel time of light across the source, Cygnus X-1 cannot be more than 1 light-millisecond, or 300 km, in diameter.

These properties suggest that Cygnus X-1 could be a black hole. Figure 13.24 is an artist's conception of this intriguing object. The X-ray-emitting region is likely an accretion disk formed as matter drawn from the visible star spirals down onto the unseen component. The rapid variability of the X-ray emission indicates that the unseen component must be very compact—a neutron star or a black hole. Its large mass, well above the neutron star mass limit discussed earlier, implies the latter.

A few other black hole candidates are known. For example, LMC X-3—the third X-ray source ever discovered in the Large Magellanic Cloud, a small galaxy orbiting our own (see Section 15.2)—is an invisible object that, like Cygnus X-1, orbits a bright companion star. LMC X-3's visible companion seems to be distorted into the shape of an egg by the unseen object's intense gravitational pull. Reasoning similar to that applied to Cygnus X-1 leads to the conclusion that LMC X-3 has a mass nearly 10 times that of the Sun, making it too massive to be anything but a black hole. The X-ray binary system A0620-00 has been found to contain an invisible compact object of mass 3.8 times the mass of the Sun. In total, about 20 known objects may turn out to be black holes; Cygnus X-1, LMC X-3, and A0620-00 probably have the strongest claims.

(a)

B-star
HDE226868

This is a background star

1.5 arc minutes

R I V U X G

(b)

The X-ray source is coincident with the visible star.

|← 1 pc →|

R I V U X G

▲ **FIGURE 13.23 Cygnus X-1** (a) The brightest star in this photograph (called HDE226868) is a member of a binary system whose unseen companion, called Cygnus X-1, is a leading black hole candidate. The system lies approximately 2000 pc away. (b) A *Chandra* X-ray image of the field of view outlined by the square in part (a). The X-rays are not coming from the star. Rather, they come from the black hole companion, which orbits so close to the star that their roughly 0.1 AU separation cannot be seen on this (roughly 1 pc) scale. Neither the star nor Cygnus X-1 are nearly as large as these images suggest. Both are so bright that their light spills over into the surrounding image. *(Harvard-Smithsonian Center for Astrophysics; CXO)*

(MA) ANIMATION / VIDEO
Black Hole and Companion Star

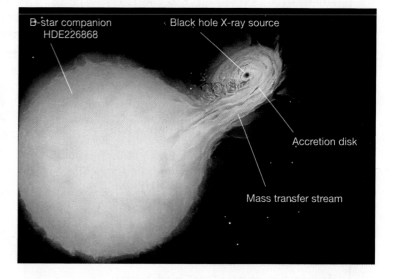

B-star companion
HDE226868

Black hole X-ray source

Accretion disk

Mass transfer stream

▶ **FIGURE 13.24 Black Hole INTERACTIVE** Artist's conception of a binary system containing a large, bright, visible star and an invisible, X-ray-emitting black hole. (Compare Figure 12.14). This painting is based on data obtained from many observations of Cygnus X-1. *(L. Chaisson)*

◀ **FIGURE 13.25 Intermediate-Mass Black Holes?**   X-ray observations (inset) of the center of the starburst galaxy M82 reveal a collection of bright sources that may be the result of matter accreting onto intermediate-mass black holes. The black holes are probably young, have masses between 100 and 1000 times the mass of the Sun, and lie relatively far from the center of M82. *(Subaru; NASA)*

5000 pc

Suspected
black hole

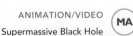

Black Holes in Galaxies

Perhaps the strongest evidence for black holes comes not from binary systems in our own Galaxy but from observations of the centers of many galaxies (including our own), where astronomers have found that stars and gas are moving extremely rapidly, orbiting some very massive, unseen object. Masses inferred from Newton's laws range from millions to billions of times the mass of the Sun. ∞ *(More Precisely 1-1)* The leading (and at present, the only) explanation is that these objects are black holes. We will return to these observations, and the question of how such *supermassive black holes* might have formed, in the next three chapters.

In 2000, X-ray astronomers reported evidence for a long-sought but elusive missing link between "stellar-mass" black holes in binaries and the supermassive black holes in the hearts of galaxies. Figure 13.25 shows an unusual looking galaxy called M82, currently the site of an intense and widespread burst of star formation (see Chapter 15). The inset shows a *Chandra* image of the innermost few thousand parsecs of M82, revealing a number of bright X-ray sources close to—but *not* at—the center of the galaxy. Their spectra and X-ray luminosities suggest that they may be accreting compact objects with masses ranging from 100 to almost 1000 times the mass of the Sun. If confirmed, they will be the first *intermediate-mass black holes* ever observed.

Too large to be remnants of normal stars and too small to warrant the "supermassive" label, these objects present a puzzle to astronomers. One possible origin is suggested by an apparent association between some of the X-ray emitters in M82 and elsewhere and dense, young star clusters. Theorists speculate that collisions between high-mass stars in the congested cores of these clusters might lead to the runaway growth of an extremely massive and very unstable star, which then collapses to form an intermediate-mass black hole. However, both the observations and this explanation remain controversial. Figure 13.26 shows the current best candidate for a "nearby" star cluster harboring an intermediate-mass black hole—the globular cluster G1 in the Andromeda galaxy. ∞ *(Sec. 2.1)* The peculiar orbits of stars near the center of this cluster suggest a black hole of mass 20,000 times that of the Sun, and observations of the cluster in both radio and X-rays are consistent with theoretical expectations of the emission from such a massive object in the cluster's core.

10 pc

▲ **FIGURE 13.26 Black Hole Host?**   The star cluster G1 is the most massive globular cluster in the Andromeda galaxy, but the stars near its center do not move as expected if the cluster's mass is as smoothly distributed as its light. Instead, the observations suggest that an intermediate-mass black hole resides at the cluster's center. *(NASA)*

CONCEPT CHECK

How do astronomers "see" black holes?

THE BIG QUESTION

Are black holes for real? Decades ago, many astronomers regarded black holes as a kind of cosmic cop-out—the last resort of researchers who couldn't decipher the bizarre phenomena they observed. But now, improved observations really do point toward stellar remnants that are denser, brighter, and much more puzzling than the already peculiar neutron stars. We have to conclude that, strange as they are, black holes have been detected in our Galaxy and beyond. Perhaps someday future generations of space travelers will visit Cygnus X-1 or the center of our Galaxy and (carefully!) test these conclusions firsthand. Still, might there be other weird collapsed remnants— quark stars, perhaps—that stop short of being genuine black holes?

CHAPTER REVIEW

SUMMARY

LO1 A core-collapse supernova may leave behind an ultracompressed ball of material called a **neutron star** (p. 362). This is the remnant of the inner core that rebounded and blew the rest of the star apart. Neutron stars are extremely dense and, at formation, are predicted to be extremely hot, strongly magnetized, and rapidly rotating. They cool down, lose much of their magnetism, and slow down as they age.

LO2 According to the **lighthouse model** (p. 364), neutron stars, because they are magnetized and rotating, send regular bursts of electromagnetic energy into space. The beams are produced by charged particles confined by the strong magnetic fields. When we can see the beams from Earth, we call the source neutron star a **pulsar** (p. 362). The pulse period is the rotation period of the neutron star.

LO3 A neutron star that is a member of a binary system can draw matter from its companion, forming an accretion disk, which is usually a strong source of X-rays. As gas builds up on the star's surface, it eventually becomes hot enough to fuse hydrogen. When hydrogen burning starts on a neutron star, it does so explosively, and an **X-ray burster** (p. 366) results. The rapid rotation of the inner part of the accretion disk causes the neutron star to spin faster as new gas arrives on its surface. The eventual result is a very rapidly rotating neutron star—a **millisecond pulsar** (p. 366). Many millisecond pulsars are found in the hearts of old globular clusters. They cannot have formed recently, and must have been spun up by interactions with other stars. Careful analysis of the radiation received has shown that some millisecond pulsars are orbited by planet-sized objects.

LO4 **Gamma-ray bursts** (p. 368) are very energetic flashes of gamma rays that occur about once a day and are distributed uniformly over the entire sky. In some cases, their distances have been measured, placing them at very large distances and implying that they are extremely luminous. The leading theoretical models for these

explosions involve the violent merger of neutron stars in a distant binary system, or the recollapse and subsequent violent explosion following a "failed" supernova in a very massive star.

LO5 Einstein's special theory of relativity deals with the behavior of particles moving at speeds comparable to the speed of light. It agrees with Newton's theory at low velocities, but makes many very different predictions for high-speed motion. All of its predictions have been repeatedly verified by experiment. The modern replacement for Newtonian gravity is Einstein's **general theory of relativity** (p. 373), which describes gravity in terms of the warping, or bending, of space by the presence of mass. The more mass, the greater the warping. All particles—including photons—respond to that warping by moving along curved paths.

LO6 The upper limit on the mass of a neutron star is about three solar masses. Beyond that mass, the star can no longer support itself against its own gravity, and it collapses to form a **black hole** (p. 373), a region of space from which nothing can escape. The most massive stars, after exploding in a supernova, form black holes rather than neutron stars. Conditions in and near black holes can only be described by general relativity. The radius at which the escape speed from a collapsing star equals the speed of light is called the **Schwarzschild radius** (p. 374). The surface of an imaginary sphere centered on the collapsing star and having a radius equal to the star's Schwarzschild radius is called the **event horizon** (p. 374).

LO7 To a distant observer, light leaving a spaceship that is falling into a black hole would be subject to **gravitational redshift** (p. 381) as the light climbed out of the hole's intense gravitational field. At the same time, a clock on the spaceship would show **time dilation** (p. 381)—the clock would appear to slow down as the ship approached the event horizon. The observer would never see the ship reach the surface of the hole. Once within the event horizon, no known force can prevent a collapsing star from contracting all the way to a pointlike **singularity** (p. 382), at which point both the density and the gravitational field of the star become infinite. This prediction of relativity theory has yet to be proved. Singularities are places where the known laws of physics break down.

LO8 Once matter falls into a black hole, it can no longer communicate with the outside. However, on its way in, it can form an accretion disk and emit X-rays.

The best place to look for a black hole is in a binary system in which one component is a compact X-ray source. Cygnus X-1, a well-studied X-ray source in the constellation Cygnus, is a long-standing black-hole candidate. Studies of orbital motions imply that some binaries contain compact objects too massive to be neutron stars, leaving black holes as the only alternative. There is also substantial evidence for more massive black holes residing in or near the centers of many galaxies, including our own.

MasteringAstronomy®   For instructor-assigned homework go to www.masteringastronomy.com

Problems labeled **POS** explore the process of science. **VIS** problems focus on reading and interpreting visual information. **LO** connects to the introduction's numbered Learning Outcomes.

REVIEW AND DISCUSSION

1. **LO1** How does the way in which a neutron star forms determine some of its most basic properties?

2. What would happen to a person standing on the surface of a neutron star?

3. **LO2** Why aren't all neutron stars seen as pulsars?

4. What are X-ray bursters?

5. **LO3** What is the favored explanation for the rapid spin rates of millisecond pulsars?

6. Why do you think astronomers were surprised to find a pulsar with a planetary system?

7. Why do astronomers think there are two basically different types of gamma-ray burst?

8. What does it mean to say that the measured speed of a light beam is independent of the motion of the observer?

9. **LO6** Use your knowledge of escape speed to explain why black holes are said to be "black."

10. What is an event horizon?

11. **LO5 POS** Why is it so difficult to test the predictions of the theory of general relativity? Describe two tests of the theory.

12. **LO7** What would happen to someone falling into a black hole?

13. **LO8 POS** What makes Cygnus X-1 a good black hole candidate?

14. **POS** Imagine that you had the ability to travel at will through the Milky Way Galaxy. Explain why you would discover many more neutron stars than those known to observers on Earth. Where would you be most likely to find these objects?

15. **POS** Do you think that planet-sized objects discovered in orbit around a pulsar should be called planets? Why or why not?

CONCEPTUAL SELF-TEST: TRUE OR FALSE?/MULTIPLE CHOICE

1. Newly formed neutron stars have extremely strong magnetic fields. (T/F)

2. All millisecond pulsars are now, or once were, members of binary star systems. (T/F)

3. The fact that gamma-ray bursts are so distant means that they must be very energetic events. (T/F)

4. All things, except light, are attracted by gravity. (T/F)

5. According to general relativity, space is warped, or curved, by matter. (T/F)

6. Although visible light cannot escape from a black hole, high-energy radiation, like gamma rays, can. (T/F)

7. Thousands of black holes have now been identified in our Galaxy. (T/F)

8. A neutron star is about the same size as (a) a school bus; (b) a U.S. city; (c) the Moon; (d) Earth.

9. The most rapidly "blinking" pulsars are those that (a) spin fastest; (b) are oldest; (c) are most massive; (d) are hottest.

10. The X-ray emission from a neutron star in a binary system comes mainly from (a) the hot surface of the neutron star itself; (b) heated material in an accretion disk around the neutron star; (c) the neutron star's magnetic field; (d) the surface of the companion star.

11. Gamma-ray bursts are observed to occur (a) mainly near the Sun; (b) throughout the Milky Way Galaxy; (c) approximately uniformly over the entire sky; (d) near pulsars.

12. If the Sun were magically to turn into a black hole of the same mass, (a) Earth would start to spiral inward; (b) Earth's orbit would remain unchanged; (c) Earth would fly off into space; (d) Earth would be torn apart by the black hole's gravity.

13. The best place to search for black holes is in a region of space that (a) is dark and empty; (b) has recently lost some stars; (c) has strong X-ray emission; (d) is cooler than its surroundings.

14. **VIS** According to Figure 13.3 (Pulsar Model), the beam of a pulsar is emitted (a) along the rotation axis; (b) in the equatorial plane; (c) from the magnetic poles; (d) in a single direction in space.

15. **VIS** Figure 13.10 (Gamma-Ray Bursts) shows that gamma-ray bursts (a) are all very similar; (b) sometimes last for less than a second; (c) are confined to the Galactic plane; (d) come from great distances.

PROBLEMS

The number of squares preceding each problem indicates its approximate level of difficulty.

1. ■ The angular momentum of a spherical body is proportional to the body's angular speed times the square of its radius. ∞ *(More Precisely 4-1)* Using the law of conservation of angular momentum, estimate how fast a collapsed stellar core would spin if its initial spin rate was one revolution per day and its radius decreased from 10,000 km to 10 km.

2. ■ What would your mass be if you were composed entirely of neutron star material of density 3×10^{10} (Assume that your average density is 1000 kg/m^2.) Compare this with the mass of (a) the Moon; (b) a typical 1-km diameter asteroid.

3. ■ Calculate the surface gravity (relative to Earth's gravity of) and the escape speed of a 1.4-solar-mass neutron star with a radius of 10 km. What would be the escape speed of a 1-solar-mass object with a radius of 3 km? ∞ *(More Precisely 5-1)*

4. ■■ Use the radius–luminosity–temperature relation to calculate the luminosity of a 10-km-radius neutron star for temperatures of 10^5 K, 10^7 K and 10^9 K. What do you conclude about the visibility of neutron stars? Could the coolest of them be plotted on our H–R diagram?

5. ■■ A gamma-ray detector of area 0.5 m^2 observing a gamma-ray burst records photons having total energy 10^{-8} joules. If the burst occurred 1000 Mpc away, calculate the total amount of energy it released (assuming that the energy was emitted equally in all directions). How would this figure change if the burst occurred 10,000 pc away instead, in the halo of our Galaxy? What if it occurred within the Oort cloud of our own solar system, at a distance of 50,000 AU?

6. ■■ An unstable elementary particle is known to decay into other particles in 2 μs, as measured in a laboratory where the particle is at rest. A beam of such particles is accelerated to a speed of 99.99 percent of the speed of light. How long do the particles in the beam take to decay, in the laboratory frame of reference?

7. ■ Supermassive black holes are thought to exist in the centers of some galaxies. What would be the Schwarzschild radii of black holes of 1 million and 1 billion solar masses? How does the first black hole compare in size with the Sun? How does the second compare in size with the solar system?

8. ■■ Calculate the tidal acceleration on a 2-m-tall human falling feet first into a 1-solar-mass black hole—that is, compute the difference in the accelerations (forces per unit mass) on his head and his feet just before his feet cross the event horizon. Repeat the calculation for a 1-million-solar-mass black hole and for a 1-billion-solar-mass black hole (see the previous question). Compare these accelerations with the acceleration due to gravity on Earth.

9. ■■■ Endurance tests suggest that the human body cannot withstand stress greater than about 10 times the acceleration due to gravity on Earth's surface. At what distance from a 1-solar-mass black hole would the human in the previous question be torn apart? Calculate the minimum mass of a black hole for which an infalling human could just reach the event horizon intact.

10. ■■ Using the data given in the text (assume the upper limit on the stated range for the black hole mass), calculate the orbital separation of Cygnus X-1 and its B-type stellar companion. ∞ *(Sec. 1.4)*

ACTIVITIES

Collaborative

1. We can't easily observe a black hole, or make one to study, but theorists have lots to say about their properties! The text focuses on the simplest possible type of black hole—the uncharged, non-rotating *Schwarzchild* black hole—but there is a large body of literature on charged and spinning black holes, too. Charged black holes don't figure much in astronomy—the universe is electrically neutral on macroscopic scales—but rotating *Kerr* black holes are in fact very important. Divide your group in two and research online the properties of Schwarzchild and Kerr black holes. You'll probably find more information on the simpler Schwarzchild case, but some persistence will yield a lot on the rotating Kerr case, too. Combine your research to make a joint presentation on the similarities and differences between the two types. Focus on properties such as the event horizon, the singularity, and the orbits of light and matter near the hole. How fast can a black hole rotate? Which kind of black hole is thought to be most common in nature?

Individual

1. Find the ninth-magnitude companion to Cygnus X-1, the sky's most famous black hole candidate. Because none of us can see X-rays, no sign of anything unusual can be seen. Still, it's fun to gaze toward this region of the heavens and contemplate Cygnus X-1's powerful energy emission and strange properties. Even without a telescope, it is easy to locate the region of the heavens where Cygnus X-1 resides. The constellation Cygnus contains a recognizable star pattern called the Northern Cross. The star in the center of the crossbar is called Sadr. The star at the bottom of the cross is called Albireo. Approximately midway between Sadr and Albireo lies the star Eta Cygni. Cygnus X-1 is located slightly less than 0.5° from this star. Whether or not you are using a telescope, sketch what you see.

Galaxies and the Universe

Our home Galaxy—the Milky Way—is a thousand times bigger than anything we have considered thus far in this book. Yet this vast system, more than 100,000 light-years across and home to 100 billion stars, is but one of billions of galaxies strewn across the vast expanse of deep space.

Part 4 explores the properties of galaxies, the glorious "building blocks" of the universe. Collectively they, trace out truly gargantuan patterns that top the hierarchy of material structures in the cosmos. On these huge scales, we enter the realm of fundamental inquiry that has preoccupied philosophers since the dawn of recorded time: How did our universe begin? And how will it end?

The images here illustrate cosmic objects in Part 4, all of them at least a million million million times beyond any familiar terrestrial scale.

Active Galaxy ~10^{22} m

Normal Galaxy ~10^{21} m

Dwarf Galaxy ~10^{20} m

Large-scale Structure $\sim 10^{25}$ m

Galaxy Cluster $\sim 10^{23}$ m

A single galaxy has more stars in it than the number of people who have ever lived on Earth.

14

The Milky Way Galaxy

A Spiral in Space

Looking up on a dark, clear night, we are struck by two aspects of the night sky. The first is that the individual stars we see are roughly uniformly distributed in all directions. They all lie relatively close to us, mapping out the local galactic neighborhood within a few hundred parsecs of the Sun. But this is only a local impression. Ours is a rather provincial view. Beyond those nearby stars, the second thing we notice is a fuzzy band of light—the Milky Way—stretching across the heavens. From the Northern Hemisphere, this band is most easily visible in the summertime, arcing high above the horizon. Its full extent forms a great circle that encompasses the entire celestial sphere. This is the insider's view of the galaxy in which we live, the blended light of countless distant stars. As we consider much larger volumes of space, on scales far, far greater than the distances between neighboring stars, a new level of organization becomes apparent as the large-scale structure of the Milky Way Galaxy is revealed.

THE BIG PICTURE

Our Milky Way Galaxy is just one among nearly a hundred billion other galaxies in the observable universe—a hundred billion galaxies! For astronomers, the Milky Way plays much the role for galaxies as the Sun does for stars. Our understanding of galaxies throughout the cosmos rests squarely on our knowledge of the size, scale, structure, composition, and dynamics of our own Galactic home.

LEARNING OUTCOMES

Studying this chapter will enable you to:

LO1 Describe the overall structure of the Milky Way Galaxy, and say how the various regions differ from one another.

LO2 Explain the importance of variable stars in determining the size and shape of our Galaxy.

LO3 Compare and contrast the orbital motions of stars in different regions of the Galaxy.

LO4 Interpret the differences between disk and halo stars in terms of our current understanding of how our Galaxy formed.

LO5 Present some possible explanations for the spiral arms observed in our own and many other galaxies.

LO6 Explain what studies of Galactic rotation reveal about the size and mass of our Galaxy, and discuss the possible nature of dark matter.

LO7 Describe the evidence for a supermassive black hole at the center of our Galaxy.

◄ Stars cluster into gigantic assemblages called galaxies, of which our Milky Way Galaxy is just one among roughly a hundred billion others. Galaxies such as this one, known as M74 and shown here in true color, in turn contain roughly a hundred billion stars. As we now enter the realm of very big dimensions, the graceful winding arms of this majestic spiral galaxy resemble a curving staircase sweeping across some 100,000 light-years of space. Its size, shape, and mass approximate those of our Galaxy, which has never been photographed in its full grandeur because we live inside it. If this were our Galaxy, the Sun would reside in one of its spiral arms, about two-thirds of the way out from the center. (NASA)

MasteringAstronomy®

Visit www.masteringastronomy.com for quizzes, animations, videos, interactive figures, and self-guided tutorials.

14.1 Our Parent Galaxy

A **galaxy** is a gargantuan collection of stellar and interstellar matter—stars, gas, dust, neutron stars, black holes—isolated in space and held together by its own gravity. Astronomers are aware of literally billions of galaxies beyond our own. The particular galaxy we happen to inhabit is known as the **Milky Way Galaxy,** or just "the Galaxy," with a capital "G."

Our Sun lies in a part of the Galaxy known as the **Galactic disk,** an immense, circular, flattened region containing most of our Galaxy's luminous stars and interstellar matter (and almost everything we have studied so far in this book). Figure 14.1 illustrates how, viewed from within, the Galactic disk appears as a band of light—the *Milky Way*—stretching across our nighttime sky. Seen from Earth, the center of the disk (indeed, of the entire Galaxy) lies in the direction of the constellation Sagittarius. As indicated in the figure, if we look in a direction away from the Galactic disk (red arrows), relatively few stars lie in our field of view. However, if our line of sight happens to lie within the disk (white and blue arrows), we see so many stars that their light merges into a continuous blur.

Paradoxically, although we can study individual stars and interstellar clouds near the Sun in great detail, our location within the Galactic disk makes deciphering our Galaxy's large-scale structure from Earth a very difficult task—a little like trying to unravel the layout of paths, bushes, and trees in a city park without being able to leave one particular park bench. In many directions, the interpretation of what we see is inconclusive, with foreground objects obscuring our view of what lies beyond. As a result, astronomers who study the Milky Way Galaxy are often guided in their efforts by comparisons with more distant, but much more easily observable, galaxies. There is every reason to suppose that *all* the features of our Galaxy described in this chapter are shared by billions of other galaxies across the universe.

Figure 14.2 shows three galaxies thought to resemble our own in overall structure. Figure 14.2(a) is the Andromeda Galaxy, the nearest major galaxy to the Milky Way Galaxy, lying about 800 kpc (about 2.5 million light-years) away. Andromeda's apparent elongated shape is a result of the angle at which we happen to view it. In fact, this galaxy, like our own, consists of a thin circular *galactic disk* of matter that fattens to a **galactic bulge** at the center. The disk and bulge are embedded in a roughly spherical ball of faint old stars known as the **galactic halo.** These three basic galactic regions are indicated on the figure (the halo stars are too faint to see here; see Figure 14.9 for a clearer view). Figures 14.2(b) and (c) show images of two other galaxies—one seen face-on, the other edge-on—that illustrate these points more clearly.

CONCEPT CHECK

Why do we see the Milky Way as a band of light across the sky?

This is an artist's conception of our Milky Way from afar . . .

Galactic disk
Galactic bulge
Galactic center
Earth
8 kpc

(a)

. . . and this is a real image of the Milky Way seen from inside, looking along the white arrow in part (a).

(b)

R I V U X G

◀ **FIGURE 14.1 Galactic Plane**   (a) Gazing from Earth toward the Galactic center (white arrow) in this artist's conception, we see myriad stars stacked up within the thin band of light known as the Milky Way. In the opposite direction (blue arrow), we see little of our Galaxy, much as when looking perpendicular to the disk (red arrows) where far fewer stars can be seen. (b) A real optical view of the sky in the direction of the white arrow shows the fuzzy (mostly white and "milky") band or disk of our Milky Way Galaxy (see Fig 11.1). (*Axel Mellinger*)

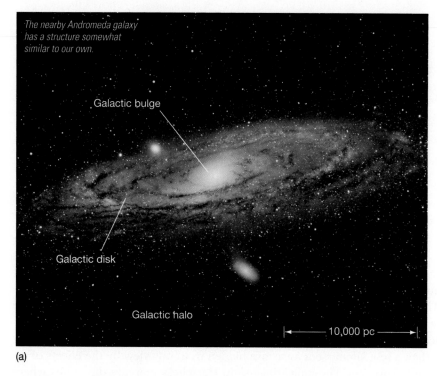

The nearby Andromeda galaxy has a structure somewhat similar to our own.

Galactic bulge

Galactic disk

Galactic halo

|←——— 10,000 pc ———→|

(a)

Another nearby galaxy, seen face on . . .

(b)

. . . and another, seen edge on.

(c)

▲ **FIGURE 14.2 Disk Galaxies**   (a) The Andromeda galaxy closely resembles the overall layout of our Milky Way Galaxy. Its disk and bulge are clearly visible in this image, which is about 30,000 pc across. (b) This galaxy, named M101 and seen nearly face-on, is similar in its overall structure to our own Milky Way Galaxy and Andromeda. (c) The galaxy NGC 4565 is oriented edge-on, allowing us to see clearly its disk and central bulge. *(R. Gendler; NASA)*

R I V U X G

14.2 Measuring the Milky Way

Before the 20th century, astronomers' conception of the cosmos differed markedly from the modern view. The growth in our knowledge of our Galaxy, as well as the realization that there are many other distant galaxies similar to our own, has gone hand in hand with the development of the cosmic distance scale.

Star Counts

In the late 18th century, long before the distances to any stars were known, English astronomer William Herschel tried to estimate the shape of our Galaxy simply by counting how many stars he could see in different directions in the sky. Assuming that all stars were of about equal brightness, he concluded that the Galaxy was a somewhat flattened, roughly disk-shaped collection of stars lying in the plane of the Milky Way, with the Sun at its *center* (Figure 14.3). Subsequent refinements to this approach led to essentially the same picture. Herschel was unable to estimate the *size* of the Galaxy by this method, but early in the 20th century, with improved knowledge of the properties of stars, some researchers estimated the dimensions of this "Galaxy" as about 10 kpc in diameter by 2 kpc thick.

   In fact, the Milky Way Galaxy is several tens of kiloparsecs across, and the Sun lies far from the center. The flaw in the above reasoning is that the observations were made in the visible part of the electromagnetic spectrum, and astronomers failed to take into account the absorption of visible light by the then unknown interstellar gas and dust. ∞ *(Sec. 11.1)* Only in the 1930s did

▶ FIGURE 14.3 Herschel's Galaxy Model   Eighteenth-century English astronomer William Herschel constructed this "map" of the Galaxy by counting the number of stars he saw in different directions in the sky. Our Sun (marked by the yellow dot) appears to lie near the center of the distribution. The long axis of the diagram roughly parallels the plane of the disk.

Early models of the Galaxy placed the Sun near the center of a relatively small system.

Sun

◀—————— 3 kpc ——————▶

(a)

(b)

(c)

▲ FIGURE 14.4 Variable Stars   Light curves of (a) the pulsating variable star RR Lyrae, which has a period of less than a day, and (b) of the Cepheid variable star WW Cygni, having a period of several days. (c) This same Cepheid is shown here (boxed) on successive nights, near its maximum and minimum brightness; two photos, one from each night, were superposed and then slightly displaced. (*Harvard College Observatory*)

astronomers begin to realize the true extent and importance of the interstellar medium.

The apparent falloff in the density of stars with distance in the plane of the Milky Way is not a real thinning of their numbers in space, but simply a consequence of the murky environment in the Galactic disk. Objects in the disk lying more than a few kiloparsecs away are hidden from our view by the effects of interstellar absorption. The long "fingers" in Herschel's map are directions where the obscuration happens to be a little less severe than in others.

The falloff in density perpendicular to the disk is real. Radiation coming to us from above or below the plane of the Galaxy, where there is less gas and dust along the line of sight, arrives on Earth relatively unscathed. There is still some patchy obscuration, but the Sun happens to lie in a location where the view out of the disk is largely unimpeded by nearby interstellar clouds.

Observations of Variable Stars

An important by-product of the laborious effort to catalog stars around the turn of the 20th century was the systematic study of **variable stars.** These are stars whose luminosities change significantly over relatively short periods of time—some quite erratically, others more regularly. Only a small fraction of stars fall into this category, but those that do are of great astronomical significance.

We have encountered several examples of variable stars in earlier chapters. Often, the variability is the result of membership in a binary system. Eclipsing binaries and novae are cases in point. ∞ *Secs 10.7 and 12.3* However, sometimes the variability is an intrinsic property of the star itself. Particularly important to Galactic astronomy are the **pulsating variable stars,** which vary cyclically in luminosity in very characteristic ways. Note that these stars have *nothing* whatsoever to do with pulsars, which represent an entirely different stage of stellar evolution. ∞ *Sec. 13.2* Two types of pulsating variable stars that have played central roles in revealing both the true extent of our Galaxy and the distances to our galactic neighbors are the **RR Lyrae** and **Cepheid** variables. (Following long-standing astronomical practice, the names come from the first star of each class discovered—in this case the variable star labeled RR in the constellation Lyra and the variable star Delta Cephei, the fourth brightest star in the constellation Cepheus.)

RR Lyrae and Cepheid variable stars are recognizable by the characteristic shapes of their light curves. RR Lyrae stars all pulsate in essentially similar ways (Figure 14.4a), with only small differences in period (the time from peak to peak) from one to another. Observed periods range from about 0.5 to 1 day. Cepheid variables also pulsate in distinctive ways (the regular "sawtooth" pattern in Figure 14.4b), but different Cepheids can have very different pulsation periods, ranging from about 1 to 100 days. In either case, the stars can be recognized and identified *just by observing the variations in the light they emit.*

Pulsating variable stars are normal stars experiencing a brief (perhaps a few million years) period of instability as a natural part of stellar evolution. The conditions necessary to cause pulsations are not found in main-sequence stars. However, they do occur in post-main-sequence stars as they expand and cool on their way to becoming red giants. On the H–R diagram, these unstable stars are found in a region called the *instability strip* (Figure 14.5). When a star's temperature and luminosity place it in this strip, the star becomes internally unstable, and both its temperature and its radius vary in a regular way, causing the pulsations we observe. Cepheid variables are high-mass, high-luminosity stars evolving across the upper part of the H–R diagram. ∞ *(Sec. 12.4)* The less-luminous RR Lyrae variables are lower-mass horizontal-branch (core helium fusion) stars that happen to lie within the lower portion of the instability strip. ∞ *(Sec. 12.2)*

A New Yardstick

Pulsating variable stars are important to Galactic astronomy because once we recognize a star as being of the RR Lyrae or Cepheid type, we can infer its luminosity, and that in turn allows us to measure its distance. Comparing the star's (known) luminosity with its (observed) apparent brightness yields an estimate of its distance, by the inverse-square law. ∞ *(Sec. 10.2)* In this way, astronomers can use pulsating variables as a means of determining distances, both within our own Galaxy and far beyond.

How do we infer a variable star's luminosity? For RR Lyrae variables, this is simple. All such stars have basically the same luminosity (averaged over a complete pulsation cycle)—about 100 times that of the Sun. For Cepheids, we use a correlation between average luminosity and pulsation period, known as the **period–luminosity relationship.** In 1908, Henrietta Leavitt of Harvard College Observatory (see *Discovery 14-1*) discovered that Cepheids that vary slowly—that is, have long periods—have high luminosities, while short-period Cepheids have low luminosities. Figure 14.6 illustrates the period–luminosity relationship for Cepheids found within a thousand parsecs or so of Earth. Astronomers can plot such a diagram for nearby Cepheid variables because they can measure their distances, and hence their luminosities, using stellar or spectroscopic parallax. ∞ *Secs. 10.1 and 10.6* Thus, a simple measurement of a Cepheid variable's pulsation period immediately tells us its luminosity—we just read it off the graph in Figure 14.6. (The roughly constant luminosities of the RR Lyrae variables are also indicated in the figure.)

This distance-measurement technique works well provided the variable star can be clearly identified and its pulsation period measured. With Cepheids, this method allows astronomers to estimate distances out to about 25 million parsecs, more than enough to take us to the nearest galaxies. Indeed, the existence of galaxies beyond our own was first established in the late 1920s, when American astronomer Edwin Hubble observed Cepheids in the Andromeda galaxy and thereby succeeded in measuring its distance. The less luminous RR Lyrae stars are not as easily detectable as Cepheids, so their usable range is not as great. However, they are much more common, so within this limited range, they are actually more useful than Cepheids.

We began our cosmic distance ladder in Chapter 1 with radar ranging in the solar system. ∞ *(Sec. 1.3)* In Chapter 10, we extended it to include stellar and

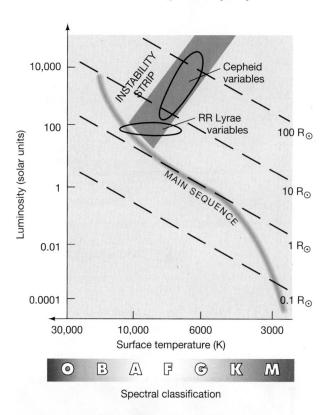

▲ **FIGURE 14.5 Variable Stars on the H–R Diagram** Pulsating variable stars are found in the instability strip of the H–R diagram. As a high-mass star evolves through the strip, it becomes a Cepheid variable. Low-mass horizontal-branch stars in the instability strip are RR Lyrae variables.

(MA) ANIMATION / VIDEO
Cepheid Variable Star in Distant Galaxy

CONCEPT CHECK

Can variable stars be used to map out the structure of the Galactic disk?

▲ **FIGURE 14.6 Period–Luminosity Relationship** A plot of pulsation period versus average absolute brightness (that is, luminosity) for a group of Cepheid variable stars. The two properties are tightly correlated. The pulsation periods of some RR Lyrae variables are also shown.

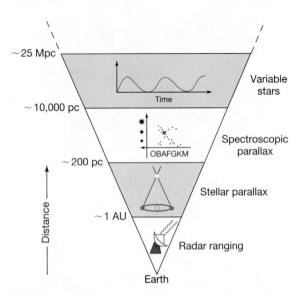

▲ FIGURE 14.7 **Variable Stars on Distance Ladder**
INTERACTIVE   Application of the period–luminosity relationship for Cepheid variable stars allows estimates of distances out to about 25 Mpc with reasonable accuracy.

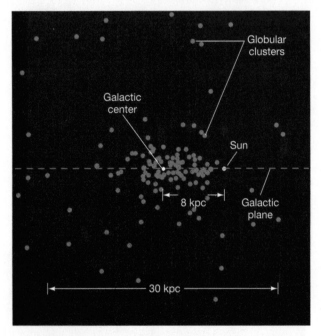

▲ FIGURE 14.8 **Globular Cluster Distribution**   Our Sun does not coincide with the center of the very large collection of globular clusters (indicated by the pink dots). Instead, more globular clusters are found in one direction than in any other. The Sun resides closer to the edge of the collection, which measures roughly 30 kpc across. The globular clusters outline the true distribution of stars in the Galactic halo.

PROCESS OF SCIENCE CHECK

Go to a library or the Web and research the Great Debate in 1920 between Harlow Shapley and Heber Curtis on the size of our Galaxy and scale of the universe. Which viewpoint was ultimately proved right, and what were the consequences for our understanding of the cosmos?

---

spectroscopic parallax. ∞ *(Secs. 10.1, 10.6)* Figure 14.7 expands it further by adding variable stars as a fourth method of determining distance.

The Size and Shape of Our Galaxy

Many RR Lyrae variables are found in globular clusters, those tightly bound swarms of old, reddish stars we first met in Chapter 11. ∞ *(Sec. 11.6)* Early in the 20th century, the American astronomer Harlow Shapley used observations of RR Lyrae stars to make two very important discoveries about the Galactic globular cluster system. First, he showed that most globular clusters reside at great distances—many thousands of parsecs—from the Sun. Second, by measuring the direction and distance of each cluster, he determined their three-dimensional distribution in space (Figures 14.8 and 14.9). In this way, Shapley demonstrated that the globular clusters map out a truly gigantic, and roughly *spherical*, volume of space, now known to be about 30 kpc across. However, the center of the distribution lies nowhere near our Sun. It is located nearly 8 kpc away from us, in the constellation Sagittarius.

In a brilliant intellectual leap, Shapley realized that the distribution of globular clusters maps out the true extent of stars in the Milky Way Galaxy—the region that we now call the Galactic halo. The hub of this vast collection of matter, 8 kpc from the Sun, is the **Galactic center.** As illustrated in Figure 14.9, we live in the suburbs of this huge ensemble, in the Galactic disk—the thin sheet of young stars, gas, and dust that cuts through the center of the halo. Since Shapley's time, astronomers have identified many individual stars—that is, stars not belonging to any globular cluster—within the Galactic halo.

Shapley's bold interpretation of the globular clusters as defining the overall structure of our Galaxy was an enormous step forward in human understanding of our place in the universe. Five hundred years ago Earth was considered the center of all things. Copernicus argued otherwise, demoting our planet to an undistinguished place far from the center of the solar system. In Shapley's time the prevailing view was that our Sun was the center not only of the Galaxy but also of the universe. Shapley showed otherwise. With his observations of globular clusters, he simultaneously increased the size of our Galaxy by almost a factor of 10 over earlier estimates and banished our parent Sun to its periphery, virtually overnight!

14.3 Galactic Structure

Based on optical, infrared, and radio studies of our Galaxy, Figure 14.9 illustrates the very different spatial distributions of the disk, bulge, and halo components of the Milky Way Galaxy.

Mapping the Galaxy

The extent of the halo in Figure 14.9 is based largely on optical observations of globular clusters and other halo stars. However, as we have seen, optical techniques can cover only a small portion of the dusty Galactic disk. Much of our knowledge of the structure of the disk on larger scales is based on radio observations, particularly of the 21-cm radio emission line produced by atomic hydrogen. ∞ *(Sec. 11.3)* Because long-wavelength radio waves are largely unaffected by interstellar dust, and hydrogen is by far the most abundant element in interstellar space, the 21-cm signals are strong enough that virtually the entire disk can be observed in this way.

14.1 DISCOVERY

Early Computers

A large portion of the early research in observational astronomy focused on monitoring stellar luminosities and analyzing stellar spectra. Much of this pioneering work was done using photographic methods. What is not so well known is that most of the labor was accomplished by women. Around the turn of the 20th century, a few dozen dedicated women—assistants at the Harvard College Observatory—created an enormous database by observing, sorting, measuring, and cataloging photographic information that helped form the foundation of modern astronomy. Some of them went far beyond their duties in the lab to make many of the basic astronomical discoveries often taken for granted today.

This 1910 photograph shows several of those women carefully examining star images and measuring variations in luminosity or wavelengths of spectral lines. In the cramped quarters of the Harvard Observatory, these women inspected image after image to collect a vast body of data on hundreds of thousands of stars. Note the plot of stellar luminosity changes pasted on the wall at the left. The cyclical pattern is so regular that it likely belongs to a Cepheid variable. Known as "computers" (there were no electronic devices then), these women were paid 25 cents an hour.

In 1880 these workers began a survey of the skies that would continue for half a century. Their first major accomplishment was a catalog of the brightnesses and spectra of tens of thousands of stars, published in 1890 under the direction of Williamina Fleming. On the basis of this compilation, several of these women

(Harvard College Observatory)

made fundamental contributions to astronomy. In 1897 Antonia Maury undertook the most detailed study of stellar spectra to that time, enabling Hertzsprung and Russell independently to develop what is now called the H–R diagram. In 1898 Annie Cannon proposed the spectral classification system (described in Chapter 10) that is now the international standard for categorizing stars. ∞ *(Sec. 10.5)* In 1908 Henrietta Leavitt discovered the period–luminosity relationship for Cepheid variable stars, which later allowed astronomers to recognize our Sun's true position in our Galaxy, as well as our Galaxy's true place in the universe.

According to the radio studies, the center of the gas distribution coincides roughly with the center of the globular cluster system, lying about 8 kpc from the Sun. In fact, this figure is derived most accurately from radio observations of Galactic gas. The densities of both stars and gas in the disk decline quite rapidly beyond about 15 kpc from the Galactic center (although some radio-emitting gas has been observed out to at least 50 kpc).

Perpendicular to the Galactic plane, the disk in the vicinity of the Sun is "only" about 300 pc thick, or about 1/100 of the 30-kpc Galactic diameter. Don't be fooled, though. Even if you could travel at the speed of light, it would take you 1000 years to traverse the

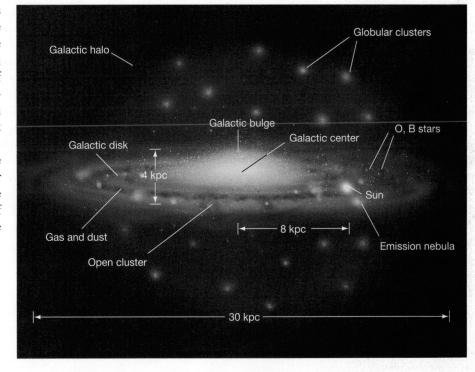

▶ **FIGURE 14.9 Stellar Populations in Our Galaxy**
INTERACTIVE   Artist's conception of a (nearly) edge-on view of the Milky Way Galaxy, showing schematically the distributions of young blue stars, open clusters, old red stars, and globular clusters. (The brightness and size of our Sun are greatly exaggerated for clarity.)

▲ **FIGURE 14.10 Infrared View of the Milky Way Galaxy**
INTERACTIVE A wide-angle infrared image of the disk and bulge of the Milky Way Galaxy, as observed by the Two Micron All Sky Survey. Compare Figure 14.2(c). *(Atlas Image)*

INTERACTIVE **MA**

Stellar Populations in a Star Cluster

DATA POINTS

Galactic Structure

More than one-third of students had difficulty describing and distinguishing the different regions of our Galaxy. Some points to remember:

- The disk is the youngest part of our Galaxy. It is circular and highly flattened, and it is the site of ongoing star formation.

- The halo is the oldest part of our Galaxy. It is approximately spherical and consists primarily of stars formed long ago.

- The bulge contains both young and old stars. It is not as flattened as the disk but flatter than the halo.

- Stars in the disk move in roughly circular, planar orbits around the Galactic center; stars in the bulge and halo also orbit the Galactic center but move in 3 dimensions in all directions, often traveling far above or below the disk.

thickness of the Galactic disk. The disk may be thin compared with the Galactic diameter, but it is huge by human standards.

Also shown in Figure 14.9 is our Galaxy's central bulge, measuring roughly 6 kpc across in the plane of the Galactic disk by 4 kpc perpendicular to that plane. Obscuration by interstellar dust makes it difficult to study the bulge at optical wavelengths. However, at longer wavelengths, which are less affected by interstellar matter, a much clearer picture emerges (compare Figure 14.10 with Figures 14.1b and 14.2c). Detailed measurements of the motion of gas and stars in and near the bulge imply that it is actually football shaped, about half as wide as it is long, with the long axis of the football lying in the Galactic plane. (In this case, our Galaxy is probably of the "barred-spiral" type—see Section 15.1.)

Stellar Populations

Aside from their shapes, the three components of the Galaxy—disk, bulge, and halo—have other properties that distinguish them from one another. First, the halo contains essentially *no* gas or dust—just the opposite of the disk and bulge, in which interstellar matter is common. Second, there are clear differences in the *colors* of the stars in the disk, bulge, and halo. Stars in the Galactic bulge and halo are distinctly redder than those found in the disk. Observations of other spiral galaxies also show this trend—the blue-white tint of the disk and the yellowish coloration of the bulge are evident in Figures 14.2(a) and (c).

All the bright, blue stars visible in our night sky are part of the Galactic disk, as are the young open star clusters and star-forming regions. In contrast, the cooler, redder stars—including those found in the old globular clusters—are more uniformly distributed throughout the disk, bulge, and halo. Galactic disks appear bluish because main-sequence O- and B-type blue supergiants are very much brighter than G-, K-, and M-type dwarfs, even though the dwarfs are present in far greater numbers.

The explanation for the marked difference in stellar content between disk and halo is that, whereas the gas-rich Galactic disk is the site of ongoing star formation and thus contains stars of all ages, all the stars in the Galactic halo are *old*. The absence of dust and gas in the halo means that no new stars are forming there, and star formation apparently ceased long ago—at least 10 billion years in the past, judging from the ages of the globular clusters. ∞ *(Sec. 12.6)* The gas density is very high in the inner part of the Galactic bulge, making this region the site of vigorous ongoing star formation, and both very old and very young stars mingle there. The bulge's gas-poor outer regions have properties more similar to those of the halo. As we will see in Chapter 15, All these statements are equally true for other spiral galaxies—young, bright stars are always found in the disk and inner bulge because the interstellar medium is densest there.

Spectroscopic studies indicate that halo stars are far less abundant in heavy elements (that is, elements heavier than helium) than are stars in the disk. Each successive cycle of star formation and evolution enriches the interstellar medium with the products of stellar evolution, leading to a steady increase in heavy elements with time. ∞ *(Sec. 12.7)* Thus, the scarcity of these elements in halo stars is consistent with the view that the halo formed long ago.

Astronomers often refer to young disk stars as *Population I* stars and old halo stars as *Population II* stars. The idea of two stellar "populations" dates back to the

▶ FIGURE 14.11 **Orbital Motion in the Galactic Disk** Stars and interstellar clouds in the neighborhood of the Sun show systematic Doppler motions, implying that the disk of the Galaxy spins in a well-ordered way. These four Galactic quadrants are drawn to intersect not at the Galactic center, but at the Sun, the location from which observations are made. Because the Sun orbits faster than stars and gas at larger radii, it moves away from material at top left and gains on that at top right, resulting in the Doppler shifts indicated. Likewise, stars and gas in the bottom left quadrant are gaining on us, while material at bottom right is pulling away.

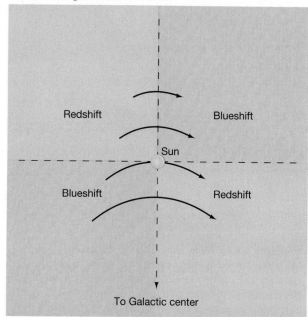

The curved arrows denote the speed of the disk material, which is greater closer to the center.

1930s, when the differences between disk and halo stars first became clear. It represents an oversimplification, as there is actually a continuous variation in stellar ages throughout the Milky Way Galaxy and not a simple division of stars into two distinct "young" and "old" categories. Nevertheless, the terminology is still widely used.

Orbital Motion

Are the internal motions of our Galaxy's members chaotic and random, or are they part of some gigantic "traffic pattern"? The answer depends on our perspective. The motion of stars and clouds we see on small scales (within a few tens of parsecs of the Sun) seems random, but on larger scales (hundreds or thousands of parsecs) the motion is much more orderly.

As we look around the Galactic disk in different directions, a clear pattern of motion emerges (see Figure 14.11). Radiation received from stars and interstellar gas clouds in the upper-right and the lower-left quadrants of Figure 14.11 is generally *blueshifted*. At the same time, radiation from stars and gas sampled in the upper-left and lower-right quadrants tends to be *redshifted*. In other words, some regions of the Galaxy (in the blueshifted directions) are approaching the Sun, while others (the redshifted ones) are receding from us. Careful study of the orbits of stars and gas clouds near the Sun leads us to the conclusion that the entire Galactic disk is *rotating* about the Galactic center. In the vicinity of the Sun, 8 kpc from the center, the orbital speed is about 220 km/s, so material takes about 225 million years to complete one circuit. At other distances from the center, the rotation period is different—shorter closer to the center, longer at greater distances—the Galactic disk rotates not as a solid object but *differentially*.

This picture of orderly circular orbital motion about the Galactic center applies only to the Galactic disk. The old globular clusters in the halo and the faint, reddish individual stars in both the halo and the bulge do *not* share the disk's well-defined rotation. Instead, as shown in Figure 14.12, their orbits are oriented randomly. Although they do orbit the Galactic center, these stars move in all directions, their paths filling an entire three-dimensional volume rather than a nearly two-dimensional disk.

At any given distance from the Galactic center, bulge or halo stars move in their orbits at speeds comparable to the disk's rotational speed at that radius. However, unlike stars in the disk, bulge and halo stars orbit in *all* directions, not just in a roughly circular path confined to a narrowly defined plane. Their orbits carry these stars repeatedly through the disk and out the other side. (They don't collide with stars in the disk because interstellar distances are huge compared with the diameters of individual stars—a star or even an entire star cluster passes through the disk almost as though it wasn't there—see *Discovery 15-1*.)

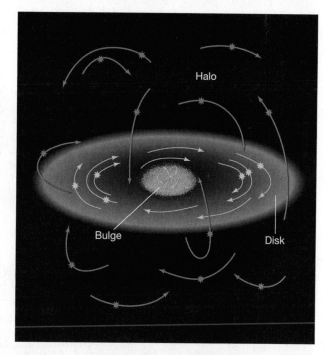

▲ FIGURE 14.12 **Stellar Orbits in Our Galaxy** Stars in the Galactic disk (blue curves) move in orderly, circular orbits about the Galactic center. In contrast, halo stars (orange curves) move randomly around the center. The orbit of a typical halo star takes it high above the Galactic disk, then down through the disk plane, then out the other side and far below the disk. Orbital properties of bulge stars are intermediate between those of the disk and halo stars.

CONCEPT CHECK

Why do astronomers regard the disk and halo as distinctly different components of our Galaxy?

14.4 Formation of the Milky Way

Table 14.1 compares some key properties of the three basic components of the Galaxy. Is there some evolutionary scenario that can naturally account for the structure we see today? The answer is yes, and it takes us all the way back to the birth of our Galaxy, more than 10 billion years ago. For simplicity we confine our discussion here to the Galactic disk and halo. In many ways the bulge is intermediate in its properties between these two extremes.

Figure 14.13 illustrates the current view of our Galaxy's evolution. Somewhat like the star formation scenario outlined in Chapter 11 discussed earlier, it starts from an extended cloud of pregalactic gas. ∞ *(Sec. 11.4)* When the first Galactic stars and globular clusters formed, the gas in our Galaxy had not yet accumulated into a thin disk. Instead, it was spread out over an irregular region of space spanning many tens of kiloparsecs. When the first stars formed, they were distributed throughout this volume (Figure 14.13b). Their distribution today (the Galactic halo) reflects that fact—it is an imprint of their birth. Most astronomers think that the very first stars formed even earlier, in smaller systems that later merged to create our Galaxy (Figure 14.13a; see also ∞ *Section 16.4*). Many more stars were likely born during the mergers themselves, as interstellar gas clouds collided and collapsed. The present-day halo would look much the same in either case.

Since those early times, rotation has flattened the gas in our Galaxy into a thin disk (Figure 14.13c). Physically, this process is similar to the flattening of the solar nebula during the formation of the solar system, except on a vastly larger scale. ∞ *(Sec. 4.3)* Star formation in the halo ceased billions of years ago when the raw materials—interstellar gas and dust—fell to the Galactic plane. Ongoing star formation in the disk gives it its bluish tint, but the halo's short-lived blue stars died long ago leaving only the long-lived red stars that give it its characteristic pinkish glow (Figure 14.13d). The Galactic halo is ancient; the disk is full of youthful activity.

Studies of the composition of stars in the Galactic disk suggest that the infall of halo gas is still going on today. The best available models of star formation and stellar evolution predict that the fraction of heavy elements in disk stars should be significantly *greater* than is actually observed, unless the gas in the disk is steadily being "diluted" by fresh gas arriving from the halo at a rate of perhaps 5–10 solar masses per year. ∞ *(Sec. 12.5)* Accumulated over billions of years, this amounts to a significant fraction of the total mass of the disk (see Section 14.6).

This theory also explains the chaotic orbits of the halo stars. When the halo developed, the irregularly shaped Galaxy was rotating very slowly, so there was no strongly preferred direction in which matter tended to move. As a result, halo stars were free to travel along nearly any path once they formed (or when their parent systems merged). As the Galactic disk formed, however, conservation of angular momentum caused it to spin more rapidly. Stars forming from the gas and dust of the disk inherit its rotational motion and so move on well-defined circular orbits.

▼ **FIGURE 14.13 Milky Way Galaxy Formation**   (a) The Milky Way Galaxy likely formed by a merger of several smaller systems. (b) Early on, our Galaxy was irregularly shaped, with gas distributed throughout its volume. When stars formed during this stage, their orbits carried them throughout an extended three-dimensional volume surrounding the newborn Galaxy. (c) In time, the gas and dust fell to the Galactic plane and formed a spinning disk. The stars that had already formed were left behind in the halo. (d) New stars forming in the disk inherit its overall rotation and so orbit the Galactic center on ordered, circular orbits.

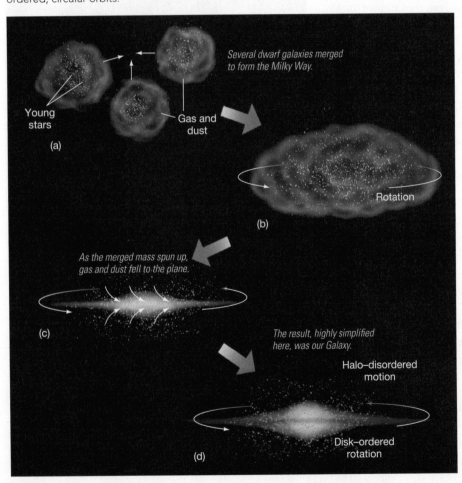

TABLE 14.1 Overall Properties of the Galactic Disk, Halo, and Bulge

Galactic Disk	Galactic Halo	Galactic Bulge
Highly flattened	roughly spherical—mildly flattened	somewhat flattened—elongated in the plane of the disk (football-shaped)
Contains both young and old stars	contains old stars only	contains both young and old stars; more old stars at greater distances from the center
Contains gas and dust	contains no gas and dust	contains gas and dust, especially in the inner regions
Site of ongoing star formation	no star formation during the last 10 billion years	ongoing star formation in the inner regions
Gas and stars move in circular orbits in the Galactic plane	stars have random orbits in three dimensions	stars have largely random orbits, but with some net rotation about the Galactic center
Spiral arms (Sec. 14.5)	little discernible substructure; globular clusters, tidal streams (Sec. 14.3)	ring of gas and dust near center; central Galactic nucleus (Sec. 14.7)
Overall white coloration, with blue spiral arms	reddish in color	yellow-white

In principle, the structure of our Galaxy bears witness to the conditions that created it. In practice, however, the sheer complexity of the system we inhabit means that the early stages of the Milky Way are still quite poorly understood. We will return to the subject of galaxy formation in Chapters 15 and 16.

CONCEPT CHECK

Why are there no young halo stars?

14.5 Galactic Spiral Arms

Radio studies provide perhaps the best direct evidence that we live in a spiral Galaxy. As illustrated in Figure 14.14, radio observations of interstellar gas, repeated in all directions in the Galactic plane, allow astronomers to map out its distribution and paint a detailed picture of the Galactic disk.

The Galaxy's differential rotation means that gas clouds at different distances are moving at different speeds with respect to Earth, so the radiation received from them is Doppler shifted by different amounts. Astronomers use all available data, coupled with knowledge of Newtonian mechanics, to construct a mathematical model of the rotation of stars and gas throughout the Galactic disk. ∞ *(Sec. 2.7)* The model allows us to turn a measured velocity into a distance along the line of sight. As in so many areas of astronomy, theory and observations complement one another—the data refine the theoretical model, while the model in turn provides the framework needed to understand and interpret further observations. ∞ *(Sec. 0.5)*

Figure 14.15 is an artist's conception (based on observational data) of the appearance of our Galaxy as seen from far above the disk. The figure clearly shows our Galaxy's **spiral arms,** pinwheel-like structures originating close to the Galactic bulge and extending outward throughout much of the Galactic disk. One of these arms wraps around a large part of

▼ **FIGURE 14.14 Gas in the Galactic Disk** Because the disk of our Galaxy is rotating differentially (inside faster than the outside), signals from different clumps of hydrogen matter along any line of sight are Doppler shifted by different amounts. Repeated observations in many different directions allow astronomers to map out the distribution of gas in our Galaxy.

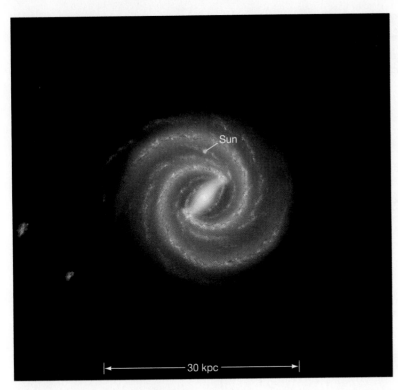

▲ FIGURE 14.15 **Milky Way Spiral Structure** An artist's conception of our Milky Way Galaxy seen face-on, based on radio and infrared maps of stars, gas, and dust in the Galactic disk. The illustration is painted from the perspective of an observer 100 kpc above the Galactic plane, showing the spiral arms emanating from a bar whose length is twice its width. Everything is drawn to scale (except for the oversized yellow dot near the top, which represents our Sun). The two small blotches to the left are dwarf galaxies, called the Magellanic Clouds, which are studied in Chapter 15. *(Adapted from JPL)*

INTERACTIVE **MA**

Spiral Arms and Star Formations

CONCEPT CHECK

Why can't spiral arms simply be clouds of gas and young stars orbiting the Galactic center?

the disk and contains our Sun. Notice, incidentally, the scale markers on Figures 14.8, 14.9, and 14.15. The Galactic globular-cluster distribution (shown in Figure 14.8), the luminous stellar component of the disk (Figure 14.9), and the known spiral structure (Figure 14.15) all have roughly the same diameter—about 30 kpc. This diameter is fairly typical of spiral galaxies observed elsewhere in the universe.

The spiral arms in our Galaxy are made up of much more than just interstellar gas and dust. Studies of the Galactic disk within a kiloparsec or so of the Sun indicate that young stellar and prestellar objects—emission nebulae, O- and B-type stars, and recently formed open clusters—are also distributed in a spiral pattern that closely follows the distribution of interstellar clouds. The obvious conclusion is that the spiral arms are the part of the Galactic disk where star formation takes place. The brightness of the young stellar objects just listed is the main reason that the spiral arms of other galaxies are easily seen from afar (Figure 14.2b).

A central problem in understanding spiral structure is explaining how that structure persists over long periods of time. The basic issue is simple: differential rotation makes it impossible for any large-scale structure "tied" to the disk material to survive. Figure 14.16 shows how a spiral pattern always consisting of the same group of stars and gas clouds would necessarily "wind up" and disappear within a few hundred million years. How then do the Galaxy's spiral arms retain their structure over long periods of time despite differential rotation? A leading explanation for the existence of spiral arms is that they are **spiral density waves**—coiled waves of gas compression that move through the Galactic disk, squeezing clouds of interstellar gas and triggering the process of star formation as they go. ∞ *(Sec. 11.4)* Like traffic slowing as it passes an obstacle on the highway (see *Discovery 14-2*), Galactic gas slows down and becomes denser as it passes through the wave. The spiral arms we observe are defined by the denser-than-normal clouds of gas thus created and by the new stars formed as a result of the spiral waves' passage.

This explanation of spiral structure avoids the problem of differential rotation because the wave pattern is not tied to any particular piece of the Galactic disk. The spirals we see are merely patterns moving through the disk, not matter being transported from place to place. The density wave moves through the collection of stars and gas making up the disk just as a sound wave moves through air or an ocean wave passes through water, compressing different parts of the disk at different times. Even though the rotation rate of the disk material varies with distance from the Galactic center, the wave itself remains intact, defining the Galaxy's spiral arms.

Over much of the visible portion of the Galactic disk (within about 15 kpc of the center), the spiral wave pattern is predicted to rotate *more slowly* than the stars and gas. Thus, as shown in Figure 14.17, Galactic material catches up with the wave, is temporarily slowed down and compressed by the wave's gravitational pull as it passes through, then continues on its way.

As the density wave enters the arm from behind, the gas is compressed and forms stars. Dust lanes mark the regions of highest-density gas. The most prominent stars—the bright O- and B-type blue giants—live for only a short time, so emission nebulae and young star clusters are found only within the arms, near their birth sites, just ahead of the dust lanes. Their brightness emphasizes the spiral structure. Further downstream, ahead of the spiral arms, we see mostly older stars and star clusters, which have had enough time since their formation to pull ahead of the wave. Over millions of years their random individual motions, superimposed on the overall rotation around the Galactic center, distort and

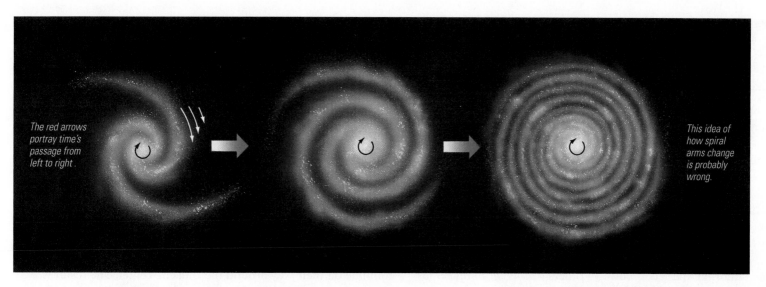

The red arrows portray time's passage from left to right.

This idea of how spiral arms change is probably wrong.

▲ FIGURE 14.16 **Differential Galactic Rotation** INTERACTIVE
The disk of our Galaxy rotates differentially, as shown by the small white arrows that represent the angular speed of the disk. If spiral arms were somehow tied to the material of the Galactic disk, such uneven rotation would cause the spiral pattern to wind up and disappear in a few hundred million years. Spiral arms would be too short-lived to be consistent with the numbers of spiral galaxies we observe today.

eventually destroy their original spiral configuration, and they become part of the general disk population.

Note, incidentally, that although the spirals shown in Figure 14.17 have two arms each, astronomers are not completely certain how many arms make up the spiral structure in our Galaxy. The theory makes no strong predictions on this point. As illustrated in Figure 14.15, the best available data suggest that our Galaxy has two major arms.

An alternative possibility is that the formation of stars drives the waves, instead of the other way around. Imagine a row of newly formed massive stars somewhere in the disk. The emission nebula created when these stars form, and the supernovae when they die, send shock waves through the surrounding gas, possibly triggering new star formation. ∞ *(Sec. 12.7)* Thus, as illustrated in Figure 14.18(a), the formation of one group of stars provides the mechanism for the creation of more stars. Computer simulations suggest that it is possible for the "wave" of star formation thus created to take on the form of a partial spiral and for this pattern to persist for some time. However, this process, known as **self-propagating star formation,** can produce only pieces of spirals, as are seen in some galaxies (Figure 14.18b). It apparently cannot produce the galaxywide spiral arms seen in other galaxies and present in our own. It may well be that there is more than one process at work in the spectacular spirals we see.

▶ FIGURE 14.17 **Spiral Density Waves**
Density-wave theory holds that the spiral arms seen in our own and many other galaxies are waves of gas compression and star formation moving through the material of the galactic disk. Gas enters an arm from behind, is compressed, and forms stars. The spiral pattern is delineated by dust lanes, regions of high gas density, and newly formed bright stars. The inset at right shows the spiral galaxy NGC 1566, which displays many of the features just described. *(AURA)*

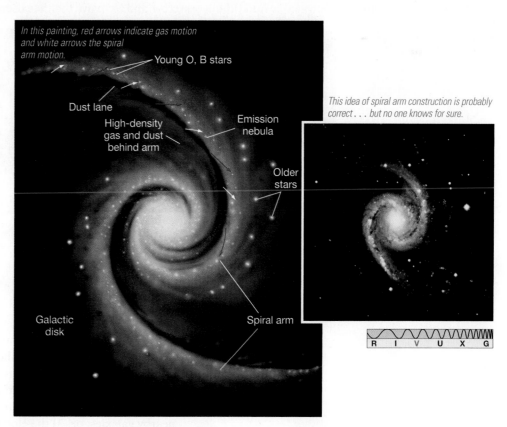

In this painting, red arrows indicate gas motion and white arrows the spiral arm motion.

Young O, B stars

Dust lane

High-density gas and dust behind arm

Emission nebula

Older stars

Galactic disk

Spiral arm

This idea of spiral arm construction is probably correct . . . but no one knows for sure.

R I V U X G

14.2 DISCOVERY

Density Waves

In the late 1960s, American astrophysicists C. C. Lin and Frank Shu proposed a way in which spiral arms in the Galaxy could persist for many Galactic rotations. They argued that the arms themselves contain no "permanent" matter. They should not be viewed as assemblages of stars, gas, and dust moving intact through the disk because such structures would quickly be destroyed by differential rotation. Instead, as described in the text, a spiral arm should be envisaged as a *density wave*—a wave of alternating compression and expansion sweeping through the Galaxy.

A wave in water builds up material temporarily in some places (crests) and lets it down in others (troughs). Similarly, as Galactic matter encounters a spiral density wave, the matter is compressed to form a region of higher than normal density. The matter enters the wave, is temporarily slowed down (by gravity) and compressed as it passes through, then continues on its way. This compression triggers the formation of new stars and nebulae. In this way, the spiral arms are formed and reformed repeatedly, without wrapping up.

The accompanying figure illustrates the formation of a density wave in a more familiar context—a traffic jam triggered by the presence of a repair crew moving slowly down the road. Cars slow down temporarily as they approach the crew, then speed up again as they pass the worksite and continue on their way. The result observed by a traffic helicopter flying overhead is a region of high traffic density concentrated around the work crew and moving with it. An observer on the side of the road, however, sees that the jam never contains the same cars for very long. Cars constantly catch up to the bottleneck, move slowly through it, then speed up again, only to be replaced by more cars arriving from behind.

The traffic jam is analogous to the region of high stellar density in a Galactic spiral arm. Just as the traffic density wave is not tied to any particular group of cars, the spiral arms are not attached to any particular piece of disk material. Stars and gas enter a spiral arm, slow down for a while, then exit the arm and continue on their way around the Galactic center. The result is a moving region of high stellar and gas density, involving different parts of the disk at different times. Notice also that, just as in our Galaxy, the traffic jam wave moves more slowly than, and independently of, the overall traffic flow.

We can extend our traffic analogy a little further. Most drivers are well aware that the effects of such a tie-up can persist long after the road crew has stopped work and gone home for the night. Similarly, spiral density waves can continue to move through the disk even after the disturbance that originally produced them has long since subsided. According to spiral density wave theory, this is precisely what has happened in the Milky Way Galaxy. Some disturbance in the past—an encounter with a satellite galaxy, perhaps, or the effect of the central bar—produced a wave that has been moving through the Galactic disk ever since.

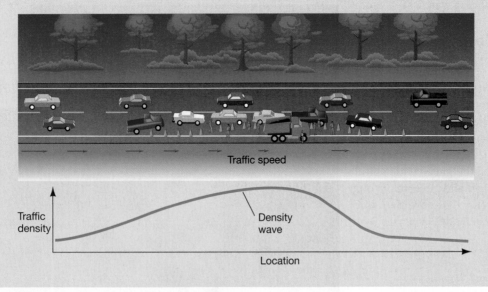

An important question, but one not answered by either of the two theories just described, is, Where do these spirals come from? What was responsible for generating the density wave in the first place or for creating the line of newborn stars whose evolution drives the advancing spiral arm? Scientists speculate that (1) instabilities in the gas near the galactic bulge, (2) the gravitational effects of nearby galaxies, or (3) the elongated shape of the bulge itself may have had a big

Again, time unfolds left to right, creating new stars over and over . . .

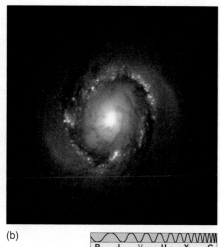

(a)

(b)

▲ **FIGURE 14.18 Self-Propagating Star Formation**   (a) In this theory of the formation of spiral arms, the shock waves produced by the formation and later evolution of a group of stars provide the trigger for new rounds of star formation. (b) This process may well be responsible for the partial spiral arms seen in some galaxies, such as NGC 4314, shown here in true color. Its distinct blue appearance derives from the vast numbers of young stars that pepper its ill-defined spiral arms. *(R. Gendler)*

enough influence on the disk to get the process going. The fact is that we still don't know for sure how galaxies—including our own—acquire such beautiful spiral arms.

14.6 The Mass of the Milky Way Galaxy

We can measure our Galaxy's mass by studying the motions of gas clouds and stars in the Galactic disk. Recall from Chapter 1 that Kepler's third law (as modified by Newton) connects the orbital period, orbit size, and masses of any two objects in orbit around one another:

$$\text{total mass (solar masses)} = \frac{\text{orbit size (astronomical units)}^3}{\text{orbit period (years)}^2}$$

The distance from the Sun to the Galactic center is about 8 kpc, and the Sun's orbital period is 225 million years. Plugging these numbers into the equation, we find a mass of almost 10^{11} solar masses—100 *billion* times the mass of our Sun.

But what mass have we just measured? When we performed the analogous calculation in the case of a planet orbiting the Sun, there was no ambiguity: Neglecting the planet's mass, the result of our calculation was the mass of the Sun. ∞ *(Sec. 1.4)* However, the Galaxy's matter is not concentrated at the Galactic center (as the Sun's mass is concentrated at the center of the solar system). Instead, Galactic matter is distributed over a large volume of space. What portion of the Galaxy's mass controls the Sun's orbit? Isaac Newton answered this question three centuries ago: The Sun's orbital period is determined by the portion of the Galaxy that lies *within the orbit of the Sun* (Figure 14.19). This is the mass computed from the above equation.

▼ **FIGURE 14.19 Weighing the Galaxy**   The orbital speed of a star or gas cloud moving around the Galactic center is determined only by the mass of the Galaxy lying inside the orbit (within the gray-shaded sphere). Thus, to measure the Galaxy's total mass, we must observe objects orbiting at large distances from the center.

Dark Matter

To determine the mass of the Galaxy on larger scales, we must measure the orbital motion of stars and gas at greater distances from the Galactic center. Astronomers have found that the most effective way to do this is to make radio observations of gas in the Galactic disk, because radio waves are relatively unaffected by interstellar absorption and allow us to probe to great distances. On the basis of these studies, radio astronomers have determined our Galaxy's rotation rate at various distances from the Galactic center. The resultant plot of rotation speed versus distance from the center (Figure 14.20) is called the Galactic **rotation curve.**

Using the Galactic rotation curve, we can now repeat our earlier calculation to compute the total mass that lies within any distance from the Galactic center. We find, for example, that the mass within about 15 kpc from the center—the volume defined by the globular clusters and the known spiral structure—is roughly 2×10^{11} solar masses, about twice the mass contained within the Sun's orbit. Does the distribution of matter in the Galaxy end at 15 kpc, where the luminosity drops off sharply? Surprisingly, the answer is no.

If all of the mass of the Galaxy were contained within the edge of the visible structure, Newton's laws of motion predict that the orbital speed of stars and gas beyond 15 kpc would decrease with increasing distance from the Galactic center, just as the orbital speeds of the planets diminish as we move outward from the Sun. The dashed line in Figure 14.20 indicates how the rotation curve should look in that case. However, the actual rotation curve is quite different. Far from declining at larger distances, it *rises* slightly out to the limits of our measurement capabilities. This implies that the amount of mass contained within successively larger radii continues to grow beyond the orbit of the Sun, apparently out to a distance of at least 50 kpc.

According to the equation above, the amount of mass within 50 kpc is approximately 6×10^{11} solar masses. Since 2×10^{11} solar masses lies within 15 kpc of the Galactic center, we have to conclude that at least twice as much mass lies *outside* the luminous part of our Galaxy—the part made up of stars, star clusters, and spiral arms—as lies inside!

Based on these observations, astronomers conclude that the luminous portion of the Milky Way Galaxy—the region outlined by the globular clusters and by the spiral arms—is merely the "tip of the Galactic iceberg." Our Galaxy is in reality very much larger. The luminous region is surrounded by an extensive, invisible **dark halo,** which dwarfs the inner halo of stars and globular clusters and extends well beyond the 15-kpc radius once thought to represent the limit of our Galaxy. But what is this dark halo made of? We do not detect enough stars or interstellar matter to account for the mass that our computations tell us must be there. We are inescapably drawn to the conclusion that most of the mass in our Galaxy (and, as we will see, of all galaxies) exists in the form of invisible **dark matter,** whose gravitational effects we can measure and quantify, but whose precise nature we do not understand.

The term *dark* here does not refer just to matter undetectable in visible light. The material has (so far) escaped detection at *all* electromagnetic wavelengths, from radio to gamma rays. Only by its gravitational pull do we know of its existence. Dark matter is not hydrogen gas (atomic or molecular), nor is it made up of ordinary stars. Given the amount of matter that must be accounted for, we would have been able to detect it with present-day equipment if it were in either of those forms. Its composition and its consequences for the

▲ FIGURE 14.20 **Galaxy Rotation Curve**   The rotation curve for the Milky Way Galaxy plots rotation speed against distance from the Galactic center. The dashed curve is the rotation curve expected if the Galaxy "ended" abruptly at a radius of 15 kpc, the limit of most of the known spiral structure. The fact that the red curve does not follow this dashed line, but instead stays well above it, indicates that there must be additional unseen matter beyond that radius.

evolution of galaxies and the universe are among the most important questions in astronomy today.

Many candidates have been suggested for this dark matter, although none is proven. Among the strongest "stellar" contenders are brown dwarfs and white dwarfs. ∞ *(Secs. 11.5, 12.3)* These objects could in principle exist in great numbers throughout the Galaxy, yet would be exceedingly hard to see. Note that the ultimate "dark" candidates—black holes—are not thought to contribute much to the dark matter total, simply because the massive stars that produce them are too rare. Only about 1 star in 10,000 ends up as a black hole.

A radically different alternative is that the dark matter is made up of exotic *subatomic particles* that pervade the entire universe. In order to account for the properties of dark matter, these particles must have mass (to produce the observed gravitational effects) but otherwise interact hardly at all with "normal" matter (because otherwise we would be able to see them).[1] Many astrophysicists think that such particles could have been produced in abundance during the very earliest moments of our universe. If they survived to the present day, there might be enough of them to account for all the dark matter we think must be out there. These ideas are hard to test, however, because such particles would be very difficult to detect. Several detection experiments have been attempted, so far without success.

A few astronomers have proposed a very different explanation for the "dark matter problem," suggesting that its resolution may lie not in the nature of dark matter, but rather in a modification to Newton's law of gravity that increases the gravitational force on very large (galactic) scales, doing away with the need for dark matter in the first place. The vast majority of scientists do *not* accept this view, but the very fact that it has been proposed—and is being seriously discussed in (some) scientific circles—underscores our current level of uncertainty. Dark matter is one of the great unsolved mysteries in astronomy today.

The Search for Stellar Dark Matter

Researchers have gained insight into the distribution of stellar dark matter by using a key element of Albert Einstein's theory of general relativity—the prediction that a beam of light can be deflected by a gravitational field, which has already been verified in the case of starlight passing close to the Sun. ∞ *(More Precisely 13-2)* Although this effect is small, it has the potential for making otherwise invisible stellar objects observable from Earth. Here's how.

Imagine looking at a distant star as a faint foreground object (such as a brown dwarf or a white dwarf) happens to cross your line of sight. As illustrated in Figure 14.21, the intervening object deflects a little more starlight than usual toward you, resulting in a temporary, but quite substantial, *brightening* of the distant star. Because the effect is in some ways like the focusing of light by a lens, this process is known as **gravitational lensing.** The foreground object is referred to as a *gravitational lens*. The amount of brightening and the duration of the effect depend on the mass, distance, and speed of the lensing object. Typically, the apparent brightness of the background star increases by a factor of 2 to 5 for a period of several weeks. Thus, even though the foreground object cannot be seen directly, its effect on the light of the background star makes it detectable.

Of course, the probability of one star passing almost directly in front of another, as seen from Earth, is extremely small. However, by observing millions of

SELF-GUIDED TUTORIAL
Gravitational Lensing

---

[1]One class of candidate particles satisfying these requirements has been dubbed *W*eakly *I*nteracting *M*assive *P*articles, or WIMPs. Not to be outdone, astronomers searching for the more mundane stellar dark matter have labeled them *MA*ssive *C*ompact *H*alo *O*bjects, or MACHOs. Who says astronomers don't have a sense of humor?

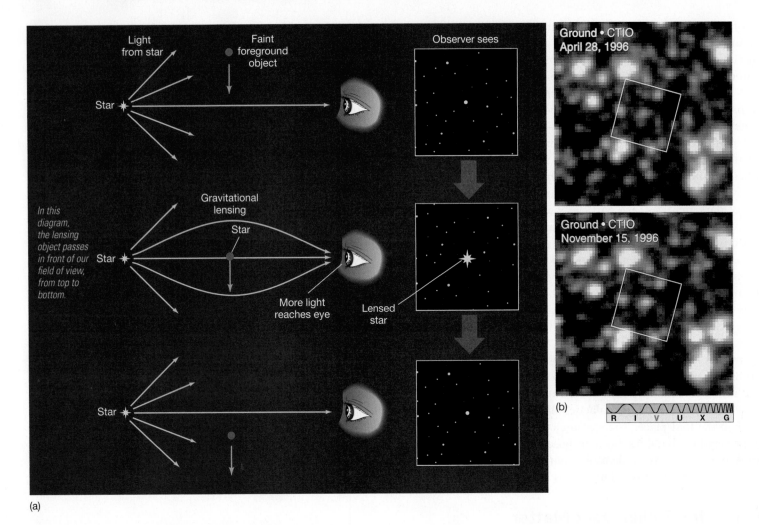

(a)

▲ **FIGURE 14.21 Gravitational Lensing** (a) Gravitational lensing by a faint foreground object (such as a brown dwarf) can temporarily cause a background star to brighten significantly, providing a means of detecting otherwise invisible stellar dark matter. (b) These two images show the brightening of a star during a lensing event, this one implying that a massive, but unseen, object passed in front of the unnamed star at the center of the two boxes imaged 6 months apart. *(AURA)*

CONCEPT CHECK

In what sense is dark matter "dark"?

stars every few days over a period of years (using automated telescopes and high-speed computers to reduce the burden of coping with so much data), astronomers have so far seen thousands of events. The observations are consistent with lensing by low-mass white dwarfs and suggest that such stars could account for at least half—but apparently not all—of the Galactic dark matter inferred from dynamical studies.

Bear in mind, though, that the identity of the dark matter is not necessarily an "all-or-nothing" proposition. It is quite possible that more than one type of dark matter exists. For example, most of the dark matter in the inner (visible) parts of galaxies could be in the form of low-mass stars, while the dark matter farther out might be primarily in the form of exotic particles. We will return to this perplexing problem in later chapters.

14.7 The Galactic Center

Theory predicts that the Galactic bulge, and especially the region close to the Galactic center, should be densely populated with billions of stars. However, we are unable to see this region of our Galaxy—the interstellar medium in the Galactic disk shrouds what otherwise would be a stunning view. Figure 14.22 shows the (optical) view we do have of the region of the Milky Way toward the Galactic center, in the general direction of the constellation Sagittarius.

Galactic Activity

With the help of infrared and radio techniques we can peer more deeply into the central regions of our Galaxy than we can by optical means. Infrared observations (Figure 14.22 inset) show that the innermost parsec of our Galaxy harbors a dense cluster containing roughly 1 million stars. That's a stellar density about 100 *million* times greater than in our solar neighborhood, high enough that stars must experience frequent close encounters and even collisions.

Figure 14.23(a) is an infrared view of part of Figure 14.22, with the Galactic plane now horizontal. On this scale, infrared radiation has been detected from what appear to be huge clouds rich in dust. Radio observations indicate a ring of molecular gas nearly 400 pc across, containing hundreds of thousands of solar masses of material and rotating around the Galactic center at about 100 km/s. Its origin is unclear, although researchers think that the gravity of the Galaxy's central rotating bar may deflect gas from farther out into the dense central regions.

Higher-resolution radio studies show much more structure. Figure 14.23(b) shows a region called Sagittarius A. (The name just means that it is the brightest radio source in Sagittarius.) It lies at the center of the boxed region in Figures 14.22 and 14.23(a) and, we think, at the center of our Galaxy. On a scale of about 25 pc, extended filaments can be seen. Their presence suggests to many astronomers that strong magnetic fields operate near the center, creating structures similar in appearance to (but much larger than) those observed on the active Sun.

On even smaller scales (Figure 14.23c), *Chandra* observations indicate an extended region of hot X-ray-emitting gas, apparently associated with a supernova remnant, in addition to many other individual bright X-ray sources. Within that region lies a rotating ring of star-forming molecular gas only a few parsecs across, with streams of matter spiraling inward toward the center (Figure 14.23d).

What could cause all this activity? An important clue comes from the Doppler broadening of infrared spectral lines emitted from the central swirling whirlpool of gas. The extent of the broadening indicates that the gas is moving very rapidly. In order to keep this gas in orbit, whatever is at the center must be extremely massive—several million solar masses. Given the twin requirements of large mass and small size, a leading contender is a *supermassive black hole*. The strong magnetic fields are thought to be generated within the accretion disk around the black hole as matter spirals inward and may act as "particle accelerators," creating the extremely high-energy particles detected on Earth as *cosmic rays*. ∞ *(Sec. 13.3)*

INTERACTIVE
Spiral Galaxy M81

▼ FIGURE 14.22 **Galactic Center INTERACTIVE** Photograph of stellar and interstellar matter in the direction of the Galactic center. The field is roughly 20°, top to bottom, and is a continuation of the bottom part of Figure 11.5. The overlaid box indicates the location of the center of our Galaxy. The inset shows an adaptive-optics infrared view of the dense stellar cluster surrounding the Galactic center, whose very core is indicated by the twin arrows. *(AURA; ESO)*

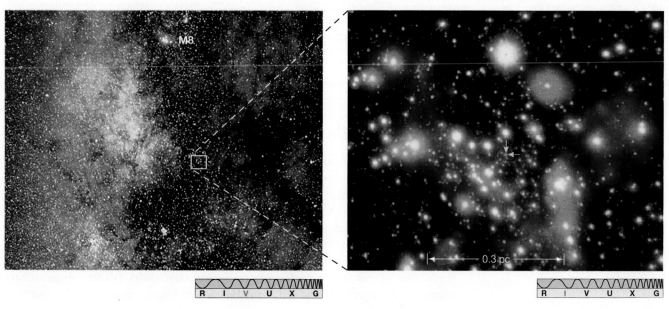

R I V U X G

R I V U X G

▲ **FIGURE 14.23 Galactic Center Close-up INTERACTIVE** (a) An infrared image of the region around the center of our Galaxy shows many bright stars packed into a relatively small volume. (b) The central portion of our Galaxy, as observed in the radio part of the spectrum, shows a region about 100 pc across surrounding the Galactic center (which lies within the orange-yellow region at bottom right). The long-wavelength radio emission cuts through the Galaxy's dust, providing a view of matter in the immediate vicinity of the Galaxy's center. (c) This *Chandra* X-ray image shows the relation of a hot supernova remnant (red) and Sgr A*, the suspected black hole at the very center of our Galaxy. (d) The spiral pattern of radio emission arising from Sagittarius A itself suggests a rotating ring of matter only a few parsecs across. All images are false color, since they lie outside the visible spectrum. *(SST; NRAO; NASA)*

The Central Black Hole

At the very center of our Galaxy, at the heart of Sagittarius A, lies a remarkable object with the odd-sounding name **Sgr A*** (pronounced "saj A star"). By the standards of the active galaxies to be studied in Chapter 15, this compact **Galactic nucleus** is not particularly energetic. Still, radio observations made during the past two decades, along with more recent X- and gamma-ray observations, suggest that it is nevertheless a pretty violent place. Its total energy output (at all wavelengths) is estimated to be W, which is more than a million times that of the Sun. Very-long-baseline interferometry (VLBI) observations using radio telescopes arrayed from Hawaii to Massachusetts imply that Sgr

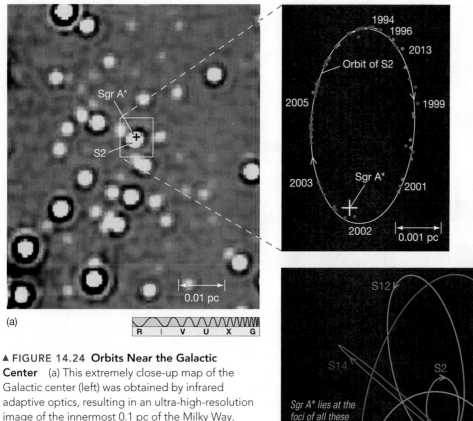

▲ FIGURE 14.24 Orbits Near the Galactic Center (a) This extremely close-up map of the Galactic center (left) was obtained by infrared adaptive optics, resulting in an ultra-high-resolution image of the innermost 0.1 pc of the Milky Way. The inset shows the orbit of the innermost star in the frame, labeled S2, between 1992 and 2003. The solid line shows the best-fitting orbit for S2 around a black hole of 4 million solar masses, located at Sgr A* (marked with a cross). (b) The orbits of several stars near the Galactic center share a common focus—Sgr A*. *(ESO)*

A* cannot be much larger than 10 AU, and it is probably a good deal smaller than that. ∞ *(Sec. 3.4)* This size is consistent with the view that the energy source is a massive black hole. Figure 14.24 is perhaps the strongest evidence to date supporting the black hole picture. It shows a high-resolution infrared image of an 0.04-pc (8000 AU) field near the Galactic center, centered on Sgr A*. Using advanced adaptive-optics techniques on the Keck telescope and the VLT, two teams of researchers from the United States and Europe have created the first-ever diffraction-limited (0.05 resolution) images of the region. ∞ *(Sec. 3.3)*

Remarkably, the image quality is good enough that the *proper motions* of several of the stars—their orbits around the Galactic center—can clearly be seen. The inset shows a series of observations of one of the brightest stars—called S2—over a 10-year period. The motion is consistent with an orbit around a massive object at the location of Sgr A*, in accordance with Newton's laws of motion. ∞ *(Sec. 1.4)* The solid curve on the figure shows the elliptical orbit that best fits the observations. It corresponds to a 15-year orbit with a semimajor axis of 950 AU, corresponding (from Kepler's third law, as modified by Newton) to a

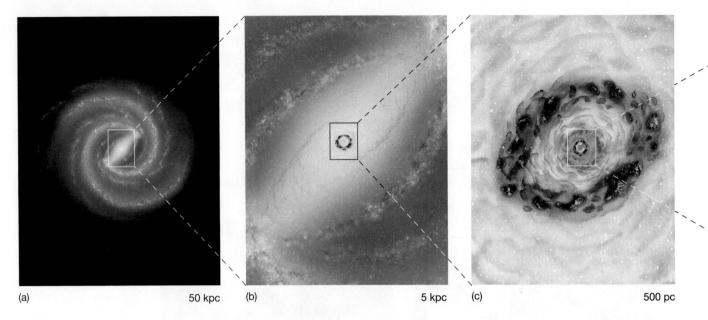

(a)   50 kpc   (b)   5 kpc   (c)   500 pc

▲ FIGURE 14.25 **Galactic Center Zoom**   This series of artist's conceptions of the Galactic center depicts each frame increasing in resolution by a factor of 10. Frame (a) is the same scene as Figure 14.15. Frame (f) is a rendition of a vast whirlpool within the innermost 0.5 parsec of our Galaxy. The data imaged in Figure 14.23 do not closely match these artistic renderings because the Figure 14.23 view is parallel to the Galactic disk—along the line of sight from the Sun to the Galactic center—whereas these six paintings portray a simplified view perpendicular to the disk, while progressively zooming down onto that disk.

central mass of 4 million solar masses. The small size of the central object is clearly demonstrated by the orbit of another star (S16), whose extremely eccentric orbit brings it within 45 AU of the center. Note that, even with this large mass, if Sgr A* is indeed a genuine black hole, the size of its event horizon is still only 0.08 AU. ∞ *(Sec. 13.5)* Such a small region, 8 kpc away, is currently unresolvable, although radio astronomers are hopeful that improving VLBI techniques will allow them to "see" the event horion and study the surrounding accretion disk within the next decade.

Figure 14.25 places these findings into a simplified perspective—all frames are artist's conceptions, but are based on real data. Each frame is centered on the Galaxy's core, with an increase in resolution of a factor of 10 from one frame to the next. Frame (a) renders the Galaxy's overall shape, as shown in Figure 14.15. This frame measures about 50 kpc across. Frame (b) spans a distance of 5 kpc and is nearly filled by the Galactic bar and the great circular sweep of the innermost spiral arm. Moving in to a 500-pc span, frame (c) depicts the 400-pc ring of gas mentioned earlier and some young dense star clusters, evidence of recent star formation near the Galactic center. The dark blobs represent giant molecular clouds, the pink patches emission nebulae associated with star formation within them.

In frame (d), at 50 pc, a pinkish (thin, warm) region of ionized gas surrounds the reddish (thicker, warmer) heart of the Galaxy. The energy responsible for this vast ionized region comes from frequent supernovae and other violent activity in the Galactic center. The central star cluster (diluted in

Violent stellar activity in the Galactic center is driving high-speed flows of gas out of the Galactic core.

|← 5000 pc →|

R I V U X G

◀ FIGURE 14.26 **Galactic Outflow**   Radio observations reveal a huge outflow of gas (blue) from the Galactic center, here superimposed on an optical image of the Galactic disk, most likely driven by supernovae in the dense Galactic core.

(d)   50 pc   (e)   5 pc   (f)   0.5 pc

the painting for clarity) and the surrounding star-forming ring can also be seen. Recent multiwavelength observations reveal that this intense stellar activity has blasted huge (10 kpc long) magnetized jets of high-energy particles out of the Galactic center, roughly perpendicular to the disk (Figure 14.26). The total energy in the jets exceeds that of a typical supernova by about a factor of a million. Astronomers suspect that similar events are occurring at the centers of many other galaxies. Also shown in frame (d) are numerous young dense star clusters, further evidence of recent bursts of star formation near the Galactic center. Frame (e), spanning 5 pc, depicts the ring in more detail as well as the central cluster, along with the tilted, spinning whirlpool of hot (10^4 K) gas surrounding the center of our Galaxy. The innermost part of this gigantic whirlpool is shown in frame (f), in which a swiftly spinning, white-hot disk of gas with temperatures in the millions of kelvins nearly engulfs the central black hole (represented here as a black dot). Two rings of stars, possibly the remains of disrupted star clusters, have also been detected, as roughly sketched in frame (f). The black hole itself, and the stellar orbits shown in Figure 14.24, are far too small to be pictured on this scale.

The last decade has seen an explosion in our knowledge of the innermost few parsecs of our Galaxy, and astronomers are working hard to decipher the clues hidden within its invisible radiation. Still, we are only now beginning to appreciate the full complexity of this strange realm deep in the heart of the Milky Way.

CONCEPT CHECK

What is the most likely explanation of the energetic events observed at the Galactic center?

THE BIG QUESTION

How big is the Milky Way? How well do we know the size, shape, and mass of our colossal home in the cosmos? In recent years, astronomers have reexamined our system of stars, gas, and dark matter: Its total mass has been upgraded by nearly a factor of ten, its extended halo might reach halfway to the nearest galaxy, and its dark matter outweighs its normal matter by at least a factor of five. Even so, have we seriously underestimated the scale of this grand system?

CHAPTER REVIEW

SUMMARY

LO1 A **galaxy** (p. 392) is a huge collection of stellar and interstellar matter isolated in space and bound together by its own gravity. Because we live within it, the **Galactic disk** (p. 392) of our **Milky Way Galaxy** (p. 392) appears as a broad band of light across the sky—the Milky Way. Near the center, the Galactic disk thickens into the **Galactic bulge** (p. 392). The disk is surrounded by a roughly spherical **Galactic halo** (p. 392) of old stars and star clusters. Our Galaxy, like many others visible in the sky, is a spiral galaxy. Disk and halo stars differ in their spatial distributions, ages, colors, and orbital motion. The luminous portion of our Galaxy has a diameter of about 30 kpc. In the vicinity of the Sun, the Galactic disk is about 300 pc thick.

LO2 The halo can be studied using **variable stars** (p. 394), whose luminosity changes with time. Two types of pulsating variable stars of particular importance to astronomers are **RR Lyrae variables** (p. 394) and **Cepheid variables** (p. 394). All RR Lyrae stars have roughly the same luminosity. For Cepheids, the luminosity can be determined using the **period–luminosity relationship** (p. 395). Knowing the luminosity, astronomers apply the inverse-square law to find the distance. The brightest Cepheids can be seen at distances of millions of parsecs, extending the cosmic distance ladder well beyond our own Galaxy. In the early 20th century, Harlow Shapley used RR Lyrae stars to determine the distances to many of the Galaxy's globular clusters and found that they have a roughly spherical distribution in space, but the center of the sphere lies far from the Sun—it is close to the **Galactic center** (p. 396), about 8 kpc away.

LO3 Stars and gas within the Galactic disk move on roughly circular orbits around the Galactic center. Stars in the halo and bulge move on largely random three-dimensional orbits that pass repeatedly through the disk plane but have no preferred orientation.

LO4 The halo lacks gas and dust, so no stars are forming there. All halo stars are old. The gas-rich disk is the site of current star formation and contains many young stars. Halo stars appeared early on, before the Galactic disk took shape, when there was still no preferred orientation for their orbits. As the gas and dust formed a rotating disk, stars that formed in the disk inherited its overall spin and so moved on circular orbits in the Galactic plane, as they do today.

LO5 Radio observations clearly reveal the extent of our Galaxy's **spiral arms** (p. 401), regions of the densest interstellar gas where star formation is taking place. The spirals cannot be "tied" to the disk material, as differential rotation would have wound them up long ago. Instead, they may be **spiral density waves** (p. 402) that move through the disk, triggering star formation as they pass by. Alternatively, the spirals may arise from **self-propagating star formation** (p. 403), when shock waves produced by the formation and evolution of one generation of stars trigger the formation of the next.

LO6 The Galactic **rotation curve** (p. 406) plots the orbital speed of matter in the disk against distance from the Galactic center. By applying Newton's laws of motion, astronomers can determine the mass of the Galaxy. The Galactic mass continues to increase beyond the radius defined by the globular clusters and the spiral structure we observe, indicating that our Galaxy has an invisible **dark halo** (p. 406). The **dark matter** (p. 406) making up this dark halo is of unknown composition. Candidates include low-mass stars and exotic subatomic particles. Recent attempts to detect stellar dark matter have used the fact that a faint foreground object can occasionally pass in front of a more distant star, deflecting the star's light and causing its apparent brightness to increase temporarily. This deflection is called **gravitational lensing** (p. 407).

LO7 Astronomers working at infrared and radio wavelengths have uncovered evidence for energetic activity within a few parsecs of the Galactic center. The leading explanation is that a black hole roughly 4 million times more massive

than the Sun resides there. The hole lies at the center of a dense star cluster containing millions of stars, which is in turn surrounded by a star-forming disk of molecular gas. The observed activity is thought to be powered by accretion onto the black hole, as well as by supernova explosions in the cluster surrounding it.

MasteringAstronomy®   For instructor-assigned homework go to www.masteringastronomy.com

Problems labeled **POS** explore the process of science. **VIS** problems focus on reading and interpreting visual information. **LO** connects to the introduction's numbered Learning Outcomes.

REVIEW AND DISCUSSION

1. **LO1 POS** What do globular clusters tell us about our Galaxy and our place within it?

2. How are Cepheid variables used in determining distances?

3. How far away can we use Cepheids to measure distance?

4. **LO2** What important discoveries were made early in this century using RR Lyrae variables?

5. Why can't optical astronomers easily study the center of our Galaxy?

6. Why are radio studies often more useful in the study of Galactic structure than observations made in visible light?

7. **LO3** Contrast the motions of disk and halo stars.

8. **LO5 POS** Explain why Galactic spiral arms are thought to be regions of recent and ongoing star formation.

9. Describe what happens to interstellar gas as it passes through a spiral density wave.

10. What is self-propagating star formation?

11. **LO4** What do halo stars tell us about the history of our Galaxy?

12. What does a galaxy's rotation curve tell us about its total mass?

13. **LO6 POS** What evidence is there for dark matter in the Galaxy?

14. Describe some candidates for Galactic dark matter.

15. **LO7 POS** Why do astronomers think that a supermassive black hole lies at the center of the Milky Way Galaxy?

CONCEPTUAL SELF-TEST: TRUE OR FALSE?/MULTIPLE CHOICE

1. Herschel's attempt to map the Milky Way by counting stars led to an inaccurate estimate of the Galaxy's size because he was unaware of absorption by interstellar dust. (T/F)

2. Cepheid variables can be used to determine the distances to the nearest galaxies. (T/F)

3. Globular clusters trace out the structure of the Galactic disk. (T/F)

4. The Galactic halo contains as much gas and dust as the disk. (T/F)

5. The Galactic disk contains only old stars. (T/F)

6. Stars and gas in the Galactic disk move in roughly circular orbits around the Galactic center. (T/F)

7. We can use 21-cm radiation to study molecular clouds. (T/F)

8. The most likely explanation of the high-speed motion of stars and gas near the Galactic center is that they are orbiting a supermassive black hole. (T/F)

9. In the Milky Way Galaxy, our Sun is located (a) near the Galactic center; (b) about halfway out from the center; (c) at the outer edge; (d) in the halo.

10. A telescope searching for newly formed stars would make the most discoveries if it were pointed (a) directly away from the Galactic center; (b) perpendicular to the Galactic disk; (c) within a spiral arm; (d) between spiral arms.

11. The first stars that formed in the Milky Way now (a) have chaotic orbits; (b) orbit in the Galactic plane; (c) orbit near the Galactic center; (d) orbit in the same direction as the Milky Way spins.

12. Stars in the outermost regions of the Milky Way Galaxy (a) are youngest; (b) orbit faster than astronomers would expect based on the Galactic mass we can see; (c) are more likely to explode as supernovae; (d) are more luminous than other stars.

13. Most of the mass of the Milky Way exists in the form of (a) stars; (b) gas; (c) dust; (d) dark matter.

14. VIS In Figure 14.6 (Period–Luminosity Relationship), a Cepheid variable star with luminosity 1000 times that of the Sun has a pulsation period of roughly (a) 1 day; (b) 3 days; (c) 10 days; (d) 50 days.

15. VIS Figure 14.20 (Galaxy Rotation Curve) tells us that (a) the Galaxy rotates like a solid body; (b) far from the center, the Galaxy rotates more slowly than we would expect based on the light we see; (c) far from the center, the Galaxy rotates more rapidly than we would expect based on the light we see; (d) there is no matter beyond about 15 kpc from the Galactic center.

PROBLEMS

The number of squares preceding each problem indicates its approximate level of difficulty.

1. ■ At one time some astronomers claimed that the Andromeda "nebula" (Figure 14.2a) was a star-forming region lying within our Galaxy. Calculate the angular diameter of a prestellar nebula of radius 100 AU, lying 100 pc from Earth. How close to Earth would the nebula have to lie for it to have the same angular diameter (roughly 6°) as Andromeda?

2. ■ What is the greatest distance at which an RR Lyrae star of absolute magnitude 0 could be seen by a telescope capable of detecting objects as faint as 20th magnitude? ∞ *(Sec. 10.2)*

3. ■ A typical Cepheid variable is 100 times brighter than a typical RR Lyrae star. On average, how much farther away than RR Lyrae stars can Cepheids be used as distance-measuring tools?

4. ■■ An astronomer using *HST* can just see a star with solar luminosity at a distance of 100,000 pc. The brightest Cepheids are 30,000 times more luminous than the Sun. If the Sun's absolute magnitude is 5, calculate the absolute magnitudes of these bright Cepheids. Neglecting interstellar absorption, how far away can *HST* see them?

5. ■■ Calculate the proper motion (in arc seconds per year) of a globular cluster with a transverse velocity (relative to the Sun) of 200 km/s and a distance of 3 kpc. Might this motion be measurable? ∞ *(Sec. 10.1)*

6. ■■ Calculate the total mass of the Galaxy lying within 20 kpc of the Galactic center if the rotation speed at that radius is 240 km/s.

7. ■■■ Using the data presented in Figure 14.20, calculate how long it takes the Sun to "lap" stars orbiting 15 kpc from the Galactic center.

8. ■ A two-arm spiral density wave is moving through the Galactic disk. At the 8-kpc radius of the Sun's orbit around the Galactic center, the wave's speed is 120 km/s, and the Galactic rotation speed is 220 km/s. How many times has the Sun passed through a spiral arm since the Sun formed 4.6 billion years ago?

9. ■■ Given the data in the previous question and the fact that O-type stars live at most 10 million years before exploding as supernovae, calculate the maximum distance at which an O-type star (orbiting at the Sun's distance from the Galactic center) can be found from the density wave in which it formed.

10. ■■■ Material at an angular distance of 0.2 from the Galactic center is observed to have an orbital speed of 1200 km/s. If the Sun's distance to the Galactic center is 8 kpc and the material's orbit is circular and is seen edge-on, calculate the radius of the orbit and the mass of the object around which the material is orbiting.

ACTIVITIES

Collaborative

1. Construct your own electronic version of the Messier catalog, listing the names, types, and coordinates of all 110 Messier objects. (You should be able to find all the needed information on the Web, although you may end up combining information from more than one source.) Use a spreadsheet to plot the celestial coordinates—right ascension and declination—of all objects. *(Sec. 0.1)* Color-code the objects to distinguish among emission nebulae, young star clusters, older open star clusters, globular clusters, and galaxies. What do you notice about the distributions of these objects on the sky? It might help to sketch the approximate location of the Milky Way (more research!) on the plot. Why are the young objects mainly found in the plane of the Milky Way. What about the globular clusters? Why do you think the galaxies apparently avoid the Galactic plane? If you have time, add deep-sky objects from other catalogs to your plot.

Individual

1. If you are far from city lights, look for a hazy band of light arching across the sky. This is our edgewise view of the Milky Way Galaxy. The Galactic center is located in the direction of the constellation Sagittarius, highest in the sky in summer, but visible from spring through fall. Look at the band making up the Milky Way and sketch what you see. Look for dark regions and faint fuzzy spots in the Milky Way; note their positions on your sketch. Add the major constellations for reference. Compare your sketch with a map of the Milky Way in a star atlas. Can you identify the dark regions and faint fuzzy spots?

1. Observe the Andromeda galaxy, M31. It's the most distant object visible to the naked eye, but don't expect to see anything like Figure 14.2(a)! To the unaided eye, from all but the darkest sites, only its nucleus will be visible, looking like a slightly fuzzy star. To locate M31, find Polaris, the pole star, and the constellations Cassiopeia and Andromeda. Follow a line from Polaris through the second "V" in the "W" of Cassiopeia, and continue south. That line will pass through M31 before you reach the northern arc of stars in Andromeda. Use binoculars or a wide-angle eyepiece to view the galaxy and its disk. Switch to higher magnification to view the nucleus and the small satellite galaxies M32, just to its south, and M110, to the northwest.

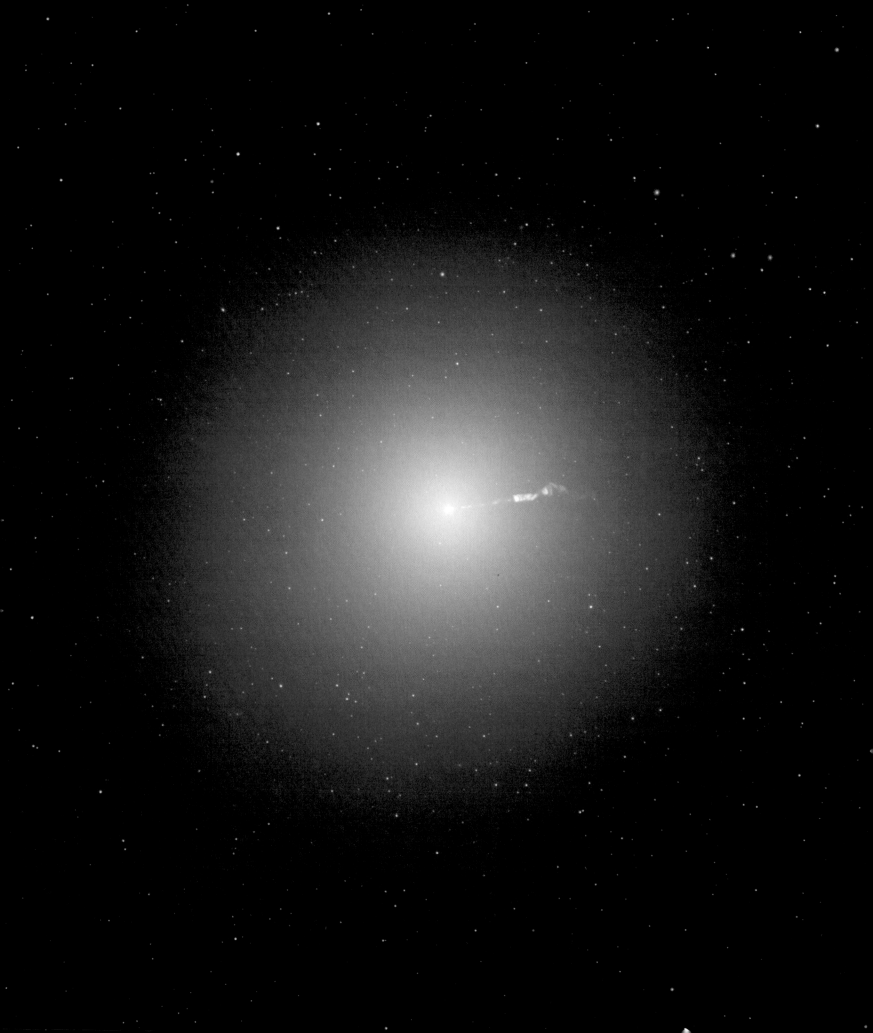

15

Normal and Active Galaxies

Building Blocks of the Universe

As our field of view expands to truly cosmic scales, the focus of our studies shifts dramatically. Planets become inconsequential, stars themselves mere points of hydrogen consumption. Now entire galaxies become the "atoms" from which the universe is built—distant realms completely unknown to scientists just a century ago. We know of literally millions of galaxies beyond our own. Most are smaller than the Milky Way, some comparable in size, a few much larger. Many are sites of explosive events far more energetic than anything ever witnessed in our own Galaxy. All are vast, gravitationally bound assemblages of stars, gas, dust, dark matter, and radiation separated from us by almost incomprehensibly large distances. The light we receive tonight from the most distant galaxies was emitted long before Earth existed. By studying the properties of galaxies and the violence that ensues when they collide, we gain insight into the history of our Galaxy and the universe in which we live.

THE BIG PICTURE

Light collected tonight from the most distant galaxies was emitted by those objects long before Earth even formed. Racing for billions of years across the darkened realms of the cosmos, a minute fraction of their radiation is now intercepted by our telescopes and spacecraft. Captured in the many images of this book, that radiation tells us not only about the properties of faraway galaxies, but also about the history of our Galaxy and the universe in which we live.

◄ The giant elliptical galaxy M87 lies some 17 megaparsecs from Earth, at the center of the Virgo galaxy cluster. Its full diameter is about 30 kpc—not much different from the Milky Way—although only the innermost 5 kpc is shown here. However, with several trillion stars, more than 10,000 globular clusters, and a 4 billion-solar-mass black hole at its center, M87 dwarfs our Galaxy in all other respects. Accretion of stars onto the central supermassive black hole has created a high-speed jet of matter leaving the galaxy's central nucleus at nearly the speed of light. The jet, nucleus, and surrounding galaxy can all be seen in this visible-light *Hubble Space Telescope* image. *(NASA/ESA)*

MasteringAstronomy®

Visit www.masteringastronomy.com for quizzes, animations, videos, interactive figures, and self-guided tutorials.

LEARNING OUTCOMES

Studying this chapter will enable you to:

LO1 List the basic properties and main types of normal galaxies.

LO2 Explain the distance-measurement techniques that enable astronomers to map the universe beyond the Milky Way.

LO3 Describe how galaxies are observed to clump into clusters.

LO4 State Hubble's law, and explain how it is used to determine distances to the most remote objects in the observable universe.

LO5 Specify how active galaxies differ from normal galaxies, and describe some of their basic features.

LO6 Describe some types of active galaxies.

LO7 Explain what drives the central engine thought to power all active galaxies.

15.1 Hubble's Galaxy Classification

Figure 15.1 shows a vast expanse of space lying about 150 million parsecs from Earth. Almost every patch or point of light in this figure is a galaxy—several hundred can be seen in just this one photograph. Images of galaxies look distinctly nonstellar. They have fuzzy edges, and many are quite elongated—not at all like the sharp, point-like images normally associated with stars. Some of the blobs of light in Figure 15.1 are spiral galaxies like the Milky Way Galaxy and Andromeda. Others, however, are definitely not spirals—no disks or spiral arms can be seen. Even when we take into account their different orientations in space, galaxies do *not* all look the same.

American astronomer Edwin Hubble was the first to categorize galaxies in a comprehensive way. Working with the then recently completed 2.5-m optical telescope on Mount Wilson in California in 1924, he classified the galaxies he saw into four basic types—*spirals, barred spirals, ellipticals,* and *irregulars*—solely on the basis of their visual appearance. Many modifications and refinements have been incorporated over the years, but the basic **Hubble classification scheme** is still widely used today.

▼ **FIGURE 15.1 Galaxy Cluster** This is Abell S0740, a collection of many galaxies, each consisting of hundreds of billions of stars. Although dominated by the large elliptical galaxy at the upper right, the system contains galaxies of many different types. The spiral galaxy at lower left is comparable in size to the Milky Way. *(NASA)*

(a) M81   Type Sa   (b) M51   Type Sb   (c) M101   Type Sc

R I V U X G

▲ **FIGURE 15.2 Spiral Galaxy Shapes** Variation in shape among spiral galaxies. As we progress from type Sa to Sb to Sc, the bulges become smaller, while the spiral arms tend to become less tightly wound. *(R. Gendler; NASA)*

Spirals

We saw several examples of **spiral galaxies** in Chapter 14 previously—for example, our own Milky Way Galaxy and our neighbor Andromeda. All galaxies of this type contain a flattened galactic disk in which spiral arms are found, a central galactic bulge with a dense nucleus, and an extended halo of faint, old stars. *(Sec. 14.3)* The stellar density (that is, the number of stars per unit volume) is greatest in the galactic nucleus, at the center of the bulge. ∞ *(Sec. 14.7)*

In Hubble's scheme, a spiral galaxy is denoted by the letter S and classified as type a, b, or c according to the size of its central bulge: Type Sa galaxies have the largest bulges, type Sc the smallest (Figure 15.2). The tightness of the spiral pattern is quite well correlated with the size of the bulge (although the correspondence is not perfect). Type Sa spiral galaxies tend to have tightly wrapped, almost circular, spiral arms. Type Sb galaxies typically have more open spiral arms, while type Sc spirals often have loose, poorly defined spiral structure. The arms also tend to become more "knotty," or clumped, in appearance as the spiral pattern becomes more open. Based on all available evidence, the Milky Way is a barred spiral galaxy, most likely of type SBb. ∞ *(Sec. 14.5)*

The bulges and halos of spiral galaxies contain large numbers of reddish old stars and globular clusters, similar to those observed in our own Galaxy and in Andromeda. ∞ *(Sec. 14.1)* Most of the light from spirals, however, comes from younger A- through G-type stars in the disk, giving these galaxies an overall whitish glow. As in the Milky Way, the disks of typical spiral galaxies are rich in gas and dust. The 21-cm radio radiation emitted by spirals betrays the presence of interstellar gas, and obscuring dust lanes are clearly visible in many systems (see Figures 15.2b and c). Stars are forming within the spiral arms, which contain numerous emission nebulae and newly formed O- and B-type stars. ∞ *(Sec. 14.5)* The arms appear bluish because of the bright blue O- and B-type stars there. Type Sc galaxies contain the most interstellar gas and dust, Sa galaxies the least.

Most spirals are not seen face-on as in Figure 15.2. Many are tilted with respect to our line of sight, making their spiral structure hard to discern. However, we do not need to see spiral arms to classify a galaxy as a spiral. The presence of the disk, with its gas, dust, and newborn stars, is sufficient. For example, the galaxy shown in Figure 15.3 is classified as a spiral because of the clear line of obscuring dust seen along its midplane. That dust is evident most clearly in the inset infrared image, which was acquired by the *Spitzer Space Telescope*.

A variation of the spiral category in Hubble's classification scheme is the **barred-spiral galaxy**. Barred spirals differ from ordinary spirals mainly by the presence of an elongated "bar" of stellar and interstellar matter passing through the center and extending beyond the bulge, into the disk. The spiral arms project from near the ends of the bar rather than from the bulge (as they do in normal spirals). Barred spirals are designated by the letters SB and are subdivided, like the ordinary spirals, into categories SBa, SBb, and SBc, depending on the size of the bulge. Again like ordinary spirals, the tightness of the spiral pattern is correlated with bulge size. Figure 15.4 shows the variation among barred-spiral galaxies. In the case of the SBc category, it is often hard to tell where the bar ends and the spiral arms begin.

▲ **FIGURE 15.3 Sombrero Galaxy** The Sombrero Galaxy (M104), a spiral system seen edge-on, has a dark band composed of interstellar gas and dust. The large size of this galaxy's central bulge marks it as type Sa, even though its spiral arms cannot be seen from our perspective. The inset shows this galaxy in the infrared part of the spectrum, highlighting its dust content in false-colored pink. *(NASA)*

Ellipticals

Unlike the spirals, **elliptical galaxies** have no spiral arms and, in most cases, no obvious galactic disk—in fact, other than a dense central nucleus, they often exhibit little internal structure of any kind. As with spirals, the stellar density increases sharply in the nucleus. Denoted by the letter E, these systems are subdivided according to how elliptical they appear on the sky. The most circular are designated E0, slightly flattened systems are labeled E1, and so on, all the way to the most elongated ellipticals, of type E7 (Figure 15.5).

Note that an elliptical galaxy's Hubble type depends both on its intrinsic three-dimensional shape *and* on its orientation relative to the line of sight. A spherical galaxy, a cigar-shaped galaxy seen end-on, and a disk-shaped galaxy seen face-on, would all appear to be circular on the sky and be classified as E0. It can be difficult to decipher a galaxy's true shape solely from its visual appearance.

▼ **FIGURE 15.4 Barred-Spiral Galaxy Shapes** Variation in shape among barred-spiral galaxies from SBa to SBc is similar to that for the spirals in Figure 15.2, except that here the spiral arms begin at either end of a bar through the galactic center. In frame (c), the bright star is a foreground object in our own Galaxy. *(NASA; ESO; Gemini)*

(a) NGC 1300   Type SBa   (b) NGC 1365   Type SBb   (c) NGC 6872   Type SBc

▶ **FIGURE 15.5 Elliptical Galaxy Shapes** (a) The E1 elliptical galaxy IC 2006 is nearly circular in appearance. (b) NGC 1132 is slightly more elongated and classified as E4. Both galaxies lack spiral structure, and neither shows evidence of interstellar dust or gas, although each has an extensive X-ray halo of hot gas that extends far beyond the visible portion of the galaxy. (c) M110 is a dwarf elliptical companion to the much larger Andromeda galaxy. *(NASA/ESA; NOAO)*

There is a large range in both the size and the number of stars contained in elliptical galaxies. The largest elliptical galaxies are much larger than our Milky Way Galaxy. These *giant ellipticals* can range up to hundreds of thousands of megaparsecs across and contain tens of trillions of stars (see the chapter-opening image on p. 418 of the E0 giant elliptical M87). At the other extreme, *dwarf ellipticals* may be as small as 1 kpc in diameter and contain fewer than a million stars. Their many differences suggest to astronomers that giant and dwarf ellipticals represent distinct galaxy classes, with quite dissimilar formation histories and stellar content. The dwarfs are by far the most common type of ellipticals, outnumbering their brighter counterparts by about 10 to 1. However, most of the *mass* that exists in the form of elliptical galaxies is contained in the larger systems.

Lack of spiral arms is not the only difference between spirals and ellipticals. Most ellipticals also contain little or no cool gas and dust and display no evidence of young stars or ongoing star formation. Like the halo of our Galaxy, ellipticals are composed mostly of old, reddish, low-mass stars. Again like the halo of our Galaxy, the orbits of stars in ellipticals are disordered, exhibiting little or no overall rotation; objects move in all directions, not in regular, circular paths as in our Galaxy's disk. Ellipticals differ from our Galaxy's halo in at least one important respect, however: X-ray observations reveal large amounts of very *hot* (several million kelvin) interstellar gas distributed throughout their interiors, often extending well beyond the visible portions of the galaxies.

Some giant ellipticals are exceptions to many of these general statements, as they contain disks of gas and dust in which stars are forming. Astronomers think that these systems may be the results of mergers between gas-rich galaxies (see Chapter 16). Indeed, galactic collisions may have played an important role in determining the appearance of many of the systems we observe today.

Intermediate between the E7 ellipticals and the Sa spirals in the Hubble classification is a class of galaxies that show evidence of a thin disk and a flattened bulge yet contain neither loose gas nor spiral arms. Two such objects are shown in Figure 15.6. These galaxies are called **S0 galaxies** (S-zero) if no bar is evident and **SB0 galaxies** if a bar is present. They are also known as *lenticular* galaxies, because of their lens-shaped appearance. They look a little like spirals whose dust and gas have been stripped away, leaving behind just a stellar disk. Observations in recent years have shown that many normal elliptical galaxies have faint disks within them, like the S0 galaxies. As with the S0s, the origin of these disks is uncertain, but some researchers suspect that S0s and ellipticals may be closely related.

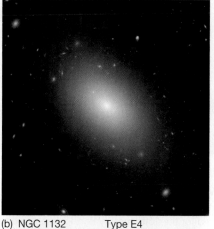

(a) IC 2006   Type E1

(b) NGC 1132   Type E4

(c) M110   Type E5

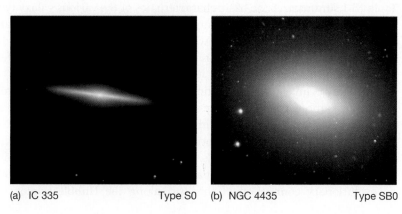

(a) IC 335   Type S0   (b) NGC 4435   Type SB0

◀ **FIGURE 15.6 S0 Galaxies** (a) S0 (or lenticular) galaxies contain a disk and a bulge, but no interstellar gas and no spiral arms. Their properties are intermediate between E7 ellipticals and Sa spirals. (b) SB0 galaxies are similar to S0 galaxies, except for a bar of stellar material extending beyond the central bulge. *(Palomar/Caltech)*

(b)

(c)

R   I   U   X   G

(a)

▲ **FIGURE 15.7 Magellanic Clouds**   The Magellanic Clouds are prominent features of the night sky in the Southern Hemisphere. Named for the 16th-century Portuguese explorer Ferdinand Magellan, whose around-the-world expedition first brought word of these fuzzy patches of light to Europe, these dwarf irregular galaxies orbit our Galaxy and accompany it on its trek through the cosmos. (a) The Clouds' relationship to one another in the southern sky reveals both the Small (b) and the Large (c) Clouds to have distorted, irregular shapes. *(Mount Stromlo & Siding Spring Observatories)*

Irregulars

The final galaxy class identified by Hubble is a catch-all category—**irregular galaxies**—so named because their visual appearance excludes them from the other categories just discussed. Irregulars tend to be rich in interstellar matter and young, blue stars, but they lack any regular structure, such as well-defined spiral arms or central bulges. Irregular galaxies tend to be smaller than spirals but somewhat larger than dwarf ellipticals. They typically contain between 10^8 and 10^{10} stars. The smallest are called *dwarf irregulars*. As with elliptical galaxies, the dwarf type is the most common. Dwarf ellipticals and dwarf irregulars occur in approximately equal numbers and together make up the vast majority of galaxies in the universe. They are often found close to a larger "parent" galaxy.

Irregular galaxies are divided into two subclasses—Irr I and Irr II. The Irr I galaxies often look like misshapen spirals. Figure 15.7 shows the **Magellanic Clouds**, a famous pair of Irr I galaxies that orbit the Milky Way Galaxy. They are shown to proper scale in Figure 14.15. Studies of Cepheid variables within the Clouds show them to be approximately 50 kpc from the center of our Galaxy. ∞ *(Sec. 14.2)* The Large Cloud contains about 6 billion solar masses of material and is a few kiloparsecs across. Both Magellanic Clouds contain lots of gas, dust, and blue stars (as well as the well-documented supernova discussed in *Discovery 12-1*), indicating ongoing star formation. Both also contain many old stars and several old globular clusters, so we know that star formation has been taking place there for a very long time.

The much rarer Irr II galaxies (Figure 15.8), in addition to their irregular shapes, have other peculiarities, often exhibiting a distinctly explosive or filamentary appearance. Their appearance once led astronomers to suspect that violent events had occurred within them. However, it now seems more likely that in some (but probably not all) cases, we are seeing the result of a close encounter or collision between two previously "normal" systems.

The Hubble Sequence

Table 15.1 summarizes the basic characteristics of the various galaxy types. When he first developed his classification scheme, Hubble arranged the galaxy types into the "tuning fork" diagram shown in Figure 15.9. The variation in galaxy types across the diagram, from ellipticals to spirals to irregulars, is often referred to as the **Hubble sequence.**

Hubble's primary aim in creating this diagram was to indicate similarities in appearance among galaxies, but he also regarded the tuning fork as an evolutionary sequence from left to right, with E0 ellipticals evolving into flatter ellipticals and S0 systems that ultimately formed disks and spiral arms. Indeed, Hubble's terminology referring to ellipticals as "early-type" galaxies and spirals as "late-type" is still widely used today. But as far as modern astronomers can tell, there is *no* evolutionary connection of this sort along the Hubble sequence. Isolated

(a) AM0644-741

R I V U X G

(b) NGC 1569

R I V U X G

▲ **FIGURE 15.8 Irregular Galaxy Shapes** (a) The strangely shaped galaxy AM0644-741 is probably plunging headlong into a group of several other galaxies (two of them shown here), causing huge rearrangements of its stars, gas, and dust. (b) The galaxy NGC 1569 seems to show an explosive appearance, probably the result of a recent galaxywide burst of star formation. *(NASA)*

normal galaxies do *not* evolve from one type to another. Spirals are not ellipticals with arms, nor are ellipticals spirals that have somehow shed their star-forming disks. In short, astronomers know of no simple parent–child relationship among Hubble types.

However, the key word here is *isolated.* As we will see, there is now strong observational evidence indicating that collisions and tidal interactions *between* galaxies are commonplace, and these may be the main physical processes driving galaxy evolution. We will return to this important subject in Chapter 16.

ANIMATION/VIDEO
The Evolution of Galaxies

CONCEPT CHECK

In what ways are large spirals such as the Milky Way and Andromeda not representative of galaxies as a whole?

▲ **FIGURE 15.9 Galactic "Tuning Fork"** The placement of the four basic types of galaxies—ellipticals, spirals, barred spirals, and irregulars—in Hubble's "tuning-fork" diagram is suggestive, but this galaxy classification scheme has no known physical meaning.

TABLE 15.1   Basic Galaxy Properties by Type

	Spiral/Barred Spiral (S/SB)	Elliptical (E)[1]	Irregular (Irr)
Shape and structural properties	Highly flattened disk of stars and gas containing spiral arms and thickening to central bulge. SB galaxies have an elongated central "bar" of stars and gas.	No disk. Stars smoothly distributed through an ellipsoidal volume. No obvious substructure other than a dense central nucleus.	No obvious structure. Irr II galaxies often have "explosive" appearance.
Stellar content	Disks contain both young and old stars; halos consist of old stars only.	Contain old stars only.	Contain both young and old stars.
Gas and dust	Disks contain substantial amounts of gas and dust; halos contain little of either.	Contain hot X-ray-emitting gas, little or no cool gas and dust.	Very abundant in gas and dust.
Star formation	Ongoing star formation in spiral arms.	No significant star formation during the last 10 billion years.	Vigorous ongoing star formation.
Stellar motion	Gas and stars in disk move in circular orbits around the galactic center; halo stars have random orbits in three dimensions.	Stars have random orbits in three dimensions.	Stars and gas have very irregular orbits.

[1]As noted in the text, some giant ellipticals appear to be the result of mergers between gas-rich galaxies and are exceptions to many of the statements listed here.

15.2 The Distribution of Galaxies in Space

Now that we have seen some of their basic properties, let us ask how galaxies are spread throughout the universe beyond the Milky Way. Galaxies are not distributed uniformly in space. Rather, they tend to clump into still larger agglomerations of matter. As always in astronomy, our understanding hinges on knowing how far away an object lies. We therefore begin by looking more closely at the means used by astronomers to measure distances to galaxies.

Extending the Distance Scale

Astronomers estimate that some 40 billion galaxies brighter than our own exist in the observable universe. Some reside close enough for the Cepheid variable technique to work—astronomers have detected and measured the periods of Cepheids in galaxies as far away as 25 Mpc. ∞ *(Sec. 14.2)* However, some galaxies contain no Cepheid stars (can you think of some reasons why this might be?), and in any case most known galaxies lie much farther away than 25 Mpc. Cepheid variables in very distant galaxies simply cannot be observed well enough, even through the world's most sensitive telescopes, to allow us to measure their apparent brightness and periods. To extend our distance-measurement ladder, we must find some new object to study.

One way in which researchers have tackled this problem is by using **standard candles**—bright, easily recognizable astronomical objects whose luminosities are confidently known. The basic idea is very simple. Once an object is identified as a standard candle—by its appearance or by the shape of its light curve, say—its luminosity can be estimated. Comparison of the luminosity with the apparent brightness then gives the object's distance, and hence the distance, to the galaxy in which it resides. ∞ *(Sec. 10.2)* Note that, apart from the way in which the luminosity is determined, the Cepheid variable technique relies on identical reasoning—the measured period tells us the luminosity, and the inverse-square law then tells us the distance.

PROCESS OF SCIENCE CHECK

What are some of the problems in measuring accurately the distances to faraway galaxies?

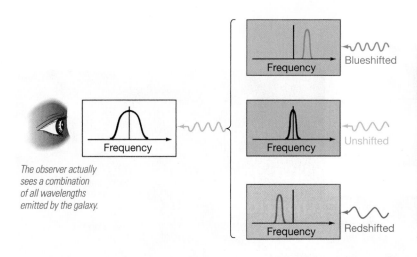

The observer actually sees a combination of all wavelengths emitted by the galaxy.

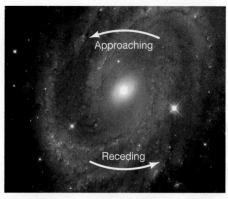

▲ **FIGURE 15.10 Galaxy Rotation**   A galaxy's rotation causes some of the radiation it emits to be blueshifted and some redshifted. From a distance, when all the radiation from the galaxy is combined into a single beam and analyzed spectroscopically, the redshifted and blueshifted components produce a broadening of the galaxy's spectral lines. The amount of broadening is a direct measure of the rotation speed of the galaxy, such as NGC 4603 shown here. *(NASA)*

To be most useful, a standard candle must (1) have a well-defined luminosity so that the uncertainty in estimating its brightness is small and (2) be bright enough to be seen at great distances. Currently, Cepheid variable stars are usable for distance measurement "only" out to about 25 Mpc. In recent years, planetary nebulae and Type I supernovae have proved particularly reliable as standard candles. ∞ *(Secs. 12.3, 12.5)* The latter have remarkably consistent peak luminosities and are bright enough to be identified and measured out to distances of more than a gigaparsec (1000 Mpc).

An important alternative to standard candles was discovered in the 1970s, when astronomers found a close correlation between the rotational speeds and the luminosities of spiral galaxies within a few tens of megaparsecs of the Milky Way Galaxy. Rotation speed is a measure of a spiral galaxy's total mass, so it is perhaps not surprising that this property should be related to luminosity. ∞ *(Sec. 14.5)* What *is* surprising, though, is how tight the correlation is. The **Tully–Fisher relation**, as it is now known (after its discoverers), allows us to obtain a remarkably accurate estimate of a spiral galaxy's luminosity simply by observing how fast it rotates (Figure 15.10). As usual, comparing the galaxy's (true) luminosity with its (observed) apparent brightness yields its distance.

The Tully–Fisher relation can be used to measure distances to spiral galaxies out to about 200 Mpc, beyond which the line broadening in Figure 15.10 becomes increasingly difficult to measure accurately. A somewhat similar connection, relating line broadening to a galaxy's *diameter,* exists for elliptical galaxies. Once the galaxy's diameter and angular diameter are known, its distance can be computed from elementary geometry. ∞ *(More Precisely 4-1)* These methods bypass many of the standard candles often used by astronomers and so provide independent means of determining distances to faraway objects.

As indicated in Figure 15.11, standard candles and the Tully–Fisher relation form the fifth and sixth rungs of our cosmic distance ladder, introduced in Chapter 1 and expanded in Chapters 10 and 14. ∞ *(Secs. 1.4, 10.6, 14.2)* In fact, they stand for perhaps a dozen or so related but separate techniques that astronomers have employed in their quest to map out the universe on large scales. Just as with the lower rungs, we calibrate the properties of these new techniques using distances measured by more local means. In this way, the distance-measurement process "bootstraps" itself to greater and greater distances. However, at the same time, the errors and uncertainties in each step accumulate, so the distances to the farthest objects are the least well known.

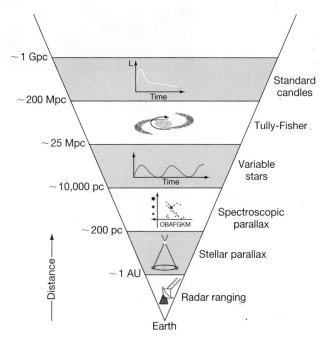

▲ **FIGURE 15.11 Extragalactic Distance Ladder INTERACTIVE** An inverted pyramid summarizes the distance techniques used to study different realms of the universe. The techniques shown in the bottom four layers—radar ranging, stellar parallax, spectroscopic parallax, and variable stars—take us as far as the nearest galaxies. To go farther, we must use other techniques—the Tully–Fisher relation and the use of standard candles, among other methods—based on distances determined by the four lowest techniques.

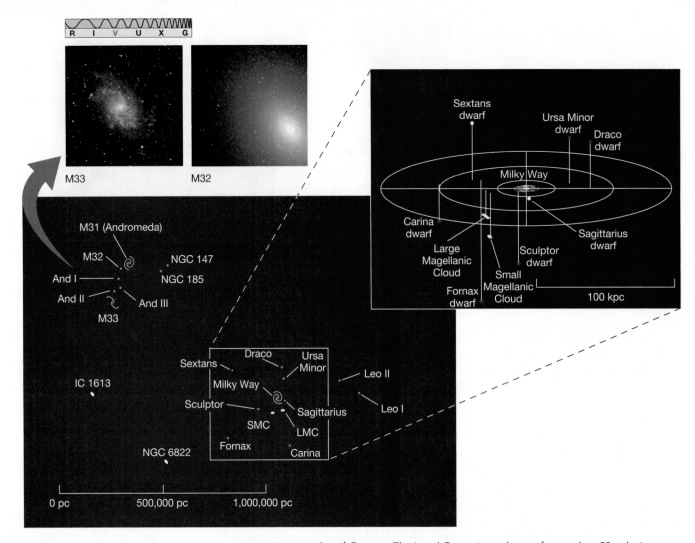

M33   M32

▲ **FIGURE 15.12 Local Group** The Local Group is made up of more than 50 galaxies within approximately 1 Mpc of our Milky Way Galaxy. Only a few are spirals; most are dwarf-elliptical or irregular galaxies, only some of which are shown here. Spirals are colored blue, ellipticals pink, and irregulars white, all of them depicted roughly to scale. The map inset (top right) shows the Milky Way in relation to some of its satellite galaxies. The photographic insets (top left) show two well-known neighbors of the Andromeda galaxy (M31): the spiral galaxy M33 and the dwarf elliptical galaxy M32 (also visible in Figure 14.2a, a larger-scale view of the Andromeda system). *(M. BenDaniel; NASA)*

Clusters of Galaxies

Figure 15.12 sketches the locations of all the known major astronomical objects within about 1 Mpc of the Milky Way. Our Galaxy appears with its dozen or so satellite galaxies—including the two Magellanic Clouds discussed earlier and a small companion (labeled "Sagittarius" in the figure) lying almost within our own Galactic plane. The Andromeda galaxy, lying 800 kpc from us, is also shown, surrounded by satellites of its own. Two of Andromeda's galactic neighbors are shown in the inset to the figure. M33 is a spiral, while M32 is a dwarf elliptical, easily seen in Figure 14.2(a) to the bottom right of Andromeda's central bulge.

All told, some 55 galaxies are known to populate our Galaxy's neighborhood. Three of them (the Milky Way, Andromeda, and M33) are spirals; the remainder are dwarf irregulars and dwarf ellipticals. Together, these galaxies form the **Local Group**—a new level of structure in the universe above the scale of our Galaxy. As

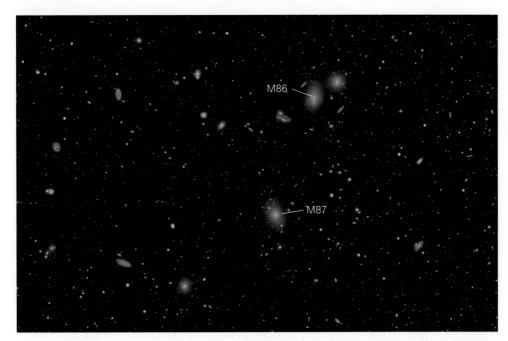

◀ **FIGURE 15.13 Virgo Cluster**   In the central region of the Virgo Cluster of galaxies, about 17 Mpc from Earth, many large spiral and elliptical galaxies can be seen. The inset shows several galaxies surrounding the giant elliptical M86. An even bigger elliptical galaxy, M87 noted at bottom, will be discussed later in this chapter. *(M. BenDaniel; AURA)*

indicated in Figure 15.12, the Local Group's diameter is a little over 1 Mpc. The Milky Way Galaxy and Andromeda are by far its largest members, and most of the smaller galaxies are gravitationally bound to one or other of them. The combined gravity of the galaxies in the Local Group binds them together, like stars in a star cluster, but on a scale a millionfold larger. More generally, a group of galaxies held together by their mutual gravitational attraction is called a **galaxy cluster.**

Moving beyond the Local Group, the next large concentration of galaxies we come to is the *Virgo Cluster* (Figure 15.13), named for the constellation in which it is found. It lies some 18 Mpc from the Milky Way. The Virgo Cluster does not contain a mere 45 galaxies, however. Rather, it houses more than 2500 galaxies, bound by gravity into a tightly knit group about 3 Mpc across.

Wherever we look in the universe we find galaxies, and most galaxies are members of galaxy clusters. Small clusters, such as the Local Group, contain only a few galaxies and are quite irregular in shape. Large, "rich" clusters such as Virgo contain thousands of individual galaxies distributed fairly smoothly in space. Abell S0740, shown in Figure 15.1, is another example of a rich cluster. It lies approximately 100 Mpc away. Figure 15.14 is a long-exposure photograph of a much more distant rich cluster, residing roughly 2 billion parsecs from Earth. A sizeable minority of galaxies (possibly as many as 40 percent) are not members of any cluster. They are apparently isolated systems, moving alone through intercluster space.

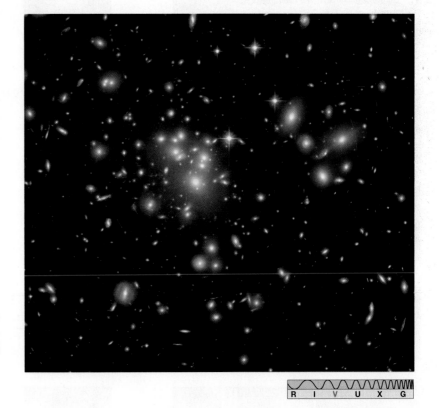

R I V U X G

▶ **FIGURE 15.14 Distant Galaxy Cluster**   The galaxy cluster Abell 1689 contains huge numbers of galaxies and resides roughly 2 billion parsecs from Earth. Virtually every patch of light in this photograph is a separate galaxy. With the most powerful telescopes, astronomers can now discern, even at this great distance, spiral structure in some of the galaxies, as well as many galaxies colliding—some tearing matter from one another, others merging into single systems. *(NASA)*

The dynamics of galaxy clusters resemble a bunch of buzzing fireflies trapped within a jar moving away from us.

15.3 Hubble's Law

Now that we have seen some basic properties of galaxies across the universe, let's turn our attention to the large-scale *motions* of galaxies and galaxy clusters. Within a galaxy cluster, individual galaxies move more or less randomly. You might expect that, on even larger scales, the clusters themselves would also have random, disordered motion—some clusters moving this way, some that, but this is not the case. On the largest scales, galaxies and galaxy clusters alike move in a very *ordered* way.

Universal Recession

In 1912 American astronomer Vesto M. Slipher, working under the direction of Percival Lowell, discovered that virtually every spiral galaxy he observed had a redshifted spectrum—they were *receding* from our Galaxy. ⚭ *(Sec. 2.7)* In fact, with the exception of a few nearby systems, *every* known galaxy is part of a general motion away from us in all directions. Individual galaxies that are not part of galaxy clusters are steadily receding. Galaxy clusters, too, have an overall recessional motion, although their individual member galaxies move randomly with respect to one another. (Consider a jar full of fireflies that has been thrown into the air. The fireflies within the jar, like the galaxies within the cluster, have random motions due to their individual whims, but the jar as a whole, like the galaxy cluster, has some directed motion as well.)

Figure 15.15 shows the optical spectra of several galaxies, arranged in order of increasing distance from the Milky Way Galaxy. These spectra are redshifted, indicating that the galaxies are receding, and the extent of the redshift increases from top to bottom in the figure. There is a connection between Doppler shift and distance: the greater the distance, the greater the redshift. This trend holds for nearly all galaxies in the universe. (Two galaxies within our Local Group, including Andromeda, and a few galaxies in the Virgo Cluster display blueshifts and so are moving toward us, but this results from their local motions within their parent clusters. Recall the fireflies in the jar.)

Figure 15.16(a) plots recessional velocity against distance for the galaxies shown in Figure 15.15. A similar plot for more galaxies within about 1 billion parsecs of Earth is shown in Figure 15.16(b). Plots like these were first made by Edwin Hubble in the 1920s and now bear his name—*Hubble diagrams*. The data points fall close to a straight line, indicating that the rate at which a galaxy recedes is *directly proportional* to its distance from us. This rule is called **Hubble's law**. We could construct such a diagram for any group of galaxies, provided we could determine their distances and velocities. The universal recession described by the Hubble diagram is called the *Hubble flow*.

The recessional motions of the galaxies prove that the cosmos is not steady and unchanging on the largest scales. The universe (actually, *space itself*—see Sec. 17.2) is expanding. But let's be clear on just *what* is expanding and what is not. Hubble's law does not mean that humans, Earth, the solar system, or even individual galaxies and galaxy clusters are physically increasing in size. These groups of atoms, rocks, planets, stars, and galaxies are held together by their own internal forces and are not

▼ **FIGURE 15.15 Galaxy Spectra** Optical spectra (left) of several galaxies shown on the right. Both the extent of the redshift (denoted by the horizontal red arrows) and the distance from the Milky Way Galaxy to each galaxy (numbers in center column) increase from top to bottom. The vertical yellow arrows denote a pair of dark absorption lines in the observed spectra. The many white vertical lines at the top and bottom of each spectrum are laboratory references. *(Adapted from Palomar/Caltech)*

Radial velocities in km/s	Distance in megaparsecs	Cluster galaxy in
1210	17	Virgo
15,000	210	Ursa Major
21,600	310	Corona Borealis
39,300	560	Bootes
61,200	870	Hydra

R I V U X G

themselves getting bigger. Only the largest scales in the universe—the distances separating the galaxy clusters—are expanding.

To distinguish recessional redshift from redshifts caused by motion *within* an object—for example, galaxy orbits within a cluster or explosive events in a galactic nucleus—astronomers refer to the redshift resulting from the Hubble flow as the **cosmological redshift**. Objects that lie so far away that they exhibit a large cosmological redshift are said to be at *cosmological distances*—distances comparable to the scale of the universe itself.

Hubble's law has some fairly dramatic implications. If nearly all galaxies show recessional velocities according to Hubble's law, then doesn't that mean that they all started their journey from a single point? If we could run time backward, wouldn't all the galaxies fly back to this one point, perhaps the site of some explosion in the remote past? The answer is yes—but not in the way you might expect! In Chapter 17 we will explore the ramifications of the Hubble flow for the past and future evolution of our universe. For now, however, we set aside its cosmic implications and use Hubble's law simply as a convenient distance-measuring tool.

Hubble's Constant

The constant of proportionality between recessional velocity and distance in Hubble's law is known as **Hubble's constant**, denoted by the symbol H_0. The data shown in Figure 15.16 then obey the equation:

$$\text{recessional velocity} = H_0 \times \text{distance}.$$

The value of Hubble's constant is the slope of the straight line—recessional velocity divided by distance—in Figure 15.16(b). Reading the numbers off the graph, this comes to roughly 70,000 km/s divided by 1000 Mpc, or 70 km/s/Mpc (kilometers per second per megaparsec, the most commonly used unit for H_0). Astronomers continually strive to refine the accuracy of the Hubble diagram and the resulting estimate of H_0 because Hubble's constant is one of the most fundamental quantities of nature—it specifies the rate of expansion of the entire cosmos.

Hubble's original value for H_0 was about 500 km/s/Mpc, far higher than the currently accepted value. This overestimate was due almost entirely to errors in the cosmic distance scale at the time, particularly the calibrations of Cepheid variables and standard candles. The measured value dropped rapidly as various observational errors were recognized and resolved and distance measurements became more reliable. Published estimates of H_0 entered the "modern" range (within, say, 20 percent of the current value) in roughly the mid-1960s.

As measurement techniques have continued to improve, the uncertainty in the result has steadily decreased to the point that now, early in the 21st century, all leading measurements of H_0, determined by a variety of different techniques—Tully–Fisher measurements, studies of Cepheid variables in the Virgo cluster, and observations of standard candles, such as Type I supernovae—are all remarkably consistent with one another. We will adopt a rounded-off value of $H_0 = 70$ km/s/Mpc (a choice roughly in the middle of all recent results and also in line with some precise cosmological measurements to be discussed in Chapter 17) as the best current estimate of Hubble's constant for the remainder of the text.

The Top of the Distance Ladder

Using Hubble's law, we can derive the distance to a remote object simply by measuring the object's recessional velocity and dividing by Hubble's constant. Hubble's law thus tops our inverted pyramid of distance-measurement techniques (Figure 15.17). This seventh method simply assumes that Hubble's law holds. If this assumption is correct, then Hubble's law enables us to measure great

(a)

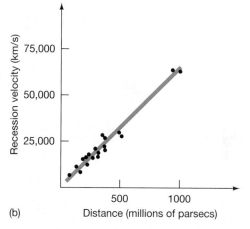

(b)

▲ **FIGURE 15.16 Hubble's Law INTERACTIVE** Plots of recessional velocity against distance (a) for the galaxies shown in Figure 15.15 and (b) for numerous other galaxies within about 1 billion pc of Earth.

DATA POINTS

Hubble's Law

More than half of all students had difficulty describing and using Hubble's Law. Some points to remember:

- Hubble's Law is an observed relation between a galaxy's distance and the speed with which it moving away from us—on average, the more distant a galaxy, the greater its recession velocity.

- If we plot galaxy distances horizontally and velocities vertically, Hubble's law means that galaxies lie approximately on a straight line on the graph. The slope of that line is Hubble's constant, a direct measure of the expansion of the universe.

- Hubble's law allows us to measure the distance to a galaxy simply by measuring its recession velocity. Dividing the velocity (in km/s) by Hubble's constant (in km/s/Mpc) gives us the distance in megaparsecs.

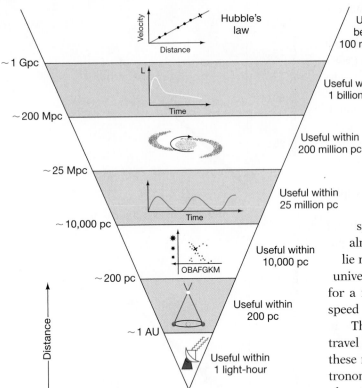

▲ FIGURE 15.17 Cosmic Distance Ladder INTERACTIVE
Hubble's law tops the hierarchy of distance-measurement techniques. It is used to find the distances of astronomical objects all the way out to the limits of the observable universe.

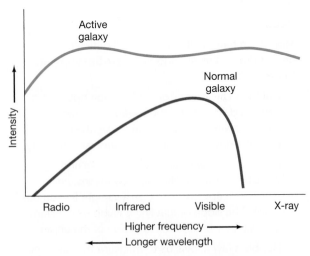

▲ FIGURE 15.18 Galaxy Energy Spectra The energy emitted by a normal galaxy differs significantly from that emitted by an active galaxy. This plot illustrates the general spread of radiation intensity for all galaxies of a particular type and does not represent any one individual galaxy.

distances in the universe—so long as we can obtain an object's spectrum, we can determine how far away it is.

The precise astronomical definition of *redshift* is the fractional increase in the wavelength of a beam of radiation due to the source's motion away from us. For many nearby objects, recession velocities are small compared to the speed of light, and the redshifts are correspondingly low—just a few percent. ∞ *(Sec. 2.7)* However, more distant objects can have recessional motions that are substantial fractions of the speed of light. The most distant objects thus far observed in the universe—some young galaxies and quasars—have redshifts of around 8, meaning that their radiation has been stretched in wavelength not by a few percent, but *ninefold*. Their ultraviolet spectral lines are shifted all the way into the infrared part of the spectrum! Such huge redshifts correspond to recessional velocities of more almost 98 percent of the speed of light. Hubble's law implies that such objects lie more than 9000 Mpc away from us, as close to the limits of the observable universe as astronomers have yet been able to probe. (See *More Precisely 15-1,* for a fuller discussion of redshifts and recession velocities comparable to the speed of light.)

The speed of light is finite. It takes time for light or any kind of radiation to travel from one point in space to another. The radiation that we now see from these most distant objects originated long ago. Incredibly, the radiation that astronomers now detect was emitted roughly billion years ago, well before our planet, our Sun, and perhaps even our Galaxy came into being.

15.4 Active Galactic Nuclei

The galaxies described in Section 15.1—those falling into the various Hubble classes—are generally referred to as **normal galaxies.** As we have seen, their luminosities range from a million or so times that of the Sun for dwarf ellipticals and irregulars to more than a trillion solar luminosities for the largest giant ellipticals. For comparison, in round numbers, the luminosity of the Milky Way Galaxy is 2×10^{10} solar luminosities, or roughly 10^{37} W.

In these last two sections we focus our attention on "bright" galaxies, conventionally taken to mean galaxies with luminosities more than about 10^{10} times the solar value. By this measure, our Galaxy is bright, but not abnormally so.

Galactic Radiation

A substantial fraction of bright galaxies—perhaps as many as 40 percent—don't fit well into the "normal" category. Their *spectra* differ significantly from those of their normal cousins, and their *luminosities* can be extremely large. Known collectively as **active galaxies,** they are of great interest to astronomers. The brightest among them are the most energetic objects known in the universe, and all may represent an important, if intermittent, phase of galactic evolution (see Section 16.4). At optical wavelengths, active galaxies often *look* like normal galaxies—familiar components such as disks, bulges, stars, and dark dust lanes can be identified. At other wavelengths, however, their unusual properties are much more apparent.

As illustrated schematically in Figure 15.18, most of a normal galaxy's energy is emitted in or near the visible portion of the electromagnetic spectrum, much like the radiation from stars. Indeed, to a large extent, the light we see from a normal galaxy *is* just the accumulated light of its many component stars (once the effects of interstellar dust are taken into account), each described approximately by a blackbody curve. ∞ *(Sec. 2.4)* By contrast the radiation from active galaxies does *not* peak in the visible portion of the spectrum. Most active galaxies do emit substantial amounts of visible radiation, but far more of their energy is emitted at invisible wavelengths, both longer and shorter than those in the visible range. Put another way, the radiation from active galaxies is *inconsistent* with what we would expect if it were the combined radiation of billions of stars. Their radiation is said to be *nonstellar.*

Many luminous galaxies with nonstellar emission are known to be **starburst galaxies**—previously normal systems currently characterized by widespread episodes of star formation, most likely as a result of interactions with a neighbor. The irregular galaxy NGC 1569 shown in Figure 15.8(b) is a prime example. We will study these important systems and their role in galaxy evolution in Chapter 16. For purposes of this text, however, we will use the term "active galaxy" to mean a system whose abnormal activity is related to violent events occurring in or near the galactic nucleus. The nuclei of such systems are called **active galactic nuclei**.

Astronomers have identified and cataloged a bewildering array of systems falling into the "active" category. For example, Figure 15.19 shows an active galaxy exhibiting both nuclear activity and widespread star formation, with a blue-tinted ring of newborn stars surrounding an extended 1-kpc-wide core of intense emission. Rather than attempting to describe the entire "zoo" of active galaxies, we will confine ourselves to three basic types: the energetic *Seyfert galaxies* and *radio galaxies,* and the even more luminous *quasars.* Their properties will allow us to identify and discuss features common to active galaxies in general.

The association of galactic activity with the galactic nucleus is reminiscent of the discussion of the center of the Milky Way in Chapter 14. ∞ *(Sec. 14.7)* In our Galaxy, it seems that the activity in the nucleus is associated with the central *supermassive black hole,* whose presence is inferred from observations of stellar orbits in the innermost fraction of a parsec. As we will see, most astronomers think that basically the same thing occurs in the nuclei of active galaxies and that "normal" and "active" galaxies may differ principally in the degree to which the nonstellar nuclear component of the radiation outshines the light from the rest of the galaxy. This is a powerful unifying theme for understanding the evolution of galaxies, and we will return to it in Chapter 16. For the remainder of this chapter, we concentrate on describing the properties of active galaxies and the black holes that power them.

Seyfert Galaxies

In 1943 Carl Seyfert, an American optical astronomer studying spiral galaxies from Mount Wilson Observatory, discovered the type of active galaxy that now bears his name. **Seyfert galaxies** are a class of astronomical objects whose properties lie between normal galaxies and the most energetic active galaxies known.

Superficially, Seyferts resemble normal spiral galaxies (Figure 15.20). Indeed, the stars in a Seyfert's galactic disk and spiral arms produce about the same amount of visible radiation as the stars in a normal spiral galaxy. However, most of a Seyfert's energy is emitted from the galactic nucleus the center of the overexposed white patch in Figure 15.20. The nucleus of a Seyfert galaxy is some 10,000 times brighter than the center of our own Galaxy. In fact, the brightest Seyfert nuclei are 10 times more energetic than the *entire* Milky Way.

Some Seyferts produce radiation spanning a broad range in wavelengths, from the infrared all the way through ultraviolet and even X-rays. However, the majority (about 75 percent) emit most of their energy in the infrared. Scientists think that much of the high-energy radiation in these Seyferts is absorbed by dust in or near the nucleus, then reemitted as infrared radiation.

▲ **FIGURE 15.19 Active Galaxy**   This image of the galaxy NGC 7742 resembles a fried egg, with a ring of blue star-forming regions surrounding a very bright yellow core that spans about 1 kpc. This active galaxy combines star formation with intense emission from its central nucleus and lies roughly 24 Mpc away. *(NASA)*

CONCEPT CHECK

The energy emission from an active galactic nucleus does not resemble a blackbody curve. Why is this important?

▲ **FIGURE 15.20 Seyfert Galaxy**   NGC 1566, a beautiful Sb/c spiral, lies about 13 Mpc away in the southern constellation Dorado. Its bright, compact nucleus also identifies it as a Seyfert galaxy—in fact, one of the brightest known. This infrared/optical/ultraviolet composite image was obtained by *HST* in 2014. *(NASA/ESA)*

▲ FIGURE 15.21 Seyfert Time Variability This graph illustrates the irregular variations of the luminosity of Seyfert galaxy 3C 84 over three decades. These observations were made in the radio part of the electromagnetic spectrum; the optical and X-ray luminosities vary as well. *(NRAO)*

Seyfert spectral lines have many similarities to those observed toward the center of our own Galaxy. ∞ *(Sec. 14.7)* The lines are very broad, most likely indicating rapid (5000 km/s or more) internal motion within the nuclei. ∞ *(Sec. 2.8)* In addition, the energy emission often varies in time (Figure 15.21). A Seyfert's luminosity can double or halve within a fraction of a year. These rapid fluctuations in luminosity lead us to conclude that the source of energy emissions must be quite compact—simply put, an object cannot "flicker" in less time than radiation takes to cross it. ∞ *(Sec. 13.4)* The emitting region must therefore be less than 1 light-year across—an extraordinarily small region, considering the amount of energy emanating from it.

Taken together, the rapid time variability and large radio and infrared luminosities observed in Seyfert galaxies imply violent nonstellar activity in their nuclei. As just mentioned, this activity is probably similar in *nature* to processes occurring at the center of our own Galaxy, but its *magnitude* is thousands of times greater than the comparatively mild events within our own Galaxy's heart. ∞ *(Sec. 14.7)*

Radio Galaxies

As the name suggests, **radio galaxies** are active galaxies that emit large amounts of energy in the radio portion of the electromagnetic spectrum. They differ from Seyferts not only in the wavelengths at which they radiate, but also in both the appearance and the extent of their emitting regions.

Figure 15.22(a) shows the radio galaxy Centaurus A, which lies about 4 Mpc from Earth. Almost none of this galaxy's radio emission comes from a compact nucleus. Instead, the energy is released from two huge extended regions called **radio lobes**—roundish clouds of gas spanning about half a megaparsec, lying well beyond the visible galaxy.[1] Undetectable in visible light, the radio lobes of radio galaxies are truly enormous. From end to end, they typically span more than 10 times the size of the Milky Way Galaxy, comparable in scale to the entire Local Group.

---

[1]The term *visible galaxy* is commonly used to refer to those components of an active galaxy that emit visible "stellar" radiation, as opposed to the nonstellar and invisible "active" component of the galaxy's emission.

(a)

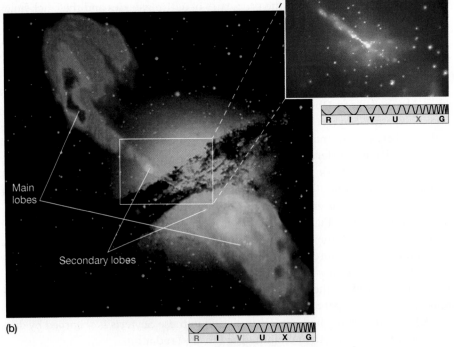

Main lobes

Secondary lobes

(b)

▲ FIGURE 15.22 Centaurus A Radio Lobes Radio galaxies, such as Centaurus A (a), often have giant radio-emitting lobes (b) extending a million light-years or more beyond the central galaxy. The lobes cannot be imaged in visible light and must be observed with radio telescopes; they are shown here in false color, with decreasing intensity from red to yellow to green to blue. The inset is a *Chandra* X-ray image of one of the lobes, showing that the jets in the inner parts of the lobes also emit higher-energy radiation. *(ESO; NRAO; SAO)*

► FIGURE 15.23 **Cygnus A** (a) A visible-light image of Cygnus A appears to show two galaxies in collision. (b) On a much larger scale, it displays radio-emitting lobes (mapped in blue) on both sides of the visible image. The galaxy in (a) is about the size of the small dot at the center of (b). The distance from one lobe to the other is approximately a million light-years. (*NOAO; NRAO*)

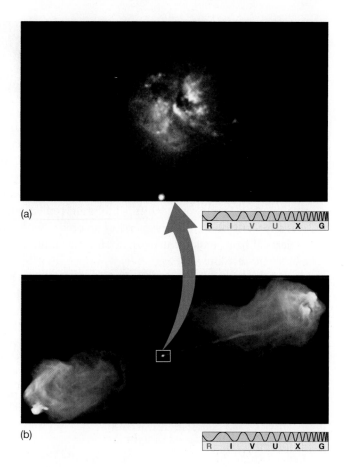

(a)

(b)

Figure 15.22(b) shows the relationship between the galaxy's visible, radio, and X-ray emission. In visible light, Centaurus A is apparently a large E2 galaxy some 500 kpc in diameter, bisected by an irregular band of dust. Centaurus A is a member of a small cluster of galaxies, and numerical simulations suggest that this peculiar galaxy is probably the result of a collision between an elliptical galaxy and a smaller spiral galaxy about 500 million years ago (see Chapter 16). The radio lobes are roughly symmetrically placed, jutting out from the center of the visible galaxy, roughly perpendicular to the dust lane, suggesting that they consist of material ejected in opposite directions from the galactic nucleus. This conclusion is strengthened by the presence of a pair of smaller secondary lobes closer to the visible galaxy and by the presence of a roughly 1-kpc-long jet of matter in the galactic center, all aligned with the main lobes (and marked in the figure).

If the material was ejected from the nucleus at close to the speed of light, then Centaurus A's outer lobes were created a few hundred million years ago, quite possibly around the time of the elliptical/spiral merger thought to be responsible for the galaxy's odd optical appearance. The secondary lobes were expelled more recently. Apparently, some violent process at the center of Centaurus A—most probably triggered by the collision—started up around that time and has been intermittently firing jets of matter out into intergalactic space ever since.

Centaurus A is a relatively low-luminosity source that happens to lie very close to us, astronomically speaking, making it particularly easy to study. Figure 15.23 shows a much more powerful emitter, called Cygnus A, lying roughly 250 Mpc from Earth. The high-resolution radio map in Figure 15.23(b) clearly shows two narrow, high-speed jets joining the radio lobes to the center of the visible galaxy (the dot at the center of the radio image). Notice also that, as with Centaurus A, Cygnus A is a member of a small group of galaxies, and the optical image (Figure 15.23a) appears to show a galaxy collision in progress.

The radio lobes of the brightest radio galaxies (such as Cygnus A) emit roughly 10 times more energy than the Milky Way Galaxy at all wavelengths, coincidentally about the same as the most luminous Seyfert nuclei. However, despite their names, radio galaxies actually radiate far more energy at shorter wavelengths. Their total energy output can be 100 times or more greater than their radio emission. Most of this energy comes from the nucleus of the visible galaxy. Having total luminosities up to 1000 times that of the Milky Way Galaxy, bright radio galaxies are among the most energetic objects known in the universe. Their radio emission is important to astronomers because it lets them study in detail the connection between the small-scale nucleus and the large-scale radio lobes.

Not all radio galaxies have obvious radio lobes. Figure 15.24 shows a *core-dominated* radio galaxy, most of whose energy is emitted from a small central nucleus (which radio astronomers refer to as the *core*) less than 1 pc across. Weaker radio emission comes from an extended region surrounding the nucleus. Probably all radio galaxies have jets and lobes, but what we observe depends on our perspective. As illustrated in Figure 15.25, when we view the radio galaxy from the side we see the jets and lobes. However, if a weak jet fails to escape from the parent galaxy, or if we view the jet almost head-on—in other words, looking *through* the lobe—then we see a core-dominated system.

▲ FIGURE 15.24 **Core-Dominated Radio Galaxy** As shown by this radio contour map of the galaxy M86, the radio emission comes from a bright central nucleus, which is surrounded by an extended region of less intense emission. The radio map is superimposed on an optical image of the galaxy and some of its neighbors, a wider-field version of which was shown previously in Figure 15.13. (*Harvard-Smithsonian Center for Astrophysics*)

15.1 MORE PRECISELY

Relativistic Redshifts and Look-Back Time

When discussing very distant objects, astronomers usually talk about their redshifts rather than their distances. Indeed, it is very common for researchers to speak of an event occurring "at" a certain redshift—meaning that the light received today from that event is redshifted by the specified amount. Of course, because of Hubble's law, redshift and distance are equivalent to one another. However, redshift is the preferred quantity because it is a directly observable property of an object, whereas distance is derived from redshift using Hubble's constant, whose value is not accurately known.

The redshift of a beam of light is, by definition, the *fractional* increase in its wavelength resulting from the recessional motion of the source. ∞ *(Sec. 2.7)* A redshift of 1 corresponds to a *doubling* of the wavelength. Using the formula for the Doppler shift given previously, the redshift of radiation received from a source moving away from us with speed v is

$$\text{redshift} = \frac{\text{observed wavelength} - \text{true wavelength}}{\text{true wavelengh}}$$
$$= \frac{\text{recessional velocity, } v}{\text{speed of light, } c}.$$

Let's illustrate this with two examples, rounding the speed of light, c, to 300,000 km/s. A galaxy at a distance of 100 Mpc has a recessional speed (by Hubble's law) of 70 km/s/Mpc × 100 Mpc = 7000 km/s. Its redshift therefore is 7000 km/s ÷ 300,000 km/s = 0.023. Conversely, an object that has a redshift of 0.05 has a recessional velocity of and hence a distance of 0.05 × 300,000 km/s = 15,000 km/s and hence a distance of 15,000 km/s ÷ 70 km/s/Mpc = 210 Mpc.

Unfortunately, while the foregoing equation is correct for low speeds, it does not take into account the effects of relativity. As we saw in Chapter 13, the rules of everyday physics have to be modified when speeds begin to approach the speed of light, and the formula for the Doppler shift is no exception. ∞ *(More Precisely 13-1)* Our formula is valid for speeds much less than the speed of light, but when $v = c$, the redshift is not 1, as the equation suggests, but is in fact *infinite*. Radiation received from an object moving away from us at nearly the speed of light is redshifted to almost infinite wavelength.

Thus, do not be alarmed to find that many galaxies and quasars have redshifts greater than 1. This does not mean that they are receding faster than light! It simply means that their recessional speeds are relativistic—comparable to the speed of light—and the preceding simple formula is not applicable. Table 15.2 presents a conversion chart relating redshift, recession speed, and present distance. The column headed "v/c" gives recession velocities based on the Doppler effect, taking relativity properly into account (but see Section 17.2 for a more correct interpretation of the redshift). All values are based on reasonable assumptions and are usable even for $v \approx c$. We take Hubble's constant to be 70 km/s/Mpc and assume a flat universe (see Section 17.3 for

details) in which matter (mostly dark) contributes just under one-third of the total density. The conversions in the table are used consistently throughout this text.

Because the universe is expanding, the "distance" to a galaxy is not very well defined—do we mean the distance when the galaxy emitted the light we see today or the present distance (as listed in the table), even though we do not see the galaxy as it is today, or is some other measure more appropriate? Largely because of this ambiguity, astronomers prefer to work in terms of a quantity known as the *look-back time* (last column of Table 15.2), which is simply how long ago an object emitted the radiation we see today. While astronomers talk frequently about redshifts and sometimes about look-back times, they hardly ever talk of distances to high-redshift objects.

For nearby sources, the look-back time is numerically equal to the distance in light-years—the light we receive tonight from a galaxy at a distance of 100 million light-years was emitted 100 million years ago. However, for more distant objects, the look-back time and the present distance in light-years differ because of the expansion of the universe, and the divergence increases dramatically with increasing redshift (see Section 17.2). For example, a galaxy now located 15 billion light-years from Earth was much closer to us when it emitted the light we now see. Consequently, its light has taken considerably less than 15 billion years—in fact, a little less than 10 billion years—to reach us.

TABLE 15.2 Redshift, Distance, and Look-Back Time

Redshift	v/c	Present Distance (Mpc)	Present Distance (10^6 light-years)	Look-Back Time (millions of years)
0.000	0.000	0	0	0
0.010	0.010	42	137	137
0.025	0.025	105	343	338
0.050	0.049	209	682	665
0.100	0.095	413	1,350	1,290
0.250	0.220	999	3,260	2,920
0.500	0.385	1,880	6,140	5,020
1.000	0.600	3,320	10,800	7,730
2.000	0.800	5,250	17,100	10,300
3.000	0.882	6,460	21,100	11,500
4.000	0.923	7,310	23,800	12,100
5.000	0.946	7,940	25,900	12,500
6.000	0.960	8,420	27,500	12,700
8.000	0.976	9,150	29,800	13,000
10.000	0.984	9,660	31,500	13,200
20.000	0.995	11,000	35,900	13,500
100.000	1.000	12,900	42,200	13,700
∞	1.000	14,600	47,500	13,700

Our precise location with respect to the jet can also radically affect the *type* of radiation we see. The theory of relativity tells us that radiation emitted by particles moving close to the speed of light is strongly concentrated, or beamed, in the direction of motion. ∞ *(More Precisely 13-1)* As a result, if the observer in Figure 15.25 happens to be directly in line with the beam, the radiation received will be both very intense and Doppler shifted toward short wavelengths. ∞ *(Sec. 2.7)* The resulting object is called a **blazar.** Much of the luminosity of the several hundred known blazars is received in the form of X-rays or gamma rays.

Jets are common in active galaxies of all types. Figure 15.26 presents several images of the giant elliptical galaxy M87, a prominent member of the Virgo Cluster (see the chapter-opening image on p. 418 and Figure 15.13). Detailed study of the central regions of this fairly normal looking E0 galaxy (Figure 15.26a) reveal that its core contains a long (2 kpc) thin jet of matter ejected from the galactic center at nearly the speed of light. Computer enhancement shows that it is made up of a series of distinct "blobs" more or less evenly spaced along its length, suggesting that the material was ejected during bursts of activity. The jet has also been imaged in the radio, infrared (Figure 15.26b), and X-ray regions of the spectrum.

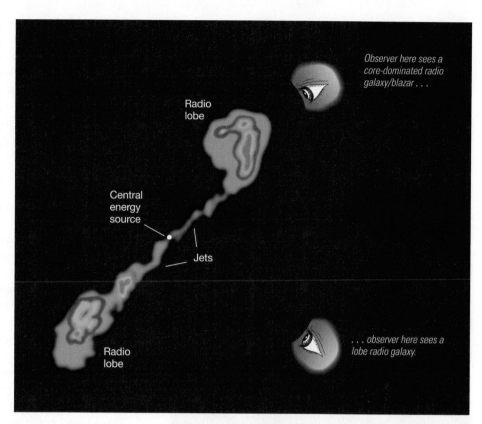

▲ FIGURE 15.25 **Radio Galaxy** A central energy source produces high-speed jets of matter that interact with intergalactic gas to form radio lobes. The system may appear to us as either radio lobes or a core-dominated radio galaxy, depending on our location with respect to the jets and lobes.

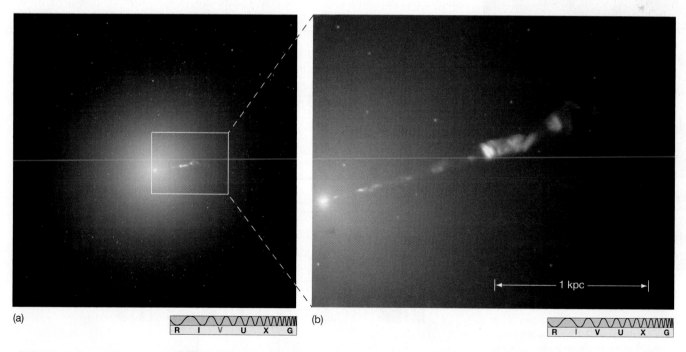

(a)

(b)

▲ FIGURE 15.26 **M87 Jet INTERACTIVE**   (a) This *Hubble* image of the innermost 5 kpc of the giant elliptical M87 shows an intriguing jet of matter emerging from the galactic nucleus. (b) An infrared image of M87's jet, showing more detail. *(NASA/ESA)*

▼ **FIGURE 15.27 Quasar 3C 273** (a) The bright quasar 3C 273 displays a luminous jet of matter, but the main body of the quasar is starlike in appearance. The spikes are a camera artifact. (b) The jet extends for about 30 kpc and emits radiation across the electromagnetic spectrum, here in X-rays, as seen by *Chandra.* (NASA/ESA)

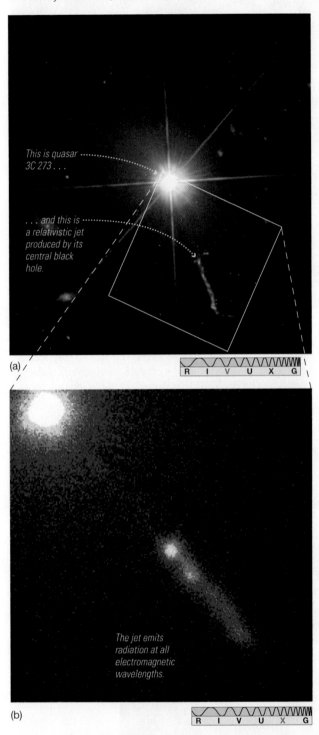

This is quasar 3C 273 . . .

. . . and this is a relativistic jet produced by its central black hole.

(a)

R I V U X G

The jet emits radiation at all electromagnetic wavelengths.

(b)

R I V U X G

▶ **FIGURE 15.28 Quasar Spectrum** Optical spectrum of the distant quasar 3C 273. (This is a negative, as the lines are actually in emission.) Notice the redshift of the three hydrogen spectral lines marked as Hβ, Hγ, and Hδ. ∞ *(Sec. 2.6)* The widths of these lines imply rapid internal motion within the quasar. *(Adapted from Palomar/Caltech)*

Quasars

In the early days of radio astronomy, many radio sources were detected for which no corresponding visible object was known. By 1960 several hundred such sources were listed in the *Third Cambridge Catalog*, and astronomers were scanning the skies in search of visible counterparts to these radio sources. Their job was made difficult both by the low resolution of the radio observations (which meant that the observers did not know exactly where to look) and by the faintness of these objects at visible wavelengths.

In 1960 astronomers detected what appeared to be a faint blue star at the location of the radio source 3C 48 (the 48th object on the third *Cambridge* list) and obtained its spectrum. Containing many unknown broad emission lines, the unusual spectrum defied interpretation. 3C 48 remained a unique curiosity until 1962, when another similar-looking, and similarly mysterious, faint blue object with "odd" spectral lines was discovered and identified with the radio source 3C 273 (Figure 15.27).

The following year saw a breakthrough when astronomers realized that the strongest unknown lines in 3C 273's spectrum were simply familiar spectral lines of hydrogen redshifted by a very unfamiliar amount—about 16 percent, corresponding to a recession velocity of 48,000 km/s. Figure 15.28 shows the spectrum of 3C 273, with prominent emission lines and their redshifts marked. Once the nature of the strange spectral lines was known, astronomers quickly found a similar explanation for the spectrum of 3C 48. Its 37 percent redshift implied that it is receding from Earth at almost one-third the speed of light.

These huge speeds mean that neither of these two objects can possibly be members of our Galaxy. In fact, their large redshifts mean that they are very far away indeed. Applying Hubble's law (with our adopted value of the Hubble constant $H_0 = 70$ km/s/Mpc), we obtain distances of 650 Mpc for 3C 273 and 1400 Mpc for 3C 48. (See *More Precisely 15-1* and Chapter 17 for an account of how these distances are determined and an explanation of what these large redshifts really mean.)

However, this explanation of the unusual spectra created a greater mystery. A simple calculation using the inverse-square law reveals that despite their unimpressive optical appearance these faint "stars" are in fact the brightest known objects in the universe! ∞ *(Sec. 10.2)* 3C 273, for example, has a luminosity of about 10^{40} W, comparable to 20 trillion Suns or 1000 Milky Way Galaxies. More generally, quasars range in luminosity from around 10^{38} W—about the same as the brightest Seyferts—up to nearly 10^{42} W. A value of 10^{40} W (comparable to the luminosity of a bright radio galaxy) is fairly typical.

Hβ   Hγ   Hδ

Redshift

This is the observed spectrum of 3C 273.

Red

Blue

This is a lab comparison spectrum.

Hβ   Hγ Hδ

600 nm   500 nm   400 nm

Clearly not stars (with such enormous luminosities), these objects became known as *quasi-stellar radio sources* (quasi-stellar means "starlike"), or **quasars.** Because we now know that not all such highly redshifted, starlike objects are strong radio sources, the term **quasi-stellar object** (or QSO) is more common today. However, the name quasar persists, and we will continue to use it here. More than 200,000 quasars are now known, and the numbers are increasing rapidly as large-scale surveys probe deeper and deeper into space. The distance to the *closest* quasar is 240 Mpc; the farthest lies some 9000 Mpc away. Most quasars lie well over 1000 Mpc from Earth. Since light travels at a finite speed, these faraway objects represent the universe as it was in the distant past. Most quasars date back to much earlier periods of galaxy formation and evolution, rather than more recent times. The prevalence of these energetic objects at great distances tells us that the universe was once a much more violent place than it is today.

Quasars share many properties with Seyferts and radio galaxies. Their radiation is nonstellar and may vary irregularly in brightness over periods of months, weeks, days, or (in some cases) even hours. Some show evidence of jets and extended emission features. Note the luminous jet in 3C 273 (Figure 15.27), reminiscent of the jet in M87 and extending nearly 30 kpc from the center of the quasar. Figure 15.29 shows a quasar with radio lobes similar to those in Cygnus A (Figure 15.23b). Quasars have been observed in all parts of the electromagnetic spectrum, but many emit most of their energy in the infrared. About 10–15 percent of quasars (called "radio-loud" quasars) also emit substantial amounts of energy at radio wavelengths, presumably as a result of unresolved jets.

Astronomers once distinguished between active galaxies and quasars on the basis of their appearance, spectra, and distance from us, but today most astronomers think that quasars are in fact just the intensely bright nuclei of distant active galaxies lying too far away for the galaxies themselves to be seen. (Figure 16.18 presents *Hubble Space Telescope* observations of two relatively nearby quasars in which the surrounding galaxies are clearly visible.)

15.5 The Central Engine of an Active Galaxy

The present consensus among astronomers is that, despite their differences in appearance and luminosity, Seyferts, radio galaxies, and quasars share a common energy-generation mechanism. As a class, active galactic nuclei have some or all of the following properties:

1. They have *high luminosities*, generally greater than the 10^{37} W characteristic of a bright normal galaxy.
2. Their energy emission is mostly *nonstellar*—it cannot be explained as the combined radiation of even trillions of stars.
3. Their energy output can be highly *variable*, implying that it is emitted from a small central nucleus much less than a parsec across.
4. They often exhibit *jets* and other signs of explosive activity.
5. Their optical spectra may show broad emission lines, indicating *rapid internal motion* within the energy-producing region.
6. Often the activity appears to be associated with *interactions* between galaxies.

The principal questions then are: How can such vast quantities of energy arise from these relatively small regions of space? Why is the radiation nonstellar? And what is the origin of the jets and extended radio-emitting lobes? We first consider how the energy is *produced,* then turn to the question of how it is actually *emitted* into intergalactic space.

PROCESS OF SCIENCE CHECK

How did the determination of quasar distances change astronomers' understanding of these objects?

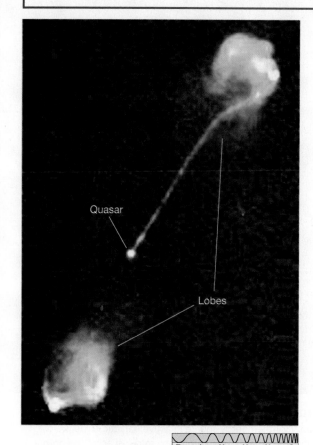

▲ **FIGURE 15.29 Quasar Jets**   This radio image of the quasar 3C 175, which is some 3000 Mpc away, shows radio jets feeding faint radio lobes. The lobes span approximately a million light-years—comparable in size to the radio galaxies discussed earlier. *(NRAO)*

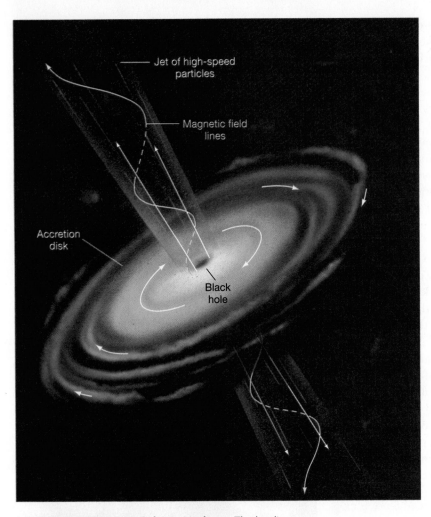

▲ FIGURE 15.30 Active Galactic Nucleus The leading theory for the energy source in active galactic nuclei holds that these objects are powered by material accreting onto a supermassive black hole. As matter spirals toward the hole, it heats up, producing large amounts of energy. High-speed jets of gas may be ejected perpendicular to the accretion disk, forming the jets and lobes seen in many active objects. Magnetic fields generated in the disk by charged matter in motion are carried by the jets out to the radio lobes, where they play a crucial role in producing the observed radiation.

ANIMATION/VIDEO (MA)
Cosmic Jets

CONCEPT CHECK

How does accretion onto a supermassive black hole power the energy emission from the extended radio lobes of a radio galaxy?

Energy Production

As illustrated in Figure 15.30, the leading model for the central engines of active galaxies is a scaled-up version of the process powering X-ray binaries in our Galaxy and the activity in the Galactic center—accretion of gas onto a supermassive black hole, releasing huge amounts of energy as the matter sinks onto the central object. ∞ *(Secs. 13.3, 13.8, 14.7)* In order to power the brightest active galaxies, theory suggests that the black holes involved must be *billions* of times more massive than the Sun.

As with this model's smaller-scale counterparts, infalling gas forms an accretion disk and spirals down toward the black hole. It is heated to high temperatures by friction within the disk and emits large amounts of radiation as a result. In this case, however, the origin of the accreted gas is not a binary companion, as in stellar X-ray sources, but rather entire stars and clouds of interstellar gas, most likely diverted into the galactic center by an encounter with another galaxy, that come too close to the hole and are torn apart by its strong gravity.

Accretion is extremely efficient at converting infalling mass (in the form of gas) into energy (in the form of electromagnetic radiation). As much as 10 or 20 percent of the total mass–energy of the infalling matter can be radiated away before it crosses the hole's event horizon and is lost forever. ∞ *(Sec. 13.5)* Since the total mass–energy of a star like the Sun—the mass times the speed of light squared—is about 2×10^{47} J, it follows that the 10^{38} W luminosity of a bright active galaxy can be produced by the consumption of "only" 1 solar mass of gas per decade by a billion-solar-mass black hole. More or less luminous active galaxies would require correspondingly more or less fuel. To power a 10^{40}-W quasar, which is 100 times brighter, the black hole consumes 100 times more fuel, or 10 stars per year. The central black hole of a 10^{36}-W Seyfert galaxy would devour only one Sun's worth of material every thousand years.

Even a billion-solar-mass black hole has a radius of only 3×10^9 km, or 10^{-4} pc—about 20 AU—so the part of the accretion disk responsible for most of the emission is much less than 1 pc across. ∞ *(Sec. 13.5)* Instabilities in the disk cause fluctuations in the energy released, leading to the variability observed in many objects. The broadening of the spectral lines observed in the nuclei of many active galaxies results from the rapid orbital motion of the gas in the black hole's intense gravity.

Jets appear to be a common feature of accretion flows, large and small, and consist of material (mainly protons and electrons) blasted out into space—and completely out of the visible portion of the galaxy—from the inner regions of the disk. They are most likely formed by strong magnetic fields produced within the accretion disk itself, which accelerate charged particles to nearly the speed of light and eject them parallel to the disk's rotation axis. Figure 15.31 shows a *Hubble Space Telescope* image of a disk of gas and dust at the core of the radio galaxy NGC 4261 in the Virgo Cluster. Consistent with the theory just described, the disk is perpendicular to the huge jets emanating from the galaxy's center.

Figure 15.32 shows further evidence for this model in the form of imaging and spectroscopic data from the center of M87, suggesting a rapidly rotating disk of matter orbiting the galaxy's center, again perpendicular to the jet. Measurements of the gas velocity on opposite sides of the disk indicate that the mass within a few parsecs of the center is approximately 3×10^9 solar masses—we assume that this is the mass of the central black hole.

► FIGURE 15.31 **Giant Elliptical Galaxy** (a) A combined optical/radio image of the giant elliptical galaxy NGC 4261, in the Virgo Cluster, shows a white visible galaxy at center, from which blue-orange (false color) radio lobes extend for about 60 kpc. (b) A close-up of the galaxy's nucleus reveals a 100-pc-diameter disk surrounding a bright hub thought to harbor a black hole. *(NRAO; NASA)*

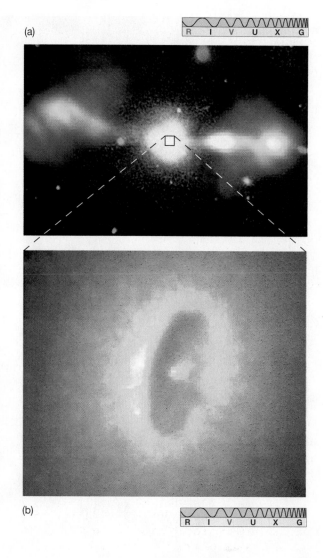

Energy Emission

The radiation emitted by the hot accretion disk surrounding a supermassive black hole spans a broad range of wavelengths, from infrared through X-rays, corresponding to the broad range of gas temperatures in the disk. This accounts for the spectra of some active galactic nuclei, but as mentioned earlier, it appears that in many cases the high-energy radiation emitted from the accretion disk itself is "reprocessed"—that is, absorbed and reemitted at infrared wavelengths—by dust surrounding the nucleus before eventually reaching our detectors.

Many researchers think that the most likely site of this reprocessing is in a rather fat, donut-shaped ring of gas and dust surrounding the inner accretion disk where the energy is actually produced. As illustrated in Figure 15.33, if our line of sight to the black hole does not intersect the dusty donut, we see the "bare" energy source, emitting large amounts of high-energy radiation. If the donut intervenes, we see instead large amounts of infrared radiation reradiated from the dust. The structure of the donut itself is quite uncertain and may in reality bear little resemblance to the rather regular-looking ring shown in the figure. Some astronomers prefer a different explanation: If quasars, like stars are born shrouded in dense gas, then the difference between "reprocessed" and "bare" systems may be an evolutionary one, as the radiation from the obscured disk disperses its surrounding cocoon and bursts into view. ∞ *(Sec. 11.2)*

A different reprocessing mechanism operates in many jets and radio lobes. It involves the *magnetic fields* possibly produced within the accretion disk and transported by the jets into intergalactic space (Figure 15.30). As sketched in Figure 15.34(a), whenever a charged particle (here an electron) encounters a magnetic field, the particle tends to spiral around the magnetic field lines. We have already encountered this idea on smaller scales in Earth's magnetosphere and at the center of our own Galaxy. ∞ *(Sec.5.7, 9.4, 13.2, 14.7)*

As the particles whirl around, they emit electromagnetic radiation. ∞ *(Sec. 2.2)* The radiation produced in this way—called **synchrotron radiation**, after the type

► FIGURE 15.32 **M87 Disk** Both images and spectra of M87 support the idea of a rapidly whirling accretion disk at this galaxy's heart. (a) The central region of M87, similar to that shown in Figure 15.26(c), shows its bright nucleus and jet. (b) A magnified view of the nucleus implies a spiral swarm of stars, gas, and dust. (c) Spectral-line features observed on opposite sides of the nucleus show contrasting red and blue Doppler shifts, implying that matter on one side is coming toward us while matter on the other side is moving away. Apparently, an accretion disk spins perpendicular to the jet, and at its center is a black hole having some three billion times the mass of the Sun. *(NASA)*

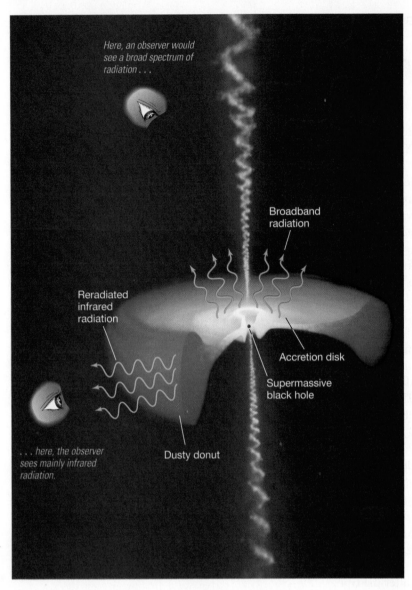

Here, an observer would see a broad spectrum of radiation . . .

Broadband radiation

Reradiated infrared radiation

Accretion disk

Supermassive black hole

Dusty donut

. . . here, the observer sees mainly infrared radiation.

◀ **FIGURE 15.33 Dusty Donut**   The accretion disk surrounding a massive black hole, drawn here with some artistic license, consists of hot gas at many different temperatures (hottest nearest the center). When viewed from above or below, the disk radiates a broad spectrum of electromagnetic energy extending into the X-ray band. However, the dusty infalling gas that ultimately powers the system is thought to form a fat, donut-shaped region outside the accretion disk (shown here in red), which effectively absorbs much of the high-energy radiation reaching it, reemitting it mainly in the form of cooler, infrared radiation. When the accretion disk is viewed from the side, strong infrared emission is observed. (Compare with Figure 15.25).

of particle accelerator in which it was first observed—is *nonthermal* in nature—the emitted radiation is not described, even approximately, by a blackbody curve. Instead, its intensity decreases with increasing frequency, as shown in Figure 15.34(b). This is just what is needed to explain the overall spectrum of radiation from many active galaxies. Observations of the radiation received from the jets and radio lobes of active galaxies are completely consistent with this process.

Eventually, the jet is slowed and stopped by the intergalactic medium, the flow becomes turbulent, and the magnetic field grows tangled. The result is a gigantic radio lobe emitting virtually all of its energy in the form of synchrotron radiation. Thus, even though the radio *emission* comes from an enormously extended volume of space that dwarfs the visible galaxy, the *source* of the energy is still the accretion disk—a billion billion times smaller in volume than the radio lobe—lying at the galactic center. The jets transport energy from the nucleus, where it is generated, into the lobes, where it is finally radiated into space.

The existence of the inner lobes of Centaurus A and the blobs in M87's jet imply that the formation of a jet may be an intermittent process (or, as in the case of the Seyferts discussed earlier, may not occur at all). Many nearby active galaxies (like Centaurus A) appear to have been "caught in the act" of interacting with another galaxy, suggesting that the fuel supply can be "turned on" by a companion.

We will explore further the connection between normal and active galaxies in Chapter 16, where we turn to the subject of *galaxy evolution*.

▶ **FIGURE 15.34 Nonthermal Radiation**   (a) Charged particles, especially fast-moving electrons (red), emit synchrotron radiation (blue) while spiraling in a magnetic field (black). Thermal and synchrotron (nonthermal) radiation vary differently with frequency. Thermal radiation, described by a blackbody curve, peaks at a frequency that depends on the temperature of the source. By contrast, nonthermal synchrotron radiation (b) is more intense at low frequencies and is independent of the temperature of the emitting object. (Compare with Figure 15.18.)

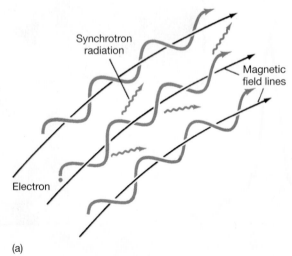

Synchrotron radiation

Magnetic field lines

Electron

(a)

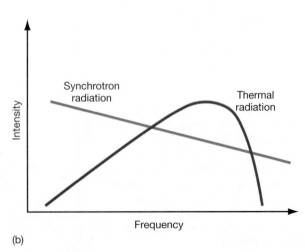

Synchrotron radiation

Thermal radiation

Intensity

Frequency

(b)

THE BIG QUESTION

Galactic research lags stellar research by about 50 years. That's because galaxies were discovered only in the 20th century, and we are still learning about them. How did they form, and how do they evolve? Those are the biggest questions regarding galaxies, and they will not be answered until more and better data accumulate, especially regarding the most distant systems. Will the much larger galaxy surveys now on the horizon help solve these important issues?

CHAPTER REVIEW

SUMMARY

LO1 The **Hubble classification scheme** (p. 420) divides galaxies into several classes, depending on their appearance. **Spiral galaxies** (p. 421) have flattened disks, central bulges, and spiral arms. Their halos consist of old stars, whereas the gas-rich disks are the sites of ongoing star formation. **Barred-spiral galaxies** (p. 422) contain an extended "bar" of material projecting beyond the central bulge. **Elliptical galaxies** (p. 422) have no disk and contain little or no cool gas or dust, although very hot interstellar gas is observed. In most cases, they consist entirely of old stars. They range in size from dwarf ellipticals, which are much less massive than the Milky Way Galaxy, to giant ellipticals, which may contain trillions of stars. **S0 and SB0 galaxies** (p. 423) are intermediate in their properties between ellipticals and spirals. **Irregular galaxies** (p. 424) are galaxies that do not fit into either of the other categories. Some may be the result of galaxy collisions or close encounters. Many irregulars are rich in gas and dust and are the sites of vigorous star formation.

LO2 Astronomers often use **standard candles** (p. 426) as distance-measuring tools. These are objects that are easily identifiable and whose luminosities lie in some reasonably well-defined range. Comparing luminosity and apparent brightness, astronomers determine the distance using the inverse-square law. An alternative approach is the **Tully–Fisher relation** (p. 427), an empirical correlation between rotational velocity and luminosity in spiral galaxies.

LO3 The Milky Way, Andromeda, and several other smaller galaxies form the **Local Group** (p. 428), a small **galaxy cluster** (p. 429). Galaxy clusters consist of a collection of galaxies orbiting one another, bound together by their own gravity. The nearest large galaxy cluster to the Local Group is the Virgo Cluster.

LO4 Distant galaxies are observed to be receding from the Milky Way at speeds proportional to their distances from us. This relationship is called **Hubble's law** (p. 430). The constant of proportionality in the law is **Hubble's constant** (p. 431). Its value is approximately 70 km/s/Mpc. Astronomers use Hubble's law to determine distances to the most remote objects in the universe simply by measuring their redshifts and converting the corresponding speed directly to a distance. The redshift associated with the Hubble expansion is called the **cosmological redshift** (p. 431).

LO5 **Active galaxies** (p. 432) are much more luminous than normal galaxies and have nonstellar spectra, emitting most of their energy outside the visible part of the electromagnetic spectrum. Often the nonstellar activity suggests rapid internal motion and is associated with a bright central **active galactic nucleus** (p. 433). Many active galaxies have high-speed, narrow jets of matter shooting out from their central nuclei. Often the jets transport energy from the nucleus, where it is generated, to enormous **radio lobes** (p. 434) lying far beyond the visible portion of the galaxy, where it is radiated into space. The jets often appear to be made up of distinct "blobs" of gas, suggesting that the process generating the energy is intermittent.

LO6 A **Seyfert galaxy** (p. 433) looks like a normal spiral, except that the Seyfert has an extremely bright central **galactic nucleus** (p. 433). Spectral lines from Seyfert nuclei are very broad, indicating rapid internal motion, and rapid variability implies that the source of the radiation is much less than 1 light-year across. **Radio galaxies** (p. 434) emit large amounts of energy in the radio part of the spectrum. The corresponding visible galaxy is usually elliptical. **Quasars** (p. 439), or **quasi-stellar objects** (p. 439), are the most luminous objects known. In visible light they appear starlike, and their spectra are usually substantially redshifted. All quasars are very distant, indicating that we see them as they were in the distant past.

LO7 The generally accepted explanation for the observed properties of all active galaxies is that the energy is generated by accretion of galactic gas onto a supermassive (million- to billion-solar-mass) black hole lying in the galactic center. The small size of the accretion disk explains the compact extent of the emitting region, and the high-speed orbit of gas in the black hole's intense gravity accounts for the rapid motion observed. Much of the high-energy radiation emitted from the disk is absorbed by a fat, donut-shaped region outside the disk and reemitted in the form of infrared radiation. Typical active galaxy luminosities require the consumption of about 1 solar mass of material every few years. Some of the infalling matter is blasted out into space, producing magnetized jets that create and feed the radio lobes. Charged particles spiraling around the magnetic field lines produce **synchrotron radiation** (p. 441), whose spectrum is consistent with the nonstellar radiation observed in radio galaxies and jets.

MasteringAstronomy®   For instructor-assigned homework go to www.masteringastronomy.com

Problems labeled **POS** explore the process of science. **VIS** problems focus on reading and interpreting visual information. **LO** connects to the introduction's numbered Learning Outcomes.

REVIEW AND DISCUSSION

1. **LO1** Distinguish among the various types of spiral galaxies.

2. Compare an elliptical galaxy with the halo of the Milky Way.

3. **LO2** Describe the four rungs in the distance-measurement ladder involved in determining the distance to a galaxy lying 5 Mpc away.

4. How does the Tully–Fisher relation allow astronomers to measure the distances to galaxies?

5. **LO3** What is the Virgo Cluster?

6. **LO4** What is Hubble's law, and how is it used by astronomers to measure distances to galaxies?

7. What is the most likely range of values for Hubble's constant?

8. Why do astronomers prefer to speak in terms of redshifts rather than distances to faraway objects?

9. **LO5** Name two differences between normal and active galaxies.

10. **LO6 POS** What is the evidence that the radio lobes of some active galaxies consist of material ejected from the galaxy's nucleus?

11. **POS** How do we know that the energy-emitting regions of many active galaxies must be very small?

12. **LO7** Briefly describe the leading model for the central engine of an active galaxy.

13. **POS** How do astronomers account for the differences in the spectra observed from active galaxies?

14. **POS** How is the process of synchrotron emission related to observations of active galaxies?

15. How do we know that quasars are extremely luminous?

CONCEPTUAL SELF-TEST: TRUE OR FALSE?/MULTIPLE CHOICE

1. Most galaxies are spirals like the Milky Way. (T/F)

2. Most elliptical galaxies contain only old stars. (T/F)

3. Irregular galaxies, although small, often have lots of star formation taking place in them. (T/F)

4. Type I supernovae can be used to determine distances to galaxies. (T/F)

5. Most galaxies are receding from the Milky Way Galaxy. (T/F)

6. Hubble's law can be used to determine distances to the farthest objects in the universe. (T/F)

7. Active galaxies can emit thousands of times more energy than our own Galaxy. (T/F)

8. Using the method of standard candles, we could, in principle, estimate the distance to a campfire if we knew (a) the number of logs used; (b) the fire's temperature; (c) the length of time the fire has been burning; (d) the type of wood used in the fire.

9. Within 30 Mpc of the Sun, there are about (a) 3 galaxies; (b) 30 galaxies; (c) a few thousand galaxies; (d) a few million galaxies.

10. Hubble's law states that (a) more distant galaxies are younger; (b) the greater the distance to a galaxy, the greater is the galaxy's redshift; (c) most galaxies are found in clusters; (d) the greater the distance to a galaxy, the fainter the galaxy appears.

11. If the light from a galaxy fluctuates in brightness very rapidly, the region producing the radiation must be (a) very large; (b) very small; (c) very hot; (d) rotating very rapidly.

12. Quasar spectra (a) are strongly redshifted; (b) show no spectral lines; (c) look like the spectra of stars; (d) contain emission lines from unknown elements.

13. Active galaxies are very luminous because they (a) are hot; (b) contain black holes in their cores; (c) are surrounded by hot gas; (d) emit jets.

14. VIS If the galaxy in Figure 15.10 (Galaxy Rotation) were smaller and spinning more slowly, then the figure should be redrawn to show (a) a greater blueshift; (b) a greater redshift; (c) a narrower combined line; (d) a larger combined amplitude.

15. VIS According to Figure 15.18 (Galaxy Energy Spectra), active galaxies (a) emit most of their energy at long wavelengths; (b) emit very little energy at high frequencies; (c) emit large amounts of energy at all wavelengths; (d) emit most of their energy in the visible part of the spectrum.

PROBLEMS

The number of squares preceding each problem indicates its approximate level of difficulty.

1. ■ The Andromeda galaxy lies 800 kpc from our Galaxy and is approaching it with a radial velocity of 266 km/s. Neglecting both the transverse component of the velocity and the effect of gravity in accelerating the motion, estimate when the two galaxies will collide.

2. ■■ A Cepheid variable star in the Virgo Cluster has an apparent magnitude of 26 and is observed to have a pulsation period of 20 days. Use these numbers and Figure 14.6 to estimate the distance to the Virgo Cluster. The absolute magnitude of the Sun is +4.85.

3. ■ A supernova 1 billion times more luminous than the Sun is used as a standard candle to measure the distance to a faraway galaxy. From Earth, the supernova appears as bright as the Sun would appear from a distance of 10 kpc. What is the distance to the galaxy?

4. ■ According to Hubble's law, with, with H_0 = 70 km/s/Mpc, what is the recessional speed of a galaxy at a distance of 200 Mpc? How far away is a galaxy whose recessional speed is 4000 km/s? How would these answers change if (a) H_0 = 50 km/s/Mpc, (b) H_0 = 80 km/s/Mpc?

5. ■■ If H_0 = 70 km/s/Mpc, how long will it take for the distance from the Milky Way Galaxy to the Virgo Cluster to double? How long for the distance to the Coma Cluster (100 Mpc) to double?

6. ■■ A quasar is observed to have luminosity fluctuations and broadened emission lines indicating a speed of 5000 km/s within 0.1 light-years of its center. Assuming circular orbits, use Kepler's laws to estimate the mass within this radius. ∞ (Sec. 14.6)

7. ■ Assuming a jet speed of 75 percent of the speed of light, calculate the time taken for material in Cygnus A's jet to cover the 250 kpc between the galaxy's nucleus and its radio-emitting lobes.

8. ■■ Assuming the upper end of the efficiency range indicated in the text, calculate how much energy an active galaxy would generate if it consumed one Earth mass of material every day. Compare this value with the luminosity of the Sun.

9. ■ Using the data presented in the text, calculate the orbital speed of material orbiting at a distance of 0.25 pc from the center of M87.

10. ■■■ Calculate the amount of energy received per unit area per unit time that would be observed at Earth from a 10^{37}-W Seyfert nucleus located at the Galactic center, 8 kpc away, neglecting the effects of absorption by interstellar dust. Using the data presented in Appendix 3, Table 3, compare this with the energy received from Sirius A, the brightest star in the night sky. From what you know about active galaxy energy emission, is it reasonable to ignore interstellar absorption?

ACTIVITIES

Collaborative

1. Observe the Virgo Cluster of galaxies. An 8-inch telescope is the optimal size for this project. The constellation Virgo is visible from the United States during much of fall through spring. To locate the cluster, first find the constellation Leo. The eastern part of Leo is composed of a distinct triangle of stars, Denebola (β), Chort (θ), and Zosma (δ). Go from Chort to Denebola in a straight line east; continue on the same distance as between the two stars, and you will be approximately at the center of the Virgo Cluster. Look for the following Messier objects, the brightest galaxies in the cluster: M49, M58, M59, M60, M84, M86, M87, M89, and M90. Examine each galaxy for unusual features; some have very bright nuclei. Sketch or photograph what you see, and construct your own pictorial catalog of the brightest galaxies in Virgo.

Individual

1. In the previous exercise, you were given directions for finding the Virgo Cluster of galaxies. M87, in the central part of this cluster, is the nearest core–halo radio galaxy. M87 has coordinates RA = 12^h 30.8^m, dec = +12° 24′. At magnitude 8.6 it should not be difficult to find with an 8-inch telescope. Its distance is roughly 18 Mpc. Describe its nucleus; compare what you see with other nearby ellipticals in the Virgo Cluster.

2. 3C 273 is the nearest and brightest quasar, but that does not mean it is easy to find! Its coordinates are RA = 12^h 29.2^m, dec = +2° 03′. It is located in the southern part of the Virgo Cluster, but it is not associated with it. At magnitude 12–13 (again, it is variable), it may require a 10- or 12-inch telescope to see, but try it first with an 8-inch. It should appear as a very faint star. The significance of seeing this object is that it is 640 Mpc distant. The light you are seeing left this object over 2 billion years ago! It is the most distant object observable with a small telescope.

16

Galaxies and Dark Matter

The Large-Scale Structure of the Cosmos

On scales much larger than even the largest galaxy clusters, new levels of structure are revealed, and a humbling new reality emerges. We may be star stuff, the product of countless cycles of stellar evolution, but we are not the stuff of the cosmos. The universe in the large is composed of matter fundamentally different from the familiar atoms and molecules that compose our bodies, our planet, our star and galaxy, and all the luminous matter we observe in the heavens. Only its gravity announces the presence of this strange kind of matter. By comparing and classifying the properties of galaxies near and far, astronomers have begun to understand their evolution. By mapping out their distribution in space, astronomers trace out the immense realms of the universe. Points of light in the uncharted darkness, the galaxies remind us that our position in the universe is no more special than that of a boat adrift at sea.

THE BIG PICTURE

Galaxies are among the grandest, most beautiful objects in the universe—each one a colossal collection of hundreds of billions of stars held together loosely by gravity. Galaxies dominate our view of deep space—they seem to be everywhere—yet they represent just a tiny fraction of all matter in the cosmos. Vast quantities of unseen cosmic material—dark matter—actually account for most of the mass in the universe.

◄ Galaxies are among the grandest, most beautiful objects in the universe—each one a colossal collection of hundreds of billions of stars held together loosely by gravity. Some are bright and splendid, like the two big ones in this image. Others are dim and distant, like several that appear smaller in the background. This pair of galaxies, nearly 300 million light-years away and known collectively as Arp 273, is in the process of colliding over millions of years. Notice the rose-like shape of the top galaxy, caused by the gravitational pull of the bottom one, and the swath of clusters of young blue stars glowing like jewels. Mergers and acquisitions are apparently common among galaxies, but astronomers still don't understand well how the galaxies originated long ago. *(STScI)*

16.1 Dark Matter in the Universe

In Chapter 14 we saw how measurements of the orbital velocities of stars and gas in our Galaxy reveal the presence of an extensive *dark-matter halo* surrounding the galaxy we see. ∞ *(Sec. 14.6)* Do other galaxies have similar dark halos? And what evidence do we have for dark matter on larger scales? To answer these questions, we need to know the masses of galaxies and galaxy clusters.

How can we measure the masses of such large systems? Surely, we can neither count all their stars nor estimate their interstellar content very well. Galaxies are just too complex to take direct inventory of their material makeup. Instead, we must rely on indirect techniques. Despite their enormous sizes, galaxies and galaxy clusters obey the same physical laws as do the planets in our own solar system. To determine galaxy masses, we turn as usual to Newton's law of gravity.

Galaxy Masses

Astronomers can determine the masses of some spiral galaxies by measuring their *rotation curves*, which plot rotation speed versus distance from the galactic center (Figure 16.1a). ∞ *(Sec. 14.6)* The mass within any given radius then follows directly from Newton's laws. Rotation curves for a few nearby spirals are shown in Figure 16.1(b). They imply masses ranging from about 10^{11} to 5×10^{11} solar masses within about 25 kpc of the center—comparable to the results obtained for our own Galaxy using the same technique. For galaxies too distant for detailed rotation curves to be drawn, the overall rotation speed can still be inferred from the broadening of spectral lines, as described in the previous chapter. ∞ *(Sec. 15.2)*

This approach is useful for measuring the mass lying within about 50 kpc of a galaxy's center (the extent of the electromagnetic emission from stellar and interstellar material). To probe farther from the centers of galaxies, astronomers turn to *binary* galaxy systems (Figure 16.2a), whose components may lie hundreds of kiloparsecs apart. The orbital period of such a system is typically billions of years, far too long for the orbit to be accurately measured. However, by estimating the period and semimajor axis using the information available—the line-of-sight velocities and angular separation of the components—an approximate total mass can be derived. ∞ *(Sec. 1.4)*

Galaxy masses obtained in this way are fairly uncertain, but by combining many such measurements, astronomers can obtain quite reliable *statistical* information about galaxy masses. Most normal spirals (the Milky Way Galaxy included) and large ellipticals contain between 10^{11} and 10^{12} solar masses of material. Irregular galaxies often contain less mass, about 10^8 to 10^{10} times that of the Sun. Dwarf ellipticals and dwarf irregulars can contain as little as 10^6 or 10^7 solar masses of material.

We can use another statistical technique to derive the combined mass of all the galaxies within a galaxy cluster. As depicted in Figure 16.2(b), each galaxy within a cluster moves relative to all other cluster members, and we can estimate the cluster's mass simply by asking how massive it must be in order to bind its galaxies gravitationally. Typical cluster masses obtained in this way lie in the range of 10^{13}–10^{14} solar masses. Notice that this calculation gives us no information whatsoever about the masses of individual galaxies. It tells us only about the *total* mass of the entire cluster.

INTERACTIVE (MA)
Rotation Curve for a Merry-Go-Round

ANIMATION/VIDEO (MA)
Dark Matter

(a)

The Milky Way's rotation curve (see Fig. 14.20) is marked in red for comparison.

NGC 4984
NGC 4378
NGC 3145
NGC 1620
NGC 7664

Rotational speed (km/s) — 50, 150, 250, 350
Distance from galactic center (kpc) — 0, 5, 10, 15, 20, 25

(b)

◀ **FIGURE 16.1 Galaxy Rotation Curves INTERACTIVE**   (a) Orbital velocities can be measured at different distances from the center of a disk galaxy, as illustrated here for M64, the "Evil Eye" galaxy, some 5 Mpc distant. (b) Rotation curves for some nearby spiral galaxies indicate masses of a few hundred billion times the mass of the Sun. *(NASA)*

Visible Matter and Dark Halos

The rotation curves of the spiral galaxies shown in Figure 16.1(b) remain flat (that is, do not decline and may even rise slightly) far beyond the visible images of the galaxies themselves, implying that these galaxies—and perhaps all spiral galaxies—have invisible dark halos similar to that surrounding the Milky Way. ∞ *(Sec. 14.5)* Spiral galaxies seem to contain from 3 to 10 times more mass than can be accounted for in the form of visible matter. Studies of elliptical galaxies suggest similarly large dark halos surrounding those galaxies, too.

Astronomers find even greater discrepancies between visible light and total mass when they study galaxy clusters. Calculated cluster masses range from 10 to nearly 100 times the mass suggested by the light emitted by individual cluster galaxies. Put another way, a lot more mass is needed to bind galaxy clusters than we can see. Thus, the problem of dark matter exists not just in our own Galaxy, but also in other galaxies and, to an even greater degree, in galaxy clusters, too. In that case, we are compelled to accept the fact that *upward of 90 percent of the matter in the universe is dark*. As noted in Chapter 14, this matter is not just dark in the visible portion of the spectrum—it is undetected at *any* electromagnetic wavelength. ∞ *(Sec. 14.5)*

The dark matter in clusters is not simply the accumulation of dark matter within individual galaxies. Even including the galaxies' dark halos, we still cannot account for all the dark matter in galaxy clusters. As we look on larger and larger scales, we find that a larger and larger fraction of the matter in the universe is dark.

Intracluster Gas

In addition to the luminous matter observed within the cluster galaxies themselves, astronomers also have evidence for large amounts of *intracluster gas—* superhot (more than 10 million K), diffuse intergalactic matter filling the space among the galaxies. Satellites orbiting above Earth's atmosphere have detected X-rays from hot gas in many clusters. ∞ *(Sec. 3.5)* Figure 16.3 shows false-color X-ray images of two such systems. In each case, the X-ray–emitting region is centered on, and comparable in size to, the visible cluster image.

Further evidence for intracluster gas can be found in the appearance of the radio lobes of some active galaxies. ∞ *(Sec. 15.4)* In some systems, known as *head–tail* radio galaxies, the lobes seem to form a "tail" behind the main part of the galaxy. For example, the lobes of radio galaxy NGC 1265, shown in Figure 16.4, appear to be "swept back" by some onrushing wind, and, indeed, this is the most likely explanation for the galaxy's appearance. If NGC 1265 were at rest, it would be just another double-lobe source, perhaps quite similar to Centaurus A (see Figure 15.22). However, the galaxy is traveling through the intergalactic medium of its parent galaxy cluster (known as the Perseus Cluster), and the outflowing matter forming the lobes tends to be left behind as NGC 1265 moves.

How much gas do these observations reveal? At least as much matter—and in a few cases substantially more—exists within clusters in the form of hot gas as is visible in the form of stars. This is a lot of material, but it still doesn't solve the dark-matter problem. To account for the total masses of galaxy clusters implied by dynamical studies, we would have to find from 10 to 100 times more mass in gas than in stars.

Why is the intracluster gas so hot? Simply because its particles are bound by gravity and hence are moving at speeds comparable to those of the galaxies in the cluster—1000 km/s or so. Since temperature is just a measure of the speed at which the gas particles move, this speed translates (for protons, which make up the bulk of the mass) to a temperature of 40 million kelvins. ∞ *(More Precisely 5-1)*

(a)

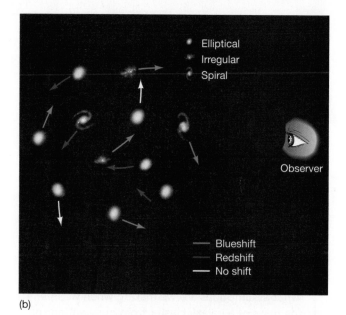

(b)

▲ **FIGURE 16.2 Galaxy Masses** (a) In a binary galaxy system, galaxy masses can be estimated by observing the orbit of one galaxy about the other. (b) The mass of a galaxy cluster can be estimated by observing the motion of many galaxies in the cluster and then estimating how much mass is needed to prevent the cluster from flying apart.

▶ **FIGURE 16.3 Galaxy Cluster X-Ray Emission**   The galaxy cluster Abell 1689 lies some 700 million parsecs from Earth. Its X-ray emission, shown here in blue, is superimposed on a *Hubble* image of the galaxies themselves, and shows clearly how the entire cluster is filled with hot, X-ray emitting gas. *(NASA)*

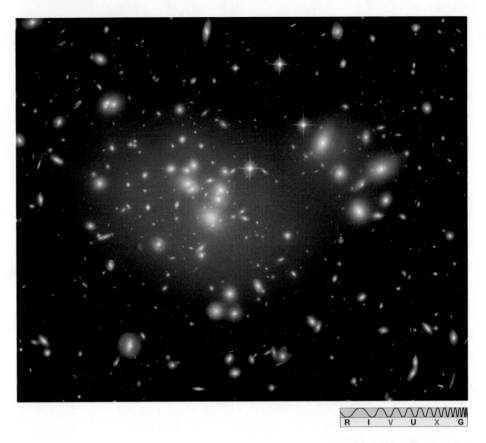

CONCEPT CHECK

What assumptions are we making when we infer the mass of a galaxy cluster from observations of the spectra of its constituent galaxies?

Where did the gas come from? There is so much of it that it could not have been expelled from the galaxies themselves. Instead, astronomers think that it is mainly *primordial*—gas that has been around since the universe began and that never became part of a galaxy. However, the intracluster gas does contain some heavy elements—carbon, nitrogen, and so on—implying that at least some of it is material ejected from galaxies after enrichment by stellar evolution. ∞ *(Sec. 12.7)* Just how this occurred remains a mystery.

The swept-back lobes of this radio galaxy imply that it is moving through the background cluster gas, roughly as indicated.

(a)   (b)

▶ **FIGURE 16.4 Head–Tail Radio Galaxy**   (a) Radiograph, in false color, of the head–tail radio galaxy NGC 1265. (b) The same radio data, in contour form, superimposed on the optical image of the galaxy. *(NRAO; Palomar/Caltech)*

16.2 Galaxy Collisions

Contemplating the congested confines of a rich galaxy cluster (such as those shown in Figures 15.1 and 15.13) with thousands of member galaxies orbiting within a few megaparsecs, we might expect that collisions among galaxies would be common. ∞ *(Sec. 15.2)* Gas particles collide in our atmosphere, and hockey players collide in the rink. Do galaxies in clusters collide too? The answer is yes.

Figure 16.5 apparently shows the aftermath of a bull's-eye collision between a small galaxy (perhaps one of the two at the right, although that is by no means certain) and the larger galaxy at the left. The result is the Cartwheel Galaxy, about 150 Mpc from Earth, its halo of young stars resembling a vast ripple in a pond. The ripple is most likely a density wave created by the passage of the smaller galaxy through the disk of the larger one. ∞ *(Sec. 14.5)* The disturbance is now spreading outward from the region of impact, creating new stars as it goes.

Figure 16.6 shows an example of a close encounter that hasn't (yet) led to an actual collision. Two spiral galaxies are apparently passing each other like majestic ships in the night. The larger and more massive galaxy at the bottom is called NGC 2207; the smaller one at the top is IC 2163. Analysis of this image suggests that IC 2163 is now swinging past NGC 2207 in a counterclockwise direction, having made a close approach some 40 million years ago. The two galaxies seem destined to undergo further close encounters, as IC 2163 apparently does not have enough energy to escape the gravitational pull of NGC 2207. In roughly a billion years, these two galaxies will probably merge into a single, massive galaxy.

These examples illustrate how galaxy interactions—close encounters or actual collisions—can have dramatic consequences for the galaxies involved, especially on their interstellar gas. The rapidly varying gravitational forces during the interaction compress the gas, often resulting in a galaxy-wide episode of star formation. The result is called a **starburst galaxy**. Figure 16.7 shows several more starburst systems in the (relatively) nearby universe.

No human will ever witness an entire galaxy collision, for it lasts many millions of years. However, modern computers can follow the event in a matter of hours. Large-scale simulations modeling in detail the gravitational interactions among stars and gas, and incorporating the best available models of gas dynamics, allow astronomers to better understand the effects of a collision on the galaxies involved and even estimate the eventual outcome of the interaction. The

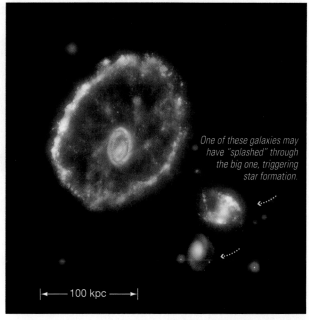

One of these galaxies may have "splashed" through the big one, triggering star formation.

|◄——— 100 kpc ———►|

▲ **FIGURE 16.5 Cosmic Splash** The Cartwheel Galaxy (left) may have resulted from a collision (possibly with one of the smaller galaxies at right) that has led to an expanding ring of star formation moving outward through the galactic disk. This is a false color composite image combining four spectral bands: infrared in red (from *Spitzer*), optical in green (from *Hubble*), ultraviolet in blue (from *Galex*), and X-ray in purple (from *Chandra*). *(NASA)*

◄ **FIGURE 16.6 Galaxy Encounter** This encounter between two spiral galaxies, NGC 2207 (left) and IC 2163 (right), has already led to bursts of star formation in each. Eventually the two will merge, but not for a billion years or so. *(NASA)*

|◄—— 10 kpc ——►|

R I V U X G

R I V U X G

R I V U X G

▲ **FIGURE 16.7 Starburst Galaxies**   (a) This interacting galaxy pair (IC 694, at left, and NGC 3690) shows starbursts now under way in both galaxies—hence the bluish tint. (b) The galaxy NGC 1313 is experiencing a widespread burst of star formation, although the origin of that activiity is unclear. It currently seems to have no neighbors, but its asymmetric appearance and distorted halo outside the frame shown here suggest that it may have absorbed a smaller companion in the relatively recent past. (c) The peculiar (Irr II) active galaxy NGC 1275 contains a system of long filaments that seem to be exploding outward into space. The system was most likely formed by the collision of two galaxies. *(W. Keel; NASA)*

INTERACTIVE   **MA**®

Colliding Galaxies

ANIMATION/VIDEO   **MA**®

Galaxy Collision I

computer simulation shown in Figure 16.8(b) began with two colliding spiral galaxies, not so different from those shown in Figure 16.6, but the details of the original structure have been largely obliterated by the collision. Notice the similarity to the real image of NGC 4038/4039 (Figure 16.8a), the so-called Antennae galaxies, which show extended tails, as well as double galactic centers only a few hundred parsecs across. Star formation induced by the collision is clearly traced by the blue light from thousands of young, hot stars and star clusters. The simulations indicate that, as with the galaxies in Figure 16.6, ultimately the two galaxies will merge into one.

Galaxies in clusters collide because their sizes are comparable to the distances between them—they simply don't have much room to roam around without bumping into one another. The large dark-matter halos surrounding many, if not all, galaxies are crucial to this process. These halos extend far beyond the visible galaxies, making them much larger than their optical appearance would suggest, greatly increasing the chances of interactions and mergers. Many researchers think that most galaxies in clusters have been strongly influenced by collisions, often in the relatively recent past.

Collisions seen in these real images at left can be studied in computer simulations like that at right.

Such simulations demonstrate the crucial role played by dark-matter halos during galaxy interactions.

▶ **FIGURE 16.8 Galaxy Collision**   (a) The long tidal "tails" (black-and-white image at left) mark the final plunge of the "Antennae" galaxies a few tens of millions of years ago. Strings of young, bright "super star clusters" (magnified color image at center) were caused by violent shock waves produced in the gas disks of the two colliding galaxies. (b) A computer simulation of the encounter shows many of the same features as the real object at left. *(AURA; NASA; J. Barnes)*

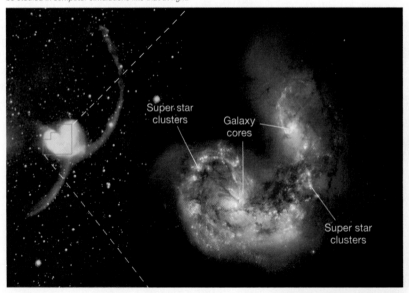

Super star clusters

Galaxy cores

Super star clusters

(a)

R I V U X G

3000 pc

(b)

Incidentally, we don't have to go far afield to find an example of an impending galaxy collision in a small cluster. Our nearest large neighbor, the Andromeda galaxy (see Figure 14.2), is currently approaching us at a velocity of about 120 km/s. In a few billion years it will collide with the Milky Way, and we will have an opportunity to see for ourselves up close what a galaxy collision is really like!

Curiously, although a collision may wreak havoc on the large-scale structure of the galaxies involved, it has essentially no effect on the individual stars they contain. The stars within each galaxy just glide past one another. In contrast to galaxies in the cluster, the stars in a galaxy are so small compared with the distances between them, that when two galaxies collide, the star population merely doubles for a time, and the stars continue to have so much space that they do not run into each other. Collisions can rearrange the stellar and interstellar contents of each galaxy, often producing a spectacular burst of star formation that may be visible to enormous distances, but from the point of view of the stars, it's clear sailing.

16.3 Galaxy Formation and Evolution

With Hubble's law as our guide to distances in the universe, and armed now with knowledge of the distribution of dark matter on galactic and larger scales, let's finally turn to the question of how galaxies came to be the way they are. Can we explain the different galaxy types we see? Astronomers know of no simple evolutionary connections among the various categories in the Hubble classification scheme. ∞ *(Sec. 15.1)* To answer the question, then, we must understand how galaxies formed.

Unfortunately, compared with the theories of star formation and stellar evolution, the theory of galaxy formation is still very much in its infancy. Galaxies are much more complex than stars, they are harder to observe, and the observations are harder to interpret. More important, stars almost never *collide* with one another (apart from situations where stellar evolution causes binary components to merge), so most stars in galaxies evolve in isolation. ∞ *(Sec. 12.3)* Galaxies, however, may suffer numerous collisions during their lives, making it much harder to decipher their pasts. Indeed, collisions like those described in the previous section blur the distinction between formation and evolution, making it hard to separate one from the other.

Nevertheless, some general ideas have gained widespread acceptance, and we can offer some insights into the processes responsible for the galaxies we see. We first describe a general scenario for how large galaxies form from smaller ones, then discuss how galaxies change in time due to both internal stellar evolution and external influences. Finally, we consider how the galaxy types in Hubble's classification fit in to this broad picture.

Mergers and Acquisitions

The seeds of galaxy formation were sown in the very early universe, when small density fluctuations in the primordial matter began to grow (see Section 17.7). Our discussion here begins with these "pregalactic" blobs of gas already formed. The masses of these fragments were quite small—only a few million solar masses, comparable to the masses of the smallest present-day dwarf galaxies, which may in fact be remnants of this early time. Most astronomers think that galaxies grew by repeated *merging* of smaller objects, as illustrated in Figure 16.9(a). Contrast this with the process of star formation, in which a large cloud fragments into smaller pieces that eventually become stars. ∞ *(Sec. 11.3)*

Theoretical evidence for this picture of **hierarchical merging** is provided by computer simulations of the early universe, which clearly show merging taking place. Further strong support comes from observations that indicate that galaxies

(c)

(b)

R I V U X G

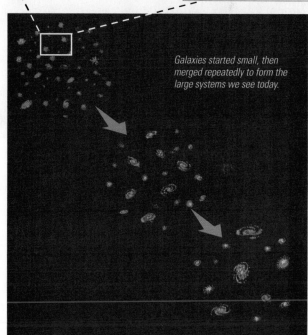

Galaxies started small, then merged repeatedly to form the large systems we see today.

(a)

▲ **FIGURE 16.9 Galaxy Formation** (a) The best current theory of galaxy formation holds that large systems were built up from smaller ones through collisions and mergers, as shown schematically in the drawing at bottom. (b) This photograph, one of the deepest ever taken of the universe, provides "fossil evidence" for hundreds of galaxy shards and fragments, up to 5000 Mpc distant. (c) Enlargements of selected portions of (b) reveal rich (billion-star) "star clusters," all lying within a relatively small volume of space (about 1 Mpc across). Such pregalactic fragments might be about to merge to form a galaxy. The events pictured took place about 10 billion years ago. *(NASA)*

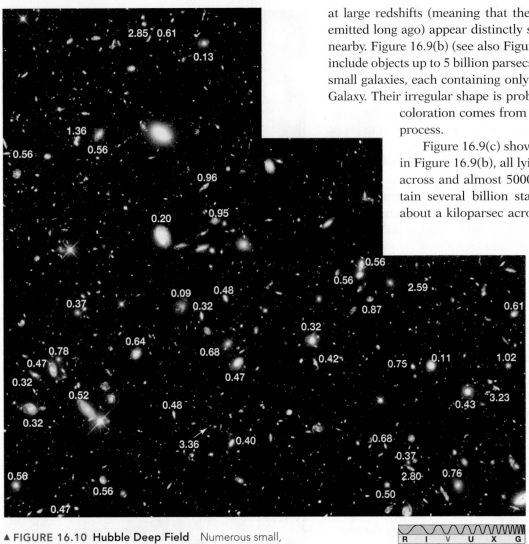

▲ **FIGURE 16.10 Hubble Deep Field** Numerous small, irregularly shaped young galaxies can be seen in this very deep optical image. Known as the Hubble Deep Field-North, this image, made with an exposure of approximately 100 hours, captured objects as faint as 30th magnitude. Redshift measurements (as denoted by the superposed values observed at the Keck Observatory in Hawaii) indicate that some of these galaxies lie billions of megaparsecs from Earth. ∞ *(Sec 15.3)* The field of view is about 2 arc minutes across, or less than 1 percent of the area subtended by the full Moon. *(NASA; Keck)*

at large redshifts (meaning that they are very distant and the light we see was emitted long ago) appear distinctly smaller and more irregular than those found nearby. Figure 16.9(b) (see also Figure 16.10) shows some of these images, which include objects up to 5 billion parsecs away. The vague bluish patches are separate small galaxies, each containing only a few percent of the mass of the Milky Way Galaxy. Their irregular shape is probably the result of galaxy mergers; the bluish coloration comes from young stars that formed during the merging process.

Figure 16.9(c) shows more detailed views of some of the objects in Figure 16.9(b), all lying in the same region of space, about 1 Mpc across and almost 5000 Mpc from Earth. Each blob seems to contain several billion stars spread throughout a distorted spheroid about a kiloparsec across. Their decidedly bluish tint suggests that active star formation is already under way. We see them as they were nearly 10 billion years ago, a group of young galaxies possibly poised to merge into one or more larger objects.

Hierarchical merging provides the conceptual framework for all modern studies of galaxy evolution. It describes a process that began billions of years ago and continues (albeit at a greatly reduced rate) to the present day, as galaxies continue to collide and merge. By studying how galaxy properties vary with distance, and hence look-back time, astronomers try to piece together the merger history of the universe. ∞ *(More Precisely 15-1)*

Figure 16.10 is a remarkable *Hubble* image showing billions of years of galaxy evolution in a single tiny patch of the sky. The large, bright galaxies with easily discernible Hubble types are mostly (according to their redshifts) relatively nearby objects. They are seen here against a backdrop of small, faint, irregular galaxies lying much farther away. The size and appearance of these distant galaxies compared with those in the foreground strongly support the basic idea that galaxies grew by mergers and were smaller and less regular in the past.

Evolution and Interaction

Left alone, a galaxy will evolve slowly and fairly steadily as interstellar clouds of gas and dust are turned into new generations of stars and main sequence stars evolve into giants and ultimately into compact remnants—white dwarfs, neutron stars, and black holes. ∞ *(Secs. 11.3, 13.5)* The galaxy's overall color, composition, and appearance change in a more or less predictable way as the cycle of stellar evolution recycles and enriches the galaxy's interstellar matter. ∞ *(Sec. 12.7)*

If the galaxy is an elliptical, lacking interstellar gas, it will tend to become fainter and redder in time as its more massive stars burn out and are not replaced. ∞ *(Sec. 12.6)* For a gas-rich galaxy, such as a spiral or irregular, hot, bright stars will lend a bluish coloration to the overall light for as long as gas remains available to form them. As in our own Galaxy, the star-forming lifetime of a spiral disk may be prolonged by infall of fresh gas from the galaxy's surroundings. ∞ *(Sec. 14.4)*

But many galaxies—perhaps most—are not alone. They reside in small groups and clusters, and as we have just seen, their orderly "internal" evolution can be significantly complicated by external events—close encounters, mergers, and the accumulation of smaller satellite galaxies over extended periods of time. As described in Section 16.2, these interactions can rearrange a galaxy's internal structure and trigger sudden, intense bursts of star formation. Encounters may also divert fuel to a central black hole, powering violent activity in some galactic nuclei. ∞ *(Sec. 15.5)* Thus, starbursts and nuclear activity are key indicators of interactions and mergers between galaxies.

Careful studies of starburst galaxies and active galactic nuclei indicate that most such encounters probably took place long ago—in galaxies having redshifts greater than about 1, meaning that we see them as they were roughly 10 billion years ago (see Table 15.2). The galaxy interactions observed in the local universe are the extension of this same basic process into the present day. Figure 16.11 presents a graphical summary of these (mostly) ancient events.

The many different types and masses of galaxies lead to an almost bewildering variety of possible interactions. Here we highlight just a few. Consider first two galaxies orbiting one another—a binary galaxy system. As they orbit, the galaxies interact with each other's dark halos, one galaxy stripping halo material from the other by tidal forces. The freed matter is redistributed between the galaxies or is entirely lost from the binary system. In either case, the interaction changes the orbits of the galaxies, causing them to spiral toward one another and eventually to merge.

If one galaxy of the pair happens to have a much lower mass than the other, the process is colloquially termed *galactic cannibalism*. Such cannibalism might explain why supermassive galaxies are often found at the cores of rich galaxy clusters. Having dined on their companions, they now lie at the center of the cluster, waiting for more food to arrive. Figure 16.12 is a remarkable combination of images that has apparently captured this process at work.

We can also find examples of galactic cannibalism closer to home. Figure 16.13(a) illustrates how the disruption of a dwarf galaxy by the Milky Way leaves behind a **tidal stream** of stripped stars, all with similar orbits and composition, still following the orbital path of their parent galaxy. The small Sagittarius dwarf galaxy (see Figure 15.12) is well on its way to suffering such a fate, and theory suggests that the Magellanic Clouds (Figure 15.7) will eventually meet the same end. ∞ *(Sec. 15.2)* Astronomers have discovered numerous tidal streams in the halo of the Milky Way. ∞ *(Sec. 14.4)* Figure 16.13(b) is a wide-angle Sloan Digital Sky Survey mosaic (see Discovery 16-1) of roughly half of the northern sky, looking in the direction away from the Galactic center. It shows several streams of stars crossing the field of view. The most prominent streams (marked) represent two orbits of the Sagittarius dwarf over the past 500 million years. Their locations in the sky are consistent with the measured orbital properties of the Sagittarius galaxy, which currently lies in the opposite direction, as seen from Earth.

Now consider two interacting disk galaxies, one a little smaller than the other but each having a mass comparable to the Milky Way Galaxy. As shown in the computer-generated frames of Figure 16.14, the smaller galaxy can substantially distort the larger one, causing spiral arms to appear where none existed before. The entire event requires several hundred million years—a span of evolution that a supercomputer can model in minutes. The final frame of Figure 16.14 looks remarkably similar to the double galaxy shown in Figure 15.2(b), demonstrating how the two galaxies might have interacted millions of years ago and how spiral arms might have been created or enhanced as a result.

What if the colliding galaxies are comparable in size and mass? Computer simulations reveal that such a merger can destroy a spiral galaxy's disk, creating a galaxywide starburst episode. ∞ *(Sec. 15.4)* The violence of the merger and the

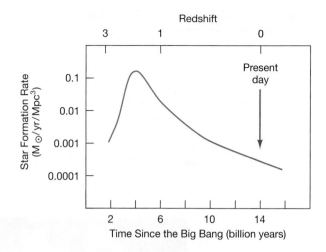

▲ FIGURE 16.11 **Galaxies Build and Stars Form** Observations of the luminosities of many different galaxies at various distances from us show that the star formation rate in the universe peaked a few billion years after the Big Bang. That epoch marked the time at which galaxies were growing most rapidly, as smaller galaxies merged to form the largest systems.

INTERACTIVE
Starburst Galaxy M82

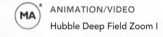

ANIMATION/VIDEO
Hubble Deep Field Zoom I

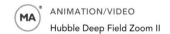

ANIMATION/VIDEO
Hubble Deep Field Zoom II

► FIGURE 16.12 **Galactic Cannibalism**   Most galaxies probably developed by means of a "bottom up" scenario that formed large systems by merging smaller building blocks. This dramatic image captures the process in action about 10 billion years ago, just a few billion years after the Big Bang. The larger image shows part of a young cluster containing hundreds of galaxies, with a radio galaxy catalogued MRC 1138-262 and nicknamed the "Spiderweb" galaxy at its center. The inset shows more clearly dozens of small galaxies about to merge into a single huge object—in fact one of the most massive galaxies known. *(NASA/ESA)*

▼ FIGURE 16.13 **Tidal Streams in the Milky Way**
(a) Breakup and dispersal of the contents of an incoming galaxy companion captured by our Milky Way. Eventually, the smaller galaxy dissolves, much as other dwarf companion galaxies were probably consumed by our Galaxy long ago. (b) This view of the outer part of the Milky Way shows innumerable stars that have been torn from disrupted satellite galaxies (colors indicate distance, with blue being closest). Several tidal streams are evident, the biggest one at center showing two orbits of the enormous, arching death spiral of the Sagittarius dwarf galaxy. *(V. Belokurov; SDSS)*

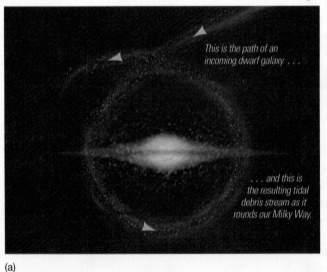

This is the path of an incoming dwarf galaxy . . .

. . . and this is the resulting tidal debris stream as it rounds our Milky Way.

(a)

Yellow arrows indicate motions of stars within the two streams.

Sagittarius streams

(b)

▲ FIGURE 16.14 Galaxy Interaction Galaxies can change their shapes long after their formation. In this computer-generated sequence, two galaxies closely interact over several hundred million years. The smaller galaxy, in red, has gravitationally disrupted the larger galaxy, in blue, changing it into a spiral galaxy. Compare the result of this supercomputer simulation with Figure 15.2(b), a photograph of M51 and its small companion. *(J. Barnes & L. Hernquist)*

effects of subsequent supernovae eject much of the remaining gas into intergalactic space, creating the hot intracluster gas noted in Section 16.1. Once the burst of star formation has subsided, the resulting object looks very much like an elliptical galaxy. The elliptical's hot X-ray halo is the last vestige of the original spiral's disk. ∞ *(Sec. 15.1)* The merging galaxies in Figure 16.8 and the irregular galaxies in Figure 16.7 may be examples of this phenomenon in progress. The blue blobs in some of these images are young star clusters formed during the starburst, and the explosive appearance of especially Figure 16.7(c) suggests that we are witnessing gas and dust being ejected.

Making the Hubble Sequence

If galaxies form and evolve by repeated mergers, can we account for the Hubble sequence and, specifically, differences between spirals and ellipticals? ∞ *(Sec. 15.1)* The details are still far from certain, but remarkably, the answer now seems to be a qualified yes. Collisions and close encounters are random events and do not represent a genuine evolutionary sequence linking all galaxies. However, computer simulations suggest a plausible way in which the observed Hubble types might have arisen, starting from a universe populated only by irregular, gas-rich galaxy fragments.

As we have just seen, the simulations reveal that "major" mergers—collisions between large galaxies of comparable size—tend to destroy galactic disks, effectively turning spirals into ellipticals. On the other hand, "minor" mergers, in which a small galaxy interacts with, and ultimately is absorbed by, a larger one, generally leave the larger galaxy intact, with more or less the same Hubble type as it had before the merger. This is the most likely way for large spirals to grow—in particular, our own Galaxy probably formed in such a manner.

Supporting evidence for this general picture comes from observations that spiral galaxies are relatively rare in regions of high galaxy density, such as the central regions of rich galaxy clusters. This is consistent with the view that their fragile disks are easily destroyed by collisions, which are more common in dense galactic environments. Spirals also seem to be more common at larger redshifts (that is, in the past), implying that their numbers are decreasing with time, presumably also as the result of collisions. However, nothing in this area of astronomy is clear-cut. Astronomers know of numerous isolated elliptical galaxies in low-density regions of the universe that are hard to explain as the result of mergers.

In principle, the starbursts associated with galaxy mergers leave their imprint on the star-formation history of the universe in a way that can be correlated with the properties of galaxies. As a result, studies of star formation in distant galaxies have become a very important way of testing and quantifying the details of the entire hierarchical merger scenario.

CONCEPT CHECK

Why would we expect the evolution of galaxies in voids to differ from that of galaxies in clusters?

PROCESS OF SCIENCE CHECK

In what ways can astronomers test the predictions of the hierarchical merger scenario?

16.4 Black Holes in Galaxies

Now let's ask how quasars and active galaxies fit into the framework of galaxy evolution just described. The fact that quasars are more common at great distances from us demonstrates that they were much more prevalent in the past than they are today. ∞ *(Sec. 15.4)* Quasars have been observed with redshifts exceeding 7, and some observers have reported galaxy redshifts of more than 10, so the process must have started at least 13 billion years ago (see Table 15.2). However, most quasars have redshifts between 2 and 3, corresponding to an epoch some 2 billion years later. Most astronomers agree that quasars represent an early stage of galaxy evolution—an "adolescent" phase of development, prone to frequent flare-ups and outbursts before settling into more steady "adulthood." This view is reinforced by the fact that the same black hole energy-generation mechanism can account for the luminosities of quasars, active galaxies, and the central regions of normal galaxies like our own.

Black Hole Masses

In Chapter 15 we saw the standard model of active galactic nuclei accepted by most astronomers—accretion of gas onto a supermassive black hole. ∞ *(Sec. 15.5)* We also saw that a large fraction of all "bright" galaxies exhibit activity of some sort, even though in many cases it represents only a small fraction of the galaxy's total energy output. This suggests that these galaxies may also harbor central black holes with the potential of far greater activity under the right circumstances. Our own Galaxy is a case in point. ∞ *(Sec. 14.7)* The 3- to 4-million-solar-mass black hole at the center of the Milky Way is not currently active, but if fresh fuel were supplied (say, by a star or molecular cloud coming too close to the hole's intense gravitational field), it might well become a (relatively weak) active galactic nucleus.

In recent years, astronomers have found that many bright normal galaxies have supermassive black holes at their centers. Figure 16.15 presents perhaps the most compelling evidence for a supermassive black hole at the center of a normal galaxy. Using the Very Long Baseline Array, a continent-wide interferometer comprising 10 radio telescopes, a U.S.–Japanese team has achieved an angular resolution hundreds of times better than that attainable with the *Hubble* telescope. ∞ *(Sec. 3.4)* The observations reveal a group of molecular clouds swirling in an organized fashion about the galaxy's center. Doppler measurements indicate a slightly warped, spinning disk centered precisely on the galaxy's heart. The rotation speeds imply the presence of more than 40 million solar masses packed into a region less than 0.2 pc across.

Similar evidence exists for supermassive black holes in the nuclei of several dozen bright galaxies—some normal, some active—within a few tens of megaparsecs of the Milky Way. Some observers would go so far as to say that in every case where a galaxy has been surveyed and a black hole *could* have been detected, given the resolution and the sensitivity of the observations, a black hole *has* in fact been found. It is a small step to the remarkable conclusion that *every bright galaxy—active or not—contains a central supermassive black hole.* This unifying principle connects our theories of normal and active galaxies in a fundamental way.

Astronomers have also found a correlation between the masses of the central black holes and the properties of the galaxy in which they reside. As illustrated in Figure 16.16, the largest black holes tend to be found in the most massive galaxies (as measured by the mass of the bulge). The reason for this correlation is not fully understood, but most astronomers take it to mean that the evolution of normal and active galaxies must be very closely connected, as we now discuss.

|←— 0.2 pc —→|

|←— 2 kpc —→|

R I V U X G

▲ **FIGURE 16.15 Galactic Black Hole** A network of radio telescopes has probed the core of the spiral galaxy NGC 4258, shown here in the light of mostly hydrogen emission. Within the innermost region (inset) a disk of Doppler-shifted molecular clouds (designated by red, green, and blue dots) obey Kepler's third law perfectly, apparently revealing a huge black hole at the center of the disk. *(J. Moran)*

The Quasar Epoch

Where did the supermassive black holes in galaxies come from? To be honest, the processes whereby the first billion-solar-mass black holes formed early in the history of the universe are not fully understood. The accretion rates needed to form the oldest known massive systems challenge current theoretical models. However, the accretion responsible for the energy emission also naturally accounts for the mass of the black hole—only a few percent of the infalling mass is converted to energy; the rest is trapped forever in the black hole once it crosses the event horizon. ∞ *(Sec. 13.5)*. Simple estimates suggest that the accretion rates needed to power the quasars are generally consistent with black-hole masses inferred by other means.

Since the brightest known quasars devour about a thousand solar masses of material every year, it is unlikely that they could maintain their luminosity for very long—even a million years would require a billion solar masses, enough to account for the most massive black holes known. ∞ *(Sec. 15.4)* This suggests that a typical quasar spends only a fairly short amount of time in its highly luminous phase—perhaps a few million years—before running out of fuel. Thus, most quasars were relatively brief events that occurred long ago.

To make a quasar, we need a black hole and enough fuel to power it. While fuel was abundant at early times in the universe's history in the form of gas and newly formed stars, black holes were not. They had yet to form, most probably by the same basic stellar evolutionary processes we saw in Chapter 12, although the details are still not known. ∞ *(Sec. 12.4)* The building blocks of the supermassive black holes that would ultimately power the quasars may well have been relatively small black holes having masses tens or perhaps a few hundreds of times the mass of the Sun. These small black holes sank to the center of their still-forming parent galaxy and merged to form a single, more massive black hole.

As galaxies merged, so, too, did their central black holes, and eventually supermassive (1-million- to 1-billion-solar mass) black holes existed in the centers of many young galaxies. Some supermassive black holes may have formed directly by the gravitational collapse of the dense central regions of a protogalactic fragment or perhaps by accretion or a rapid series of mergers in a particularly dense region of the universe. These events resulted in the earliest (redshift 6–7) quasars known, shining brightly 13 billion years ago. However, in most cases, the mergers took longer—roughly another 2 billion years. By then (at redshifts between 2 and 3, roughly 11 billion years ago), many supermassive black holes had formed, and there was still plenty of merger-driven fuel available to power them. This was the height of the "quasar epoch" in the universe.

Until recently, astronomers were confident that black holes would merge when their parent galaxies collided, but they had no direct evidence of the process—no image of two black holes "caught in the act." In 2002, the *Chandra* X-ray observatory discovered a binary black hole—two supermassive objects, each having a mass a few tens of millions of times that of the Sun—in the center of the ultraluminous starburst galaxy NGC 6240, itself the product of a galaxy merger some 30 million years ago. Figure 16.17 shows optical and X-ray views of the system. The black holes are the two blue-white objects near the center of the (false-color) X-ray image. Orbiting just 1000 pc apart, they are losing energy through interactions with stars and gas and are predicted to merge in about 400 million years. Astronomers now know of several binary black holes in relatively nearby galaxies, caught in the act of spiraling together on their way to merging. NGC 6240 lies just 120 Mpc from Earth, so we are far from seeing a quasar merger in the early universe. Nevertheless, astronomers think that events similar to this must have occurred countless times billions of years ago, as galaxies collided and quasars blazed.

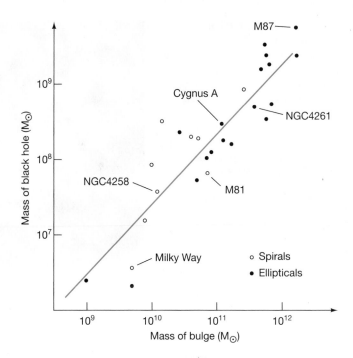

▲ FIGURE 16.16 **Black Hole Masses**   Observations of nearby normal and active galaxies show that the mass of the central black hole is closely correlated with the mass of the galactic bulge. The straight line is the best fit to the data points for many galaxies, implying a black hole mass of $^1/_{200}$ the mass of the bulge. *(L. Ferrarese)*

▶ **FIGURE 16.17 Binary Black Hole** These (a) optical (*Hubble*) and (b) X-ray (*Chandra*) images of the starburst galaxy NGC 6420 show two supermassive black holes (the blue-white objects near the center of the X-ray image) orbiting about 1 kpc apart. Theoretical estimates imply that they will merge in about 400 million years, releasing an intense burst of gravitational radiation. ∞ *(Discovery 13-1) (NASA)*

(a)

(b)

3000 pc

R I V U X G

Distant galaxies are generally much fainter than their bright quasar cores. As a result, until quite recently, astronomers were hard-pressed to discern any galactic structure in quasar images. Since the mid-1990s, several groups of astronomers have used the *Hubble Space Telescope* to search for the "host" galaxies of moderately distant quasars. After removing the bright quasar core from the *HST* images and carefully analyzing the remnant light, the researchers have reported that in every case studied—several dozen quasars so far—a host galaxy can be seen enveloping the quasar. Figure 16.18 shows some of the longest quasar exposures ever taken. Even without sophisticated computer processing, the hosts are clearly visible.

As we saw in Chapter 15, the connection between active galaxies and galaxy clusters is well established, and many relatively nearby quasars are also known to be members of clusters. ∞ *(Sec. 15.4)* The link is less clear-cut for the most distant quasars, however, simply because they are so far away that other cluster members are very faint and extremely hard to see. However, as the number of known quasars continues to increase, evidence for quasar clustering

▶ **FIGURE 16.18 Quasar Host Galaxies** These long-exposure images of distant quasars show the young host galaxies in which the quasars reside, lending support to the idea that quasars represent an early, very luminous phase of galactic evolution. The quasar at the left is the best example, having the catalog name PG0052 + 251 and residing roughly 690 Mpc from Earth. *(NASA)*

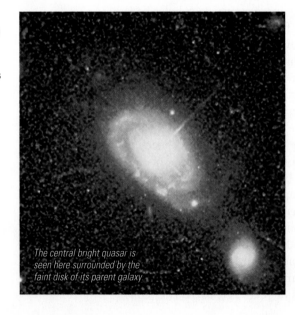

The central bright quasar is seen here surrounded by the faint disk of its parent galaxy.

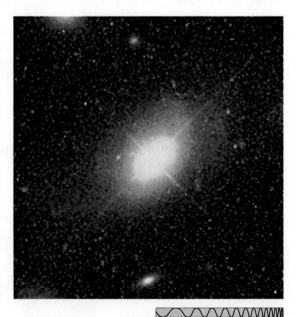

R I V U X G

(and presumably, therefore, for quasar membership in young galaxy clusters) mounts. Thus, as best we can tell, quasar activity—and, in fact, galaxy activity of *all* sorts—is intimately related to interactions and collisions in galaxy clusters.

This connection also suggests a possible way in which the growth of black holes might be tied to the growth of their parent galaxies. Astronomers speculate that a process called **quasar feedback**, in which some fraction of the quasar's enormous energy output is absorbed by the surrounding galactic gas, might help explain the correlation of black hole and bulge masses shown in Figure 16.16. According to this view, which is appealing but by no means certain, the absorbed energy expels the gas from the galaxy, simultaneously shutting down both galactic star formation and the quasar's own fuel supply, thus tying the growth of the central black hole to the formation of new stars in the bulge.

Active and Normal Galaxies

Early on, frequent mergers may have replenished the quasar's fuel supply, extending its luminous lifetime. However, as the merger rate declined, these systems spent less and less of their time in the "bright" phase. The rapid decline in the number of bright quasars roughly 10 billion years ago marks the end of the quasar epoch. Today, the number of quasars has dropped virtually to zero (recall that the nearest lies hundreds of megaparsecs away). ∞ *(Sec. 15.4)*

Large black holes do not simply vanish. If a galaxy contained a bright quasar 10 billion years ago, the black hole responsible for all that youthful activity must still be present in the center of the galaxy today. We see some of these black holes as active galaxies. The remainder reside dormant in normal galaxies all around us. In this view, *the difference between an active galaxy and a normal one is mainly a matter of fuel supply.* When the fuel runs out and a quasar shuts down, its central black hole remains behind, its energy output reduced to a relative trickle. The black holes at the hearts of normal galaxies are quiescent, awaiting another interaction to trigger a new active outburst. Occasionally, two nearby galaxies may interact, causing a flood of new fuel to be directed toward the central black hole of one or both. The engine starts up for a while, giving rise to the nearby active galaxies—Seyferts, radio galaxies, and others—we observe.

Should this general picture be correct, it follows that many relatively nearby galaxies (but probably *not* our own Milky Way, whose central black hole is even now only a paltry 3–4 million solar masses) must once have been brilliant quasars. ∞ *(Sec. 14.7)* Perhaps some alien astronomer, thousands of megaparsecs away, is at this very moment observing the progenitor of M87 in the Virgo Cluster—seeing it as it was billions of years ago—and is commenting on its enormous luminosity, nonstellar spectrum, and high-speed jets and wondering what exotic physical process can possibly account for its violent activity! ∞ *(Sec. 14.4)*

Finally, Figure 16.19 suggests some possible (but unproven) evolutionary connections among quasars, active galaxies, and normal galaxies. If the largest black holes reside in the most massive galaxies, and also tend to power the brightest active galactic nuclei, then we would expect that the most luminous nuclei should reside in the largest galaxies, which probably came into being via "major" mergers of other large galaxies. Since the products of such mergers are elliptical galaxies, we have a plausible explanation of why the brightest active galaxies— the radio galaxies—should be associated with large ellipticals. ∞ *(Sec. 15.4)* The path to spiral galaxies would have entailed a series of mergers involving smaller galaxies, resulting in the less violent Seyferts along the way.

PROCESS OF SCIENCE CHECK

How does the existence of supermassive black holes in normal galaxies fit in with theories of active galaxy evolution?

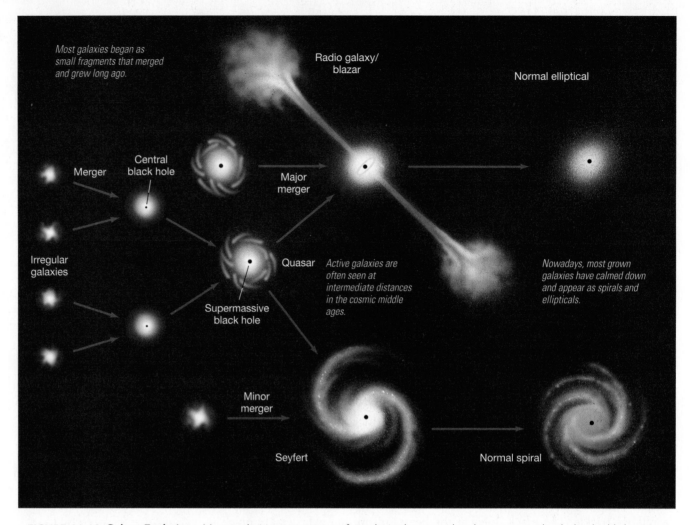

▲ **FIGURE 16.19 Galaxy Evolution**   Most evolutionary sequences for galaxies began with galaxy mergers that led to highly luminous quasars, after which they decreased in violence through the radio and Seyfert galaxies, eventually resulting in normal ellipticals and spirals. The central black holes that powered the early activity remain at later times, but many of them have run out of fuel.

Active Galaxies and the Scientific Method

When active galactic nuclei—and especially quasars—were first discovered, their extreme properties defied conventional explanation. Initially, the idea of super-massive (million- to billion-solar mass) black holes in galaxies was just one of several competing, and very different, hypotheses advanced to account for the enormous luminosities and small sizes of those baffling objects. Some astronomers suggested instead that the energy source might be multiple supernova explosions, others that perhaps some exotic form of matter–antimatter annihilation was taking place. A few went so far as to propose that these inexplicable objects demanded an even more radical explanation: that Hubble's law was incorrect and that the large redshifts of quasars had some other, unknown cause.

However, as observational evidence mounted, the other hypotheses were abandoned one by one, and massive black holes in galactic nuclei became first the leading, and eventually the standard, theory of active galaxies. As often happens in science, a theory once itself considered extreme is now the accepted explanation. Far from threatening the laws of physics, as some astronomers once feared, active galaxies are now an integral part of our understanding of how galaxies form and evolve. The synthesis of studies of normal and active galaxies, galaxy formation, and large-scale structure is one of the great triumphs of extragalactic astronomy.

CONCEPT CHECK

Does *every* galaxy have the potential for activity?

► FIGURE 16.20 Local Supercluster   More than 4500 galaxies are plotted here in the vicinity of the Virgo cluster, and several prominent galaxy clusters are labeled. The diagram depicts the Virgo Supercluster roughly as we see it from our own Galaxy, which is located approximately 20 Mpc (two grid squares) above the page. Notice the supercluster's irregular, elongated shape. (B. Tully and S. Levy)

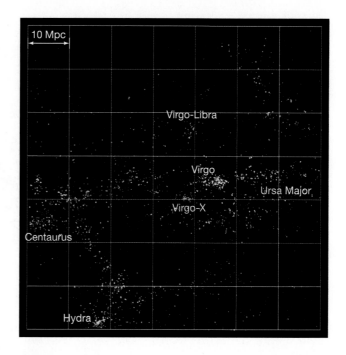

16.5 The Universe on Very Large Scales

Many galaxies, including our own, are members of galaxy clusters—megaparsec-sized structures held together by their own gravity. ∞ (Sec. 15.2) Our own small cluster is called the Local Group. Figure 16.20 shows the locations of the Virgo Cluster, the closest "large" cluster, and of several other well-defined clusters in our cosmic neighborhood. The region displayed is about 70 Mpc across. Each point in the figure represents an entire galaxy whose distance has been determined by one of the methods described in Chapter 15.

Clusters of Clusters

Do galaxy clusters top the cosmic hierarchy, or does the universe have even larger groupings of matter? Most astronomers have concluded that the galaxy clusters are themselves clustered, forming titanic agglomerations of matter known as **superclusters**.

Together, the galaxies and clusters shown in Figure 16.20 form the *Local Supercluster*, also known as the Virgo Supercluster. Aside from the Virgo Cluster itself, it contains the Local Group and numerous other clusters lying within about 20–30 Mpc of Virgo. Most of the galaxies depicted in the figure are fairly large spirals and ellipticals; the fainter irregulars and dwarfs are not included in the diagram. Figure 16.21 shows a three-dimensional rendering of an even wider view, illustrating the Virgo Supercluster (near the center) relative to other "nearby" galaxy superclusters within a vast imaginary rectangle roughly 100 Mpc on its short side.

All told, the Local Supercluster is about 40–50 Mpc across, contains some 10^{15} solar masses of material (several tens of thousands of galaxies), and is very irregular in shape. The Local Supercluster is significantly elongated perpendicular to the line joining the Milky Way to Virgo, with its center lying near the Virgo Cluster. By now it should come as no surprise that the Local Group is *not* found at the heart of the Local Supercluster—we live far off in the periphery, about 18 Mpc from the center.

Redshift Surveys

The farther we peer into deep space, the more galaxies, clusters of galaxies, and superclusters we see. Is there structure on even larger scales? To answer this question, astronomers use Hubble's law to map out the distribution of galaxies in the universe and have developed indirect techniques to probe the dark recesses of intergalactic space.

Figure 16.22 shows part of an early survey of the universe performed by astronomers at Harvard University in the 1980s. Using Hubble's law as a distance indicator, the team systematically mapped out the locations of galaxies within about 200 Mpc of the Milky Way in a series of wedge-shaped "slices," each 6° thick, starting in the northern sky. The first slice (Figure 16.22) covered a region of the sky, which happens to lie in a direction almost perpendicular to our Galaxy's plane. Because redshift is used as the primary distance indicator, these studies are known as *redshift surveys*.

ANIMATION/VIDEO
Cluster Merger

▲ FIGURE 16.21 Virgo Supercluster in 3-D   The elongated shape of the Virgo Supercluster (left) is mapped relative to other neighboring galaxy superclusters within about 100 Mpc of the Milky Way (within the Local Group near center). Individual galaxies are not shown; rather, smoothed contour plots outline galaxy clusters, each named or numbered by its most prominent member. (M. Hudson)

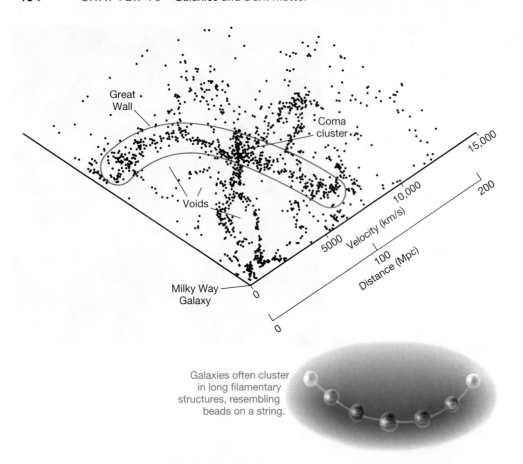

Galaxies often cluster in long filamentary structures, resembling beads on a string.

▲ **FIGURE 16.22 Galaxy Survey** A "slice" of the universe, covering 1732 galaxies out to a distance of about 200 Mpc, shows that galaxies and clusters are not randomly distributed on large scales. Instead, they appear to have a filamentary structure, surrounding vast, nearly empty voids. The large Coma cluster of galaxies near the center of the slice can also be seen in Figure 16.21. *(Harvard-Smithsonian Center for Astrophysics)*

The most striking feature of maps such as this is that the distribution of galaxies on very large scales is decidedly nonrandom. The galaxies appear to be arranged in a network of strings, or filaments, surrounding large, relatively empty regions of space known as **voids**. Voids account for some 50 percent of the total volume of the nearby universe, but only 5–10 percent of the mass. The biggest voids measure some 100 Mpc across. The most likely explanation for the voids and filamentary structure in Figure 16.22 is that the galaxies and galaxy clusters are spread across the surfaces of vast "bubbles" in space. The voids are the interiors of these gigantic bubbles. The galaxies seem to be distributed like beads on strings only because of the way our slice of the universe cuts through the bubbles. Like suds on soapy water, these bubbles fill the entire universe. The densest clusters and superclusters lie in regions where several bubbles meet. The elongated shape of the Virgo Supercluster (see Figure 16.20) is a local example of this same filamentary structure.

Most theorists think that this "frothy" distribution of galaxies, and in fact all structure on scales larger than a few megaparsecs, traces its origin directly to conditions in the very earliest stages of the universe (Chapter 17). As such, studies of large-scale structure are vital to our efforts to understand the origin and nature of the cosmos itself. The idea that the filaments are the intersection of the survey slice with much larger structures (the bubble surfaces) was confirmed when the next three slices of the survey, lying above and below the first, were completed. The region of Figure 16.22 indicated by the red outline was found to continue through both the other slices. This extended sheet of galaxies, which has come to be known as the *Great Wall*, measures at least 70 Mpc (out of the plane of the page) by 200 Mpc (across the page). It is one of the largest known structures in the universe.

Figure 16.23 shows a more recent redshift survey, considerably larger than the one presented in Figure 16.22. This survey includes nearly 24,000 galaxies within about 750 Mpc of the Milky Way. Numerous voids and "Great Wall-like" filaments can be seen (some are marked), but apart from the general fall-off in numbers of galaxies at large distances—basically because the more distant galaxies are harder to see due to the inverse-square law—there is no obvious evidence for any structures on scales larger than about 200 Mpc. Careful statistical analysis and even larger surveys (see *Discovery* 16-1) confirm this impression. Apparently, voids and walls represent the largest structures in the universe. We will return to the far-reaching implications of this fact in Chapter 17.

Quasar Absorption Lines

How can we probe the structure of the universe on very large scales? As we have seen, much of the matter is dark, and even the "luminous" component is so faint that it is hard to detect at large distances. One way to study large-scale structure is to take advantage of the great distances, point-like appearance, and large luminosities of quasars, in much the same way as astronomers use

CONCEPT CHECK

What do observations of distant quasars tell us about the structure of the universe closer to home?

bright stars to probe the interstellar medium near the Sun. ∞ *(Secs. 11.1, 15.3)* Since quasars are so far away, light traveling from a quasar to Earth has a pretty good chance of passing through or near something "interesting" en route. By analyzing quasar images and spectra, it is possible to piece together a partial picture of the intervening space.

In addition to their own strongly redshifted spectra, many quasars also show additional absorption features that are redshifted by substantially *less* than the lines from the quasar itself. For example, the quasar PHL 938 has an emission-line redshift of 1.954, placing it at a distance of some 5200 Mpc, but it also shows absorption lines having redshifts of just 0.613. These are interpreted as arising from intervening gas that is much closer to us (only about 2200 Mpc away) than the quasar itself. Most probably this gas is part of an otherwise invisible galaxy lying along the line of sight. Quasar spectra, then, afford astronomers a means of probing previously undetected parts of the universe.

The absorption lines of atomic hydrogen are of particular interest, since hydrogen makes up so much of all matter in the cosmos. Specifically, hydrogen's ultraviolet (122 nm) "Lyman-alpha" line, associated with transitions between the ground and first excited states, is often used in this context. ∞ *(Sec. 2.6)* As illustrated in Figure 16.24, when astronomers observe the spectrum of a high-redshift quasar, they typically see a "forest" of absorption lines starting at the (redshifted) wavelength of the quasar's own Lyman-alpha emission line and extending to shorter wavelengths. These lines are interpreted as Lyman-alpha absorption features produced by gas clouds in foreground structures—galaxies, clusters, and so on—giving astronomers crucial information about the distribution of matter along the line of sight.

Quasar light thus explores an otherwise invisible component of cosmic gas. In principle, every intervening cloud of atomic hydrogen leaves its own characteristic imprint on the quasar's spectrum, in a form that lets us explore the distribution of matter in the universe. By comparing these *Lyman-alpha forests* with the results of simulations, astronomers hope to refine many key elements of the theories of galaxy formation and the evolution of large-scale structure.

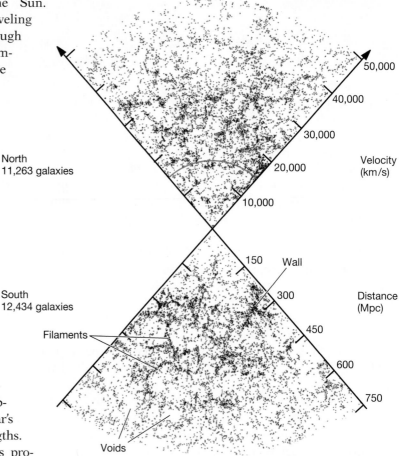

▲ **FIGURE 16.23 The Universe on Larger Scales** This large-scale galaxy survey, carried out at the Las Campanas Observatory in Chile, consists of 23,697 galaxies within about 1000 Mpc, in two 80° (wide) × 4.5° (thick) wedges of the sky. Many voids and "walls" on scales of up to 100–200 Mpc are evident, but nothing much larger. For scale, the extent of the survey shown in Figure 16.22 is marked as a thin blue arc in the northern sky here.

MA INTERACTIVE
Quasar Spectrum due to Absorption from Distant Clouds

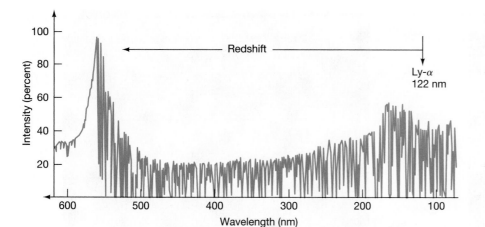

◄ **FIGURE 16.24 Absorption Line "Forest"** The huge number of absorption lines in the spectrum of quasar QSO 1422 + 2309 are the ultraviolet 122 nm Lyman-alpha lines from hundreds of clouds of foreground hydrogen gas, each redshifted by a slightly different amount (but less than the quasar itself). The peak at left marks the Lyman-alpha emission line from the quasar itself, emitted at 122 nm but redshifted here to 564 nm, in the visible range.

16.1 DISCOVERY

The Sloan Digital Sky Survey

Many of the photographs used in this book—not to mention most of the headline-grabbing imagery found in the popular media—come from large, high-profile, and usually very expensive, instruments such as NASA's *Hubble Space Telescope* and the European Southern Observatory's Very Large Telescope in Chile. ∞ *(Secs. 3.3, 3.4)* Their spectacular views of deep space have revolutionized our view of the universe, yet a less well-known, considerably cheaper, but no less ambitious, project begun in 2000 may, in the long run, have every bit as great an impact on astronomy and our understanding of the cosmos.

The Sloan Digital Sky Survey (SDSS) was originally a 5-year project, but it has been extended numerous times, most recently until 2020. It was designed to systematically map out a quarter of the entire sky on a scale and at a level of precision never before attempted. It cataloged almost 1 billion celestial objects, recording their apparent brightnesses at five different colors (wavelength ranges) spread across the optical part of the spectrum. In addition, spectroscopic follow-up observations determined redshifts and hence distances to 2 million galaxies and 500,000 quasars. These data have been used to construct even more detailed redshift surveys than those described in the text and to probe the structure of the universe on very large scales. The sensitivity of the survey was such that it could detect bright galaxies like our own out to distances of more than 1 billion parsecs. Very bright objects, such as quasars and young starburst galaxies, were detectable almost throughout the entire observable universe.

The first figure shows the Sloan Survey telescope, a special-purpose 2.5-m instrument sited in Apache Point Observatory, near Sunspot, New Mexico. The telescope is not space-based, does not employ active or adaptive optics, and cannot probe as deeply (i.e., far) into space as do larger instruments. How could it possibly compete with these other systems? The answer is that, unlike most other large telescopes in current use, where hundreds or even thousands of observers share the instrument and compete for its time, the SDSS telescope was designed specifically for the purpose of the survey. It has a wide field of view and was dedicated to the task, carrying out observations of the sky on *every* clear night during the duration of the project.

The survey used this single instrument night after night, applying tight quality controls on which nights' data are actually incorporated into the survey. Nights with poor seeing or other problematic conditions were discarded and the observations repeated. The end product was a database of exceptionally high quality and uniformity spanning an enormous volume of space—a monumental achievement and an indispensable tool for cosmology. The survey field of view covers much of the sky away from the Galactic plane in the north, together with a large swath of the sky around the Galactic south pole.

Archiving images and spectra on millions of galaxies produces a lot of data. The total amount of data gathered by the survey's 120-megapixel camera comprises more than 100 *trillion* bytes of high-quality data—far more than the entire Library of Congress! All of the survey data have now been released to the public. The second figure shows a nearby galaxy cluster with a giant elliptical at its center—one of hundreds of thousands of images that make up the full dataset. Among recent highlights, SDSS has detected the largest known structure in the universe, observed the most distant known galaxies and quasars, and has been instrumental in pinning down the key observational parameters describing our universe (see Chapter 17).

SDSS impacts astronomy in areas as diverse as the large-scale structure of the universe, the origin and evolution of galaxies, the nature of dark matter, the structure of the Milky Way, and the properties and distribution of interstellar matter and exoplanetary systems. Its uniform, accurate, and detailed database is likely to be used by generations of scientists for decades to come, and its success has spawned several much larger follow-up surveys; the most ambitious is due to become operational in 2021.

(SDSS; R. Lupton)

► **FIGURE 16.25 Twin Quasar**   This twin quasar (designated AC114 and located about 2 billion parsecs away) is not two separate objects at all. Instead, the two large "blobs" (at upper left and lower right) are images of the same object, created by a gravitational lens. The lensing galaxy itself is probably not visible in this image—the two objects near the center of the frame are thought to be unrelated galaxies in a foreground cluster. *(NASA)*

Two images of the same background galaxy, formed by an unseen foreground lens

Quasar "Mirages"

In 1979 astronomers were surprised to discover what appeared to be a binary quasar—two quasars with exactly the same redshift and very similar spectra, separated by only a few arc seconds on the sky. Remarkable as the discovery of such a binary would have been, the truth about this pair of quasars turned out to be even more amazing. Closer study of the quasars' radio emission revealed that they were *not* two distinct objects. Instead, they were two separate images of the *same* quasar! Optical views of such a *twin quasar* are shown in Figure 16.25.

What could produce such a "doubling" of a quasar image? The answer is gravitational lensing—the deflection and focusing of light from a background object by the gravity of some foreground body (Figure 16.26). In Chapter 14 we saw how lensing by compact objects in the halo of the Milky Way Galaxy may amplify the light from a distant star, allowing astronomers to detect otherwise invisible stellar dark matter. ∞ *(Sec. 14.6)* In the case of quasars, the idea is the same, except that the foreground lensing object is an entire galaxy or galaxy cluster and the deflection of the light is so great (a few arc seconds) that several separate images of the background object may be formed, as shown in Figure 16.27. About two dozen such gravitational lenses are known.

The existence of these multiple images provides astronomers with a number of useful observational tools. First, lensing by a foreground galaxy tends to amplify the light of the quasar, as just mentioned, making it easier to observe. At the same time, *microlensing* by individual stars within the galaxy may cause large fluctuations in the quasar's brightness, allowing astronomers to study the galaxy's stellar content.

Second, because the light rays forming the images usually follow paths of different lengths, there is often a time delay, ranging from days to years, between them. This delay provides advance notice of interesting events, such as sudden changes in the quasar's brightness—if one image flares up, in time the other(s) will too, giving astronomers a second chance to study the event. The time delay also allows astronomers to determine the distance to the lensing galaxy. This method provides an alternative means of measuring Hubble's constant that is independent of any of the techniques discussed previously. The average value of H_0 reported by workers using this approach is consistent with the 70 km/s/Mpc value used throughout this text.

▼ **FIGURE 16.26 Gravitational Lens INTERACTIVE**   When light from a distant object passes close to a galaxy or cluster of galaxies along the line of sight, the image of the background object (here, the quasar) can sometimes be split into two or more separate images (A and B). The foreground object is a gravitational lens.

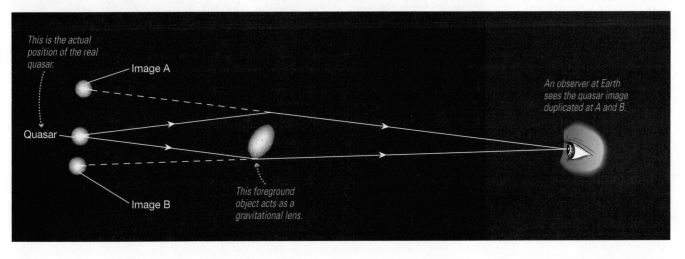

This is the actual position of the real quasar.

Image A

Quasar

An observer at Earth sees the quasar image duplicated at A and B.

Image B

This foreground object acts as a gravitational lens.

▶ **FIGURE 16.27 Einstein Cross**
(a) The Einstein Cross, a multiply-imaged quasar, spans only a couple of arc seconds, showing four separate images of the same quasar produced by the galaxy at the center. (b) This 2015 *Hubble* image shows four images (marked in the inset) of a supernova in a distant background galaxy some 3 billion parsecs away, due to lensing by a foreground galaxy and the cluster in which it resides. *(NASA)*

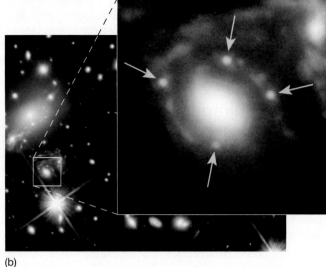

(a) (b)

R I V U X G

DATA POINTS

Scales in the Universe

More than 50 percent of students had difficulty ranking the scale of various objects in the universe. Some key points to bear in mind:

■ Planets are smaller—and usually much smaller—than stars, but stars are in turn much smaller than the planetary systems that surround them. A typical planetary system is around 100 astronomical units across.

■ Stars are born in star clusters, which are typically a few parsecs across, much larger than any planetary system orbiting a star.

■ Stars and star clusters form galaxies, which might be anywhere from a few thousand to 100,000 parsecs across—much larger than stars, planetary systems, or star clusters.

■ Galaxies themselves form larger groupings called galaxy clusters, typically a few million parsecs (a few megaparsecs) in size.

■ On even larger scales, galaxies and clusters form superclusters, tens to hundreds of megaparsecs across.

■ All of these scales are tiny compared to the size of the observable universe, which is roughly 10 billion parsecs, or 10,000 megaparsecs.

Mapping Dark Matter

Astronomers have extended the ideas first learned from studies of quasars to use lensing of any distant object to probe the universe. Distant faint irregular galaxies—the raw material of the universe, if current theories are correct (see Section 16.4)—are of particular interest here, as they are far more common than quasars; thus, they provide much better coverage of the sky. By studying the lensing of background quasars and galaxies by foreground galaxy clusters, astronomers can obtain a better understanding of the distribution of dark matter on large scales.

Figures 16.28(a) and (b) show how the images of faint, background galaxies are bent into arcs by the gravity of a foreground galaxy cluster. The degree of bending allows the total mass of the cluster (including the dark matter) to be measured. The (mostly blue) loop- and arc-shaped features visible in Figure 16.28(b) are multiple images of a single, distant (unseen) spiral or ring-shaped galaxy, lensed by the foreground galaxy cluster (the yellow-red blobs in the image).

It is even possible to reconstruct the foreground dark mass distribution by carefully analyzing the distortions of the background objects, providing a means of tracing out the distribution of matter on scales far larger than have previously been possible Figure 16.29(a) is an optical image of a (hard to see) foreground galaxy

▶ **FIGURE 16.28 Galaxy Cluster Lensing**
INTERACTIVE (a) This spectacular example of gravitational lensing shows more than a hundred faint arcs from very distant galaxies. The wispy pattern spread across the foreground galaxy cluster (A2218, about a billion parsecs distant) is caused by A2218's gravitational field, which deflects the light from background galaxies and distorts their appearance. (b) Another galaxy cluster, known only by its catalog name 0024 + 1654 and residing some 1.5 billion pc away, shows reddish-yellow blobs that are mostly normal elliptical galaxies and bluish loop-like features that are images of a single background galaxy. *(NASA)*

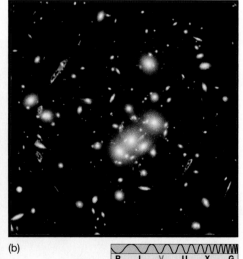

(a) (b)

R I V U X G   R I V U X G

(a)

(b)

▲ **FIGURE 16.29 Dark Matter Map**   By measuring distortions in the images of background objects, astronomers can make maps of dark matter in the universe. Analysis of an optical view (a) of a region of the sky containing a small galaxy cluster (the clump of yellowish galaxies near the center of the frame indicated by the arrow) reveals the distribution of dark matter (b) in and near the visible cluster and on the same scale as (a). *(J. A. Tyson, Bell Labs; Alcatel-Lucent; NOAO)*

cluster set against a background of much fainter distant galaxies. Figure 16.29(b) is the reconstructed dark-matter image, revealing the presence of dark mass many megaparsecs from the cluster center, extending far beyond the visible galaxies. Evidently, the dark matter fraction is even greater on supercluster scales than it is on the scales of individual galaxies or galaxy clusters. Notice also the elongated structure of the dark-matter distribution, reminiscent of the Virgo Supercluster and filamentary structure seen in large-scale galaxy surveys. ∞ *(Sec. 14.5)*

Astronomers have used these techniques to obtain what may be the first direct observational evidence for dark matter. Figure 16.30 shows combined optical

ANIMATION/VIDEO
Simulation of Gravitational Lens in Space

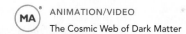

ANIMATION/VIDEO
The Cosmic Web of Dark Matter

The arrows indicate the directions in which the two clusters are now moving, subsequent to what might have been the most energetic collision in the universe since the Big Bang.

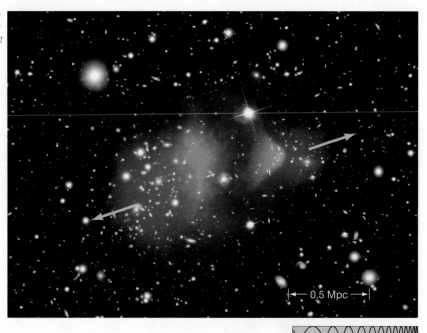

◄ **FIGURE 16.30 Cluster Collision**
INTERACTIVE   Clusters of galaxies must also occasionally collide, as is apparently the case here for this combined cluster with the innocuous name 1E 0657-56 and the nickname "bullet cluster." This is a composite image of a region about 1 billion parsecs away, showing optical light from the galaxies themselves in white and X-ray emission from the hot intracluster gas in red. By contrast, the blue color represents the inferred dark matter within the two large clusters that is distinctly displaced from their normal matter. *(NOAO/NASA)*

and X-ray images of a distant galaxy cluster called 1E 0657-56. The fuzzy red region shows the location of the hot X-ray–emitting gas in the system, the dominant *luminous* component of the mass. The blue regions indicate where most of the mass actually lies, as determined from lensing studies of background galaxies. Note that the bulk of the mass is *not* found in the form of hot gas, implying that the dark matter is distributed differently from the "normal" matter in the cluster.

The explanation for this odd state of affairs is that we are witnessing a collision between two clusters. Each initially contained hot gas and dark matter distributed throughout the cluster, but when the two collided, the pressure of each gas cloud effectively stopped the other, leaving the gas behind in the middle as the galaxies and dark matter moved on. This separation between the gas and the dark matter directly contradicts some alternative theories of gravity that have been invoked to avoid the "dark matter problem" in galaxies and clusters and may prove to be a crucial piece of evidence in our understanding of large-scale structure in the universe.

THE BIG QUESTION

What is dark matter? Is it actual matter that exerts a gravitational pull but is impossible to detect by electromagnetic means, or does it imply that something is badly wrong with our theoretical understanding of the way gravity works on very large scales? Dark matter—and dark energy, too (see Chapter 17)—represent the foremost scientific conundrums in astronomy today, and whoever solves them will become immediately famous. How will science resolve these cosmic mysteries, and what new mysteries will arise as a result?

CHAPTER REVIEW

SUMMARY

LO1 The masses of nearby spiral galaxies can be determined by studying their rotation curves. Astronomers also use studies of binary galaxies and distant galaxy clusters to obtain statistical mass estimates of the galaxies involved.

LO2 Measurements of galaxy and cluster masses reveal the presence of large amounts of dark matter. The fraction of dark matter grows as the scale under consideration increases. More than 90 percent of the mass in the universe is dark. Substantial quantities of hot X-ray–emitting gas have been detected among the galaxies in many clusters, but not enough to account for the dark matter inferred from dynamical studies.

LO3 Researchers know of no simple evolutionary sequence that links spiral, elliptical, and irregular galaxies. Most astronomers think that large galaxies formed by the merger of smaller ones and collisions and mergers of galaxies play very important roles in galactic evolution. A starburst galaxy may result when a galaxy has a close encounter or a collision with a neighbor. The strong tidal distortions caused by the encounter compress galactic gas, resulting in a widespread burst of star formation. Mergers between spirals most likely result in elliptical galaxies.

LO4 Quasars, active galaxies, and normal galaxies may represent an evolutionary sequence. When galaxies began to form and

merge, conditions may have been suitable for the formation of large black holes at their centers, and highly luminous quasars could have been the result. The brightest quasars consume so much fuel that their energy-emitting lifetimes must be quite short. As the fuel supply diminished, the quasar dimmed, and the galaxy in which it was embedded became intermittently visible as an active galaxy. At even later times, the nucleus became virtually inactive, and a normal galaxy was all that remained. Many normal galaxies have been found to contain central massive black holes, suggesting that most galaxies in clusters have the capacity for activity if they should interact with a neighbor.

LO5 Galaxy clusters themselves tend to clump together into **superclusters** (p. 463). The Virgo Cluster, the Local Group, and several other nearby clusters form the Local Supercluster. On even larger scales, galaxies and galaxy clusters are arranged

on the surfaces of enormous "bubbles" of matter surrounding vast low-density regions called **voids** (p. 464). The origin of this structure is thought to be closely related to conditions in the very earliest epochs of the universe.

LO6 Quasars can be used as probes of the universe along the line of sight. Some quasars have been observed to have double or multiple images. These result from gravitational lensing, in which the gravitational field of a foreground galaxy or galaxy cluster bends and focuses the light from the more distant quasar. Analysis of the images of distant galaxies, distorted by the gravitational effect of a foreground cluster, provides a means of determining the masses of galaxy clusters—including the dark matter—far beyond the information that optical images of the galaxies themselves afford.

MasteringAstronomy®   For instructor-assigned homework go to www.masteringastronomy.com

Problems labeled POS explore the process of science. VIS problems focus on reading and interpreting visual information. LO connects to the introduction's numbered Learning Outcomes.

REVIEW AND DISCUSSION

1. **LO1** Describe two techniques for measuring the mass of a galaxy.

2. **LO2** Why do astronomers think that galaxy clusters contain more mass than we can see?

3. Why do some clusters of galaxies emit X-rays?

4. What evidence do we have that galaxies collide with one another?

5. **LO3** Describe the role of collisions in the formation and evolution of galaxies.

6. **POS** Do you think that collisions between galaxies constitute "evolution" in the same sense as the evolution of stars? Why or why not?

7. What are starburst galaxies, and what do they have to do with galaxy evolution?

8. Why do astronomers think that quasars represent an early stage of galactic evolution?

9. **LO4 POS** Why does the theory of galaxy evolution suggest that there should be supermassive black holes at the centers of many normal galaxies?

10. What evidence do we have for supermassive black holes in galaxies?

11. How might a normal galaxy become active?

12. What is a redshift survey?

13. **LO5** What are voids? What is the distribution of galactic matter on very large (more than 100 Mpc) scales?

14. **LO6 POS** How can observations of distant quasars be used to probe the space between us and them?

15. **POS** How can astronomers "see" dark matter?

CONCEPTUAL SELF-TEST: TRUE OR FALSE?/MULTIPLE CHOICE

1. Intergalactic gas in galaxy clusters emits large amounts of energy in the form of radio waves. (T/F)

2. Distant galaxies appear to be much larger than those nearby. (T/F)

3. Collisions between galaxies are rare and have little or no effect on the stars and interstellar gas in the galaxies involved. (T/F)

4. The quasar stage of a galaxy ends because the central black hole swallows up all the matter around it. (T/F)

5. Elliptical galaxies may be formed by mergers between spirals. (T/F)

6. The fact that a typical quasar would consume an entire galaxy's worth of mass in 10 billion years suggests that quasar lifetimes are relatively long. (T/F)

7. On the largest scales, galaxies in the universe appear to be arranged on huge sheets surrounding nearly empty voids. (T/F)

8. The image of a distant quasar can be split into several images by the gravitational field of a foreground cluster. (T/F)

9. A galaxy containing substantial amounts of dark matter will (a) appear darker; (b) spin faster; (c) repel other galaxies; (d) have more tightly wound arms.

10. According to X-ray observations, the space between galaxies in a galaxy cluster is (a) completely devoid of matter; (b) very cold; (c) very hot; (d) filled with faint stars.

11. The fraction of mass of the universe made up of dark matter is (a) zero; (b) less than 10 percent; (b) roughly 50 percent; (d) more than 90 percent.

12. In current theories of galaxy evolution, quasars occur (a) early in the evolutionary sequence; (b) near the Milky Way; (c) when elliptical galaxies merge; (d) late in the evolutionary sequence.

13. Many nearby galaxies (a) may be much more active in the future; (b) contain quasars; (c) have radio lobes; (d) may have been much more active in the past.

14. VIS If light from a distant quasar did not pass through any intervening atomic hydrogen clouds, then Figure 16.24 (Absorption Line "Forest") would have to be redrawn to show (a) more absorption features; (b) fewer absorption features; (c) a single, large absorption feature; (d) more features at short wavelengths, but fewer at long wavelengths.

15. VIS If Figure 16.26 (Gravitational Lens) were to be redrawn using a more massive lensing galaxy, the quasar images would be (a) farther apart; (b) closer together; (c) fainter; (d) redder.

PROBLEMS

The number of squares preceding each problem indicates its approximate level of difficulty.

1. ■ Two galaxies orbit one another at a separation of 500 kpc. Their orbital period is estimated to be 30 billion years. Use Kepler's third law to find the total mass of the pair. ∞ *(Sec. 14.6)*

2. ■ Based on the data in Figure 16.1, estimate the mass of the galaxy NGC 4984 inside 20 kpc.

3. ■■■ A small satellite galaxy is moving in a circular orbit around a much more massive parent and just happens to be moving exactly parallel to the line of sight as seen from Earth. The recession velocities of the satellite and the parent galaxy are measured to be 6450 km/s and 6500 km/s, respectively, and the two galaxies are separated by an angle of 0.1° on the sky. Assuming $H_0 = 70$ km/s/Mpc, calculate the mass of the parent galaxy.

4. ■ Calculate the average speed of hydrogen nuclei (protons) in a gas of temperature 20 million K. ∞ *(More Precisely 5-1)* Compare this with the speed of a galaxy moving in a circular orbit of radius 1 Mpc around the center of a cluster of mass 10^{14} solar masses.

5. ■ In a galaxy collision, two similar-sized galaxies pass through each other with a combined relative velocity of 1500 km/s. If each galaxy is 100 kpc across, how long does the event last?

6. ■■ A certain quasar has a redshift of 0.25 and an apparent magnitude of 13. Using the data from Table 15.2, calculate the quasar's absolute magnitude and hence its luminosity. ∞ *(More Precisely 10-1)* Compare the apparent brightness of the quasar, viewed from a distance of 10 pc, with that of the Sun as seen from Earth.

7. ■ What are the absolute magnitude and luminosity of a quasar with a redshift of 5 and an apparent magnitude of 22?

8. ■■ Assuming an energy-generation efficiency (that is, the ratio of energy released to total mass-energy consumed) of 20 percent, calculate how long a 10^{40}-W quasar can shine if a total of 10^8 solar masses of fuel is available.

9. ■■ The spectrum of a quasar with a redshift of 0.20 contains two sets of absorption lines, redshifted by 0.15 and 0.155, respectively. If $H_0 = 70$ km/s/Mpc, estimate the distance between the intervening galaxies responsible for the two sets of lines.

10. ■■■ Light from a distant quasar is deflected through an angle of 3″ by an intervening lensing galaxy and is subsequently detected on Earth (see Figure 16.26). If Earth, the galaxy, and the quasar are all aligned, the quasar's redshift is 3.0 (see Table 15.2), and the galaxy lies midway between Earth and the quasar, calculate the minimum distance between the light ray and the center of the galaxy.

ACTIVITIES

Collaborative

1. Figure 16.10 is called the Hubble Deep Field. It contains too many galaxies for one person to easily count. Each group member should count the galaxies in a random area 2 cm × 2 cm and then determine a group average. Since the entire image is approximately 500 cm², multiply your group's average number of galaxies in a 2 cm × 2 cm area by 125 to estimate the number of galaxies in the image. How does your value compare to that of another group?

Individual

1. Look for a copy of the *Atlas of Peculiar Galaxies* by Halton Arp. It is available in book form, but it will be more convenient to find a version online. Search for examples of interacting galaxies of various types: (1) tidal interactions, (2) starburst galaxies, (3) collisions between two spirals, and (4) collisions between a spiral and an elliptical. For (1) look for galactic material pulled away from a galaxy by a neighboring galaxy. Is the latter galaxy also tidally distorted? In (2) the surest signs of starburst activity are bright knots of star formation. In what type(s) of galaxies do you find starburst activity? For (3) and (4) how do collisions differ depending on the types of galaxies involved? What typically happens to a spiral galaxy after a near miss or collision? Do ellipticals suffer the same fate?

17

Cosmology
The Big Bang and the Fate of the Universe

Our field of view now extends for billions of parsecs into space and billions of years back in time. We have asked and answered many questions about the structure and evolution of planets, stars, and galaxies. At last we are in a position to address the central issues of the biggest puzzle of all: How big is the universe? How long has it been around, and how long will it last? What was its origin, and what will be its fate? What is it made of? Many cultures have asked these questions, in one form or another, and have developed their own cosmologies—theories about the nature, origin, and destiny of the universe—to answer them. In this chapter, we see how modern scientific cosmology addresses these important issues and what it has to tell us about the universe we inhabit. After more than 10,000 years of civilization, science may be ready to provide some insight into the ultimate origin of all things.

THE BIG PICTURE

Cosmology is the study of the origin, structure, evolution, and fate of the universe on the largest scales, yet one of its key predictions is that the entire cosmos can be traced back to an extraordinarily hot and dense energy state billions of years ago. As mind-boggling as it may seem, all that we see around us apparently arose from microscopic "quantum" fluctuations that occurred a fraction of a second after the Big Bang. Ironically, the largest structures observed today are inextricably tied to the smallest scales known in physics.

◄ The universe began in a fiery expansion some 14 billion years ago, and out of this maelstrom emerged all the energy that would later form galaxies, stars, and planets. The story of the origin, evolution, and fate of all these systems comprises the subject of cosmology. This image—called the Ultra Deep Field—was taken with the Advanced Camera for Surveys aboard the *Hubble* telescope. More than a thousand galaxies are crowded into this one image, displaying many different types and shapes. In all, astronomers estimate that the observable universe contains about 100 billion such galaxies. *(NASA/ESA)*

17.1 The Universe on the Largest Scales

Galaxy surveys have revealed the existence of structures in the universe up to 200 Mpc across. ∞ *(Sec. 16.5)* Are there even larger structures? Does the hierarchy of clustering of stars into galaxies and galaxies into galaxy clusters, superclusters, voids, and filaments continue forever, on larger and larger scales? Perhaps surprisingly, given the trend we have just described, most astronomers think the answer is *no*.

We saw in Chapter 16 how astronomers use redshift surveys to construct three-dimensional maps of the universe on truly "cosmic" scales. ∞ *(Sec. 16.5)* Figure 17.1 is a map similar to those shown previously, but based on data from the most extensive redshift survey to date—the Sloan Digital Sky Survey. ∞ *(Discovery 16-1)* It extends out to a distance of almost 1000 Mpc, comparable to Figure 16.23, but because it includes much fainter galaxies, the Sloan map contains many more galaxies than the earlier figure, making structure easier to discern, particularly at large distances. Many voids and filaments can be seen on scales of a few tens of megaparsecs, but on the largest scales, apart from the general falloff in numbers of galaxies at large distances—basically because more distant galaxies are harder to see due to the inverse-square law—there is no obvious evidence for any structures on scales larger than about 200 Mpc. ∞ *(Sec. 10.2)* Careful statistical analysis of the galaxy distribution confirms this impression—even the Sloan Great Wall in Figure 17.1 can be explained statistically as a chance superposition of smaller structures.

The results of this and other large-scale surveys strongly suggest that the universe is **homogeneous** (the same everywhere) on scales greater than a few hundred megaparsecs. In other words, if we took a huge cube—300 Mpc on a side, say—and placed it anywhere in the universe, its overall contents would look much the same no matter where it was centered. The universe also appears to be **isotropic** (the same in all directions) on these scales. Excluding directions that

CONCEPT CHECK

In what sense is the universe homogeneous and isotropic?

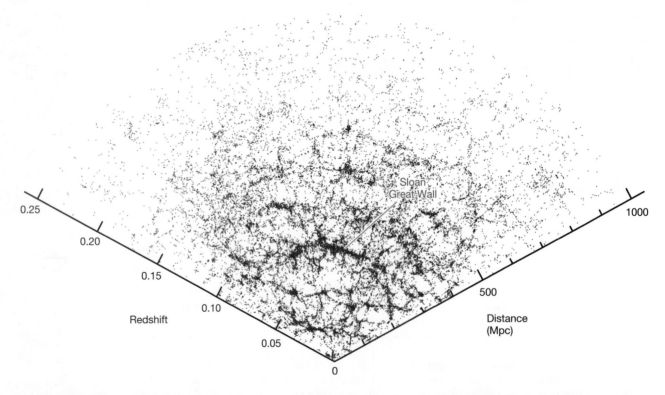

▲ **FIGURE 17.1 Galaxy Survey**   This map of the universe is drawn using data from the Sloan Digital Sky Survey. ∞ *(Discovery 16-1)* It shows the locations of 66,976 galaxies lying within 12 degrees of the celestial equator and extending to a distance of almost 1000 Mpc (or a redshift of about 0.25). The largest known structure in the universe, the Sloan Great Wall, stretches nearly 300 Mpc across the center of the frame. *(SDSS Collaboration)*

are obscured by our Galaxy, we count roughly the same number of galaxies per square degree in any patch of the sky we choose to observe, provided we look deep (far) enough that local inhomogeneities don't distort our sample.

In the science of **cosmology**—the study of the structure and evolution of the entire universe—researchers generally assume that the universe is homogeneous and isotropic on sufficiently large scales. No one knows if these assumptions are precisely correct, but we can at least say that they are consistent with current observations. In this chapter we simply assume that they hold. The twin assumptions of homogeneity and isotropy are known as the **cosmological principle**.

The cosmological principle has some very far-reaching implications. For example, it implies that there can be no *edge* to the universe because that would violate the assumption of homogeneity. And there can be no *center* because then the universe would not look the same in all directions from any noncentral point, violating the assumption of isotropy. This is the familiar Copernican principle expanded to truly cosmic proportions—not only are we not central to the universe, but *no one* can be central, because *the universe has no center*!

17.2 The Expanding Universe

Every time you go outside at night and notice that the sky is dark, you are making a profound cosmological observation. Here's why.

Olbers's Paradox

Let's assume for a moment that, in addition to being homogeneous and isotropic, the universe is also infinite in spatial extent and unchanging in time. On average, then, the universe is uniformly populated with galaxies filled with stars. In that case, as illustrated in Figure 17.2, when you look up at the night sky, your line of sight must *eventually* encounter a star. The star may lie at an enormous distance, in some remote galaxy, but the laws of probability dictate that, sooner or later, any line drawn outward from Earth will run into a bright stellar surface.

Faraway stars appear fainter than those nearby because of the inverse-square law, but they are also much more numerous—the number of stars we see at any given distance in fact increases as the *square* of the distance (just consider the area of a sphere of increasing radius). ∞ *(Sec. 10.2)* Thus the diminishing brightnesses of distant stars are exactly balanced by their increasing numbers, and stars at all distances contribute equally to the total amount of light received on Earth.

This fact has a dramatic implication: No matter where you look, the sky should be as bright as the surface of a star—the entire night sky should be as brilliant as the surface of the Sun! The obvious difference between this prediction and the actual appearance of the night sky is known as **Olbers's paradox**, after the 19th-century German astronomer Heinrich Olbers, who popularized the idea.

So why is the sky dark at night? Given that the universe is homogeneous and isotropic, then one (or both) of the other two assumptions must be false. Either the universe is finite in extent, or it evolves in time. In fact, the answer involves a little of each and is intimately tied to the behavior of the universe on the largest scales.

The Birth of the Universe

We have seen that all the galaxies in the universe are rushing away from us as described by Hubble's law:

$$\text{recession velocity} = H_0 \times \text{distance},$$

where we take Hubble's constant H_0 to be 70 km/s/Mpc. ∞ *(Sec. 15.3)* We have used this relationship as a convenient means of determining the distances to galaxies and quasars, but it is much more than that.

In a forest, every line of sight eventually intersects a tree, much as Olbers's paradox suggests for stars seen from Earth.

▲ **FIGURE 17.2 Olbers's Paradox**   If the universe were homogeneous, isotropic, infinite in extent, and unchanging, then any line of sight from Earth should eventually meet a star and the entire night sky should be bright. This obvious contradiction of the facts is known as Olbers's paradox.

Assuming that all velocities have remained constant in time, we can ask how long it has taken for any given galaxy to reach its present distance from us. The answer follows from Hubble's law. The time taken is simply the distance traveled divided by the velocity:

$$\text{time} = \frac{\text{distance}}{\text{velocity}}$$

$$= \frac{\text{distance}}{H_0 \times \text{distance}} \qquad \text{(using Hubble's law for the velocity)}$$

$$= \frac{1}{H_0}.$$

For $H_0 = 70$ km/s/Mpc, this time is about 14 billion years. Notice that it is *independent* of the distance—galaxies twice as far away are moving twice as fast, so the time they took to cross the intervening distance is the same.

Hubble's law therefore implies that, at some time in the past—14 billion years ago, according to the foregoing simple calculation (see also Section 17.4)—*all* the galaxies in the universe lay right on top of one another. In fact, astronomers think that *everything* in the universe—matter and radiation alike—was confined at that instant to a single point of enormously high temperature and density. Then the universe began to expand at a furious rate, its density and temperature falling rapidly as the volume increased. This stupendous, unimaginably violent event, involving literally everything in the cosmos, is known as the **Big Bang**. Astronomers often refer to the superhot, expanding universe just after the Big Bang as the primeval fireball. It marked the beginning of the universe.

The Big Bang provides the resolution of Olbers's paradox. Whether the universe is actually finite or infinite in extent is irrelevant, at least as far as the appearance of the night sky is concerned. The key point is that we see only a *finite* part of it—the region lying within roughly 14 billion light-years of us. What lies beyond is unknown—its light has not had time to reach us.

Where Was the Big Bang?

Now we know *when* the Big Bang occurred. Is there any way of telling *where*? Astronomers think that the universe is the same everywhere, yet we have just seen that the observed recession of the galaxies described by Hubble's law suggests that all the galaxies expanded from a point at some time in the past. Wasn't that point, then, different from the rest of the universe, violating the assumption of homogeneity expressed in the cosmological principle? The answer is a definite *no*!

To understand why there is no center to the expansion, we must make a great leap in our perception of the universe. If the Big Bang was simply an enormous explosion that spewed matter out into space, ultimately to form the galaxies we see, then the foregoing reasoning would be quite correct. The universe would have a center and an edge, and the cosmological principle would not apply. But the Big Bang was *not* an explosion in an otherwise featureless, empty universe. The only way that we can have Hubble's law *and* retain the cosmological principle is to realize that the Big Bang involved the entire universe—not just the matter and radiation within it, but the universe *itself*. In other words, the galaxies are not flying apart into the rest of the universe. The universe itself is expanding. Like raisins in a loaf of raisin bread that move apart as the bread expands in an oven, the galaxies are just along for the ride.

Let's reconsider some of our earlier statements in this new light. We now recognize that Hubble's law describes the expansion of the universe itself. Apart from small-scale individual random motions, galaxies are not moving with respect to the fabric of space. The component of the galaxies' motion that makes up the Hubble flow is really an expansion of space itself. The expanding universe remains homogeneous at all times. There is no "empty space" beyond the galaxies into which they rush. At the time of the Big Bang, the galaxies did not reside at a point located at some well-defined place within the universe. The *entire universe* was a point. The Big Bang happened *everywhere* at once.

To illustrate these ideas, imagine an ordinary balloon with coins taped to its surface, as shown in Figure 17.3. (Better yet, do the experiment yourself.) The coins represent galaxies, and the two-dimensional surface of the balloon represents the "fabric" of our three-dimensional universe. The cosmological principle applies here because every point on the balloon looks pretty much the same as every other. Imagine yourself as a resident of one of the coin "galaxies" on the leftmost balloon, and note your position relative to your neighbors. As the balloon inflates (that is, as the universe expands), the other galaxies recede from you and more distant galaxies recede more rapidly.

Regardless of which galaxy you chose to consider, you would see all the other galaxies receding from you, but nothing is special or peculiar about this fact. There is no center to the expansion and no position that can be identified as the location from which the universal expansion began. Everyone sees an overall expansion described by Hubble's law, with the same value of Hubble's constant in all cases. Now imagine letting the balloon deflate, running the "universe" backward from the present time to the moment of the Big Bang. *All* the galaxies (coins) would arrive at the same place at the same time at the instant the balloon reached zero size, but no single point on the balloon could be said to be *the* place where that occurred.

Coins on the surface of an expanding balloon . . .

. . . obey Hubble's law as they recede from one another.

▲ **FIGURE 17.3 Receding Galaxies INTERACTIVE** Coins taped to the surface of a spherical balloon recede from each other as the balloon inflates (left to right). Similarly, galaxies recede from each other as the universe expands. As the coins separate, the distance between any two of them increases, and the rate of increase of this distance is proportional to the distance between them.

PROCESS OF SCIENCE CHECK

Why does Hubble's law imply a Big Bang?

The Cosmological Redshift

Up to now, we have explained the cosmological redshift of galaxies as a Doppler shift, a consequence of their motion relative to us. ∞ *(Sec. 15.3)* However, we have just argued that the galaxies are not moving with respect to the universe, in which case the Doppler interpretation cannot be correct. The true explanation is that, as a photon moves through space, its wavelength is influenced by the expansion of the universe. Think of the photon as being attached to the expanding fabric of space, so its wavelength expands along with the universe, as illustrated in Figure 17.4. Although it is common in astronomy to refer to the cosmological redshift in terms of recessional velocity, bear in mind that, strictly speaking, this is not the right thing to do. The cosmological redshift is a consequence of the changing size of the universe and is *not* related to velocity at all.

The redshift of a photon measures the amount by which the universe has expanded since that photon was emitted. For example, when we measure the light from a quasar to have a redshift of 7, it means that the observed wavelength is 8 times (1 plus the redshift) greater than the wavelength at the time of emission, which means that the light was emitted at a time when the universe was just one-eighth its present size (and we are observing the quasar as it was at that time). ∞ *(More Precisely 15-1)* In general, the larger a photon's redshift, the smaller the universe was at the time the photon was emitted and so the longer ago that emission occurred. Because the universe expands with time and redshift is related to that expansion, cosmologists routinely use redshift as a convenient means of expressing time.

Radiation also shifts with cosmic expansion.

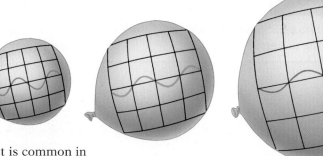

▲ **FIGURE 17.4 Cosmological Redshift INTERACTIVE** As the universe expands, photons of radiation are stretched in wavelength, giving rise to the cosmological redshift.

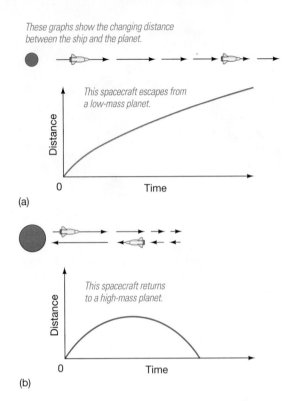

These graphs show the changing distance between the ship and the planet.

This spacecraft escapes from a low-mass planet.

(a)

This spacecraft returns to a high-mass planet.

(b)

▲ **FIGURE 17.5 Escape Speed** (a) A spacecraft (arrow) leaving a planet (blue ball) with a speed greater than the planet's escape speed follows an unbound trajectory and escapes. (b) If the launch speed is less than the escape speed, the ship eventually drops back to the planet. Its distance, as graphed, first rises and then falls.

CONCEPT CHECK

What are the two basic possibilities for the future expansion of the universe?

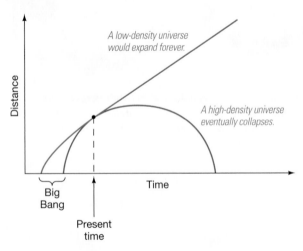

A low-density universe would expand forever.

A high-density universe eventually collapses.

Big Bang

Present time

▲ **FIGURE 17.6 Model Universes** Distance between two galaxies as a function of time in each of the two basic universes discussed in the text: a low-density universe that expands forever and a high-density cosmos that collapses. The point where the two curves touch represents the present time.

17.3 Cosmic Dynamics and the Geometry of Space

Will the universe expand forever? This fundamental question about the fate of the universe has been at the heart of cosmology since Hubble's law was first discovered. Until the late 1990s, the prevailing view among cosmologists was that the answer would most likely be found by determining the extent to which gravity would slow, and perhaps ultimately reverse, the current expansion. However, it now appears that the answer is rather more subtle—and perhaps a lot more profound in its implications—than was hitherto thought.

Two Futures

Consider a rocket ship launched from the surface of a planet. What are the likely outcomes of the its motion? According to Newtonian mechanics, there are just two basic possibilities, depending on the launch speed of the ship relative to the escape speed of the planet. ∞ *(More Precisely 5-1)* If the launch speed is high enough, it will exceed the planet's escape speed, and the ship will never return to the surface. The speed will diminish because of the planet's gravitational pull, but it will never reach zero. The spacecraft leaves the planet on an unbound trajectory, as illustrated in Figure 17.5(a). Alternatively, if the launch speed is lower than the escape speed, the ship will reach a maximum distance from the planet and then fall back to the surface. Its bound trajectory is shown in Figure 17.5(b).

Similar reasoning applies to the expansion of the universe. Imagine two galaxies at some known distance from one another, their present relative velocity given by Hubble's law. The same two basic possibilities exist for these galaxies as for our rocket ship. The distance between them can increase forever, or it can increase for a while and then start to decrease. What's more, the cosmological principle says that whatever the outcome, it must be the same for *any* two galaxies—in other words, the same statement applies to the universe *as a whole*. Thus, as illustrated in Figure 17.6, the universe has only two options: It can continue to expand forever, or the present expansion will someday stop and turn around into a contraction. The two curves in the figure are drawn so that they pass through the same point at the present time. Both are possible descriptions of the universe, given its current size and expansion rate.

What determines which of these possibilities will actually occur? In the case of a rocket ship of fixed launch speed, the *mass* of the planet determines whether or not escape will occur—a more massive planet has a higher escape speed, making it less likely that the rocket can escape. For the universe, the corresponding quantity is the *density* of the cosmos. A high-density universe contains enough mass to stop the expansion and eventually cause a recollapse. Such a universe is destined ultimately to shrink toward a superdense, superhot singularity much like the one from which it originated—a "Big Crunch." A low-density universe, conversely, will expand forever. In that case, in time, even with the most powerful telescopes, we will see no galaxies in the sky beyond the Local Group (which is not itself expanding). The rest of the observable universe will appear dark, the distant galaxies too faint to be seen.

The dividing line between these outcomes—the density corresponding to a universe in which *gravity acting alone* would be just sufficient to halt the present expansion—is called the universe's **critical density**. For $H_0 = 70$ km/s/Mpc, the critical density is about 9×10^{-27} kg/m^3. That's an extraordinarily low density—just five hydrogen atoms per cubic meter, a volume the size of a typical household closet. In more "cosmological" terms, it corresponds to about 0.1 Milky Way Galaxy (including the dark matter) per cubic megaparsec.

We will see in a moment that the separation between never-ending expansion and cosmic collapse is not nearly as straightforward as the foregoing simple reasoning would suggest. Several independent lines of evidence now indicate that gravity is *not* the only influence on the dynamics of the universe on large scales (Section 17.4). As a result, while the "futures" just described are still the only two possibilities for the long-term evolution of the universe, the distinction between them turns out to be more than just a matter of density alone. Nevertheless, the density of the universe—or, more precisely, the *ratio of the total density to the critical value*—is a vitally important quantity in cosmology.

The Shape of the Universe

We have reverted to the familiar notions of Newtonian mechanics and gravity and moved away from general relativity and the more correct concept of warped spacetime, because speaking in Newtonian terms makes our discussion of the evolution of the universe easier to understand. ∞ *(Secs. 1.4, 13.5)* However, general relativity makes some predictions about the universe that do not have a simple description within Newton's theory.

Foremost among these non-Newtonian predictions is the fact that space is curved and that the degree of curvature is determined by the total density of the cosmos. ∞ *(Sec. 13.5)* But general relativity is clear on what "density" means here. Both matter *and* energy must be taken into account, with energy (E) properly "converted" into mass (m) units via Einstein's famous relation $E = mc^2$. ∞ *(More Precisely 13-1)* The density of the universe includes not just the atoms and molecules that make up the familiar normal matter around us, but also the invisible dark matter that dominates the masses of galaxies and galaxy clusters, as well as *everything* that carries energy—photons, relativistic neutrinos, gravity waves, and anything else we can think of.

In a homogeneous universe, the curvature (on sufficiently large scales) must be the same everywhere, so there are really only three distinct possibilities for the large-scale geometry of space. General relativity tells us that the geometry of the universe depends only on the ratio of the density of the universe to the critical density (as defined in the previous section). Cosmologists call the ratio of the universe's actual density to the critical value the *cosmic density parameter* and denote it by the symbol Ω_0 ("omega nought"). In terms of this quantity, then, a universe with density equal to the critical value has $\Omega_0 = 1$, a "low-density" cosmos has Ω_0 less than 1, and a "high-density" universe has Ω_0 greater than 1.

In a high-density universe (Ω_0 greater than 1), space is curved so much that it bends back on itself and closes off, making the universe *finite* in size. Such a universe is known as a **closed universe**. It is difficult to visualize a three-dimensional volume uniformly arching back on itself in this way, but the two-dimensional version is well known: It is the surface of a sphere, like the balloon we discussed earlier. Figure 17.3, then, is the two-dimensional likeness of a three-dimensional closed universe. Like the surface of a sphere, a closed universe has no boundary, yet it is finite in extent.[1] One remarkable property of a closed universe is illustrated in Figure 17.7: A flashlight beam shining in some direction in space might eventually traverse the entire universe and return from the opposite direction.

The surface of a sphere curves, loosely speaking, "in the same direction" no matter which way we move from a given point. A sphere is said to have *positive curvature*. However, if the density of the universe is below the critical value, then the curvature of space is qualitatively quite different from that of a sphere. The two-dimensional surface corresponding to such a space curves like a saddle and is said to have *negative curvature*. Most people have seen a saddle—it curves "up" in

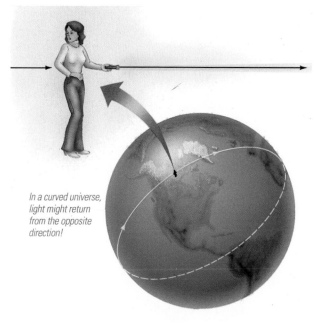

In a curved universe, light might return from the opposite direction!

▲ **FIGURE 17.7 Einstein's Curve Ball** In a closed universe, a beam of light launched in one direction might return someday from the opposite direction after circling the universe, just as motion in a "straight line" on Earth's surface will eventually encircle the globe.

---

[1] Notice that for the sphere analogy to work, we must imagine ourselves as two-dimensional "flatlanders" who cannot visualize or experience in any way the third dimension perpendicular to the sphere's surface. Flatlanders are confined to the sphere's surface, just as we are confined to the three-dimensional volume of our universe.

CONCEPT CHECK

How is the curvature of space related to the size and density of the universe?

one direction and "down" in another, but no one has ever seen a uniformly negatively curved surface, for the good reason that it cannot be constructed in three-dimensional space! It is just "too big" to fit. A low-density (Ω_0 less than 1) saddle-curved universe is infinite in extent and is usually called an **open universe**.

The intermediate case, when the density is precisely equal to the critical density (i.e., $\Omega_0 = 1$), is the easiest to visualize. This **critical universe** has no curvature. It is said to be "flat," and it is infinite in extent. In this case, and only in this case, the geometry of space on large scales is precisely the familiar Euclidean geometry taught in high school. Apart from its overall expansion, this is basically the universe that Newton knew.

17.4 The Fate of the Cosmos

Is there any way for us to determine which of the futures just described actually applies to our universe (apart from just waiting to find out)? Will the universe end as a small, dense point much like that from which it began? Or will it expand forever? And can we hope to determine the geometry of the vast cosmos we inhabit? Finding answers to these questions has been the dream of astronomers for decades. We are fortunate to live at a time when we can subject these questions to intensive observational tests and come up with definite answers—even though they aren't what most cosmologists expected! Let's begin by looking at the density of the universe (or, equivalently, the cosmic density parameter Ω_0).

The Density of the Universe

How might we determine the average density of the universe? On the face of it, it would seem simple—just measure the average mass of the galaxies residing within a large parcel of space, calculate the volume of that space, and compute the total mass density. When astronomers do this, they usually find a little less than 10^{-28} kg/m^3 in the form of luminous matter. Largely independent of whether the chosen region contains only a few galaxies or a rich galaxy cluster, the resulting density is about the same, within a factor of 2 or 3. Galaxy counts thus yield a value of Ω_0 of only a few percent. If that measure were correct, and galaxies were all that existed, then we would live in a low-density open universe destined to expand forever.

But there is a catch. Most of the matter in the universe is *dark*—it exists in the form of invisible material that has been detected only through its gravitational effect in galaxies and galaxy clusters. ∞ *(Secs. 14.6, 16.1)* We currently do not know *what* the dark matter is, but we *do* know that it is there. Galaxies may contain as much as 10 times more dark matter than luminous material, and the figure for galaxy clusters is even higher—perhaps as much as 95 percent of the total mass in clusters is invisible. Even though we cannot see it, dark matter contributes to the average density of the universe and plays its part in opposing the cosmic expansion. Including all the dark matter that is known to exist in galaxies and galaxy clusters increases the value of Ω_0 to about 0.25.

Although we can detect and quantify the effects of dark matter in galaxies and galaxy clusters, its distribution on larger scales is harder to measure. Astronomers have developed techniques to study matter on supercluster and larger scales, using gravitational lensing of distant objects and the large-scale motions of galaxies and clusters to probe the gravitational fields of cosmic clumps of invisible matter. ∞ *(Sec. 16.5)* Yet all these studies add little to the overall density. As best we can tell, there doesn't seem to be much additional dark matter "tucked away" on very large scales. Most cosmologists agree that the overall density of matter (luminous plus dark) in the universe is between 25 and 30 percent of the critical value—not enough to halt the universe's current expansion.

Cosmic Acceleration

Determining the density of the universe is an example of a *local* measurement that provides an estimate of Ω_0. But the result we obtain depends on just how local our measurement is, and there are many uncertainties in the result, especially on large scales. In an attempt to get around this problem, astronomers have devised alternative methods that rely instead on *global* measurements, covering much larger regions of the observable universe. In principle, such global tests should indicate the universe's overall density, not just its value in our cosmic neighborhood.

One such global method is based on observations of Type I (carbon detonation) supernovae. ∞ *(Sec. 12.5)* Recall that these objects are both very bright and have a remarkably narrow spread in luminosities, making them particularly useful as standard candles. ∞ *(Sec. 15.2)* They can be used as probes of the universe because, by measuring their distances (*without* using Hubble's law) and their redshifts, we can determine the rate of cosmic expansion in the distant past. Here's how the method works.

Suppose the universe is decelerating, as we would expect if gravity were slowing its expansion. Then, because the expansion rate is decreasing, objects at great distances—that is, objects that emitted their radiation long ago—should appear to be receding *faster* than Hubble's law predicts. Figure 17.8 illustrates this concept. If the universal expansion were constant in time, recessional velocity and distance would be related by the black line in the figure. (The line is not quite straight because it takes the expansion of the universe properly into account in computing the distance.) ∞ *(More Precisely 15-1)* In a decelerating universe, the velocities of distant objects should lie *above* the black curve, and the deviation from that curve is greater for a denser universe in which gravity has been more effective at slowing the expansion.

How does theory compare with reality? In the late 1990s two groups of astronomers announced the results of independent, systematic surveys of distant supernovae. Some of these supernovae are shown in Figure 17.8(b); the data are marked on Figure 17.8(a). Far from clarifying the picture of cosmic deceleration, however, these findings indicated that the expansion of the universe is not slowing, but actually accelerating! According to the supernova data, galaxies at large distances are receding *less* rapidly than Hubble's law would predict. The deviations from the decelerating curves appear small in the figure, but they are statistically very significant, and both groups report similar findings.

These observations are *inconsistent* with the standard "gravity only" Big Bang theory just described and sparked a major revision of our view of the cosmos. The measurements are difficult, and the results depend quite sensitively on just how "standard" the supernovae luminosities really are. Not surprisingly, since so much hangs on this result, the reliability of the supernova measurement technique has been the subject of intense scrutiny by cosmologists. However, no convincing argument against the method has yet been put forward, so there is no reason to believe that we are somehow being "fooled" by nature. As far as we can tell, the measurements are good, and the cosmic acceleration is real. This unexpected but crucially important result earned the leaders of the two groups who discovered it the 2011 Nobel Prize in Physics.

Dark Energy

What could cause an overall acceleration of the universe? Frankly, cosmologists don't know, although several possibilities have been suggested. Whatever it is, the mysterious cosmic force causing the universe to accelerate is neither matter nor radiation. Although it carries energy, it exerts an overall *repulsive* effect on the universe, causing empty space to expand. It has come to be known simply as **dark energy**, and it is perhaps *the* leading puzzle in astronomy today.

▲ FIGURE 17.8 **Accelerating Universe** (a) In a decelerating universe (purple and red curves), redshifts of distant objects are greater than would be predicted from Hubble's law (black curve). The reverse is true for an accelerating universe. The data points show observations of some 50 supernovae that strongly suggest cosmic expansion is accelerating. (b) The bottom frames show three Type I supernovae (marked by arrows) that exploded in distant galaxies when the universe was nearly half its current age. The top frames show the same areas prior to the explosions. (*CfA/NASA*)

17.1 DISCOVERY

Einstein and the Cosmological Constant

Even the greatest minds are fallible. The first scientist to apply general relativity to the universe was (not surprisingly) the theory's inventor, Albert Einstein. When he derived and solved the equations describing the behavior of the universe, Einstein discovered that they predicted a universe that evolved in time. But in 1917 neither he nor anyone else knew about the expansion of the universe, as described by Hubble's law, which would not be discovered for another 10 years. ∞ *(Sec. 15.3)* At the time Einstein, like most scientists, believed that the universe was static—that is, unchanging and everlasting. The discovery that there was no static solution to his equations seemed to Einstein a near-fatal flaw in his new theory.

To bring his theory into line with his beliefs, Einstein tinkered with his equations, introducing a "fudge factor" now known as the cosmological constant. The accompanying figure shows the effect of introducing a cosmological constant into the equations describing the expansion of a critical-density universe. One possible solution is a "coasting" universe, whose radius does indeed remain constant for an indefinite period of time. Einstein took this to be the static universe he expected. Instead of predicting an evolving cosmos, which would have been one of general relativity's greatest triumphs, Einstein yielded to a preconceived notion of the way the universe "should be," unsupported by observational evidence. Later, when the expansion of the universe was discovered and Einstein's equations—without the cosmological constant—were found to describe it perfectly, he declared that the cosmological constant was the biggest mistake of his scientific career.

For many researchers—Einstein included—the main problem with the cosmological constant was (and still is) the fact that it had no clear physical interpretation. Einstein introduced it to fix what he thought was a problem with his equations, but he discarded it immediately once he realized that no problem actually existed. Scientists are very reluctant to introduce unknown quantities into their equations purely to make the results "come out right." As a result, the cosmological constant fell out of favor among astronomers for many years. However, as described in the text, it has apparently been completely rehabilitated and is now identified as a leading candidate for the "dark energy" whose existence is inferred from studies of the universe on very large scales.

Attempts are underway to construct theories of dark energy that preserve the character of the cosmological constant, yet also account for its value in some natural way, perhaps somehow coupling it to the density of matter. Before we make too many sweeping statements about the role of the cosmological constant in cosmology, we should probably remember the experience of its inventor and bear in mind that—at least for now—its physical meaning remains completely unknown.

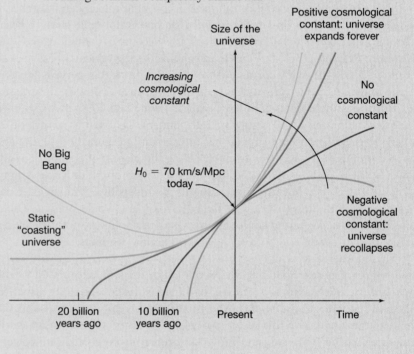

As illustrated in Figure 17.9, the repulsive effect of dark energy is proportional to the size of the universe, so it increases as the universe expands. Thus, it was negligible at early times, but today it is the major factor controlling the cosmic expansion. Furthermore, since the effect of gravity *weakens* as the expansion proceeds, it follows that once dark energy begins to dominate, gravity can never catch up, and the universe will continue to accelerate at an ever-increasing pace. Thus, by opposing the attractive force of gravity, the repulsive effect of dark energy strengthens our earlier conclusion that the universe will expand forever.

A leading candidate for dark energy is an additional "vacuum pressure" force associated with empty space and effective only on very large scales. Known simply as the **cosmological constant,** it has a long and checkered history (see *Discovery 17-1*) and has gone in and out of favor since its invention by Einstein

◄ **FIGURE 17.9 Dark Energy** The expansion of the universe is opposed by the attractive force of gravity and sped up by the repulsion due to dark energy. As the universe expands, the gravitational force weakens, whereas the force due to dark energy increases. A few billion years ago, dark energy began to dominate, and the expansion of the cosmos has been accelerating ever since.

almost a century ago. Note, however, that while models including the cosmological constant can fit the observational data, astronomers as yet have no clear physical interpretation of what it actually means—it is neither required nor explained by any known law of physics.

The fact that the present value of the repulsive dark energy force is comparable to the attractive force of gravity poses another problem to astronomers. When we calculate the evolution of a universe governed by a cosmological constant, consistent with current observations (see Figure 17.12), we find that this state of affairs was not true in the early universe (when galaxies were forming, say), nor will it be true in 10 or 20 billion years' time. In other words, the observations seem to suggest that we live at a special time in the history of the universe—a conclusion viewed with great suspicion by astronomers schooled in the Copernican principle. ∞ *(Sec. 1.1)* Cosmologists have tried to address this problem by constructing alternative theories in which the behavior of dark energy depends on the densities of matter and radiation in the universe, but for now at least, the cosmological constant seems to give just as good a fit to the observational data.

Cosmic Composition

In addition to measurements of density and acceleration, astronomers have several other means of estimating the "cosmological parameters" that describe the large-scale properties of our universe. Theoretical studies of the very early universe (Section 17.7) strongly suggest that the geometry of the universe should be precisely flat—and hence the density should be exactly critical. These ideas first became widespread in the 1980s and for many years seemed to be at odds with observations indicating a cosmic matter density of just 20–30 percent of the critical value, even taking the dark matter into account. The cosmic acceleration just discussed actually resolves that conflict, albeit at the price of introducing yet another unknown component—dark energy—into the cosmic mix.

Detailed measurements of the radiation field known to fill the entire cosmos (Sections 17.5 and 17.8) support the theoretical prediction of critical density and are also consistent with the dark energy inferred from the supernova studies. Further support comes from careful analysis of galaxy surveys such as those discussed in Section 17.1, which allow astronomers to measure the growth of large-scale structure in the universe, and in turn to constrain the value of Ω_0.

Remarkably, all the approaches just described yield consistent results! The current consensus among cosmologists is that the universe is of precisely critical

MA ANIMATION/VIDEO
Gravitational Curvature of Space

PROCESS OF SCIENCE CHECK

Why have astronomers concluded that dark energy is the major constituent of the universe?

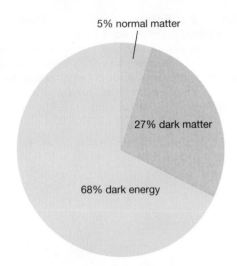

▲ FIGURE 17.10 **Composition of the Universe**   The universe today is made mostly of mysterious dark energy, accounting for nearly three-quarters of the total. Dark matter accounts for over a quarter. Normal matter amounts to only a few percent—and of that 3.6% is mostly in galactic and intergalactic gas, and only a miniscule 0.4% comprises stars, planets, and life-forms.

▲ FIGURE 17.11 **Geometry of the Universe**   The universe on the largest scales seems geometrically flat—governed by the same familiar Euclidean geometry taught in high schools.

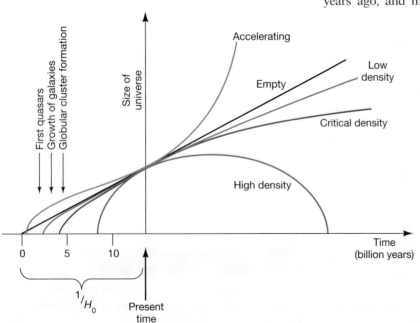

◀ FIGURE 17.12 **Cosmic Age**   The age of a universe without dark energy (represented by all three lower curves colored red, purple, and blue) is always less than $1/H_0$ and decreases for larger values of the present-day density. The existence of a repulsive cosmological constant (green curve) increases the age of the cosmos.

density, $\Omega_0 = 1$, but this density is made up of both matter (mostly dark) and dark energy (suitably converted into mass units using the relation $E = mc^2$). ∞ *(Sec. 9.5)* Note that, while dark energy *opposes* the force of gravity in its effect on cosmic expansion, it still *adds* to the total density (and Ω_0) in its effect on the curvature of space. Based on all available data, as illus- trated in Figure 17.10, the best estimate is that matter accounts for 32 percent of the total and dark energy for the remaining 68 percent. This is the assumption underlying Table 15.1 and used consistently throughout this book. ∞ *(More Precisely 15-1)*

Note that such a universe will expand forever and, the heavy machinery of general relativity and curved spacetime notwithstanding, is perfectly flat (Figure 17.11)—an irony that would no doubt have amused Newton!

The Age of the Universe

In Section 17.2, when we estimated the age of the universe from the accepted value of Hubble's constant, we made the assumption that the expansion speed of the cosmos was constant in the past. However, as we have just seen, this is a considerable oversimplification. Gravity tends to slow the universe's expansion, while dark energy acts to accelerate it. In the absence of a cosmological constant, the universe expanded faster in the past than it does today, so the assumption of a constant expansion rate leads to an overestimate of the universe's age—such a universe is younger than the 14 billion years calculated earlier. Conversely, the repulsive effect of dark energy tends to increase the age of the cosmos.

Figure 17.12 illustrates these points. It is similar to Figure 17.6, except that here we have added three curves, including one (labeled "Empty") corresponding to a constant expansion rate at the present value and another to an accelerating universe with the "best estimate" parameters given above. The age of a critical-density universe with no cosmological constant (whose curve is also added here) is about 9 billion years. A low-density, unbound universe is older than 9 billion years but still less than 14 billion years old. The age corresponding to the accelerating universe is about 14 billion years, coincidentally very close to the value calculated for constant expansion.

How does this compare with an age estimated by other means? On the basis of the theory of stellar evolution, the oldest globular clusters formed about 12 billion years ago, and most are estimated to be between 10 and 12 billion years old. ∞ *(Sec. 12.6)* This range is indicated on Figure 17.12. These ancient star clusters are thought to have formed at around the same time as our Galaxy, so they date the epoch of galaxy formation. More important, they can't be older than the universe! Figure 17.12 shows that globular cluster ages are consistent with a 14 billion-year-old cosmos, and even allow a couple of billion years for galaxies to grow, as discussed in Chapter 16. ∞ *(Sec. 16.3)* Note also that the cluster ages are *not* consistent with a critical-density universe without dark energy. This independent check of a key prediction is an important piece of evidence supporting the modern version of the Big Bang theory.

Thus, for $H_0 = 70$ km/s/Mpc, our current best guess of the history of the universe places the Big Bang at 14 billion years ago. The first quasars appeared about 13 billion years ago (at a redshift of 7), the peak quasar epoch (redshifts 2–3) occurred during the next 1 billion years, and the oldest known stars in our Galaxy formed during the 2 billion years after that. ∞ *(Sec. 16.3)* Even though astronomers do not presently understand the nature of dark energy, the good agreement between so many separate lines of reasoning has convinced many that the dark-matter/dark-energy Big Bang theory just described is the correct description of the universe.

Bear in mind as we proceed through the remainder of this chapter that the Big Bang *is* a scientific theory and as such is continually being challenged and scrutinized. ∞ *(Sec. 0.5)* It makes testable predictions about the state and history of the cosmos and must change—or fail—if these predictions are found to be at odds with observations. Astronomers aren't ready to relax just yet—the history of this subject suggests that there may be a few more unexpected twists and turns in the road before the details are finally resolved.

17.5 The Early Universe

We now turn from studies of the far future of our universe to the quest to understand its distant past. Just how far back in time can we probe? Is there any way to study the universe beyond the most remote quasar? How close can we come to perceiving directly the edge of time, the very origin of the universe?

The Cosmic Microwave Background

A partial answer to these questions was discovered by accident in 1964, during an experiment designed to improve America's telephone system. As part of a project to identify and eliminate unwanted interference in satellite communications, Arno Penzias and Robert Wilson, scientists at Bell Telephone Laboratories in New Jersey, were studying the Milky Way's emission at microwave (radio) wavelengths, using the horn-shaped antenna shown in Figure 17.13. In their data they noticed a bothersome background "hiss"—a little like the background static on an AM radio station. Regardless of where and when they pointed their antenna, the hiss persisted. Never diminishing or intensifying, the weak signal was detectable at any time of the day and on any day of the year, apparently filling all of space.

Eventually, after discussions with colleagues at Bell Labs and theorists at nearby Princeton University, the two experimentalists realized that the origin of the mysterious static was nothing less than the fiery creation of the universe itself. The radio hiss that Penzias and Wilson detected is now known as the **cosmic microwave background**. Their discovery won them the 1978 Nobel Prize in Physics.

In fact, theorists had predicted the existence and general properties of the microwave background well before its discovery. They reasoned that, in addition to being extremely dense, the early universe must also have been very *hot*, and so, shortly after the Big Bang, the universe must have been filled with extremely high-energy thermal radiation—gamma rays of very short wavelength. As illustrated in Figure 17.14, this intensely hot primordial radiation has been redshifted by the expansion of the universe from gamma rays, through optical and infrared wavelengths, all the way into the radio (microwave) range of the electromagnetic spectrum. ∞ *(Sec. 2.4)* The radiation observed today is the "fossil remnant" of the primeval fireball that existed at those very early times. The discovery of the microwave background was thus a powerful piece of evidence in support of the Big Bang theory of the universe and was instrumental in convincing many astronomers of the basic correctness of the idea.

Why do astronomers think the universe will expand forever?

▲ FIGURE 17.13 **Microwave Background Discoverers** This "sugarscoop" antenna was built to communicate with Earth-orbiting satellites but was used by Robert Wilson (right) and Arno Penzias to discover the 2.7-K cosmic background radiation. *(Alcatel-Lucent)*

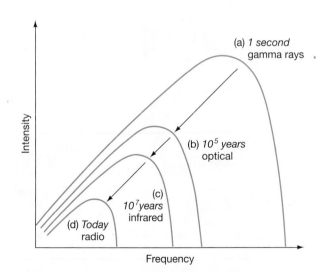

▲ FIGURE 17.14 **Cosmic Blackbody Curves**   Theoretically derived blackbody curves for the entire universe (a) 1 second after the Big Bang, (b) 100,000 years later, (c) 10 million years after that, and (d) today, approximately 14 billion years after the Big Bang.

▲ FIGURE 17.15 Microwave Background Spectrum The intensity of the cosmic background radiation, as measured by the *COBE* satellite, agrees well with theory. The curve is the best fit to the data, corresponding to a temperature of 2.725 K. Experimental errors in this remarkably accurate observation are smaller than the dots representing the data points.

CONCEPT CHECK

When was the cosmic microwave background formed? Why is it still a blackbody curve?

Researchers at Princeton confirmed the existence of the microwave background and estimated its temperature at about 3 K. However, this part of the electromagnetic spectrum happens to be very difficult to observe from the ground, and it was 25 years before astronomers could demonstrate conclusively that the radiation was indeed described by a blackbody curve. ∞ *(Sec. 2.3)* In 1989 the *Cosmic Background Explorer* satellite—*COBE* for short—measured the intensity of the microwave background at wavelengths straddling the peak, from half a millimeter to about 1 cm. The results are shown in Figure 17.15. The solid line is the blackbody curve that best fits the *COBE* data. The near-perfect fit corresponds to a universal temperature of just over 2.7 K.

A striking aspect of the cosmic microwave background is its high degree of *isotropy*. When we correct for Earth's motion through space (which causes the microwave background to appear a little hotter than average in front of us and slightly cooler behind, by the Doppler effect), the intensity of the radiation is virtually constant (to about one part in 10^5) from one direction in the sky to another.

Remarkably, the cosmic microwave background contains more energy than has been emitted by all the stars and galaxies that have ever existed in the history of the universe. Stars and galaxies, though very intense sources of radiation, occupy only a small fraction of space. When their energy is averaged out over the volume of the entire cosmos, it falls short of the energy in the microwave background by at least a factor of 10. For our present purposes, then, the cosmic microwave background is the only significant form of radiation in the universe.

Radiation in the Universe

What role does radiation play in the evolution of the universe on large scales? We can begin to address this question by expressing the energy in the microwave background as an equivalent density, calculating the number of photons in a cubic meter of space, and then converting the total energy of these photons into a mass using the relation $E = mc^2$. ∞ *(Sec. 16.6)* When we do this, we find an equivalent density for the microwave background of about 5×10^{-21} kg/m³—much less than the current value of the cosmic density, which as we have just seen (Figure 17.10), is roughly 9×10^{-27} kg/m³, virtually all in the form of dark energy and matter. Thus, *at the present moment*, the densities of both dark energy and matter in the universe far exceed the density of radiation.

Has this always been the case? Have dark energy and matter always dominated the cosmos? To answer this we must understand how the densities of dark energy, matter, and radiation change as the universe expands. According to current theory, dark energy is a property of space itself, and its density remains *constant* as the universe evolves. In contrast, the densities of matter and radiation both *decrease*, as the expansion dilutes the numbers of atoms and photons alike, but the radiation is further diminished in energy by the cosmological redshift, so its equivalent density falls even faster than the density of matter as the universe grows. Thus, as illustrated in Figure 17.16, even

◄ FIGURE 17.16 Changing Dominance As the universe expanded, the numbers of both matter particles and photons per unit volume decreased. The photons were additionally reduced in energy by the cosmological redshift. As a result, the density of radiation fell faster than the density of matter as the universe grew, and that radiation dominated matter at early times, before the crossover point. Today, dark energy dominates both matter and radiation.

though dark energy dominates today, it was *unimportant* in the early universe. Furthermore, even though the present density of radiation is much less than that of matter, there must have been a time in the past when they were equal, and before that time, radiation was the *main* constituent of the cosmos. At early times the universe was **radiation-dominated**.

17.6 Formation of Nuclei and Atoms

According to the Big Bang theory, at the very earliest times the cosmos consisted entirely of radiation. During the first minute or so of the universe's existence, temperatures were high enough that individual photons of radiation had sufficient energy to transform themselves into matter in the form of elementary particles. This period saw the creation of all the basic building blocks of matter we know today—both "normal" matter, such as protons, neutrons, and electrons, and the "exotic" particles that may compose the dark matter in galaxies and galaxy clusters. ∞ *(Sec. 14.5)* Since then, matter has evolved, clumping together into more and more complex structures forming the nuclei, atoms, planets, stars, galaxies, and large-scale structure we know today, but no *new* matter has been created. *Everything we see around us formed out of radiation as the early universe expanded and cooled.*

Helium Formation

As the temperature of the universe fell below about 900 million K, roughly 2 minutes after the Big Bang, conditions became favorable for protons and neutrons to fuse to form deuterium.[2] Prior to that time, deuterium nuclei were broken apart by high-energy gamma rays as quickly as they were created. Once the deuterium was at last able to survive the radiation background, other fusion reactions quickly converted it into heavier elements, especially helium-4. In just a few minutes, most of the available neutrons were consumed, leaving a universe whose "normal" matter content was primarily hydrogen and helium, with a trace of deuterium. Figure 17.17 illustrates some of the reactions responsible for helium formation. Contrast it with Figure 9.25, which depicts how helium is formed today in the cores of main-sequence stars such as the Sun. ∞ *(Sec. 9.5)* The production of elements heavier than hydrogen by nuclear fusion shortly after the Big Bang is called **primordial nucleosynthesis**.

This early burst of fusion did not last long. Conditions in the cooling, thinning universe became less and less conducive to further fusion reactions as time went by. For all practical purposes, primordial nucleosynthesis stopped at helium-4. Detailed calculations show that, by about 15 minutes after the Big Bang, the cosmic elemental abundance was set, and helium accounted for approximately one-quarter of the total mass of normal matter in the universe. The remaining 75 percent was hydrogen. It would be almost a billion years before nuclear reactions in stars would change these figures.

Astronomers find that no matter where they look, and no matter how low a star's supply of heavy elements may be, there is a minimum amount of helium—a little less than 25 percent by mass—in all stars. This is the case even in stars containing very few heavy elements, indicating that the helium already existed when those ancient stars formed. The accepted explanation is that this base level of helium is *primordial*—created during the early, hot epochs of the universe, before any stars had formed, as just described. The striking agreement between the theoretical calculation and actual observations of helium in stars is another key point in favor of the Big Bang theory.

---

[2]Recall from Chapter 9 that deuterium is simply a heavy form of hydrogen. Its nucleus contains one proton and one neutron. ∞ *(Sec. 9.5)*

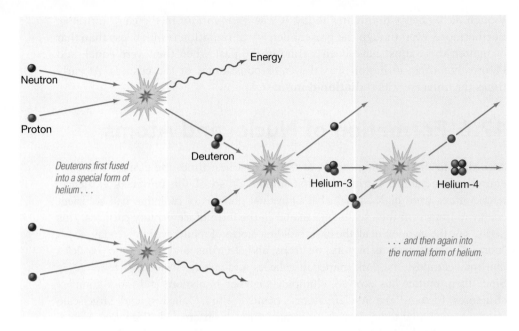

▲ **FIGURE 17.17 Helium Formation**   Some of the reaction sequences that led to the formation of helium in the early universe.

Nucleosynthesis and the Composition of the Universe

Although most deuterium was quickly burned into helium as soon as it formed, a small amount was left over when the primordial nuclear reactions ceased. Nucleosynthesis calculations indicate that the amount of deuterium remaining is a very sensitive indicator of the present-day density of matter in the universe: The *denser* the universe is today, the more matter (protons and neutrons) there was at those early times to react with the deuterium and the less deuterium was left when nucleosynthesis ended. Comparison between theoretical calculations of the production of deuterium in the early universe and the observed abundance of deuterium in stars and the interstellar medium implies a present-day density of at most 3 or 4 percent of the critical value.

But there is an important qualification. Primordial nucleosynthesis depends only on the presence of protons and neutrons in the early universe. Thus, measurements of the abundance of helium and deuterium tell us only about the density of "normal" matter—matter made up of protons and neutrons—in the cosmos. This finding has a momentous implication for the overall composition of the universe. Since the total matter density is about one-third the critical value, if the density of normal matter is only a few percent of this density, then we are forced to conclude that not only is most of the matter in the universe dark, but most of the dark matter is *not* composed of protons and neutrons (see Figure 17.10). The bulk of the matter in the universe apparently exists in the form of elusive subatomic particles whose nature we do not fully understand and whose very existence has yet to be conclusively demonstrated in laboratory experiments. ∞ *(Sec. 14.5)*

For the sake of brevity, we will adopt from here on the convention that the term *dark matter* refers only to these unknown particles and not to "stellar" dark matter, such as brown or white dwarfs, which are made of normal matter.

The Formation of Atoms

When the universe was tens of thousands of years old, matter (consisting of electrons, protons, helium nuclei, and dark matter) began to dominate over radiation.

CONCEPT CHECK

How do we know that most of the dark matter in the universe is not of "normal" composition?

INTERACTIVE
Creation of the Cosmic Microwave Background

The temperature at that time was several tens of thousands of kelvins—too hot for atoms of hydrogen to exist, although some helium ions may already have formed. During the next few hundred thousand years, the universe expanded by another factor of 10, the temperature dropped to a few thousand kelvins, and electrons and nuclei combined to form neutral atoms. By the time the temperature had fallen to 3000 K, the universe consisted of atoms, photons, and dark matter.

The period during which nuclei and electrons combined to form atoms is often called the epoch of **decoupling**, for it was during this period that the radiation background parted company with normal matter. At early times, when matter was ionized, the universe was filled with large numbers of free electrons that interacted frequently with electromagnetic radiation of all wavelengths. A photon could not travel far before encountering an electron and scattering off it. The universe was *opaque* to radiation. However, when the electrons combined with nuclei to form atoms of hydrogen and helium, only certain wavelengths of radiation—those corresponding to the spectral lines of those atoms—could interact with matter. ∞ *(Sec. 2.6)* Light of all other wavelengths could travel virtually forever without being absorbed. The universe became nearly *transparent*. After that time, as the universe expanded, the radiation simply cooled, eventually to become the microwave background we observe today.

The microwave photons now detected on Earth have been traveling through the universe ever since they decoupled. According to the models that best fit the observational data, their last interaction with matter (at the epoch of decoupling) occurred when the universe was about 400,000 years old and roughly 1100 times smaller (and hotter) than it is today—that is, at a redshift of 1100. As illustrated in Figure 17.18, the epoch of atom formation created a kind of "photosphere" in the universe, often referred to as the *surface of last scattering*, completely surrounding Earth at a distance of approximately 14,000 Mpc, corresponding to a redshift of 1100. ∞ *(More Precisely 15-1)* On our side of the photosphere—that is, since decoupling—the universe is transparent. On the far side—before decoupling—it was opaque. Thus, by observing the microwave background, we are probing conditions in the universe almost all the way back in time to the Big Bang itself, in much the same way as studying sunlight tells us about the surface layers of the Sun.

17.7 Cosmic Inflation

In the late 1970s, cosmologists were confronted with two nagging problems that had no easy explanation within the standard Big Bang model. Their resolution has caused theorists to completely rethink their views of the very early universe.

The Horizon and Flatness Problems

The first problem is known as the **horizon problem**. Imagine observing the microwave background in two opposite directions on the sky, as illustrated in Figure 17.19. As we have just seen, in doing so we are actually observing two distant regions of the universe, marked A and B on the figure, where the radiation background last interacted with matter. The fact that the background is known to be *isotropic* to high accuracy means that regions A and B must have had very similar densities and temperatures at the time the radiation we see left them.

The problem is, within the Big Bang theory as just described, there is no particular reason *why* these regions should have been so similar. At the time in question, a few hundred thousand years after the Big Bang, regions A and B were separated by many megaparsecs, and no signal—sound waves, light rays, or anything else—had had time to travel from one to the other. In cosmological

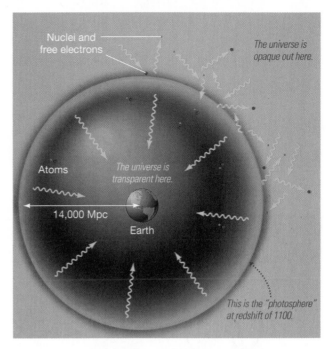

▲ **FIGURE 17.18 Radiation–Matter Decoupling** When atoms formed, the universe became transparent to radiation. Thus, observations of the cosmic background radiation reveal conditions in the universe around a time when the redshift was 1100 and the temperature was less than about 3000 K. For an explanation of how we can see a region of space 14,000 Mpc (46 billion light-years) away when the universe is just 14 billion years old, see *More Precisely 15-1*.

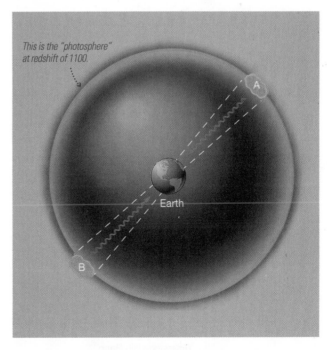

▲ **FIGURE 17.19 Horizon Problem** The isotropy of the microwave background indicates that regions A and B in the universe were very similar to each other when the radiation we now observe left them, but there has not been enough time since the Big Bang for them ever to have interacted with one another. Why then should they look the same?

▲ **FIGURE 17.20 Flatness Problem**   If the universe deviates even slightly from critical density, that deviation grows rapidly in time. For the universe to be as close to critical as it is today, it must have differed from the critical density in the past by only a tiny amount.

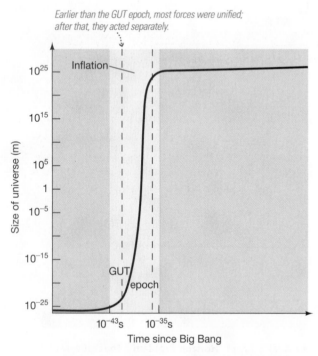

▲ **FIGURE 17.21 Cosmic Inflation**   During the period of inflation, the universe expanded enormously in a very short time. Afterward, it resumed its earlier "normal" expansion rate, except that the size of the cosmos had become about 10^{50} times bigger than it was before inflation.

terminology, each region lay outside the other's *horizon*. But with no possibility of information or energy exchange between these regions, there was no way for them to "know" that they should look the same. The only alternative is that they simply started off looking alike—an assumption that cosmologists are very unwilling to make.

The second problem with the standard Big Bang model is called the **flatness problem**. Whatever the exact value of the cosmic density, it is quite close to the critical value. In terms of spacetime curvature, the universe is remarkably close to being *flat*. But again, there is no good reason why the universe should have formed with a density very close to critical. Why not a millionth or a million times that value? Furthermore, as can be seen in Figure 17.20, a universe that starts off close to, but not exactly on, the critical curve soon deviates greatly from it, so if the universe is close to critical now, it must have been *extremely* close to critical in the past. For example, even if Ω_0 were as low as 0.1 today, the departure from the critical density at the time of nucleosynthesis would have been only change: 1 part in 10^{15} (a thousand trillion).

These observations constitute "problems" because cosmologists want to be able to explain the present condition of the universe, not just accept it "as is." They would prefer to resolve the horizon and flatness problems in terms of physical processes that could have taken a universe with no special properties and caused it to evolve into the cosmos we now see. The resolution of both problems takes us back in time even earlier than nucleosynthesis or the formation of any of the elementary particles we know today—back, in fact, almost to the instant of the Big Bang itself.

The Epoch of Inflation

In the 1970s and 1980s, theoretical physicists succeeded in combining, or *unifying*, the three nongravitational forces in the universe—electromagnetism and the strong and weak forces—into a single all-encompassing "superforce." ∞ *(More Precisely 9-1)* A general prediction of the **Grand Unified Theories** (GUTs) that describe this superforce is that the three forces are unified and indistinguishable from one another *only* at enormously high energies—corresponding to temperatures in excess of 10^{28} K. At lower temperatures, the superforce splits into three, revealing its separate electromagnetic, strong, and weak characters.

In the 1980s, cosmologists discovered that Grand Unified Theories had some remarkable implications for the very early universe. Less than 10^{-34} s after the Big Bang, parts of the cosmos could have temporarily found themselves in a very odd and unstable condition in which empty space—the very fabric of the universe—acquired **vacuum energy**, in essence becoming *excited* above its normal equilibrium state. Esoteric though this sounds, these regions are of direct interest to us—if theorists are correct, we live in one!

The appearance of vacuum energy had dramatic consequences. For a short while, as illustrated in Figure 17.21, the extra energy caused the affected region to expand at an enormously accelerated rate. (The effect is similar in concept but vastly greater in magnitude than the cosmic acceleration associated with the cosmological constant, discussed in *Discovery 17-1*.) The vacuum energy density remained almost constant as the region grew, and the expansion *accelerated* with time while this out-of-equilibrium condition persisted. In fact, the size of the region doubled many times over. This period of unchecked cosmic expansion is known as the **epoch of inflation**.

Eventually, the vacuum returned to its normal "true" state and inflation stopped. Regions of normal space began to appear within the false vacuum and rapidly spread to include the entire expanding region. The whole episode lasted a mere 10^{-32} s, but during that time the patch of the universe that had become unstable swelled in size by the incredible factor of about 10^{50}. After the inflationary phase, the universe once again resumed its (relatively) leisurely expansion, but a number of important changes had occurred that would have far-reaching ramifications for the evolution of the cosmos.

The original theory of inflation was developed in the early 1980s and associated the inflationary period (as in Figure 17.21) with the end of the so-called, *GUT epoch,* as temperatures fell below 10^{28} K and the basic forces of nature became reorganized—a little like a gas liquefying or water freezing as the temperature drops. However, researchers have since realized that conditions suitable for inflation could have occurred under many different circumstances—and possibly *many times*—during the evolution of the early universe, making inflation a rather generic prediction of current theoretical models.

Implications for the Universe

Inflation provides a natural solution for the horizon and flatness problems. The horizon problem is solved because inflation took regions of the universe that had already had time to communicate with one another—and so had established similar physical properties—and then dragged them far apart, well out of communication range. In effect, the universe expanded much faster than the speed of light during the inflationary epoch, so what was once within the horizon now lies far beyond it. (Relativity restricts matter and energy to speeds less than the speed of light, but imposes no such limit on the universe as a whole.) Regions A and B in Figure 17.19 have been out of contact with one another since 10^{-32} s after the Big Bang, but they *were* in contact before then. Their properties are the same today because they were the same long ago, before inflation separated them.

Figure 17.22 illustrates how inflation solves the flatness problem. Imagine that you are a 1-mm-long ant sitting on the surface of an expanding balloon. When the balloon is just a few centimeters across, you can easily see that its surface is curved because its circumference is comparable to your own size. However, as the balloon expands, the curvature becomes less pronounced. When the balloon is a few kilometers across, your "ant-sized" patch of the surface looks quite flat, just as Earth's surface looks flat to us. Now imagine that the balloon expands 100 trillion trillion trillion trillion times. Your local patch of the surface would be completely indistinguishable from a perfectly flat plane. Thus, any curvature the universe may have had before inflation has been expanded so much that space is now perfectly flat on all scales we can ever hope to observe.

This resolution to the flatness problem—the universe appears close to being flat because the universe *is* in fact precisely flat, to very high accuracy—has another very important consequence. Because the universe is geometrically flat, relativity tells us the total density must be exactly equal to the critical value: $\Omega_0 = 1$ (Section 17.3). This is the key result that led us to conclude in Section 17.4 that dark energy—whatever it is—must dominate total density of the universe. Not only is most matter dark (Section 17.6), but *most of the cosmic density isn't made up of matter at all.*

Even though inflation solves the horizon and flatness problems in a quite convincing way, for nearly two decades after it was first proposed the theory was resisted by many astronomers. The main reason was that its prediction of $\Omega_0 = 1$ was at odds with the growing evidence that the density of matter in the universe was no more than 30 percent or so of the critical value. Many cosmologists had in fact considered the possibility

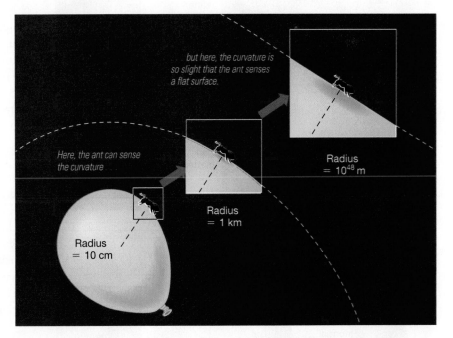

▲ **FIGURE 17.22 Inflation and the Flatness Problem**
Inflation solves the flatness problem by enormously expanding a curved surface, here represented by the surface of a balloon. To an ant on the surface, the balloon looks virtually flat when the expansion is complete.

that a cosmological constant might provide a way to account for the remaining 70 percent of the cosmic density, but without independent corroboration, a conclusive case could not be made. That is why the supernova observations were so important—by demonstrating acceleration in the cosmic expansion rate, they established independent evidence for the effects of dark energy and in doing so reconciled inflation with the discrepant observations.

Physicists will probably never create in terrestrial laboratories conditions even remotely similar to those that existed in the universe during the inflationary epoch. Creation of a false vacuum is (safely) beyond our reach. Nevertheless, cosmic inflation seems to be a natural consequence of many Grand Unified Theories, and it explains two otherwise intractable problems within the Big Bang theory. For these reasons, despite the lack of direct evidence for the process, inflation theory has become an integral part of modern cosmology. It makes definite, testable predictions about the large-scale geometry and structure of the universe that are critically important to current theories of galaxy formation. As we will see in the next section, astronomers are now subjecting these predictions to rigorous scrutiny.

17.8 Large-Scale Structure in the Universe

Most cosmologists think that the present large-scale structure in the universe grew from small *inhomogeneities*—slight deviations from perfectly uniform density—that existed in the early universe. Denser than average clumps of matter contracted under the influence of gravity and merged with other clumps, eventually reaching the point where stars began to form and luminous galaxies appeared. ∞ *(Sec. 16.3)*

The intense background radiation in the early universe effectively prevented clumps of normal matter from forming or growing. Just as in a star, the pressure of the radiation stabilized the clumps against the inward pull of their own gravity. ∞ *(Sec. 12.1)* Instead, the first structures to appear did so in the *dark* matter that makes up the bulk of matter in the universe. While the true nature of dark matter remains uncertain, its defining property is the fact that it interacts only very weakly with both normal matter and radiation. Consequently, clumps of dark matter were unaffected by the radiation background and began to grow as soon as matter first began to dominate the cosmic density.

Thus, as illustrated in Figure 17.23, dark matter determined the overall distribution of mass in the universe and clumped to form the observed large-scale structure. Then, at later times, normal matter was drawn by gravity into the regions of highest density, eventually forming galaxies and galaxy clusters. This picture explains why so much dark matter is found outside the visible galaxies. The luminous material is strongly concentrated near the density peaks and dominates the dark matter there, but the rest of the universe is essentially devoid of normal matter. Like foam on the crest of an ocean wave, the universe we see is only a tiny fraction of the total.

ANIMATION/VIDEO **MA**
Cosmic Structure

The plots below are graphical representations of structure growth.

(a) Time = 1 second

(b) Time = 1000 years

(c) Time = 10^8 years

The maps above show what those structures might look like on the sky.

◄ **FIGURE 17.23 Structure Formation**   (a) The very early universe was a mixture of (mostly) dark and normal matter. (b) A few thousand years after the Big Bang, the dark matter began to clump. (c) Eventually, the dark matter formed large structures (represented here by the two high-density peaks) into which normal matter flowed, ultimately to form the galaxies seen today.

Figure 17.24 shows the results of a supercomputer simulation of a "best-bet" universe consisting of 5 percent normal matter, 27 percent dark matter, and 68 percent dark energy, in the form of a cosmological constant (see Figure 17.10). ∞ *(Sec. 17.3)* The similarities with real observations of cosmic structure shown in Figures 16.22, 16.23, and 17.1, are striking, and more detailed statistical analysis shows that these models in fact agree extremely well with reality. Although calculations like this cannot prove that these models are the correct description of the universe, the agreement in detail between simulations and reality strongly favor the cosmological constant/dark-matter model of the cosmos.

Dark matter does not interact directly with photons, so its growing density fluctuations did not directly affect the radiation background. However, as just described, those fluctuations were coupled to normal matter by gravity, with the result that, before decoupling, the cosmos was filled with tiny fluctuations in the density of normal matter. These density fluctuations in turn caused temperature fluctuations in the radiation field, and those features were "imprinted" in the microwave background when matter and radiation finally parted company at a redshift of 1100. Thus a key prediction of dark-matter models is that there should be tiny "ripples" in the microwave background—temperature variations of only a few parts per million from place to place on the sky.

In 1992, after almost 2 years of careful observation, the *COBE* team (Section 17.5) announced that the expected ripples had been detected. The temperature fluctuations were only 30–40 millionths of a kelvin, but they were there. The *COBE* data were limited by relatively low (roughly 7°) resolution, but combined with computer simulations such as that shown in Figure 17.24, they predicted a present-day structure that is consistent with the superclusters, voids, filaments, and Great Walls we see around us. Detailed analysis also supported the key prediction of inflation theory—that the universe is of exactly critical density and hence spatially flat. For these reasons, the *COBE* observations rank alongside the discovery of the microwave background itself in terms of their importance to the field of cosmology. The lead investigators of the *COBE* program won the 2006 Nobel Prize in physics for their groundbreaking work.

Subsequent missions have radically improved our view of the microwave background, confirming and extending the *COBE* results. NASA's *Wilkinson Microwave Anisotropy Probe* (*WMAP*) operated from 2001 until 2009. Its angular resolution was roughly 20′–30′, some 20 times finer than that of *COBE*, allowing extraordinarily detailed measurements of many cosmological parameters to be made. More recently, the European Space Agency's *Planck* mission, launched in 2009, refined

◄ **FIGURE 17.24 Structure Simulated** Three views of a (present-day) 100 × 100 × 100-Mpc cube in a simulated dark-matter universe with $\Omega_0 = 1$. The cube expands as the universe expands, so that it encloses the same material at all times. The three frames show structures progressively growing from small-density fluctuations in the early universe, displayed at three different times after the Big Bang. *(V. Springel)*

DATA POINTS

Structure in the Universe

Forty-two percent of students had difficulty describing the order in which different kinds of structure formed in the early universe. Some points to remember:

- Matter formed as the expanding universe became cool enough to allow first elementary particles (protons and neutrons), then nuclei to exist—all within the first few minutes after the Big Bang.

- The first atoms formed hundreds of thousands of years later, when the background radiation temperature dropped below a few thousand kelvins.

- Large-scale structure began to form in the dark matter component of the cosmos at very early times, not long after nuclei formed, but the radiation background prevented normal matter from clumping into stars and galaxies until long after atoms formed.

- Stars, galaxies, and quasars began to form a few hundred million years after the Big Bang, as cool gas flowed into regions of high dark matter density.

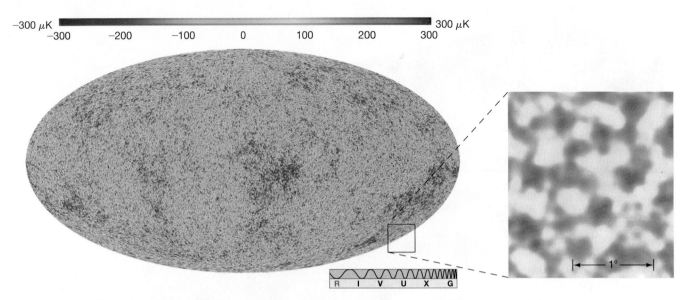

−300 µK   300 µK

−300   −200   −100   0   100   200   300

R   I   V   U   X   G

▲ **FIGURE 17.25 Early Structure INTERACTIVE** The entire microwave sky, as seen by the *Planck* spacecraft. The inset shows a comparably high-resolution map of a small patch of sky, obtained by the ground-based *Cosmic Background Imager* instrument at 30 GHz (1-cm wavelength). The bright blobs are slightly denser-than-average regions of the universe at an age of roughly 400,000 years; they will eventually contract to form clusters of galaxies. *(ESA; CBI)*

WMAP's observations by a further factor of 3 in resolution and 10 in sensitivity, making still more precise measurements of the microwave sky. Figure 17.25 shows an all-sky map of temperature fluctuations in the microwave background, based on *Planck* observations. The inset shows a smaller-scale (2° wide) high-resolution image returned by Cosmic Background Imager, a ground-based microwave telescope located high in the Chilean Andes. Each of the irregular high-temperature regions in these images represents a slightly denser than average clump of dark matter that will one day collapse to form a supercluster-sized clump of galaxies. According to the *Planck* data, we are seeing the universe here when it was less than 400,000 years old, yet the future development of the cosmos is already imprinted in these tiny ripples.

Both maps in Figure 17.25 show temperature fluctuations of a few hundred microkelvins, with a characteristic angular scale of about 1°—pretty much what your eye sees from one "blob" to the next and confirmed by more quantitative means. This scale is related to conditions in the early universe in a very important way: It corresponds to the maximum distance a sound wave can travel between the end of inflation and the time of decoupling. Sound waves were produced every time an overdense clump of dark matter began to collapse under its own gravity, eventually to form a galaxy cluster or supercluster. Normal matter and radiation were also slightly compressed by the contracting clump, but the radiation "pushed back" on the matter, causing it to expand rapidly outward in a shell, as illustrated in Figure 17.26.

The weak gravity of the dark matter clump was far too weak to prevent the shell from escaping, so it continued to expand until the epoch of decoupling, at which time the background radiation stopped interacting with normal matter and the shell stalled. The combined imprints of many overlapping shells on the microwave background sets the scale we see in Figure 17.25. Cosmologists can compute this scale for any given set of cosmological assumptions, allowing direct comparison between observations and theory. The observed fluctuations are in excellent agreement with the theoretical prediction for a universe with $\Omega_0 = 1$ and having the composition indicated in Figure 17.10.

Remarkably, these ideas also connect the microwave background to the observed distribution of galaxies at later times. Because the expanding shell just discussed itself represents a denser-than-average part of the universe, it too will tend to attract more matter, and it will eventually form galaxies of its own. The

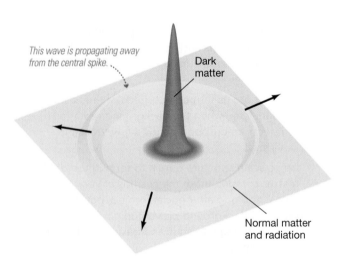

This wave is propagating away from the central spike.

Dark matter

Normal matter and radiation

▲ **FIGURE 17.26 Acoustic Oscillation** This sketch is a two-dimensional rendering of a three-dimensional wave of normal matter pushed by radiation away from a clump of dark matter in the early universe. In reality, many such sound waves would have propagated like this all over the sky wherever there were concentrations of dark matter.

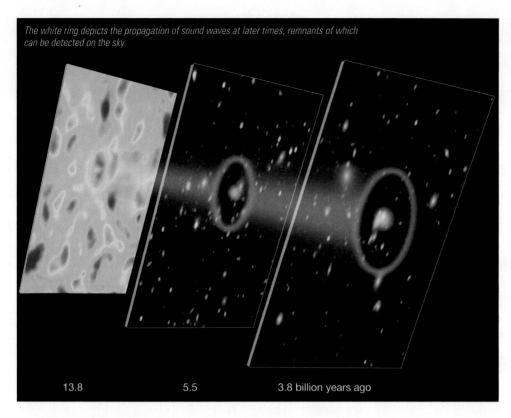

The white ring depicts the propagation of sound waves at later times, remnants of which can be detected on the sky.

13.8   5.5   3.8 billion years ago

◀ **FIGURE 17.27 Acoustic Remnants**   The record of expanding sound waves in the early universe allows astronomers to retrace cosmic history. This simulation shows how the small-density variations of the early universe (left) grew to become the clusters, walls, and filaments seen in more recent times. *(Z. Rostomian/SDSS)*

result is that *every* dark matter region that forms a galaxy or galaxy cluster is expected to have associated with it a secondary shell of galaxies. This feature is imprinted on the galaxy distribution throughout the universe and at all redshifts. Figure 17.27 shows how such a ripple grows with the expansion of the universe; today, its radius would be about 150 Mpc. It forms a "standard yardstick," telling us precisely the scale of the universe at different times in the past, and represents an alternative powerful means of probing the expansion of the universe, *independent* of the supernova studies earlier in this chapter.

Of course, every density fluctuation will give rise to a similar wave, so what we actually observe is a large number of shells overlapping and mixed together. Nevertheless, astronomers can infer statistically the existence of these shells, and that has been a major goal of the *Sloan Digital Sky Survey*, which includes data from more than a million galaxies. ∞ *(Discovery 15-1)* The separations of those galaxies, again measured statistically, are completely consistent with the predictions of the process just described.

The first decade of the 21st century saw the basic parameters of the universe measured (even if not yet fully understood) to an accuracy only dreamed of just a few years ago. It now appears that the second decade is well on track to explore the nature of dark matter and dark energy with unprecedented precision.

ANIMATION/VIDEO
Ripples in the Early Universe

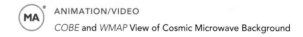

ANIMATION/VIDEO
COBE and *WMAP* View of Cosmic Microwave Background

CONCEPT CHECK

Why is dark matter important to galaxy formation? How do observations of the microwave background support this view?

THE BIG QUESTION

What is the origin of the universe, and what will be its ultimate fate? Did it actually have an origin, or has it existed forever? Can astronomers ever hope to answer such fundamental questions? No one knows the answers, but for the first time in recorded history, human beings are using logic, rationality, and some very sophisticated (and expensive) experimental equipment to try to address them. Might we be on the cusp of answering some of the deepest queries any human has ever asked?

CHAPTER REVIEW

SUMMARY

LO1 On scales larger than a few hundred megaparsecs, the universe appears roughly **homogeneous** (p. 476; the same everywhere) and **isotropic** (p. 476; the same in all directions). In **cosmology** (p. 477)—the study of the universe as a whole—researchers usually assume that the universe is homogeneous and isotropic on large scales. This assumption is known as the **cosmological principle** (p. 477).

LO2 If the universe were homogeneous, isotropic, infinite, and unchanging, then the night sky would be bright because any line of sight would eventually intercept a star. The fact that the night sky is instead dark is called **Olbers's paradox** (p. 477). Its resolution is that we see only a finite part of the universe—the region within which light has had time to reach us since the universe formed. Tracing the observed motions of galaxies back in time implies that, about 14 billion years ago, the universe consisted of a single point that expanded rapidly in the **Big Bang** (p. 478). Space itself was compressed to a point at that instant—the Big Bang happened everywhere at once. The cosmological redshift occurs as a photon's wavelength is "stretched" by cosmic expansion. The extent of the observed redshift is a direct measure of the expansion of the universe since the photon was emitted.

LO3 There are only two possible outcomes to the current expansion: Either the universe will expand forever, or it will recollapse to a point. The **critical density** (p. 480) is the density of matter needed for gravity to overcome the present expansion and cause the universe to collapse. The total density of the universe (including matter, radiation, and dark energy) determines the geometry of the universe on the largest scales, as described by general relativity. In a high (greater-than-critical) density universe, the curvature of space is sufficiently large that the universe "bends back" on itself and is finite in extent, somewhat like the surface of a sphere. Such a universe is said to be a **closed universe** (p. 481). A low-density **open universe** (p. 482) is infinite in extent and has a "saddle-shaped" geometry. The **critical universe** (p. 482) has a density precisely equal to the critical value and is spatially flat.

LO4 Luminous and dark matter account for only 25–30 percent of the critical cosmic density. Observations of distant supernovae suggest that the expansion of the universe may be accelerating,

possibly driven by the effects of a force commonly called **dark energy** (p. 483). One candidate for this dark energy is the **cosmological constant** (p. 484), a repulsive force that may exist throughout all space. Its physical nature is unknown. Other, independent observations are consistent with the idea that the universe is flat—that is, of exactly critical density—with (mostly dark) matter making up 32 percent of the total and dark energy making up the rest. Such a universe will expand forever.

LO5 The **cosmic microwave background** (p. 487) is isotropic blackbody radiation that fills the entire universe. Its present temperature is about 3 K. Its existence is evidence that the universe expanded from a hot, dense state. As the universe has expanded, the initially high-energy radiation has been redshifted to lower and lower temperatures. At the present time, the densities of dark energy and matter in the universe both greatly exceed the equivalent mass density of radiation. The density of dark energy remains constant as the universe expands, but the matter density was much greater in the past, when the universe was smaller. However, because radiation is redshifted as the universe expands, the density of radiation was greater still. The early universe was **radiation-dominated** (p. 489).

LO6 All of the hydrogen in the universe is primordial, formed from radiation as the hot early universe expanded and cooled. Most of the helium observed in the universe today is also primordial, created by fusion between protons and neutrons a few minutes after the Big Bang. This is known as **primordial nucleosynthesis** (p. 489). Other, heavier elements were formed much later, in the cores of stars. Detailed studies of this process indicate that "normal" matter can account for at most 3 percent of the critical density. By the time the universe was about 1100 times smaller than it is today, the temperature had become low enough for the first atoms to form. At that time, the (then visible) radiation background **decoupled** (p. 491) from the matter. The photons that now make up the microwave background have been traveling freely through space ever since.

LO7 According to modern **Grand Unified Theories** (p. 492), the three nongravitational forces of nature first began to display their separate characters about 10^{-34} s after the Big Bang. A brief period of rapid cosmic expansion called the **epoch of inflation** (p. 492) ensued, during which the size of the universe increased by a factor of about 10^{50}. The **horizon problem** (p. 491) is the fact that, according to the standard (that is, noninflationary) Big Bang model, there is no good reason for widely separated parts of the universe to be as similar as they are. Inflation solves the horizon problem by taking a small homogeneous patch of the early universe and expanding it enormously. Inflation also solves the **flatness problem** (p. 492), which is the fact that there is no obvious reason why the density of the universe is so close to critical. Inflation implies that the universe is spatially flat and hence that the cosmic density is exactly critical.

LO8 The large-scale structure observed in the universe today formed when density inhomogeneities in the dark matter clumped and grew to create the "skeleton" of the structure now observed. Normal matter then flowed into the densest regions of space, eventually forming the galaxies we now see. "Ripples" in the microwave background, discovered by the *COBE* satellite, are the imprint of these early inhomogeneities on the radiation field and lend strong support to the inflationary prediction that we live in a flat, critical-density universe. Detailed observations of the microwave background, combined with studies of large-scale structure in the cosmos, provide detailed information on the basic cosmological parameters of the universe.

MasteringAstronomy®   For instructor-assigned homework go to www.masteringastronomy.com

Problems labeled **POS** explore the process of science. **VIS** problems focus on reading and interpreting visual information. **LO** connects to the introduction's numbered Learning Outcomes.

REVIEW AND DISCUSSION

1. **LO1** What evidence do we have that there is no structure in the universe on very large scales? How large is "very large"?

2. What is the cosmological principle?

3. What is Olbers's paradox? How is it resolved?

4. **LO2** Explain how an accurate measure of Hubble's constant can lead to an estimate of the age of the universe.

5. Why isn't it correct to say that the expansion of the universe involves galaxies flying outward into empty space?

6. Where did the Big Bang occur?

7. How does the cosmological redshift relate to the expansion of the universe?

8. What is Ω_0? What is the current best estimate of its value, and what does this imply about the geometry of the universe?

9. **LO4 POS** Why are observations of distant supernovae so important to cosmology?

10. **LO5** What is the significance of the cosmic microwave background?

11. **LO6** Why do all stars, regardless of their supply of heavy elements, seem to contain at least one-quarter helium by mass?

12. When did the universe become transparent to radiation?

13. **LO7 POS** What is cosmic inflation? How does inflation solve the horizon problem? The flatness problem?

14. **POS** What is the connection between dark matter and the formation of large-scale structure in the universe?

15. **LO8 POS** What prediction of inflation theory has since been verified by observations of the microwave background?

CONCEPTUAL SELF-TEST: TRUE OR FALSE/MULTIPLE CHOICE

1. Deep surveys of the universe indicate that the largest structures in space are about 50 Mpc in size. (T/F)

2. If the universe had an edge, that fact would violate the assumption of isotropy in the cosmological principle. (T/F)

3. Hubble's law implies that the universe will expand forever. (T/F)

4. The cosmological redshift is a direct measure of cosmic expansion. (T/F)

5. The cosmic microwave background is the highly redshifted radiation of the early Big Bang. (T/F)

6. Observations suggest that the density of the universe is made up mostly of dark matter. (T/F)

7. The microwave background radiation last interacted with matter around the time of decoupling. (T/F)

8. The theory of inflation predicts that the density of the universe is exactly equal to the critical density. (T/F)

9. Olbers's paradox is resolved by (a) the finite size of the universe; (b) the finite age of the universe; (c) light from distant galaxies being redshifted so we can't see them; (d) the fact that there is an edge to the universe.

10. The galactic distances used to measure the acceleration of the universe are determined by observations of (a) trigonometric parallax; (b) line broadening; (c) Cepheid variable stars; (d) exploding white dwarfs.

11. On the basis of our current best estimate of the present mass density of the universe, astronomers think that (a) the universe is finite in extent and will expand forever; (b) the universe is finite in extent and will eventually collapse; (c) the universe is

infinite in extent and will expand forever; (d) the universe is infinite in extent and will eventually collapse.

12. The age of the universe is estimated to be (a) less than Earth's age; (b) the same as the age of the Sun; (c) the same as the age of the Milky Way Galaxy; (d) greater than the age of the Milky Way Galaxy.

13. The horizon problem in the standard Big Bang model is solved by having the universe (a) accelerate; (b) inflate rapidly early in its existence; (c) have tiny, but significant fluctuations in temperature; (d) be geometrically flat.

14. VIS The data points in Figure 17.8 (Accelerating Universe) (a) prove that the universe is not expanding; (b) imply that the expansion is decelerating faster than expected; (c) allow a measurement of Hubble's constant; (d) indicate that the redshifts of distant galaxies are greater than would be expected if gravity alone were acting.

15. VIS According to Figure 17.16 (Changing Dominance), when the universe was 10 years old, the density of radiation was (a) much greater than that of matter; (b) much less than that of matter; (c) comparable to that of matter; (d) unknown.

PROBLEMS

The number of squares preceding each problem indicates its approximate level of difficulty.

1. ■ What is the greatest distance at which a galaxy survey sensitive to objects as faint as 20th magnitude could detect a galaxy as bright as the Milky Way (absolute magnitude –20)? ∞ *(More Precisely 10-1)*

2. ■■ If the entire universe were filled with Milky Way–like galaxies, with an average density of 0.1 per cubic megaparsec, calculate the total number of galaxies observable by the survey in the previous question if it covered the entire sky.

3. ■■ Eight galaxies are located at the corners of a cube. The present distance from each galaxy to its nearest neighbor is 10 Mpc, and the entire cube is expanding according to Hubble's law, with $H_0 = 70$ km/s/Mpc. Calculate the recession velocity (magnitude *and* direction) of one corner of the cube relative to the opposite corner.

4. ■ According to the standard Big Bang theory (neglecting any effects of cosmic acceleration), what is the *maximum* possible age of the universe if $H_0 = 50$ km/s/Mpc? 70 km/s/Mpc? 80 km/s/Mpc?

5. ■ For a Hubble constant of 70 km/s/Mpc, the critical density is 9×10^{-27} kg/m^3. (a) How much mass does this correspond to within a volume of 1 cubic astronomical unit? (b) How large a cube would be required to enclose 1 Earth mass of material?

6. ■■ The Virgo Cluster is observed to have a recessional velocity of 1200 km/s. Assuming $H_0 = 70$ km/s/Mpc and a critical-density universe, calculate the total mass contained within a sphere centered on Virgo and just enclosing the Milky Way Galaxy. What is the escape speed from the surface of this sphere? ∞ *(More Precisely 5-1)*

7. ■■ (a) What is the present peak wavelength of the cosmic microwave background? Calculate the size of the universe relative to the present size when the radiation background peaked in (b) the infrared, at 10 μm, (c) in the ultraviolet, at 100 nm, and (d) in the gamma ray, at 1 nm.

8. ■■ What were the matter density and the equivalent mass density of the cosmic radiation field when the universe was one-thousandth its present size? Assume critical density, and don't forget the cosmological redshift.

9. ■ What was the distance at the time of decoupling between the points that would eventually become the center of the Milky Way and the center of a nearby galaxy cluster, if the present distance to the cluster is 100 Mpc?

10. ■■■ The "blobs" evident in the inset to Figure 17.25 are about 20′ (arcminutes). If those blobs represent clumps of matter around the time of decoupling (redshift = 1100), and assuming Euclidean geometry, estimate the size of these clumps at the time of decoupling. ∞ *(More Precisely 15-1)*

ACTIVITIES

Collaborative

1. Make a model of a two-dimensional universe and use it to examine Hubble's law. Find a balloon that will expand into a nice, large sphere. Blow it up about halfway, and mark dots all over its surface; the dots will represent galaxies. Each group member should choose one dot as his/her home galaxy. Using a measuring tape, measure the distances to various other galaxies, numbering the dots so you will not confuse them later. Now blow the balloon up to full size and measure the distances again. Calculate the change in the distances for each galaxy; this is a measure of its velocity (change in position/ change in time; the time is the same for all and is arbitrary). Plot their velocities versus their new distances as in Figure 15.16 or 17.8. Do you get a straight-line correlation, that is, a "Hubble" law? Does it matter which dot you choose as home? Demonstrate this to your class.

Individual

1. Isotropy is the extent to which things look the same in every direction. Considering buildings, geographic features, and similar objects within a few miles of your current location, is your local universe isotropic? If not, is there any scale on which isotropy applies, even approximately?

2. Write a paper on the philosophical differences between living in an open, closed, or flat universe. Are there aspects of any of these three possibilities that are hard to accept? Do you have a preference?

3. Go to your library or online and read about the *steady-state universe*, which enjoyed some measure of popularity in the 1950s and 1960s. How did it differ from the standard Big Bang model? What were its main assumptions and predictions, and why did its proponents regard it as superior to the Big Bang? Why is the steady-state model not widely accepted today—what key predictions did it make that were inconsistent with observations?

18

Life in the Universe

Are We Alone?

Are we unique? Is life on our planet the only example of life in the universe? If so, then what would be the implications of that momentous discovery? If not, then how and where should we look for other intelligent beings? These are difficult questions, for the subject of extraterrestrial life is one for which we have no data, but they are important questions, with profound implications for the human species. Earth is the only place in the universe where we know for certain that life exists. In this final chapter we take a look at how humans evolved on Earth and then consider whether those evolutionary steps might have happened elsewhere. Having done that, we assess the likelihood of our having Galactic neighbors and consider how we might learn about them if they exist.

THE BIG PICTURE

Earth is the only place in the universe where we know for certain that life exists. Despite the likelihood of life elsewhere in the universe, we have no clear evidence for it. None of the thousands of known extrasolar planets has yet shown any sign of life, intelligent or otherwise. Even so, astronomers keep watchful eyes on the sky, constantly aware and hopeful that evidence for extraterrestrial intelligence (ETI) might emerge at any moment.

◄ Is there intelligent life elsewhere in the universe? We saw in Chapter 4 how astronomers are amassing growing evidence of potentially Earth-like planets orbiting other stars. Perhaps, as this fanciful artist's conception suggests, Earth-like moons may orbit many giant exoplanets, providing important alternative sites for the development of extraterrestrial life. Still, despite blockbuster movies, science-fiction novels, and many claims of extraterrestrial contact, astronomers have so far found no evidence for life, intelligent or otherwise, anywhere else in the universe. Nevertheless, the search goes on.

18.1 Cosmic Evolution

In our study of the universe, we have been careful to avoid any inference or conclusion that places Earth in a special place in the cosmos. This Copernican principle has been our essential guide in helping define our place in the "big picture." ∞ *(Sec. 1.1)* But when discussing the topic of life in the universe we face a problem: Ours is the only planet we know of on which life and intelligence have evolved, making it hard for any discussion of intelligent life *not* to treat humans as special cases.

Accordingly, in this final chapter we must adopt a decidedly different approach. We first describe the chain of events leading to the only technologically proficient, intelligent civilization we know—us. Then, we try to assess the likelihood of finding and communicating with intelligent life elsewhere in the cosmos.

Life in the Universe

With this human-centered viewpoint clearly evident, Figure 18.1 identifies seven major phases in the history of the universe: *particulate, galactic, stellar, planetary, chemical, biological,* and *cultural* evolution. Taken together, these evolutionary stages make up the grand sweep of **cosmic evolution**—the continuous transformation of matter and energy that has led to the appearance of life and civilization on Earth. The first four phases represent, in reverse order, the contents of this book up to this point. We now expand our field of view beyond astronomy to include the other three.

From the Big Bang to the evolution of intelligence and culture, the universe has evolved from simplicity to complexity. We are the result of an incredibly complex chain of events that spanned billions of years. Were those events random, making us unique, or are they in some sense *natural*, so that a technological civilization like ours is inevitable? Put another way, are we alone in the universe, or are we just one among countless other intelligent life-forms?

Before trying to answer these important questions, we need a working definition of *life*. The distinction between the living and the nonliving is not as obvious as we might at first think. Generally speaking, scientists regard the following as characteristics of living organisms: (1) they can *react* to their

► **FIGURE 18.1 Arrow of Time** Some highlights of cosmic history, as it relates to the emergence of life on Earth, are noted along this arrow of time, from the beginning of the universe to the present. At the bottom of the arrow are seven "windows" outlining the major phases of cosmic evolution: evolution of primal energy into elementary particles; of atoms into galaxies and stars; of stars into heavy elements; of elements into solid rocky planets; of those elements into the molecular building blocks of life; of those molecules into life itself; and of advanced life-forms into intelligence, culture, and technological civilization. *(D. Berry)*

environment and can often heal themselves when damaged; (2) they can *grow* by taking in nourishment from their surroundings; (3) they can *reproduce*, passing along some of their own characteristics to their offspring; and (4) they have the capacity for genetic change and can therefore *evolve* from generation to generation to adapt to a changing environment.

These rules are not hard and fast, and a good working definition of life is elusive. Stars, for example, react to the gravity of their neighbors, grow by accretion, generate energy, and "reproduce" by triggering the formation of new stars, but no one would suggest that they are alive. For example, a virus is inert, often crystalline in form, when isolated from living organisms, but once inside a living system, it exhibits all the properties of life, seizing control of the cell and using the cell's own genetic machinery to grow and reproduce (Figure 18.2). Most researchers think that the distinction between living and nonliving is more one of structure and complexity than a simple checklist of rules.

The general case in favor of extraterrestrial life is summed up in what are sometimes called the *assumptions of mediocrity*: (1) because life on Earth depends on just a few basic molecules, (2) because the elements that make up these molecules are (to a greater or lesser extent) common to all stars, and (3) if the laws of science we know apply to the entire universe (as we have supposed throughout this book), then—given sufficient time—life must have originated elsewhere in the cosmos. The opposing view maintains that intelligent life on Earth is the product of a series of extremely fortunate accidents—astronomical, geological, chemical, and biological events unlikely to have occurred anywhere else in the universe. The purpose of this chapter is to examine some of the scientific arguments for and against these viewpoints.

▲ **FIGURE 18.2 Virus** This electron microscope image shows the H1N1 (Swine Flu) virus that killed thousands of people worldwide in 2009. Viruses exist in the gray area between the living and the nonliving—they are lifeless when isolated from living organisms, but once inside a living system, they take on all the properties of life. A virus comes alive by transferring its genetic material into a living cell, seizing control and establishing itself as the new master of chemical activity, often with devastating results to the host organism. *(Centers for Disease Control)*

Life on Earth

What information do we have about the earliest stages of planet Earth? Unfortunately, not very much. Geological hints about the first billion years or so were largely erased by violent surface activity as volcanoes erupted and meteorites bombarded our planet; subsequent erosion by wind and water has seen to it that little evidence has survived to the present day. However, the early Earth was probably barren, with shallow lifeless seas washing upon grassless, treeless continents. Outgassing from our planet's interior through volcanoes, fissures, and geysers produced an atmosphere rich in hydrogen, nitrogen, and carbon compounds and poor in free oxygen. As Earth cooled, ammonia, methane, carbon dioxide, and water formed. The stage was set for the appearance of life.

The surface of the young Earth was a very violent place. Natural radioactivity, lightning, volcanism, solar ultraviolet radiation, and meteoritic impacts all provided large amounts of energy that eventually shaped the ammonia, methane, carbon dioxide, and water into more complex molecules known as **amino acids** and **nucleotide bases**—organic (carbon-based) molecules that are the building blocks of life as we know it. Amino acids build *proteins*, and proteins control *metabolism*, the daily utilization of food and energy by means of which organisms stay alive and carry out their vital activities. Sequences of nucleotide bases form *genes*—parts of the DNA molecule—that direct the synthesis of proteins and hence determine the characteristics of organisms (Figure 18.3). These same genes

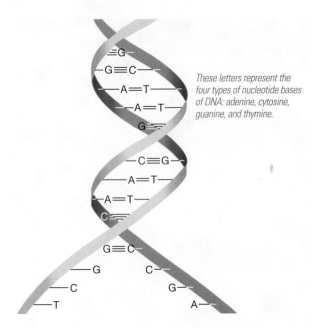

These letters represent the four types of nucleotide bases of DNA: adenine, cytosine, guanine, and thymine.

▲ FIGURE 18.3 DNA Molecule   The DNA (deoxyribonucleic acid) molecule contains all the genetic information needed for a living organism to reproduce and survive. Often consisting of literally tens of billions of individual atoms, its double-helix structure allows it to "unzip," exposing its internal structure to control the creation of proteins needed for a cell to function. The ordering of its constituent parts is unique to each individual organism.

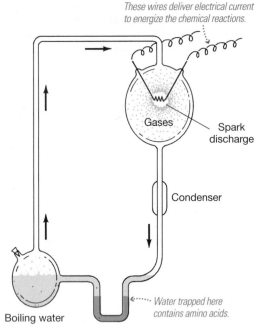

These wires deliver electrical current to energize the chemical reactions.

Gases

Spark discharge

Condenser

Water trapped here contains amino acids.

Boiling water

▲ FIGURE 18.4 Urey-Miller Experiment   This chemical apparatus is designed to make complex biochemical molecules by energizing a mixture of simple chemicals. Gases (ammonia, methane, carbon dioxide, water vapor) are placed in the upper bulb to simulate Earth's primordial atmosphere and then zapped by spark-discharge electrodes akin to lightening. After about a week, amino acids and other complex molecules emerge in the trap at the bottom, which simulates the primordial oceans into which heavy molecules produced in the overlying atmosphere would have fallen.

also carry an organism's hereditary characteristics from one generation to the next. In all living creatures on Earth—from bacteria to amoebas to humans—genes mastermind life, while proteins maintain it.

The first experimental demonstration that complex molecules could have evolved naturally from simpler ingredients found on the primitive Earth came in 1953. Scientists Harold Urey and Stanley Miller, using laboratory equipment somewhat similar to that shown in Figure 18.4, took a mixture of the materials thought to be present on Earth long ago—a "primordial soup" of water, methane, carbon dioxide, and ammonia—and energized it by passing an electrical discharge ("lightning") through it. After a few days, they analyzed their mixture and found that it contained many of the same amino acids found today in all living things on Earth. About a decade later, scientists succeeded in constructing nucleotide bases in a similar manner. These experiments have been repeated in many different forms, with more realistic mixtures of gases and a variety of energy sources, but always with the same basic outcomes.

None of these experiments has ever produced a living organism, and the molecules created are far less complex than the strand of DNA illustrated in Figure 18.3. Nevertheless, they demonstrate conclusively that "biological" molecules can be synthesized by strictly nonbiological means, using raw materials available on the early Earth. More advanced experiments have fashioned protein-like blobs (Figure 18.5a) that behave to some extent like true biological cells. They resist dissolution in water and tend to cluster into small droplets called *microspheres*—a little like oil globules floating on the surface of water. The walls of these laboratory-made droplets permit the inward passage of small molecules, which then combine within the droplet to produce more complex molecules too large to pass back out through the walls. As the droplets "grow," they tend to "reproduce," forming smaller droplets.

Can we consider these proteinlike microspheres to be alive? Almost certainly not. The microspheres contain many of the basic ingredients needed to form life but are not life itself. They lack the hereditary molecule DNA. However, they do have similarities to ancient cells found in the fossil record (Figure 18.5b). Thus, while no actual living cells have yet been created "from scratch," many biochemists feel that many key steps in *chemical evolution*—the chain of events leading from simple nonbiological molecules to the building blocks of life itself—have now been demonstrated.

An Interstellar Origin?

Recently, a dissenting view has emerged. Some scientists argue that there may not have been sufficient energy available to power the chemical reactions in Earth's primitive atmosphere, and there may not have been enough raw material for the reactions to occur at a significant rate in any case. Instead, they suggest, most of the organic material that combined to form the first living cells was produced in *interstellar space* and subsequently arrived on Earth in the form of interplanetary dust and meteors that did not burn up during their descent through the atmosphere.

To test this hypothesis, NASA researchers carried out their own version of the Urey-Miller experiment, exposing an icy mixture of water, methanol, ammonia, and carbon monoxide—representative of many interstellar grains—to ultraviolet radiation to simulate the energy from a nearby newborn star. As shown in Figure 18.6, when they later placed the irradiated ice in water and examined the results, they found that it had formed droplets, surrounded by membranes, containing complex organic molecules. No amino acids, proteins, or DNA were found in the mix, but the results, repeated numerous times, clearly show that even the harsh, cold vacuum of interstellar space can be a suitable medium for the formation of complex molecules and primitive cellular structures.

▶ **FIGURE 18.5 Chemical Evolution** (a) These carbon-rich, protein-like droplets contain as many as a billion amino acid molecules in a liquid sphere. The droplets can "grow," and parts of them can separate from their "parent" droplet to become new individual droplets (as at A, B, C). (b) This photograph shows primitive fossils that display concentric spheres or walls connected by smaller spheroids; they were found in sediments radioactively dated to be about 2 billion years old. (c) For comparison, modern blue-green algae are shown, on approximately the same scale. *(S. Fox/E. Barghoorn)*

These three photographs, taken through a microscope, show structures on the scale of 1 micrometer, which equals 1/10,000 of a centimeter.

2.5 μm

(a)

Interstellar molecular clouds are known to contain many complex molecules, and large amounts of organic material were detected on comets Halley and Hale-Bopp when they last visited the inner solar system. ∞ *(Sec. 4.2)* We saw earlier that many scientists think cometary impacts early in our planet's history were responsible for most of Earth's water. ∞ *(Sec. 4.3)* Perhaps those comets carried complex molecules, too.

In addition, a small fraction of meteorites that survive the plunge to Earth's surface (including perhaps the controversial "Martian meteorite" discussed in Chapter 6) have been shown to contain organic compounds. ∞ *(Discovery 6-2)* Figure 18.7 shows a particularly well-studied example, which fell near Murchison, Australia, in 1969. Located soon after crashing to the ground, this meteorite contained 12 of the amino acids normally found in living cells. Such discoveries strongly suggest that complex molecules can form in an interplanetary or interstellar environment and could have reached Earth's surface unscathed after their fiery descent.

The moderately large molecules found in meteorites and in interstellar clouds are our only evidence that chemical evolution has occurred elsewhere in the universe. However, most researchers regard this organic matter as prebiotic—matter that could eventually lead to life, but that has not yet done so. Nevertheless, the hypothesis that organic matter is constantly raining down on Earth from space in the form of interplanetary debris is quite plausible. Whether this was the *primary* means by which those molecules appeared on Earth remains unclear.

Diversity and Culture

However the basic materials appeared on Earth, we know that life *did* appear. The fossil record chronicles how life on Earth became widespread and diversified over time. The study of fossil remains shows the initial appearance about

10 μm

(b)　　　(c)

ANIMATION/VIDEO
Icy Organics in Planet-Forming Disc

10 μm

◀ **FIGURE 18.6 Interstellar Globules** These oily, hollow droplets rich in organic molecules were made by exposing a freezing mixture of primordial matter to harsh ultraviolet radiation. When immersed in water, the larger ones display cell-like membrane structure. Although they are not alive, they bolster the idea that life on Earth could have come from space. *(NASA)*

|←——→| 10 µm
R I V U X G

▲ **FIGURE 18.7 Murchison Meteorite** The Murchison meteorite contains relatively large amounts of amino acids and other organic material, indicating that chemical evolution of some sort has occurred beyond our own planet. In this magnified view of a fragment from the meteorite, an arrow points to a microscopic sphere of organic matter. *(Harvard-Smithsonian Center for Astrophysics)*

CONCEPT CHECK

Has chemical evolution been verified in the laboratory?

3.5 billion years ago of simple one-celled organisms, such as blue-green algae. These were followed about 2 billion years ago by more complex one-celled creatures like the amoeba. Multicellular organisms did not appear until about 1 billion years ago, after which a wide variety of increasingly complex organisms flourished—among them insects, reptiles, mammals, and humans. Figure 18.8 illustrates some of the key developments in the evolution of life on our planet.

The fossil record leaves no doubt that biological organisms have changed over time—all scientists accept the reality of *biological evolution.* As conditions on Earth have shifted and Earth's surface has evolved, those organisms that could best take advantage of their new surroundings succeeded and thrived—often at the expense of those organisms that could not make the necessary adjustments and consequently became extinct.

In the opinion of many anthropologists, intelligence, like any other highly advantageous trait, is strongly favored by natural selection. The social cooperation that went with coordinated hunting efforts was an important competitive advantage that developed as brain size increased. Perhaps most important of all was the development of language. Experience and ideas, stored in the brain as memories, could be passed down from one generation to the next. A new kind of evolution had begun, namely, *cultural evolution*—the changes in the ideas and behavior of society. Our more recent ancestors created, within about 10,000 years, the entirety of human civilization.

To put all this into historical perspective, let's imagine the entire lifetime of Earth to be 46 years rather than 4.6 billion years. On this scale, we have no reliable record of the first decade of our planet's existence. Life originated at least 35 years ago, when Earth was about 10 years old. Our planet's middle age is still largely a mystery. Not until about 6 years ago did abundant life flourish throughout Earth's oceans. Life came ashore about 4 years ago, and plants and animals mastered the land only about 2 years ago. Dinosaurs reached their peak about 1 year ago, only to die suddenly about 4 months later. ∞ *(Discovery 4-1)* Humanlike apes changed into apelike humans only last week, and the latest ice ages occurred only a few days ago. *Homo sapiens*—our species—did not emerge until about 4 hours ago. Agriculture was invented within the last hour, and the Renaissance—along with all of modern science—is just 3 minutes old!

18.2 Life in the Solar System

Simple one-celled life-forms reigned supreme on Earth for most of our planet's history. It took time—a great deal of time—for life to emerge from the oceans, to evolve into simple plants, to continue to evolve into complex animals, and to develop intelligence, culture, and technology. Have those (or similar) events occurred elsewhere in the solar system? Let's try to assess what little evidence we have on the subject.

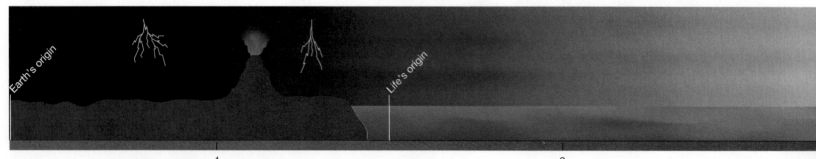

Earth's origin

Life's origin

4

3

Life as We Know It

"Life as we know it" is generally taken to mean carbon-based life that originated in a liquid-water environment—in other words, life on Earth. We must always be careful in our studies not to be misled by this "Earth-centric" viewpoint—it is all too easy to attribute human or Earth-based characteristics to other life-forms—but since we have no extraterrestrial examples to guide us, at the very least this is a reasonable starting point. Thus we begin by asking, Might Earth-like life exist (or have existed) on some other body in our planetary system?

The Moon and Mercury lack liquid water, protective atmospheres, and magnetic fields, and so are subjected to fierce bombardment by solar ultraviolet radiation, the solar wind, meteoroids, and cosmic rays. Simple molecules could not survive in such hostile environments. On the other hand, Venus has far too much protective atmosphere. Its dense, dry, scorchingly hot atmospheric blanket effectively rules it out as a possible abode for life.

The jovian planets have no solid surfaces, and most of their moons (except for volcanic Io) have frozen surfaces far too cold to support Earth-like life. One possible exception is Saturn's moon Titan. With its thick atmosphere of methane, ammonia, and nitrogen gases, liquid hydrocarbon lakes, and apparent geological activity, Titan might conceivably be a place where surface life could have arisen, although the most recent results from the *Cassini-Huygens* mission suggest that the environment there would be hostile to anything remotely familiar to us. ∞ *(Sec. 8.2)*

A more promising scenario comes from the discovery that four jovian moons—Jupiter's Europa and Ganymede, and Saturn's Titan and Enceladus—may contain significant amounts of liquid water in their interiors. ∞ *(Sec. 8.1, 8.2)* This possibility has fueled speculation about the development of life within these bodies, making them prime candidates for future exploration. Europa in particular is high on the priority lists of both NASA and the European Space Agency. Again, conditions in or on these moons are far from ideal by Earthly standards, but, as we discuss below, scientists are finding more and more examples of terrestrial organisms that thrive in extreme environments once regarded as uninhabitable.

The planet most likely to harbor life (or, more likely, to have harbored it in the past) still seems to be Mars. The red planet is harsh by Earth standards—liquid water is scarce, the atmosphere is thin, and the lack of magnetism and an ozone layer allows high-energy solar particles and radiation to reach the surface unabated. But the Martian atmosphere was once thicker, and the surface warmer and much wetter. ∞ *(Sec. 6.8)* There is strong photographic evidence from orbiters such as *Viking* and *Mars Global Surveyor* for flowing and standing water on Mars in the distant (and perhaps even relatively recent) past, and in 2004, the European *Mars Express* orbiter confirmed the long-hypothesized presence of water ice at the Martian poles, while NASA's *Opportunity Rover* reported strong geological evidence

▼ FIGURE 18.8 **Life On Earth** This simplified timeline of the origin and evolution of life on our planet begins at the far left with the origin of Earth about 4.6 billion years ago and extends linearly to the present at right. Notice how life-forms most familiar to us—fish, reptiles, mammals—emerged relatively recently in the history of Earth.

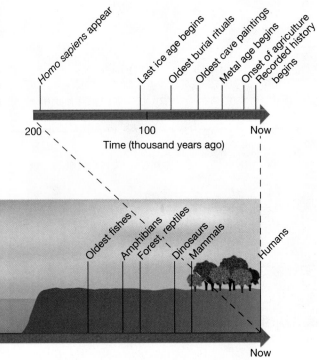

Which solar system bodies (other than Earth) are the leading candidates in the search for extraterrestrial life, and why?

▼ FIGURE 18.9 **Hydrothermal Vents** (a) A two-person submarine (the *Alvin*, partly seen at bottom) took this picture of a hot spring, or "black smoker"—one of many along the mid-ocean ridge in the eastern Pacific Ocean. As hot water rich in sulfur pours out of the top of the vent's tube (near center), black clouds billow forth, providing a strange environment for many life-forms thriving near the vent. The inset shows a close-up of the vent base, where extremophilic life thrives, including as seen here giant red tube worms and huge crabs. (b) Extremophiles also thrive in inhospitable environments on Earth's surface, such as here in the Yellowstone National Park. *(WHOI)*

that the region around its landing site was once "drenched" with water for an extended period. The *Viking* landers found no evidence of life, although they landed on the safest Martian terrain, not in the most interesting regions, such as near the moist polar caps, where scientists consider life to be more likely. ∞ *(Discovery 6-2)* *Mars Pathfinder* also surveyed part of the Martian surface, again without finding any evidence of present or past life. The recent *Mars Phoenix* mission was not equipped to test for life directly, although it did confirm the presence of water in the soil near the Martian north pole. ∞ *(Sec. 6.6)* The chemical composition of the polar soil seems less Earth-like than was at first thought, but it remains to be seen how big an obstacle this poses to the chances of life on the planet.

Some scientists think that a different type of biology might be operating, or might have operated, on the Martian surface. They suggest that Martian microbes capable of digesting oxygen-rich compounds in the Martian soil could explain the *Viking* results. This speculation will be greatly strengthened if recent announcements of fossilized bacteria in meteorites originating on Mars are confirmed. ∞ *(Discovery 6-2)* The consensus among biologists and chemists today is that Mars does not house any life similar to that on Earth, but a solid verdict regarding present or past life on Mars will likely not be reached until we have thoroughly explored our intriguing neighbor.

Life in the Extreme

In considering the emergence of life under adversity, we should perhaps not be too quick to rule out an environment based solely on its extreme properties. Figure 18.9 (a) shows a very hostile environment on a deep-ocean floor, where hydrothermal vents spew forth boiling water from vertical tubes a few meters tall. The conditions are unlike anything on our planet's surface, yet life thrives in an environment rich in sulfur, poor in oxygen, and completely dark. Similar hot springs might conceivably exist on alien worlds, raising the possibility of life-forms with much greater diversity over a much wider range of conditions than those known to us on Earth.

In recent years scientists have discovered many instances of **extremophiles**—life-forms that have adapted to live in extreme environments. The superheated vents in Figure 18.9 (a) are one example, but extremophiles have also been found in frigid lakes buried millions of years ago beneath the Antarctic glaciers; in the dark, oxygen-poor and salt-rich floor of the Mediterranean Sea; in the mineral-rich acidic environment surrounding Yellowstone's volcanic Grand Prismatic Spring (Figure 18.9b); and even in the hydrogen-rich volcanic darkness far below Earth's crust. Often these organisms have evolved to create the energy they need by purely chemical means, rather than relying on sunlight. These environments may present conditions comparable to those found on Mars, Europa, or Titan, suggesting that life as we know it might be able to thrive in these hostile, alien worlds.

(a)

|← 50 m →|

(b)

18.3 Intelligent Life in the Galaxy

With humans apparently the only intelligent life in the solar system, we must broaden our search for extraterrestrial intelligence to other stars, perhaps even other galaxies. At such distances, though, we have little hope of detecting life with current equipment. Instead, we must ask, How likely is it that life in any form—carbon-based, silicon-based, water-based, ammonia-based, or something we cannot even dream of—exists? Let's look at some numbers to develop statistical estimates of the probability of life elsewhere in the universe.

The Drake Equation

An early approach to this statistical problem is known as the **Drake equation**, after Frank Drake, the U.S. astronomer who pioneered this analysis:

| number of technological, intelligent civilizations now present in the Galaxy | = | rate of star formation, averaged over the lifetime of the Galaxy | × | fraction of stars having planetary systems | × | average number of habitable planets within those planetary systems | × | fraction of those habitable planets on which life arises | × | fraction of those life-bearing planets on which intelligence evolves | × | fraction of those intelligent-life planets that develop technological society | × | average lifetime of a technologically competent civilization |

Of course, several of the factors in this equation are largely a matter of opinion. We do not have nearly enough information to determine—even approximately—all of them, so the Drake equation cannot give us a hard-and-fast answer. Its real value is that it subdivides a large and very difficult question into smaller pieces that we can attempt to answer separately. Figure 18.10 illustrates how, as our requirements become more and more stringent, only a small fraction of star systems in the Milky Way Galaxy have the advanced qualities specified by the combination of factors on the right-hand side of the equation.

Let's examine the factors in the equation one by one and make some educated guesses about their values. Bear in mind, though, that if you ask two scientists for their best estimates of any given one, you will likely get two very different answers!

Rate of Star Formation

We can estimate the average number of stars forming each year in the Galaxy simply by noting that at least 100 billion stars now shine in it. Dividing this number by the roughly 10-billion-year lifetime of the Galaxy, we obtain a formation rate of 10 stars per year. This may be an overestimate because astronomers think that fewer stars are forming now than formed during earlier epochs, when more interstellar gas was available. ∞ *(Sec. 16.3)* However, we do know that stars are forming today, and our estimate does not include stars that formed in the past and have since died, so our value of 10 stars per year is probably reasonable when averaged over the lifetime of the Milky Way Galaxy.

Fraction of Stars with Planetary Systems

Many astronomers regard planet formation as a natural result of the star-formation process. If the condensation theory is correct, and if there is nothing special about our Sun, as we have argued throughout this book, we would expect many stars to have a planetary system of some sort. ∞ *(Sec. 4.3)*

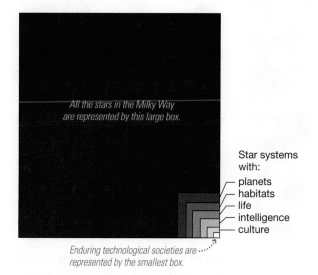

All the stars in the Milky Way are represented by this large box.

Star systems with:
- planets
- habitats
- life
- intelligence
- culture

Enduring technological societies are represented by the smallest box.

▲ **FIGURE 18.10 Drake Equation** Of all the star systems in our Milky Way Galaxy, progressively fewer and fewer have each of the qualities typical of a long-lasting technological society.

As observational techniques have improved over the past two decades, these expectations have been borne out, and there is now overwhelming evidence for planets orbiting hundreds of other stars. ∞ *(Sec. 4.4)* The first planets discovered were much larger than Earth and mostly moved on eccentric or "hot" orbits, but as we saw in Chapter 4, these were the only planets that could have been detected with the instruments available at the time. As detection technology has advanced, more and more planets with masses comparable to Earth have been discovered, to the point that, since 2010, several Earth-sized planets, some of them on roughly Earth-like orbits, have been confirmed. These observations are at the very edge of current capabilities, and many astronomers expect the numbers of "Earth-like" planets to grow rapidly as new detectors come on-line.

Only about 10 percent of the nearby stars surveyed to date have been found to have planets. However, most researchers think that this is a significant underestimate of the true fraction due to observational limitations and selection effects. ∞ *(Sec. 4.4)* Thus, accepting the condensation theory and its consequences, and without being either too conservative or naively optimistic, we assign a value near 1 to this factor—that is, we think that essentially all stars have planetary systems of some sort.

Number of Habitable Planets per Planetary System

What determines the feasibility of life on a given planet? Temperature is perhaps the single most important factor. However, the possibility of catastrophic external events, such as cometary impacts or even distant supernovae, must also be considered. ∞ *(Discovery 4-1, Sec. 12.4)*

As discussed previously (Chapter 4), the surface temperature of a planet depends on two things: the planet's distance from its parent star and the thickness of its atmosphere. ∞ *Sec. 4.4* Planets with a nearby (but not too close) parent star and some atmosphere (though not too thick) should be reasonably warm, like Earth or Mars. Objects far from the star and with no atmosphere, like Pluto, will surely be cold by our standards. And planets too close to the star and with a thick atmosphere, like Venus, will be very hot indeed.

Figure 18.11 illustrates how a three-dimensional *stellar habitable zone* of "comfortable" temperatures surrounds every star. ∞ *(Sec. 4.4)* (The zones are indicated as rings in this two-dimensional figure.) The habitable zone represents the range of distances within which a planet of mass and composition similar to those of Earth would have a surface temperature between the freezing and boiling points of water. (Our Earth-based bias is plainly evident here!) The hotter the star, the larger this zone. For example, an A- or an F-type star has a rather large habitable zone. G-, K-, and M-type stars have successively smaller zones. O- and B-type stars are not considered here because they are not expected to last long enough for life to develop, even if they do have planets.

Three planets—Venus, Earth, and Mars—reside in or near the habitable zone surrounding our Sun. Venus is too hot because of its thick atmosphere and proximity to the Sun. Mars is a little too cold because its atmosphere is too thin and it is too far from the Sun. But if the orbits of Venus and Mars were swapped—not inconceivable, since chance played such a large role in the formation of the terrestrial planets—then both of these nearby planets might conceivably have evolved surface conditions resembling those on Earth. ∞ *(Secs. 4.3, 6.8)* In that case our solar system would have had three habitable planets instead of one. Perhaps just as important, proximity to a giant planet may also render the interior of a moon (such as Europa or Titan) habitable, the planet's tidal heating making up for the lack of sunlight. ∞ *(Sec. 8.1)* Sheltered by its parent planet's gravity, such a moon might be largely immune to the habitable limitations just described for planets.

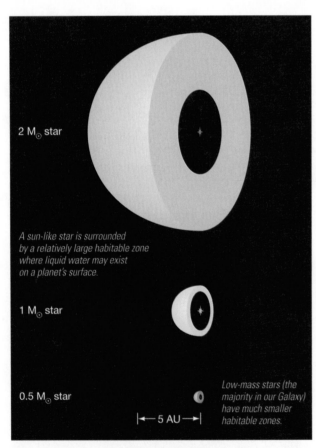

2 M⊙ star

A sun-like star is surrounded by a relatively large habitable zone where liquid water may exist on a planet's surface.

1 M⊙ star

0.5 M⊙ star

|← 5 AU →|

Low-mass stars (the majority in our Galaxy) have much smaller habitable zones.

▲ **FIGURE 18.11 Stellar Habitable Zones** Hot stars have bigger habitable zones than cool ones. For a G-type star like the Sun, the zone extends from about 0.8 AU to 2 AU. For a hotter F-type star, the range is 1.2 to 2.8 AU. For a cool M-type star, only Earth-like planets orbiting between about 0.02 and 0.06 AU would be habitable.

▶ **FIGURE 18.12 Galactic Habitable Zone**   Some regions of the Galaxy are more conducive to life than others. Too far from the Galactic center, there may not be enough heavy elements for terrestrial planets to form or technological society to evolve. Too close, the radiative or gravitational effects of nearby stars may render life impossible. The result is a ring-shaped habitable zone, colored here in green, although its full extent is uncertain.

A planet moving on a "habitable" orbit may still be rendered uninhabitable by external events. Many scientists think that the outer planets in our own solar system are critical to the habitability of the inner worlds, both by stabilizing their orbits and by protecting them from cometary impacts, deflecting would-be impactors away from the inner part of the solar system. The theory presented in Chapter 4 suggests that a star with inner terrestrial planets on stable orbits would probably also need the jovian worlds to safeguard their survival. ∞ *(Sec. 4.3)* However, observations of extrasolar planets are not yet sufficiently refined to determine the fraction of stars having "outer planet" systems like our own. ∞ *(Sec. 4.4)*

Other external forces may also influence a planet's survival. Some researchers have suggested that there is a *Galactic habitable zone* for stars in general, outside of which conditions are unfavorable for life (see Figure 18.12). Far from the Galactic center, the star formation rate is low and few cycles of star formation have occurred, so there are insufficient heavy elements to form terrestrial planets or populate them with technological civilizations. ∞ *(Sec. 12.7)* Too close, and the radiation from bright stars and supernovae in the crowded inner part of the Galaxy might be detrimental to life. More important, the gravitational effects of nearby stars may send frequent showers of comets from the Oort cloud into the inner regions of a planetary system, impacting the terrestrial planets and terminating any chain of evolution that might lead to intelligent life.

Thus, to estimate the number of habitable planets per planetary system, we must first take inventory of how many stars of each type shine in the Galactic habitable zone, then calculate the sizes of their stellar habitable zones and estimate the number of planets likely to be found there. In doing so, we eliminate almost all of the stars around which planets have so far been observed, and presumably a similar fraction of stars in general. We also exclude the majority of binary star systems: Given the observed properties of binaries in our Galaxy, "habitable" planetary orbits in binary systems would be unstable in many cases, as illustrated in Figure 18.13, so there would not be time for life to develop.

The scant observational evidence currently available on Earth-like planets in habitable orbits suggests that only a few percent of the known planetary systems contain a habitable planet. ∞ *(Sec. 4.4)* However, because these planets are so close to the limits of detectability with current equipment, many astronomers think the true fraction will turn out to be significantly higher. Potentially habitable jovian moons may increase the fraction still further. However, many uncertainties remain. The inner and outer radii of the Galactic habitable zone are not known with any certainty, and the simple fact is that we still have insufficient data about most stars to make any strong statement about habitable worlds in their planetary systems.

Taking the many uncertainties into account as best we can, we assign a value of 1/10 to this factor in our equation. In other words, we think that, on average, there is 1 potentially habitable planet for every 10 planetary systems that might exist in our Galaxy. Single F-, G-, and K-type stars are the best candidates.

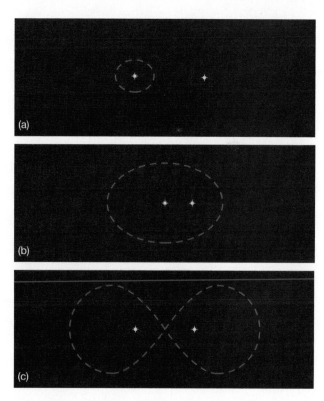

▲ **FIGURE 18.13 Binary Star Planets**   In binary star systems, planets are restricted to only a few kinds of orbits that are gravitationally stable. (a) This orbit is stable only if the planet lies very close to its parent star, so the gravity of the other star is negligible. (b) A planet circulating at a great distance about both stars in an elliptical orbit is stable only if it lies far from both stars. (c) Another possible path interweaves between the two stars in a figure-eight pattern.

Fraction of Habitable Planets on Which Life Arises

The number of possible combinations of atoms is incredibly large. If the chemical reactions that led to the complex molecules that make up living organisms occurred completely at random, then it is extremely unlikely that those molecules could have formed at all. In that case, life is extraordinarily rare, this factor is close to 0, and we are probably alone in the Galaxy, perhaps even in the entire universe.

However, laboratory experiments (like the Urey-Miller experiment described earlier) seem to suggest that certain chemical combinations are strongly favored over others—that is, the reactions are not random. Of the billions upon billions of basic organic groupings that could possibly occur on Earth from the random combination of all sorts of simple atoms and molecules, only about 1500 actually do occur. Furthermore, these 1500 organic groups of terrestrial biology are made from only about 50 simple "building blocks" (including the amino acids and nucleotide bases mentioned earlier). This suggests that molecules critical to life may not be assembled by pure chance. Apparently, additional factors are at work at the microscopic level. If a relatively small number of chemical "evolutionary tracks" is likely to exist, then the formation of complex molecules—and hence, we assume, life—becomes much more likely, given sufficient time.

To assign a very low value to this factor in the equation is to say that life arises randomly and rarely. Assigning a value close to 1 implies that we think life is inevitable given the proper ingredients, a suitable environment, and a long enough period of time. No easy experiment can distinguish between these extreme alternatives, and there is little or no middle ground. To many researchers, the discovery of life—past or present—on Mars, Europa, Titan, or any other object in our solar system would convert the appearance of life from an unlikely miracle to a virtual certainty throughout the Galaxy. In the absence of any objective evidence, we will simply take the optimistic view and adopt a value of 1.

Fraction of Life-Bearing Planets on Which Intelligence Arises

As with the evolution of life, the appearance of a well-developed brain is a very unlikely event if only random chance is involved. However, biological evolution through natural selection is a mechanism that generates apparently highly improbable results by singling out and refining useful characteristics. Organisms that profitably use adaptations can develop more complex behavior, and complex behavior provides organisms with the *variety* of choices needed for more advanced development.

One school of thought maintains that given enough time, intelligence is inevitable. In this view, assuming that natural selection is a universal phenomenon, at least one organism on a planet will always rise to the level of "intelligent life." If this is correct, then the fifth factor in the Drake equation equals or nearly equals 1.

Others argue that there is only one known case of intelligence, and that case is life on Earth. For 2.5 billion years—from the start of life about 3.5 billion years ago to the first appearance of multicellular organisms about 1 billion years ago—life did not advance beyond the one-celled stage. Life remained simple and dumb, but it survived. If this latter view is correct, then the fifth factor in our equation is very small, and we are faced with the depressing prospect that humans may be the smartest form of life anywhere in the Galaxy. As with the previous factor, we will be optimistic and adopt a value of 1 here.

Fraction of Planets on Which Intelligent Life Develops and Uses Technology

To evaluate the sixth factor of our equation, we need to estimate the probability that intelligent life eventually develops technological competence. Should the rise of technology be inevitable, this factor is close to 1, given long enough periods of time. If it is not inevitable—if intelligent life can somehow "avoid" developing technology (as dolphins on Earth seem to have done)—then this factor could be much less than 1. The latter possibility envisions a universe possibly teeming with civilizations, but very few among them ever becoming technologically competent. Perhaps only one managed it—ours.

Again, it is difficult to decide conclusively between these two views. We don't know how many (if any) prehistoric Earth cultures either failed to develop technology or rejected its use. We do know that tool-using societies arose independently at several places on Earth, including Mesopotamia, India, China, Egypt, Mexico, and Peru. Because so many of these ancient "technological" cultures originated *independently* at about the same time, we conclude that the chances are good that some sort of technological society will inevitably develop, given some basic intelligence and enough time. Accordingly, we take this factor to be close to 1.

Average Lifetime of a Technological Civilization

The reliability of these estimates declines markedly from the first to the last factor of the Drake equation. Our knowledge of astronomy allows us to make a reasonably good stab at the first two, but even the third is hard to evaluate, and the values adopted for the last few are more wishful thinking than science. The final factor is totally unknown. There is only one example of such a civilization that we are aware of—humans on planet Earth. Our own civilization has presently survived in its technological state for only about 100 years, and how long we will be around before a human-made catastrophe or a planetwide natural disaster ends it all is impossible to tell. ∞ *(Discovery 4-1)*

One thing is certain: If the correct value for *any one factor* in the equation is very small—and we have just seen at least two for which this could well be the case, optimistic choices notwithstanding—then few technological civilizations now exist in the Galaxy. If the pessimistic view of the development of life or of intelligence is correct, then we are unique, and that is the end of our story. However, if both life and intelligence are inevitable consequences of chemical and biological evolution, as many scientists think, and if intelligent life always becomes technological, then we can plug the higher, more optimistic values into the Drake equation. In that case, combining our estimates for the other six factors (and noting that, conveniently, $10 \times 1 \times 1/10 \times 1 \times 1 \times 1 = 1$), we can say

$$\begin{matrix} \text{number of technological} \\ \text{intelligent civilizations} \\ \text{now present in the Milky} \\ \text{Way Galaxy} \end{matrix} = \begin{matrix} \text{average lifetime of a} \\ \text{technologically} \\ \text{competent civilization,} \\ \text{in years.} \end{matrix}$$

Thus, if advanced civilizations typically survive for 1000 years, there should be 1000 of them currently in existence scattered throughout the Galaxy. If they survive for a million years, on average, we would expect there to be a million of them in the Galaxy, and so on.

Note that even setting aside language and cultural issues, the sheer size of the Galaxy presents a significant hurdle to communication between technological civilizations. The minimum requirement for a two-way conversation is that we can send a signal and receive a reply in a time shorter than our own lifetime. If the lifetime is short, then civilizations are literally few and far between—small in number, according to the Drake equation, and scattered over the vastness of the Milky Way—and the distances between them (in light-years) are much greater than their lifetimes (in years). In that case, two-way communication, even at the speed of light, will be impossible. However, as the lifetime increases, the distances get smaller as the Galaxy becomes more crowded, and the prospects improve.

Taking into account the size, shape, and distribution of stars in the Galactic disk (why do we exclude the halo?) and under the assumptions made above, we find that unless the life expectancy of a civilization is *at least* a few thousand years, it is unlikely to have time to communicate with even its nearest neighbor.

18.4 The Search for Extraterrestrial Intelligence

Let's continue our optimistic assessment of the prospects for life and assume that civilizations enjoy a long stay on their parent planet once their initial technological "teething problems" are past. In that case, they may be plentiful in the Galaxy. How might we become aware of their existence?

Meeting Our Neighbors

Let's assume that the average lifetime of a technological civilization is 1 million years—only 1 percent of the reign of the dinosaurs, but 100 times longer than our civilization has survived thus far and 10,000 times longer than human society has so far been in the "technological" state. The Drake equation then tells us that there are 1 million such civilizations in the Galaxy, and we can then estimate the average distance between these civilizations to be some 30 pc, or about 100 light-years. Thus, any two-way communication with our neighbors will take at least 200 years (100 years for the message to reach the planet and another 100 years for the reply to travel back to us)—a long time by human standards, but comfortably less than the lifetime of the civilization.

Might we hope to visit our neighbors by developing the capability of traveling far outside our solar system? This may never be a practical possibility. At a speed of 50 km/s, the speed of the current fastest space probes, the round-trip to even the nearest Sun-like star, Alpha Centauri, would take about 50,000 years. The journey to the nearest technological neighbor (assuming a distance of 100 pc) and back would take 1 million years—the entire lifetime of our civilization! Interstellar travel at these speeds is clearly not feasible. Speeding up our ships to near the speed of light would reduce the travel time, but this is far beyond our present technology.

Actually, our civilization has already launched some interstellar probes, although they have no specific stellar destination. Figure 18.14 is a reproduction of a plaque mounted on board the *Pioneer 10* spacecraft launched in the mid-1970s and now well beyond the orbit of Pluto, on its way out of the solar system. Similar information was also included aboard the *Voyager* probes launched in 1978. While these spacecraft would be incapable of reporting back to Earth the news that they had encountered an alien culture, scientists hope that the civilization on the other end would be able to unravel most of its contents using the universal language of mathematics. The caption to Figure 18.14 notes how the aliens might discover from where and when the *Pioneer* and *Voyager* probes were launched.

Radio Searches

A cheaper and much more practical alternative is to try to make contact with extraterrestrials using electromagnetic radiation, the fastest known means of

PROCESS OF SCIENCE CHECK

If most of the factors are largely a matter of opinion, how does the Drake equation assist astronomers in refining their search for extraterrestrial life?

▲ FIGURE 18.14 *Pioneer 10* **Plaque**   This replica of a plaque mounted on board the *Pioneer 10* spacecraft shows a scale drawing of the craft, a man, and a woman; a diagram of the hydrogen atom undergoing a change in energy (top left); a starburst pattern representing various pulsars and the frequencies of their radio waves that can be used to estimate when the craft was launched (middle left); and a depiction of the solar system, showing that the spacecraft departed the third planet from the Sun and passed the fifth planet on its way into outer space (bottom). *(NASA)*

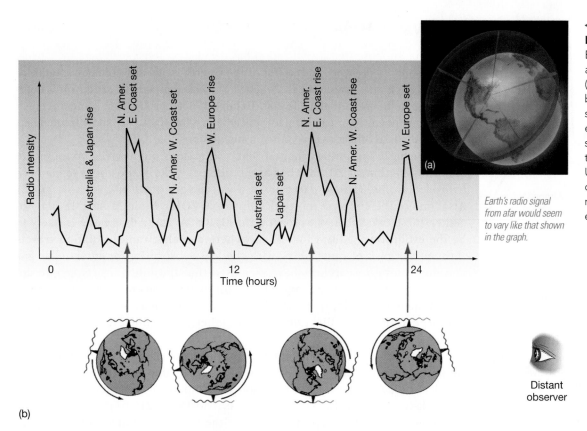

(b)

► **FIGURE 18.15 Earth's Radio Leakage** Radio radiation leaks from Earth into space because of the daily activities of our technological civilization. (a) Most radio and television transmitters broadcast their energy parallel to Earth's surface, sending a great "sheet" of electromagnetic radiation into interstellar space. (b) Because the great majority of transmitters are clustered in the eastern United States and western Europe, a distant observer would detect blasts of radiation from Earth as our planet rotates each day.

Earth's radio signal from afar would seem to vary like that shown in the graph.

transferring information from one place to another. Because light and other short-wavelength radiation are heavily scattered while moving through dusty interstellar space, long-wavelength radio radiation seems to be the best choice. ∞ *(Sec. 2.3)* We do not attempt to broadcast to all nearby candidate stars, however—that would be far too expensive and extremely inefficient. Instead, radio telescopes on Earth listen *passively* for radio signals emitted by other civilizations.

In what direction should we aim our radio telescopes? The answer to this question at least is fairly easy. On the basis of our earlier reasoning, we should target all F-, G-, and K-type stars in our vicinity. But are extraterrestrials broadcasting radio signals? If they are not, this search technique will obviously fail. If they are, how do we distinguish their artificially generated radio signals from signals emitted naturally by interstellar gas clouds? This depends on whether the signals are produced deliberately or are simply "waste radiation" escaping from a planet.

Consider first how "waste" radio wavelengths originating on Earth would look to extraterrestrials. Figure 18.15 shows the pattern of radio signals we broadcast into space. From the viewpoint of a distant observer, the spinning Earth emits a bright flash of radio radiation every few hours as the most technological regions of our planet rise or set. In fact, Earth is now a more intense radio emitter than the Sun. Our radio and television transmissions race out into space and have been doing so since the invention of these technologies nearly seven decades ago. Another civilization as advanced as ours might have constructed devices capable of detecting this radiation. If any sufficiently advanced (and sufficiently interested) civilization resides within about 70 light-years (20 pc) of Earth, then we have already announced our presence to them.

Of course, it may very well be that, having discovered cable and fiber-optics technology, most civilizations' indiscriminate transmissions cease after a few decades. In that case, radio silence becomes the hallmark of intelligence, and we must find an alternative means of locating our neighbors.

DATA POINTS

Searching for Extraterrestrial Life

Nearly 60 percent of students had difficulty describing how an electromagnetic signal might suggest extraterrestrial life. Some points to remember:

■ The presence of large amounts of nitrogen and oxygen in a planet's atmosphere might suggest the presence of Earth-like life.

■ Many natural cosmic sources produce steady or regularly varying electromagnetic emission.

■ A regularly modulated complex signal might indicate leakage of radio or other communications from a planet's surface.

■ A repeating complex signal might indicate a deliberate attempt at communication, particularly if it were transmitted in the electromagnetic "water hole," where natural background emission is minimized.

CONCEPT CHECK

What assumptions go into the identification of the water hole as a likely place to search for extraterrestrial signals?

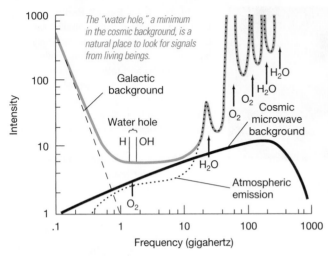

The "water hole," a minimum in the cosmic background, is a natural place to look for signals from living beings.

▲ **FIGURE 18.16 Water Hole** The "water hole" is bounded by the natural emission of the hydrogen (H) atom at 21-cm wavelength and the hydroxyl (OH) molecule at 18-cm wavelength. ∞ *Sec. 11.3* The topmost solid (blue) curve sums the natural emissions of our Galaxy (dashed line) and Earth's atmosphere (dotted line). This sum is minimized near the water hole frequencies, and thus all intelligent civilizations might conduct their interstellar communications within this quiet "electromagnetic oasis."

(a)

(b)

The Water Hole

Now let us suppose that a civilization has decided to assist searchers by actively broadcasting its presence to the rest of the Galaxy. At what frequency should we listen for such an extraterrestrial beacon? The electromagnetic spectrum is enormous; the radio domain alone is vast. To hope to detect a signal at some unknown radio frequency is like searching for a needle in a haystack. Are some frequencies more likely than others to carry alien transmissions?

Some basic arguments suggest that civilizations might well communicate at a wavelength near 20 cm. The basic building blocks of the universe, namely, hydrogen atoms, radiate naturally at a wavelength of 21 cm. ∞ *(Sec. 11.2)* Also, one of the simplest molecules, hydroxyl (OH), radiates near 18 cm. Together, these two substances form water (H_2O). Arguing that water is likely to be the evolution medium for life anywhere and that radio radiation travels through the disk of our Galaxy with the least absorption by interstellar gas and dust, some researchers have proposed that the interval between 18 cm and 21 cm is the best wavelength range for civilizations to transmit or listen. Called the **water hole**, this radio interval might serve as an "oasis" where all advanced Galactic civilizations would gather to conduct their electromagnetic business.

This water-hole frequency interval is only a guess, of course, but its use is supported by other arguments. Figure 18.16 shows the water hole's location in the electromagnetic spectrum and plots the amount of natural emission from our Galaxy and from Earth's atmosphere. The 18- to 21-cm range lies within the quietest part of the spectrum, where the Galactic "static" from stars and interstellar clouds happens to be minimized. Furthermore, the atmospheres of typical planets are also expected to interfere least at these wavelengths. Thus, the water hole seems like a good choice for the frequency of an interstellar beacon, although we cannot be sure of this reasoning until contact is actually achieved.

A few radio searches are now in progress at frequencies in and around the water hole. One of the most sensitive and comprehensive projects in the ongoing search for extraterrestrial intelligence (known to many by its acronym SETI) is now underway with the Allen Telescope Array (Figure 18.17a). This collection of many small dishes is currently searching millions of channels simultaneously in the 1- to 3-GHz range. Figure 18.17(b) shows what a typical narrowband, 1-Hz signal—a potential "signature" of an intelligent transmission—would look like on a computer monitor. However, this observation was merely a test to detect the weak, redshifted radio signal emitted by the *Pioneer 10* robot, now receding into the outer realm of our solar system—a sign of intelligence, but one that we put there. Nothing resembling an extraterrestrial signal has yet been detected.

The space surrounding all of us could be, right now, flooded with radio signals from extraterrestrial civilizations. If only we knew the proper direction and frequency, we might be able to make one of the most startling discoveries of all time. The result would provide whole new opportunities to study the cosmic evolution of energy, matter, and life throughout the universe.

◄ **FIGURE 18.17 Project Phoenix** (a) This array of small radio dishes at the SETI Institute in California is designed to search for extraterrestrial intelligent signals. (b) A typical recording of an alien signal—here, as a test, the Doppler-shifted broadcast from the *Pioneer 10* spacecraft, now well beyond the Kuiper belt—shows a diagonal line across the computer monitor, in contrast to the random noise in the background. *(S. Shostak/SETI Institute)*

THE BIG QUESTION

Humans have long wondered if planets might orbit the innumerable stars observed in the universe. The answer, we now know, is yes. We have further wondered if there might be intelligent beings on some of those planets. Not knowing the answer, the search for extraterrestrial life continues. That is perhaps the grandest of all unsolved questions in astronomy: Do alien beings reside beyond Earth?

CHAPTER REVIEW

SUMMARY

LO1 **Cosmic evolution** (p. 504) is the continuous process that has led to the appearance of galaxies, stars, planets, and life on Earth. Living organisms may be characterized by their ability to react to their environment; to grow by taking in nutrition from their surroundings; to reproduce, passing along some of their own characteristics to their offspring; and to evolve in response to a changing environment. Organisms that can best take advantage of their new surroundings succeed at the expense of those organisms that cannot make the necessary adjustments. Intelligence is strongly favored by natural selection.

LO2 Powered by natural energy sources, reactions between simple molecules in the oceans of the primitive Earth may have led to the formation of **amino acids** (p. 505) and **nucleotide bases** (p. 505), the basic molecules of life. Amino acids build proteins, which control metabolism, while sequences of nucleotide bases make up DNA, the genetic blueprint of a living organism. Alternatively, some of these complex molecules may have been formed in interstellar space and then delivered to Earth by meteors or comets.

LO3 The best hope for life beyond Earth in the solar system is the planet Mars, although no direct evidence for life—current or extinct—has yet been found. Some of the icy moons of the outer planets—Jupiter's Europa and Ganymede and Saturn's Titan and Enceladus—may also be possibilities for life of some sort. Conditions on those frozen bodies are harsh

by terrestrial standards, although **extremophiles** (p. 510) on Earth have been found to thrive in hostile environments in which life had previously been thought impossible.

LO4 The **Drake equation** (p. 511) provides a means of estimating the probability of other intelligent life in the Galaxy. The astronomical terms in the equation are the Galactic star-formation rate, the likelihood of planets, and the number of habitable planets. Chemical and biological terms are the probability of life appearing and the probability that it subsequently develops intelligence. Cultural and political terms are the probability that intelligence leads to technology and the lifetime of a technological civilization. Taking an optimistic view of the development of life and intelligence leads to the conclusion that the total number of technologically competent civilizations in the Galaxy is approximately equal to the lifetime of a typical civilization, expressed in years.

LO5 A technological civilization would probably "announce" itself to the universe by the radio and television signals it emits into space. Observed from afar, our planet would appear as a radio source with a 24-hour period, as different regions of the planet rise and set. The **water hole** (p. 518) is a region in the radio range of the electromagnetic spectrum, near the 21-cm line of hydrogen and the 18-cm line of hydroxyl, where natural emissions from the Galaxy happen to be minimal. Many researchers regard this as the best part of the spectrum for communications purposes.

MasteringAstronomy®   For instructor-assigned homework go to www.masteringastronomy.com

Problems labeled POS explore the process of science. VIS problems focus on reading and interpreting visual information. LO connects to the introduction's numbered Learning Outcomes.

REVIEW AND DISCUSSION

1. **LO1** Why is life difficult to define?

2. What is chemical evolution?

3. What is the Urey-Miller experiment? What important organic molecules were produced in this experiment?

4. **LO2** What are the basic ingredients from which biological molecules formed on Earth?

5. **POS** How do we know anything at all about the early episodes of life on Earth?

6. What is the role of language in cultural evolution?

7. Where else, besides Earth, have organic molecules been found?

8. **LO3** Where—besides the planet Mars—might we find signs of life in our solar system?

9. **POS** Do we know whether Mars ever had life at any time during its past? What argues in favor of the position that it may once have harbored life?

10. **POS** What is generally meant by "life as we know it"? What other forms of life might be possible?

11. **LO4** How many of the terms in the Drake equation are known with any-degree of certainty? Which factor is least well-known?

12. What is the relationship between the average lifetime of Galactic civilizations and the possibility of our someday communicating with them?

13. How would Earth appear, at radio wavelengths, to extraterrestrial astronomers?

14. What are the advantages of using radio waves for communication over interstellar distances?

15. **LO5** What is the water hole? What advantage does it have over other parts of the radio spectrum?

CONCEPTUAL SELF-TEST: TRUE OR FALSE?/MULTIPLE CHOICE

1. The definition of life requires only that, to be considered "alive," you must be able to reproduce. (T/F)

2. Laboratory experiments have created living cells from nonbiological molecules. (T/F)

3. Dinosaurs lived on Earth for more than a thousand times longer than human civilization has existed to date. (T/F)

4. The *Viking* landers on Mars discovered microscopic evidence of life but found no fossil evidence. (T/F)

5. The development of life and intelligence on Earth are extremely unlikely if chance is the only evolutionary factor involved. (T/F)

6. We have no direct evidence for Earth-like planets orbiting other stars. (T/F)

7. Our civilization has already launched probes into interstellar space and broadcast our presence to our neighbors. (T/F)

8. The chemical elements that form the basic molecules needed for life are found (a) in the cores of Sun-like stars; (b) commonly throughout the cosmos; (c) only on planets that have liquid water; (d) only on Earth.

9. Fossil records of early life-forms on Earth suggest that life began about (a) 6000 years ago; (b) 65 million years ago; (c) 3.5 billion years ago; (d) 14 billion years ago.

10. The least well-known factor in the Drake equation is (a) the rate of star formation; (b) the fraction of stars having planetary systems; (c) the average lifetime of a technologically competent civilization; (d) the diameter of the Milky Way Galaxy.

11. We don't expect life on planets orbiting B-type stars because the star (a) has too much gravity; (b) is too short-lived for life to evolve; (c) is at too low a temperature to sustain life; (d) would have only gas giant planets.

12. NASA's Space Shuttle orbits Earth at about 17,500 mph. If it traveled to the next Sun-like star at that speed, the trip would take at least (a) 1 week; (b) 1 decade; (c) 1 century; (d) 100 millennia.

13. The strongest radio-wavelength emission in the solar system comes from (a) human-made signals from Earth; (b) the Sun; (c) the Moon; (d) Jupiter.

14. **VIS** From the data shown in Figure 18.11 (Stellar Habitable Zones), the habitable zone surrounding a main sequence K-type star (a) cannot be determined; (b) extends from roughly 1 to 2 AU from the star; (c) is larger than that of an F-type star; (d) is larger than that of an M-type star.

15. **VIS** If Figure 18.15 (Earth's Radio Leakage) were to be redrawn for a planet spinning twice as fast, the new jagged line would be (a) unchanged; (b) taller; (c) stretched out horizontally; (d) compressed horizontally.

PROBLEMS

The number of squares preceding each problem indicates its approximate level of difficulty.

1. ■ As described in the text, imagine that Earth's 4.6-billion-year age is compressed to just 46 years. What would be your age, in seconds? How long ago (in seconds) was the end of World War II, the Declaration of Independence, Columbus's discovery of the New World? How long ago did the extinction of the dinosaurs occur (in days)?

2. ■■■ According to the inverse-square law, a planet receives energy from its parent star at a rate proportional to the star's luminosity and inversely proportional to the square of the planet's distance from the star. ∞ *(Sec. 10.2)* According to Stefan's law, the rate at which the planet radiates energy into space is proportional to the fourth power of its surface temperature. ∞ *(Sec. 2.4)* In equilibrium, the two rates are equal. Based on this information and given the fact that (taking into account the greenhouse effect) the Sun's habitable zone extends from 0.6 AU to 1.5 AU, estimate the extent of the habitable zone surrounding a K-type main-sequence star of luminosity one-tenth the luminosity of the Sun.

3. ■ Based on the numbers presented in the text, and assuming an average lifetime of 5 billion years for suitable stars, estimate the total number of habitable planets in the Galaxy.

4. ■■ A planet orbits one component of a binary star system at a distance of 1 AU (see Figure 18.13a). If both stars have the same mass and their orbit is circular, estimate the minimum distance between the stars for the tidal force due to the companion not to exceed a "safe" 0.01 percent of the gravitational force between the planet and its parent star.

5. ■ Suppose that each of the "fraction" terms in the Drake equation turns out to have a value of 1/10, that stars form at an average rate of 20 per year, and that each star with a planetary system has exactly one habitable planet orbiting it.

Estimate the present number of technological civilizations in the Milky Way Galaxy if the average lifetime of a civilization is (a) 100 years, (b) 10,000 years, (c) 1 million years.

6. ■■■ Adopting the estimate from the text that the number of technological civilizations in the Milky Way Galaxy is equal to the average lifetime of a civilization, in years, it follows that the distance to our nearest neighbor decreases as the average lifetime increases. Assuming that civilizations are uniformly spread over a two-dimensional Galactic disk of radius 15 kpc, and all have the same lifetime, calculate the *minimum* lifetime for which a two-way radio communication with our nearest neighbor would be possible before our civilization ends. Repeat the calculation for a round-trip personal visit, using current-technology spacecraft that travel at 50 km/s.

7. ■ How fast would a spacecraft have to travel in order to complete the trip from Earth to Alpha Centauri (a distance of 1.3 pc) and back in less than an average human lifetime (80 years, say)?

8. ■ Assuming that there are 10,000 FM radio stations on Earth, each transmitting at a power level of 50 kW, calculate the total radio luminosity of Earth in the FM band. Compare this value with the roughly 10^6 W radiated by the Sun in the same frequency range.

9. ■ Convert the water hole's wavelengths to frequencies. For practical reasons, any search of the water hole must be broken up into channels, much as you find on a television, except these channels are very narrow in radio frequency, about 100 Hz wide. How many channels must astronomers search in the water hole?

10. ■ There are 20,000 stars within 100 light-years that are to be searched for radio communications. How long will the search take if 1 hour is spent looking at each star? What if 1 day is spent per star?

ACTIVITIES

Collaborative

1. As a group, compose a paragraph everyone agrees with that defines life. It should clearly show that rocks are not alive and that plants are alive. According to your definition, are stars alive? What about viruses? Compare and contrast your group's definition with that from another group.

2. Independently, each person in your group should estimate the average lifetime of a technologically competent civilization, as used in the Drake equation. Explain the variation in your estimates.

3. If your group was appointed to "speak for Earth" upon establishing communication with an extraterrestrial world, what would you say? What questions would you ask, and what aspects of our planet would you choose to present? Write your group's speech and annotate it with explanations of why you chose to say this.

Individual

1. It has been suggested that if extraterrestrial life is discovered, it will have a profound effect on human culture. Interview as

many people as you can and ask the following two questions: (1) Do you think that extraterrestrial life exists? (2) Why? From your results, try to decide whether the discovery of extraterrestrial life would indeed profoundly affect life on Earth.

2. Conduct another poll separately, or do it at the same time as the first one. Ask the following question: What one question would you like to ask an extraterrestrial life-form in a radio communication? How many responses do you receive that indicate "Earth-centered" thinking? How many responses suggest a lack of understanding of how alien an extraterrestrial life-form might be? Does this change your conclusion from the first project in any way?

3. The Drake equation should be able to "predict" at least one civilization in our Galaxy: us. Try changing the values of various factors so that you end up with at least one. What do these various combinations of factors imply about how life arises and develops? Are there some combinations that just don't make any sense?

Scientific Notation

The objects studied by astronomers range in size from the smallest particles to the largest expanse of matter we know—the entire universe. Subatomic particles have sizes of about 0.000000000000001 meter, while galaxies typically measure some 1,000,000,000,000,000,000,000 meters across. The most distant known objects in the universe lie on the order of 100,000,000,000,000,000,000,000,000 meters from Earth.

Obviously, writing all those zeros is both cumbersome and inconvenient. More important, it is also very easy to make an error—write down one zero too many or too few and your calculations become hopelessly wrong! To avoid this, scientists always write large numbers using a shorthand notation in which the number of zeros following or preceding the decimal point is denoted by a superscript power, or *exponent*, of 10. The exponent is simply the number of places between the first significant (nonzero) digit in the number (reading from left to right) and the decimal point. Thus 1 is 10^0, 10 is 10^1, 100 is 10^2, 1000 is 10^3, and so on. For numbers less than 1, with zeros between the decimal point and the first significant digit, the exponent is negative: 0.1 is 10^{-1}, 0.01 is 10^{-2}, 0.001 is 10^{-3}, and so on. Using this notation we can shorten the number describing subatomic particles to 10^{-15} meter and write the number describing the size of a galaxy as 10^{21} meters.

More complicated numbers are expressed as a combination of a power of 10 and a multiplying factor. This factor is conventionally chosen to be a number between 1 and 10, starting with the first significant digit in the original number. For example, 150,000,000,000 meters (the distance from Earth to the Sun, in round numbers) can be more concisely written at 1.5×10^{11} meters, 0.000000025 meters as 2.5×10^{-8} meter, and so on. The exponent is simply the number of places the decimal point must be moved *to the left* to obtain the multiplying factor.

Some other examples of scientific notation are:

- the approximate distance to the Andromeda Galaxy = 2,500,000 light-years = 2.5×10^6 light-years

- the size of a hydrogen atom = 0.00000000005 meter = 5×10^{-11} meter

- the diameter of the Sun = 1,392,000 kilometers = 1.392×10^6 kilometers

- the U.S. national debt (as of June 1, 2015) = \$18,152,841,000.00 = \$18.152841 trillion = 1.8152841×10^{13} dollars.

In addition to providing a simpler way of expressing very large or very small numbers, this notation also makes it easier to do basic arithmetic. The rule for multiplication of numbers expressed in this way is simple: Just multiply the factors and add the exponents. Similarly for division, divide the factors and subtract the exponents. Thus, 3.5×10^{-2} multiplied by 2.0×10^3 is simply $(3.5 \times 2.0) \times 10^{-2+3} = 7.0 \times 10^1$—that is, 70. Again, 5×10^6 divided by 2×10^4 is just $(5/2) \times 10^{6-4}$, or $2.5 \times 10^2 (= 250)$. Applying these rules to unit conversions, 200,000 nanometers is $200,000 \times 10^{-9}$ meter (since 1 nanometer = 10^{-9} meter; see Appendix 2), or $(2 \times 10^5) \times 10^{-9}$ meter, or $2 \times 10^{5-9} = 2 \times 10^{-4}$ meter = 0.2 mm. Verify these rules yourself with a few examples of your own. The advantages of this notation when considering astronomical objects will soon become obvious.

Scientists often use "rounded-off" versions of numbers, both for simplicity and for ease of calculation. For example, we will usually write the diameter of the Sun as 1.4×10^6 kilometers, instead of the more precise number given earlier. Similarly, the diameter of Earth is 12,756 kilometers, or 1.2756×10^4 kilometers, but for "ballpark" estimates, we really don't need so many digits, and the more approximate number 1.3×10^4 kilometers will suffice. Very often, we perform rough calculations using only the first one or two significant digits in a number, and that may be all that is necessary to make a particular point. For example, to support the statement, "The Sun is much larger than Earth," we need only say that the ratio of the two diameters is roughly 1.4×10^6 divided by 1.3×10^4. Since 1.4/1.3 is close to 1, the ratio is approximately $10^6/10^4 = 10^2$, or 100. The essential fact here is that the ratio is much larger than 1; calculating it to greater accuracy (to get 109.13) would give us no additional *useful* information. This technique of stripping away the arithmetic details to get to the essence of a calculation is very common in astronomy, and we use it frequently throughout this text.

Astronomical Measurement

Astronomers use many different kinds of units in their work, simply because no single system of units will do. Rather than the *Système Internationale* (SI), or meter-kilogram-second (MKS), metric system used in most high school and college science classes, many professional astronomers still prefer the older centimeter-gram-second (CGS) system. However, astronomers also commonly introduce new units when convenient. For example, when discussing stars, the mass and radius of the Sun are often used as reference points. The solar mass, written as $M_\odot$, is equal to 2.0×10^{33} g, or 2.0×10^{30} kg (since 1 kg = 1000 g). The solar radius, $R_\odot$, is equal to 700,000 km, or 7.0×10^8 m (1 km = 1000 m). The subscript $\odot$ always stands for Sun. Similarly, the subscript $\oplus$ always stands for Earth. In this book, we try to use the units that astronomers commonly use in any given context, but we also give the "standard" SI equivalents where appropriate.

Of particular importance are the units of length astronomers use. On small scales, the *angstrom* ($1\text{Å} = 10^{-10}$ m $= 10^{-8}$ cm), the *nanometer* (1 nm $= 10^{-9}$ m $= 10^{-7}$ cm), and the *micron* (1 μm $= 10^{-6}$ m $= 10^{-4}$ cm) are used. Distances within the solar system are usually expressed in terms of the *astronomical unit* (AU), the mean distance between Earth and the Sun. One AU is approximately equal to 150,000,000 km, or 1.5×10^{11} m. On larger scales, the *light-year* (1 ly $= 9.5 \times 10^{15}$ m $= 9.5 \times 10^{12}$ km) and the *parsec* (1 pc $= 3.1 \times 10^{16}$ m $= 3.1 \times 10^{13}$ km $= 3.3$ ly) are commonly used. Still larger distances use the regular prefixes of the metric system: *kilo* for one thousand and *mega* for one million. Thus 1 kiloparsec (kpc) $= 10^3$ pc $= 3.1 \times 10^{19}$ m, 10 megaparsecs (Mpc) $= 10^7$ pc $= 3.1 \times 10^{23}$ m, and so on.

Astronomers use units that make sense within a context, and as contexts change, so do the units. For example, we might measure densities in grams per cubic centimeter (g/cm^3), in atoms per cubic meter (atoms/m^3), or even in solar masses per cubic megaparsec ($M_\odot$/Mpc^3), depending on the circumstances. The important thing to know is that once you understand the units, you can convert freely from one set to another. For example, the radius of the Sun could equally well be written as $R_\odot = 6.96 \times 10^8$ m, or 6.96×10^{10} cm, or 109 $R_\oplus$, or 4.65×10^{-3} AU, or even 7.36×10^{-8} ly—whichever happens to be most useful. Some of the more common units used in astronomy, and the contexts in which they are most likely to be encountered, are listed below.

Length:

1 angstrom (Å)	$= 10^{-10}$ m	atomic physics,
1 nanometer (nm)	$= 10^{-9}$ m	spectroscopy
1 micron (μm)	$= 10^{-6}$ m	interstellar dust and gas
1 centimeter (cm)	$= 0.01$ m	in widespread use
1 meter (m)	$= 100$ cm	throughout all
1 kilometer (km)	$= 1000$ m $= 10^5$ cm	astronomy
Earth radius ($R_\oplus$)	$= 6378$ km	planetary astronomy
Solar radius ($R_\odot$)	$= 6.96 \times 10^8$ m	solar system, stellar
1 astronomical unit (AU)	$= 1.496 \times 10^{11}$ m	evolution
1 light-year (ly)	$= 9.46 \times 10^{15}$ m $= 63{,}200$ AU	galactic astronomy,
1 parsec (pc)	$= 3.09 \times 10^{16}$ m $= 206{,}000$ AU $= 3.26$ ly	stars and star clusters
1 kiloparsec (kpc)	$= 1000$ pc	galaxies, galaxy
1 megaparsec (Mpc)	$= 1000$ kpc	clusters, cosmology

Mass:

1 gram (g)		in widespread use in
1 kilogram (kg)	$= 1000$ g	many different areas
Earth mass ($M_\oplus$)	$= 5.98 \times 10^{24}$ kg	planetary astronomy
Solar mass ($M_\odot$)	$= 1.99 \times 10^{30}$ kg	"standard" unit for all mass scales larger than Earth

Time:

1 second (s)		in widespread use throughout astronomy
1 hour (h)	$= 3600$ s	planetary and stellar
1 day (d)	$= 86{,}400$ s	scales
1 year (yr)	$= 3.16 \times 10^7$ s	virtually all processes occurring on scales larger than a star

Tables

TABLE 1 Some Useful Constants and Physical Measurements[1]

astronomical unit	$1\ \text{AU} = 1.496 \times 10^8\ \text{km}\ (1.5 \times 10^8\ \text{km})$
light-year	$1\ \text{ly} = 9.46 \times 10^{12}\ \text{km}\ (10^{13}\ \text{km; about 6 trillion miles})$
parsec	$1\ \text{pc} = 3.09 \times 10^{13}\ \text{km} = 3.3\ \text{ly}$
speed of light	$c = 299{,}792.458\ \text{km/s}\ (3 \times 10^5\ \text{km/s})$
Stefan-Boltzmann constant	$\sigma = 5.67 \times 10^{-8}\ \text{W/m}^2 \cdot \text{K}^4$
gravitational constant	$G = 6.67 \times 10^{-11}\ \text{N m}^2/\text{kg}^2$
mass of Earth	$M_\oplus = 5.97 \times 10^{24}\ \text{kg}\ (6 \times 10^{24}\ \text{kg; 6000 billion tons})$
radius of Earth	$R_\oplus = 6378\ \text{km}\ (6500\ \text{km})$
mass of the Sun	$M_\odot = 1.99 \times 10^{30}\ \text{kg}\ (2 \times 10^{30}\ \text{kg})$
radius of the Sun	$R_\odot = 6.96 \times 10^5\ \text{km}\ (7 \times 10^5\ \text{km})$
luminosity of the Sun	$L_\odot = 3.90 \times 10^{26}\ \text{W}\ (4 \times 10^{26}\ \text{W})$
effective temperature of the Sun	$T_\odot = 5778\ \text{K}\ (5800\ \text{K})$
Hubble constant	$H_0 \approx 70\ \text{km/s/Mpc}$
mass of an electron	$m_e = 9.11 \times 10^{-31}\ \text{kg}$
mass of a proton	$m_p = 1.67 \times 10^{-27}\ \text{kg}$

[1] The rounded-off values used in the text are shown above in parentheses.

Conversions Between Common English and Metric Units

English	Metric
1 inch	= 2.54 centimeters (cm)
1 foot (ft)	= 0.3048 meters (m)
1 mile	= 1.609 kilometers (km)
1 pound (lb)	= 453.6 grams (g) or 0.4536 kilograms (kg) (on Earth)

TABLE 2 Main-Sequence Stellar Properties by Spectral Class

Spectral Class	Typical Surface Temperature	Color	Mass*	Luminosity*	Lifetime*	Familiar Examples
	(K)		($M_\odot$)	($L_\odot$)	(10^6 yr)	
O	>30,000	Electric Blue	>20	>100,000	<2	Mintaka (O9)
B	20,000	Blue	7	500	140	Spica (B2)
A	10,000	White	3	60	500	Vega (A0) Sirius (A1)
F	7000	Yellow-white	1.5	7	2000	Procyon (F5)
G	6000	Yellow	1.0	1.0	10,000	Sun (G2) Alpha Centauri (G2)
K	4000	Orange	0.8	0.3	30,000	Epsilon Eridani (K2)
M	3000	Red	0.1	0.00006	16,000,000	Proxima Centauri (M5) Barnard's Star (M5)

*Approximate values for stars of solar composition

TABLE 3A Planetary Data: Orbital Properties

Planet	Semimajor Axis (AU)	(10⁶ km)	Sidereal Period (tropical years)	Mean Orbital Speed (km/s)	Orbital Eccentricity	Inclination to the Ecliptic (degrees)
Mercury	0.39	57.9	0.241	47.9	0.206	7.00
Venus	0.72	108.2	0.612	35.0	0.007	3.39
Earth	1.00	149.6	1.00	29.8	0.017	0.01
Mars	1.52	227.9	1.88	24.1	0.093	1.85
Jupiter	5.20	778.4	11.86	13.1	0.048	1.31
Saturn	9.54	1427	29.42	9.65	0.054	2.49
Uranus	19.19	2871	83.75	6.80	0.047	0.77
Neptune	30.07	4498	163.7	5.43	0.009	1.77

TABLE 3B Planetary Data: Physical Properties

Planet	Equatorial Radius (km)	(Earth = 1)	Mass (kg)	(Earth = 1)	Mean Density (kg/m³)	Sidereal Rotation Period (solar days)[1]
Mercury	2440	0.38	3.30×10^{23}	0.055	5430	58.6
Venus	6052	0.95	4.87×10^{24}	0.82	5240	−243.0
Earth	6378	1.00	5.97×10^{24}	1.00	5520	0.9973
Mars	3394	0.53	6.42×10^{23}	0.11	3930	1.026
Jupiter	71,492	11.21	1.90×10^{27}	317.8	1330	0.41
Saturn	60,268	9.45	5.68×10^{26}	95.2	690	0.44
Uranus	25,559	4.01	8.68×10^{25}	14.5	1270	−0.72
Neptune	24,766	3.88	1.02×10^{26}	17.1	1640	0.67

Planet	Axial Tilt (degrees)	Surface Gravity (Earth = 1)	Escape Speed (km/s)	Surface Temperature (K)[2]	Number of Moons[3]
Mercury	0.0	0.38	4.2	100 to 700	0
Venus	177.4	0.91	10.4	730	0
Earth	23.5	1.00	11.2	290	1
Mars	24.0	0.38	5.0	180 to 270	2
Jupiter	3.1	2.53	60	124	16
Saturn	26.7	1.07	36	97	18
Uranus	97.9	0.91	21	58	27
Neptune	29.6	1.14	24	59	13

[1]A negative sign indicates retrograde rotation.

[2]Temperature is effective temperature for jovian planets.

[3]Moons more than 10 km in diameter

TABLE 4 The 20 Brightest Stars in Earth's Night Sky

Name	Star	Spectral Type* A	B	Parallax (arc seconds)	Distance (pc)	Apparent Visual Magnitude* A	B
Sirius	α CMa	A1V	wd†	0.379	2.6	−1.44	+8.4
Canopus	α Car	F0Ib–II		0.010	96	−0.62	
Arcturus	α Boo	K2III		0.089	11	−0.05	
Rigel Kentaurus (Alpha Centauri)	α Cen	G2V	K0V	0.742	1.3	−0.01	+1.4
Vega	α Lyr	A0V		0.129	7.8	+0.03	
Capella	α Aur	GIII	M1V	0.077	13	+0.08	+10.2
Rigel	β Ori	B8Ia	B9	0.0042	240	+0.18	+6.6
Procyon	α CMi	F5IV–V	Wd†	0.286	3.5	+0.40	+10.7
Betelgeuse	α Ori	M2Iab		0.0076	130	+0.45	
Achernar	α Eri	B5V		0.023	44	+0.45	
Hadar	α Cen	B1III	?	0.0062	160	+0.61	+4
Altair	α Aq1	A7IV–V		0.194	5.1	+0.76	
Acrux	α Cru	B1IV	B3	0.010	98	+0.77	+1.9
Aldebaran	α Tau	K5III	M2V	0.050	20	+0.87	+13
Spica	α Vir	B1V	B2V	0.012	80	+0.98	2.1
Antares	α Sco	M1Ib	B4V	0.005	190	+1.06	+5.1
Pollux	β Gem	K0III		0.097	10	+1.16	
Formalhaut	α PsA	A3V	?	0.130	7.7	+1.17	+6.5
Deneb	α Cyg	A2Ia		0.0010	990	+1.25	
Mimosa	β Cru	B1IV		0.0093	110	+1.25	

Name	Visual Luminosity* (Sun = 1) A	B	Absolute Visual Magnitude[1] A	B	Proper Motion (arc seconds/yr)	Transverse Velocity (km/s)	Radial Velocity (km/s)
Sirius	22	0.0025	+1.5	+11.3	1.33	16.7	−7.6‡
Canopus	4.1×10^4		−5.5		0.02	9.1	20.5
Arcturus	110		−0.3		2.28	119	−5.2
Rigel Kentaurus	1.6	0.45	+4.3	+5.7	3.68	22.7	−24.6
Vega	50		+0.6		0.34	12.6	−13.9
Capella	130	0.01	−0.5	+9.6	0.44	27.1	30.2‡
Rigel	4.1×10^4	110	−6.7	−0.3	0.00	1.2	20.7
Procyon	7.2	0.0006	+2.7	+13.0	1.25	20.7	−3.2
Betelgeuse	9700		−5.1		0.03	18.5	21.0
Achernar	1100		−2.8		0.10	20.9	19
Hadar	1.3×10^4	560	−5.4	−2.0	0.04	30.3	−12
Altair	11		+2.2		0.66	16.3	−26.3
Acrux	4100	2200	−4.2	−3.5	0.04	22.8	−11.2
Aldebaran	150	0.002	−0.6	+11.5	0.20	19.0	54.1
Spica	2200	780	−3.5	−2.4	0.05	19.0	1.0
Antares	1.1×10^4	290	−5.3	−1.3	0.03	27.0	−3.2
Pollux	31		+1.1		0.62	29.4	3.3
Formalhaut	17	0.13	+1.7	+7.1	0.37	13.5	6.5
Deneb	2.6×10^5		−8.7		0.003	14.1	−4.6
Mimosa	3200		−3.9		0.05	26.1	−

* Energy output in the visible part of the spectrum; A and B columns identify individual components of binary star systems.

† "wd" stands for "white dwarf."

‡ Average value of variable velocity

TABLE 5 The 20 Nearest Stars

Name	Spectral Type* A	B	Parallax (arc seconds)	Distance (pc)	Apparent Visual Magnitude* A	B
Sun	G2V				−26.74	
Proxima Centauri	M5		0.772	1.30	+11.01	
Alpha Centauri	G2V	K1V	0.742	1.35	−0.01	+1.35
Barnard's Star	M5V		0.549	1.82	+9.54	
Wolf 359	M8V		0.421	2.38	+13.53	
Lalande 21185	M2V		0.397	2.52	+7.50	
UV Ceti	M6V	M6V	0.387	2.58	+12.52	+13.02
Sirius	A1V	wd†	0.379	2.64	−1.44	+8.4
Ross 154	M5V		0.345	2.90	+10.45	
Ross 248	M6V		0.314	3.18	+12.29	
ε Eridani	K2V		0.311	3.22	+3.72	
Ross 128	M5V		0.298	3.36	+11.10	
61 Cygni	K5V	K7V	0.294	3.40	+5.22	+6.03
ε Indi	K5V		0.291	3.44	+4.68	
Grm 34	M1V	M6V	0.290	3.45	+8.08	+11.06
Luyten 789-6	M6V		0.290	3.45	+12.18	
Procyon	F5IV–V	wd†	0.286	3.50	+0.40	+10.7
Σ 2398	M4V	M5V	0.285	3.55	+8.90	+9.69
Lacaille 9352	M2V		0.279	3.58	+7.35	
G51-15	MV		0.278	3.60	+14.81	

Name	Visual Luminosity* (Sun = 1) A	B	Absolute Visual Magnitude* A	B	Proper Motion (arc seconds/yr)	Transverse Velocity (km/s)	Radial Velocity (km/s)
Sun	1.0		+4.83				
Proxima Centauri	5.6×10^{-5}		+15.4		3.86	23.8	−16
Alpha Centauri	1.6	0.45	+4.3	+5.7	3.68	23.2	−22
Barnard's Star	4.3×10^{-4}		+13.2		10.34	89.7	−108
Wolf 359	1.8×10^{-5}		+16.7		4.70	53.0	+13
Lalande 21185	0.0055		+10.5		4.78	57.1	−84
UV Ceti	5.4×10^{-4}	0.00004	+15.5	+16.0	3.36	41.1	+30
Sirius	22	0.0025	+1.5	+11.3	1.33	16.7	−8
Ross 154	4.8×10^{-4}		+13.3		0.72	9.9	−4
Ross 248	1.1×10^{-4}		+14.8		1.58	23.8	−81
ε Eridani	0.29		+6.2		0.98	15.3	+16
Ross 128	3.6×10^{-4}		+13.5		1.37	21.8	−13
61 Cygni	0.082	0.039	+7.6	+8.4	5.22	84.1	−64
ε Indi	0.14		+7.0		4.69	76.5	−40
Grm 34	0.0061	0.00039	+10.4	+13.4	2.89	47.3	+17
Luyten 789-6	1.4×10^{-4}		+14.6		3.26	53.3	−60
Procyon	7.2	0.00055	+2.7	+13.0	1.25	2.8	−3
Σ 2398	0.0030	0.0015	+11.2	+11.9	2.28	38.4	+5
Lacaille 9352	0.013		+9.6		6.90	117	+10
G51-15	1.4×10^{-5}		+17.0		1.26	21.5	—

* A and B columns identify individual components of binary star systems.

† "wd" stands for "white dwarf."

Glossary

Key terms that are boldface in the text are followed by a page reference in the Glossary.

A

A ring One of three Saturnian rings visible from Earth. The A ring is farthest from the planet and is separated from the B ring by the Cassini Division. (p. 238)

aberration of starlight Small shift in the observed direction to a star, caused by Earth's motion perpendicular to the line of sight.

absolute brightness The apparent brightness a star would have if it were placed at a standard distance of 10 parsecs from Earth.

absolute magnitude The apparent magnitude a star would have if it were placed at a standard distance of 10 parsecs from Earth. (p. 284)

absolute zero The lowest possible temperature that can be obtained; all thermal motion ceases at this temperature.

absorption line Dark line in an otherwise continuous bright spectrum, where light within one narrow frequency range has been removed. (p. 57)

abundance Relative amounts of different elements in a gas.

acceleration The rate of change of velocity of a moving object. (p. 38)

accretion Gradual growth of bodies, such as planets, by the accumulation of other, smaller bodies. (p. 123)

accretion disk Flat disk of matter spiraling down onto the surface of a neutron star or black hole. Often, the matter originated on the surface of a companion star in a binary star system. (p. 341)

active galactic nucleus Region of intense emission at the center of an active galaxy responsible for virtually all of the galaxy's nonstellar luminosity. (p. 433)

active galaxies The most energetic galaxies, which can emit hundreds or thousands of times more energy per second than the Milky Way, mostly in the form of long-wavelength nonthermal radiation. (p. 432)

active optics Collection of techniques used to increase the resolution of ground-based telescopes. Minute modifications are made to the overall configuration of an instrument as its temperature and orientation change; used to maintain the best possible focus at all times. (p. 84)

active region Region of the photosphere of the Sun surrounding a sunspot group, which can erupt violently and unpredictably. During sunspot maximum, the number of active regions is also a maximum. (p. 266)

active Sun The unpredictable aspects of the Sun's behavior, such as sudden explosive outbursts of radiation in the form of prominences and flares.

adaptive optics Technique used to increase the resolution of a telescope by deforming the shape of the mirror's surface under computer control while a measurement is being taken; used to undo the effects of atmospheric turbulence. (p. 84)

aerosol Suspension of liquid or solid particles in air.

alpha particle A helium-4 nucleus.

alpha process Process occurring at high temperatures, in which high-energy photons split heavy nuclei to form helium nuclei.

ALSEP Acronym for Apollo Lunar Surface Experiments Package.

altimeter Instrument used to determine altitude.

amino acids Organic molecules that form the basis for building the proteins that direct metabolism in living creatures. (p. 505)

Amor asteroid Asteroid that crosses only the orbit of Mars.

amplitude The maximum deviation of a wave above or below zero point. (p. 47)

angstrom Distance unit equal to 0.1 nanometers, or one 10-billionth of a meter.

angular diameter Angle made between the top (or one edge) of an object, the observer and the bottom (or opposite edge) of the object.

angular distance Angular separation between two objects as seen by some observer.

angular momentum Tendency of an object to keep rotating; proportional to the mass, radius, and rotation speed of the body.

angular resolution The ability of a telescope to distinguish between adjacent objects in the sky. (p. 81)

annular eclipse Solar eclipse occurring at a time when the Moon is far enough away from Earth that it fails to cover the disk of the Sun completely, leaving a ring of sunlight visible around its edge. (p. 16)

antiparallel Configuration of the electron and proton in a hydrogen (or other) atom when their spin axes are parallel but the two rotate in opposite directions.

antiparticle A particle of the same mass but opposite in all other respects (e.g., charge) to a given particle; when a particle and its antiparticle come into contact, they annihilate and release energy in the form of gamma rays.

aphelion The point on the elliptical path of an object in orbit about the Sun that is most distant from the Sun. (p. 35)

Apollo asteroid *See* Earth-crossing asteroid.

apparent brightness The brightness that a star appears to have, as measured by an observer on Earth. (p. 282)

apparent magnitude The apparent brightness of a star, expressed using the magnitude scale. (p. 284)

association Small grouping of (typically 100 or less) bright stars, spanning up to a few tens of parsecs across, usually rich in very young stars. (p. 324)

Assumption of Mediocrity Statements suggesting that the development of life on Earth did not require any unusual circumstances, suggesting that extraterrestrial life may be common.

asteroid One of thousands of very small members of the solar system orbiting the Sun between the orbits of Mars and Jupiter. Often referred to as "minor planets." (p. 107)

asteroid belt Region of the solar system, between the orbits of Mars and Jupiter, in which most asteroids are found. (p. 108)

asthenosphere Layer of Earth's interior, just below the lithosphere, over which the surface plates slide.

astrology Pseudoscience that purports to use the positions of the planets, sun, and moon to predict daily events and human destiny.

astronomical unit (AU) The average distance of Earth from the Sun. Precise radar measurements yield a value for the AU of 149,603,500 km. (p. 35)

astronomy Branch of science dedicated to the study of everything in the universe that lies above Earth's atmosphere. (p. 6)

asymptotic giant branch Path on the Hertzsprung–Russell diagram corresponding to the changes that a star undergoes after helium burning ceases in the core. At this stage, the carbon core shrinks and drives the expansion of the envelope, and the star becomes a swollen red giant for a second time.

aten asteroid Earth-crossing asteroid with semimajor axis less than 1 AU.

atmosphere Layer of gas confined close to a planet's surface by the force of gravity. (p. 140)

atom Building block of matter, composed of positively charged protons and neutral neutrons in the nucleus surrounded by negatively charged electrons. (p. 60)

atomic epoch Period after decoupling when the first simple atoms and molecules formed.

aurora Event that occurs when atmospheric molecules are excited by incoming charged particles from the solar wind, then emit energy as they fall back to their ground states. Aurorae generally occur at high latitudes, near the north and south magnetic poles. (p. 160)

autumnal equinox Date on which the Sun crosses the celestial equator moving southward, occurring on or near September 22. (p. 12)

B

B ring One of three Saturnian rings visible from Earth. The B ring is the brightest of the three and lies just past the Cassini Division, closer to the planet than the A ring. (p. 238)

background noise Unwanted light in an image, from unresolved sources in the telescope's field of view, scattered light from the atmosphere, or instrumental "hiss" in the detector itself.

barred-spiral galaxy Spiral galaxy in which a bar of material passes through the center of the galaxy, with the spiral arms beginning near the ends of the bar. (p. 422)

basalt Solidified lava; an iron-magnesium-silicate mixture.

baseline The distance between two observing locations used for the purposes of triangulation measurements. The larger the baseline, the better the resolution attainable. (p. 19)

belt Dark, low-pressure region in the atmosphere of a jovian planet, where gas flows downward. (p. 209)

Big Bang Event that cosmologists consider the beginning of the universe, in which all matter and radiation in the entire universe came into being. (p. 478)

Big Crunch Point of final collapse of a bound universe.

binary asteroid Asteroids that have a partner in orbit around it.

binary pulsar Binary system in which both components are pulsars.

binary-star system A system that consists of two stars in orbit about their common center of mass, held together by their mutual gravitational attraction. Most stars are found in binary-star systems. (p. 295)

biological evolution Change in a population of biological organisms over time.

bipolar flow Jets of material expelled from a protostar perpendicular to the surrounding protostellar disk.

blackbody curve The characteristic way in which the intensity of radiation emitted by a hot object depends on frequency. The frequency at which the emitted intensity is highest is an indication of the temperature of the radiating object. Also referred to as the Planck curve. (p. 53)

black dwarf The endpoint of the evolution of an isolated, low-mass star. After the white dwarf stage, the star cools to the point where it is a dark "clinker" in interstellar space. (p. 339)

black hole A region of space where the pull of gravity is so great that nothing—not even light—can escape. A possible outcome of the evolution of a very massive star. (p. 373)

blazar Particularly intense active galactic nucleus in which the observer's line of sight happens to lie directly along the axis of a high-speed jet of particles emitted from the active region. (p. 437)

blue giant Large, hot, bright star at the upper-left end of the main sequence on the Hertzsprung–Russell diagram. Its name comes from its color and size. (p. 290)

blueshift Motion-induced changes in the observed wavelength from a source that is moving toward us. Relative approaching motion between the object and the observer causes the wavelength to appear shorter (and hence bluer) than if there were no motion at all.

blue straggler Star found on the main sequence of the Hertzsprung–Russell diagram, but which should already have evolved off the main sequence, given its location on the diagram; thought to have formed from mergers of lower mass stars.

blue supergiant The very largest of the large, hot, bright stars at the uppermost-left end of the main sequence on the Hertzsprung–Russell diagram. (p. 290)

Bohr model First theory of the hydrogen atom to explain the observed spectral lines. This model rests on three ideas: that there is a state of lowest energy for the electron, that there is a maximum energy beyond which the electron is no longer bound to the nucleus, and that within these two energies the electron can only exist in certain energy levels. (p. 60)

bok globule Dense, compact cloud of interstellar dust and gas on its way to forming one or more stars.

boson Particle that exerts or mediates forces between elementary particles in quantum physics.

bound trajectory Path of an object with launch speed low enough that it cannot escape the gravitational pull of a planet.

brown dwarf Remnants of fragments of collapsing gas and dust that did not contain enough mass to initiate core nuclear fusion. Such objects are then frozen somewhere along their pre-main-sequence contraction phase, continually cooling into compact dark objects. Because of their small size and low temperature, they are extremely difficult to detect observationally. (p. 322)

brown oval Feature of Jupiter's atmosphere that appears only at latitudes near 20 degrees N, this structure is a long-lived hole in the clouds that allows us to look down into Jupiter's lower atmosphere. (p. 211)

C

C ring One of three Saturnian rings visible from Earth. The C ring lies closest to the planet and is relatively thin compared to the A and B rings. (p. 238)

caldera Crater that forms at the summit of a volcano.

capture theory (Moon) Theory suggesting that the Moon formed far from Earth but was later captured by it.

carbonaceous asteroid The darkest, or least reflective, type of asteroid, containing large amounts of carbon.

carbon-based molecule Molecule containing atoms of carbon.

carbon-detonation supernova *See* Type I supernova. (p. 346)

cascade Process of deexcitation in which an excited electron moves down through energy states one at a time.

Cassegrain telescope A type of reflecting telescope in which incoming light hits the primary mirror and is then reflected upward toward the prime focus, where a secondary mirror reflects the light back down through a small hole in the main mirror into a detector or eyepiece. (p. 74)

Cassini Division A relatively empty gap in Saturn's ring system, discovered in 1675 by Giovanni Cassini. It is now known to contain a number of thin ringlets. (p. 238)

cataclysmic variable Collective name for novae and supernovae.

catalyst Something that causes or helps a reaction to occur, but is not itself consumed as part of the reaction.

catastrophic theory A theory that invokes statistically unlikely accidental events to account for observations.

celestial coordinates Pair of quantities—right ascension and declination—similar to longitude and latitude on Earth, used to pinpoint locations of objects on the celestial sphere. (p. 8)

celestial equator The projection of Earth's equator onto the celestial sphere. (p. 8)

celestial mechanics Study of the motions of bodies, such as planets and stars, that interact via gravity.

celestial pole Projection of Earth's North or South pole onto the celestial sphere.

celestial sphere Imaginary sphere surrounding Earth to which all objects in the sky were once considered to be attached. (p. 7)

Celsius Temperature scale in which the freezing point of water is 0 degrees and the boiling point of water is 100 degrees.

center of mass The "average" position in space of a collection of massive bodies, weighted by their masses. For an isolated system this point moves with constant velocity, according to Newtonian mechanics. (p. 40)

centigrade *See* Celsius.

centripetal force (literally "center seeking") Force directed toward the center of a body's orbit.

centroid Average position of the material in an object; in spectroscopy, the center of a spectral line.

Cepheid variable Star whose luminosity varies in a characteristic way, with a rapid rise in brightness followed by a slower decline. The period of a Cepheid variable star is related to its luminosity, so a determination of this period can be used to obtain an estimate of the star's distance. (p. 394)

Chandrasekhar mass Maximum possible mass of a white dwarf.

chaotic rotation Unpredictable tumbling motion that nonspherical bodies in eccentric orbits, such as Saturn's satellite Hyperion, can exhibit. No amount of observation of an object rotating chaotically will ever show a well-defined period.

charge-coupled device (CCD) An electronic device used for data acquisition; composed of many tiny pixels, each of which records a buildup of charge to measure the amount of light striking it. (p. 78)

chemical bond Force holding atoms together to form a molecule.

chemosynthesis Analog of photosynthesis that operates in total darkness.

chromatic aberration The tendency for a lens to focus red and blue light differently, causing images to become blurred.

chromosphere The Sun's lower atmosphere, lying just above the visible atmosphere. (p. 254)

circumnavigation Traveling all the way around an object.

cirrus High-level clouds composed of ice or methane crystals.

closed universe Geometry that the universe as a whole would have if the density of matter is above the critical value. A closed universe is finite in extent and has no edge, like the surface of a sphere. It has enough mass to stop the present expansion and will eventually collapse. (p. 481)

CNO cycle Chain of reactions that converts hydrogen into helium using carbon, nitrogen, and oxygen as catalysts.

cocoon nebula Bright infrared source in which a surrounding cloud of gas and dust absorb ultraviolet radiation from a hot star and reemits it in the infrared.

cold dark matter Class of dark-matter candidates made up of very heavy particles, possibly formed in the very early universe.

collecting area The total area of a telescope capable of capturing incoming radiation. The larger the telescope, the greater its collecting area and the fainter the objects it can detect. (p. 79)

collisional broadening Broadening of spectral lines due to collisions between atoms, most often seen in dense gases.

color index A convenient method of quantifying a star's color by comparing its apparent brightness as measured through different filters. If the star's radiation is well described by a blackbody spectrum, the ratio of its blue intensity (B) to its visual intensity (V) is a measure of the object's surface temperature.

color–magnitude diagram A way of plotting stellar properties, in which absolute magnitude is plotted against color index.

coma An effect occurring during the formation of an off-axis image in a telescope. Stars whose light enters the telescope at a large angle acquire comet-like tails on their images. The brightest part of a comet, often referred to as the "head." (p. 112)

comet A small body, composed mainly of ice and dust, in an elliptical orbit about the Sun. As it comes close to the Sun, some of its material is vaporized to form a gaseous head and extended tail. (p. 107)

common envelope Outer layer of gas in a contact binary.

comparative planetology The systematic study of the similarities and differences among the planets, with the goal of obtaining deeper insight into how the solar system formed and has evolved in time.

composition The mixture of atoms and molecules that make up an object.

condensation nuclei Dust grains in the interstellar medium that act as seeds around which other material can cluster. The presence of dust was very important in causing matter to clump during the formation of the solar system. (p. 123)

condensation theory Currently favored model of solar system formation that combines features of the old nebular theory with new information about interstellar dust grains, which acted as condensation nuclei. (p. 122)

conjunction Orbital configuration in which a planet lies in the same direction as the Sun, as seen from Earth.

conservation of mass and energy *See* law of conservation of mass and energy.

constellation A human grouping of stars in the night sky into a recognizable pattern. (p. 6)

constituents *See* composition.

contact binary A binary star system in which both stars have expanded to fill their Roche lobes, and the surfaces of the two stars merge. The binary system now consists of two nuclear burning stellar cores surrounded by a continuous common envelope.

continental drift The movement of the continents around Earth's surface.

continuous spectra Spectra in which the radiation is distributed over all frequencies, not just a few specific frequency ranges. A prime example is the blackbody radiation emitted by a hot, dense body. (p. 56)

convection Churning motion resulting from the constant upwelling of warm fluid and the concurrent downward flow of cooler material to take its place. (p. 144)

convection cell Circulating region of upwelling hot fluid and sinking cooler fluid in convective motion. (p. 144)

convection zone Region of the Sun's interior, lying just below the surface, where the material of the Sun is in constant convection motion. This region extends into the solar interior to a depth of about 20,000 km. (p. 254)

co-orbital satellites Satellites sharing the same orbit around a planet.

Copernican principle The removal of Earth from any position of cosmic significance.

Copernican revolution The realization, toward the end of the 16th century, that Earth is not at the center of the universe. (p. 31)

core The central region of Earth, surrounded by the mantle. (p. 256) The central region of any planet or star. (p. 256)

core-accretion theory Theory that the jovian planets formed when icy protoplanetary cores became massive enough to capture gas directly from the solar nebula. *See* gravitational instability theory.

core-collapse supernova *See* Type II supernova.

core hydrogen burning The energy-burning stage for main-sequence stars, in which the helium is produced by hydrogen fusion in the central region of the star. A typical star spends up to 90 percent of its lifetime in hydrostatic equilibrium brought about by the balance between gravity and the energy generated by core hydrogen burning. (p. 332)

cornea (eye) The curved transparent layer covering the front part of the eye. (p. 171)

corona One of numerous large, roughly circular regions on the surface of Venus, thought to have been caused by upwelling mantle material causing the planet's crust to bulge outward (plural, *coronae*). (p. 178)

The tenuous outer atmosphere of the Sun, which lies just above the chromosphere and, at great distances, turns into the solar wind. (p. 246)

coronal hole Vast regions of the Sun's atmosphere where the density of matter is about 10 times lower than average. The gas there streams freely into space at high speeds, escaping the Sun completely. (p. 268)

coronal mass ejection Giant magnetic "bubble" of ionized gas that separates from the rest of the solar atmosphere and escapes into interplanetary space. (p. 267)

corpuscular theory Early particle theory of light.

cosmic density parameter Ratio of the universe's actual density to the critical value corresponding to zero curvature.

cosmic distance scale Collection of indirect distance-measurement techniques that astronomers use to measure distances in the universe. (p. 18)

cosmic evolution The collection of the seven major phases of the history of the universe, namely particulate, galactic, stellar, planetary, chemical, biological, and cultural evolution. (p. 504)

cosmic microwave background The almost perfectly isotropic radio signal that is the electromagnetic remnant of the Big Bang. (p. 487)

cosmic ray Very energetic subatomic particle arriving at Earth from elsewhere in the Galaxy.

cosmological constant Quantity originally introduced by Einstein into general relativity to make his equations describe a static universe. Now one of several candidates for the repulsive "dark energy" force responsible for the observed cosmic acceleration. (p. 484)

cosmological distance Distance comparable to the scale of the universe.

cosmological principle Two assumptions that make up the basis of cosmology, namely that the universe is homogeneous and isotropic on sufficiently large scales. (p. 477)

cosmological redshift The component of the redshift of an object that is due only to the Hubble flow of the universe. (p. 431)

cosmology The study of the structure and evolution of the entire universe. (p. 477)

cosmos The universe.

coudé focus Focus produced far from the telescope using a series of mirrors. Allows the use of heavy and/or finely tuned equipment to analyze the image.

crater Bowl-shaped depression on the surface of a planet or moon, resulting from a collision with interplanetary debris. (p. 156)

crescent Appearance of the Moon (or a planet) when less than half of the body's hemisphere is visible from Earth.

crest Maximum departure of a wave above its undisturbed state.

critical density The cosmic density corresponding to the dividing line between a universe that recollapses and one that expands forever. (p. 480)

critical universe Universe in which the density of matter is exactly equal to the critical density. The universe is infinite in extent and has zero curvature. The expansion will continue forever, but will approach an expansion speed of zero. (p. 482)

crust Layer of Earth that contains the solid continents and the seafloor. (p. 140)

C-type asteroid *See* carbonaceous asteroid.

cultural evolution Change in the ideas and behavior of a society over time.

current sheet Flat sheet on Jupiter's magnetic equator where most of the charged particles in the magnetosphere lie due to the planet's rapid rotation.

D

D ring Collection of very faint, thin rings, extending from the inner edge of the C ring down nearly to the cloud tops of Saturn. This region contains so few particles that it is completely invisible from Earth. (p. 240)

dark dust cloud A large cloud, often many parsecs across, which contains gas and dust in a ratio of about 1012 gas atoms for every dust particle. Typical densities are a few tens or hundreds of millions of particles per cubic meter. (p. 311)

dark energy Generic name given to the unknown cosmic force field thought to be responsible for the observed acceleration of the Hubble expansion. (p. 483)

dark halo Region of a galaxy beyond the visible halo where dark matter is believed to reside. (p. 406)

dark matter Term used to describe the mass in galaxies and clusters whose existence we infer from rotation curves and other techniques, but that has not been confirmed by observations at any electromagnetic wavelength. (p. 406)

dark matter particle Particle undetectable at any electromagnetic wavelength, but can be inferred from its gravitational influence.

daughter/fission theory Theory suggesting that the Moon originated out of Earth.

declination Celestial coordinate used to measure latitude above or below the celestial equator on the celestial sphere. (p. 8)

decoupling Event in the early universe when atoms first formed, after which photons could propagate freely through space. (p. 491)

deferent A construct of the geocentric model of the solar system that was needed to explain observed planetary motions. A deferent is a large circle encircling Earth, on which an epicycle moves. (p. 29)

degree Unit of angular measure. There are 360 degrees in one complete circle.

density A measure of the compactness of the matter within an object, computed by dividing the mass of the object by its volume. Units are kilograms per cubic meter (kg/m^3), or grams per cubic centimeter (g/cm^3). (p. 105)

detached binary Binary system where each star lies within its respective Roche lobe.

detector noise Readings produced by an instrument even when it is not observing anything; produced by the electronic components within the detector itself.

deuterium A form of hydrogen with an extra neutron in its nucleus.

deuterium bottleneck Period in the early universe between the start of deuterium production and the time when the universe was cool enough for deuterium to survive.

deuteron An isotope of hydrogen in which a neutron is bound to the proton in the nucleus. Often called "heavy hydrogen" because of the extra mass of the neutron. (p. 271)

differential rotation The tendency for a gaseous sphere, such as a jovian planet or the Sun, to rotate at a different rate at the equator than at the poles. More generally, a condition where the angular speed varies with location within an object. (p. 207)

differentiation Variation in the density and composition of a body, such as Earth, with low-density material on the surface and higher-density material in the core. (p. 150)

diffraction The ability of waves to bend around corners. The diffraction of light establishes its wave nature. (p. 48)

diffraction grating Sheet of transparent material with many closely spaced parallel lines ruled on it, designed to separate white light into a spectrum.

diffraction-limited resolution Theoretical resolution that a telescope can have due to diffraction of light at the telescope's aperture. Depends on the wavelength of radiation and the diameter of the telescope's mirror. (p. 81)

direct motion *See* prograde motion.

distance modulus Difference between the apparent and absolute magnitude of an object; equivalent to distance, by the inverse-square law.

diurnal motion Apparent daily motion of the stars caused by Earth's rotation. (p. 9)

DNA Deoxyribonucleic acid, the molecule that carries genetic information and determines the characteristics of a living organism.

Doppler effect Any motion-induced change in the observed wavelength (or frequency) of a wave. (p. 64)

double-line spectroscopic binary Binary system in which spectral lines of both stars can be distinguished and seen to shift back and forth as the stars orbit one another.

double-star system System containing two stars in orbit around one another.

Drake equation Expression that gives an estimate of the probability that intelligence exists elsewhere in the Galaxy, based on a number of supposedly necessary conditions for intelligent life to develop. (p. 511)

dust grain An interstellar dust particle, roughly 10^{-7} m in size, comparable to the wavelength of visible light. (p. 305)

dust lane A lane of dark, obscuring interstellar dust in an emission nebula or galaxy. (p. 309)

dust tail The component of a comet's tail that is composed of dust particles. (p. 113)

dwarf Any star with radius comparable to, or smaller than, that of the Sun (including the Sun itself). (p. 289)

dwarf elliptical Elliptical galaxy as small as 1 kiloparsec across, containing only a few million stars.

dwarf galaxy Small galaxy containing a few million stars.

dwarf irregular Small irregular galaxy containing only a few million stars.

dwarf planet A body that orbits the Sun, is massive enough that its own gravity has caused its shape to be approximately spherical, but is insufficiently massive to have cleared other bodies from "the neighborhood" of its orbit. (p. 107)

dynamo theory Theory that explains planetary and stellar magnetic fields in terms of rotating, conducting material flowing in an object's interior. (p. 161)

E

E ring A faint ring, well outside the main ring system of Saturn, which was discovered by *Voyager* and is believed to be associated with volcanism on the moon Enceladus. (p. 241)

Earth-crossing asteroid An asteroid whose orbit crosses that of Earth. Earth-crossing asteroids are also called Apollo asteroids, after the first asteroid of this type discovered. (p. 108)

earthquake A sudden dislocation of rocky material near Earth's surface. (p. 149)

eccentricity A measure of the flatness of an ellipse, equal to the distance between the two foci divided by the length of the major axis. (p. 35)

eclipse Event during which one body passes in front of another, so that the light from the occulted body is blocked. (p. 14)

eclipse season Times of the year when the Moon lies in the same plane as Earth and Sun, so that eclipses are possible.

eclipse year Time interval between successive orbital configurations in which the line of nodes of the Moon's orbit points toward the Sun.

eclipsing binary Rare binary star system that is aligned in such a way that from Earth we observe one star pass in front of the other, eclipsing the other star. (p. 300)

ecliptic The apparent path of the Sun, relative to the stars on the celestial sphere, over the course of a year. (p. 10)

effective temperature Temperature of a blackbody of the same radius and luminosity as a given star or planet.

ejecta (planetary) Material thrown outward by a meteoroid impact.

ejecta (stellar) Material thrown into space by a nova or supernova.

electric field A field extending outward in all directions from a charged particle, such as a proton or an electron. The electric field determines the electric force exerted by the particle on all other charged particles in the universe; the strength of the electric field decreases with increasing distance from the charge according to an inverse-square law. (p. 48)

electromagnetic energy Energy carried in the form of rapidly fluctuating electric and magnetic fields.

electromagnetic radiation Another term for light, electromagnetic radiation transfers energy and information from one place to another. (p. 46)

electromagnetic spectrum The complete range of electromagnetic radiation, from radio waves to gamma rays, including the visible spectrum. All types of electromagnetic radiation are basically the same phenomenon, differing only by wavelength, and all move at the speed of light. (p. 50)

electromagnetism The union of electricity and magnetism, which do not exist as independent quantities but are in reality two aspects of a single physical phenomenon. (p. 49)

electron An elementary particle with a negative electric charge; one of the components of the atom. (p. 48)

electron degeneracy pressure The pressure produced by the resistance of electrons to further compression once they are squeezed to the point of contact.

electrostatic force Force between electrically charged objects.

electroweak force Unification of the weak electromagnetic forces.

element Matter made up of one particular atom. The number of protons in the nucleus of the atom determines which element it represents. (p. 63)

elementary particle Technically, a particle that cannot be subdivided into component parts; however, the term is also often used to refer to particles such as protons and neutrons, which are themselves made up of quarks.

ellipse Geometric figure resembling an elongated circle. An ellipse is characterized by its degree of flatness, or eccentricity, and the length of its long axis. In general, bound orbits of objects moving under gravity are elliptical. (p. 34)

elliptical galaxy Category of galaxy in which the stars are distributed in an elliptical shape on the sky, ranging from highly elongated to nearly circular in appearance. (p. 422)

elongation Angular distance between a planet and the Sun.

emission line Bright line in a specific location of the spectrum of radiating material, corresponding to emission of light at a certain frequency. A heated gas in a glass container produces emission lines in its spectrum. (p. 57)

emission nebula A glowing cloud of hot interstellar gas. The gas glows as a result of one or more nearby young stars that ionize the gas. Since the gas is mostly hydrogen, the emitted radiation falls predominantly in the red region of the spectrum, because of the hydrogen-alpha emission line. (p. 308)

emission spectrum The pattern of spectral emission lines produced by an element. Each element has its own unique emission spectrum. (p. 57)

empirical Discovery based on observational evidence (rather than from theory).

Encke Gap A small gap in Saturn's A ring. (p. 238)

energy flux Energy per unit area per unit time radiated by a star (or recorded by a detector).

epicycle A construct of the geocentric model of the solar system that was necessary to explain observed planetary motions. Each planet rides on a small epicycle whose center in turn rides on a larger circle (the deferent). (p. 29)

epoch of inflation Short period of unchecked cosmic expansion early in the history of the universe. During inflation, the universe swelled in size by a factor of about 10^{50}. (p. 492)

equinox See autumnal equinox, vernal equinox. (p. 12)

escape speed The speed necessary for one object to escape the gravitational pull of another. Anything that moves away from a gravitating body with more than the escape speed will never return. (p. 140)

euclidean geometry Geometry of flat space.

event horizon Imaginary spherical surface surrounding a collapsing star with radius equal to the Schwarzschild radius, within which no event can be seen, heard, or known about by an outside observer. (p. 374)

evolutionary theory A theory that explains observations in a series of gradual steps, explainable in terms of well-established physical principles.

evolutionary track A graphical representation of a star's life as a path on the Hertzsprung–Russell diagram. (p. 319)

excited state State of an atom when one of its electrons is in a higher energy orbital than the ground state. Atoms can become excited by absorbing a photon of a specific energy or by colliding with a nearby atom. (p. 59)

exposure time Time spent gathering light from a source. (p. 79)

extinction The dimming of starlight as it passes through the interstellar medium.

extrasolar planet Planet orbiting a star other than the Sun. (p. 119)

extremophilic Adjective describing organisms that can survive in very harsh environments. (p. 510)

eyepiece Secondary lens through which an observer views an image. This lens is often chosen to magnify the image.

F

F ring Faint narrow outer ring of Saturn, discovered by *Pioneer 11* in 1979. The F ring lies just inside the Roche limit of Saturn and was found by *Voyager 1* to be made up of several ring strands apparently braided together. (p. 240)

Fahrenheit Temperature scale in which the freezing point of water is 32 degrees and the boiling point of water is 212 degrees.

false color Representation of an image in which color does not represent true visual color, but rather an invisible wavelength of radiation, or some other property, such as temperature. (p. 87)

false vacuum Region of the universe that remained in the "unified" state after the strong and electroweak forces separated; one possible cause of cosmic inflation at very early times.

fault line Dislocation on a planet's surface, often indicating the boundary between two plates.

field line Imaginary line indicating the direction of an electric or magnetic field.

fireball Large meteor that burns up brightly and sometimes explosively in Earth's atmosphere.

firmament Old-fashioned term for the heavens (i.e., the sky).

flare Explosive event occurring in or near an active region on the Sun. (p. 267)

flatness problem One of two conceptual problems with the standard Big Bang model, which is that there is no natural way to explain why the density of the universe is so close to the critical density. (p. 492)

fluidized ejecta The ejecta blankets around some Martian craters, which apparently indicate that the ejected material was liquid at the time the crater formed.

fluorescence Phenomenon in which an atom absorbs energy, then radiates photons of lower energy as it cascades back to the ground state; in astronomy, often produced as ultraviolet photons from a hot young star that are absorbed by a neutral gas, causing some of the gas atoms to become excited and give off an optical (red) glow.

flyby Unbound trajectory of a spacecraft around a planet or other body.

focal length Distance from a mirror or the center of a lens to the focus.

focus One of two special points within an ellipse, whose separation from each other indicates the eccentricity. (p. 34)

In a bound orbit, planets orbit in ellipses with the Sun at one focus. (p. 34)

forbidden line A spectral line seen in emission nebulae, but not seen in laboratory experiments because, under laboratory conditions, collisions kick the electron in question into some other state before emission can occur.

force Action on an object that causes its momentum to change. The rate at which the momentum changes is numerically equal to the force. (p. 37)

fragmentation The breaking up of a large object into many smaller pieces (for example, as the result of high-speed collisions between planetesimals and protoplanets in the early solar system). (p. 124)

Fraunhofer lines The collection of over 600 absorption lines in the spectrum of the Sun, first categorized by Joseph Fraunhofer in 1812.

frequency The number of wave crests passing any given point per unit time. (p. 47)

full When the full hemisphere of the Moon or a planet can be seen from Earth.

full Moon Phase of the Moon in which it appears as a complete circular disk in the sky. (p. 14)

fusion See nuclear fusion.

G

G ring Faint, narrow ring of Saturn, discovered by *Pioneer 11* and lying just outside the F ring.

galactic bulge Thick distribution of warm gas and stars around the galactic center. (p. 392)

galactic cannibalism A galaxy merger in which a larger galaxy consumes a smaller one.

galactic center The center of the Milky Way, or any other galaxy. The point about which the disk of a spiral galaxy rotates. (p. 396)

galactic disk An immense, circular, flattened region containing most of a galaxy's luminous stars and interstellar matter. (p. 392)

galactic epoch Period from 100 million to 3 billion years after the Big Bang when large agglomerations of matter (galaxies and galaxy clusters) formed and grew.

galactic habitable zone Region of a galaxy in which conditions are conducive to the development of life.

galactic halo Region of a galaxy extending far above and below the galactic disk, where globular clusters and other old stars reside. (p. 392)

galactic nucleus Small, central, high-density region of a galaxy. Many galactic nuclei are thought to contain supermassive black holes, and almost all the radiation from active galaxies is generated within the nucleus. (pp. 410, 419)

galactic rotation curve Plot of rotation speed versus distance from the center of a galaxy.

galactic year Time taken for objects at the distance of the Sun (about 8 kpc) to orbit the center of the Galaxy, roughly 225 million years.

galaxy Gravitationally bound collection of a large number of stars. The Sun is a star in the Milky Way Galaxy. (pp. 6, 392)

galaxy cluster A collection of galaxies held together by their mutual gravitational attraction. (p. 429)

Galilean moons The four large moons of Jupiter discovered by Galileo Galilei.

Galilean satellites The four brightest and largest moons of Jupiter (Io, Europa, Ganymede, Callisto), named after Galileo Galilei, the 17th-century astronomer who first observed them.

gamma ray Region of the electromagnetic spectrum, far beyond the visible spectrum, corresponding to radiation of very high frequency and very short wavelength. (p. 46)

gamma-ray burst Object that radiates tremendous amounts of energy in the form of gamma rays, possibly due to the collision and merger of two neutron stars initially in orbit around one another. (p. 368)

gamma-ray spectrograph Spectrograph designed to work at gamma-ray wavelengths. Used to map the abundances of certain elements on the Moon and Mars.

gaseous Composed of gas.

gas-exchange experiment Experiment to look for life on Mars. A nutrient broth was offered to Martian soil specimens. If there were life in the soil, gases would be created as the broth was digested.

gene Sequence of nucleotide bases in the DNA molecule that determines the characteristics of a living organism.

general theory of relativity Theory proposed by Einstein to incorporate gravity into the framework of special relativity (p. 373).

geocentric model A model of the solar system that holds that Earth is at the center of the universe and all other bodies are in orbit around it. The earliest theories of the solar system were geocentric. (p. 29)

giant A star with a radius between 10 and 100 times that of the Sun. (p. 289)

giant elliptical Elliptical galaxy up to a few megaparsecs across, containing trillions of stars.

gibbous Appearance of the Moon (or a planet) when more than half (but not all) of the body's hemisphere is visible from Earth.

globular cluster Tightly bound, roughly spherical collection of hundreds of thousands, and sometimes millions, of stars spanning about 50 parsecs. Globular clusters are distributed in the halos around the Milky Way and other galaxies. (p. 324)

gluon Particle that exerts or mediates the strong force in quantum physics.

gradient Rate of change of some quantity (e.g., temperature or composition) with respect to location in space.

Grand Unified Theories Class of theories describing the behavior of the single force that results from unification of the strong, weak, and electromagnetic forces in the early universe. (p. 492)

granite Igneous rock, containing silicon and aluminum, that makes up most of Earth's crust.

granulation Mottled appearance of the solar surface, caused by rising (hot) and falling (cool) material in convective cells just below the photosphere. (p. 259)

gravitational force Force exerted on one body by another due to the effect of gravity. The force is directly proportional to the masses of both bodies involved and inversely proportional to the square of the distance between them. (p. 38)

gravitational instability theory Theory that the jovian planets formed directly from the solar nebula via instabilities in the gas leading to gravitational contraction. *See* core-accretion theory.

gravitational lensing The effect induced on the image of a distant object by a massive foreground object. Light from the distant object is bent into two or more separate images. (p. 407)

gravitational radiation Radiation resulting from rapid changes in a body's gravitational field.

gravitational redshift A prediction of Einstein's general theory of relativity. Photons lose energy as they escape the gravitational field of a massive object. Because a photon's energy is proportional to its frequency, a photon that loses energy suffers a decrease in frequency, which corresponds to an increase, or redshift, in wavelength. (p. 381)

graviton Particle carrying the gravitational field in theories attempting to unify gravity and quantum mechanics.

gravity The attractive effect that any massive object has on all other massive objects. The greater the mass of the object, the stronger its gravitational pull. (p. 38)

gravity assist Using gravity to change the flight path of a satellite or spacecraft.

gravity wave Gravitational counterpart of an electromagnetic wave.

great attractor A huge accumulation of mass in the relatively nearby universe (within about 200 Mpc of the Milky Way).

Great Dark Spot Prominent storm system in the atmosphere of Neptune observed by *Voyager 2*, near the equator of the planet. The system was comparable in size to Earth. (p. 215)

Great Red Spot A large, high-pressure, long-lived storm system visible in the atmosphere of Jupiter. The Red Spot is roughly twice the size of Earth. (p. 215)

great wall Extended sheet of galaxies measuring at least 200 megaparsecs across; one of the largest known structures in the universe.

greenhouse effect The partial trapping of solar radiation by a planetary atmo-sphere, similar to the trapping of heat in a greenhouse. (p. 147)

greenhouse gas Gas (such as carbon dioxide or water vapor) that efficiently absorbs infrared radiation.

ground state The lowest energy state that an electron can have within an atom. (p. 60)

GUT epoch Period when gravity separated from the other three forces of nature.

gyroscope System of rotating wheels that allows a spacecraft to maintain a fixed orientation in space.

H

habitable zone Three-dimensional zone of comfortable temperature (corresponding to liquid water) that surrounds every star. (p. 133)

half-life The amount of time it takes for half of the initial amount of a radioactive substance to decay into something else.

Hayashi track Evolutionary track followed by a protostar during the final premain-sequence phase before nuclear fusion begins.

heat Thermal energy, the energy of an object due to the random motion of its component atoms or molecules.

heat death End point of a bound universe, in which all matter and life are destined to be incinerated.

heavy element In astronomical terms, any element heavier than hydrogen and helium.

heliocentric model A model of the solar system that is centered on the Sun, with Earth in motion about the Sun. (p. 31)

helioseismology The study of conditions far below the Sun's surface through the analysis of internal "sound" waves that repeatedly cross the solar interior. (p. 250)

helium-burning shell Shell of burning helium gas surrounding a nonburning stellar core of carbon ash.

helium capture The formation of heavy elements by the capture of a helium nucleus. For example, carbon can form heavier elements by fusion with other carbon nuclei, but it is much more likely to occur by helium capture, which requires less energy.

helium flash An explosive event in the post-main-sequence evolution of a low-mass star. When helium fusion begins in a dense stellar core, the burning is explosive in nature. It continues until the energy released is enough to expand the core, at which point the star achieves stable equilibrium again. (p. 335)

helium precipitation Mechanism responsible for the low abundance of helium of Saturn's atmosphere. Helium condenses in the upper layers to form a mist that rains down toward Saturn's interior, just as water vapor forms into rain in the atmosphere of Earth.

helium shell flash Condition in which the helium-burning shell in the core of a star cannot respond to rapidly changing conditions within it, leading to a sudden temperature rise and a dramatic increase in nuclear reaction rates.

Hertzsprung–Russell (H–R) diagram A plot of luminosity against temperature (or spectral class) for a group of stars. (p. 289)

hierarchical merging Widely accepted galaxy-formation scenario in which galaxies formed as relatively small objects in the early universe and subsequently collided and merged to form the large galaxies observed today. The present Hubble type of a galaxy depends on the sequence of merger events in the galaxy's past. (p. 453)

high-energy astronomy Astronomy using X- or gamma-ray radiation rather than optical radiation.

high-energy telescope Telescope designed to detect X- and gamma-ray radiation. (p. 94)

highlands Relatively light-colored regions on the surface of the Moon that are elevated several kilometers above the maria. Also called terrae. (p. 155)

high-mass star Star with a mass more than 8 times that of the Sun; progenitor of a neutron star or black hole.

HI region Region of space containing primarily neutral hydrogen.

HII region Region of space containing primarily ionized hydrogen.

homogeneity Assumed property of the universe such that the number of galaxies in an imaginary large cube of the universe is the same no matter where in the universe the cube is placed. (p. 477)

horizon problem One of two conceptual problems with the standard Big Bang model, which is that some regions of the universe that have very similar properties are too far apart to have exchanged information within the age of the universe. (p. 491)

horizontal branch Region of the Hertzsprung–Russell diagram where post-main-sequence stars again reach hydrostatic equilibrium. At this point, the star is burning helium in its core and fusing hydrogen in a shell surrounding the core. (p. 335)

hot dark matter A class of candidates for the dark matter in the universe, composed of lightweight particles such as neutrinos, much less massive than the electron.

hot Jupiter A massive, gaseous planet orbiting very close to its parent star. (p. 130)

hot longitudes (Mercury) Two opposite points on Mercury's equator where the Sun is directly overhead at perihelion.

Hubble classification scheme Method of classifying galaxies according to their appearance, developed by Edwin Hubble. (p. 420)

Hubble diagram Plot of galactic recession velocity versus distance; evidence for an expanding universe.

Hubble flow Universal recession described by the Hubble diagram and quantified by Hubble's law.

Hubble sequence Hubble's arrangement of elliptical, S0, and spiral galaxies, often used to classify galaxies but not thought to represent a true evolutionary sequence in any sense. (p. 424)

Hubble's constant The constant of proportionality that gives the relation between recessional velocity and distance in Hubble's law. (p. 431)

Hubble's law Law that relates the observed velocity of recession of a galaxy to its distance from us. The velocity of recession of a galaxy is directly proportional to its distance away. (p. 430)

hydrocarbon Molecules consisting solely of hydrogen and carbon.

hydrogen envelope An invisible sheath of gas engulfing the coma of a comet, usually distorted by the solar wind and extending across millions of kilometers of space. (p. 112)

hydrogen shell burning Fusion of hydrogen in a shell that is driven by contraction and heating of the helium core. Once hydrogen is depleted in the core of a star, hydrogen burning stops and the core contracts due to gravity, causing the temperature to rise, heating the surrounding layers of hydrogen in the star, and increasing the burning rate there. (p. 334)

hydrosphere Layer of Earth that contains the liquid oceans and accounts for roughly 70 percent of Earth's total surface area. (p. 140)

hydrostatic equilibrium Condition in a star or other fluid body in which gravity's inward pull is exactly balanced by internal forces due to pressure. (p. 255)

hyperbola Curve formed when a plane intersects a cone at a small angle to the axis of the cone.

hypernova Explosion where a massive star undergoes core collapse and forms a black hole and a gamma-ray burst. *See* supernova.

I

igneous Rocks formed from molten material.

image The optical representation of an object produced when the object is reflected or refracted by a mirror or lens. (p. 72)

impact theory (Moon) Combination of the capture and daughter theories, suggesting that the Moon formed after an impact that dislodged some of Earth's mantle and placed it in orbit.

inertia The tendency of an object to continue moving at the same speed and in the same direction, unless acted upon by a force. (p. 38)

inferior conjunction Orbital configuration in which an inferior planet (Mercury or Venus) lies closest to Earth.

inflation *See* epoch of inflation.

infrared Region of the electromagnetic spectrum just outside the visible range, corresponding to light of a slightly longer wavelength than red light. (p. 46)

infrared telescope Telescope designed to detect infrared radiation. Many such telescopes are designed to be lightweight so that they can be carried above (most of) Earth's atmosphere by balloons, airplanes, or satellites. (p. 90)

infrared waves Electromagnetic radiation with wavelength in the infrared part of the spectrum. (p. 46)

inhomogeneity Deviation from perfectly uniform density; in cosmology, inhomogeneities in the universe are ultimately due to quantum fluctuations before inflation.

inner core The central part of Earth's core, believed to be solid and composed mainly of nickel and iron. (p. 150)

instability strip Part of the Hertzsprung–Russell diagram where pulsating post-main-sequence stars are found.

intensity A basic property of electromagnetic radiation that specifies the amount or strength of the radiation. (p. 51)

intercloud medium Superheated bubbles of hot gas extending far into interstellar space.

intercrater plains Regions on the surface of Mercury that do not show extensive cratering but are relatively smooth. (p. 175)

interference The ability of two or more waves to interact in such a way that they either reinforce or cancel each other. (p. 48)

interferometer Collection of two or more telescopes working together as a team, observing the same object at the same time and at the same wavelength. The effective diameter of an interferometer is equal to the distance between its outermost telescopes. (p. 88)

interferometry Technique in widespread use to dramatically improve the resolution of radio and infrared maps. Several telescopes observe the object simultaneously, and a computer analyzes how the signals interfere with each other. (p. 88)

intermediate-mass black hole Black hole having a mass 100–1000 times greater than the mass of the Sun.

interplanetary matter Matter in the solar system that is not part of a planet or moon—cosmic "debris."

interstellar dust Microscopic dust grains that populate space between the stars, having their origins in the ejected matter of long-dead stars.

interstellar gas cloud A large cloud of gas found in the space among the stars.

interstellar medium The matter between stars, composed of two components, gas and dust, intermixed throughout all of space. (p. 304)

intrinsic variable Star that varies in appearance due to internal processes (rather than, say, interaction with another star).

inverse-square law The law that a field follows if its strength decreases with the square of the distance. Fields that follow the inverse-square law decrease rapidly in strength as the distance increases, but never quite reach zero. (p. 39)

Io plasma torus Doughnut-shaped region of energetic ionized particles emitted by the volcanoes on Jupiter's moon Io and swept up by Jupiter's magnetic field.

ion An atom that has lost one or more of its electrons. (p. 60)

ionization state Term describing the number of electrons missing from an atom: I refers to a neutral atom, II refers to an atom missing one electron, and so on.

ionosphere Layer in Earth's atmosphere above about 100 km where the atmosphere is significantly ionized and conducts electricity. (p. 144)

ion tail Thin stream of ionized gas that is pushed away from the head of a comet by the solar wind. It extends directly away from the Sun. Often referred to as a plasma tail. (p. 113)

irregular galaxy A galaxy that does not fit into any of the other major categories in the Hubble classification scheme. (p. 424)

isotopes Nuclei containing the same number of protons but different numbers of neutrons. Most elements can exist in several isotopic forms. A common example of an isotope is deuterium, which differs from normal hydrogen by the presence of an extra neutron in the nucleus.

isotropic Assumed property of the universe such that the universe looks the same in every direction. (p. 478)

J

jet stream Relatively strong winds in the upper atmosphere channeled into a narrow stream by a planet's rotation. Normally refers to a horizontal, high-altitude wind.

joule The SI unit of energy.

jovian planet One of the four giant outer planets of the solar system, resembling Jupiter in physical and chemical composition. (p. 108)

K

Kelvin scale Temperature scale in which absolute zero is at 0 K; a change of 1 kelvin is the same as a change of 1 degree Celsius.

Kelvin-Helmholtz contraction phase Evolutionary track followed by a star during the protostar phase.

Kepler's laws of planetary motion Three laws, based on precise observations of the motions of the planets by Tycho Brahe that summarize the motions of the planets about the Sun.

kinetic energy Energy of an object due to its motion.

Kirchhoff's laws Three rules governing the formation of different types of spectra. (p. 59)

Kirkwood gaps Gaps in the spacings of orbital semimajor axes of asteroids in the asteroid belt, produced by dynamical resonances with nearby planets, especially Jupiter.

Kuiper belt A region in the plane of the solar system outside the orbit of Neptune where most short-period comets are thought to originate. (p. 115)

Kuiper belt object Small icy body orbiting in the Kuiper belt. (p. 112)

L

labeled-release experiment Experiment to look for life on Mars. Radioactive carbon compounds were added to Martian soil specimens. Scientists looked for indications that the carbon had been eaten or inhaled.

Lagrangian point One of five special points in the plane of two massive bodies orbiting one another, where a third body of negligible mass can remain in equilibrium.

lander Spacecraft that lands on the object it is studying.

laser ranging Method of determining the distance to an object by firing a laser beam at it and measuring the time taken for the light to return.

lava dome Volcanic formation formed when lava oozes out of fissures in a planet's surface, creating the dome, and then withdraws, causing the crust to crack and subside.

law of conservation of mass and energy A fundamental law of modern physics that states that the sum of mass and energy must always remain constant in any physical process. In fusion reactions, the lost mass is converted into energy, primarily in the form of electromagnetic radiation. (p. 270)

laws of planetary motion Three laws derived by Kepler describing the motion of the planets around the Sun. (p. 34)

leap year Year in which an additional day is inserted into the calendar in order to keep the calendar year synchronized with Earth's orbit around the Sun.

lens Optical instrument made of glass or some other transparent material, shaped so that, as a parallel beam of light passes through it, the rays are refracted so as to pass through a single focal point.

lens (eye) The part of the eye that refracts light onto the retina.

lepton Greek word for light, referring in particle physics to low-mass particles such as electrons, muons, and neutrinos that interact via the weak force.

lepton epoch Period when the light elementary particles (leptons) were in thermal equilibrium with the cosmic radiation field.

lidar Light Detection and Ranging—a device that uses laser-ranging to measure distance.

light See electromagnetic radiation.

light curve The variation in brightness of a star with time. (p. 295)

light element In astronomical terms, hydrogen and helium.

light-gathering power Amount of light a telescope can view and focus, proportional to the area of the primary mirror.

lighthouse model The leading explanation for pulsars. A small region of the neutron star, near one of the magnetic poles, emits a steady stream of radiation that sweeps past Earth each time the star rotates. The period of the pulses is the star's rotation period. (p. 364)

light-year The distance that light, moving at a constant speed of 300,000 km/s, travels in 1 year. One light-year is about 10 trillion kilometers.

limb Edge of a lunar, planetary, or solar disk.

linear momentum Tendency of an object to keep moving in a straight line with constant velocity; the product of the object's mass and velocity.

line of nodes Intersection of the plane of the Moon's orbit with Earth's orbital plane.

line of sight technique Method of probing interstellar clouds by observing their effects on the spectra of background stars.

lithosphere Earth's crust and a small portion of the upper mantle that make up Earth's plates. This layer of Earth undergoes tectonic activity.

Local Bubble The particular low-density intercloud region surrounding the Sun.

Local Group The small galaxy cluster that includes the Milky Way Galaxy. (p. 428)

local supercluster Collection of galaxies and clusters centered on the Virgo cluster. See also supercluster.

logarithm The power to which 10 must be raised to produce a given number.

logarithmic scale Scale using the logarithm of a number rather than the number itself; commonly used to compress a large range of data into more manageable form.

look-back time Time in the past when an object emitted the radiation we see today.

low-mass star Star with a mass less than 8 times that of the Sun; progenitor of a white dwarf.

luminosity One of the basic properties used to characterize stars, luminosity is defined as the total energy radiated by a star each second, at all wavelengths. (p. 255)

luminosity class A classification scheme that groups stars according to the width of their spectral lines. For a group of stars with the same temperature, luminosity class differentiates between supergiants, giants, main-sequence stars, and subdwarfs. (p. 294)

lunar dust See regolith.

lunar eclipse Celestial event during which the moon passes through the shadow of Earth, temporarily darkening its surface. (p. 14)

lunar phase The appearance of the Moon at different points along its orbit.

Lyman-alpha forest Collection of lines in an object's spectra starting at the redshifted wavelength of the object's own Lyman-alpha emission line and extending to shorter wavelengths; produced by gas in galaxies along the line of sight.

M

macroscopic Large enough to be visible by the unaided eye.

Magellanic Clouds Two small irregular galaxies that are gravitationally bound to the Milky Way Galaxy. (p. 424)

magnetic field Field that accompanies any changing electric field and governs the influence of magnetized objects on one another. (p. 49)

magnetic poles Points on a planet where the planetary magnetic field lines intersect the planet's surface vertically.

magnetism The presence of a magnetic field.

magnetometer Instrument that measures magnetic field strength.

magnetopause Boundary between a planet's magnetosphere and the solar wind.

magnetosphere A zone of charged particles trapped by a planet's magnetic field, lying above the atmosphere. (p. 140)

magnitude scale A system of ranking stars by apparent brightness, developed by the Greek astronomer Hipparchus. Originally, the brightest stars in the sky were categorized as being of first magnitude, while the faintest stars visible to the naked eye were classified as sixth magnitude. The scheme has since been extended to cover stars and galaxies too faint to be seen by the unaided eye. Increasing magnitude means fainter stars, and a difference of five magnitudes corresponds to a factor of 100 in apparent brightness. (p. 284)

main sequence Well-defined band on the Hertzsprung–Russell diagram on which most stars are found, running from the top left of the diagram to the bottom right. (p. 290)

main-sequence turnoff Special point on the Hertzsprung–Russell diagram for a cluster, indicative of the cluster's age. If all the stars in the cluster are plotted, the lower-mass stars will trace out the main sequence up to the point where stars begin to evolve off the main sequence toward the red giant branch. The point where stars are just beginning to evolve off is the main-sequence turnoff. (p. 351)

major axis The long axis of an ellipse.

mantle Layer of Earth just interior to the crust. (p. 140)

marginally bound universe Universe that will expand forever but at an increasingly slow rate.

maria Relatively dark-colored and smooth regions on the surface of the Moon (singular: mare). (p. 155)

mass A measure of the total amount of matter contained within an object. (p. 38)

mass function Relation between the component masses of a single-line spectroscopic binary.

massive compact halo object (MACHO) Collective name for "stellar" candidates for dark matter, including brown dwarfs, white dwarfs, and low-mass red dwarfs.

mass–luminosity relation The dependence of the luminosity of a main-sequence star on its mass. The luminosity increases roughly as the mass raised to the third power.

mass–radius relation The dependence of the radius of a main-sequence star on its mass. The radius rises roughly in proportion to the mass.

mass transfer Process by which one star in a binary system transfers matter onto the other.

mass-transfer binary See semidetached binary.

matter Anything having mass.

matter–antimatter annihilation Reaction in which matter and antimatter annihilate to produce high energy gamma rays. See also antiparticle.

matter-dominated universe A universe in which the density of matter exceeds the density of radiation. The present-day universe is matter dominated.

matter era (Current) era following the radiation era when the universe is larger and cooler and matter is the dominant constituent of the universe.

Maunder minimum Lengthy period of solar inactivity that extended from 1645 to 1715.

mean solar day Average length of time from one noon to the next, taken over the course of a year—24 hours.

medium Material through which a wave, such as sound, travels.

meridian An imaginary line on the celestial sphere through the north and south celestial poles, passing directly overhead at a given location.

mesosphere Region of Earth's atmosphere lying between the stratosphere and the ionosphere, 50–80 km above Earth's surface. (p. 144)

Messier object Member of a list of "fuzzy" objects compiled by astronomer Charles Messier in the 18th century.

metabolism The daily utilization of food and energy by which organisms stay alive.

metallic Composed of metal or metal compounds.

metamorphic Rocks created from existing rocks exposed to extremes of temperature or pressure.

meteor Bright streak in the sky, often referred to as a "shooting star," resulting from a small piece of interplanetary debris entering Earth's atmosphere and heating air molecules, which emit light as they return to their ground states. (p. 115)

meteorite Any part of a meteoroid that survives passage through the atmosphere and lands on the surface of Earth. (p. 116)

meteoroid Chunk of interplanetary debris prior to encountering Earth's atmo-sphere. (p. 107)

meteoroid swarm Pebble-sized cometary fragments dislodged from the main body, moving in nearly the same orbit as the parent comet. (p. 116)

meteor shower Event during which many meteors can be seen each hour, caused by the yearly passage of Earth through the debris spread along the orbit of a comet.

microlensing Gravitational lensing by individual stars in a galaxy.

micrometeoroid Relatively small chunks of interplanetary debris ranging from dust-particle size to pebble-sized fragments. (p. 116)

microsphere Small droplet of protein-like material that resists dissolution in water.

midocean ridge Place where two plates are moving apart, allowing fresh magma to well up.

Milky Way Galaxy The spiral galaxy in which the Sun resides. The disk of our Galaxy is visible in the night sky as the faint band of light known as the Milky Way. (p. 392)

millisecond pulsar A pulsar whose period indicates that the neutron star is rotating nearly 1000 times each second. The most likely explanation for these rapid rotators is that the neutron star has been spun up by drawing in matter from a companion star. (p. 366)

molecular cloud A cold, dense interstellar cloud that contains a high fraction of molecules. It is widely believed that the relatively high density of dust particles in these clouds plays an important role in the formation and preservation of the molecules. (p. 314)

molecular cloud complex Collection of molecular clouds that spans as much as 50 parsecs and may contain enough material to make millions of Sun-size stars. (p. 314)

molecule A tightly bound collection of atoms held together by the atoms' electromagnetic fields. Molecules, like atoms, emit and absorb photons at specific wavelengths. (p. 62)

molten In liquid form due to high temperatures.

moon A small body in orbit around a planet.

mosaic (photograph) Composite photograph made up of many smaller images.

M-type asteroid Asteroid containing large fractions of nickel and iron.

multiple star system Group of two or more stars in orbit around one another.

muon A type of lepton (along with the electron and tau).

N

naked singularity A singularity that is not hidden behind an event horizon.

nanobacterium Very small bacterium with diameter in the nanometer range.

nanometer One billionth of a meter.

neap tide The smallest tides, occurring when the Earth–Moon line is perpendicular to the Earth–Sun line at the first and third quarters.

nebula General term used for any "fuzzy" patch on the sky, either light or dark. (pp. 121, 306)

nebular theory One of the earliest models of solar system formation, dating back to Descartes, in which a large cloud of gas began to collapse under its own gravity to form the Sun and planets. (p. 121)

nebulosity "Fuzziness," usually in the context of an extended or gaseous astronomical object.

neon–oxygen white dwarf White dwarf formed from a low-mass star with a mass close to the "high-mass" limit, in which neon and oxygen form in the core.

neutrino Virtually massless and chargeless particle that is one of the products of fusion reactions in the Sun. Neutrinos move at close to the speed of light and interact with matter hardly at all. (p. 274)

neutrino oscillations Possible solution to the solar neutrino problem, in which the neutrino has a very tiny mass. In this case, the correct number of neutrinos can be produced in the solar core, but on their way to Earth some can "oscillate," or become transformed into other particles, and thus go undetected. (p. 274)

neutron An elementary particle with roughly the same mass as a proton, but which is electrically neutral. Along with protons, neutrons form the nuclei of atoms. (p. 63)

neutron capture The primary mechanism by which very massive nuclei are formed in the violent aftermath of a supernova. Instead of fusion of like nuclei, heavy elements are created by the addition of more and more neutrons to existing nuclei.

neutron degeneracy pressure Pressure due to the Pauli exclusion principle, arising when neutrons are forced to come into close contact.

neutronization Process occurring at high densities, in which protons and electrons are crushed together to form neutrons and neutrinos.

neutron spectrometer Instrument designed to search for water ice by looking for hydrogen.

neutron star A dense ball of neutrons that remains at the core of a star after a supernova explosion has destroyed the rest of the star. Typical neutron stars are about 20 km across and contain more mass than the Sun. (p. 362)

new Moon Phase of the moon during which none of the lunar disk is visible. (p. 14)

Newtonian mechanics The basic laws of motion postulated by Newton, which are sufficient to explain and quantify virtually all of the complex dynamical behavior found on Earth and elsewhere in the universe. (p. 37)

Newtonian telescope A reflecting telescope in which incoming light is intercepted before it reaches the prime focus and is deflected into an eyepiece at the side of the instrument. (p. 74)

nodes Two points on the Moon's orbit when it crosses the ecliptic.

nonrelativistic Speeds that are much less than the speed of light.

nonthermal spectrum Continuous spectrum not well described by a blackbody.

north celestial pole Point on the celestial sphere directly above Earth's North Pole. (p. 8)

Northern and Southern Lights Colorful displays produced when atmospheric molecules, excited by charged particles from the Van Allen belts, fall back to their ground state.

nova A star that suddenly increases in brightness, often by a factor of as much as 10,000, then slowly fades back to its original luminosity. A nova is the result of an explosion on the surface of a white-dwarf star, caused by matter falling onto its surface from the atmosphere of a binary companion. (p. 340)

nuclear binding energy Energy that must be supplied to split an atomic nucleus into neutrons and protons.

nuclear epoch Period when protons and neutrons fused to form heavier nuclei.

nuclear fusion Mechanism of energy generation in the core of the Sun, in which light nuclei are combined, or fused, into heavier ones, releasing energy in the process. (p. 270)

nuclear reaction Reaction in which two nuclei combine to form other nuclei, often releasing energy in the process. *See also* fusion.

nucleotide base An organic molecule, the building block of genes that pass on hereditary characteristics from one generation of living creatures to the next. (p. 505)

nucleus Dense, central region of an atom containing both protons and neutrons and orbited by one or more electrons. (p. 60)

The solid region of ice and dust that composes the central region of the head of a comet. (p. 112)

The dense central core of a galaxy. (p. 433)

O

obscuration Blockage of light by pockets of interstellar dust and gas.

Olbers's paradox A thought experiment suggesting that if the universe were homogeneous, infinite, and unchanging, the entire night sky would be as bright as the surface of the Sun. (p. 477)

Oort cloud Spherical halo of material surrounding the solar system out to a distance of about 50,000 AU; where most comets reside. (p. 115)

opacity A quantity that measures a material's ability to block electromagnetic radiation. Opacity is the opposite of transparency. (p. 52)

open cluster Loosely bound collection of tens to hundreds of stars, a few parsecs across, generally found in the plane of the Milky Way. (p. 324)

open universe Geometry that the universe would have if the density of matter were less than the critical value. In an open universe there is not enough matter to halt the expansion of the universe. An open universe is infinite in extent. (p. 482)

opposition Orbital configuration in which a planet lies in the opposite direction from the Sun, as seen from Earth.

optical double Chance superposition in which two stars appear to lie close together but are actually widely separated.

optical telescope Telescope designed to observe electromagnetic radiation at optical wavelengths.

orbital One of several energy states in which an electron can exist in an atom.

orbital period Time taken for a body to complete one full orbit around another.

orbiter Spacecraft that orbits an object to make observations.

organic compound Chemical compound (molecule) containing a significant fraction of carbon atoms; the basis of living organisms.

outer core The outermost part of Earth's core, believed to be liquid and composed mainly of nickel and iron. (p. 150)

outflow channel Surface feature on Mars, evidence that liquid water once existed there in great quantity; believed to be the relics of catastrophic flooding about 3 billion years ago. Found only in the equatorial regions of the planet. (p. 184)

outgassing Production of atmospheric gases (carbon dioxide, water vapor, methane, and sulphur dioxide) by volcanic activity.

ozone layer Layer of Earth's atmosphere at an altitude of 20–50 km where incoming ultraviolet solar radiation is absorbed by oxygen, ozone, and nitrogen in the atmosphere. (p. 145)

P

parallax The apparent motion of a relatively close object with respect to a more distant background as the location of the observer changes. (p. 19)

parsec The distance at which a star must lie in order for its measured parallax to be exactly 1 arc second; 1 parsec equals 206,000 AU. (p. 280)

partial eclipse Celestial event during which only a part of the occulted body is blocked from view. (p. 16)

particle A body with mass but of negligible dimension.

particle accelerator Device used to accelerate subatomic particles to relativistic speeds.

particle–antiparticle pair Pair of particles (e.g., an electron and a positron) produced by two photons of sufficiently high energy.

particle detector Experimental equipment that allows particles and antiparticles to be detected and identified.

Pauli exclusion principle A rule of quantum mechanics that prohibits electrons in dense gas from being squeezed too closely together.

penumbra Portion of the shadow cast by an eclipsing object in which the eclipse is seen as partial. (p. 15)

The outer region of a sunspot, surrounding the umbra, which is not as dark and not as cool as the central region. (p. 254)

perihelion The closest approach to the Sun of any object in orbit about it. (p. 35)

period The time needed for an orbiting body to complete one revolution about another body. (p. 35)

period–luminosity relation A relation between the pulsation period of a Cepheid variable and its absolute brightness. Measurement of the pulsation period allows the distance of the star to be determined. (p. 395)

permafrost Layer of permanently frozen water ice believed to lie just under the surface of Mars. (p. 185)

phase Appearance of the sunlit face of the Moon at different points along its orbit, as seen from Earth. (p. 14)

photodisintegration Process occurring at high temperatures, in which individual photons have enough energy to split a heavy nucleus (e.g., iron) into lighter nuclei.

photoelectric effect Emission of an electron from a surface when a photon of electromagnetic radiation is absorbed.

photoevaporation Process in which a cloud in the vicinity of a new-born hot star is dispersed by the star's radiation.

photometer A device that measures the total amount of light received in all or part of the image.

photometry Branch of observational astronomy in which the brightness of a source is measured through each of a set of standard filters.

photomicrograph Photograph taken through a microscope.

photon Individual packet of electromagnetic energy that makes up electromagnetic radiation. (p. 61)

photosphere The visible surface of the Sun, lying just above the uppermost layer of the Sun's interior and just below the chromosphere. (p. 254)

photosynthesis Process by which plants manufacture carbohydrates and oxygen from carbon dioxide and water, using chlorophyll and sunlight as the energy source.

pixel One of many tiny picture elements, organized into an array, making up a digital image.

Planck curve See blackbody curve.

Planck epoch Period from the beginning of the universe to roughly 10^{-43} second, when the laws of physics are not understood.

Planck's constant Fundamental physical constant relating the energy of a photon to its radiation frequency (color).

planet One of eight major bodies that orbit the Sun, visible to us by reflected sunlight. (p. 6)

planetary nebula The ejected envelope of a red-giant star, spread over a volume roughly the size of our solar system. (p. 335)

planetary ring system Material organized into thin, flat rings encircling a giant planet, such as Saturn.

planetesimal Term given to objects in the early solar system that had reached the size of small moons, at which point their gravitational fields were strong enough to begin influencing their neighbors. (p. 123)

plasma A gas in which the constituent atoms are completely ionized.

plate tectonics The motions of regions of Earth's lithosphere, which drift with respect to one another. Also known as continental drift. (p. 152)

plutino Kuiper belt object whose orbital period (like that of Pluto) is in a 3:2 resonance with the orbit of Neptune.

plutoid A dwarf planet orbiting beyond Neptune. (p. 246)

polarity A measure of the direction of the solar magnetic field in a sunspot. Conventionally, lines coming out of the surface are labeled "S," while those going into the surface are labeled "N." (p. 264)

polarization The alignment of the electric fields of emitted photons, which are generally emitted with random orientations.

Population I and II stars Classification scheme for stars based on the abundance of heavy elements. Within the Milky Way, Population I refers to young disk stars, and Population II refers to old halo stars.

positron Atomic particle with properties identical to those of a negatively charged electron, except for its positive charge. This positron is the antiparticle of the electron. Positrons and electrons annihilate one another when they meet, producing pure energy in the form of gamma rays. (p. 271)

prebiotic compound Molecule that can combine with others to form the building blocks of life.

precession The slow change in the direction of the rotation axis of a spinning object, caused by some external gravitational influence. (p. 12)

primary mirror Mirror placed at the prime focus of a telescope. See prime focus.

prime focus The point in a reflecting telescope where the mirror focuses incoming light to a point. (p. 72)

prime-focus image Image formed at the prime focus of a telescope.

primeval fireball Hot, dense state of the universe at very early times, just after the Big Bang. (p. 478)

primordial matter Matter created during the early, hot epochs of the universe.

primordial nucleosynthesis The production of elements heavier than hydrogen by nuclear fusion in the high temperatures and densities that existed in the early universe. (p. 489)

principle of cosmic censorship A proposition to separate the unexplained physics near a singularity from the rest of the well-behaved universe. The principle states that nature always hides any singularity, such as a black hole, inside an event horizon, which insulates the rest of the universe from seeing it.

progenitor "Ancestor" star of a given object; for example, the star that existed before a supernova explosion is the supernova's progenitor.

prograde motion Motion across the sky in the eastward direction.

prominence Loop or sheet of glowing gas ejected from an active region on the solar surface that then moves through the inner parts of the corona under the influence of the Sun's magnetic field. (p. 267)

proper motion The angular movement of a star across the sky, as seen from Earth, measured in seconds of arc per year. This movement is a result of the star's actual motion through space. (p. 282)

protein Molecule made up of amino acids that controls metabolism.

proton An elementary particle carrying a positive electric charge, a component of all atomic nuclei. The number of protons in the nucleus of an atom dictates what type of atom it is. (p. 48)

proton–proton chain The chain of fusion reactions, leading from hydrogen to helium, that powers main-sequence stars. (p. 271)

protoplanet Clump of material, formed in the early stages of solar system formation, that was the forerunner of the planets we see today. (p. 123)

protostar Stage in star formation when the interior of a collapsing fragment of gas is sufficiently hot and dense that it becomes opaque to its own radiation. The protostar is the dense region at the center of the fragment. (p. 316)

protosun The central accumulation of material in the early stages of solar system formations, the forerunner of the present-day Sun. (p. 124)

Ptolemaic model Geocentric solar system model, developed by the second century astronomer Claudius Ptolemy. It predicted with great accuracy the positions of the then known planets. (p. 29)

pulsar Object that emits radiation in the form of rapid pulses with a characteristic pulse period and duration. Charged particles, accelerated by the magnetic field of a rapidly rotating neutron star, flow along the magnetic field lines, producing radiation that beams outward as the star spins on its axis. (p. 362)

pulsating variable star A star whose luminosity varies in a predictable, periodic way. (p. 393)

P-waves Pressure waves from an earthquake that travel rapidly through liquids and solids.

pyrolific-release experiment Experiment to look for life on Mars. Radioactively tagged carbon dioxide was added to Martian soil specimens. Scientists looked for indications that the radioactive material had been absorbed.

Q

quantization The fact that light and matter on small scales behave in a discontinuous manner and manifest themselves in the form of tiny "packets" of energy, called quanta.

quantum fluctuation Temporary random change in the amount of energy at a point in space.

quantum gravity Theory combining general relativity with quantum mechanics.

quantum mechanics The laws of physics as they apply on atomic scales.

quark A fundamental matter particle that interacts via the strong force; basic constituent of protons and neutrons.

quark epoch Period when all heavy elementary particles (composed of quarks) were in thermal equilibrium with the cosmic radiation field.

quarter Moon Lunar phase in which the Moon appears as a half disk. (p. 14)

quasar Starlike radio source with an observed redshift that indicates an extremely large distance from Earth. The brightest nucleus of a distant active galaxy. (p. 439)

quasar feedback The idea that some of the energy released by a quasar in an active galactic nucleus heats and ejects the gas in the surrounding galaxy, shutting off the quasar fuel supply and suppressing star formation in the galaxy, connecting the growth of the central black hole to the properties of the host galaxy. (p. 461)

quasi-stellar object (QSO) *See* quasar.

quiescent prominence Prominence that persists for days or weeks, hovering high above the solar photosphere.

quiet Sun The underlying predictable elements of the Sun's behavior, such as its average photospheric temperature, which do not change in time.

R

radar Acronym for RAdio Detection And Ranging. Radio waves are bounced off an object, and the time taken for the echo to return indicates its distance. (p. 36)

radial motion Motion along a particular line of sight, which induces apparent changes in the wavelength (or frequency) of radiation received.

radial velocity Component of a star's velocity along the line of sight.

radian Angular measure equivalent to $180/\pi = 57.3$ degrees.

radiant Constellation from which a meteor shower appears to come.

radiation A way in which energy is transferred from place to place in the form of a wave. Light is a form of electromagnetic radiation.

radiation darkening The effect of chemical reactions that result when high-energy particles strike the icy surfaces of objects in the outer solar system. The reactions lead to a buildup of a dark layer of material.

radiation-dominated universe Early epoch in the universe, when the equivalent density of radiation in the cosmos exceeded the density of matter. (p. 489)

radiation era The first few thousand years after the Big Bang when the universe was small, dense, and dominated by radiation.

radiation zone Region of the Sun's interior where extremely high temperatures guarantee that the gas is completely ionized. Photons only occasionally interact with electrons and travel through this region with relative ease. (p. 254)

radio Region of the electromagnetic spectrum corresponding to radiation of the longest wavelengths. (p. 46)

radioactivity The release of energy by rare, heavy elements when their nuclei decay into lighter nuclei. (p. 151)

radio galaxy Type of active galaxy that emits most of its energy in the form of long-wavelength radiation. (p. 434)

radiograph Image made from observations at radio wavelengths.

radio lobe Roundish extended region of radio-emitting gas, lying well beyond the center of a radio galaxy. (p. 434)

radio telescope Large instrument designed to detect radiation from space at radio wavelengths. (p. 84)

radio wave Electromagnetic radiation with wavelength in the radio part of the spectrum (p. 46)

radius–luminosity–temperature relationship A mathematical proportionality, arising from Stefan's law, that allows astronomers to indirectly determine the radius of a star once its luminosity and temperature are known. (p. 289)

rapid mass transfer Mass transfer in a binary system that proceeds at a rapid and unstable rate, transferring most of the mass of one star onto the other.

ray The path taken by a beam of radiation.

Rayleigh scattering Scattering of light by particles in the atmosphere.

recession velocity Rate at which two objects are separating from one another.

recurrent nova Star that "goes nova" several times over the course of a few decades.

reddening Dimming of starlight by interstellar matter, which tends to scatter higher-frequency (blue) components of the radiation more efficiently than the lower-frequency (red) components. (p. 305)

red dwarf Small, cool, faint star at the lower-right end of the main sequence on the Hertzsprung–Russell diagram. (p. 290)

red giant A giant star whose surface temperature is relatively low so that it glows red. (p. 289)

red giant branch The section of the evolutionary track of a star corresponding to intense hydrogen shell burning, which drives a steady expansion and cooling of the outer envelope of the star. As the star gets larger in radius and its surface temperature cools, it becomes a red giant. (p. 334)

red giant region The upper-right corner of the Hertzsprung–Russell diagram, where red giant stars are found. (p. 291)

redshift Motion-induced change in the wavelength of light emitted from a source moving away from us. The relative recessional motion causes the wave to have an observed wavelength longer (and hence redder) than it would if it were not moving.

redshift survey Three-dimensional survey of galaxies, using redshift to determine distance.

red supergiant An extremely luminous red star. Often found on the asymptotic giant branch of the Hertzsprung–Russell diagram. (pp. 289, 342)

reflecting telescope A telescope that uses a mirror to gather and focus light from a distant object. (p. 72)

reflection nebula Bluish nebula caused by starlight scattering from dust particles in an interstellar cloud located just off the line of sight between Earth and a bright star.

refracting telescope A telescope that uses a lens to gather and focus light from a distant object. (p. 72)

refraction The tendency of a wave to bend as it passes from one transparent medium to another. (p. 72)

regolith Surface dust on the moon, several tens of meters thick in places, caused by billions of years of meteoritic bombardment.

relativistic Speeds comparable to the speed of light.

relativistic fireball Leading explanation of a gamma-ray burst in which an expanding region of superhot gas radiates in the gamma-ray part of the spectrum.

residual cap Portion of Martian polar ice caps that remains permanently frozen, undergoing no seasonal variations.

resonance Circumstance in which two characteristic times are related in some simple way, e.g., an asteroid with an orbital period exactly half that of Jupiter.

retina (eye) The back part of the eye onto which light is focused by the lens.

retrograde motion Backward, westward loop traced out by a planet with respect to the fixed stars. (p. 28)

revolution Orbital motion of one body about another, such as Earth about the Sun. (p. 9)

revolving *See* revolution.

Riemannian geometry Geometry of positively curved space (such as the surface of a sphere).

right ascension Celestial coordinate used to measure longitude on the celestial sphere. The zero point is the position of the Sun at the vernal equinox. (p. 8)

rille A ditch on the surface of the Moon where molten lava flowed in the past.

ring *See* planetary ring system.

ringlet Narrow region in Saturn's planetary ring system where the density of ring particles is high. *Voyager* discovered that the rings visible from Earth are actually composed of tens of thousands of ringlets. (p. 240)

Roche limit Often called the tidal stability limit, the Roche limit gives the distance from a planet at which the tidal force (due to the planet) between adjacent objects exceeds their mutual attraction. Objects within this limit are unlikely to accumulate into larger objects. The rings of Saturn occupy the region within Saturn's Roche limit. (p. 239)

Roche lobe An imaginary surface around a star. Each star in a binary system can be pictured as being surrounded by a teardrop-shaped zone of gravitational influence, the Roche lobe. Any material within the Roche lobe of a star can be considered to be part of that star. During evolution, one member of the binary system can expand so that it overflows its own Roche lobe and begins to transfer matter onto the other star.

rock Material made predominantly from compounds of silicon and oxygen.

rock cycle Process by which surface rock on Earth is continuously redistributed and transformed from one type into another.

rotation Spinning motion of a body about an axis. (p. 7)

rotation curve Plot of the orbital speed of disk material in a galaxy against its distance from the galactic center. Analysis of rotation curves of spiral galaxies indicates the existence of dark matter. (p. 406)

R-process "Rapid" process in which many neutrons are captured by a nucleus during a supernova explosion.

RR Lyrae variable Variable star whose luminosity changes in a characteristic way. All RR Lyrae stars have more or less the same average luminosity. (p. 394)

runaway greenhouse effect A process in which the heating of a planet leads to an increase in its atmosphere's ability to retain heat and thus to further heating, causing extreme changes in the temperature of the surface and the composition of the atmosphere. (p. 194)

runoff channel Riverlike surface feature on Mars, evidence that liquid water once existed there in great quantities. They are found in the southern highlands and are thought to have been formed by water that flowed nearly 4 billion years ago. (p. 184)

S

S0 galaxy Galaxy that shows evidence of a thin disk and a bulge, but that has no spiral arms and contains little or no gas. (p. 423)

Sagittarius A/Sgr A Strong radio source corresponding to the supermassive black hole at the center of the Milky Way.

Saros cycle Time interval between successive occurrences of the "same" solar eclipse, equal to 18 years, 11.3 days.

satellite A small body orbiting another larger body.

SB0 galaxy An S0-type galaxy whose disk shows evidence of a bar. (p. 423)

scarp Surface feature on Mercury believed to be the result of cooling and shrinking of the crust forming a wrinkle on the face of the planet. (p. 175)

Schmidt telescope Type of telescope having a very wide field of view, allowing large areas of the sky to be observed at once.

Schwarzschild radius The distance from the center of an object such that, if all the mass were compressed within that region, the escape speed would equal the speed of light. Once a stellar remnant collapses within this radius, light cannot escape and the object is no longer visible. (p. 374)

science A step-by-step process for investigating the physical world based on natural laws and observed phenomena. (p. 20)

scientific method The set of rules used to guide science, based on the idea that scientific "laws" be continually tested and modified or replaced if found inadequate. (p. 20)

scientific notation Expressing large and small numbers using power-of-10 notation.

seasonal cap Portion of Martian polar ice caps that is subject to seasonal variations, growing and shrinking once each Martian year.

seasons Changes in average temperature and length of day that result from the tilt of Earth's (or any planet's) axis with respect to the plane of its orbit. (p. 11)

secondary atmosphere The chemicals that composed Earth's atmosphere after the planet's formation, once volcanic activity outgassed chemicals from the interior. (p. 193)

sedimentary Rocks formed from the buildup of sediment.

seeing A term used to describe the ease with which good telescopic observations can be made from Earth's surface, given the blurring effects of atmospheric turbulence. (p. 83)

seeing disk Roughly circular region on a detector over which a star's pointlike images is spread, due to atmospheric turbulence. (p. 83)

seismic wave A wave that travels outward from the site of an earthquake through Earth. (p. 149)

seismology The study of earthquakes and the waves they produce in Earth's interior.

seismometer Equipment designed to detect and measure the strength of earthquakes (or quakes on any other planet).

selection effect Observational bias in which a measured property of a collection of objects is due to the way in which the measurement was made, rather than being intrinsic to the objects themselves. (p. 131)

self-propagating star formation Mode of star formation in which shock waves produced by the formation and evolution of one generation of stars triggers the formation of the next. (p. 403)

semidetached binary Binary system where one star lies within its Roche lobe but the other fills its Roche lobe and is transferring matter onto the first star.

semimajor axis One-half of the major axis of an ellipse. The semimajor axis is the way in which the size of an ellipse is usually quantified. (p. 34)

SETI Acronym for Search for Extraterrestrial Intelligence.

Seyfert galaxy Type of active galaxy whose emission comes from a very small region within the nucleus of an otherwise normal-looking spiral system. (p. 433)

shepherd satellite Satellite whose gravitational effect on a ring helps preserve the ring's shape. Examples are two satellites of Saturn, Prometheus and Pandora, whose orbits lie on either side of the F ring. (p. 241)

shield volcano A volcano produced by repeated nonexplosive eruptions of lava, creating a gradually sloping, shield-shaped low dome. Often contains a caldera at its summit. (p. 177)

shock wave Wave of matter, which may be generated by a newborn star or supernova, that pushes material outward into the surrounding molecular cloud. The material tends to pile up, forming a rapidly moving shell of dense gas.

short-period comet Comet with orbital period less than 200 years.

SI Système International d'unités, the international system of metric units used to define mass, length, time, etc.

sidereal day The time needed between successive risings of a given star. (p. 9)

sidereal month Time required for the Moon to complete one trip around the celestial sphere. (p. 14)

sidereal year The time required for the constellations to complete one cycle around the sky and return to their starting points, as seen from a given point on Earth. Earth's orbital period around the Sun is 1 sidereal year. (p. 12)

single-line spectroscopic binary Binary system in which one star is too faint for its spectrum to be distinguished, so only the spectrum of the brighter star can be seen to shift back and forth as the stars orbit one another.

singularity A point in the universe where the density of matter and the gravitational field are infinite, such as at the center of a black hole. (p. 382)

sister/coformation theory (Moon) Theory suggesting that the Moon formed as a separate object close to Earth.

soft landing Use of rockets, parachutes, or packaging to break the fall of a space probe as it lands on a planet.

solar activity Unpredictable, often violent events on or near the solar surface, associated with magnetic phenomena on the Sun. (p. 262)

solar constant The amount of solar energy reaching Earth per unit area per unit time, approximately 1400 W/m^2. (p. 255)

solar core The region at the center of the Sun, with a radius of nearly 200,000 km, where powerful nuclear reactions generate the Sun's energy output.

solar cycle The 22-year period that is needed for both the average number of spots and the Sun's magnetic polarity to repeat themselves. The Sun's polarity reverses on each new 11-year sunspot cycle. (p. 265)

solar day The period of time between the instant when the Sun is directly overhead (i.e., noon) to the next time it is directly overhead. (p. 9)

solar eclipse Celestial event during which the new Moon passes directly between Earth and the Sun, temporarily blocking the Sun's light. (p. 16)

solar interior The region of the Sun between the solar core and the photosphere.

solar maximum Point of the sunspot cycle during which many spots are seen. They are generally confined to regions in each hemisphere, between about 15 and 20 degrees latitude.

solar minimum Point of the sunspot cycle during which only a few spots are seen. They are generally confined to narrow regions in each hemisphere at about 25–30 degrees latitude.

solar nebula The swirling gas surrounding the early Sun during the epoch of solar system formation, also referred to as the primitive solar system. (p. 121)

solar neutrino problem The discrepancy between the theoretically predicted flux of neutrinos streaming from the Sun as a result of fusion reactions in the core and the flux that is actually observed. The observed number of neutrinos is only about half the predicted number. (p. 274)

solar system The Sun and all the bodies that orbit it—Mercury, Venus, Earth, Mars, Jupiter, Saturn, Uranus, Neptune, their moons, the asteroids, the Kuiper belt, and the comets. (p. 104)

solar wind An outward flow of fast-moving charged particles from the Sun. (pp. 113, 254)

solstice See summer solstice; winter solstice.

south celestial pole Point on the celestial sphere directly above Earth's South Pole. (p. 8)

spacetime Single entity combining space and time in special and general relativity.

spatial resolution The dimension of the smallest detail that can be seen in an image.

special relativity Theory proposed by Einstein to deal with the preferred status of the speed of light.

speckle interferometry Technique whereby many short-exposure images of a star are combined to make a high-resolution map of the star's surface.

spectral class Classification scheme, based on the strength of stellar spectral lines, which is an indication of the temperature of a star. (p. 287)

spectral window Wavelength range in which Earth's atmosphere is transparent.

spectrograph Instrument used to produce detailed spectra of stars. Usually, a spectrograph records a spectrum on a CCD detector, for computer analysis.

spectrometer Instrument used to produce detailed spectra of stars. Usually, a spectrograph records a spectrum on a photographic plate, or more recently, in electronic form on a computer.

spectroscope Instrument used to view a light source so that it is split into its component colors. (p. 56)

spectroscopic binary A binary-star system that appears as a single star from Earth, but whose spectral lines show back-and-forth Doppler shifts as two stars orbit one another. (p. 295)

spectroscopic parallax Method of determining the distance to a star by measuring its temperature and then determining its absolute brightness by comparing with a standard Hertzsprung–Russell diagram. The absolute and apparent brightness of the star give the star's distance from Earth. (p. 293)

spectroscopy The study of the way in which atoms absorb and emit electromagnetic radiation. Spectroscopy allows astronomers to determine the chemical composition of stars. (p. 58)

spectrum The separation of light into its component colors.

speed Distance moved per unit time, independent of direction. *See also* velocity.

speed of light The fastest possible speed, according to the currently known laws of physics. Electromagnetic radiation exists in the form of waves or photons moving at the speed of light. (p. 49)

spicule Small solar storm that expels jets of hot matter into the Sun's lower atmosphere.

spin–orbit resonance State that a body is said to be in if its rotation period and its orbital period are related in some simple way.

spiral arm Distribution of material in a galaxy forming a pinwheel-shaped design, beginning near the galactic center. (p. 401)

spiral density wave A wave of matter formed in the plane of planetary rings, similar to ripples on the surface of a pond, that wraps around the rings, forming spiral patterns similar to grooves in a record disk. Spiral density waves can lead to the appearance of ringlets. (p. 402)

A proposed explanation for the existence of galactic spiral arms, in which coiled waves of gas compression move through the galactic disk, triggering star formation.

spiral galaxy Galaxy composed of a flattened, star-forming disk component that may have spiral arms and a large central galactic bulge. (p. 421)

spiral nebula Historical name for spiral galaxies, describing their appearance.

spring tide The largest tides, occurring when the Sun, Moon, and Earth are aligned at new and full Moon.

S-process "Slow" process in which neutrons are captured by a nucleus; the rate is typically one neutron capture per year.

standard candle Any object with an easily recognizable appearance and known luminosity, which can be used in estimating distances.

Supernovae, which all have the same peak luminosity (depending on type), are good examples of standard candles and are used to determine distances to other galaxies. (p. 426)

standard solar model A self-consistent picture of the Sun, developed by incorporating the important physical processes that are believed to be important in determining the Sun's internal structure into a computer program. The results of the program are then compared with observations of the Sun, and modifications are made to the model. The standard solar model, which enjoys widespread acceptance, is the result of this process. (p. 255)

standard time System of dividing Earth's surface into 24 time zones, with all clocks in each zone keeping the same time.

star A glowing ball of gas held together by its own gravity and powered by nuclear fusion in its core. (pp. 6, 254)

starburst galaxy Galaxy in which a violent event, such as near collision, has caused an intense episode of star formation in the recent past. (p. 451)

star cluster A grouping of anywhere from a dozen to a million stars that formed at the same time from the same cloud of interstellar gas. Stars in clusters are useful to aid our understanding of stellar evolution because, within a given cluster, stars are all roughly the same age and chemical composition and lie at roughly the same distance from Earth. (p. 323)

Stefan's law Relation that gives the total energy emitted per square centimeter of its surface per second by an object of a given temperature. Stefan's law shows that the energy emitted increases rapidly with an increase in temperature, proportional to the temperature raised to the fourth power. (p. 54)

stellar epoch Most recent period, when stars, planets, and life have appeared in the universe.

stellar nucleosynthesis The formation of heavy elements by the fusion of lighter nuclei in the hearts of stars. Except for hydrogen and helium, all other elements in our universe resulted from stellar nucleosynthesis.

stellar occultation The dimming of starlight produced when a solar system object such as a planet, moon, or ring passes directly in front of a star. (p. 241)

stratosphere The portion of Earth's atmosphere lying above the troposphere, extending up to an altitude of 40–50 km. (p. 144)

string theory Theory that interprets all particles and forces in terms of particular modes of vibration of submicroscopic strings.

strong nuclear force Short-range force responsible for binding atomic nuclei together. The strongest of the four fundamental forces of nature. (p. 271)

S-type asteroid Asteroid made up mainly of silicate or rocky material.

subatomic particle Particle smaller than the size of an atomic nucleus.

subduction zone Place where two plates meet and one slides under the other.

subgiant branch The section of the evolutionary track of a star that corresponds to changes that occur just after hydrogen is depleted in the core, and core hydrogen burning ceases. Shell hydrogen burning heats the outer layers of the star, which causes a general expansion of the stellar envelope. (p. 334)

sublimation Process by which element changes from the solid to the gaseous state, without becoming liquid.

summer solstice Point on the ecliptic where the Sun is at its northernmost point above the celestial equator, occurring on or near June 21. (p. 10)

sunspot An Earth-sized dark blemish found on the surface of the Sun. The dark color of the sunspot indicates that it is a region of lower temperature than its surroundings. (p. 262)

sunspot cycle The fairly regular pattern that the number and distribution of sunspots follows, in which the average number of spots reaches a maximum every 11 or so years, then falls off to almost zero. (p. 265)

supercluster Grouping of several clusters of galaxies into a larger, but not necessarily gravitationally bound unit. (p. 463)

super-Earth An extrasolar planet with a mass between 2 and 10 times that of Earth. (p. 130)

superforce An attempt to combine the strong and electroweak forces into one single force.

supergiant A star with a radius between 100 and 1000 times that of the Sun. (p. 289)

supergranulation Large-scale flow pattern on the surface of the Sun, consisting of cells measuring up to 30,000 km across, believed to be the imprint of large convective cells deep in the solar interior. (p. 259)

superior conjunction Orbital configuration in which an inferior planet (Mercury or Venus) lies farthest from Earth (on the opposite side of the Sun).

supermassive black hole Black hole having a mass a million to a billion times greater than the mass of the Sun; usually found in the central nucleus of a galaxy.

supernova Explosive death of a star, caused by the sudden onset of nuclear burning (Type I), or an enormously energetic shock wave (Type II). One of the most energetic events of the universe, a supernova may temporarily outshine the rest of the galaxy in which it resides. (p. 344)

supernova remnant The scattered glowing remains from a supernova that occurred in the past. The Crab Nebula is one of the best-studied supernova remnants. (p. 347)

supersymmetric relic Massive particles that should have been created in the Big Bang if supersymmetry is correct.

surface gravity The acceleration due to gravity at the surface of a star or planet. (p. 140)

S-waves Shear waves from an earthquake, which can travel only through solid material and move more slowly than p-waves.

synchronous orbit State of an object when its period of rotations is exactly equal to its average orbital period. The Moon is in a synchronous orbit and so presents the same face toward Earth at all times. (p. 142)

synchrotron radiation Type of nonthermal radiation produced by high-speed charged particles, such as electrons, as they are accelerated in a strong magnetic field. (p. 441)

synodic month Time required for the Moon to complete a full cycle of phases. (p. 14)

synodic period Time required for a body to return to the same apparent position relative to the Sun, taking Earth's own motion into account; the time between one closest approach to Earth and the next (for a planet).

T

T Tauri phase star Protostar in the late stages of formation, often exhibiting violent surface activity. T Tauri phase stars have been observed to brighten noticeably in a short period of time, consistent with the idea of rapid evolution during this final phase of stellar formation. (p. 319)

tail Component of a comet that consists of material streaming away from the main body, sometimes spanning hundreds of millions of kilometers. May be composed of dust or ionized gases. (p. 112)

tau A type of lepton (along with the electron and muon).

tectonic fracture Cracks on a planet's surface, in particular on the surface of Mars, caused by internal geological activity.

telescope Instrument used to capture as many photons as possible from a given region of the sky and concentrate them into a focused beam for analysis. (p. 72)

temperature A measure of the amount of heat in an object, and an indication of the speed of the particles that comprise it. (p. 53)

tenuous Thin, having low density.

terminator The line separating night from day on the surface of the Moon or a planet.

terrae *See* highlands.

terrestrial planet One of the four innermost planets of the solar system, resembling Earth in general physical and chemical properties. (p. 106)

theoretical model An attempt to construct a mathematical explanation of a physical process or phenomenon within the assumptions and confines of a given theory. In addition to providing an explanation of the observed facts, the model generally also makes new predictions that can be tested by further observation or experimentation. (p. 20)

theories of relativity Einstein's theories, on which much of modern physics rests. Two essential facts of the theory are that nothing can travel faster than the speed of light and that everything, including light, is affected by gravity.

theory A framework of ideas and assumptions used to explain some set of observations and make predictions about the real world. (p. 20)

thermal equilibrium Condition in which new particle–antiparticle pairs are created from photons at the same rate as pairs annihilate one another to produce new photons.

thick disk Region of a spiral galaxy where an intermediate population of stars resides, younger than the halo stars but older than stars in the disk.

threshold temperature Critical temperature above which pair production is possible and below which pair production cannot occur.

tidal bulge Elongation of Earth caused by the difference between the gravitational force on the side nearest the Moon and the force on the side farthest from the Moon. The long axis of the tidal bulge points toward the Moon. More generally, the deformation of any body produced by the tidal effect of a nearby gravitating object. (p. 141)

tidal force The variation in one body's gravitational force from place to place across another body—for example, the variation of the Moon's gravity across Earth. (p. 141)

tidal locking Circumstance in which tidal forces have caused a moon to rotate at exactly the same rate at which it revolves around its parent planet, so that the moon always keeps the same face turned toward the planet. (p. 142)

tidal stability limit The minimum distance within which a moon can approach a planet before being torn apart by the planet's tidal force.

tidally locked Condition in which tidal forces have caused a body (such as a moon) to rotate at exactly the same rate at which it orbits another body so that it always keeps the same face toward the other body. (p. 142)

tides Rising and falling motion of terrestrial bodies of water, exhibiting daily, monthly, and yearly cycles. Ocean tides on Earth are caused by the competing gravitational pull of the Moon and Sun on different parts of Earth. (p. 141)

time dilation A prediction of the theory of relativity, closely related to the gravitational reshift. To an outside observer, a clock lowered into a strong gravitational field will appear to run slow. (p. 381)

time zone Region on Earth in which all clocks keep the same time, regardless of the precise position of the Sun in the sky, for consistency in travel and communications.

total eclipse Celestial event during which one body is completely blocked from view by another. (p. 16)

transit Orbital configuration where a planet (i.e., Mercury or Venus) passes between Earth and the Sun.

transition zone The region of rapid temperature increases that separates the Sun's chromosphere from the corona. (p. 254)

transverse motion Motion perpendicular to a particular line of sight, which does not result in Doppler shift in radiation received.

transverse velocity Component of star's velocity perpendicular to the line of sight.

triangulation Method of determining distance based on the principles of geometry. A distant object is sighted from two well-separated locations. The distance between the two locations and the angle between the line joining them and the line to the distant object are all that are necessary to ascertain the object's distance. (p. 18)

triple-alpha process The creation of carbon-12 by the fusion of three helium-4 nuclei (alpha particles). Helium-burning stars occupy a region of the Hertzsprung–Russell diagram known as the horizontal branch.

triple star system Three stars that orbit one another, bound together by gravity.

Trojan asteroid One of two groups of asteroids that orbit at the same distance from the Sun as Jupiter, 60 degrees ahead of and behind the planet. (p. 108)

tropical year The time interval between one vernal equinox and the next. (p. 12)

troposphere The portion of Earth's atmosphere from the surface to about 15 km. (p. 144)

trough Maximum departure of a wave below its undisturbed state.

true space motion True motion of a star, taking into account both its transverse and radial motion according to the Pythagorean theorem.

Tully-Fisher relation A relation used to determine the absolute luminosity of a spiral galaxy. The rotational velocity, measured from the broadening of spectral lines, is related to the total mass, and hence the total luminosity. (p. 427)

turnoff mass The mass of a star that is just now evolving off the main sequence in a star cluster.

21-centimeter line Spectral line in the radio region of the electromagnetic spectrum associated with a change of spin of the electron in a hydrogen atom. (p. 313)

21-centimeter radiation Radio radiation emitted when an electron in the ground state of a hydrogen atom flips its spin to become parallel to the spin of the proton in the nucleus.

twin quasar Quasar that is seen twice at different locations in the sky due to gravitational lensing.

Type I supernova One possible explosive death of a star. A white dwarf in a binary star system can accrete enough mass that it cannot support its own weight. The star collapses and temperatures become high enough for carbon fusion to occur. Fusion begins throughout the white dwarf almost simultaneously, and an explosion results. (p. 345)

Type II supernova One possible explosive death of a star, in which the highly evolved stellar core rapidly implodes and then explodes, destroying the surrounding star. (p. 345)

U

ultraviolet Region of the electromagnetic spectrum, just beyond the visible range, corresponding to wavelengths slightly shorter than blue light. (p. 46)

ultraviolet telescope A telescope that is designed to collect radiation in the ultraviolet part of the spectrum. Earth's atmosphere is partially opaque to these wavelengths, so ultraviolet telescopes are put on rockets, balloons, and satellites to get high above most or all of the atmosphere. (p. 92)

umbra Central region of the shadow cast by an eclipsing body.

The central region of a sunspot, which is its darkest and coolest part. (p. 16)

unbound An orbit that does not stay in a specific region of space, but where an object escapes the gravitational field of another. Typical unbound orbits are hyperbolic in shape.

unbound trajectory Path of an object with launch speed high enough that it can escape the gravitational pull of a planet.

uncompressed density The density a body would have in the absence of any compression due to its own gravity.

universal time Mean solar time at the Greenwich meridian.

universe The totality of all space, time, matter, and energy. (p. 6)

unstable nucleus Nucleus that cannot exist indefinitely, but rather must eventually decay into other particles or nuclei.

upwelling Upward motion of material having temperature higher than the surrounding medium.

V

vacuum energy Property of empty space when it is excited above its normal zero-energy state. Astronomers think that the temporary appearance of vacuum energy was responsible for one or more periods of inflation during the early universe. (p. 492)

Van Allen belts At least two doughnut-shaped regions of magnetically trapped, charged particles high above Earth's atmosphere. (p. 159)

variable star A star whose luminosity changes with time. (p. 394)

velocity Displacement (distance plus direction) per unit time. *See also* speed.

vernal equinox Date on which the Sun crosses the celestial equator moving northward, occurring on or near March 21. (p. 12)

visible light The small range of the electromagnetic spectrum that human eyes perceive as light. The visible spectrum ranges from about 400 to 700 nm, corresponding to blue through red light. (p. 46)

visible spectrum The small range of the electromagnetic spectrum that human eyes perceive as light. The visible spectrum ranges from about 4000 to 7000 angstroms, corresponding to blue through red light.

visual binary A binary star system in which both members are resolvable from Earth. (p. 295)

void Large, relatively empty region of the universe around which superclusters and "walls" of galaxies are organized. (p. 464)

volcano Upwelling of hot lava from below Earth's crust to the planet's surface. (p. 150)

W

wane (referring to the Moon or a planet) To shrink. The Moon appears to wane, or shrink in size, for 2 weeks after full Moon.

warm longitudes (Mercury) Two opposite points on Mercury's equator where the Sun is directly overhead at aphelion. Cooler than the hot longitudes by 150 degrees.

water hole The radio interval between 18 cm and 21 cm, the respective wavelengths at which hydroxyl (OH) and hydrogen (H) radiate, in which intelligent civilizations might conceivably send their communication signals. (p. 518)

water volcano Volcano that ejects water (molten ice) rather than lava (molten rock) under cold conditions.

watt/kilowatt Unit of power: 1 watt (W) is the emission of 1 joule (J) per second; 1 kilowatt (kW) is 1000 watts.

wave A pattern that repeats itself cyclically in both time and space. Waves are characterized by the speed at which they move, their frequency, and their wavelength. (p. 46)

wavelength The distance from one wave crest to the next, at a given instant in time. (p. 47)

wave period The amount of time required for a wave to repeat itself at a specific point in space. (p. 47)

wave theory of radiation Description of light as a continuous wave phenomenon, rather than as a stream of individual particles.

wax (referring to the Moon or a planet) To grow. The Moon appears to wax, or grow in size, for 2 weeks after new Moon.

weakly interacting massive particle (WIMP) Class of subatomic particles that might have been produced early in the history of the universe; dark-matter candidates.

weak nuclear force Short-range force, weaker than both electromagnetism and the strong force, but much stronger than gravity; responsible for certain nuclear reactions and radioactive decays. (p. 271)

weight The gravitational force exerted on you by Earth (or the planet on which you happen to be standing).

weird terrain A region on the surface of Mercury with oddly rippled features. This feature is thought to be the result of a strong impact that occurred on the other side of the planet and sent seismic waves traveling around the planet, converging in the weird region.

white dwarf A dwarf star with sufficiently high surface temperature that it glows white. (p. 289)

white dwarf region The bottom-left corner of the Hertzsprung–Russell diagram, where white dwarf stars are found. (p. 291)

white oval Light-colored region near the Great Red Spot in Jupiter's atmosphere. Like the red spot, such regions are apparently rotating storm systems. (p. 211)

Wien's law Relation between the wavelength at which a blackbody curve peaks and the temperature of the emitter. The peak wavelength is inversely proportional to the temperature, so the hotter the object, the bluer its radiation. (p. 54)

winter solstice Point on the ecliptic where the Sun is at its southernmost point below the celestial equator, occurring on or near December 21. (p. 11)

wispy terrain Prominent light-colored streaks on Rhea, one of Saturn's moons.

X

X-ray Region of the electromagnetic spectrum corresponding to radiation of high frequency and short wavelength, far beyond the visible spectrum. (p. 46)

X-ray burster X-ray source that radiates thousands of times more energy than our Sun in short bursts lasting only a few seconds. A neutron star in a binary system accretes matter onto its surface until temperatures reach the level needed for hydrogen fusion to occur. The result is a sudden period of rapid nuclear burning and release of energy. (p. 366)

X-ray nova Nova explosion detected at X-ray wavelengths.

Z

Zeeman effect Broadening or splitting of spectral lines due to the presence of a magnetic field.

zero-age main sequence The region on the Hertzsprung–Russell diagram, as predicted by theoretical models, where stars are located at the onset of nuclear burning in their cores. (p. 322)

zodiac The 12 constellations on the celestial sphere through which the Sun appears to pass during the course of a year. (p. 10)

zonal flow Alternating regions of westward and eastward flow, roughly symmetrical about the equator of Jupiter, associated with the belts and zones in the planet's atmosphere. (p. 209)

zone Bright, high-pressure region in the atmosphere of a jovian planet, where gas flows upward. (p. 209)

Answers to Check Questions
Concept Checks and Process of Science Checks

Chapter 0

1. (p. 8) (1) Because the celestial sphere provides a natural means of specifying the locations of stars on the sky. Celestial coordinates are directly related to Earth's orientation in space but are independent of Earth's rotation. (2) Distance information is lost. 2. (p. 13) Because of the tilt of Earth's rotation axis, the Sun is highest in the northern sky during northern summer, and the days are longest. 3. (p. 17) (1) The angular size of the Moon would remain the same, but that of the Sun would be halved, making it easier for the Moon to eclipse the Sun. We would expect to see total or partial eclipses, but no annular ones. (2) If the distance were halved, the angular size of the Sun would double, and total eclipses would never be seen—only partial or annular ones. 4. (p. 20) Because astronomical objects are too distant for direct ("measuring tape") measurements, we must rely on indirect means and mathematical reasoning. 5. (p. 23) A theory can never become proven "fact," because it can always be invalidated, or forced to change by a single contradictory observation. However, once a theory's predictions have been repeatedly verified by experiments over many years, it is often widely regarded as "true."

Chapter 1

1. (p. 32) In the geocentric view, retrograde motion is a real backward motion of a planet as it moves on its epicycle. In the heliocentric view, the backward motion is only apparent, caused by Earth "overtaking" the planet in its orbit. 2. (p. 31) Mainly simplicity and elegance. Both theories made testable predictions, and until Newton developed his laws, neither could explain *why* the planets move as they do. However, Copernicus's model was much simpler than the Ptolemaic version, which became more and more convoluted as observations improved. 3. (p. 33) The discovery of the phases of Venus could not be reconciled with the geocentric model. Observations of Jupiter's moons proved that some objects in the universe did not orbit Earth. 4. (p. 35) Galileo was an experimentalist; Kepler was a theorist. Galileo found observational evidence supporting the Copernican theory; Kepler found an empirical description of the motions of the planets within the Copernican picture that greatly simplified the theoretical view of the solar system. 5. (p. 36) Because those planets were unknown to Kepler. The fact that the laws apply to the outer planets then constitutes a prediction subsequently verified by observation. 6. (p. 36) Because Kepler determined the overall geometry of the solar system by triangulation using Earth's orbit as a baseline, all distances were known only relative to the scale of Earth's orbit—the astronomical unit (AU). 7. (p. 39) In the absence of any force, a planet would move in a straight line with constant velocity (Newton's first law) and therefore tends to move along the tangent to its orbital path. The Sun's gravity causes the planet to accelerate toward the Sun (Newton's second law), bending its trajectory into the orbit we observe. 8. (p. 41) Kepler's laws were accurate descriptions of planetary motion based on observations, but they contained no insight into *why* the planets orbited the Sun or why the orbits are as they are. Newtonian mechanics explained the orbits in terms of universal laws, and in addition made detailed predictions about the motion of other bodies in the cosmos—moons, comets, other stars—that were hitherto impossible to make.

Chapter 2

1. (p. 47) A wave is a way in which energy or information is transferred from place to place via a repetitive, regularly varying disturbance of some sort. A wave is characterized by its wavelength, frequency, velocity, and amplitude: the product of the first two always equals the third. 2. (p. 50) Light is an electromagnetic wave produced by accelerating charged particles. All waves on the list are characterized by their wave periods, frequencies, and wavelengths, and all carry energy and information from one location to another. Unlike waves in water or air, however, light waves require no physical medium through which to propagate. 3. (p. 53) All are electromagnetic radiation and travel at the speed of light. In physical terms, they differ only in frequency (or wavelength), although their effects on our bodies (or our detectors) differ greatly. 4. (p. 55) As the switch is turned and the temperature of the filament rises, the bulb's brightness increases rapidly, by Stefan's law, and it color shifts, by Wien's law, from invisible infrared to red to yellow to white. 5. (p. 57) They are characteristic frequencies (wavelengths) at which matter absorbs or emits photons of electromagnetic radiation. They are unique to each atom or molecule and thus provide a means of identifying the gas producing them. 6. (p. 62) Electron orbits can occur only at certain specific energies, and there is a

ground state that has the lowest possible energy. Planetary orbits can have any energy. Planets can stay in any orbit indefinitely; in an atom the electron must eventually fall to the ground state, emitting electromagnetic radiation in the process. Planets may reasonably be thought of as having specific trajectories around the Sun—there is no ambiguity as to "where" a planet is. Electrons in an atom are smeared out into an electron cloud, and we can only talk of the electron's location in probabilistic terms. 7. (p. 63) The wave theory could not explain why atoms emit and absorb radiation at specific, unique wavelengths, rather than as a continuous beam. The key new insight of quantum theory was that, on microscopic scales, the properties of both matter and radiation are *quantized*. Specifically, the energy of an electron in an atom can only take on certain definite values, unique to the type of atom, and radiation exists in the form of *photons* of specific energies. When an electron changes energy within an atom, it does so by emitting or absorbing a photon of a definite energy, and hence *color*, directly connecting observations of atomic spectra to the particular atoms producing them. 8. (p. 64) Spectral lines correspond to transitions between specific orbitals within an atom. The structure of an atom determines the energies of these orbitals, and hence the possible transitions, and thus the energies (colors) of the photons involved. 9. (p. 65) Measuring masses in astronomy usually entails measuring the orbital speed of one object—a companion star or a planet, perhaps—around another. In most cases, the Doppler effect is an astronomer's only way of making such measurements. 10. (p. 66) Because, with few exceptions, spectral analysis is the only way we have of determining the physical conditions—composition, temperature, density, velocity, etc.—in a distant object. Without spectral analysis, astronomers would know next to nothing about the properties of stars and galaxies.

Chapter 3

1. (p. 80) Because large reflecting telescopes are easier to design, build, and maintain than refracting instruments. 2. (p. 83) The need to gather as much light as possible, and the need to achieve the highest possible angular resolution. 3. (p. 84) The atmosphere absorbs some radiation arriving from space, and turbulent motion blurs the incoming rays. To reduce or overcome the effects of atmospheric absorption, instruments are placed on high mountains or in space. To compensate for atmospheric turbulence, adaptive optics techniques probe the air above the observing site and adjust the mirror surface accordingly to try to recover the undistorted image. 4. (p. 87) Radio astronomy opens up a new window on the universe, allowing astronomers to study known objects in new ways and to discover new objects that would otherwise be completely unobservable. The faintness of the sources is addressed by combining the largest available telescopes with sensitive detectors. Resolution can be greatly improved by the use of interferometry, which combines the signals from two or more separate telescopes to create the effect of a single instrument of much larger diameter. 5. (p. 93) The text presents multiwavelength observations of stars, star-forming regions, and galaxies. We must study astronomical objects at many wavelengths because most emit radiation across the entire spectrum, and much of what we know is learned from studies of invisible radiation. 6. (p. 95) Benefits: They are above the atmosphere, so they are unaffected by seeing or absorption; they can also make round-the-clock observations of the entire sky. Drawbacks: Cost/smaller size, inaccessibility, vulnerability to damage by radiation and cosmic rays.

Chapter 4

1. (p. 107) Because the two classes of planets differ in almost every physical property—orbit, mass, radius, composition, existence of rings, number of moons. 2. (p. 111) Similarities: All orbit in the inner solar system, roughly in the ecliptic plane, and are solid bodies of generally "terrestrial" composition. Differences: Asteroids are much smaller than the terrestrial planets, are generally less evolved than the planets, and move on much less regular orbits. 3. (p. 115) Most comets never come close enough to the Sun for us to see them. 4. (p. 118) Because it is thought to be much less evolved than material now found in planets, and hence it is a better indicator of conditions in the early solar system. 5. (p. 120) Because it must explain certain general features of solar system architecture, while accommodating the fact that there are exceptions to many of them. 6. (p. 127) Yes. If a star formed with a disk of matter around it, then even if it doesn't have planets like Earth, the basic processes of condensation and accretion would probably have occurred there, too. 7. (p. 134) Because the search techniques are most sensitive to

massive planets orbiting close to their parent stars, which are exactly what have been observed.

Chapter 5

1. (p. 143) A tidal force is the *variation* in one body's gravitational force from place to place across another. Tidal forces tend to deform a body rather than causing an overall acceleration, and they decrease rapidly with distance. 2. (p. 147) It raises Earth's average surface temperature above the freezing point of water, which was critical for the development of life on our planet. Should the greenhouse effect continue to strengthen, however, it may conceivably cause catastrophic climate changes on Earth. 3. (p. 150) Earthquakes and seismic waves are observed on Earth's surface. These data are combined with models of how waves move through Earth's interior to construct a model of our planet's composition and physical state. We measure Earth's mass, radius, and surface composition directly. However, the composition and temperature of our planet's interior are inferred from models. 4. (p. 152) We would have far less detailed direct (from volcanoes) or indirect (from seismic studies following earthquakes) information on our planet's interior. 5. (p. 155) Convection currents in the upper mantle cause portions of Earth's crust—plates—to slide around on the surface. As the plates move and interact, they are responsible for volcanism, earthquakes, the formation of mountain ranges and ocean trenches, and the creation and destruction of oceans and continents. 6. (p. 157) The maria are younger, denser, and are much less heavily cratered than the highlands. 7. (p. 161) It tells us that the planet has a conducting, liquid core in which the magnetic field is continuously generated. 8. (p. 163) The Moon's small size, which is the main reason that the Moon (1) has lost its initial heat, making it geologically dead today, and (2) has insufficient gravity to retain an atmosphere.

Chapter 6

1. (p. 172) The Sun's tidal force has caused Mercury's rotation period to become exactly 2/3 of the orbital period and the rotation axis to be oriented exactly perpendicular to the planet's orbit plane. A synchronous orbit is not possible because of Mercury's eccentric orbit around the Sun. 2. (p. 174) Venus's atmosphere is much more massive, hotter, and denser than that of Earth, and is composed almost entirely of carbon dioxide. 3. (p. 175) Scarps are thought to have been formed when the planet's iron core cooled and contracted, causing the crust to crack. Faults on Earth are the result of tectonic activity. 4. (p. 179) No. They are mainly shield volcanoes formed by lava upwelling through "hot spots" in the crust and are not associated with plate tectonics (which is not observed on Venus). 5. (p. 183) Mercury: improved observations of the planet's surface led to a major revision of Mercury's rotation rate. Venus: observations of the atmosphere initially suggested a relatively mild climate; however, later radio studies of the surface revealed a hot, dry, inhospitable world. Mars: with improving observations, astronomers' ideas of the surface have evolved from an arid desert irrigated by canals built by intelligent beings, to a world now dry and geologically dead, but that once had liquid water and possibly even life on its surface. 6. (p. 190) Observations reveal evidence for dried-up rivers, outflow channels, and what may be ancient oceans, strongly suggesting that the surface was once warm enough and the atmosphere thick enough for large bodies of liquid water to exist on the planet's surface. Today most of the remaining water on Mars is frozen, underground or at the poles. 7. (p. 193) Venus probably has a large liquid metal core, but it rotates too slowly for a planetary dynamo to operate. Mars rotates rapidly but lacks a liquid metal core. 8. (p. 196) Both might have climates quite similar to that on Earth, as the greenhouse effect would probably not have run away on Venus, and the water in the Martian atmosphere might not have frozen and become permafrost.

Chapter 7

1. (p. 206) Uranus's orbit was observed to deviate from a perfect ellipse, leading astronomers to try to compute the mass and location of the body responsible for those discrepancies. That body was Neptune. 2. (p. 208) The magnetic field is generated by the motion of electrically conducting liquid in the deep interior and therefore presumably shares the rotation of that region of the planet. 3. (p. 214) Like weather systems on Earth, the belts and zones are regions of high and low pressure and are associated with convective motion. However, unlike storms on Earth, they wrap all the way around the planet because of Jupiter's rapid rotation. In addition, the clouds are arranged in three distinct layers, and the bright colors are the result of cloud chemistry unlike anything operating in Earth's atmosphere. Jupiter's spots are somewhat similar to hurricanes on Earth, but they are far larger and longer-lived. 4. (p. 215) Uranus's low temperature means that the planet's clouds lie deeper in the atmosphere, making them harder to distinguish from Earth. 5. (p. 219) We have no direct observations of the interiors of these worlds. Everything we know is derived from theoretical models of their interiors, coupled with observations of their masses, radii, and surface properties. 6. (p. 221) They all combine the two

essential ingredients needed to operate a planetary dynamo: (1) rapid rotation, in all cases faster than Earth's rotation, and (2) conducting liquid interiors: metallic hydrogen in the cases of Jupiter and Saturn and "slushy" ice in the cases of Uranus and Neptune.

Chapter 8

1. (p. 230) Jupiter's gravitational field, via its tidal effect on the moons. 2. (p. 235) The haze layers high in the moon's dense atmosphere are opaque to visible light. The *Huygens* probe landed on the surface, and *Cassini*'s radar and infrared sensors can penetrate the haze. 3. (p. 237) They are all very heavily cratered—a result of the heavy bombardment that marked the clearing of the comets from the outer solar system. 4. (p. 243) The Roche limit is the radius inside of which a moon will be torn apart by tidal forces. Planetary rings are found inside the Roche limit, (most) moons outside. 5. (p. 245) They have similar masses, radii, composition, and perhaps also similar origins in the Kuiper belt. 6. (p. 247) Pluto, like Ceres, Eris, and a few other bodies in the outer solar system, orbits the Sun and is large enough for its gravity to have caused it to assume an approximately spherical shape, but it is not massive enough to have cleared its orbital path of other solar system objects. In current terminology, it is called a dwarf planet, or a plutoid.

Chapter 9

1. (p. 255) When we multiply the solar constant by the total area to obtain the solar luminosity, we are implicitly assuming that the same amount of energy reaches every square meter of the large sphere in Figure 9.3. 2. (p. 259) The energy may be carried in the form of (1) radiation, where energy travels in the form of light, and (2) convection, where energy is carried by physical motion of upwelling solar gas. 3. (p. 201) The coronal spectrum shows (1) emission lines of (2) highly ionized elements, implying high temperature. 4. (p. 270) There is a strong field, with a well-defined east-west organization, just below the surface. The field direction in the southern hemisphere is opposite that in the north. However, the details of the fields are very complex. 5. (p. 273) Because the Sun shines by nuclear fusion, which converts mass into energy as hydrogen turns into helium. 6. (p. 274) The solar neutrino problem was the fact that theory predicted that significantly more neutrinos should be detected coming from the Sun than were actually observed experimentally. Both the theory of solar nuclear fusion and the experimental apparatus on Earth were carefully checked and tested, and neither was found to be at fault. In the end, a new piece of the theory, explaining how neutrinos could change from one type to another en route from the core of the Sun to the detectors on Earth, accounted for the discrepancy and solved the problem.

Chapter 10

1. (p. 281) Because stars are so far away that their parallaxes relative to any baseline on Earth are too small to measure accurately. 2. (p. 285) Nothing—We need to know their distances before the luminosities (or absolute magnitudes) can be determined. 3. (p. 288) Even though the theory behind the classification was incorrect, it provided a very useful way to organize and categorize the large database of stellar observations, making it easy for astronomers to adapt and interpret the data when an improved understanding of atomic physics emerged some three decades later. 4. (p. 288) Because temperature controls which excited states the star's atoms and ions are in, and hence which atomic transitions are possible. 5. (p. 289) Yes, by using the radius–luminosity–temperature relationship, but only if we can find a method of determining the luminosity that doesn't depend on the inverse-square law (Sec. 10.2). 6. (p. 292) Because giants are intrinsically very luminous and can be seen to much greater distances than the more common main-sequence or white dwarfs. 7. (p. 293) All stars would be further away, but their measured spectral types and apparent brightnesses would presumably be unchanged, so their luminosities would be greater than previously thought. The main sequence would therefore move vertically upward in the H–R diagram. We would then use larger luminosities in the method of spectroscopic parallax, so distances inferred by that method would also increase. 8. (p. 294) We are assuming (1) that we can tell with some confidence whether or not an observed star lies on the main sequence, and (2) that the radius of a main-sequence star of a given spectral type is fairly well determined. Both assumptions are reasonable, although significant uncertainties exist, especially with assumption (2). 9. (p. 299) We don't. We assume that their masses are the same as similar stars found in binaries.

Chapter 11

1. (p. 307) Because the scale of interstellar space is so large that even very low densities can accumulate to a large amount of obscuring matter along the line of sight to a distant star. 2. (p. 311) Because the UV light is absorbed by hydrogen gas in the surrounding cloud, ionizing it to form the emission nebulae. The red light is H radiation, part of the visible hydrogen spectrum emitted as electrons, and protons recombine to form hydrogen atoms.

3. (p. 314) Because the cloud's main constituent, hydrogen, is very hard to observe, astronomers must use other molecules as tracers of the cloud's properties. 4. (p. 321) (1) The existence of a photosphere, meaning that the inner part of the cloud becomes opaque to its own radiation, signaling the slowing of the collapse phase. (2) Nuclear fusion in the core, and equilibrium between pressure and gravity. 5. (p. 322) No. Different parts of the main sequence correspond to stars of different masses. A typical star stays at roughly the same location on the main sequence most of its lifetime. 6. (p. 326) Because stars form at different rates—high-mass stars reach the main sequence and start disrupting the parent cloud long before lower-mass stars have finished forming.

Chapter 12

1. (p. 333) Hydrostatic equilibrium means that, if some property of the Sun changes a little, the star's structure adjusts to compensate. Small changes in the Sun's internal temperature or pressure will not lead to large changes in its radius or luminosity. 2. (p. 335) Because the nonburning inner core, unsupported by fusion, begins to shrink, releasing gravitational energy, heating the overlying layers, causing them to burn more vigorously, and increasing the luminosity. 3. (p. 341) No. It does not have a binary companion to provide mass after the Sun becomes a white dwarf, so our star will never become a nova or Type I supernova; it is too low in mass to become a Type II supernova. 4. (p. 344) Because iron cannot fuse to produce energy. As a result, no further nuclear reactions are possible, and the core's equilibrium cannot be restored. 4. (p. 347) Because the two types of supernova differ in their spectra and their light curves, making it impossible to explain them in terms of a single phenomenon. 5. (p. 354) Because a star cluster gives us a "snapshot" of stars of many different masses—but of the same age and initial composition—allowing us to directly test the predictions of the theory. 6. (p. 355) Because it is responsible for creating and dispersing all the heavy elements out of which we are made. In addition, it may also have played an important role in triggering the collapse of the interstellar cloud from which our solar system formed.

Chapter 13

1. (p. 362) No. Only Type II supernovae. According to theory, the rebounding central core of the original star becomes a neutron star. 2. (p. 364) Because (1) not all supernovae form neutron stars, (2) the emission is beamed and may not be observable from Earth, and (3) pulsars spin down and become too faint to observe after a few tens of millions of years. 3. (p. 309) Some X-ray sources are binaries containing accreting neutron stars, which may be in the process of being spun up to form millisecond pulsars. 4. (p. 372) They are energetic bursts of gamma rays, roughly uniformly distributed on the sky, occurring roughly once per day. They pose a challenge because they are very distant, and hence extremely luminous, but their energy comes from a region less than a few hundred kilometers across. 5. (p. 374) Its gravity becomes so strong that even light cannot escape, and it becomes a black hole. 6. (p. 379) In Newton's theory, gravity is a force that exists between all massive bodies. In relativity, the acceleration we call gravity is actually our perception of motion through a curved spacetime, whose curvature depends on the local density of material. Special relativity makes some quite counterintuitive predictions about the universe—for example, time dilation, length contraction, and the loss of simultaneity—which scientists initially found hard to accept. However, all of its predictions have been repeatedly verified in laboratory experiments, and many predictions of general relativity have been verified by astronomical observations. 7. (p. 382) Because the object would appear to take infinitely long to reach the event horizon, and its light would be infinitely redshifted by the time it got there. 8. (p. 384) By observing their gravitational effects on other objects, and from the X-rays emitted when matter falls into them.

Chapter 14

1. (p. 392) The Milky Way is the thin plane of the Galactic disk, seen from within. When our line of sight lies in the plane of the Galaxy, we see many stars blurring into a continuous band. In other directions we see darkness. 2. (p. 395) No, because even the brightest Cepheids are unobservable at distances of more than a kiloparsec or so through the obscuration of interstellar dust. 3. (p. 396) The debate involved the size and location of the "spiral nebulae," such as M31 in Andromeda. Today we know them as spiral galaxies, but at the time astronomers had insufficient data to determine their distances, and hence their sizes. Shapley thought that the spiral nebulae were small, relatively nearby objects contained within our Galaxy. Curtis argued that the nebulae were distant galaxies comparable in size to our own. Shapley used observations of variable stars to argue that our Galaxy is large; Curtis thought it was much smaller. In the end, neither scientist was completely right. Shapley was correct about the size of the Galaxy, but Curtis was right in his assertion that the spiral nebulae are distant galaxies like the Milky Way. The

matter was resolved a few years later when Edwin Hubble used observations of variable stars to demonstrate that many spiral nebulae did indeed reside far outside our Galaxy. 4. (p. 399) They differ greatly in their spatial distributions of stars, the types and ages of stars found in each, the amount of interstellar gas they contain, and in the orbital motions of their component stars. 5. (p. 401) Because all the gas and dust in the halo fell to the Galactic disk billions of years ago, halo star formation has long since ceased. 6. (p. 402) Because differential rotation would destroy the spiral structure within a few hundred million years. 7. (p. 407) Scientists are unwilling to suggest new physics to explain observations, and competing theories of galactic rotation curves and "missing mass" on large scales do exist. However, despite their reservations, most astronomers have concluded that dark matter, introduced as material with gravity but no interaction with electromagnetic radiation, best fits the observed facts. Several theories exist to explain how dark matter might have been formed in the early universe, and they lead to experimental tests that should some day detect dark matter if it exists. 8. (p. 408) Its emission has not been observed at any electromagnetic wavelength. Its presence is inferred from its gravitational effect on stars and gas orbiting the Galactic center. 9. (p. 413) Observations of rapidly moving stars, gas, and the variability of the radiation emitted suggest the presence of a roughly 3 million-solar-mass black hole.

Chapter 15

1. (p. 425) Most galaxies are not large spirals—the most common galaxy types are dwarf ellipticals and dwarf irregulars. 2. (p. 426) Distance-measurement techniques ultimately rely upon the existence of very bright objects whose luminosities can be inferred by other means. Such objects become increasingly hard to find and calibrate the farther we look out into intergalactic space. 3. (p. 430) It doesn't use the inverse-square law. The other methods all provide a way of determining the luminosity of a distant object, which then is converted to a distance using the inverse-square law. Hubble's law gives a direct connection between redshift and distance. 4. (p. 433) Because it means that the energy source cannot simply be the summed energy of a huge number of stars—some other mechanism must be at work. 5. (p. 439) When quasars were discovered, they were thought to be faint, relative nearby stars, or starlike objects, although their unusual spectra posed problems for astronomers. Once astronomers realized that their odd spectra actually meant that they had very large redshifts, it became clear that quasars were actually among the most distant—and hence most luminous—objects in the entire universe. 6. (p. 440) The energy is generated in an accretion disk in the central nucleus of the visible galaxy, then transported by jets out of the galaxy and into the lobes, where it is eventually emitted by the synchrotron process in the form of radio waves.

Chapter 16

1. (p. 450) First, that the galaxies are gravitationally bound to the cluster. Second, and more fundamentally, that the laws of physics as we know them in the solar system—gravity, atomic structure, the Doppler effect—all apply on very large scales, and to systems possibly containing a lot of dark matter. 2. (p. 457) Galaxies in voids are much less likely to encounter other galaxies, so their evolution will be dictated more by "internal" processes—star formation and stellar evolution—than galaxies in clusters, which will be more affected by interactions with other galaxies. 3. (p. 457) The scenario makes predictions about how different types of galaxies were assembled over the history of the universe, and this has consequences for the types, compositions, and ages of the stars they contain. Astronomers can gauge the accuracy of the model by comparing its predictions with the properties of spiral and elliptical galaxies at different redshifts (that is, at different times in the past). 4. (p. 461) Astronomers think that the first black holes formed in young galaxies early in the history of the universe. The black holes grew by accretion of stars and gas from the surrounding galaxies, as well as mergers with other black holes when their parent galaxies merged. Eventually, they form the supermassive black holes observed in normal galaxies today. The galaxy mergers were often responsible for directing additional galactic gas onto the central black holes. The connection with active galaxies is that, during the periods of accretion, the black holes emitted large amounts of energy, becoming the active galactic nuclei we see. 5. (p. 462) Probably not. There may well be galaxies that do not harbor central black holes, and in any case, only galaxies in clusters are likely to experience the encounters that trigger activity. 6. (p. 464) Light traveling from these objects to Earth is influenced by cosmic structure all along the line of sight. Light rays are deflected by the gravitational field of intervening concentrations of mass, and gas along the line of sight produces absorption lines whose redshifts tell us the distance at which each feature formed. The light received on Earth thus gives astronomers a "core sample" through the universe, from which detailed information can be extracted.

Chapter 17

1. (p. 476) When viewed on very large scales—more than 300 Mpc—the distribution of galaxies seems to be roughly the same everywhere and in all

directions. (In addition, the microwave background radiation field is observed to be isotropic to very high precision.) 2. (p. 479) Because, tracing the motion backward in time, it implies that all galaxies, and in fact everything in the entire universe, were located at a single point at the same instant in the past. The cosmological principle does not allow that point to be special, leaving a Big Bang as the only possibility. 3. (p. 480) If there is enough matter in the universe for gravity to halt and reverse the current expansion, the universe will eventually recollapse in a "Big Crunch." Otherwise, the universe will expand forever. 4. (p. 481) A low-density universe has negative curvature and is infinite in extent; a critical-density universe is spatially flat (Euclidean) and is also infinite in extent; a high-density universe has positive curvature and is finite in extent. 5. (p. 485) The observed acceleration of the expansion of the universe implies that some nongravitational force must be at work. Dark energy is our current best theory of the cause of that force. In addition, galactic and cosmological observations indicate that the universe is spatially flat, and hence has critical density, but that the density of matter (mostly dark) cannot account for the total. The amount of dark energy needed to account for the cosmic acceleration is in good agreement with the required extra density—about 70 percent of the critical value. 6. (p. 487) There doesn't seem to be enough matter to halt the collapse, and in addition, the observed cosmic acceleration suggests the existence of a large-scale repulsive force in the cosmos that also opposes recollapse. 7. (p. 488) It was formed at the time of the Big Bang. It is the electromagnetic remnant of the primeval fireball. It started off as a blackbody curve because it was the thermal emission of the hot, young cosmos. It has retained this form because the cosmological redshift has stretched all photons by the same factor, preserving the shape of the blackbody curve. 8. (p. 490) The amount of deuterium observed in the universe today implies that the present density of normal matter is at most a few percent of the critical value—much less than the density of dark matter inferred from dynamical studies. 9. (p. 444) Not directly. Inflation implies that the universe is flat and hence that the total cosmic density equals the critical value. However, the matter density seems to be only about one-third of the critical value, and the density of electromagnetic radiation (the microwave background) is a tiny fraction of the critical density. Together, these statements suggest that the remaining density is in the form of "dark energy," as described in Section 17.3. 10. (p. 497) Because dark matter is largely unaffected by the background radiation field, it was able to clump earlier than normal matter, forming the structures within which luminous matter subsequently formed the galaxies we see today. The gravity of the early dark-matter clumps caused tiny fluctuations in the temperature of the cosmic microwave background. The ripples we see today in the microwave background are the beginning of galaxy formation long ago.

Chapter 18
1. (p. 508) Not yet. The formation of complex molecules from simple ingredients by nonbiological processes has been repeatedly demonstrated, but no living cell or self-replicating molecule has ever been created. 2. (p. 510) Mars remains the most likely site because of its generally Earthlike makeup and because several lines of evidence suggest that the planet was warmer in the past and may have had liquid water on its surface. Europa and Titan also have properties that might have been conducive to the emergence of living organisms—liquid water under Europa's icy crust and an atmosphere and many complex molecules on Titan. 3. (p. 516) It breaks a complex problem up into simpler "astronomical," "biochemical," "anthropological," and "cultural" pieces, which can be analyzed separately. The analysis itself helps identify the types of stars where a search might be most fruitful. 4. (p. 517) We assume that a technological civilization will opt to work in the radio part of the spectrum, where Galactic absorption is least, and will choose a region where natural Galactic background "static" is minimized. We also assume that the 21-cm line of hydrogen will have significance to them, since hydrogen is the most common element in the universe, and that hydroxyl will also have special meaning if they evolved in a water environment as we did.

Answers to Self-Test Questions

Chapter 0
True or False?/Multiple Choice 1) F 2) F 3) T 4) F 5) T 6) T 7) F 8) b 9) b 10) c
11) a 12) c 13) d 14) b 15) a
Problems 1) Scorpius 2) 29.9 km, 1.08×10^5 km, 2.58×10^6 km 3) It would
decrease by 8 minutes 4) (a) 33 arc minutes, (b) 33 arc seconds, (c) 0.55 arc
seconds; 56.4 minutes 5) 1.0 km/s 6) It is! 7) 173 m 8) (a) 57,300 km, (b)
3.44×10^6 km, (c) 2.06×10^8 km 9) $2.9°$

Chapter 1
True or False?/Multiple Choice 1) F 2) F 3) F 4) F 5) F 6) F 7) F 8) T 9) d 10) c
11) c 12) a 13) c 14) b 15) b
Problems 1) (a) 110 km, (b) 44,000 km, (c) 370,000 km 2) 700 s 3) 0.4 degrees
retrograde 4) 3 AU, 0.333, 5.2 yr 5) 35 AU 6) Mercury's perihelion and aphelion
distances are 0.307 AU and 0.467 AU, respectively. The difference is 0.159 AU.
Alternatively, Mercury's aphelion distance is 52 percent larger than the perihelion distance. 7) 9.42×10^{-4} solar $= 1.88 \times 10^{27}$ kg 8) 9.50, 7.33, 1.49 m/s^2
9) For a 70-kg person, force $= 684$ N $= 154$ pounds; we call it the person's
weight 10) 1.7 km/s

Chapter 2
True or False?/Multiple Choice 1) F 2) T 3) T 4) F 5) T 6) T 7) T 8) a 9) d 10) a
11) b 12) c 13) d 14) a 15) d
Problems 1) 1480 m/s 2) 3 m 3) 23 Hz; radio 4) A; 3.25; 112 5) 310; 9.4 microns;
infrared 6) 6.4×10^7 W/m^2; 3.9×10^{26} W 7) 3×10^{10} 8) Second excited state:
3 possibilities: $2 \rightarrow 1$ (Hα, 656 nm), $2 \rightarrow$ ground (Lyβ, 103 nm), $1 \rightarrow$ ground
(Lyα, 122 nm); Third excited state: 6 possibilities: those already listed, plus
$3 \rightarrow 2$ (called Paschen α, 1876 nm), $3 \rightarrow 1$ (Hβ, 486 nm), $3 \rightarrow$ ground (Lyγ,
97.3 nm) 9) 137 km/s approaching 10) 1.94×10^{27} kg

Chapter 3
True or False?/Multiple Choice 1) F 2) F 3) T 4) T 5) F 6) F 7) F 8) d 9) d 10) c
11) a 12) d 13) b 14) b 15) c
Problems 1) 0.3 arc seconds; 6.8 pixels 2) 580 μm; longer than the $3 - 200$ μm
operating range 3) 6.7 minutes; 1.7 minutes 4) (a) 0.25″, (b) 0.01″ 5) 15 m 6)
160 light-years, using the formula in the text to compute the resolution 7) 3.3
minutes 8) (a) 36, (b) 0.61, (c) 0.012 light-years 9) 14.1 m, 16 m 10) (a) 0.003
arc seconds, (b) 0.005 arc seconds

Chapter 4
True or False?/Multiple Choice 1) F 2) T 3) F 4) F 5) F 6) F 7) F 8) d 9) b 10) d
11) a 12) c 13) b 14) a 15) c
Problems 1) 2×10^{21} kg, 0.03 percent of Earth's mass; 40 km 2)
1.1 AU, 2.0 AU 3) 3 kg 4) 1, 0.1, 0.01, 0.001 times Ceres's mass, or
1.6×10^{21} kg, 1.6×10^{20} kg, 1.6×10^{19} kg, 1.6×10^{18} kg. 5) (a) 11 million years, (b) 50 AU 6) The orbital period is 0.00836 years $= 3.05$ days, so
(for example) January 1, 2013, would be $14 \times 365 + 4$ (leap days) $+ 31 =$
5145 days $= 1685$ periods after discovery. 7) 1.7, 0.8 Earth gravity 8) 890 AU
9) 2×10^8; once every 2.5 years 10) 0.25 AU

Chapter 5
True or False?/Multiple Choice 1) T 2) T 3) T 4) F 5) T 6) T 7) F 8) b 9) d 10) b
11) b 12) d 13) d 14) a 15) c
Problems 1) 5.3 m/s^2 $= 0.55$ times the actual value, 8.3 km/s 2) 30 kg 3) 21 m
$= 70$ feet 4) 1.6×10^{-7} 5) (a) 8.4×10^{-3}, (b) 0.16, (c) 0.84, (d) 7.0×10^{-3} times
Earth's mass 7) 45 percent 8) 43 minutes 9) Need at least 4.8×10^5 such cra-
ters, requiring at least 4.8 trillion years; rate must have been greater by a factor
of about 1000 10) 93 m

Chapter 6
True or False?/Multiple Choice 1) T 2) F 3) T 4) T 5) T 6) T 7) F 8) b 9) b 10) d
11) c 12) b 13) b 14) a 15) c
Problems 1) 8.8 minutes 2) 4.0×10^{23} kg, (for a temperature of 700 K) or
roughly 22 percent greater than it is now 3) 4.9×10^{20} kg, 98 times more massive than Earth's atmosphere, 10^{-4} times the mass of Venus 4) 400 km/h, 250
mph 5) (a) 12,100 km, (b) yes—the smallest detectable feature would be about
20 km across, much smaller than the largest impact features 6) 197 minutes,
94 minutes 7) 28 kg, assuming I weigh 70 kg on Earth 8) 10 m/s $= 22$ mph 9)
2.7×10^{17} kg; 4.5×10^{-7} times the mass of Mars; 5.5×10^{-4} times the mass of

Venus's atmosphere 10) 16′, 2.7′; no—the minimum angular diameter of the
Sun, as seen from Mars, is 19′

Chapter 7
True or False?/Multiple Choice 1) F 2) T 3) F 4) T 5) T 6) F 7) T 8) T 9) a 10) d
11) d 12) b 13) c 14) a 15) b
Problems 1) 150 km, or 0.002 planetary radii; 1100 km, or 0.04 planetary radii
2) 2.2×10^{17} newton; 1.4×10^{21} newton, or 6400 times larger 3) 4200 s $= 1.17$
hours; 84,000 km 4) 6×10^{-5} Jupiter masses, or 0.019 Earth masses 5) — 6)
1.5 percent of the actual mass 7) 10.5 days 8) 74 K 9) 1.7×10^{25} kg (2.8 Earth
masses); 81 percent 10) 49 μm; infrared

Chapter 8
True or False?/Multiple Choice 1) T 2) T 3) T 4) T 5) F 6) T 7) T 8) T 9) F 10) c
11) c 12) b 13) d 14) c 15) b
Problems 1) 20.3 km/s, 7.6 km/s; Saturn is much more massive than Earth! 2) 2.3 Jupiter radii 3) For Io: acceleration difference $= 6.3 \times 10^{-3}$ m/s^2, surface gravity $=$
1.8 m/s^2, ratio $= 0.0035 = 0.35$ percent; For Moon: acceleration
difference $= 2.2 \times 10^{-5}$ m/s^2, surface gravity $= 1.6$ m/s^2, ratio $= 1.4 \times 10^{-5}$
4) 1.35 m/s^2 (Earth: 9.80 m/s^2); 5) 2.6 km/s Io: 36′, Europa: 18′, Ganymede:
18′, Callisto: 9′, Sun: 6′; Yes 6) 38 km radius 7) 1.1×10^{18} 8) 6.8 times farther
out 9) For a weight of 70 kg, 4.7 kg on Pluto, 2.1 kg on Charon 10) 11.1 hours;
28.700 km; 2.8 orbits

Chapter 9
True or False?/Multiple Choice 1) T 2) F 3) F 4) F 5) T 6) F 7) F 8) c 9) c 10) a
11) b 12) c 13) a 14) b 15) c
Problems 1) (a) 14,600 W/m^2; (b) 52 W/m^2 2) (a) 3000 km, (b) 1500, (c) 1/33 of
the 167-minute orbital period 3) (a) 0.29 nm (hard X-ray), (b) 29 nm (far UV),
(c) 290 nm (near UV) 4) 100 s = 17 min, comparable to the granule lifetime
5) 36 percent of the average solar value (64 percent less) 6) (a) radiation mass
loss $= 4.3$ million tons/s $\approx$ twice the mass loss due to the solar wind; (b) 30
trillion years 7) 83 days 8) 300,000 years 9) 74 billion years 10) 4×10^{28}

Chapter 10
True or False?/Multiple Choice 1) F 2) F 3) F 4) F 5) T 6) F 7) F 8) d 9) b 10) b
11) c 12) d 13) c 14) b 15) b
Problems 1) 77 pc; 0.39″ 2) 47 km/s; 56 km/s 3) (a) 80 times the luminosity of
the Sun; (b) 2 solar radii 4) B is 3 times farther away 5) A is 10 times farther
away 6) 3.3×10^{-10} W/m^2; 2.3×10^{-13} times the solar constant 7) factor of 16
trillion 8) 320 pc 9) 3.5, 2.3 solar masses 10) (a) 200 billion years, (b) 1 billion
years, (c) 100 million years

Chapter 11
True or False?/Multiple Choice 1) F 2) F 3) F 4) T 5) F 6) T 7) T 8) d 9) b 10) c
11) d 12) b 13) b 14) c 15) b
Problems 1) 1.9 grams 2) 1428.6 MHz for a wavelength of 21 cm (but note
that the precise wavelength is 21.11 cm, corresponding to frequency of 1420.4
MHz); frequency: 1428.8–1428.2 MHz, wavelength: 20.9965–21.0053 cm 3)
0.017 pc $= 3500$ AU 4) 4100, 9.0 5) escape speeds (km/s): 1.8, 1.1, 1.1, 0.80;
average molecular speeds: 13.6, 14.0, 14.6, 14.2; No 6) yes, barely: the escape
speed is 0.93 km/s, molecular speed is 0.35 km/s 7) (a) 200, (b) 2.3 solar radii
8) 8 9) 10^{-6} solar luminosities 10) 37 pc

Chapter 12
True or False?/Multiple Choice 1) T 2) T 3) F 4) T 5) F 6) T 7) T 8) F 9) a 10) b
11) a 12) b 13) b 14) c 15) d
Problems 1) 3.5×10^7 kg/m^3; 5.7×10^{-4} kg/m^3, smaller by a factor of
1.6×10^{-11}; solar central density is 150,000 kg/m^3, smaller than the white
dwarf density by a factor of 4.3×10^{-3} 2) 220 pc 3) (a) 2.9 solar masses, (b) 1.7
solar masses 4) 7.1 years, 63,000 years 5) 7300 km/s, 500,000 Earth gravities
6) 20, 3.2 Mpc; 1000 Mpc 7) 0.4 pc; no—there are no O or B stars (in fact no
stars at all) within that distance of the Sun 8) apparent magnitude is –14.1, 1.6
magnitudes (4 times) brighter than the full Moon, 9.7 magnitudes (7600 times)
brighter than Venus; yes—according to Appendix 3, several O and B stars, as
well as Betelgeuse and some other red supergiants, lie within that distance 9)
roughly 1000 km/s; not too bad an assumption—the Nebula is moving too fast
to be affected much by gravity, although it is probably slowing down as it runs
into the interstellar medium 10) 1.2×10^{44} J; 220 Earth masses

Chapter 13

True or False?/Multiple Choice 1) T 2) T 3) T 4) F 5) T 6) F 7) F 8) b 9) a 10) b 11) c 12) b 13) c 14) c 15) b

Problems 1) 1 million revolutions per day, or 11.6 revolutions per second 2) 3×10^{44} times higher—2.1×10^{16} kg for a 70-kg human; Moon's mass is greater by a factor of 3.5 million; asteroid mass is roughly 1.6×10^{12} kg (assuming spherical shape and density 3000 kg/m^3), smaller by a factor of 13,000 3) 1.9×10^{12} m/s^2 or 190 billion Earth gravities; 190,000 km/s, or 64 percent of the speed of light; 300,000 km/s 4) 1.6×10^{-5}, 1.6×10^3, 1.6×10^{11} solar luminosities; ultraviolet, X-rays, gamma rays; they might be very bright immediately after formation, but would become very faint as they cool; the coolest could be plotted at the bottom left of the diagram, as a very faint O-type "star" 5) 2.4×10^{44} J, roughly twice the Sun's lifetime energy output; 2.4×10^{34} J; 1.4×10^{25} J 6) 140 μs 7) 3 million km = 4.3 solar radii; 3 billion km = 20 AU 8) 2×10^{10} m/s^2 = 2×10^9 g; 2×10^{-3} g; 2×10^{-9} g 9) 1760 km 10) 3×10^7 km = 0.2 AU

Chapter 14

True or False?/Multiple Choice 1) T 2) T 3) F 4) F 5) F 6) T 7) F 8) T 9) b 10) c 11) a 12) b 13) d 14) b 15) c

Problems 1) 2.0″, much less than the angular diameter of Andromeda 2) 100 kpc 3) 10 times 4) − 6.2; 17 Mpc 5) 0.014″/yr; proper motion has in fact been measured for several globular clusters 6) 2.7×10^{11} solar masses 7) 570 million years 8) 18 9) 1 kpc 10) 1600 AU, 2.6×10^6 solar masses

Chapter 15

True or False?/Multiple Choice 1) F 2) T 3) T 4) T 5) T 6) T 7) T 8) a 9) c 10) b 11) b 12) a 13) b 14) c 15) c

Problems 1) 3 billion years 2) 17 Mpc, taking the luminosity of the Cepheid to be 10^4 times that of the Sun 3) 320 Mpc 4) 14,000 km/s, 57 Mpc; 10,000 km/s, 80 Mpc; 16,000 km/s, 50 Mpc 5) 14 billion years; 14 billion years 6) 1.8×10^8 solar masses 7) 1.1 million years 8) 1.3×10^{36} W = 3.2 billion solar luminosi-

ties 9) 7200 km/s 10) Seyfert: ; Sirius A: ; smaller by a factor of 120; yes, if much of the emission from the nucleus is in the infrared

Chapter 16

True or False?/Multiple Choice 1) F 2) F 3) F 4) T 5) T 6) F 7) T 8) T 9) b 10) c 11) d 12) a 13) d 14) b 15) a

Problems 1) 1.2×10^{12} solar masses 2) 5.6×10^{11} solar masses, assuming a velocity of 350 km/s 3) 9.3×10^{10} solar masses 4) 700 km/s, comparable to the 660 km/s speed of the galaxy in the cluster 5) 130 million years 6) −27.2; 6.4 trillion solar luminosities; apparent magnitude of the Sun = −26.8; so the quasar at 10 pc is 0.4 magnitudes, or a factor of 1.4 times, brighter than the Sun at 1 AU 7) − 22.5; 8.3×10^{10} solar luminosities 8) 11 million years 9) 21 Mpc 10) 48 kpc

Chapter 17

True or False?/Multiple Choice 1) F 2) F 3) F 4) T 5) T 6) F 7) T 8) T 9) a 10) d 11) c 12) d 13) b 14) c 15) a

Problems 1) 1000 Mpc 2) 4×10^8 3) 1200 km/s, directly away from the opposite corner 4) 20, 14, 12 billion years 5) (a) 30,000 tons, (b) 2.8 pc 6) 3×10^{15} solar masses; 1200 km/s 7) (a) 1.1 mm, (b) 0.0093, (c) 9.3×10^{-5}, (d) 9.3×10^{-7} 8) 9×10^{-18} kg/m^3; 5×10^{-19} kg/m^3 9) 91 kpc 10) 74 kpc

Chapter 18

True or False?/Multiple Choice 1) F 2) F 3) T 4) F 5) T 6) T 7) T 8) b 9) c 10) c 11) b 12) d 13) a 14) d 15) d

Problems 1) 6.3 seconds (for a 20-year-old reader); 18 seconds (in 2003); 72 seconds; 161 seconds; 237 days 2) 0.19–0.47 AU 3) 5 billion 4) 27 AU 5) (a) 0.2, (b) 20, (c) 2000 6) 3100 years, 3.1 million years 7) 32,000 km/s 8) 5×10^8 W, 500 times the emission of the Sun 9) 1.43×10^9 to 1.67×10^9 Hz; 2.4×10^6 channels 10) 2.3 years; 55 years

Credits

NASA **p. 247** NASA **p. 247** NASA **p. 247** NASA **p. 247** NASA **p. 247** NASA **p. 247** NASA **p. 248** NASA **p. 248** NASA

Chapter 9
Opener Solar Dynamics Observatory/NASA **9.1** National Optical Astronomy Observatories **9.3** NASA **p. 257** Solar Dynamics Observatory/NASA **9.8** The Institute for Solar Physics **9.10** Jaime Vilinga/Schneider, Glen **9.11** NASA **9.12** NASA **9.14** Palomar Observatory **9.15a** Palomar Observatory **9.15b** The Institute for Solar Physics **9.16b** NASA **9.16c** Solar Dynamics Observatory/NASA **9.20a** NASA **9.20b** NASA **9.21** NASA **9.22a** NASA **9.23a** Lockheed Martin **9.23b** Lockheed Martin **9.23c** Lockheed Martin **9.23d** Lockheed Martin **9.26a** Institute for Cosmic Ray Research of the University of Tokyo **9.26b** Institute for Cosmic Ray Research of the University of Tokyo **p. 275** NASA **p. 275** NASA **p. 275** NASA **p. 276** Institute for Cosmic Ray Research of the University of Tokyo

Chapter 10
Opener NASA **10.3a** Harvard College Observatory **10.6b** Photodisc/Getty Images **10.6c** California Institute of Technology/Palomar/Hale Observatory **10.6d** UC Regents/Lick Observatory **10.6e** Association of Universities for Research in Astronomy, Inc. **10.7a** Pere Sanz/Alamy **10.7b** NASA Jet Propulsion Laboratory **10.10a** European Space Agency **10.10b** European Space Agency **10.10c** European Space Agency **10.18a-1** Harvard College Observatory **10.18a-2** Harvard College Observatory **10.18a-3** Harvard College Observatory

Chapter 11
Opener NASA **11.1** European Southern Observatory **11.2b** European Southern Observatory **11.2c** European Southern Observatory **11.3** Joyce Photographics/Science Source **p. 306** NASA **11.4** European Southern Observatory **11.5** Harvard College Observatory **11.6** Robert Gendler/Science Source **11.7a** NASA **11.7b** NASA **11.9a** NASA **11.9b** NASA **11.9c** Association of Universities for Research in Astronomy, Inc. **11.9d** NASA **11.11a** Charles J. Cole **11.11b** Charles J. Cole **11.12** Robert Gendler/Jim Misti/Steve Mazlin/Science Source **11.13a** Royal Observatory of Belgium **11.13b** European Southern Observatory **11.15** Association of Universities for Research in Astronomy, Inc. **11.16** Five College Radio Astronomy Observatory **11.19** J. Rathborne/Astrophysical Journal **11.20a** Pere Sanz/Alamy **11.20b** NASA Jet Propulsion Laboratory **11.20c** NASA Jet Propulsion Laboratory **11.20d** James M. Moran **11.24a** NASA Goddard Space Flight Center **11.24b** NASA **11.24b (inset)** NASA **11.25a** NASA **11.25b** NASA **11.27a** NASA **11.27b** NASA **11.27c** NASA Jet Propulsion Laboratory **11.28** European Southern Observatory **11.28 (inset)** NASA Goddard Space Flight Center **11.29a** National Optical Astronomy Observatories **11.30a** ESO **11.31** NASA **11.32a** Lick Observatory Publications Office **11.32b** NASA Jet Propulsion Laboratory **p. 327** European Southern Observatory **p. 327** NASA **p. 327** NASA **p. 327** NASA **p. 327** University of Michigan Astronomy

Chapter 12
Opener NASA **12.9a** Association of Universities for Research in Astronomy, Inc. **12.9b** NASA **12.9c** Space Telescope Science Institute **12.9d** NASA **12.11** Palomar Observatory **12.12a** Association of Universities for Research in Astronomy, Inc. **12.12b** NASA Jet Propulsion Laboratory **12.13a** NASA **12.13b** NASA **12.15a** Palomar Observatory **12.15b** NASA/JPL-Caltech/Univ. of Arizona **12.15c** NASA/JPL-Caltech/Univ. of Arizona **12.18** Space Telescope Science Institute **12.19a** Association of Universities for Research in Astronomy, Inc. **12.19b** Association of Universities for Research in Astronomy, Inc. **12.20a** Hubblesite.org **12.21** NASA **12.23a** European Southern Observatory **12.23b** European Southern Observatory **p. 348** NASA **p. 348** NASA **p. 348** NASA **p. 348** NASA **p. 348** NASA **p. 348** NASA **12.24** Harvard-Smithsonian Center for Astrophysics **12.26a** National Optical Astronomy Observatories **12.27a** NASA **12.28a** NASA **12.29a** Hubblesite.org **12.30a** NASA **12.30b** NASA **12.30c** National Optical Astronomy Observatory (NOAO) **12.30d** NASA **p. 356** NASA **p. 356** NASA **p. 356** NASA **p. 357** NASA **p. 357** NASA **p. 357** NASA

Chapter 13
Opener NASA **13.1** NASA **13.4a** European Southern Observatory **13.4b** NASA **13.4c** Lick Observatory Publications Office **13.5a** NASA **13.5b** NASA **13.6** NASA **13.7a** Smithsonian Astrophysical Observatory **13.7b** NASA **13.7c** NASA **13.9** NASA **13.10** NASA **13.11a** NASA **13.11b** NASA **13.11c** European Southern Observatory **13.11d** European Southern Observatory **p. 371** NASA **13.13a** ESO **13.13b** ESO **13.14** NASA **13.23** NASA **13.24** L. J. Chaisson **13.25a** NASA Jet Propulsion Laboratory **13.25a** NASA **13.26** NASA **p. 385** NASA **p. 385** NASA **p. 386** N/A

Chapter 14
Opener NASA, ESA, and The Hubble Heritage (STScI/AURA)-ESA/Hubble Collaboration **14.1b** Lund Observatory **14.2a** Robert Gendler/Science Source **14.2b** NASA **14.2c** NASA **14.4c** Harvard College Observatory **p. 397** Harvard College Observatory **14.10** Atlas Image **14.15** NASA Jet Propulsion Laboratory **14.17** Association of Universities for Research in Astronomy, Inc. **14.18** Space Telescope Science Institute **14.21b** Association of Universities for Research in Astronomy, Inc. **14.22a** Association of Universities for Research in Astronomy, Inc. **14.22b**

European Southern Observatory **14.23a** NASA Jet Propulsion Laboratory **14.23b** National Radio Astronomy Observatory (NRAO) **14.23c** NASA **14.23d** National Radio Astronomy Observatory (NRAO) **14.24** ESO **14.25a** L. J. Chaisson **14.25b** L. J. Chaisson **14.25c** L. J. Chaisson **14.26** Optical image—A. Mellinger; radio image—E. Carretti; radio data—S-PASS team; composition—E. Bressert. **14.25d** L. J. Chaisson **14.25e** L. J. Chaisson **14.25f** L. J. Chaisson **p. 414** NASA **p. 414** Association of Universities for Research in Astronomy, Inc. **p. 415** NASA

Chapter 15
Opener Hubblesite.org **15.1** NASA/ESA/Hubblesite.org **15.2a** Robert Gendler **15.2b** Space Telescope Science Institute **15.2c** European Space Agency & NASA/Space Telescope Science Institute **15.3a** NASA/Kennedy Space Center **15.3b** NASA/Kennedy Space Center **15.4a** NASA **15.4b** R. Gendler/J E. Ovaldsen/C. Thöne/C. Feron./European Southern Observatory **15.4c** ESO **15.5a** ESA/Hubble/NASA/European Space Agency (ESA) **15.5b** Hubblesite.org **15.5c** AURA/NSF/National Optical Astronomy Observatory (NOAO) **15.6a** ESA/Hubblesite.org **15.6b** European Space Observatory **15.7a** Mount Stromlo/Siding Spring Observatories/Science Source **15.7b** Science Source **15.7c** Science Source **15.8a** NASA Goddard Space Flight Center **15.8b** NASA Goddard Space Flight Center **15.10c** NASA/Kennedy Space Center **15.12a** Matt Bendaniel **15.12b** NASA **15.13** Rogelio Andreo Bernal **15.14** NASA/Kennedy Space Center **15.15a-1** Palomar Observatory **15.15a-2** Palomar Observatory **15.15b-1** Palomar Observatory **15.15b-2** Palomar Observatory **15.15c-1** Palomar Observatory **15.15c-2** Palomar Observatory **15.15d-1** Palomar Observatory **15.15d-2** Palomar Observatory **15.15e-1** Palomar Observatory **15.15e-2** Palomar Observatory **15.19** NASA **15.20** NASA, ESA, HEIC, and The Hubble Heritage Team (STScI/AURA) **15.21** From NRAO operated by AURA, Association of Universities for Research in Astronomy **15.22a** European Southern Observatory **15.22b** Jack O.Burns **15.22c** NASA/Kennedy Space Center **15.23a** Hubblesite.org **15.23b** National Radio Astronomy Observatory (NRAO) **15.24** Association of Universities for Research in Astronomy, Inc **15.26a** NASA/ESA/Hubblesite.org **15.26b** NASA **15.27a** J. Bahcall/NASA/Hubblesite.org **15.27b** SAO/H.Marshall/NASA **15.28** Adapted from Palomar/Caltech **15.29** Halton Arp **15.31a** National Radio Astronomy Observatory (NRAO) **15.31b** NASA **15.32** NASA **p. 443** NASA/Kennedy Space Center **p. 443** Rogelio Andreo Bernal **p. 444** Halton Arp

Chapter 16
Opener Space Telescope Science Institute **16.1** NASA/Kennedy Space Center **16.3** JPL/NASA **16.4a** National Radio Astronomy Observatory (NRAO) **16.4b** Palomar Observatory **16.5** NASA **16.6** NASA **16.7a** NASA **16.7b** NASA EOS Earth Observing System **16.7c** Hubblesite.org **16.8a** Association of Universities for Research in Astronomy, Inc. **16.8b** Association of Universities for Research in Astronomy, Inc. **16.9b** NASA **16.9c** NASA **16.10** NASA **16.12a** NASA **16.12b** Hubblesite.org **16.13b** Vasily Belokurov **16.14** Based on J. Barnes & L. Hernquist **16.15** James Cornell **16.17a** NASA **16.17b** NASA/Kennedy Space Center **16.18a** NASA **16.18b** NASA **16.22** Harvard-Smithsonian Center for Astrophysics **p. 466 (left)** Sloan Digital Sky Survey **p. 466 (right)** Sloan Digital Sky Survey **16.25** NASA **16.27a** NASA **16.27b** Hubblesite/NASA **16.28a** Hubblesite/NASA **16.28b** Space Telescope Science Institute **16.29a** ALCATEL-LUCENT Technologies Bell Labs **16.29b** ALCATEL-LUCENT Technologies Bell Labs **16.30** NASA **p. 470** JPL/NASA **p. 470** NASA **p. 471** NASA

Chapter 17
Opener NASA **17.1** SDSS Collaboration **17.8a** NASA **17.8b** NASA **17.8c** NASA **17.8d** NASA **17.8e** NASA **17.8f** NASA **17.13** ALCATEL-LUCENT Technologies Bell Labs **17.25a** European Space Agency (ESA) **17.27** Harvard-Smithsonian Center for Astrophysics **p. 499** European Space Agency (ESA)

Chapter 18
Opener Walter Myers/Science Source **18.1** Dana Berry **18.2** Centers for Disease Control and Prevention **18.5a** NASA **18.5b** Elso Barghoorn **18.5c** Chaisson, Eric **18.6** NASA **18.7** Harvard-Smithsonian Center for Astrophysics **18.9a** Woods Hole Oceanographic Institute **18.9b** David Mencin **p. 519** Dana Berry **p. 519** NASA **18.17a** Seth Shostak **18.17b** SETI Institute **p. 519** Woods Hole Oceanographic Institute

Cover
Opener Solar Dynamics Observatory, NASA **PO-1a** Dr. Mitsuo Ohtsuki/Science Source **PO-1b** OMIKRON/Science Source **PO-1c** Jerry Lodriguss/Science Source **PO-1d** Joyce Photographics/Science Source **PO-1e** National Oceanic and Atmospheric Administration (NOAA) **PO-2a** NASA **PO-2b** Lick Observatory Publications Office **PO-2c** NASA Jet Propulsion Laboratory **PO-2d** NASA Jet Propulsion Laboratory **PO-2e** NASA **PO-3a** Space Telescope Science Institute **PO-3b** Space Telescope Science Institute **PO-3c** Space Telescope Science Institute **PO-3d** Space Telescope Science Institute **PO-3e** Space Telescope Science Institute **PO-4a** NASA/Kennedy Space Center **PO-4b** Space Telescope Science Institute **PO-4c** NASA Jet Propulsion Laboratory **PO-4d** NASA **PO-4e** NASA

Index

H ave you ever become lost in an unfamiliar city or state? Chances are you used two things to get around: a map and some signposts. In much the same way, these two items can help you find your way around the night sky in any season. Fortunately, in addition to the seasonal Star Charts on the following pages, the sky provides us with two major signposts. Each seasonal description will talk about the Big Dipper—a group of seven bright stars that dominates the constellation Ursa Major the Great Bear. Meanwhile, the constellation Orion the Hunter plays a key role in finding your way around the sky from late autumn until early spring.

Each chart portrays the sky as seen from near 35° north latitude at the times shown at the top of the page. Located just outside the chart are the four directions: north, south, east, and west. To find stars above your horizon, hold the map overhead and orient it so a direction label matches the direction you're facing. The stars above the map's horizon now match what's in the sky.

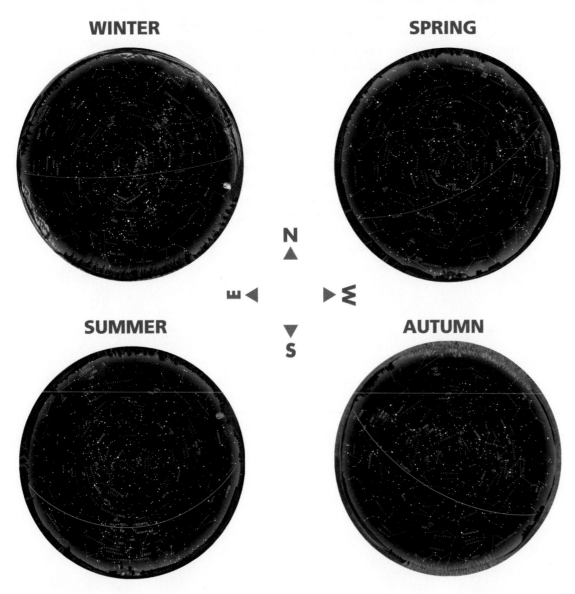

WINTER

SPRING

N
▲

E ◄   ► W

SUMMER   ▼
                S

AUTUMN

W inter finds the Big Dipper climbing the northeastern sky, with the three stars of its handle pointing toward the horizon and the four stars of its bowl standing highest. The entire sky rotates around a point near Polaris, a second-magnitude star found by extending a line from the uppermost pair of stars in the bowl across the sky to the left of the Dipper. Polaris also performs two other valuable functions: The star's altitude above the horizon equals your latitude north of the equator, and a vertical line dropped from Polaris to the horizon indicates the direction of true north.

Turn around with your back to the Dipper and you'll be facing the diamond-studded winter sky. The second great signpost in the sky, Orion the Hunter, is central to the brilliant scene. Three closely spaced, second-magnitude stars form a straight line that represents the unmistakable belt of Orion. Extending the imaginary line joining these stars to the upper right leads to Taurus the Bull and its orangish first-magnitude star, Aldebaran. Reverse the direction of your gaze to the belt's lower left and you cannot miss Sirius the Dog Star—brightest in all the heavens at magnitude –1.5.

Now move perpendicular to the belt from its westernmost star, Mintaka, and find at the upper left of Orion the red supergiant star Betelgeuse. Nearly a thousand times the Sun's diameter, Betelgeuse marks one shoulder of Orion. Continuing this line brings you to a pair of bright stars, Castor and Pollux. Two lines of fainter stars extend from this pair back toward Orion—these represent Gemini the Twins. At the northeastern corner of this constellation lies the beautiful open star cluster M35. Head south of the belt instead and your gaze will fall on the blue supergiant star Rigel, Orion's other luminary.

Above Orion and nearly overhead on winter evenings is brilliant Capella, in Auriga the Charioteer. Extending a line through the shoulders of Orion to the east leads you to Procyon in Canis Minor, the Little Dog. Once you have these principal stars mastered, using the chart to discover the fainter constellations will be a whole lot easier. Take your time, and enjoy the journey. Before leaving Orion, however, aim your binoculars at the line of stars below the belt. The fuzzy "star" in the middle is actually the glorious Orion Nebula (M42), a stellar nursery illuminated by bright, newly formed stars.

WINTER
2 A.M. on December 1; midnight on January 1; 10 P.M. on February 1

The Big Dipper, our signpost in the sky, swings high overhead during the spring and lies just north of the center of the chart. This season of rejuvenation encourages us to move outdoors with the milder temperatures, and with the new season a new set of stars beckons us.

Follow the arc of stars outlining the handle of the Dipper away from the bowl and you will land on brilliant Arcturus. This orangish star dominates the spring sky in the kite-shaped constellation Boötes the Herdsman. Well to the west of Boötes lies Leo the Lion. You can find its brightest star, Regulus, by using the pointers of the Dipper in reverse. Regulus lies at the base of a group of stars shaped like a sickle or backward question mark, which represents the head of the lion.

Midway between Regulus and Pollux in Gemini, which is now sinking in the west, is the diminutive group Cancer the Crab. Centered in this group is a hazy patch of light that binoculars reveal as the Beehive star cluster (M44).

To the southeast of Leo lies the realm of the galaxies and the constellation Virgo the Maiden. Virgo's brightest star, Spica, shines at magnitude 1.0.

During springtime, the Milky Way lies level with the horizon, and it's easy to visualize that we are looking out of the plane of our Galaxy. In the direction of Virgo, Leo, Coma Berenices, and Ursa Major lie thousands of galaxies whose light is unhindered by intervening dust in our own Galaxy. However, all these galaxies are elusive to the untrained eye and require binoculars or a telescope to be seen.

Boötes lies on the eastern border of this galaxy haven. Midway between Arcturus and Vega, the bright "summer" star rising in the northeast, is a region where no star shines brighter than second magnitude. A semicircle of stars represents Corona Borealis the Northern Crown, and adjacent to it is a large region that houses Hercules the Strongman, the fifth-biggest constellation in the sky. It is here we can find the northern sky's brightest globular star cluster, M13. A naked-eye object from a dark site, it looks spectacular when viewed through a telescope.

Returning to Ursa Major, check the second-to-last star in the Dipper's handle. Most people will see it as double, and binoculars show this easily. The pair is called Mizar and Alcor, and they lie just 0.2° apart. A telescope reveals Mizar itself to be a double. Its companion star shines at magnitude 4.0 and lies 14 arcseconds away.

SPRING
1 A.M. on March 1; 11 P.M. on April 1; 9 P.M. on May 1. Add 1 hour for daylight-saving time

The richness of the summer sky is exemplified by the splendor of the Milky Way. Stretching from the northern horizon in Perseus, through the cross-shaped constellation Cygnus overhead, and down to Sagittarius in the south, the Milky Way is packed with riches. These riches include star clusters, nebulae, double stars, and variable stars.

Let's start with the Big Dipper, our perennial signpost, which now lies in the northwest with its handle still pointing toward Arcturus. High overhead, and the first star to appear after sunset, is Vega in Lyra the Harp. Vega forms one corner of the summer triangle, a conspicuous asterism of three stars. Near Vega lies the famous double-double, Epsilon (ε) Lyrae. Two fifth-magnitude stars lie just over 3 arcminutes apart and can be split when viewed through binoculars. Each of these two stars is also double, but you need a telescope to split them.

To the east of Vega lies the triangle's second star: Deneb in Cygnus the Swan (some see a cross in this pattern). Deneb marks the tail of this graceful bird, the cross represents its outstretched wings, and the base of the cross denotes its head, which is marked by the incomparable double star Albireo. Albireo matches a third-magnitude yellow star and a fifth-magnitude blue star and offers the finest color contrast anywhere in the sky. Deneb is a supergiant star that pumps out as much light as 60,000 Suns. Also notice that the Milky Way splits into two parts in Cygnus, a giant rift caused by interstellar dust blocking starlight from beyond.

Altair, the third star of the summer triangle and the one farthest south, is the second brightest of the three. Lying 17 light-years away, it's the brightest star in the constellation Aquila the Eagle.

Frequently overlooked to the north of Deneb lies the constellation Cepheus the King. Shaped rather like a bishop's hat, the southern corner of Cepheus is marked by a compact triangle of stars that includes Delta (Δ) Cephei. This famous star is the prototype of the Cepheid variable stars used to determine the distances to some of the nearer galaxies. It varies regularly from magnitude 3.6 to 4.3 and back again with a 5.37-day period.

Hugging the southern horizon, the constellations Sagittarius the Archer and Scorpius the Scorpion lie in the thickest part of the Milky Way. Scorpius's brightest star, Antares, is a red supergiant star whose name means "rival of Mars" and derives from its similarity to the planet in both color and brightness.

SUMMER
1 A.M. on June 1; 11 P.M. on July 1; 9 P.M. on August 1. Add 1 hour for daylight-saving time

The cool nights of autumn are here to remind us the chill of winter is not far off. Along with the cool air, the brilliant stars of the summer triangle descend in the west to be replaced with a rather bland-looking region of sky. But don't let initial appearances deceive you. Hidden in the fall sky are gems equal to summertime.

The Big Dipper swings low this season, and for parts of the Southern United States it actually sets. Cassiopeia the Queen, a group of five bright stars in the shape of a "W" or "M," reaches its highest point overhead, the same spot the Big Dipper reached 6 months ago. To the east of Cassiopeia, Perseus the Hero rises high. Nestled between these two groups is the wondrous Double Cluster—NGC 869 and NGC 884—a fantastic sight in binoculars or a low-power telescope.

Our view to the south of the Milky Way is a window out of the plane of our Galaxy in the opposite direction to that visible in spring. This allows us to look at the Local Group of galaxies. Due south of Cassiopeia is the Andromeda Galaxy (M31), a fourth-magnitude smudge of light that passes directly overhead around 9 p.m. in mid-November. Farther south, between Andromeda and Triangulum, lies M33, a sprawling face-on spiral galaxy best seen in binoculars or a rich-field telescope.

The Great Square of Pegasus passes just south of the zenith. Four second- and third-magnitude stars form the square, but few stars can be seen inside of it. If you draw a line between the two stars on the west side of the square and extend it southward, you'll find first-magnitude Fomalhaut in Piscis Austrinus, the Southern Fish. Fomalhaut is the solitary bright star low in the south. Using the eastern side of the square as a pointer to the south brings you to Diphda in the large, faint constellation of Cetus, the Whale.

To the east of the Square lies the Pleiades star cluster (M45) in Taurus, which reminds us of the forthcoming winter. By late evening in October and early evening in December, Taurus and Orion have both cleared the horizon, and Gemini is rising in the northeast. In concert with the reappearance of winter constellations, the view to the northwest finds summertime's Cygnus and Lyra about to set. The autumn season is a great transition period, both on Earth and in the sky, and a fine time to experience the subtleties of these constellations.

AUTUMN
1 A.M. on September 1; 11 P.M. on October 1; 9 P.M. on November 1. Add 1 hour for daylight-saving time